FUNDAMENTALS of PLASMA PHYSICS

by

J. A. BITTENCOURT
Research Scientist and Professor
Institute for Space Research (INPE), Brazil

PERGAMON PRESS

OXFORD · NEW YORK · BEIJING · FRANKFURT
SÃO PAULO · SYDNEY · TOKYO · TORONTO

U.K.	Pergamon Press plc, Headington Hill Hall, Oxford OX3 0BW, England
U.S.A.	Pergamon Press, Inc., Maxwell House, Fairview Park, Elmsford, New York 10523, U.S.A.
PEOPLE'S REPUBLIC OF CHINA	Pergamon Press, Room 4037, Qianmen Hotel, Beijing, People's Republic of China
FEDERAL REPUBLIC OF GERMANY	Pergamon Press GmbH, Hammerweg 6, D-6242 Kronberg, Federal Republic of Germany
BRAZIL	Pergamon Editora Ltda, Rua Eça de Queiros, 346, CEP 04011, Paraiso, São Paulo, Brazil
AUSTRALIA	Pergamon Press Australia Pty Ltd., P.O. Box 544, Potts Point, N.S.W. 2011, Australia
JAPAN	Pergamon Press, 5th Floor, Matsuoka Central Building, 1-7-1 Nishishinjuku, Shinjuku-ku, Tokyo 160, Japan
CANADA	Pergamon Press Canada Ltd., Suite No. 271, 253 College Street, Toronto, Ontario, Canada M5T 1R5

First edition 1986
Reprinted with corrections 1988

British Library Cataloguing in Publication Data

Bittencourt, J.A.
Fundamentals of plasma physics.
1. Plasma (Ionized gases)
I. Title
530.4'4 QC718

ISBN 0-08-033924-7 (Hardcover)
ISBN 0-08-033923-9 (Flexicover)

Printed in Great Britain by A. Wheaton & Co. Ltd., Exeter

Contents

Preface

This text is intended as a general introduction to plasma physics and was designed with the main purpose of presenting a comprehensive, logical and unified treatment of the fundamentals of plasma physics based on the statistical kinetic theory. It should be useful primarily for advanced undergraduate and first year graduate students meeting the subject of plasma physics for the first time, and presupposes only a basic elementary knowledge of vector analysis, differential equations and complex variables, as well as courses on classical mechanics and electromagnetic theory beyond the sophomore level. Some effort has been made to make the book self-contained by including in the text developments of the fluid mechanics and kinetic theory that is needed.

Throughout the text the emphasis is on clarity rather than formality. The various derivations are explained in detail and, wherever possible, the physical interpretations are emphasized. The equations are presented in such a way that they connect together without requiring the reader to do extensive algebra to bridge the gap. The features of clarity and completeness make the book suitable for self-learning and for self-paced courses.

The structure of this book is as follows. The first chapter consists of a basic introduction to plasma physics, at a descriptive level, intended to give the reader an overall view of the subject. The motion of charged particles under the influence of specified electric and magnetic fields is treated in detail in Chapters 2, 3 and 4. In the next five chapters the basic equations necessary for an elementary description of plasma phenomena are developed. Chapter 5 introduces the concepts of phase space and distribution function, and derives the basic differential kinetic equation which governs the evolution of the distribution function in phase space. The definitions of the macroscopic variables in terms of the phase space distribution function are presented in Chapter 6 and their physical interpretations are discussed. The Maxwell-Boltzmann distribution function is introduced in Chapter 7 as the equilibrium solution of the Boltzmann equation and its kinetic properties are analyzed in some detail. In Chapter 8 the macroscopic transport equations for a plasma considered as a mixture of various interpenetrating fluids are derived, whereas the macroscopic transport equations for the whole plasma as a single conducting fluid are developed in Chapter 9. The remaining of the the book is devoted to applications of these basic equations in the description of a variety of important phenomena in plasmas. The problems of electrical conductivity and diffusion in plasmas are analyzed in Chapter 10, and other basic plasma phenomena, such as electron plasma oscillations and Debye shielding, are treated in Chapter 11. Simple applications of the magnetohydrodynamic equations, such as in plasma confinement by magnetic fields and the pinch effect, are presented in Chapters 12 and 13. The subject of wave phenomena in plasmas is organized in the next six chapters. A review of the basic concepts related to the propagation of electromagnetic waves in free space is given in Chapter 14. The propagation of very low frequency waves in a highly conducting fluid is analyzed in Chapter 15, under the title of magnetohydrodynamic waves. The various modes of wave propagation in cold and warm plasmas are considered in Chapters 16 and 17, respectively. In Chapters 18 and 19 the problems of wave propagation in hot nonmagnetized plasmas and in hot magnetized plasmas, respectively, are analyzed. Collision phenomena in plasmas and the derivation of the Boltzmann collision integral and of the Fokker-Planck collision term, are presented in Chapters 20 and 21. Finally, in Chapter 22 some applications of the Boltzmann equation to the analysis of transport phenomena in plasmas are presented.

A number of problems is provided at the end of each chapter, which illustrate additional applications of the theory and supplement the textual material. Most of the problems are designed in such a way as to provide a guideline for the student, including intermediate steps and answers in their statements.

The numbering of the equations starts over again at each section. When reference is made to an equation, using three numbers, the first number indicates the chapter and the last two numbers indicate the section and the equation, respectively. Within the same chapter the first number is omitted. Vectors are represented by an arrow above the corresponding letter and unit vectors by a circunflex above the letter.

The book contains more material than can normally be covered in one semester. This permits some freedom in the selection of topics depending on the level and desired emphasis of the course, and on the interests of the students.

In this, as in any introductory book, the topics included clearly do not cover all areas of plasma physics. No attempt was made to present the experimental aspects of the subject. Moreover, there are some important theoretical topics which are covered only very briefly and some which have been left for more advanced courses on plasma physics, such as plasma instabilities, plasma radiation, nonlinear plasma theory and plasma turbulence.

The system of units used in this text is the rationalized MKSA.

CHAPTER 1
Introduction

1. GENERAL PROPERTIES OF PLASMAS

1.1 Definition of a plasma

The term *plasma* is used to describe a wide variety of macroscopically neutral substances containing many interacting free electrons and ionized atoms or molecules, which exhibit collective behavior due to the long-range Coulomb forces. Not all media containing charged particles, however, can be classified as plasmas. For a collection of interacting charged and neutral particles to exhibit plasma behavior it must satisfy certain conditions, or *criteria*, for plasma existence. These criteria will be discussed in the next section.

The word plasma comes from the Greek and means "something molded". It was applied for the first time by Tonks and Langmuir, in 1929, to describe the inner region, remote from the boundaries, of a glowing ionized gas produced by electric discharge in a tube, the ionized gas as a whole remaining electrically neutral.

1.2 Plasma as the fourth state of matter

From a scientific point of view, matter in the known universe is often classified in terms of four states: *solid*, *liquid*, *gaseous* and *plasma*. The basic distinction between solids, liquids and gases lies in the difference between the strength of the bonds that hold their constituent particles together. These binding forces are relatively strong in a solid, weak in a liquid, and essentially almost absent in the gaseous state. Whether a given substance is found in one of these states depends on the *random kinetic energy* (thermal energy) of its atoms or molecules, that is, on its temperature. The equilibrium between this particle thermal energy and the interparticle binding forces determines the state.

By heating a solid or liquid substance the atoms or molecules acquire more thermal kinetic energy until they are able to overcome the binding potential energy. This leads to *phase transitions*, which occur at a constant temperature for a given pressure. The amount of energy required for the phase transition is called the *latent heat*.

If sufficient energy is provided, a molecular gas will gradually dissociate into an atomic gas as a result of collisions between those particles whose thermal kinetic energy exceeds the molecular binding energy. At sufficiently elevated temperatures an increasing fraction of the atoms will posses enough kinetic energy to overcome, by collisions, the binding energy of the outermost orbital electrons, and an ionized gas or plasma results. However, this transition from a gas to a plasma is not a phase transition in the thermodynamic sense, since it occurs gradually with increasing temperature.

1.3 Plasma production

A plasma can be produced by raising the temperature of a substance until a reasonably high fractional ionization is obtained. Under thermodynamic equilibrium conditions the degree of ionization and the electron temperature are closely related. This relation is given by the Saha equation (Chapter 7). Although plasmas in local thermodynamic equilibrium are found in many places in nature, as is the case for many astrophysical plasmas, they are not very common in the laboratory.

Plasmas can also be generated by ionization processes that raise the degree of ionization much above its thermal equilibrium value. There are many different methods of creating plasmas in the laboratory and, depending on the method, the plasma may have a high or low density, high or low temperature, it may be steady state or transient, stable or unstable, and so on. In what follows, a brief description is presented of the most commonly known processes of *photoionization* and *electric discharge* in gases.

In the *photoionization* process, ionization occurs by absorption of incident photons whose energy is equal to or greater than the ionization potential of the absorbing atom. The excess energy of the photon is transformed into kinetic energy of the electron-ion pair formed. For example, the ionization potential energy for the outermost electron of atomic oxygem is 13.6 eV, which can be supplied by radiation of wavelength smaller than about 910 Å i.e. in the far ultraviolet. Ionization can also be produced by X-rays or gamma-rays, which have much smaller wavelengths. The Earth's ionosphere, for example, is a natural photoionized plasma (see Section 3).

In a *gas discharge*, an electric field is applied across the ionized gas, which accelerates the free electrons to energies sufficiently high to ionize other atoms by collisions. One characteristic of this process is that the applied electric field transfers energy much more efficiently to the light electrons than to the relatively heavy ions. The electron temperature in gas discharges is therefore usually higher than the ion temperature, since the transfer of thermal energy from the electrons to the heavier particles is very slow.

When the ionizing source is turned off, the ionization decreases gradually because of recombination until it reaches an equilibrium value consistent with the temperature of the medium. In the laboratory the recombination usually occurs so fast that the plasma completely disappears in a small fraction of a second.

1.4 Particle interactions and collective effects

The properties of a plasma are markedly dependent upon the particle interactions. One of the basic features which distinguish the behavior of plasmas from that of ordinary fluids and solids is the existence of *collective effects*. Due to the long-range of electromagnetic forces, each charged particle in the plasma interacts *simultaneously* with a considerable number of other charged particles, resulting in important collective effects which are responsible for the wealth of physical phenomena that take place in a plasma.

The particle dynamics in a plasma is governed by the internal fields due to the nature and motion of the particles themselves, and by externally applied fields. The basic particle interactions are electromagnetic in character. Quantum effects are negligible, except for some cases of close collisions. In a plasma we must distinguish between charge-charge and charge-neutral interactions. A charged particle is surrounded by an electric field and interacts with the other charged particles according to the Coulomb force law, with its dependence on the inverse of the square of the separation distance. Furthermore, a magnetic field is associated with a moving charged particle, which also produces a force on other moving charges. The charged and neutral particles interact through electric polarization fields produced by distortion of the neutral particle's electronic cloud during a close passage of the charged particle. The field associated with neutral particles involves short-range forces, such that their interaction is effective only for interatomic distances sufficiently small to perturb the orbital electrons. It is appreciable when the distance between the centers of the interacting particles is of the order of their diameter, but nearly zero when they are farther apart. Its characteristics can be adequatelly described only by quantum-mechanical considerations. In many cases this interaction involves permanent or induced electric dipole moments.

A distinction can be made between *weakly ionized* and *strongly ionized plasmas* in terms of the nature of the particle interactions. In a weakly ionized plasma the charge-neutral interactions dominate over the multiple Coulomb interactions. When the degree of ionization is such that the multiple Coulomb interactions become dominant, the plasma is considered strongly ionized. As the degree of ionization increases, the Coulomb interactions become increasingly important so that in a fully ionized plasma all particles are subjected to the multiple Coulomb interactions.

1.5 Some basic plasma phenomena

The fact that some or all of the particles in a plasma are electrically charged and therefore capable of interacting with electromagnetic fields, as well as creating them, gives rise to many novel phenomena that are not present in ordinary fluids and solids. The presence of the magnetic field used, for example, in the heating and confinement of plasmas in controlled thermonuclear research greatly accentuates the novelty of plasma phenomena. To explore all features of plasma phenomena, the plasma behavior is usually studied in the presence of both electric and magnetic fields.

Because of the high electron mobility, plasmas are generally very good *electrical conductors*, as well as good *thermal conductors*. As a consequence of their high electrical conductivity they do not support electrostatic fields except, to a certain extent, in a direction normal to any magnetic field present, which inhibits the flow of charged particles in this direction.

The presence of density gradients in a plasma causes the particles to diffuse from dense regions to regions of lower density. Although the diffusion problem in nonmagnetized plasmas is somewhat similar to that which occurs in ordinary fluids, there is nevertheless a fundamental difference. Because of their lower mass, the electrons tend to diffuse faster than the ions, generating a polarization electric field as a result of charge separation. This field enhances the diffusion of the ions and decreases that of the electrons, in such a way as to make ions and electrons diffuse at approximately the same rate. This type of diffusion is called *ambipolar diffusion*. When there is an externally applied magnetic field, the diffusion of charged particles across the field lines is reduced, which indicates that strong magnetic fields are helpful in plasma confinement. The diffusion of charged particles across magnetic field lines when the diffusion coefficient is proportional to $1/B^2$, where B denotes the magnitude of the magnetic induction, is called *classical diffusion*, in contrast to *Bohm diffusion* in which the diffusion coefficient is proportional to $1/B$.

An important characteristic of plasmas is their ability to sustain a great variety of *wave phenomena*. Examples include high-frequency transverse *electromagnetic waves* and longitudinal *electrostatic plasma waves*. In the low frequency regime important wave modes in a magnetized plasma are the so called *Alfvén waves* and *magnetosonic waves*. Each of the various possible modes of wave propagation can be characterized by a *dispersion relation*, which is a functional relation between the wave frequency ω and the wave number k, and by its *polarization*. The study of waves in plasmas provides significant information on plasma properties and is very useful in plasma diagnostics.

Dissipative processes, such as collisions, produce damping of the wave amplitude. This means that energy is transferred from the wave field to the plasma particles. An essentially non-collisional mechanism of wave attenuation also exists in a plasma, the so called *Landau damping*. The mechanism responsible for Landau damping is the trapping of some plasma particles (the ones which are moving with velocities close to the wave phase velocity) in the energy potential well of the wave, the net result being the transfer of energy from the wave to the particles. It is also possible to have modes with growing amplitudes, as a result of *instabilities*, which transfer energy from the plasma particles to the wave field. Instability phenomena are important in a wide variety of physical situations involving dynamical processes in plasmas. The existence of many different types of instabilities in a plasma greatly complicates the confinement of a hot plasma in the laboratory. The study of these instabilities is of essential importance for controlled thermonuclear fusion research.

Another important aspect of plasma behavior is the emission of *radiation*. The main interest in plasma radiation lies in the fact that it can be used to measure plasma properties. The mechanisms that cause plasmas to emit or absorb radiation can be grouped into two categories: radiation from emitting atoms or molecules, and radiation from accelerated charges. At the same time that ionization is produced in a plasma, the opposite process, *recombination* of the ions and electrons to form neutral particles, is normally also ocurring. As a result of the recombination process, radiation is often emitted as the excited particles formed during recombination decay to the ground state. This radiation constitutes the *line spectra* of plasmas. On the other hand, any accelerated charged particle emits radiation. The radiation emitted whenever a charged particle is decelerated by making some kind of collisional interaction is called *bremsstrahlung*. If the charged particle remains unbound, both before and after the encounter, the process is called *free-free bremsstrahlung*. Radiation of any wavelength can be emitted or absorbed in bremsstrahlung. If the originally unbound charged particle is captured by another particle, as it emits the radiation, the process is called *free-bound radiation*. *Cyclotron radiation*, which occurs in magnetized plasmas, is due to the magnetic centripetal acceleration of the charged particles as they spiral about the magnetic field lines. *Blackbody radiation* from plasmas in thermodynamic equilibrium is important only in astrophysical plasmas, in view of the large size needed for a plasma to radiate as a blackbody.

2. CRITERIA FOR THE DEFINITION OF A PLASMA

2.1 Macroscopic neutrality

In the absence of external disturbances a plasma is *macroscopically neutral*. This means that under equilibrium with no external forces present, in a volume of the plasma sufficiently large to contain a large number of particles and yet sufficiently small compared with the characteristic lengths for variation of macroscopic parameters such as density and temperature, the net resulting electric charge is zero. In the interior of the plasma the microscopic space charge fields cancel each other and no net space charge exists over a macroscopic region.

If this macroscopic neutrality was not maintained, the potential energy associated with the resulting Coulomb forces could be enormous compared to the thermal particle kinetic energy. Consider, for example, a plasma with a charged particle number density of $10^{20} m^{-3}$ and suppose that the electron density (n_e) in a spherical volume of 10^{-3}m radius (r) were to differ by one per cent from the density of the positive ions (n_i). Denoting the ion charge by e and the electron charge by -e, the total net charge inside the sphere would be

$$q = (4\pi r^3/3)\ (n_i - n_e)e \qquad (2.1)$$

and the electric potential at the surface of the sphere would be

$$\phi = \frac{1}{4\pi\varepsilon_0}\frac{q}{r} = \frac{r^2(n_i - n_e)e}{3\varepsilon_0} \tag{2.2}$$

where ε_0 is the permittivity of free space. Plugging numerical values into (2.2) yields $\phi = 6 \times 10^3$ Volts. Recalling that 1 eV = 1.602×10^{-19} joule, we find that kT = 1 eV when T = 11 600K, where k is Boltzmann's constant (1.380×10^{-23} joule/K). Therefore, a plasma temperature of several millions of degrees Kelvin would be required to balance the electric potential energy with the average thermal particle energy.

Departures from macroscopic electrical neutrality can naturally occur only over distances in which a balance is obtained between the thermal particle energy, which tends to disturb the electrical neutrality, and the electrostatic potential energy resulting from any charge separation, which tends to restore the electrical neutrality. This distance is of the order of a characteristic length parameter of the plasma, called the *Debye length*. In the absence of external forces, the plasma cannot support departures from macroscopic neutrality over larger distances than this, since the charged particles are able to move freely to neutralize any regions of excess space charge in response to the large Coulomb forces that appear.

2.2 Debye shielding

The Debye length is an important physical parameter for the description of a plasma. It provides a measure of the distance over which the influence of the electric field of an individual charged particle (or of a surface at some non-zero potential) is felt by the other charged particles inside the plasma. The charged particles arrange themselves in such a way as to effectively shield any electrostatic fields within a distance of the order of the Debye length. This shielding of electrostatic fields is a consequence of the collective effects of the plasma particles. A calculation of the shielding distance was first performed by Debye, for an electrolyte. In Chapter 11 it will be shown that the Debye length, λ_D, is directly proportional to the square root of the temperature T and inversely proportional to the square root of the electron number density, n_e, according to

$$\lambda_D = [\varepsilon_0 kT/(n_e e^2)]^{1/2} \tag{2.3}$$

As mentioned before, the Debye length can also be regarded as a measure of the distance over which fluctuating electric potentials may appear in a plasma, corresponding to a conversion of the thermal particle kinetic energy into electrostatic potential energy.

When a boundary surface is introduced in a plasma the perturbation produced extends only up to a distance of the order of λ_D from the surface. In the neighborhood of any surface inside the plasma there is a layer of width of the order of λ_D, known as the *plasma sheath*, inside which the condition of macroscopic electrical neutrality need not be satisfied. Beyond the plasma sheath region there is the plasma region where macroscopic neutrality is maintained.

Generally λ_D is very small. For example, in a gas discharge where typical values for T and n_e are around 10^4K and 10^{16} electrons/m^3, respectively, we have $\lambda_D = 10^{-4}$m. For the Earth's ionosphere typical values can be taken as $n_e = 10^{12}$ electrons/m^3 and T = 1000K, which give $\lambda_D = 10^{-3}$m. In the interstellar plasma, on the other hand, the Debye length can be several meters long.

It is convenient to define a *Debye sphere* as a sphere inside the plasma of radius equal to λ_D. Any electrostatic fields originated outside a Debye sphere are effectively screened by the charged particles and do not contribute significantly to the electric field existing at its center. Consequently, each charge in the plasma interacts collectively only with the charges that lie inside its Debye sphere, its effect on the other charges being effectively negligible. The number of electrons N_D, inside a Debye sphere, is given by

$$N_D = \frac{4}{3}\pi\lambda_D^3 n_e = \frac{4\pi}{3}\left(\frac{\varepsilon_0 kT}{n_e^{1/3} e^2}\right)^{3/2} \tag{2.4}$$

The Debye shielding effect is a characteristic of all plasmas, although it does not occur in every media that contains charged particles. A necessary and obvious requirement for the existence of a plasma is that the physical dimensions of the system be large compared to λ_D. Otherwise there is just not sufficient space for the collective shielding effect to take place and the collection of charged particles will not exhibit plasma behavior. If L is a characteristic dimension of the plasma, the *first criterion* for the definition of a plasma is therefore

$$L \gg \lambda_D \tag{2.5}$$

Since the shielding effect is the result of the *collective* particle behavior inside a Debye sphere, it is also necessary that the number of electrons inside a Debye sphere be very large. A *second criterion* for the definition of a plasma is therefore

$$n_e \lambda_D^3 \gg 1 \tag{2.6}$$

This means that the average distance between electrons, which is roughly given by $n_e^{-1/3}$, must be very small compared to λ_D. The quantity defined by

$$g = 1/(n_e \lambda_D^3) \tag{2.7}$$

is known as the *plasma parameter* and the condition $g \ll 1$ is called the *plasma approximation*. This parameter is also a measure of the ratio of the mean interparticle potential energy to the mean plasma kinetic energy.

Note that the requirement (2.5) already implies in macroscopic charge neutrality if it is realized that deviations from neutrality can naturally occur only over distances of the order of λ_D. Nevertheless, macroscopic neutrality is sometimes considered as a *third criterion* for the existence of a plasma, although it is *not an independent one*, and can be expressed as

$$n_e = \sum_i n_i \tag{2.8}$$

2.3 The plasma frequency

An important plasma property is the stability of its macroscopic space charge neutrality. When a plasma is instantaneously disturbed from the equilibrium condition, the resulting internal space charge fields give rise to collective particle motions which tend to restore the original charge neutrality. These collective motions are characterized by a natural frequency of oscillation known as the *plasma frequency*. Since these collective oscillations are high frequency oscillations, the ions, because of their heavy mass, are to a certain extent unable to follow the motion of the electrons. The electrons oscillate collectively about the heavy ions, the necessary collective restoring force being provided by the ion-electron Coulomb attraction. The period of this natural oscillation constitutes a meaningful time scale against which the dissipative mechanisms, tending to destroy the collective electron motions, can be compared.

Consider a plasma initially uniform and at rest, and suppose that by some external means a small charge separation is produced inside it (Fig. 1). When the external disturbing force is removed instantaneously, the internal electric field resulting from charge separation collectively accelerates the electrons in an attempt to restore the charge neutrality. However, because of their inertia, the electrons keep moving beyond the equilibrium position, and an electric field is produced in the opposite direction. This sequence of movements repeats itself periodically, with a continuous transformation of kinetic energy into potential energy, and vice-versa, resulting in fast collective oscillations of the electrons about the more massive ions. On the *average* the plasma maintains its macroscopic charge neutrality.

In will be shown in Chapter 11 that the angular frequency of this collective electron oscillation, called the *(electron) plasma frequency*, is given by

$$\omega_{pe} = [n_e e^2/(m_e \varepsilon_o)]^{1/2} \tag{2.9}$$

Collisions between electrons and neutral particles tend to damp these collective oscillations and gradually diminish their amplitude. If the oscillations are to be only slightly damped, it is necessary that the electron-neutral collision frequency, ν_{en}, be smaller than the electron plasma frequency,

$$\nu_{pe} > \nu_{en} \tag{2.10}$$

where $\nu_{pe} = \omega_{pe}/2\pi$. Otherwise, the electrons will not be able to behave in an independent way, but will be forced by collisions to be in complete equilibrium with the neutrals, and the medium can be treated as a neutral gas. Eq. (2.10) constitutes therefore the *fourth criterion* for the existence of a plasma. This criterion can be alternatively written as

$$\omega\tau > 1 \tag{2.11}$$

where $\tau = 1/\nu_{en}$ represents the average time an electron travels between collisions with neutrals, and ω stands for the angular frequency of typical plasma oscillations. It implies that the average time between electron-neutral collisions must be large compared to the characteristic time during which the plasma physical parameters are changing.

Consider for example a gas containing, say, 10^{10} electrons/m³ at a temperature of 1000K, which satisfies both criteria $L >> \lambda_D$ and $n_e\lambda_D^3 >> 1$. If the density of the neutral particles (n_n) is relatively small, as in the interstellar gas for example, τ is relatively large and the electrons will behave independently, so that the medium can then be treated as a plasma. On the other hand, if n_n is many orders of magnitude greater than n_e, then the motion of the electrons will be coupled to that of the neutrals and their effect will be negligible.

The basic characteristics of various laboratory and cosmic plasmas are given in Fig. 2 in terms of their temperature T and electron density n_e, a well as of parameters which depend upon T and n_e, such as the Debye shielding distance λ_D, the electron plasma frequency ω_{pe}, and the number of electrons N_D inside a Debye sphere.

3. THE OCCURRENCE OF PLASMAS IN NATURE

With the progress made in astrophysics and in theoretical physics during this century it was realized that most of the matter in the known universe, with a few exceptions such as the surfaces of cold planets (the Earth, for example) exists as a plasma.

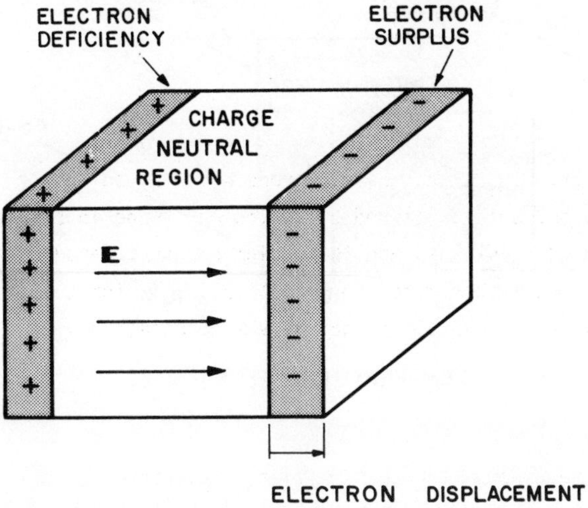

Fig. 1 - The electric field resulting from charge separation provides the force which generates the electron plasma oscillations.

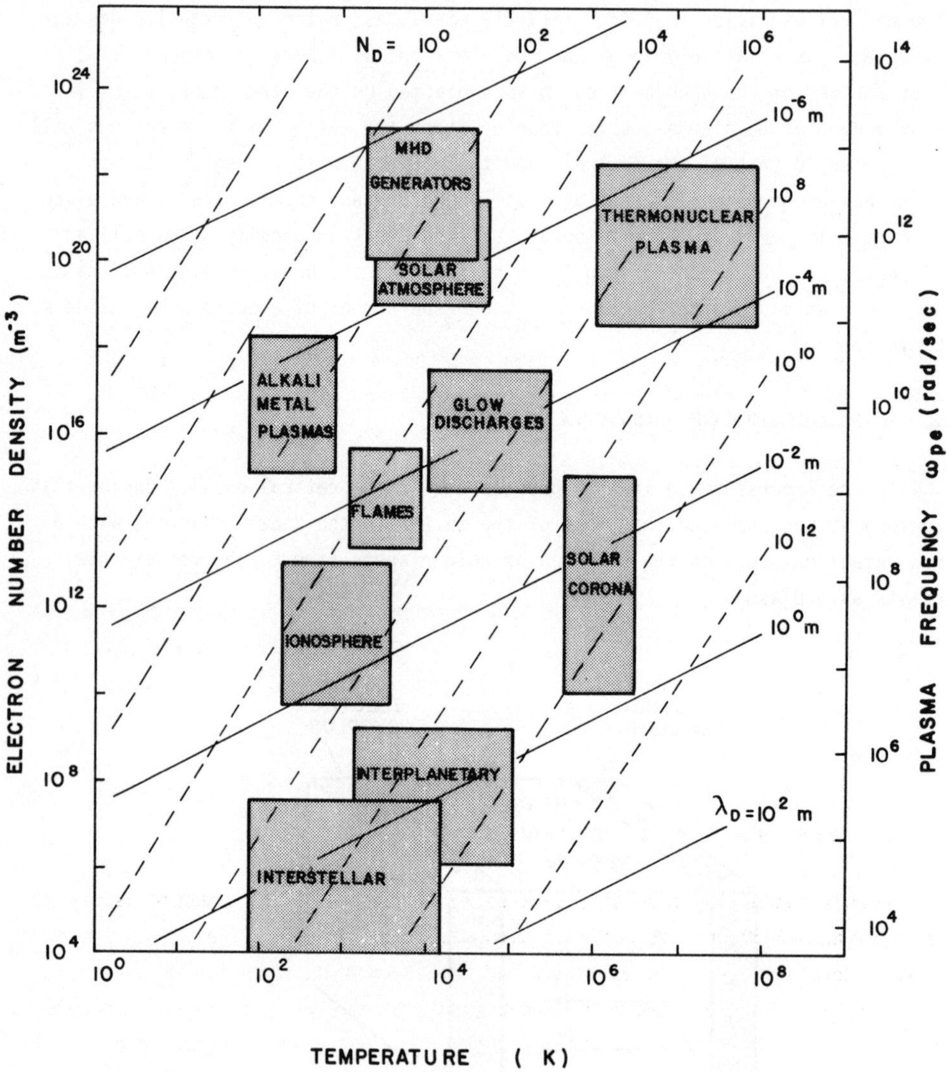

Fig. 2 - Ranges of temperature and electron density for several laboratory and cosmic plasmas and their characteristic physical parameters: Debye length λ_D, plasma frequency ω_{pe} and number of electrons N_D in a Debye sphere.

3.1 The Sun and its atmosphere

The *Sun*, which is our nearest star and upon which the existence of life on Earth fundamentally depends, is a plasma phenomenon. Its energy output is derived from thermonuclear fusion reactions of protons forming helium ions deep in its interior, where temperatures exceed 12 millions K. The high temperature of its interior and the consequent thermonuclear reactions keep the entire Sun gaseous. Due to its large mass (2×10^{30}Kg), the force of gravity is sufficient to prevent the escape of all but the most energetic particles, and of course radiation, from the hot solar plasma.

There is no sharp boundary surface to the Sun. Its visible part is known as the solar atmosphere, which is divided into three general regions or layers. The *photosphere*, with a temperature of about 6000K, comprises the visible disk, the layer in which the gases become opaque, and is a few hundred kilometers thick. Surrounding the photosphere there is a redish ring called the *chromosphere*, approximately ten thousand kilometers thick, above which flame-like proeminences rise with temperatures of the order of 100,000K. Surrounding the chromosphere there is a tenuous hot plasma, extending millions of kilometers into space, known as the *corona*. A steep temperature gradient extends from the chromosphere to the hotter corona, where the temperature exceeds 1 million K.

The Sun possesses a magnetic field which at its surface is of the order of 10^{-4} Tesla, but in the regions of *sunspots* (regions of cooler gases) the magnetic field rises to about 0.1 Tesla.

3.2 The solar wind

A highly conducting tenuous plasma called the *solar wind*, composed mainly of protons and electrons, is continuously emitted by the Sun at very high speeds into interplanetary space, as a result of the supersonic expansion of the hot solar corona. The solar magnetic field tends to remain frozen in the streaming plasma and, because of solar rotation, the field lines are carried into Archimedean spirals by the radial motion of the solar wind (Fig. 3). Typical values of the parameters in the solar wind are $n_e \cong 5 \times 10^6 m^{-3}$, $T_i \cong 10^4 K$, $T_e \cong 5 \times 10^4 K$, $B \cong 5 \times 10^{-9}$ Tesla and drift velocity 3×10^5m/s.

3.3 The magnetosphere and the Van Allen radiation belts

As the highly conducting solar wind impinges on the Earth's magnetic field it compresses the field on the sunward side and flows around it at supersonic speeds. This creates a boundary, called the *magnetopause*, which is roughly spherical on the sunward side and roughly cylindrical in the anti-sun direction (Fig. 4). The inner region, from which the solar wind is excluded and which contains the compressed Earth's magnetic field, is called the *magnetosphere*.

Inside the magnetosphere we find the *Van Allen radiation belts*, in which energetic charged particles (mainly electrons and protons) are trapped in regions where they execute complicated trajectories which spiral along the geomagnetic field lines and at the same time drift slowly around the Earth. The origin of the inner belt is ascribed to cosmic rays, which penetrate into the atmosphere and form proton-electron pairs which are then trapped by the Earth's magnetic field. The outer belt is considered to be due to and maintained by streams of plasma consisting mainly of protons and electrons which are ejected from time to time by the Sun. Depending on solar activity, particularly violent solar eruptions may occur with the projection of hot streams of plasma material into space. The separation into inner and outer belts reflects only an altitude-dependent energy spectrum, rather than two separate trapping regions.

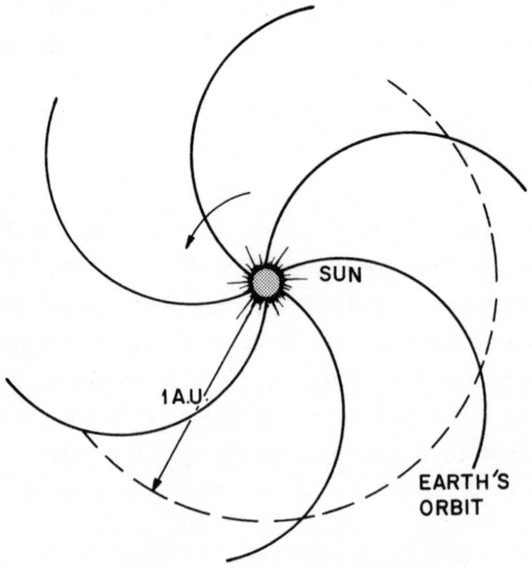

Fig. 3 - Schematic representation of the Archimedes spiral structure of the interplanetary magnetic field in the ecliptic plane.

3.4 The ionosphere

The large natural blanket of plasma in the atmosphere, which envelopes the Earth from an altitude of approximately 60 km to several thousands of kilometers, is called the *ionosphere*. The ionized particles in the ionosphere are produced during the daytime through absorption of solar extreme ultraviolet and X-ray radiation by the atmospheric species. As the ionizing

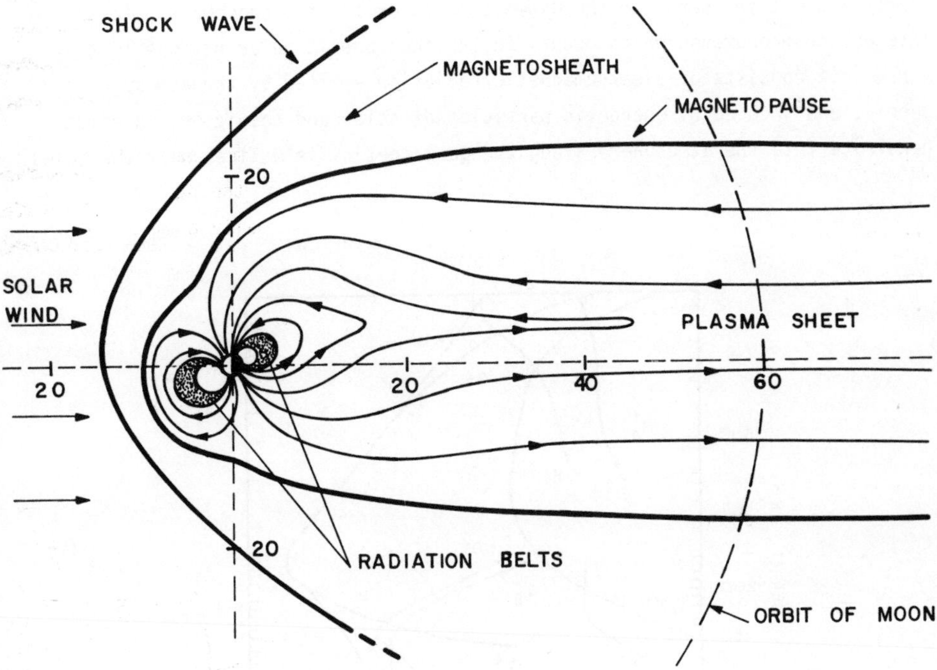

Fig. 4 - Schematic configuration of the magnetosphere in the noon-midnight plane. The dark crescents represent the regions of trapped energetic particles (Van Allen radiation belts). The turbulent region between the shock wave (bow shock) and the magnetopause is known as the magnetosheath. Geocentric distances are indicated in units of Earth radii.

radiation from the Sun penetrates deeper and deeper into the Earth's atmosphere it encounters a larger and larger density of gas particles, producing more and more electrons per unit volume. However, since radiation is absorbed in this process, there is a height where the rate of electron production reaches a maximum. Below this height the rate of electron production decreases, in spite of the increase in atmospheric density, since most of the ionizing radiation was already absorbed at the higher altitudes.

Fig. 5 provides some idea of the relative concentration and altitude distribution of the electrons and the principal positive ions, typical of the daytime ionosphere during solar minimum conditions. The Earth's magnetic field exerts a great influence on the dynamic behavior of the ionospheric plasma. An interesting phenomenon which occurs in the ionospheric polar regions is the *aurora*. It consists of electromagnetic radiation emitted by the atmospheric species and induced by energetic particles of solar and cosmic origin which penetrate into the atmosphere along the geomagnetic field lines near the poles.

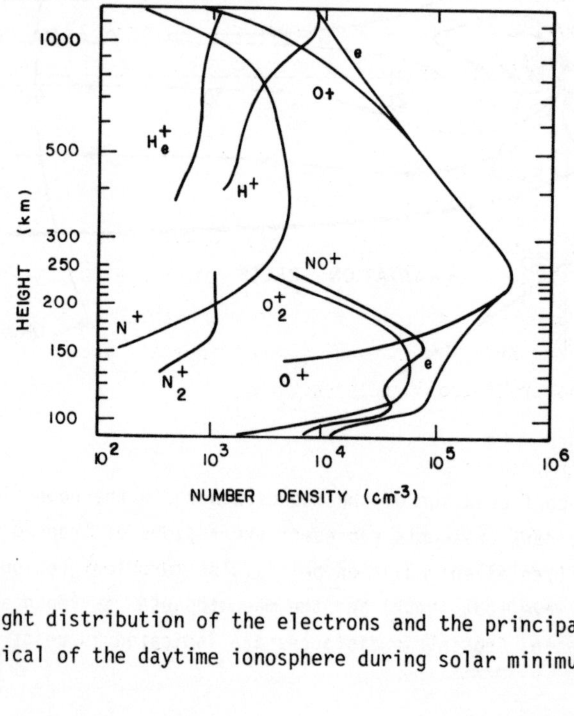

Fig. 5 - Height distribution of the electrons and the principal positive ions, typical of the daytime ionosphere during solar minimum conditions.

3.5 Plasmas beyond the solar system

Beyond the solar system we find a great variety of natural plasmas in stars, interstellar space, galaxies, intergalactic space, and far beyond to systems quite unknown before the start of astronomy from space vehicles. There we find a variety of phenomena of great cosmological and astrophysical significance: interstellar shock waves from remote supernova explosions, rapid variations of X-ray fluxes from neutron stars with densities like that of atomic nuclei, pulsing radio stars or pulsars (which are theoretically pictured as rapidly rotating neutron stars with plasmas emitting synchrotron radiation from the surface) and the plasma phenomena around the remarkable "black holes" (which are considered to be singular regions of space into which matter has collapsed, and possessing such a powerful gravitational field that nothing, whether material objects or even light itself, can escape from them).

The behavior of plasmas in the universe involves the interaction between plasmas and magnetic fields. The Crab nebula is a rich source of plasma phenomena because it contains a magnetic field. The widespread existence of magnetic fields in the universe has been demonstrated by independent measurements, and a wide range of field magnitudes has been found, varying from 10^{-9} Tesla in interstellar space to 1 Tesla on the surface of magnetic variable stars.

4. APPLICATIONS OF PLASMA PHYSICS

A wide variety of plasma experiments have been performed in the laboratory to aid in the understanding of plasmas, as well as to test and help expand plasma theory. The progress in plasma research has lead to a wide range of plasma applications. A brief exposition of some important practical applications of plasma physics is presented in this section.

4.1 Controlled thermonuclear fusion

The most important application of man-made plasmas is in the control of thermonuclear fusion reactions, which holds a vast potential for the generation of power. Nuclear fusion is the process whereby two light nuclei combine to form a heavier one, the total final mass being slightly less than the total initial mass. The mass difference (Δm) appears as energy (E) according to

Einstein's famous law $E = (\Delta m)c^2$, where c is the speed of light. The nuclear fusion reaction is the source of energy in the stars, including the Sun. The confinement of the hot plasma in this case is provided by the self-gravity of the stars.

In the nuclear fusion of hydrogen the principal reactions involve the deuterium (^2H) and tritium (^3H) isotopes of hydrogen, as follows:

$$^2H + {}^2H \Longrightarrow \begin{cases} {}^3He + {}^1n + 3.27 \text{ Mev} & (4.1a) \\ {}^3H + {}^1H + 4.03 \text{ Mev} & (4.1b) \end{cases}$$

$$^2H + {}^3H \Longrightarrow {}^4He + {}^1n + 17.58 \text{ Mev} \tag{4.1c}$$

$$^2H + {}^3He \Longrightarrow {}^4He + {}^1H + 18.34 \text{ Mev} \tag{4.1d}$$

where ^1n represents a neutron. The basic problem in achieving controlled fusion is to generate a plasma at very high temperatures (with thermal energies at least in the 10 keV range) and hold its particles together long enough for a substantial number of fusion reactions to take place. The need for high temperatures comes from the fact that, in order to undergo fusion, the positively charged nuclei must come very close together (within a distance of the order of 10^{-14}m), which requires sufficient kinetic energy to overcome the electrostatic Coulomb repulsion.

Fig. 6 presents the cross sections, as a function of the incident particle energy, for the nuclear fusion reactions of hydrogen given in (4.1). They are appreciable only for incident particles with energies above at least 10 keV. This means that the plasma must have temperatures of the order of 100 millions K. Other fusion reactions involving nuclei with larger values of the atomic number Z require even higher energies to overcome the Coulomb repulsion.

Many confinement schemes have been suggested and built which use some type of magnetic field configuration. The main experimental efforts for achieving plasma conditions for fusion can be grouped into four approaches: (1) open systems (magnetic mirrors); (2) closed systems (toruses); (3) theta pinch devices; and (4) laser-pellet fusion.

The *mirror machines* are linear devices with an axial magnetic field to keep the particles from the wall, and with magnetic mirrors (regions of converging magnetic field lines) at the ends to reduce the number of particles escaping at each end (Fig. 7). The four principal *toroidal systems* differ in the way they twist the magnetic field lines. They are the stellarators (in which the twisting of the field lines is produced by external helical conductors), the tokamaks (in which a poloidal field produced by an internal plasma current is superposed on the toroidal field), the multipoles (which have their magnetic field lines primarily in the poloidal direction and produced by internal conductors), and the Astron (in which internal relativistic particle beams modify a mirror field into a form having stable confinement regions with closed lines of force). In the *theta pinch devices* a plasma current in the azimuthal (θ) direction and a longitudinal magnetic field produce a force which compresses the cross-sectional area of the plasma. Finally, the scheme to ignite a fusion reaction using *pulsed lasers* consists in focusing converging laser beams on a small pellet of solid deuterium-tritium material producing a rapid symmetrical heating of the plasma, followed by an expansion of the heated surrounding shell and compression of the pellet core by the recoil (Fig. 8).

In addition to the heating and confinement problems attention must be given to the energy loss by radiation (predominantly electron-ion bremsstrahlung and electron cyclotron radiation). These radiation losses constitute a serious factor in maintaining a self-sustaining fusion device. To generate more energy by fusion than is required to heat and confine the plasma, and to supply the

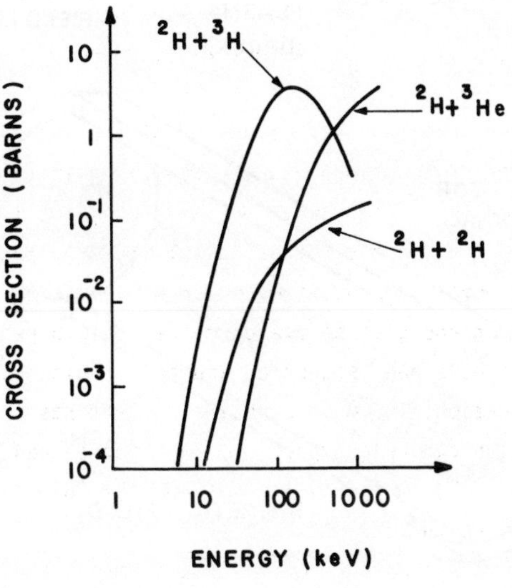

Fig. 6 - Fusion cross sections, in barns (1 barn = $10^{-28}m^2$), as a function of energy, in keV, for the hydrogen reactions given in (4.1).

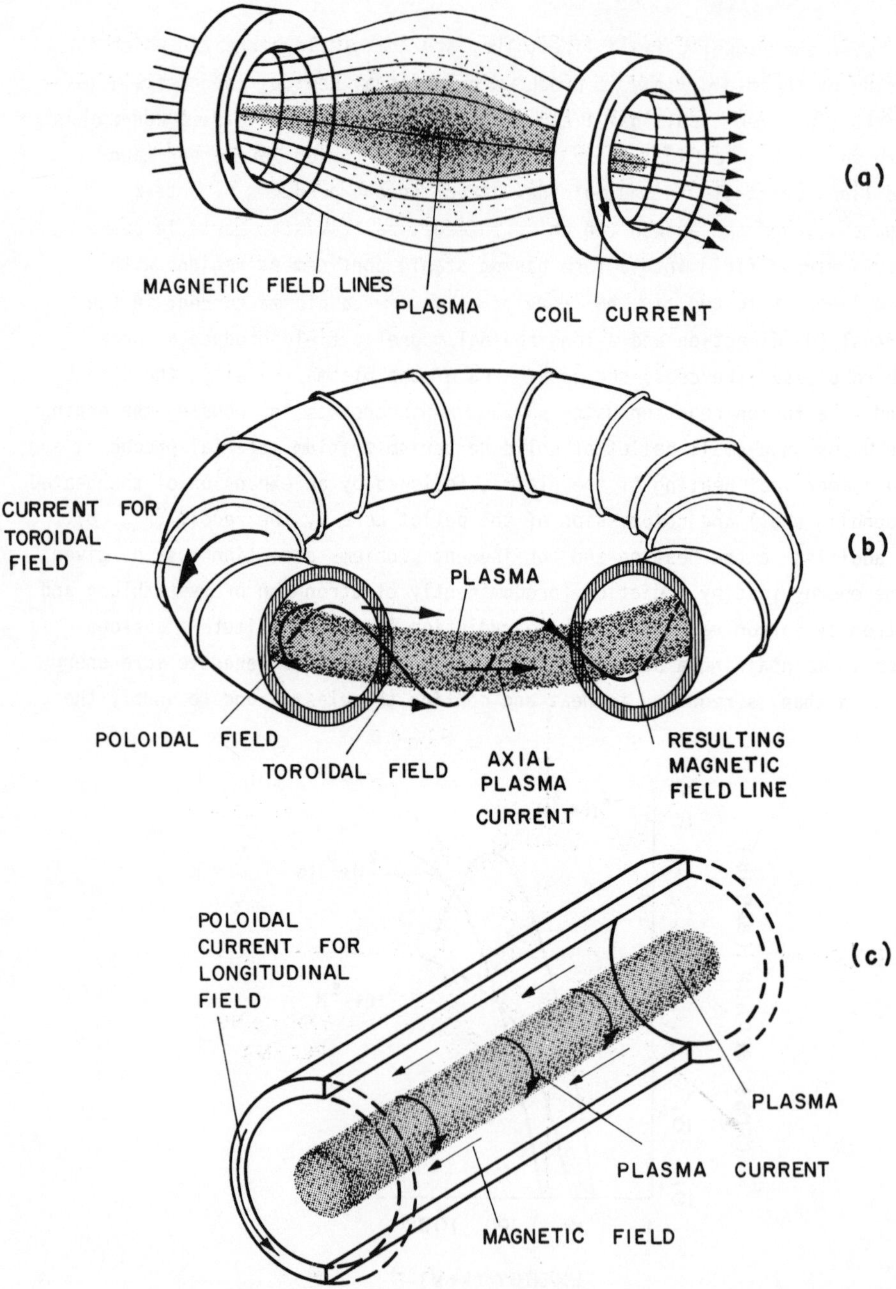

Fig. 7 - Schematic illustration showing the magnetic field configurations of some basic schemes for plasma confinement. (a) Magnetic mirror system. (b) Tokamak. (c) Linear θ-pinch.

radiation losses, a condition is imposed on the plasma density, n, and the confinement time, τ, as well as on the temperature. It turns out that the product $n\tau$ must be higher than a minimum value which, for example, is estimated to be about 10^{20} m^{-3} s for deuterium-tritium (with $T > 10^7 K$) and about 10^{22} m^{-3} s for deuterium-deuterium (with $T > 10^8 K$). This condition is known as the *Lawson criterion*. Consequently, controlled fusion can be achieved either by having a large number density of hot plasma particles confined for a short period of time, or by having a smaller number density of particles confined for a longer period of time. For this reason some fusion experiments operate in the regime of high density and short confinement time utilizing a pulsed mode of operation.

Since controlled nuclear fusion can provide an almost limitless source of energy, it is certainly one of the most important scientific challenges man faces today, and its achievement will cause an enourmous impact on our civilization.

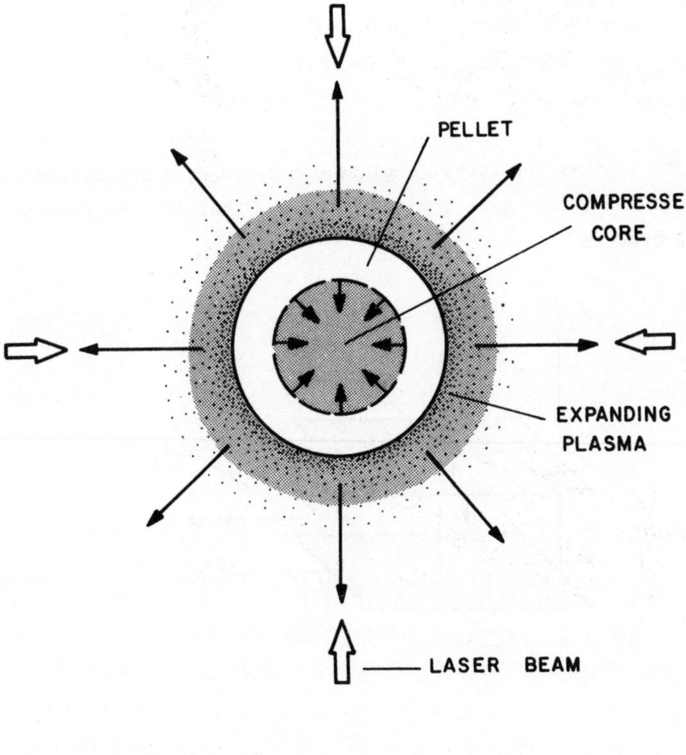

Fig. 8 - Illustrating laser-pellet fusion.

4.2 The magnetohydrodynamic generator

The magnetohydrodynamic (MHD) energy generator converts the kinetic energy of a dense plasma flowing across a magnetic field into electrical energy. While a rigorous discussion of this device becomes quite involved, its basic principle is quite simple.

Suppose (Fig. 9) that a plasma flows with velocity $\vec{u}$ (along the x direction) across an applied magnetic field $\vec{B}$ (in the y direction). The Lorentz force $q(\vec{u} \times \vec{B})$ causes the ions to drift upward (in the z direction) and the electrons downward so that if electrodes are placed in the walls of the channel and connected to an external circuit then a current density $\vec{J} = \sigma\vec{E}_{ind} = \sigma\vec{u} \times \vec{B}$ (where σ denotes the plasma conductivity and $\vec{E}_{ind}$ is the induced electric field) flows across the plasma stream in the z direction. This current density, in turn, produces a force $\vec{J} \times \vec{B}$ (in the x direction) which decelerates the flowing plasma. The net result is the conversion of some of the plasma kinetic energy entering the generator into electrical energy that can be applied to an external load. This process has the advantage that it operates without the inefficiency of a heat cycle.

4.3 Plasma propulsion

Plasma propulsion systems for rocket engines are based on a process which converts electrical energy into plasma kinetic energy, that is, the reverse of the MHD generator process.

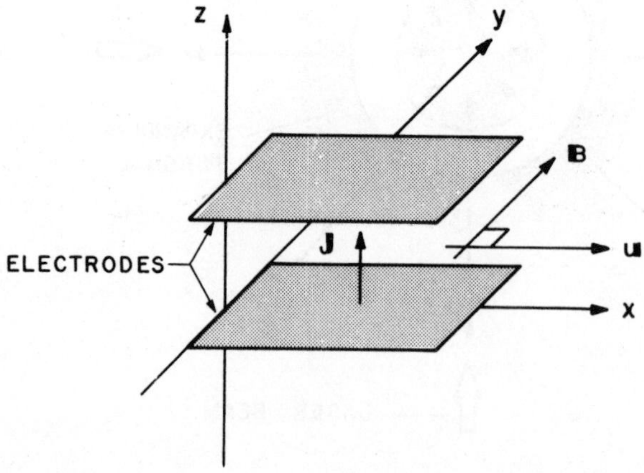

Fig. 9 - Schematic diagram illustrating the basic principle of the MHD energy generator.

The plasma rocket engine is accomplished by having both electric and magnetic fields applied perpendicular to each other, across a plasma (Fig. 10). The resulting current density $\vec{J}$ flowing in the direction of the applied $\vec{E}$ field gives rise to a $\vec{J} \times \vec{B}$ force which accelerates the plasma out of the rocket. The associated reaction force, due to conservation of momentum, accelerates the rocket in the direction opposite to the plasma flow. The ejected plasma must always be neutral, otherwise the rocket will become charged to a large potential.

An important characteristic of plasma propulsion systems is that they are capable of generating a certain amount of thrust (although small) over a very long time period, contrarily to chemical propulsion systems. Since the force the plasma rocket engine provides is too modest to overcome the Earth's gravitational field, chemical rockets must still be used as the first stage of any plasma propulsion system in order to produce the extremely high values of thrust required to leave the Earth's gravity. The plasma rocket engine is appropriate for long interplanetary and interstellar space travel.

4.4 Other plasma devices

A number of other practical applications of plasma physics should be mentioned in addition to controlled fusion, MHD energy conversion and plasma propulsion.

The *thermionic energy converter* is a device which utilizes a cesium plasma between two electrodes to convert thermal energy into electrical energy. The

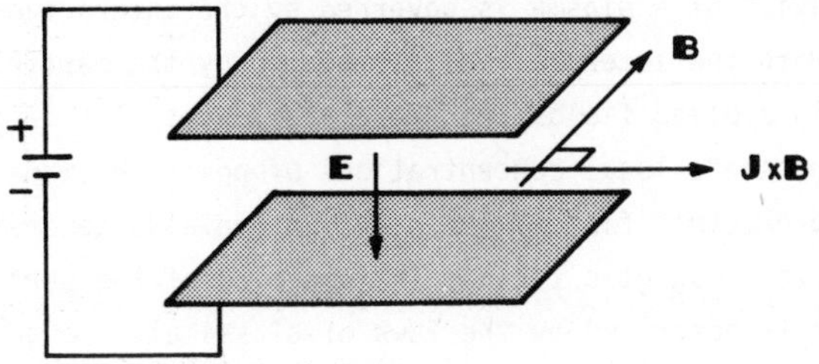

Fig. 10 - Schematic diagram illustrating the basic principle of the plasma rocket engine.

cathode is heated so that electrons are emitted from the surface, and the anode is cooled. Due to the presence of the cesium plasma, very large electrical currents can be produced at the expense of a significant fraction of the thermal energy applied to the cathode.

Examples of applications involving *gas discharges* include the ordinary fluorescent tubes and neon lights used for illumination and for signs, mercury rectifiers, spark gaps, a number of specialized tubes like the hydrogen thyratrons and the ignitrons, which are used for switching, and the arc discharges or plasma jets, which are the source of temperatures two or more times as high as the hottest gas flames and which are used in metallurgy for cutting, melting and welding metals.

Two major applications in the area of *communications* are the long-distance radio wave propagation by reflection in the ionospheric plasma and the communication with a space vehicle through the plasma layer which forms around it during the reentry period into the Earth's atmosphere.

Finally, there is the realm of *solid state plasmas*. If the usual lattice temperature is considered, it can be easily verified that solids do not satify the plasma shielding criterion $N_D >> 1$. Nevertheless, quantum mechanical effects associated with the uncertainty principle give some solids an effective electron temperature high enough to make N_D sufficiently large, so that plasma behavior can be observed. It has been demonstrated that the free electrons and holes in appropriate solid materials, particularly semiconductors, exhibit the same sort of oscillations and instabilities as gaseous plasmas. The most likely application of solid state plasmas is in electronic circuitry.

5. THEORETICAL DESCRIPTION OF PLASMA PHENOMENA

The dynamic behavior of a plasma is governed by the interactions of the plasma particles with the internal fields produced by the particle themselves and with externally applied fields. As the charged particles in a plasma move around, they can generate local concentrations of positive or negative charge, which give rise to electric fields. Their motion can also generate electric currents and therefore magnetic fields. The dynamics of the particles in a plasma is adequately described by the laws of classical (nonquantum) mechanics. Generally, the momentum of the plasma particles is high and the density low enough to keep their De Broglie wavelengths much smaller than the interparticle distance. Quantum effects turn out to be important only at very high densities and very low temperatures.

5.1 General considerations on a self-consistent formulation

The interaction of charged particles with electromagnetic fields is governed by the *Lorentz force*. For a typical particle of charge q and mass m, moving with velocity $\vec{v}$ in the presence of electric ($\vec{E}$) and magnetic induction ($\vec{B}$) fields, the equation of motion is

$$d\vec{p}/dt = q(\vec{E} + \vec{v} \times \vec{B}) \tag{5.1}$$

where $\vec{p} = m\,\vec{v}$ denotes the momentum. It is conceivable, at least in principle, to describe the dynamics of a plasma by solving the equations of motion for each particle in the plasma under the combined influence of externally applied force fields and internal fields generated by all the other plasma particles. If the total number of particles is N, we will have N nonlinear coupled differential equations of motion to solve simultaneously. A *self-consistent formulation* must be used since the fields and the trajectories of the particles are intrinsically coupled, that is, the internal fields associated with the presence and motion of the plasma particles influence their motions which, in turn, modify the internal fields.

The electromagnetic fields obey Maxwell equations

$$\vec{\nabla} \times \vec{E} = -\partial\vec{B}/\partial t \tag{5.2}$$

$$\vec{\nabla} \times \vec{B} = \mu_0\,(\vec{J} + \varepsilon_0\,\partial\vec{E}/\partial t) \tag{5.3}$$

$$\vec{\nabla} \cdot \vec{E} = \rho/\varepsilon_0 \tag{5.4}$$

$$\vec{\nabla} \cdot \vec{B} = 0 \tag{5.5}$$

where ρ, $\vec{J}$, ε_0 and μ_0 denote, respectively, the *total* charge density, the *total* electric current density, the electric permittivity and the magnetic permeability of free space. The plasma charge density and current density can be expressed, respectively, as

$$\rho_p = (1/\Delta V) \sum_i q_i \tag{5.6}$$

$$\vec{J}_p = (1/\Delta V) \sum_i q_i \vec{v}_i \tag{5.7}$$

where the summation is over all charged particles contained inside a suitably chosen small volume element ΔV. Note that since we are dealing with a discrete distribution of charges and therefore also of current densities, ρ_p and $\vec{J}_p$ should actually be expressed in terms of Dirac delta functions. If point charges are considered, the problem gets even more complicated because the fields become singular at the particle positions. However, if ΔV is chosen big enough to contain a fairly large number of particles, then (5.6) and (5.7) will give smooth functions for ρ_p and $\vec{J}_p$ which are suitable for analytical calculations.

Although this self-consistent approach is conceivable in principle, it cannot be carried out in practice. According to the laws of classical mechanics, in order to determine the position and velocity of each particle in the plasma as a function of time under the action of known forces, it is necessary to know the initial position and velocity of each particle. For a system consisting of a very large number of interacting particles these initial conditions are obviously unknown. Furthermore, in order to explain and predict the macroscopic phenomena observed in nature and in the laboratory, it is not of interest to know the detailed individual motion of each particle, since the observable macroscopic properties of a plasma are due to the average collective behavior of a large number of particles. To try to interpret the complex individual particle motions in terms of a few macroscopic observables would be a very inefficient way of making predictions. We reject, therefore, the possibility of solving simultaneously the equations of motion for a large number of interacting particles, even because the results would be of no practical interest.

5.2 Theoretical approaches

For the theoretical description of plasma phenomena there are basically four principal approaches with several different choices of approximations, each of which applies to different circumstances.

One useful approximation, known as *particle orbit theory*, consists in studying the motion of each charged particle in the presence of specified electric and magnetic fields. This approach is not really plasma theory, but rather the dynamics of a charged particle in given fields. Nevertheless it is important, since it provides some physical insight for a better understanding of the dynamical processes in plasmas. It has proven to be useful for predicting the behavior of very low density plasmas, which is determined primarily by the interaction of the particles with external fields. This is the case, for example, of the highly rarefied plasmas of the Van Allen radiation belts and the solar corona, as well as for cosmic rays, high energy accelerators and cathode ray tubes.

Since a plasma consists of a very large number of interacting particles, in order to provide a macroscopic description of plasma phenomena it is appropriate to adopt a *statistical approach.* This implies in a great reduction in the amount of information to be handled. In the *kinetic theory* statistical description it is necessary to know only the *distribution function* for the system of particles under consideration. The problem consists in solving the appropriate kinetic equations which govern the evolution of the distribution function in phase space. One example of a differential kinetic equation is the *Vlasov equation*, in which the interaction between the charged particles is described by smeared out internal electromagnetic fields consistent with the distributions of electric charge density and current density inside the plasma, and the effects of short-range correlations (close collisions) are neglected.

When collisions between the plasma particles are very frequent so that each species is able to maintain a local equilibrium distribution function, then each species can be treated as a fluid described by a local density, local macroscopic velocity and local temperature. In this case the plasma is treated as a mixture of two or more interpenetrating fluids. This theory is called *two-fluid* or *many-fluid theory*. In addition to the usual electrodynamic equations, there is a set of hydrodynamic equations expressing conservation of mass, of momentum and of energy for each particle species in the plasma.

Another approach consists in treating the whole plasma as a *single conducting fluid* using lumped macroscopic variables and their corresponding hydrodynamic conservation equations. This theory is usually referred to as the *one-fluid theory*. An appropriately simplified form of this theory, applicable to the study of very low frequency phenomena in highly conducting fluids immersed in magnetic fields, is usually referred to as the *magnetohydrodynamic* (MHD) theory.

PROBLEMS

1.1 - The interatomic or intermolecular forces are usually represented in terms of a potential energy function $V(r)$ such that $F(r) = - dV(r)/dr$. For *neutral* particles, at large internuclear distances, there is a slight attractive potential between the particles called the *van der Waals* potential (which is the long-range part of the Lennard-Jones potential). For like atoms or molecules in like states the *van der Waals* interaction potential can be represented by

$$V(r) = - C\,Ry\,(a_0/r)^6$$

where C is a constant (which depends on the type of particle), a_0 is the Bohr radius (0.529 Å) and Ry denotes the Rydberg energy unit (13.605 eV). Calculate the van der Waals force of attraction between two hydrogen molecules (for which C = 24.0), and compare with the Coulomb force between a proton and an electron at a distance $r = N\,a_0$, where $N \gg 1$.

1.2 - Consider an initially uniform plasma in which the electron and ion number densities are each equal to n. By some external means, let a one--dimensional perturbation occur such that the electrons in an infinite plane (y-z plane) are displaced by a small amount x (see Fig. P1.1).
(a) Using Gauss' law show that the electric field which appears across the perturbed plane is given by $E = (ne/\varepsilon_0)x$.
(b) Show that the equation of motion (Newton's second law) for each electron, under the action of this electric field, is

$$\frac{d^2x}{dt^2} + \left(\frac{ne^2}{m\varepsilon_0}\right)x = 0$$

Verify that this is the equation of a harmonic oscillator of frequency $\omega_{pe} = [\,ne^2/(m\varepsilon_0)]^{1/2}$.

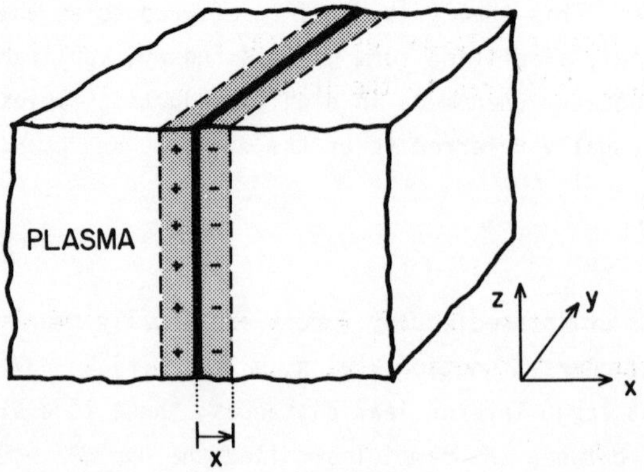

Fig. P1.1

1.3 - (a) Calculate the amount of energy released by the fusion of 1g of deuterium according to the nuclear reactions indicated in (4.1), considering as end products ^4He, ^1H and ^1n. Assume that the two possible results shown in (4.1), for the reaction $^2H + {}^2H$, occur with equal probability.
(b) How much energy can be released from the fusion of all the deuterium that exists in one liter of ordinary water? Compare this much energy with the energy obtained from the combustion of one liter of gasoline.

1.4 - Calculate the Coulomb repulsion force and the associated electric potential energy due to the Coulomb interaction of two deuterium nuclei when brought together at a distance of 10^{-14}m. What temperature must have the nuclei in a deuterium plasma, if their average thermal kinetic energy is to be equal to this electric potential energy?

1.5 - In a MHD generator a plasma of conductivity σ is driven with velocity u (in the x direction) across a magnetic field B (in the y direction). Two electrode plates, each of area A and separated by a distance d, are placed oriented parallel to the x-y plane, as shown in Fig. 9 of Chapter 1.
(a) Show that the open-circuit electric potential difference between the two electrode plates is given by ϕ = Bud.
(b) If an external load of resistance R_L is connected between the electrodes, show that the current that flows is given by

$$I = \frac{Bud}{R_L + R_P}$$

where R_P denotes the internal plasma resistance.
(c) Show that the power delivered to the load is

$$P_L = \frac{B^2u^2d^2R_L}{(R_L + R_P)^2}$$

Verify that this has a maximum ($dP_L/dR_L = 0$) when $R_L = R_P$ and show that the maximum power that can be delivered to the load is given by

$$P_{L,max} = B^2u^2d\sigma A/4$$

(d) Determine numerical results for items (a), (b) and (c) when $B = 1$ Tesla, $u = 100$ m/s, $\sigma = 100$ mho/m, $d = 0.1$ m and $A = 1$ m^2.

1.6 - Consider a rocket once it is beyond the Earth's gravitational field. Let:
v = constant velocity of the exhaust gas *relative to the rocket.*
$u(t)$ = instantaneous velocity of the rocket.
$M(t)$ = instantaneous total mass of the rocket.
$-\,dM/dt$ = constant time rate of decrease of $M(t)$, that is, the mass expelled per unit time.

(a) Verify that the equation of motion (Newton's second law) for the rocket is

$$\frac{d}{dt}[M(t)u(t)] = \frac{dM}{dt}[v - u(t)]$$

and show that the instantaneous acceleration of the rocket is

$$\frac{du}{dt} = -\frac{v}{M(t)}\frac{dM}{dt}$$

(b) Integrate the equation of motion to show that

$$u(t) = u(t_0) + v\,\ell n[M(t_0)/M(t)]$$

(c) If the rocket burns for a time interval $t - t_0 = T_0$ and if $M(t) \ll M(t_0)$, show that the initial acceleration of the rocket is

$$\left.\frac{du}{dt}\right|_{t_0} = \frac{v}{M(t_0)}\frac{[M(t_0) - M(t)]}{T_0} \cong \frac{v}{T_0}$$

(d) Calculate numerically $du/dt|_{t_0}$ and $u(t)$ for a chemical rocket with $v = 10^3$ m/s and $T_0 = 10$ s; and for a plasma propulsion system with $v = 10^4$ m/s and $T_0 = 100$ days. For the calculation of $u(t)$ consider $u(t_0) = 0$ and $M(t_0) = 10\,M(t)$.

1.7 - Using Maxwell equations (5.3) and (5.4) derive the charge conservation equation

$$\partial\rho/\partial t + \vec{\nabla} \cdot \vec{J} = 0$$

This result shows that conservation of electric charge is implied by Maxwell equations.

1.8 - From Maxwell curl equation (5.2) derive the equation

$$\vec{\nabla} \cdot \vec{B} = \text{constant}$$

Therefore, (5.5) can be considered as an *initial condition* for (5.2) since, if $\vec{\nabla} \cdot \vec{B} = 0$ at the initial time, then (5.2) implies that this condition will remain satisfied for all subsequent times.

1.9 - Using Maxwell equation derive the following energy conservation law for electromagnetic fields, known as *Poynting's theorem*,

$$\frac{\partial}{\partial t} \int_V (\varepsilon E^2/2 + \mu H^2/2)\, d^3r + \oint_S (\vec{E} \times \vec{H}) \cdot d\vec{S} = - \int_V \vec{J} \cdot \vec{E}\, d^3r$$

for a linear isotropic medium, for which $\vec{D} = \varepsilon\vec{E}$ and $\vec{B} = \mu\vec{H}$. Give the physical interpretation for each term in this equation. What are the physical dimensions of these terms?

1.10 - Consider the following Maxwell equations:

$$\vec{\nabla} \times \vec{B} = \mu_0 (\vec{J}_t + \varepsilon_0\, \partial\vec{E}/\partial t)$$

$$\vec{\nabla} \times \vec{H} = \vec{J} + \partial\vec{D}/\partial t$$

$$\vec{\nabla} \cdot \vec{E} = \rho_t/\varepsilon_0$$

$$\vec{\nabla} \cdot \vec{D} = \rho$$

For a general medium for which

$$\vec{D} = \varepsilon_0 \vec{E} + \vec{P}$$

$$\vec{B} = \mu_0 (\vec{H} + \vec{M})$$

where $\vec{P}$ is the polarization vector and $\vec{M}$ is the magnetization vector, show that the *total* electric charge density (ρ_t) and current density ($\vec{J}_t$) are given by

$$\rho_t = \rho - \vec{\nabla} \cdot \vec{P}$$

$$J_t = \vec{J} + \partial\vec{P}/\partial t + \vec{\nabla} \times \vec{M}$$

Explain why $\vec{E}$ and $\vec{B}$ are usually considered as *fundamental* fields, whereas $\vec{D}$ and $\vec{H}$ are *partial* fields.

CHAPTER 2
Charged Particle Motion in Constant and Uniform Electromagnetic Fields

1. INTRODUCTION

In this and in the following two chapters, we investigate the motion of charged particles in the presence of electric and magnetic fields known as functions of position and time. Thus, the electric and magnetic fields are assumed to be prescribed and not affected by the charged particles. This chapter, in particular, considers the fields to be constant in time and spatially uniform. This subject is considered in some detail since many of the more complex situations, considered in Chapters 3 and 4, can be treated as perturbations to this problem.

The study of the motion of charged particles in specified fields is important since it provides a good physical insight for understanding some dynamical processes in plasmas. It also permits to obtain information on some macroscopic phenomena, which are due to the collective behavior of a large number of particles. Not all of the components of the detailed microscopic particle motion contribute to macroscopic effects. It is possible, however, to isolate the components of the individual motion that contribute to the collective plasma behavior. Nevertheless, the macroscopic parameters can be obtained much more easily and conveniently from the macroscopic equations (Chapters 8 and 9).

The equation of motion for a particle of charge q, under the action of the Lorentz force $\vec{F}$ due to electric ($\vec{E}$) and magnetic induction ($\vec{B}$) fields, can be written as

$$d\vec{p}/dt = \vec{F} = q(\vec{E} + \vec{v} \times \vec{B}) \tag{1.1}$$

where $\vec{p}$ represents the momentum of the particle and $\vec{v}$ its velocity. This equation is relativistically correct if we take

$$\vec{p} = \gamma m \vec{v} \tag{1.2}$$

where m is the rest mass of the particle and γ is the so-called *Lorentz factor* defined by

$$\gamma = (1 - v^2/c^2)^{-1/2} \tag{1.3}$$

where c is the speed of light in vacuum. In the relativistic case (1.1) can also be written in the form

$$\gamma m \, d\vec{v}/dt + q(\vec{v}/c^2)\,(\vec{v} \cdot \vec{E}) = q(\vec{E} + \vec{v} \times \vec{B}) \tag{1.4}$$

noting that the time rate of change of the total relativistic energy ($U = \gamma mc^2$) is given by $dU/dt = q(\vec{v} \cdot \vec{E})$ and that $d\vec{p}/dt = d(U\vec{v}/c^2)/dt$.

In many situations of practical interest, however, the term v^2/c^2 is negligible compared to unity. For $v^2/c^2 \ll 1$ we have $\gamma \cong 1$ and m can be considered constant (independent of the particle velocity), so that (1.4) reduces to the following nonrelativistic expression

$$m \, d\vec{v}/dt = q\,(\vec{E} + \vec{v} \times \vec{B}) \tag{1.5}$$

If the velocity obtained from (1.5) does not satisfy the condition $v^2 \ll c^2$, then the corresponding result is not valid and the relativistic expression (1.4) must be used instead of (1.5). Relativistic effects become important only for highly energetic particles (a 1 MeV proton, for instance, has a velocity of 1.4×10^7 m/s, with $v^2/c^2 = 0.002$). For the situations to be considered here it is assumed that the restriction $v^2 \ll c^2$, implicit in (1.5), is not violated. Also, all radiation effects are neglected.

2. ENERGY CONSERVATION

In the absence of an electric field ($\vec{E} = 0$) the equation of motion (1.5) reduces to

$$m \, d\vec{v}/dt = q(\vec{v} \times \vec{B}) \tag{2.1}$$

Since the magnetic force is perpendicular to $\vec{v}$ it does no work on the particle. Taking the dot product of (2.1) with $\vec{v}$ and noting that $(\vec{v} \times \vec{B}) \cdot \vec{v} = 0$ for any vector $\vec{v}$, we obtain

$$m \left(\frac{d\vec{v}}{dt}\right) \cdot \vec{v} = \frac{d}{dt}(mv^2/2) = 0 \tag{2.2}$$

which shows that the particle kinetic energy ($mv^2/2$) and the magnitude of its velocity (speed v) are both constants. Therefore, a *static* magnetic field does not change the particle kinetic energy. This result is valid whatever the spatial dependence of the magnetic flux density $\vec{B}$. However, if $\vec{B}$ varies with time then, according to Maxwell equations, an electric field such that $\vec{\nabla} \times \vec{E} = -\partial\vec{B}/\partial t$ is also present which does work on the particle changing its kinetic energy.

When both *magnetostatic* and *electrostatic* fields are present, we obtain from (1.5)

$$\frac{d}{dt}(mv^2/2) = q\vec{E} \cdot \vec{v} \tag{2.3}$$

Since $\vec{\nabla} \times \vec{E} = 0$, we can express the electrostatic field in terms of the electrostatic potential according to $\vec{E} = -\vec{\nabla}\phi$, so that

$$\frac{d}{dt}(mv^2/2) = -q(\vec{\nabla}\phi) \cdot \vec{v} = -q(\vec{\nabla}\phi) \cdot \frac{d\vec{r}}{dt} = -q\frac{d\phi}{dt} \tag{2.4}$$

This result can be written as

$$\frac{d}{dt}(mv^2/2 + q\phi) = 0 \tag{2.5}$$

which shows that the sum of the kinetic energy and the electric potential energy of the particle remains constant in the presence of static electromagnetic fields. The electric potential ϕ can be considered as the potential energy per unit charge.

When the fields are time-dependent, we have $\vec{\nabla} \times \vec{E} \neq 0$ and $\vec{E}$ is not the gradient of a scalar function. But since $\vec{\nabla} \cdot \vec{B} = 0$, we can define a magnetic vector potential $\vec{A}$ by $\vec{B} = \vec{\nabla} \times \vec{A}$, and write (1.5.2) as

$$\vec{\nabla} \times \vec{E} + \partial\vec{B}/\partial t = \vec{\nabla} \times \vec{E} + \partial(\vec{\nabla} \times \vec{A})/\partial t = \vec{\nabla} \times (\vec{E} + \partial\vec{A}/\partial t) = 0 \tag{2.6}$$

Hence, we can express the electric field in the form

$$\vec{E} = -\vec{\nabla}\phi - \partial\vec{A}/\partial t \tag{2.7}$$

In this case the system is not conservative in the usual sense and there is no energy integral, but the analysis can be performed using a Lagrangian function L for a charged particle in electromagnetic fields, defined by

$$L = mv^2/2 - U \tag{2.8}$$

where U is a velocity-dependent potential energy given by

$$U = q(\phi - \vec{v} \cdot \vec{A}) \tag{2.9}$$

The energy considerations presented in this section assume that the particle energy changes only as a result of the work done by the fields. This assumption is not strictly correct since every charged particle when accelerated irradiates energy in the form of electromagnetic waves. For the situations to be considered here this effect is usually small and can be neglected.

3. UNIFORM ELECTROSTATIC FIELD

According to (1.1) the motion of a charged particle in an electric field obeys the following differential equation

$$d\vec{p}/dt = q\,\vec{E} \tag{3.1}$$

For a constant $\vec{E}$ field, (3.1) can be integrated directly giving

$$\vec{p}(t) = q\,\vec{E}\,t + \vec{p}_0 \tag{3.2}$$

where $\vec{p}_0 = \vec{p}(t = 0)$ denotes the initial particle momentum. Using the nonrelativistic expression

$$\vec{p} = m\vec{v} = m\,d\vec{r}/dt \tag{3.3}$$

and performing a second integration in (3.2), we obtain the following expression for the particle position as a function of time

$$\vec{r}(t) = (q\vec{E}/2m)\, t^2 + \vec{v}_0 t + \vec{r}_0 \tag{3.4}$$

where $\vec{r}_0$ denotes the particle initial position and $\vec{v}_0$ its initial velocity. Therefore, the particle moves with a constant acceleration, $q\vec{E}/m$, in the direction of $\vec{E}$ if $q > 0$, and in the opposite direction if $q < 0$. In a direction perpendicular to the electric field there is no acceleration and the particle state of motion remains unchanged.

4. UNIFORM MAGNETOSTATIC FIELD

4.1 Formal solution of the equation of motion

For a particle of charge q and mass m, moving with velocity $\vec{v}$ in a region of space where there is only a magnetic induction $\vec{B}$, the equation of motion is

$$m\, d\vec{v}/dt = q(\vec{v} \times \vec{B}) \tag{4.1}$$

It is convenient to separate $\vec{v}$ in components parallel ($v_\parallel$) and perpendicular ($v_\perp$) to the magnetic field,

$$\vec{v} = \vec{v}_\parallel + \vec{v}_\perp \tag{4.2}$$

as indicated in Fig. 1. Substituting (4.2) into (4.1) and noting that $\vec{v}_\parallel \times \vec{B} = 0$ we obtain

$$d\vec{v}_\parallel/dt + d\vec{v}_\perp/dt = (q/m)\,(\vec{v}_\perp \times \vec{B}) \tag{4.3}$$

Since the term $\vec{v}_\perp \times \vec{B}$ is perpendicular to $\vec{B}$, the parallel component equation can be written as

$$d\vec{v}_\parallel/dt = 0 \tag{4.4}$$

and the perpendicular component equation as

$$d\vec{v}_\perp/dt = (q/m)\ (\vec{v}_\perp \times \vec{B}) \tag{4.5}$$

Eq. (4.4) shows that the particle velocity along $\vec{B}$ does not change and is equal to the particle initial velocity. For motion in the plane perpendicular to $\vec{B}$, we can write (4.5) in the form

$$d\vec{v}_\perp/dt = - \vec{v}_\perp \times \vec{\omega}_c = \vec{\omega}_c \times \vec{v}_\perp \tag{4.6}$$

where $\vec{\omega}_c$ is a vector defined by

$$\vec{\omega}_c = - q\vec{B}/m = (|q|B/m)\ \hat{\omega}_c = \omega_c\ \hat{\omega}_c \tag{4.7}$$

Thus, $\vec{\omega}_c$ points in the direction of $\vec{B}$ for a negatively charged particle $(q < 0)$ and in the opposite direction for a positively charged particle $(q > 0)$. Its magnitude ω_c is always positive $(\omega_c = |q|\ B/m)$. The unit vector $\hat{\omega}_c$ points along $\vec{\omega}_c$.

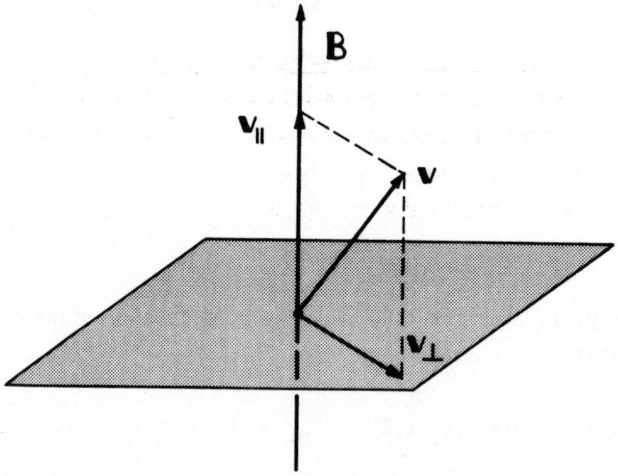

Fig. 1 - Decomposition of the velocity vector into components parallel ($\vec{v}_\parallel$) and perpendicular ($\vec{v}_\perp$) to the magnetic field.

Since $\vec{\omega}_c$ is constant and (from conservation of kinetic energy) $v_\perp = |\vec{v}_\perp|$ is also constant, (4.6) shows that the particle acceleration is constant in magnitude and its direction is perpendicular to both $\vec{v}_\perp$ and $\vec{B}$. Thus, this acceleration corresponds to a rotation of the velocity vector $\vec{v}_\perp$ in the plane perpendicular to $\vec{B}$ with the constant angular velocity $\vec{\omega}_c$. We can integrate (4.6) directly, noting that $\vec{\omega}_c$ is constant and taking $\vec{v}_\perp = d\vec{r}_c/dt$, to obtain

$$\vec{v}_\perp = \vec{\omega}_c \times \vec{r}_c \tag{4.8}$$

where the vector $\vec{r}_c$ is interpreted as the particle position vector with respect to a point G (the center of gyration) in the plane perpendicular to $\vec{B}$ which contains the particle. Since the particle speed $v_\perp$ is constant, the magnitude r_c of the position vector is also constant. Therefore, (4.8) shows that the velocity $\vec{v}_\perp$ corresponds to a rotation of the position vector $\vec{r}_c$ about the point G in the plane perpendicular to $\vec{B}$ with constant angular velocity $\vec{\omega}_c$. The component of the motion in the plane perpendicular to $\vec{B}$ is therefore a circle of radius r_c. The instantaneous center of gyration of the particle (the point G at the distance r_c from the particle) is called the *guiding center*. This circular motion about the guiding center is illustrated in Fig. 2.

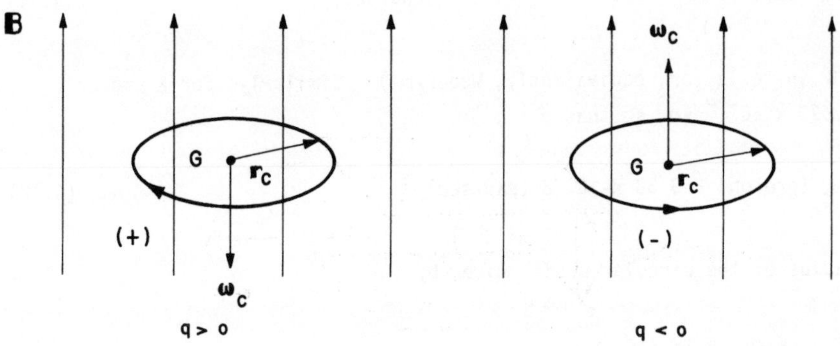

Fig. 2 - Circular motion of a charged particle about the guiding center G in a uniform magnetostatic field.

Note that according to the definition of $\vec{\omega}_c$, given in (4.7), $\hat{\omega}_c$ always points in the same direction as the particle angular momentum vector $(\vec{r}_c \times \vec{p})$, irrespective of its charge.

The resulting trajectory of the particle is given by the superposition of a uniform motion along $\vec{B}$ (with the constant velocity $\vec{v}_\parallel$) and a circular motion in the plane normal to $\vec{B}$ (with the constant speed $v_\perp$). Hence, the particle describes a helix (Fig. 3). The angle between the direction of motion of the particle and $\vec{B}$ is called the *pitch angle*,

$$\alpha = \sin^{-1}(v_\perp/v) = \tan^{-1}(v_\perp/v_\parallel) \tag{4.9}$$

where v is the total speed of the particle ($v^2 = v_\parallel^2 + v_\perp^2$). When $v_\parallel = 0$ but $v_\perp \neq 0$, we have $\alpha = \pi/2$ and the particle trajectory is a circle in the plane normal to $\vec{B}$. On the other hand, when $v_\perp = 0$ but $v_\parallel \neq 0$, we have $\alpha = 0$ and the particle moves along $\vec{B}$ with the velocity $\vec{v}_\parallel$.

The magnitude of the angular velocity,

$$\omega_c = |q|\, B/m \tag{4.10}$$

is known as the *angular frequency of gyration*, or *gyrofrequency*, or *cyclotron frequency* or *Larmor frequency*. For an electron $|q| = 1.602 \times 10^{-19}$ Coulombs and $m = 9.109 \times 10^{-31}$ kg, so that

$$\omega_c \text{ (electron)} = 1.76 \times 10^{11} B \qquad \text{(rad/sec)} \tag{4.11}$$

with B in Tesla (or, equivalently, Weber/m²). Similarly, for a proton $m = 1.673 \times 10^{-27}$ kg, so that

$$\omega_c \text{ (proton)} = 9.58 \times 10^{7}\, B \text{ (rad/sec)} \tag{4.12}$$

The radius of the circular orbit, given by

$$r_c = v_\perp/\omega_c = mv_\perp/|q|\, B \tag{4.13}$$

is called the *radius of gyration*, or *gyroradius*, or *cyclotron radius*, or *Larmor radius*. It is important to note that ω_c is directly proportional to B. Consequently, as B increases, the gyrofrequency increases and the radius decreases. Also, the smaller the particle mass the larger will be its gyrofrequency and the smaller its gyroradius. Multiplying (4.13) by B gives

$$Br_c = mv_\perp/|q| = p_\perp/|q| \tag{4.14}$$

which shows that the magnitude of $\vec{B}$ times the gyroradius of the particle is equal to its momentum per unit charge. This quantity is often called the *magnetic rigidity*.

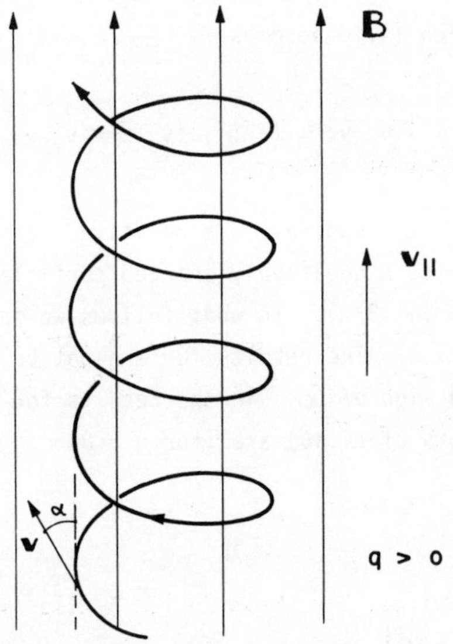

Fig. 3 - Helicoidal trajectory of a positively charged particle in a uniform magnetostatic field.

4.2 Solution in Cartesian coordinates

The treatment presented so far in this section has not been related to any particular frame of reference. Consider now a Cartesian coordinate system (x,y,z) such that $\vec{B} = B\hat{z}$. In this case, the cross product between $\vec{v}$ and $\vec{B}$ can be written as

$$\vec{v} \times \vec{B} = \begin{vmatrix} \hat{x} & \hat{y} & \hat{z} \\ v_x & v_y & v_z \\ 0 & 0 & B \end{vmatrix} = B(v_y \hat{x} - v_x \hat{y}) \tag{4.15}$$

and the equation of motion (4.1) becomes

$$d\vec{v}/dt = (qB/m)\,(v_y \hat{x} - v_x \hat{y}) = (\pm \omega_c)\,(v_y \hat{x} - v_x \hat{y}) \tag{4.16}$$

The (+) sign in front of ω_c applies to a positively charged particle ($q > 0$) and the (-) sign to a negatively charged particle ($q < 0$), since ω_c is always positive ($\omega_c = |q|\,B/m$). In what follows we shall consider a *positively charged particle*. The results for a negative charge can be obtained by changing the sign of ω_c in the results for the positive charge.

The Cartesian components of (4.16) are (for $q > 0$)

$$dv_x/dt = \omega_c v_y \tag{4.17}$$

$$dv_y/dt = -\,\omega_c v_x \tag{4.18}$$

$$dv_z/dt = 0 \tag{4.19}$$

The last of these equations gives $v_z(t) = v_z(0) = v_\parallel$, which is the initial value of the velocity component parallel to $\vec{B}$. To obtain the solution of (4.17) and (4.18) we take the derivative of (4.17) with respect to time and substitute the result into (4.18), getting

$$d^2v_x/dt^2 + \omega_c^2 v_x = 0 \tag{4.20}$$

This is the homogeneous differential equation for a harmonic oscillator of frequency ω_c, whose solution is

$$v_x(t) = v_\perp \sin(\omega_c t + \theta_o) \qquad (4.21)$$

where $v_\perp$ is the constant speed of the particle in the (x,y) plane (normal to $\vec{B}$) and θ_o is a constant of integration which depends on the relation between the initial velocities $v_x(0)$ and $v_y(0)$, according to

$$\tan(\theta_o) = v_x(0)/v_y(0) \qquad (4.22)$$

To determine $v_y(t)$ we substitute (4.21) in the left-hand side of (4.17), obtaining

$$v_y(t) = v_\perp \cos(\omega_c t + \theta_o) \qquad (4.23)$$

Note that $v_x^2 + v_y^2 = v_\perp^2$. The equations for the components of $\vec{v}$ can be further integrated with respect to time, yielding

$$x(t) = -(v_\perp/\omega_c)\cos(\omega_c t + \theta_o) + X_o \qquad (4.24)$$

$$y(t) = (v_\perp/\omega_c)\sin(\omega_c t + \theta_o) + Y_o \qquad (4.25)$$

$$z(t) = v_\parallel t + z_o \qquad (4.26)$$

where we have defined

$$X_o = x_o + (v_\perp/\omega_c)\cos\theta_o \qquad (4.27)$$

$$Y_o = y_o - (v_\perp/\omega_c)\sin\theta_o \qquad (4.28)$$

The vector $\vec{r} = x_o\hat{x} + y_o\hat{y} + z_o\hat{z}$ gives the initial particle position. From (4.24) and (4.25) we see that

$$(x - X_o)^2 + (y - Y_o)^2 = (v_\perp/\omega_c)^2 = r_c^2 \qquad (4.29)$$

The particle trajectory in the plane normal to $\vec{B}$ is therefore a circle with center at (X_0, Y_0) and radius equal to $(v_\perp/\omega_c)$. The motion of the point $[X_0, Y_0, z(t)]$, at the instantaneous center of gyration, corresponds to the *trajectory of the guiding center*. Thus, the guiding center moves with constant velocity $\vec{v}_\parallel$ along $\vec{B}$.

In the (x, y) plane, the argument $\phi(t)$, defined by

$$\phi(t) = \tan^{-1} [(y - Y_0)/(x - X_0)] = - (\omega_c t + \theta_0); \ \phi_0 = - \theta_0 \tag{4.30}$$

decreases with time for a positively charged particle. For a magnetic field pointing toward the observer, a positive charge describes a circle in the clockwise direction. For a negatively charged particle ω_c must be replaced by $- \omega_c$ in the results of this sub-section. Hence, (4.30) shows that for a negative charge $\phi(t)$ increases with time and the particle moves in a circle in the counterclockwise direction, as shown in Fig. 4. The resulting particle motion is a cylindrical helix of constant pitch angle. Fig. 5 shows the parameters of the helix with reference to a Cartesian coordinate system.

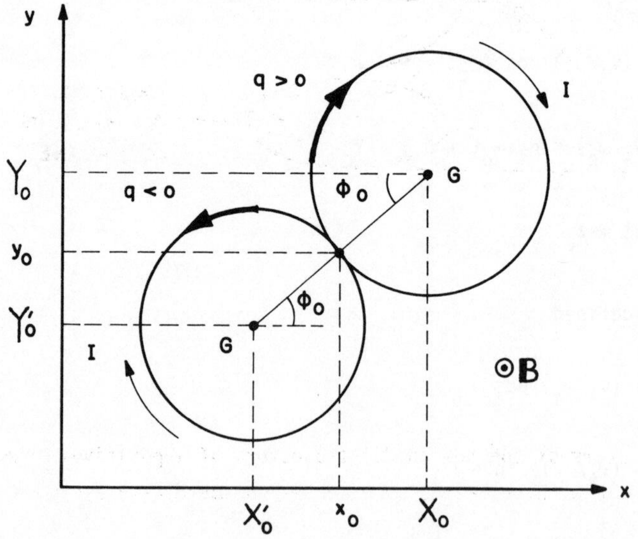

Fig. 4 - Circular trajectory of a charged particle in a uniform and constant $\vec{B}$ field (directed out of the paper), and the direction of the associated electric current.

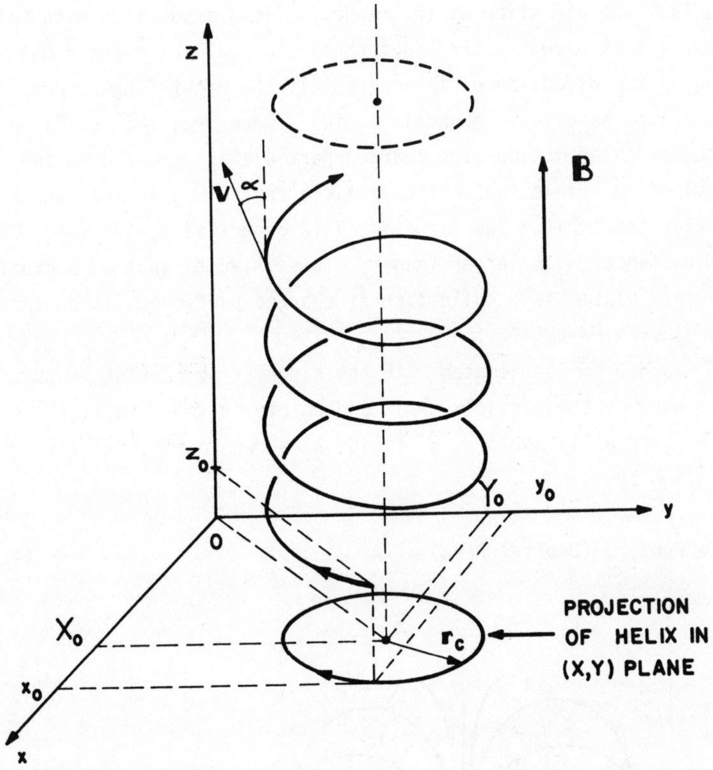

Fig. 5 - Parameters of the helicoidal trajectory of a positively charged particle with reference to a Cartesian coordinate system.

4.3 Magnetic moment

To the circular motion of a charged particle in a magnetic field there is associated a circulating electric current I. This current flows in the clockwise direction for a $\vec{B}$ field pointing toward the observer (Fig. 4). From Ampère's law, the direction of the magnetic field associated with this circulating current is given by the right-hand rule i.e. with the right thumb pointing in the direction of the current I, the right fingers curl in the direction of the associated magnetic field. Therefore, the $\vec{B}$ field produced by the circular motion of a charged particle is *opposite* to the externally applied $\vec{B}$ field *inside* the particle orbit, but in the same direction outside the orbit. The magnetic field generated by the ring current I, at distances much larger than r_c, is similar to that of a dipole (Fig. 6). Since a plasma is a collection of charged particles, it possesses therefore *diamagnetic properties*.

The *magnetic moment* $\vec{m}$ associated with the circulating current is normal to the area A bounded by the particle orbit and points in the direction opposite to the externally applied $\vec{B}$ field, as shown in Fig. 7. Its magnitude is given by

$$|\vec{m}| = (\text{Current}) \cdot (\text{Orbital Area}) = IA \tag{4.31}$$

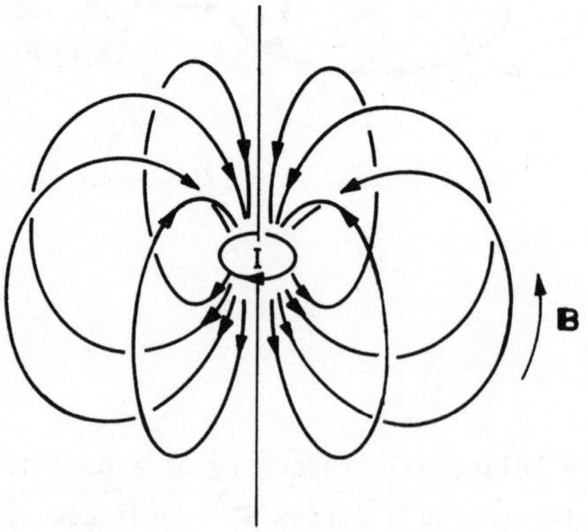

Fig. 6 - The magnetic field generated by a small ring current is that of a magnetic dipole.

The circulating current corresponds to a flow of charge and is given by

$$I = |q|/T_c = |q|\ \omega_c/2\pi \tag{4.32}$$

where $T_c = 2\pi/\omega_c$ is the period of the particle orbit, known as the *cyclotron period* or *Larmor period*. The magnitude of $\vec{m}$ is therefore

$$|\vec{m}| = (|q|\ \omega_c/2\pi)\ (\pi r_c^2) = |q|\ \omega_c\ r_c^2/2 \tag{4.33}$$

Using the relations $\omega_c = |q|\ B/m$ and $r_c = v_\perp/\omega_c$, (4.33) becomes

$$|\vec{m}| = m\ v_\perp^2/2B = W_\perp/B \tag{4.34}$$

where $W_\perp$ denotes the part of the particle kinetic energy associated with the transverse velocity $v_\perp$. Thus, in vector form,

$$\vec{m} = -\ (W_\perp/B^2)\ \vec{B} \tag{4.35}$$

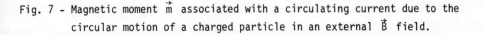

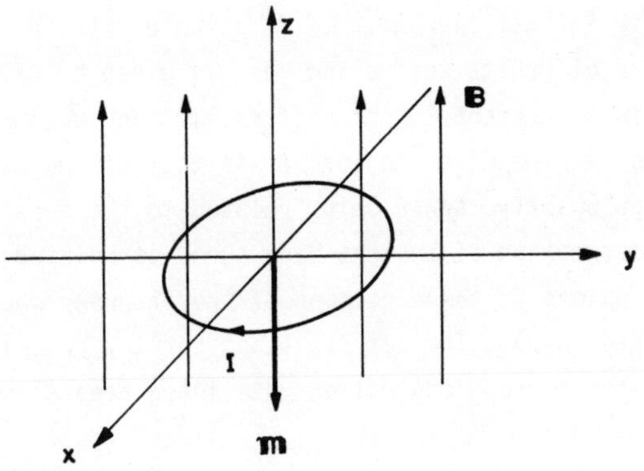

Fig. 7 - Magnetic moment $\vec{m}$ associated with a circulating current due to the circular motion of a charged particle in an external $\vec{B}$ field.

4.4 Magnetization current

Consider now a collection of charged particles, positive and negative in equal numbers (in order to have no internal macroscopic electrostatic fields), instead of just one single particle. For instance, consider the case of a low-density plasma in which the particle collisions can be neglected (collisionless plasma). The condition for this is that the average time between collisions be much greater than the cyclotron period. This condition is fulfilled for many space plasmas, for example. For a collisionless plasma in an external magnetic field, the magnetic moments due to the orbital motion of the charged particles act together, giving rise to a resultant magnetic field which may be strong enough to appreciably change the externally applied $\vec{B}$ field. The mean magnetic field produced by the orbital motion of the charged particles can be determined from the net electric current density associated with their motion.

To calculate the resultant electric current density, let us consider a macroscopic volume containing a large number of particles. Let S be an element of area in this volume, bounded by the curve C (Fig. 8-a). Orbits such as (1), which encircle the bounded surface only once, contribute to the resultant current, whereas orbits such as (2), which cross the surface twice, do not contribute to the net current. If $d\vec{\ell}$ is an element of arc along the curve C, the number of orbits encircling $d\vec{\ell}$ is given by $n\vec{A}.d\vec{\ell}$, where n is the number of orbits of current I per unit volume, and $\vec{A}$ is the vector area bounded by each orbit. The direction of $\vec{A}$ is that of the normal to the orbital area A, the positive sense being related to the sense of circulation in the way the linear motion of a right-hand screw is related to its rotary motion. Thus, $\vec{A}$ points in the direction of the observer when I flows counterclockwise (Fig. 8-b). The net resultant current crossing S is therefore given by the current encircling $d\vec{\ell}$ integrated along the curve C,

$$I_n = \oint_C I\, n\, \vec{A} \,.\, d\vec{\ell} \tag{4.36}$$

Since $\vec{m} = I\,\vec{A}$, the magnetic moment per unit volume, $\vec{M}$, (also called the *magnetization* vector) is given by

$$\vec{M} = n\,\vec{m} = n\,I\,\vec{A} \tag{4.37}$$

Hence, (4.36) can be written as

$$I_n = \oint_C \vec{M} \cdot d\vec{\ell} = \int_S (\vec{\nabla} \times \vec{M}) \cdot d\vec{S} \tag{4.38}$$

where we have applied Stoke's theorem. We may define an average *magnetization current density*, $\vec{J}_M$, crossing the surface S, by

$$I_n = \int_S \vec{J}_M \cdot d\,\vec{S} \tag{4.39}$$

Consequently, from (4.38) and (4.39) we obtain the magnetization current density as

$$\vec{J}_M = \vec{\nabla} \times \vec{M} \tag{4.40}$$

where, from (4.37) and (4.35),

$$\vec{M} = n\,\vec{m} = -\,(nW_\perp/B^2)\,\vec{B} \tag{4.41}$$

and $nW_\perp$ denotes the kinetic energy per unit volume, associated with the transverse particle velocity.

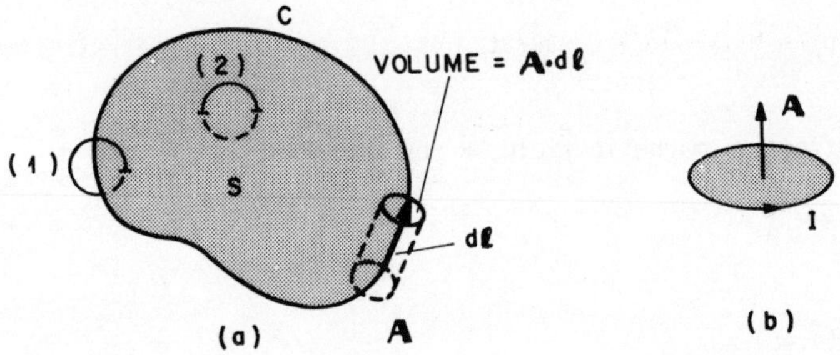

Fig. 8 - (a) Electric current orbits crossing the surface element S bounded by the curve C, in a macroscopic volume containing a large number of particles. (b) Positive direction of the area vector $\vec{A}$.

The charge density ρ_M associated with the magnetization current density $\vec{J}_M$ can be deduced from the equation of continuity,

$$\partial \rho_M / \partial t + \vec{\nabla} \cdot \vec{J}_M = 0 \tag{4.42}$$

Since $\vec{J}_M = \vec{\nabla} \times \vec{M}$ and since for any vector $\vec{a}$, $\vec{\nabla} \cdot (\vec{\nabla} \times \vec{a}) = 0$, it follows that the charge density ρ_M is constant.

In the following Maxwell equation

$$\vec{\nabla} \times \vec{B} = \mu_0 \, (\vec{J} + \varepsilon_0 \, \partial \vec{E} / \partial t) \tag{4.43}$$

we can separate the total current density $\vec{J}$ into two parts: a magnetization current density $\vec{J}_M$ and a current density $\vec{J}'$ due to other sources,

$$\vec{J} = \vec{J}_M + \vec{J}' \tag{4.44}$$

Expressing $\vec{J}_M$ in terms of $\vec{M}$, through (4.40), and substituting in (4.43), we obtain

$$\vec{\nabla} \times \vec{B} = \mu_0 \, (\vec{\nabla} \times \vec{M} + \vec{J}' + \varepsilon_0 \, \partial \vec{E} / \partial t) \tag{4.45}$$

which can be rearranged as

$$\vec{\nabla} \times \left(\vec{B} / \mu_0 - \vec{M} \right) = \vec{J}' + \varepsilon_0 \, \partial \vec{E} / \partial t \tag{4.46}$$

Defining an effective magnetic field $\vec{H}$ by the relation

$$\vec{B} = \mu_0 \, (\vec{H} + \vec{M}) \tag{4.47}$$

we can write (4.46) as

$$\vec{\nabla} \times \vec{H} = \vec{J}' + \varepsilon_0 \, \partial \vec{E} / \partial t \tag{4.48}$$

Thus, the effective magnetic field $\vec{H}$ is related to the current due to other sources $\vec{J}'$, in the way $\vec{B}$ is related to the total current $\vec{J}$. Eqs. (4.40) and (4.47) constitute the basic relations for the classical treatment of magnetic materials.

A simple linear relation between $\vec{B}$ and $\vec{H}$ exists when $\vec{M}$ is proportional to $\vec{B}$ or $\vec{H}$,

$$\vec{M} = \chi_m \vec{H} \tag{4.49}$$

where the constant χ_m is called the *magnetic susceptibility* of the medium. However, for a plasma we have seen that $M \propto 1/B$ [see (4.41)], so that the relation between $\vec{H}$ and $\vec{B}$ (or $\vec{M}$) is *not* linear. For this reason it is generally not convenient to treat a plasma as a magnetic medium.

5. UNIFORM ELECTROSTATIC AND MAGNETOSTATIC FIELDS

5.1 Formal solution of the equation of motion

We consider now the motion of a charged particle in the presence of both electric and magnetic fields which are constant in time and spatially uniform. The nonrelativistic equation of motion is

$$m\, d\vec{v}/dt = q\,(\vec{E} + \vec{v} \times \vec{B}) \tag{5.1}$$

Taking components parallel and perpendicular to $\vec{B}$,

$$\vec{v} = \vec{v}_\perp + \vec{v}_\parallel \tag{5.2}$$

$$\vec{E} = \vec{E}_\perp + \vec{E}_\parallel \tag{5.3}$$

we can resolve (5.1) into two component equations

$$m\, d\vec{v}_\parallel/dt = q\,\vec{E}_\parallel \tag{5.4}$$

$$m\, d\vec{v}_\perp/dt = q\,(\vec{E}_\perp + \vec{v}_\perp \times \vec{B}) \tag{5.5}$$

Eq. (5.4) is similar to (3.1) and represents a motion with constant acceleration $q\,\vec{E}_\parallel/m$ along the $\vec{B}$ field. Hence, according to (3.2) and (3.4),

$$\vec{v}_\parallel(t) = (q/m)\,\vec{E}_\parallel t + \vec{v}_\parallel(0) \tag{5.6}$$

$$\vec{r}_{\parallel}(t) = (q\,\vec{E}_{\parallel}/2m)\,t^2 + \vec{v}_{\parallel}(0)t + \vec{r}_{\parallel}(0) \tag{5.7}$$

To solve (5.5) it is convenient to separate $\vec{v}_{\perp}$ into two components

$$\vec{v}_{\perp}(t) = \vec{v}_{\perp}'(t) + \vec{v}_E \tag{5.8}$$

where $\vec{v}_E$ is a constant velocity in the plane normal to $\vec{B}$. Hence, $\vec{v}_{\perp}'$ represents the velocity of the particle as seen by an observer in a frame of reference moving with the constant velocity $\vec{v}_E$. Substituting (5.8) into (5.5), and writing the component of the electric field perpendicular to $\vec{B}$ in the form (see Fig. 9)

$$\vec{E}_{\perp} = -\,[(\vec{E}_{\perp} \times \vec{B})/B^2] \times \vec{B} \tag{5.9}$$

we obtain

$$d\vec{v}_{\perp}'/dt = q\,[\,\vec{v}_{\perp}' + \vec{v}_E - (\vec{E}_{\perp} \times \vec{B})/B^2\,] \times \vec{B} \tag{5.10}$$

This equation shows that in a coordinate system moving with the constant velocity

$$\vec{v}_E = (\vec{E}_{\perp} \times \vec{B})/B^2 \tag{5.11}$$

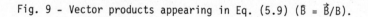

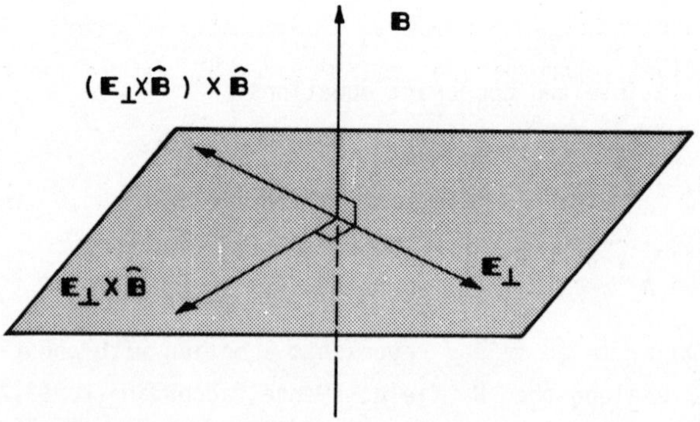

Fig. 9 - Vector products appearing in Eq. (5.9) ($\hat{B} = \vec{B}/B$).

the particle motion in the plane normal to $\vec{B}$ is governed entirely by the magnetic field, according to

$$m\, d\vec{v}'_{\perp}/dt = q\,(\vec{v}'_{\perp} \times \vec{B}) \tag{5.12}$$

Thus, in this frame of reference, the electric field component $\vec{E}_{\perp}$ is transformed away, whereas the magnetic field is left unchanged. Eq. (5.12) is identical to (4.5) and implies that in the reference system moving with the constant velocity $\vec{v}_E$, given by (5.11), the particle describes a circular motion at the cyclotron frequency ω_c with radius r_c,

$$\vec{v}_{\perp} = \vec{\omega}_c \times \vec{r}_c \tag{5.13}$$

The results obtained so far indicate that the resulting particle motion is described by a superposition of a circular motion in the plane normal to $\vec{B}$, with a uniform motion with the constant velocity $\vec{v}_E$ perpendicular to both $\vec{B}$ and $\vec{E}_{\perp}$, plus a uniform acceleration $q\vec{E}_{\parallel}/m$ along $\vec{B}$. The particle velocity can be expressed in vector form, independently of a coordinate system, as

$$\vec{v}(t) = \vec{\omega}_c \times \vec{r}_c + \vec{E}_{\perp} \times \vec{B}/B^2 + (q/m)\,\vec{E}_{\parallel} t + \vec{v}_{\parallel}(0) \tag{5.14}$$

The first term in the right hand side of (5.14) represents the cyclotron circular motion, and the following ones represent, respectively, the drift velocity of the guiding center (perpendicular to both $\vec{E}_{\perp}$ and $\vec{B}$), the constant acceleration of the guiding center along $\vec{B}$, and the initial velocity parallel to $\vec{B}$.

Note that the velocity $\vec{v}_E$ is independent of the mass and the sign of the charge and therefore is the same for both positive and negative particles. It is usually called the *plasma drift velocity*. Since $\vec{E}_{\parallel} \times \vec{B} = 0$, (5.11) can also be written as

$$\vec{v}_E = \vec{E} \times \vec{B}/B^2 \tag{5.15}$$

The resulting motion of the particle in the plane normal to $\vec{B}$ is, in general, a cycloid, as shown in Fig. 10. The physical explanation for this cycloidal motion is as follows. The electric force $q\vec{E}_{\perp}$, acting simultaneously with the magnetic force, accelerates the particle so as to increase or decrease its velocity, depending on the relative direction of the particle motion with respect to the direction of $\vec{E}_\perp$ and on the charge sign. According to (4.13) the radius of gyration increases with velocity and hence the radius of curvature of the particle path will vary under the action of $\vec{E}_\perp$. This results in a cycloidal trajectory with a net drift in the direction perpendicular to both $\vec{E}$ and $\vec{B}$. Different trajectories are obtained, depending on the initial conditions and on the magnitude of the applied electric and magnetic fields.

The ions are much more massive than the electrons and therefore the Larmor radius for ions is correspondingly greater and the Larmor frequency correspondingly smaller than for electrons. Consequently, the arcs of cycloid for ions are greater than for electrons, but there is a larger number of arcs of cycloid per second for electrons, such that the drift velocity is the same for both species.

In a collisionless plasma the drift velocity does not imply in an electric current, since both positive and negative particles move together. When collisions between charged and neutral particles are important, this drift gives rise to an electric current, since the ion-neutral collision frequency is greater than the electron-neutral collision frequency, causing the ions to

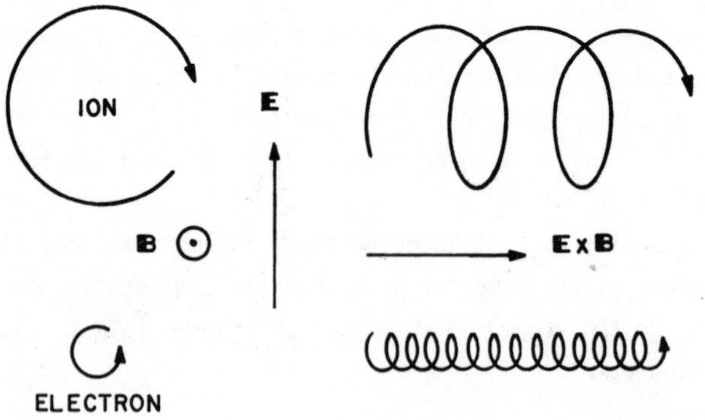

Fig. 10 - Cycloidal trajectories described by ions and electrons in crossed electric and magnetic fields. The electric field $\vec{E}$ acting together with the magnetic flux density $\vec{B}$ gives rise to a drift velocity in the direction $\vec{E} \times \vec{B}$.

move slower than the electrons. This current is normal to both $\vec{E}$ and $\vec{B}$, and is in the direction opposite to $\vec{v}_E$. It is known as the *Hall current*.

5.2 Solution in Cartesian coordinates

Let us choose a Cartesian coordinate system with the z axis pointing in the direction of $\vec{B}$, so that

$$\vec{B} = B\,\hat{z} \tag{5.16}$$

$$\vec{E} = E_x\,\hat{x} + E_y\,\hat{y} + E_z\,\hat{z} \tag{5.17}$$

Using (4.15), the equation of motion (5.1) can be written as

$$d\vec{v}/dt = (q/m)\,[(E_x + v_y\,B)\,\hat{x} + (E_y - v_x\,B)\,\hat{y} + E_z\,\hat{z}] \tag{5.18}$$

As before, we consider, in what follows, a *positive charge*. The results for a negative charge can be obtained by changing the sign of ω_c in the results for the positive charge.

The z component of (5.18) can be integrated directly and gives the same results expressed in (5.6) and (5.7). For the x and y components, we first take the derivative of dv_x/dt with respect to time and substitute the expression for dv_y/dt, which gives

$$d^2v_x/dt^2 + \omega_c^2\,v_x = \omega_c^2\,E_y/B \tag{5.19}$$

This is the inhomogeneous differential equation for a harmonic oscillator of frequency ω_c. Its solution is given by the sum of the solution of the homogeneous equation [given in (4.21)] with a particular solution (which is clearly E_y/B). Thus,

$$v_x(t) = v_\perp'\,\sin\,(\omega_c t + \theta_0) + E_y/B \tag{5.20}$$

where $v_\perp'$ and θ_0 are integration constants. The solution for $v_y(t)$ can be obtained by substituting (5.20) directly into (5.18). Hence,

$$v_y(t) = (dv_x/dt)/\omega_c - E_x/B = v_\perp'\,\cos\,(\omega_c t + \theta_0) - E_x/B \tag{5.21}$$

Therefore, the velocity components $v_x(t)$ and $v_y(t)$, in the plane perpendicular to $\vec{B}$, oscillate at the cyclotron frequency ω_c with amplitude $v_\perp'$. This motion is superposed to a constant drift velocity $\vec{v}_E$ given by

$$\vec{v}_E = (E_y/B)\,\hat{x} - (E_x/B)\,\hat{y} \tag{5.22}$$

This expression corresponds to (5.11) when $\vec{B} = B\hat{z}$.

One more integration of (5.20) and (5.21) gives the particle trajectory in the (x,y) plane

$$x(t) = -\,(v_\perp'/\omega_c)\cos(\omega_c t + \theta_0) + (E_y/B)\,t + X_0 \tag{5.23}$$

$$y(t) = (v_\perp'/\omega_c)\sin(\omega_c t + \theta_0) - (E_x/B)\,t + Y_0 \tag{5.24}$$

where X_0 and Y_0 are defined according to (4.27) and (4.28), but with $v_\perp$ replaced by $v_\perp'$.

In summary, the motion of a charged particle in uniform electrostatic and magnetostatic fields consists of three components:

(a) A constant acceleration $q\,\vec{E}_\parallel/m$ along the $\vec{B}$ field. If $\vec{E}_\parallel = 0$, the particle moves along $\vec{B}$ with its initial velocity.

(b) A rotation about the direction of $\vec{B}$ at the cyclotron frequency $\omega_c = |q|\,B/m$ and radius $r_c = v_\perp'/\omega_c$.

(c) An electromagnetic drift velocity $\vec{v}_E = (\vec{E} \times \vec{B})/B^2$, perpendicular to both $\vec{B}$ and $\vec{E}$.

6. DRIFT DUE TO AN EXTERNAL FORCE

If some additional force $\vec{F}$ (gravitational force or inertial force, if the motion is considered in a noninertial system, for example) is present, the equation of motion (1.5) must be modified to include this force,

$$m\,d\vec{v}/dt = q\,(\vec{E} + \vec{v} \times \vec{B}) + \vec{F} \tag{6.1}$$

The effect of this force is, in a formal sense, analogous to the effect of the electric field. We assume here that $\vec{F}$ is uniform and constant. In analogy with the drift velocity $(\vec{E} \times \vec{B})/B^2$, the drift produced by the force $\vec{F}$ having a component normal to $\vec{B}$ is given by

$$\vec{v}_F = \vec{F} \times \vec{B}/qB^2 \tag{6.2}$$

In the case of a uniform gravitational field, for example, we have $\vec{F} = m\vec{g}$, where $\vec{g}$ is the acceleration due to gravity, and the drift velocity is given by

$$\vec{v}_g = (m/q)\ \vec{g} \times \vec{B}/B^2 \tag{6.3}$$

This drift velocity depends on the ratio m/q and therefore it is in opposite directions for particles of opposite charge (Fig. 11). We have seen that in a coordinate system moving with the velocity $\vec{v}_E = (\vec{E} \times \vec{B})/B^2$, the electric field component $\vec{E}_\perp$ is transformed away leaving the magnetic field unchanged. The gravitational field however cannot, in this context, be transformed away.

In a collisionless plasma, associated with the gravitational drift velocity there is an electric current density, $\vec{J}_g$, in the direction of $\vec{g} \times \vec{B}$, which can be expressed as

$$\vec{J}_g = (1/\Delta V) \sum_i q_i\ \vec{v}_{gi} \tag{6.4}$$

where the summation is over all charged particles contained in a suitably chosen small volume element ΔV. Using (6.3) we obtain

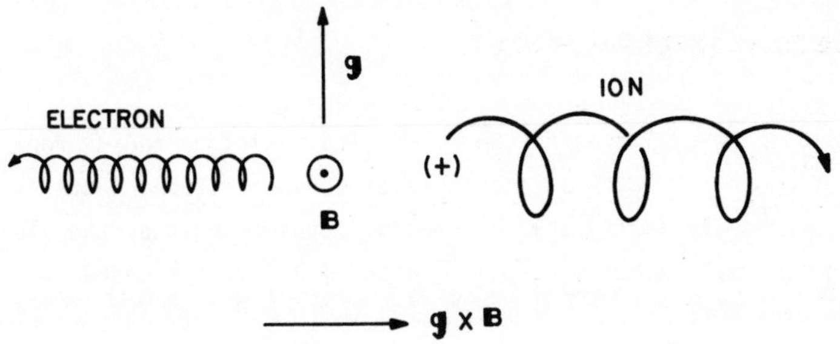

Fig. 11 - The drift of a gyrating particle in crossed gravitational and magnetic fields.

$$\vec{J}_g = \left[(1/\Delta V) \sum_i m_i\right] (\vec{g} \times \vec{B})/B^2 = \rho_m (\vec{g} \times \vec{B})/B^2 \tag{6.5}$$

where ρ_m denotes the total mass density of the charged particles.

A comment on the validity of (6.2) is appropriate here. Since we have used the nonrelativistic equation of motion, there is a limitation on the magnitude of the force $\vec{F}$ in order that (6.2) be applicable. The magnitude of the transverse drift velocity is given by

$$v_D = F_\perp/(qB) \tag{6.6}$$

Hence, for the nonrelativistic equation of motion to be applicable we must have

$$F_\perp/(qB) << c \tag{6.7}$$

or, if $\vec{F}$ is due to an electrostatic field $\vec{E}$,

$$E_\perp/B << c \tag{6.8}$$

For a magnetic field of 1 Tesla for example, (6.2) may be used as long as $E_\perp$ is much less than 10^8 Volts/m. If these conditions are not satisfied, the problem becomes a relativistic one. Although the relativistic equations of motion can be integrated exactly for constant $\vec{B}$, $\vec{E}$ and $\vec{F}$, we shall not analyze this problem here. It is left as an exercise for the reader.

PROBLEMS

2.1 - Calculate the cyclotron frequency and the cyclotron radius for:
(a) An electron in the Earth's ionosphere at 300km altitude, where the magnetic flux density $B \cong 0.5 \times 10^{-4}$ Tesla, considering that the electron moves at the thermal velocity $(kT/m)^{1/2}$ with $T = 1000$ K, where k is Boltzmann's constant.
(b) A 50 MeV proton in the Earth's inner Van Allen radiation belt at about 1.5 R_E (where $R_E = 6370$km is the Earth's radius) from the center of the Earth in the equatorial plane, where $B \cong 10^{-5}$ Tesla.
(c) A 1 MeV electron in the Earth's outer Van Allen radiation belt at about 4 R_E from the center of the Earth in the equatorial plane, where $B \cong 10^{-9}$

Tesla.

(d) A proton in the solar wind with a streaming velocity of 100km/sec, in a magnetic flux density $B \cong 10^{-9}$ Tesla.

(e) A 1 MeV proton in the solar atmosphere, in the region of a sunspot, in which $B = 0.1$ Tesla.

2.2 - For an electron and an oxygen ion O^+ in the Earth's ionosphere, at 300km altitude in the equatorial plane, where $B \cong 0.5 \times 10^{-4}$ Tesla calculate:

(a) The gravitational drift velocity $\vec{v}_g$.

(b) The gravitational current density $\vec{J}_g$, considering $n_e = n_i = 10^{12}\ m^{-3}$.

Assume that $\vec{g}$ is perpendicular to $\vec{B}$.

2.3 - Consider a particle of mass m and charge q moving in the presence of constant and uniform electromagnetic fields given by $\vec{E} = \hat{y}\, E_0$ and $\vec{B} = \hat{z}\, B_0$. Assuming that initially ($t = 0$) the particle is at rest at the origin of a Cartesian coordinate system, show that it moves on the cycloid

$$x(t) = (E_0/B_0)\ \{t - [\sin(\omega_c t)]/\omega_c\}$$

$$y(t) = (E_0/B_0)\ [1 - \cos(\omega_c t)]/\omega_c$$

Plot the trajectory of the particle in the $z = 0$ plane for $q > 0$ and $q < 0$, and consider the cases when $v_\perp > v_E$, $v_\perp = v_E$ and $v_\perp < v_E$, where $v_\perp$ denotes the particle cyclotron motion velocity and v_E is the electromagnetic drift velocity.

2.4 - In general the trajectory of a charged particle in crossed electric and magnetic fields is a cycloid. Show that, if $\vec{v} = \hat{x}\, v_0$, $\vec{B} = \hat{z}\, B$ and $\vec{E} = \hat{y}\, E$, then for $v_0 = E/B$ the path is a straight line. Explain how this situation can be exploited to design a mass spectrometer.

2.5 - Derive the relativistic equation of motion in the form (1.4), starting from (1.1) and the relation (1.2).

2.6 - Write down, in vector form, the relativistic equation of motion for a charged particle in the presence of a uniform magnetostatic field $\vec{B} = \hat{z}\, B_0$, and show that its Cartesian components are given by

$$\frac{d}{dt}[v_x/(1-\beta^2)^{1/2}] = (qB_o/m)\, v_y$$

$$\frac{d}{dt}[v_y/(1-\beta^2)^{1/2}] = -(qB_o/m)\, v_x$$

$$\frac{d}{dt}[v_z/(1-\beta^2)^{1/2}] = 0$$

were $\beta = v/c$. Show that the velocity and trajectory of the charged particle are given by the same formulas as in the nonrelativistic case, but with ω_c replaced by $(|q|B_o/m)\,(1-\beta^2)^{1/2}$.

2.7 - Study the motion of a relativistic charged particle in the presence of crossed electric ($\vec{E}$) and magnetic ($\vec{B}$) fields which are constant in time and uniform in space. What coordinate transformation must be made in order to transform away the transversal electric field? Derive equations for the velocity and trajectory of the charged particle.

CHAPTER 3

Charged Particle Motion in Nonuniform Magnetostatic Fields

1. INTRODUCTION

When the fields are spatially nonuniform, or when they vary with time, the integration of the equation of motion (2.1.1) can be a mathematical problem of great difficulty. In this case, since the equation of motion is nonlinear, the theory may become extremely involved and a rigorous analytic expression for the charged particle trajectory cannot, in general, be obtained in closed form. Complicated and tedious numerical methods of integration must be used in order to obtain all the details of the motion.

There is one particularly important case, however, in which it becomes possible to obtain an approximate, but otherwise general solution without recourse to numerical integration, if the details of the particle motion are not of interest. This is the case when the magnetic field is strong and *slowly varying* in both *space* and *time*, and when the electric field is weak. In a wide variety of situations of interest the fields are *approximately* constant and uniform, at least on the distance and time scales seen by the particle during one gyration about the magnetic field. This is the case for many laboratory plasmas, including those of relevance to the problem of controlled thermonuclear reactions, and also for a great number of astrophysical plasmas.

In this chapter we investigate the motion of a charged particle in a static magnetic field *slightly* inhomogeneous in space. The word slightly here means that the spatial variation of the magnetic field inside the particle orbit is small compared with the magnitude of $\vec{B}$. In other words, we shall consider only magnetostatic fields whose spatial change in a distance of the order of the Larmor radius, r_c, is much smaller than the magnitude of the field itself.

To specify more quantitatively this assumption, let δB represent the spatial change in the magnitude of $\vec{B}$ in a distance of the order of r_c, that is, $\delta B = r_c |\vec{\nabla} B|$, where $\vec{\nabla} B$ is the gradient of the magnitude of $\vec{B}$. It is assumed therefore that $\delta B \ll B$. Consequently, in what follows we limit our discussion to problems where the deviations from uniformity are small and solve for the trajectory only in the first order approximation. The analysis of the motion of charged particles in stationary fields based on this approximation is often referred to as the *first order orbit theory*. This theory was first used systematically by the Swedish scientist Alfvén, and it is also known as the *Alfvén approximation* or the *guiding center approximation*.

The concept of guiding center is of great utility in the development of this theory. We have seen that in a uniform magnetic field the particle motion can be regarded as a superposition of a circular motion about the direction of $\vec{B}$, with a motion of the guiding center along $\vec{B}$. In the case of a nonuniform $\vec{B}$ field, satisfying the condition (1.2), the value of $\vec{B}$ at the particle position differs only slightly from its value at the guiding center. The component of the particle motion, in a plane normal to the field line that passes through the guiding center instantaneous position, will still be nearly circular (Fig.1). However, due to the spatial variation of $\vec{B}$, we expect in this case a gradual drift of the guiding center across $\vec{B}$, as well as a gradual change of its velocity along $\vec{B}$.

The rapid gyrations of the charged particle about the direction of $\vec{B}$ are not usually of great interest and it is convenient to eliminate them from the equations of motion, and focus attention on the guiding center motion. In the motion of the guiding center, the small oscillations (of amplitudes small compared with the cyclotron radius) occurring during one gyration period may be averaged out, since they represent the effect of perturbations due to the spatial variation of the magnetic field. The problem is thus reduced to the calculation of the average values over one gyration period (and not the instantaneous values) of the guiding center *transverse drift velocity* and *parallel acceleration*.

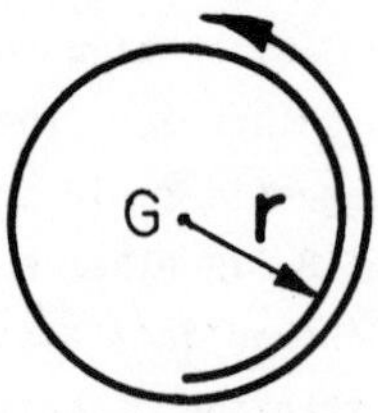

Fig. 1 - The motion of a charged particle in a magnetostatic field slightly inhomogeneous is nearly circular.

2. SPATIAL VARIATION OF THE MAGNETIC FIELD

Any of the three components of the magnetic flux density, $\vec{B} = B_x\hat{x} + B_y\hat{y} + B_z\hat{z}$, may vary with respect to the three coordinates x, y, and z. Consequently, nine parameters are needed to completely specify the spatial variation of $\vec{B}$. These parameters can be conveniently represented by the dyad (or tensor) $\vec{\nabla}\vec{B}$, which can be written in matrix form as

$$\vec{\nabla}\vec{B} = (\hat{x}\ \hat{y}\ \hat{z}) \begin{pmatrix} \partial B_x/\partial x & \partial B_y/\partial x & \partial B_z/\partial x \\ \partial B_x/\partial y & \partial B_y/\partial y & \partial B_z/\partial y \\ \partial B_x/\partial z & \partial B_y/\partial z & \partial B_z/\partial z \end{pmatrix} \begin{pmatrix} \hat{x} \\ \hat{y} \\ \hat{z} \end{pmatrix} \tag{2.1}$$

Of these nine components only eight are independent, since the following Maxwell equation

$$\vec{\nabla} \cdot \vec{B} = \partial B_x/\partial x + \partial B_y/\partial y + \partial B_z/\partial z = 0 \tag{2.2}$$

shows that only two of the divergence terms are independent.

If in the region where the particle is moving the condition $\vec{J} = 0$ is satisfied, then other restrictions exist in the number of independent components of $\vec{\nabla}\vec{B}$ since, under these circumstances, the relation $\vec{\nabla} \times \vec{B} = 0$ holds. This means that in regions where there are no electric currents $\vec{B}$ can be written as the gradient of a *scalar magnetic potential*,

$$\vec{B} = \vec{\nabla}\phi_m \tag{2.3}$$

where the magnetic potential ϕ_m satisfies the Laplace equation

$$\nabla^2\phi_m = 0 \tag{2.4}$$

In regions where an electric current density exists, we have $\nabla \times \vec{B} = \mu_o \vec{J}$ and we cannot define a scalar magnetic potential ϕ_m as indicated. The number of independent components of $\vec{\nabla}\vec{B}$ cannot, in this case, be reduced without knowing the current density $\vec{J}$.

Let us consider a Cartesian coordinate system such that at the origin the magnetic field is in the z direction,

$$\vec{B}(0,0,0) = \vec{B}_0 = B_0\hat{z} \qquad (2.5)$$

The nine components of $\vec{\nabla}\vec{B}$ can be conveniently grouped in four categories:

(a) Divergence terms: $\partial B_x/\partial x$, $\partial B_y/\partial y$, $\partial B_z/\partial z$ (2.6a)

(b) Gradient terms: $\partial B_z/\partial x$, $\partial B_z/\partial y$ (2.6b)

(c) Curvature terms: $\partial B_x/\partial z$, $\partial B_y/\partial z$ (2.6c)

(d) Shear terms: $\partial B_x/\partial y$, $\partial B_y/\partial x$ (2.6d)

2.1 Divergence terms

We shall initially discuss the magnetic field line geometry corresponding to the *divergence* terms of $\vec{\nabla}\vec{B}$. The presence of a small variation of the component B_z in the z direction (i.e. $\partial B_z/\partial z \neq 0$), implies that at least one of the terms $\partial B_x/\partial x$ or $\partial B_y/\partial y$ is also present, as can be seen from (2.2). It is of great utility to make use here of the concept of *magnetic flux lines* which, at any point, are parallel to the $\vec{B}$ field at that point and whose density at each point is proportional to the local magnitude of $\vec{B}$. To determine the differential equation of a line of force, let

$$d\vec{s} = dx\,\hat{x} + dy\,\hat{y} + dz\,\hat{z} \qquad (2.7)$$

be an element of arc along the magnetic field line. Then, we must have

$$d\vec{s} \times \vec{B} = 0 \qquad (2.8)$$

since $d\vec{s}$ is parallel to $\vec{B}$, which gives by expansion of the cross product,

$$dx/B_x = dy/B_y = dz/B_z \qquad (2.9)$$

Since we are focusing attention only on the divergence terms of $\vec{B}$, and in the region of interest the field is considered to be mainly in the z direction, we may expand B_x and B_y in a Taylor series about the origin as follows (see Fig. 2)

$$B_x(x_1,0,0) = B_x(0,0,0) + \left(\frac{\partial B_x}{\partial x}\right) x_1 = \left(\frac{\partial B_x}{\partial x}\right) x_1 \qquad (2.10)$$

$$B_y(0,y_1,0) = B_y(0,0,0) + \left(\frac{\partial B_y}{\partial y}\right) y_1 = \left(\frac{\partial B_y}{\partial y}\right) y_1 \tag{2.11}$$

where the second and higher order terms were neglected. Note that at the origin $B_x = B_y = 0$. Therefore, the magnetic field line crossing the $z = 0$ plane at the point $(x_1,y_1,0)$, when projected on the $x - z$ plane $(y = 0)$ and on the $y - z$ plane $(x = 0)$, satisfies the following differential equations, respectively,

$$\frac{dx}{dz} = \frac{B_x}{B_z} = \frac{1}{B_z}\left(\frac{\partial B_x}{\partial x}\right) x_1 \qquad (y=0) \tag{2.12}$$

$$\frac{dy}{dz} = \frac{B_y}{B_z} = \frac{1}{B_z}\left(\frac{\partial B_y}{\partial y}\right) y_1 \qquad (x=0) \tag{2.13}$$

These equations show that the field lines converge or diverge in the $x - z$ plane or in the $y - z$ plane, depending on the sign of the divergence terms of $\vec{B}$. Fig.3 illustrates the field line geometry when $\partial B_x/\partial x$ and $\partial B_y/\partial y$ are positive.

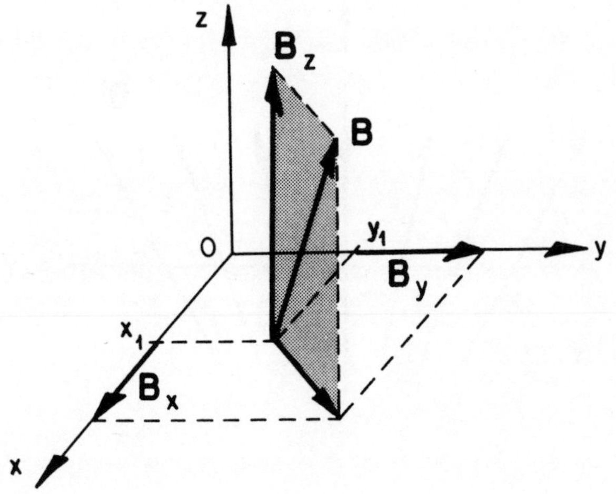

Fig. 2 - The magnetic field components B_x and B_y at the points $(x_1,0,0)$ and $(0,y_1,0)$, near the origin.

2.2 Gradient and curvature terms

The following vector field has a *gradient* in the x direction,

$$\vec{B} = B_z \hat{z} = B_0 (1 + \alpha x) \hat{z} \tag{2.14}$$

(see Fig. 4). We must note however that in a region where $\vec{J} = 0$ this vector field does not satisfy the Maxwell equation $\vec{\nabla} \times \vec{B} = 0$, so that we must add to (2.14) a term of *curvature*, given by $B_x \hat{x} = B_0 \alpha z \hat{x}$. Therefore, a magnetic field having gradient and curvature, and which satisfies $\vec{\nabla} \times \vec{B} = 0$, is

$$\vec{B} = B_0 [\alpha z \hat{x} + (1 + \alpha x) \hat{z}] \tag{2.15}$$

The geometry of the magnetic field lines corresponding to this equation is indicated in Fig. 5.

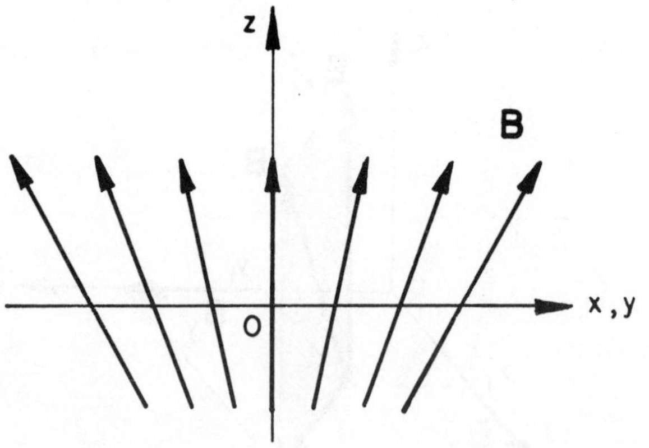

Fig. 3 - Geometry of the magnetic field lines corresponding to the divergence terms $\partial B_x/\partial x$ or $\partial B_y/\partial y$, when they are positive.

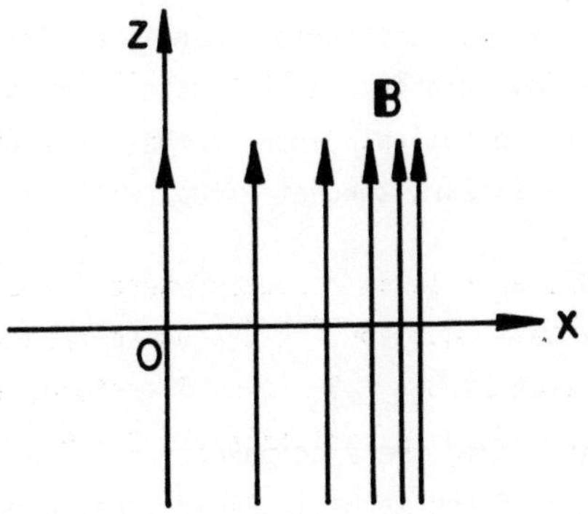

Fig. 4 - Geometry of the magnetic field lines when $\vec{B}$ has a gradient in the x direction, according to Eq. (2.14). This field geometry does not satisfy $\vec{\nabla} \times \vec{B} = 0$.

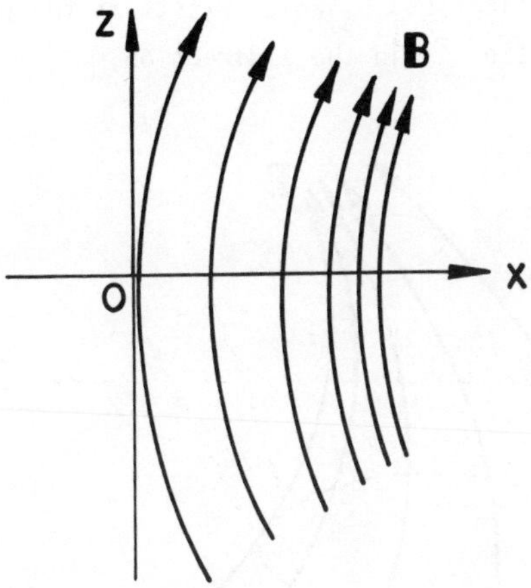

Fig. 5 - Geometry of the magnetic field lines corresponding to (2.15), with gradient and curvature terms.

Generally, all terms corresponding to divergence, gradient and curvature are simultaneously present. Fig. 6 illustrates a $\vec{B}$ field having divergence, gradient and curvature. An example is the Earth's magnetic field (Fig. 1.4). Later in this section we will investigate the effects of each of these terms separately on the charged particle motion. Since in the first order approximation the equations are linear, the net effect will be the sum of each one of them.

2.3 Shear terms

The *shear terms* of (2.6) enter into the z component of $\vec{\nabla} \times \vec{B}$, that is, into $\vec{B} \cdot (\vec{\nabla} \times \vec{B})$, and cause twisting of the magnetic field lines about each other. They do not produce any first order drifts, although the shape of the orbit can be slightly changed. They do not give rise to any particularly interesting effect on the motion of charged particles and will not be considered any further.

3. EQUATION OF MOTION IN THE FIRST ORDER APPROXIMATION

We consider that the magnetic field $\vec{B}_0$ which exists at the origin in the guiding center coordinate system is in the z direction,

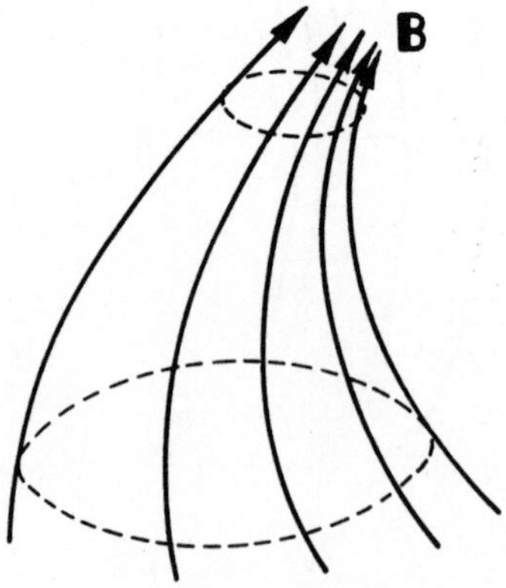

Fig. 6 - Schematic representation of a magnetic field having divergence, gradient and curvature terms.

$$\vec{B}(0,0,0) \equiv \vec{B}_0 = B_0\hat{z} \tag{3.1}$$

The particle motion in the neighborhood of the origin can be described by considering only a linear approximation to the magnetic field near the origin. Let $\vec{r}$ be the momentary position vector of the particle in the guiding center coordinate system (Fig. 1). In the region of interest (near the origin) the magnetic field can be expressed by a Taylor expansion,

$$\vec{B}(\vec{r}) = \vec{B}_0 + \vec{r} \cdot (\vec{\nabla}\vec{B}) + \ldots \tag{3.2}$$

where the derivatives of $\vec{B}$ are to be calculated at the origin. Note that actually the instantaneous position of the particle guiding center changes slightly during one period of rotation, while the origin is kept fixed during this time.

Since we are assuming that the spatial variation of $\vec{B}$ in a distance of the order of the Larmor radius is much smaller than the magnitude of $\vec{B}$ itself, the higher order terms of (3.2) have been neglected. The condition

$$\delta B = |\vec{r} \cdot (\vec{\nabla}\vec{B})| \ll |\vec{B}_0| \tag{3.3}$$

is clearly met (section 1). Thus, the magnetic field at the particle position differs only slightly from that existing at the guiding center. The first order term $\vec{r} \cdot (\vec{\nabla}\vec{B})$ can be written explicitly as

$$\vec{r} \cdot (\vec{\nabla}\vec{B}) = (\vec{r} \cdot \vec{\nabla})\,\vec{B} = (x\frac{\partial}{\partial x} + y\frac{\partial}{\partial y} + z\frac{\partial}{\partial z})\,\vec{B} = (x\frac{\partial B_x}{\partial x} + y\frac{\partial B_x}{\partial y} + z\frac{\partial B_x}{\partial z})\,\hat{x} + \\ + (x\frac{\partial B_y}{\partial x} + y\frac{\partial B_y}{\partial y} + z\frac{\partial B_y}{\partial z})\,\hat{y} + (x\frac{\partial B_z}{\partial x} + y\frac{\partial B_z}{\partial y} + z\frac{\partial B_z}{\partial z})\,\hat{z} \tag{3.4}$$

where the partial derivatives are to be calculated at the origin. Substituting (3.2) into the equation of motion (2.1.5), with $\vec{E} = 0$, gives

$$m\, d\vec{v}/dt = q\,(\vec{v} \times \vec{B}_0) + q\,\vec{v} \times [\vec{r} \cdot (\vec{\nabla}\vec{B})] \tag{3.5}$$

The last term in the right-hand side is of first-order compared to the first one. The particle velocity can be written as a superposition

$$\vec{v} = \vec{v}^{(0)} + \vec{v}^{(1)} = d\vec{r}^{(0)}/dt + d\vec{r}^{(1)}/dt \tag{3.6}$$

where $\vec{v}^{(1)}$ is a first-order perturbation

$$|\vec{v}^{(1)}| << |\vec{v}^{(0)}| \tag{3.7}$$

and $\vec{v}^{(0)}$ is the solution of the zero-order equation

$$m\, d\vec{v}^{(0)}/dt = q\, (\vec{v}^{(0)} \times \vec{B}_0) \tag{3.8}$$

which has already been discussed in section 4, Chapter 2. Neglecting second-order terms we can write, therefore,

$$\vec{v} \times [\vec{r} \cdot (\vec{\nabla}\vec{B})] = \vec{v}^{(0)} \times [\vec{r}^{(0)} \cdot (\vec{\nabla}\vec{B})] \tag{3.9}$$

The equation of motion (3.5) becomes, under these approximations,

$$m\, d\vec{v}/dt = q\, (\vec{v} \times \vec{B}_0) + q\, \vec{v}^{(0)} \times [\vec{r}^{(0)} \cdot (\vec{\nabla}\vec{B})] \tag{3.10}$$

The second term in the right-hand side constitutes the force term of (2.6.1). This additional force, however, is not constant but it depends on the *instantaneous* particle position. Thus, small oscillations occur during one period of gyration. Since we are interested in the smoothed motion of the guiding center we shall eliminate these small oscillations by averaging this force term over one gyration period. Therefore, in what follows we will be involved in calculating the average value over one gyration period of the force term $q\, \vec{v}^{(0)} \times [\vec{r}^{(0)} \cdot (\vec{\nabla}\vec{B})]$, which will allow us to determine the parallel acceleration of the guiding center and its transverse drift velocity using (2.6.2).

4. AVERAGE FORCE OVER ONE GYRATION PERIOD

Consider initially the case when the particle initial velocity along $\vec{B}$ is zero, so that the particle path differs but little from a circle. In a uniform magnetic field this would be equivalent to observe the particle motion in a coordinate system moving with the guiding center velocity $\vec{v}_{\parallel}$. However, when the field lines are bent, a coordinate system gliding along $\vec{B}$ is not an inertial system. The curvature of the field lines give rise to inertial forces and therefore to a *curvature drift* of the particle. This effect will be investigated later in section 7. For the moment we will assume that the field lines are not curved and that the coordinate system moves with velocity $\vec{v}_{\parallel}$.

Under the conditions indicated above, the zero-order variables, $\vec{v}^{(0)}$ and $\vec{r}^{(0)}$, are seen to be situated in the (x, y) plane. The force term

$$\vec{F} = q\,\vec{v}^{(0)} \times [\vec{r}^{(0)} \cdot (\vec{\nabla}\vec{B})] \tag{4.1}$$

can be separated in a component along $\vec{B}_0$ (z axis), $\vec{F}_{\parallel}$, and a component normal to $\vec{B}_0$ (x,y plane), $\vec{F}_{\perp}$. Using a local cylindrical coordinate system (r,Θ,z) with the z axis pointing along $\vec{B}_0$ at the origin (see Fig.7), we have

$$\vec{r}^{(0)} \cdot (\vec{\nabla}\vec{B}) = r^{(0)}\,(\partial\vec{B}/\partial r) \tag{4.2}$$

Of the three components of $\vec{B} = B_r\hat{r} + B_\Theta\hat{\Theta} + B_z\hat{z}$, $B_\Theta\hat{\Theta}$ is parallel to $\vec{v}^{(0)}$ and therefore gives no contribution to $\vec{F}$, while $B_r\hat{r}$ contributes to $\vec{F}_{\parallel}$ and $B_z\hat{z}$ contributes to $\vec{F}_{\perp}$. Hence, from (4.1) and (4.2),

$$\vec{F}_{\parallel} = q\,(\vec{v}^{(0)} \times \hat{r})\,r^{(0)}\,\partial B_r/\partial r = |q|\,v^{(0)}\,r^{(0)}\,(\partial B_r/\partial r)\,\hat{z} \tag{4.3}$$

$$\vec{F}_{\perp} = q\,(\vec{v}^{(0)} \times \hat{z})\,r^{(0)}\,\partial B_z/\partial r = -\,|q|\,v^{(0)}\,r^{(0)}\,(\partial B_z/\partial r)\,\hat{r} \tag{4.4}$$

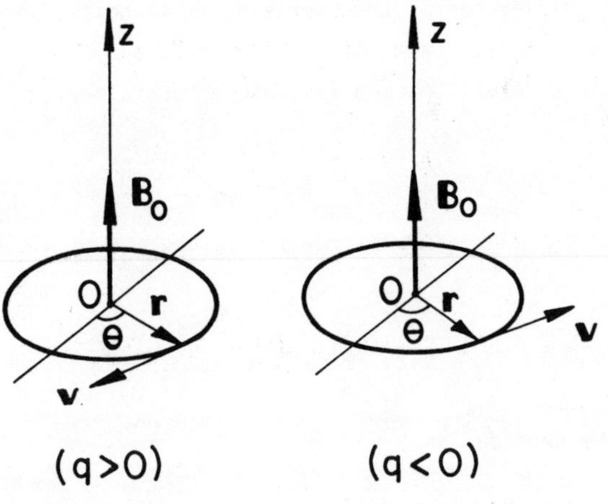

Fig. 7 - Local cylindrical coordinate system with the z axis pointing in the direction of the field $\vec{B}_0$ at the origin.

Note that if $q > 0$ we have $\vec{v}^{(0)} \times \hat{r} = v^{(0)}\hat{z}$, whereas if $q < 0$ we have $\vec{v}^{(0)} \times \hat{r} = - v^{(0)}\hat{z}$. Now, $r^{(0)}$ is the cyclotron radius corresponding to $\vec{B}_0$,

$$r^{(0)} = v^{(0)}/\omega_c = mv^{(0)}/(|q|B_0) \tag{4.5}$$

and using the expression for the magnitude of the magnetic moment (2.4.34), we can write (4.3) and (4.4) as

$$\vec{F}_{\parallel} = 2\,|\vec{m}|(\partial B_r/\partial r)\,\hat{z} \tag{4.6}$$

$$\vec{F}_{\perp} = -\,2\,|\vec{m}|(\partial B_z/\partial r)\,\hat{r} \tag{4.7}$$

These results apply to both positively and negatively charged particles.

The average values of $\vec{F}_{\parallel}$ and $\vec{F}_{\perp}$ over one gyration period are given by

$$< \vec{F}_{\parallel} > = 2\,|\vec{m}|\,\hat{z}\left[\frac{1}{2\pi}\oint \left(\frac{\partial B_r}{\partial r}\right) d\theta\right] = 2\,|\vec{m}|\,\hat{z} < \left(\frac{\partial B_r}{\partial r}\right) > \tag{4.8}$$

$$< \vec{F}_{\perp} > = -\,2\,|\vec{m}|\left[\frac{1}{2\pi}\oint \left(\frac{\partial B_z}{\partial r}\right)\hat{r}\, d\theta\right] = -\,2\,|\vec{m}| < \hat{r}\left(\frac{\partial B_z}{\partial r}\right) > \tag{4.9}$$

(4.8) produces the guiding center parallel acceleration, while (4.9) is responsible for the guiding center transverse drift velocity. The first one is the result of the *divergence* terms of $\vec{B}$, and the second one of the *gradient* terms. We proceed now to evaluate each force term separately.

4.1 Parallel force

Note that from $\vec{\nabla} \cdot \vec{B} = 0$ we have, in cylindrical coordinates,

$$\frac{1}{r}\frac{\partial}{\partial r}(r\,B_r) + \frac{1}{r}\frac{\partial}{\partial \theta}(B_\theta) + \frac{\partial}{\partial z}(B_z) = 0 \tag{4.10}$$

The first term can be expanded as

$$\frac{1}{r}\frac{\partial}{\partial r}(r\,B_r) = \frac{\partial B_r}{\partial r} + \frac{B_r}{r} \tag{4.11}$$

Since at $r = 0$ we have $B_r = 0$, and since near the origin B_r changes only very slightly with r, we can take

$$B_r/r = \partial B_r/\partial r \tag{4.12}$$

Consequently, from (4.12) and (4.11),

$$\left(\frac{\partial B_r}{\partial r}\right) = -\frac{1}{2}\left[\frac{1}{r}\left(\frac{\partial B_\Theta}{\partial \Theta}\right) + \left(\frac{\partial B_z}{\partial z}\right)\right] \tag{4.13}$$

Hence, taking the average over one gyration period,

$$< \left(\frac{\partial B_r}{\partial r}\right) > = -\frac{1}{2} < \frac{1}{r}\left(\frac{\partial B_\Theta}{\partial \Theta}\right) > - \frac{1}{2} < \left(\frac{\partial B_z}{\partial z}\right) > \tag{4.14}$$

Now, since B is single-valued,

$$< \frac{1}{r}\left(\frac{\partial B_\Theta}{\partial \Theta}\right) > = \frac{1}{2\pi} \oint \frac{1}{r}\left(\frac{\partial B_\Theta}{\partial \Theta}\right) d\Theta = 0 \tag{4.15}$$

Furthermore, since $\partial B_z/\partial z$ is a very slowly varying function inside the particle orbit, it can be taken outside the integral sign, so that we have approximately,

$$< \left(\frac{\partial B_z}{\partial z}\right) > = \frac{1}{2\pi} \oint \left(\frac{\partial B_z}{\partial z}\right) d\Theta = \frac{\partial B_z}{\partial z} = \frac{\partial B}{\partial z} \tag{4.16}$$

It is justifiable to replace B_z by B in (4.16), since all the spatial variations of the magnetic field in the region of interest are very small. Therefore, we have finally from (4.14), (4.15) and (4.16),

$$< \left(\frac{\partial B_r}{\partial r}\right) > = -\frac{1}{2}\left(\frac{\partial B}{\partial z}\right) \tag{4.17}$$

Using this result, the parallel force (4.8) becomes

$$< \vec{F}_\parallel > = - |\vec{m}| (\partial B/\partial z)\, \hat{z} = - |\vec{m}| (\vec{\nabla} B)_\parallel \tag{4.18}$$

or, equivalently,

$$< \vec{F}_\parallel > = (\vec{m} \cdot \vec{\nabla})\, B\, \hat{z} = - (|\vec{m}|/B)\, [(\vec{B} \cdot \vec{\nabla})\, \vec{B}]_\parallel \tag{4.19}$$

since $\vec{m} = - |\vec{m}| \hat{z} = - |\vec{m}| \vec{B}/B$, and where the derivatives are evaluated at the origin.

4.2 Perpendicular force

It is convenient to consider a two-dimensional Cartesian coordinate system (x,y) in the perpendicular plane, such that $x = r \cos \Theta$ and $y = r \sin \Theta$ (see Fig. 8). Hence,

$$\hat{r} = \cos \Theta \, \hat{x} + \sin \Theta \, \hat{y} \tag{4.20}$$

$$\frac{\partial}{\partial r} = \frac{dx}{dr}\frac{\partial}{\partial x} + \frac{dy}{dr}\frac{\partial}{\partial y} = \cos \Theta \frac{\partial}{\partial x} + \sin \Theta \frac{\partial}{\partial y} \tag{4.21}$$

Therefore, we obtain

$$< \hat{r} \left(\frac{\partial B_z}{\partial r}\right) > = <(\cos \Theta \, \hat{x} + \sin \Theta \, \hat{y}) \left[\cos \Theta \left(\frac{\partial B_z}{\partial x}\right) + \sin \Theta \left(\frac{\partial B_z}{\partial y}\right)\right]> =$$

$$= < \cos^2 \Theta \left(\frac{\partial B_z}{\partial x}\right) \hat{x} > + < \sin \Theta \cos \Theta \left(\frac{\partial B_z}{\partial x}\right) \hat{y} > +$$

$$+ < \cos \Theta \sin \Theta \left(\frac{\partial B_z}{\partial y}\right) \hat{x} > + < \sin^2 \Theta \left(\frac{\partial B_z}{\partial y}\right) \hat{y} > \tag{4.22}$$

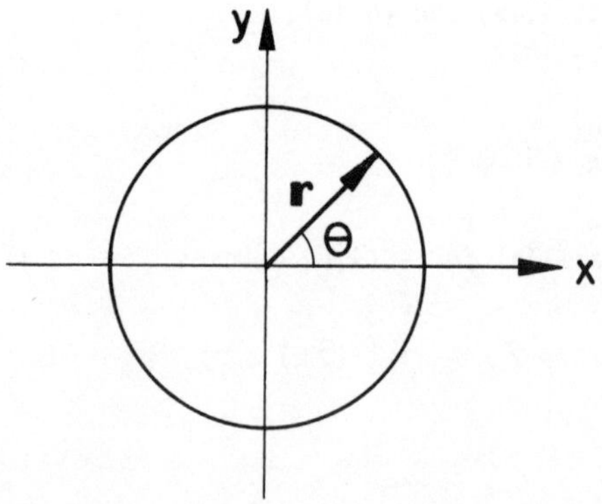

Fig. 8 - Two-dimensional coordinate system in the perpendicular plane, used in the evaluation of $< \vec{F}_\perp >$.

Next we approximate $(\partial B_z/\partial x)$ by $(\partial B/\partial x)$, and the same for the y-derivative, since these terms are slowly varying functions inside the particle orbit, so that we can take them outside the integral sign. Noting that $\langle \sin\Theta \cos\Theta \rangle = 0$ and $\langle \cos^2\Theta \rangle = \langle \sin^2\Theta \rangle = 1/2$, we obtain

$$\langle \hat{r} \left(\frac{\partial B_z}{\partial r}\right) \rangle = \frac{1}{2} \left(\frac{\partial B}{\partial x}\right) \hat{x} + \frac{1}{2} \left(\frac{\partial B}{\partial y}\right) \hat{y} \tag{4.23}$$

Substituting this result into (4.9), yields

$$\langle \vec{F}_\perp \rangle = - |\vec{m}| \, [(\partial B/\partial x)\, \hat{x} + (\partial B/\partial y)\, \hat{y}] = - |\vec{m}| \, (\vec{\nabla} B)_\perp \tag{4.24}$$

4.3 Total average force

We proceed now to write down a general expression for the *total average force* $\langle \vec{F} \rangle = \langle \vec{F}_\parallel \rangle + \langle \vec{F}_\perp \rangle$. From (4.18) and (4.24) we have

$$\langle \vec{F} \rangle = - |\vec{m}| \, (\vec{\nabla} B)_\parallel - |\vec{m}| \, (\vec{\nabla} B)_\perp = - |\vec{m}| \, \vec{\nabla} B \tag{4.25}$$

Alternatively, we can use the vector identity

$$(\vec{\nabla} \times \vec{B}) \times \vec{B} = (\vec{B} \cdot \vec{\nabla}) \vec{B} - \vec{\nabla} (B^2/2) \tag{4.26}$$

and write (4.25) in the form

$$\langle \vec{F} \rangle = - (|\vec{m}|/B) \, [(\vec{B} \cdot \vec{\nabla}) \vec{B} - (\vec{\nabla} \times \vec{B}) \times \vec{B}] \tag{4.27}$$

Since $\vec{m} = - |\vec{m}| \, \vec{B}/B$, we have

$$\langle \vec{F} \rangle = (\vec{m} \cdot \vec{\nabla}) \vec{B} + \vec{m} \times (\vec{\nabla} \times \vec{B}) \tag{4.28}$$

This is the usual expression for the force acting on a small ring current immersed in a magnetic field with spatial variation. The first term on the right-hand side of (4.28) alone gives the force acting on a magnetic dipole.

5. GRADIENT DRIFT

From (2.6.2) and (4.24) we see that $\langle \vec{F}_\perp \rangle$ causes the guiding center to drift with the velocity

$$\vec{v}_G = \frac{\langle \vec{F}_\perp \rangle \times \vec{B}}{q\,B^2} = -\frac{|\vec{m}|}{q}\,\frac{(\vec{\nabla} B) \times \vec{B}}{B^2} \tag{5.1}$$

This gradient drift is perpendicular to $\vec{B}$ and to the field gradient, and its direction depends on the charge sign. Thus, positive and negative charges drift in opposite directions, giving rise to an electric current (Fig. 9).

The physical reason for this gradient drift can be seen as follows. Since the Larmor radius of the particle orbit decreases as the magnetic field increases, the radius of curvature of the orbit is smaller in the regions of stronger $\vec{B}$ field. The positive ions gyrate in the clockwise direction for $\vec{B}$ pointing towards the observer (Fig. 9), while the electrons gyrate in the counterclockwise direction, so that the positive ions drift to the left and the electrons to the right.

In the case of a collisionless plasma, associated with this gradient drift across $\vec{B}$ there is a *magnetization current density* $\vec{J}_G$, given by

$$\vec{J}_G = (1/\Delta V) \sum_i q_i\,\vec{v}_{Gi} \tag{5.2}$$

where the summation is over all charged particles contained in a suitably chosen element of volume ΔV. From (5.1) and (5.2),

$$\vec{J}_G = -[(1/\Delta V) \sum_i |\vec{m}_i|]\,(\vec{\nabla} B) \times \vec{B}/B^2 \tag{5.3}$$

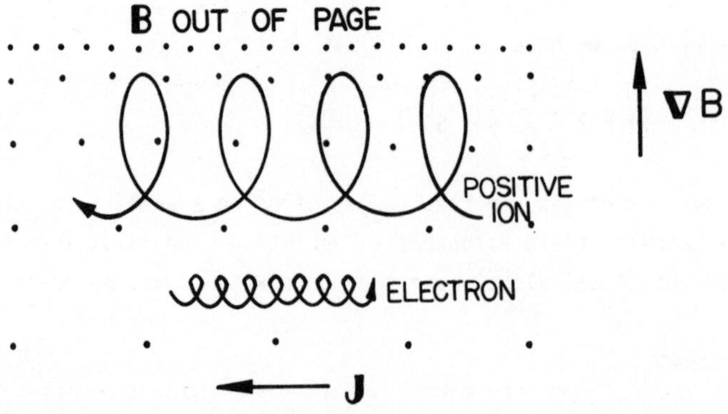

Fig. 9 - Charged particle drifts due to a $\vec{B}$ field gradient perpendicular to $\vec{B}$.

6. PARALLEL ACCELERATION OF THE GUIDING CENTER

The expression (4.18) for $< \vec{F}_{\parallel} >$ shows that, when the magnetic field has a longitudinal variation (i.e. convergence or divergence of the field lines along the z direction) as shown in Fig. 3, an axial force along z accelerates the particle in the direction of decreasing magnetic field, irrespective of whether the particle is positively or negatively charged. This is illustrated in Fig. 10. There are several important consequences of this repulsion of gyrating charges from a region of converging magnetic field lines, which we proceed to discuss.

6.1 Invariance of the orbital magnetic moment and magnetic flux

Using (4.18), the component of the equation of motion along $\vec{B}$ can be written as

$$m\ (dv_{\parallel}/dt)\ \hat{z} = < \vec{F}_{\parallel} > = -\ |\vec{m}|\ (\partial B/\partial z)\ \hat{z} \qquad (6.1)$$

If we multiply both sides of this equation by $v_{\parallel} = dz/dt$, we obtain (replacing $|\vec{m}|$ by $W_{\perp}/B$)

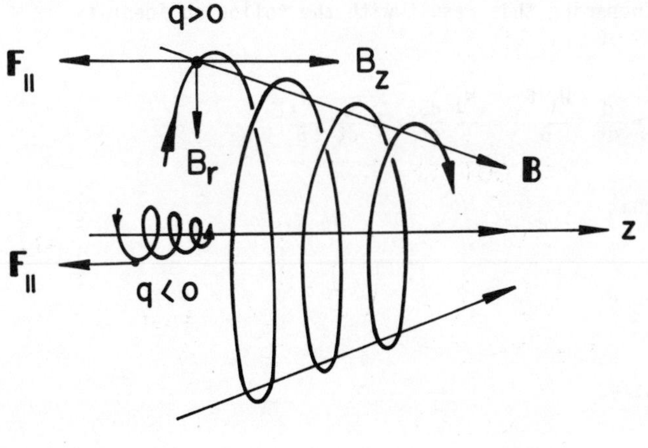

Fig. 10 - Repulsion of gyrating charges from a region of converging magnetic field lines.

$$m\, v_{\parallel} \frac{dv_{\parallel}}{dt} \equiv \frac{d}{dt}\left(\frac{1}{2} m\, v_{\parallel}^2\right) = -\frac{W_{\perp}}{B} \frac{\partial B}{\partial z} \frac{dz}{dt} \tag{6.2}$$

where $W_{\perp} = m\, v_{\perp}^2/2$ denotes the part of the kinetic energy of the particle associated with its transverse velocity. Since the total kinetic energy of a charged particle in a magnetostatic field is constant,

$$W_{\parallel} + W_{\perp} = \text{constant} \tag{6.3}$$

it follows that

$$\frac{d}{dt}(W_{\perp}) = -\frac{d}{dt}(W_{\parallel}) = -\frac{d}{dt}\left(\frac{1}{2} m\, v_{\parallel}^2\right) \tag{6.4}$$

Therefore, from (6.2) and (6.4),

$$\frac{d}{dt}(W_{\perp}) = \frac{W_{\perp}}{B} \frac{\partial B}{\partial z} \frac{dz}{dt} = \frac{W_{\perp}}{B} \frac{dB}{dt} \tag{6.5}$$

where dB/dt represents the rate of change of B as seen by the particle as it moves in the spatially varying magnetic field (i.e. in the particle frame of reference). Comparing this result with the following identity

$$\frac{d}{dt}(W_{\perp}) = \frac{d}{dt}\left(\frac{W_{\perp}\, B}{B}\right) = \frac{W_{\perp}}{B} \frac{dB}{dt} + B \frac{d}{dt}\left(\frac{W_{\perp}}{B}\right) \tag{6.6}$$

we conclude that

$$\frac{d}{dt}\left(\frac{W_{\perp}}{B}\right) = 0 \tag{6.7}$$

or, equivalently,

$$|\vec{m}| = \frac{W_{\perp}}{B} = \text{constant} \tag{6.8}$$

Therefore, as the particle moves into regions of converging or diverging $\vec{B}$ its cyclotron radius changes, but the magnetic moment remains constant. This constancy of the particle magnetic moment holds only within the approximation used, that is, when the spatial variation of $\vec{B}$ inside the particle orbit is small compared with the magnitude of $\vec{B}$. Consequently, the *orbital magnetic moment* is said to be an *adiabatic invariant*. It is usually referred to as the *first adiabatic invariant*.

The magnetic flux, Φ_m, enclosed by one orbit of the particle is given by

$$\Phi_m = \int \vec{B} \cdot d\vec{S} = \pi\, r_c^2\, B = \pi \frac{m^2\, v_\perp^2}{q^2\, B^2} B = \frac{2\pi m}{q^2} \left(\frac{W_\perp}{B}\right) \tag{6.9}$$

Therefore,

$$\frac{d}{dt}(\Phi_m) = \frac{2\pi m}{q^2} \frac{d}{dt} |\vec{m}| = 0 \tag{6.10}$$

in view of the invariance of $|\vec{m}|$. Hence, as the charged particle moves in a region of converging $\vec{B}$ field the particle will orbit with increasingly smaller radius, so that the *magnetic flux enclosed by the orbit* remains *constant*.

6.2 Magnetic mirror effect

As a consequence of the *adiabatic invariance* of $|\vec{m}|$ and Φ_m, as the particle moves into a region of converging magnetic field lines its transverse kinetic energy $W_\perp$ increases, while its parallel kinetic energy $W_\parallel$ decreases, in order to keep $|\vec{m}|$ and the total energy constant. Ultimately, if the $\vec{B}$ field becomes strong enough, the particle velocity in the direction of increasing $\vec{B}$ field may come to zero and then be reversed. After reversion the particle is speeded up in the direction of decreasing field, while its transverse velocity diminishes. Thus, the particle is *reflected* from the region of converging magnetic field lines. This phenomenon is called the *magnetic mirror effect* and is the basis for one of the primary schemes of plasma confinement.

When two coaxial magnetic mirrors are considered, as illustrated in Fig. 11, the charged particles may be reflected by the magnetic mirrors and travel back and forth in the space between them, being trapped. This trapping region has been called a *magnetic bottle* and it is used in laboratory for plasma confinement.

The trapping in a magnetic mirror system is not perfect, however. The effectiveness of the coaxial magnetic mirror system in trapping charged particles can be measured by the *mirror ratio* B_m/B_0, where B_m is the intensity of the magnetic field at the point of reflection (where the pitch angle of the particle is $\pi/2$) and B_0 is the intensity of the magnetic field at the center of the magnetic bottle.

Consider a charged particle having a pitch angle α_0 at the center of the magnetic bottle. If v is the particle speed, which in a static magnetic field remains constant, the constancy of the magnetic moment $|\vec{m}| = W_\perp/B$ leads to

$$m\, v^2 (\sin^2\alpha)/2B = m\, v^2 (\sin^2\alpha_0)/2B_0 \tag{6.11}$$

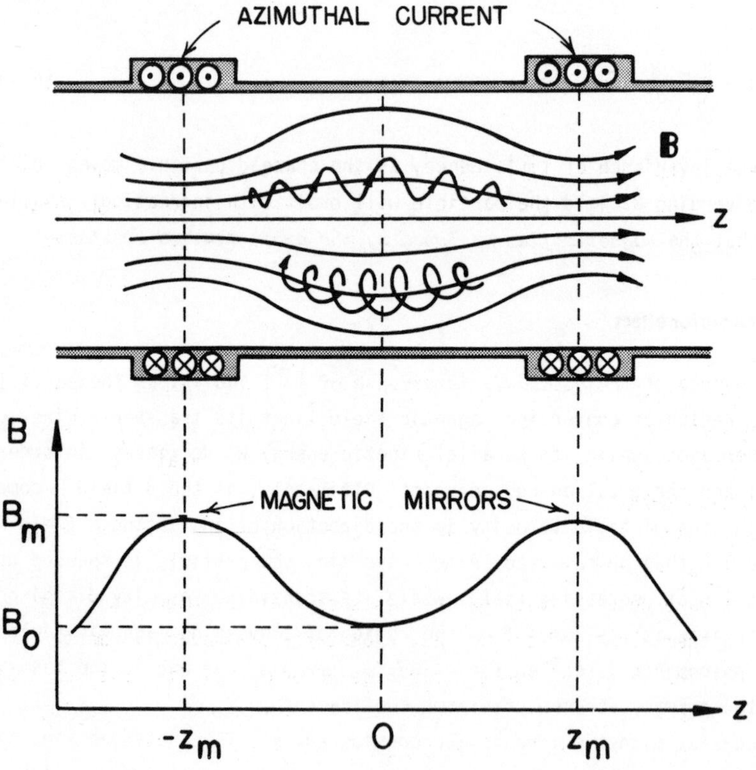

Fig. 11 - Schematic diagram showing the arrangement of coils to produce two coaxial magnetic mirrors facing each other for plasma confinement, and the relative intensity of the magnetic field.

where α is the particle pitch angle at a position where the magnetic field intensity is B. Thus, at any point inside the magnetic bottle, for this particle,

$$[\sin^2 \alpha(z)]/B(z) = (\sin^2 \alpha_0)/B_0 \tag{6.12}$$

Suppose now that this particle is reflected at the "throat" of the mirror, that is, $\alpha = \pi/2$ for $B(z) = B_m$. Therefore, from (6.12),

$$(\sin^2 \alpha_0)/B_0 = 1/B_m \tag{6.13}$$

This means that a particle having a pitch angle α_0, given by,

$$\alpha_0 = \sin^{-1} \sqrt{B_0/B_m} = \sin^{-1}(v_\perp/v)_0 \tag{6.14}$$

at the center of the bottle, is reflected at a point where the intensity of the field is B_m. Therefore, for a magnetic bottle with a fixed mirror ratio B_m/B_0, the plasma particles having a pitch angle at the center *greater* than α_0, as given by (6.14), will be reflected *before* the ends of the magnetic bottle. On the other hand, if the pitch angle of the particle at the center is less than α_0, its pitch angle will never reach the value $\pi/2$, which implies that at the ends of the bottle the particle has a non-vanishing parallel velocity and hence escapes through the ends of the mirror system. There is therefore a *loss cone*, a bi-cone of angle α_0 with its vertex at the center, as shown in Fig. 12, where particles which have velocity vectors with a pitch angle falling inside it are not trapped. The loss cone is determined by the mirror ratio B_m/B_0 according to (6.14).

Devices that have no ends, with geometries such that the magnetic field lines close on themselves, offer many advantages for plasma confinement. *Toroidal* geometries (Fig. 13) for example have no ends, but it turns out that confinement of a plasma inside a toroidal magnetic field does not provide a plasma equilibrium situation, because of the radial inhomogeneity of the field. In this case a poloidal magnetic field is normally superposed on the toroidal field, resulting in helical field lines (as in the Tokamak). The major problem in the confinement schemes, however, is that instabilities and small fluctuations from the desired configuration are always present, which lead to a rapid escape of the particles from the magnetic bottle. This instability problem is a fundamental one, and it is likely to occur in any conceivable magnetic confinement scheme.

A good example of a natural magnetic bottle is the Earth's magnetic field, which traps charged particles of solar and cosmic origin. These charged particles trapped in the Earth's magnetic field constitute the so called *Van Allen radiation belts*. As shown in Fig. 14, the geomagnetic field near the Earth is approximately that of a dipole, with the field lines converging towards the north and south magnetic poles. The electrons and protons spiral in almost helical paths along the field lines towards the magnetic poles, where

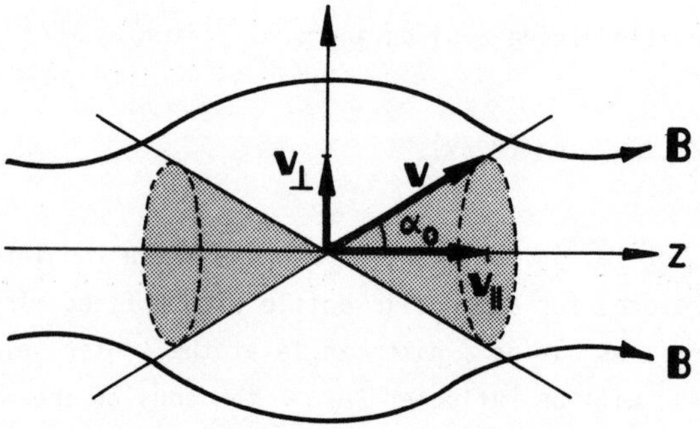

Fig. 12 - The loss cone in a coaxial magnetic mirror system.

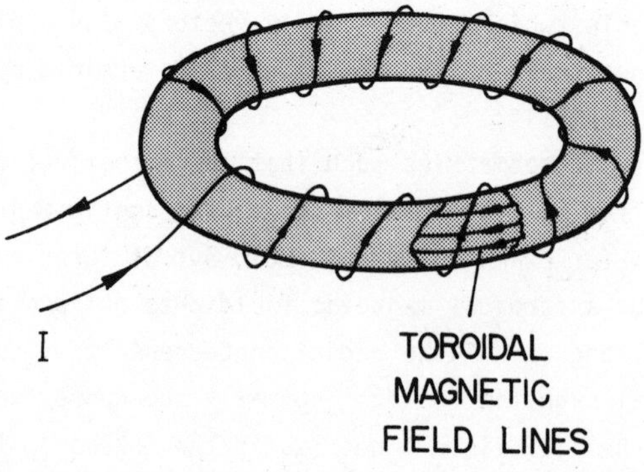

Fig. 13 - Magnetic field with toroidal geometry.

they are reflected. These particles bounce back and forth between the poles. In addition to this bouncing motion, the charged particles in the Van Allen radiation belts are also subjected to a gradient drift and a curvature drift, to be discussed later in this chapter.

6.3 The longitudinal adiabatic invariant

Consider a particle trapped between two magnetic mirrors and bouncing between them. Suppose that the separation distance between the two mirrors changes very slowly with time as compared to the bounce period. With the periodic motion of the particle between the two magnetic mirrors (whose separation varies slowly with time) there is associated an adiabatic invariant called the *longitudinal adiabatic invariant*, defined by the integral

$$J = \oint \vec{v} \cdot d\vec{\ell} = \oint v_{\parallel} \, d\ell \tag{6.15}$$

taken over one period of oscillation of the particle back and forth between the mirror points.

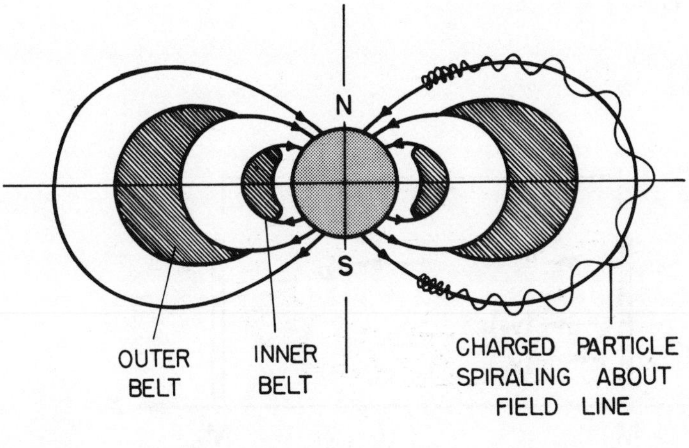

Fig. 14 - Dipole approximation of the Earth's magnetic field. The distance of the Van Allen radiation belts from the center of the Earth, at the equator, is about 1.5 Earth radii for the high-energy protons and about 3 to 4 Earth radii for the high-energy electrons.

For a simple proof of the adiabatic invariance of J, consider the idealized situation illustrated in Fig. 15, where the existing $\vec{B}$ field in z direction is uniform in space, except near the points M_1 and M_2 where the field increases to form the two mirrors separated by a distance L. The mirror M_1 approaches the other with velocity

$$v_m = - \frac{dL}{dt} \tag{6.16}$$

the negative sign being due to the fact that L decreases with time. It is assumed that this velocity is much smaller than the longitudinal component of the particle velocity, that is, $v_m \ll v_\parallel$. Thus, the distance moved by the mirror M_1 during one period of oscillation of the particle is small compared to the distance L between the mirrors.

Further, since B is assumed to be uniform throughout the space between the mirrors (except near the ends), the longitudinal particle speed $v_\parallel$ may be taken to be constant in the space between the mirrors. Neglecting the small end effects at the two mirrors, we can take

$$J = \int_0^{2L} v_\parallel \, d\ell = 2 \, v_\parallel \, L \tag{6.17}$$

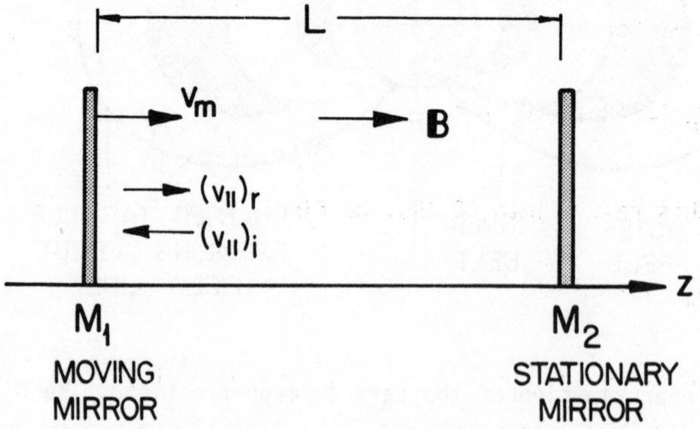

Fig. 15 - Schematic representation of a system of two coaxial magnetic mirrors approaching each other.

The time rate of change of J is

$$dJ/dt = 2\, v_{\parallel}\, (dL/dt) + 2L\, (dv_{\parallel}/dt) = -\, 2\, v_{\parallel}\, v_m + 2L\, (dv_{\parallel}/dt) \qquad (6.18)$$

where use was made of (6.16). To calculate $dv_{\parallel}/dt$, we set

$$dv_{\parallel}/dt = \Delta v_{\parallel}/\Delta t = \Delta v_{\parallel}/(2L/v_{\parallel}) \qquad (6.19)$$

where $\Delta v_{\parallel}$ denotes the change in the particle speed $v_{\parallel}$ on reflection from the moving mirror, and $\Delta t = 2L/v_{\parallel}$ is the period of oscillation between the mirrors. In order to find $\Delta v_{\parallel}$, it is convenient to transform to a coordinate system moving with the magnetic mirror M_1, at the speed v_m. Let us denote this moving coordinate system by a prime and the incident and reflected particle speeds by subscripts i and r, respectively. Thus,

$$(v_{\parallel})'_i = (v_{\parallel})_i + v_m \qquad (6.20)$$

$$(v_{\parallel})'_r = (v_{\parallel})_r - v_m \qquad (6.21)$$

which gives for the change in the particle speed, in one reflection,

$$\Delta v_{\parallel} = (v_{\parallel})_r - (v_{\parallel})_i = 2\, v_m \qquad (6.22)$$

since in the moving coordinate system $(v_{\parallel})'_i = (v_{\parallel})'_r$ with only their directions reversed. Therefore, (6.19) becomes

$$dv_{\parallel}/dt = 2\, v_m/(2L/v_{\parallel}) = v_m\, v_{\parallel}/L \qquad (6.23)$$

On substituting this result into (6.18), we find

$$dJ/dt = d(2v_{\parallel}\, L)/dt = 0 \qquad (6.24)$$

which shows that J is an *adiabatic invariant*. This quantity is also referred to as the *second adiabatic invariant*.

The parallel kinetic energy of a charged particle trapped between the two mirrors is (using $J = 2v_{\parallel}L$)

$$W_{\parallel} = mv_{\parallel}^2/2 = m\, J^2/(8L^2) \qquad (6.25)$$

which increases rapidly as L decreases. The italian physicist Fermi suggested this process as a mechanism for the acceleration of charged particles in order to explain the origin of high energy cosmic rays. Fermi proposed that two stellar clouds moving towards each other, and having a magnetic field greater than in the space between them, may trap and accelerate the cosmic charged particles. There is a limit, however, in the particle longitudinal speed increase, since the direction of the particle velocity at the center of the mirror system may eventually enter the loss cone and escape through the ends of the system. It should be noted that a magnetic mirror moving towards a stationary one involves in fact time-varying $\vec{B}$ fields and consequently electric fields, which can lead to a change in the particle kinetic energy.

7. CURVATURE DRIFT

So far the effects associated with the curvature of the magnetic field lines have not been considered. As stated previously, a $\vec{B}$ field with only curvature terms does not satisfy the equation $\vec{\nabla} \times \vec{B} = 0$, so that in practice the gradient and the curvature drifts will always be present simultaneously. In the first order orbit theory the effects corresponding to each of the components of $\vec{\nabla}\vec{B}$ are additive.

We investigate now the effect of the curvature terms $\partial B_x/\partial z$ and $\partial B_y/\partial z$ [referred in (2.6c)] on the motion of a charged particle. We will assume that these terms are so small that the radius of curvature of the magnetic field lines is very large compared to the particle cyclotron radius. Let us introduce a *local* coordinate system gliding along the magnetic field line with the particle longitudinal velocity $\vec{v}_\parallel$. Since this is not an inertial system because of the curvature of the field lines, a centrifugal force will be present. This local coordinate system can be specified by the orthogonal set of unit vectors $\hat{B}$, $\hat{n}_1$ and $\hat{n}_2$, where $\hat{B}$ is along the field line, $\hat{n}_1$ is along the principal normal to the field line, and $\hat{n}_2$ is along the binormal to the curved magnetic field line, as indicated in Fig. 16.

The centrifugal force $\vec{F}_C$ acting on the particle as seen from this noninertial system, is given by

$$\vec{F}_C = - (m\, v_\parallel^2/R)\, \hat{n}_1 \tag{7.1}$$

where R denotes the local radius of curvature of the magnetic field line and $v_\parallel$ is the particle instantaneous longitudinal speed. From (2.6.2) the curvature drift associated with this force is

$$\vec{v}_c = (\vec{F}_c \times \vec{B})/(q\, B^2) = -\,(\hat{n}_1 \times \vec{B})\, m\, v_\parallel^2/(RqB^2) \tag{7.2}$$

To express the unit vector $\hat{n}_1$ in terms of the unit vector $\hat{B}$ along the magnetic field line, we let ds represent an element of arc along the field line subtending an angle $d\phi$,

$$ds = R\, d\phi \tag{7.3}$$

If $d\hat{B}$ denotes the change in $\hat{B}$ due to the displacement ds (see Fig. 16), then $d\hat{B}$ is in the direction of $\hat{n}_1$ and its magnitude is

$$|d\hat{B}| = |\hat{B}|\; d\phi = d\phi \tag{7.4}$$

Consequently,

$$d\hat{B} = \hat{n}_1\, d\phi \tag{7.5}$$

Dividing this equation by (7.3) side by side, gives

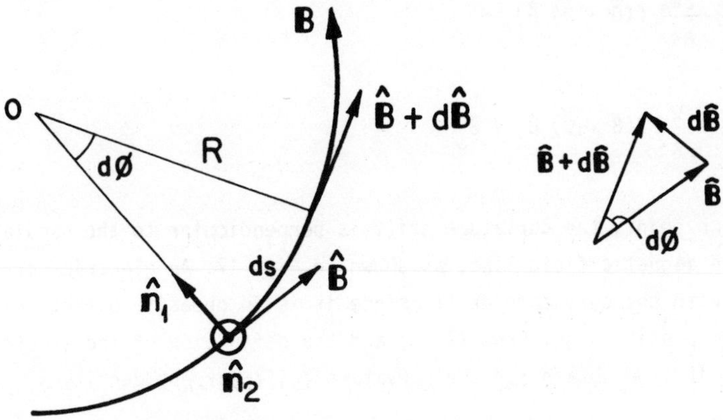

Fig. 16 - Curved magnetic field line showing the unit vector $\hat{B}$ along the field line, the principal normal $\hat{n}_1$, and the binormal $\hat{n}_2$, at an arbitrary point ($\hat{n}_1 \times \hat{n}_2 = \hat{B}$). The local radius of curvature is R.

$$d\hat{B}/ds = \hat{n}_1/R \tag{7.6}$$

The derivative d/ds along $\vec{B}$ may be written as $(\hat{B} \cdot \vec{\nabla})$, so that (7.6) becomes

$$\hat{n}_1/R = (\hat{B} \cdot \vec{\nabla})\,\hat{B} \tag{7.7}$$

Incorporating this result into equation (7.1), we obtain

$$\vec{F}_C = -\,m\,v_{\parallel}^2\,(\hat{B} \cdot \vec{\nabla})\,\hat{B} \tag{7.8}$$

This force is obviously perpendicular to the magnetic field $\vec{B}$, since it is in the $-\,\hat{n}_1$ direction [see (7.7)], and gives rise to a curvature drift whose velocity is

$$\vec{v}_C = -\,\frac{m\,v_{\parallel}^2}{q\,B^2}\,[(\hat{B} \cdot \vec{\nabla})\,\hat{B}] \times \vec{B} \tag{7.9}$$

Since $\vec{B} = B\hat{B}$ and writing $W_{\parallel} = m\,v_{\parallel}^2/2$ for the longitudinal kinetic energy of the particle, (7.8) and (7.9) can be written, respectively, as

$$\vec{F}_C = -\,\frac{2W_{\parallel}}{B^2}\,[(\vec{B} \cdot \vec{\nabla})\,\vec{B}]_{\perp} \tag{7.10}$$

$$\vec{v}_C = -\,\frac{2\,W_{\parallel}}{q\,B^4}\,[(\vec{B} \cdot \vec{\nabla})\,\vec{B}] \times \vec{B} \tag{7.11}$$

Thus, at each point, the curvature drift is perpendicular to the osculating plane of the magnetic field line, as shown in Fig. 17. An electric current is associated with the curvature drift, since it is in opposite directions for particles of opposite sign. From (7.11) and the definition of the electric current density, we obtain for the curvature drift current density

$$\vec{J}_C = -\,2\,[(1/\Delta V)\sum_i W_{i\parallel}]\,[\hat{B} \cdot \vec{\nabla})\hat{B}] \times \vec{B}/B^2 \tag{7.12}$$

where the summation extends over all charged particles contained in the small volume element ΔV.

8. COMBINED GRADIENT-CURVATURE DRIFT

The curvature drift and the gradient drift always appear together and both point in the same direction, since the term $\vec{\nabla}B$ points in the direction opposite to $\vec{F}_C$ (see Fig. 5). These two drifts can therefore be added up to form the combined *gradient-curvature drift*. Thus, from (5.1) and (7.11),

$$\vec{v}_{GC} = \vec{v}_G + \vec{v}_C = -\frac{m\,v_\perp^2}{2\,q\,B^3}(\vec{\nabla}B) \times \vec{B} - \frac{m\,v_\parallel^2}{q\,B^4}[(\vec{B}\cdot\vec{\nabla})\vec{B}] \times \vec{B} \tag{8.1}$$

When volume currents are not present (in a vacuum field, for example) so that $\vec{\nabla} \times \vec{B} = 0$, the vector identity (4.26) allows the expression (8.1) to be written in the compact form

$$\vec{v}_{GC} = -\frac{m}{q\,B^4}\left(v_\parallel^2 + \frac{v_\perp^2}{2}\right)\left(\vec{\nabla}\,\frac{B^2}{2}\right) \times \vec{B} \tag{8.2}$$

In the Earth's magnetosphere near the equatorial plane both the curvature and the gradient drifts (B decreases with altitude) cause the positively charged particles to slowly drift westward and the negative ones eastward, resulting in an east to west current, known as the *ring current*. Fig. 18

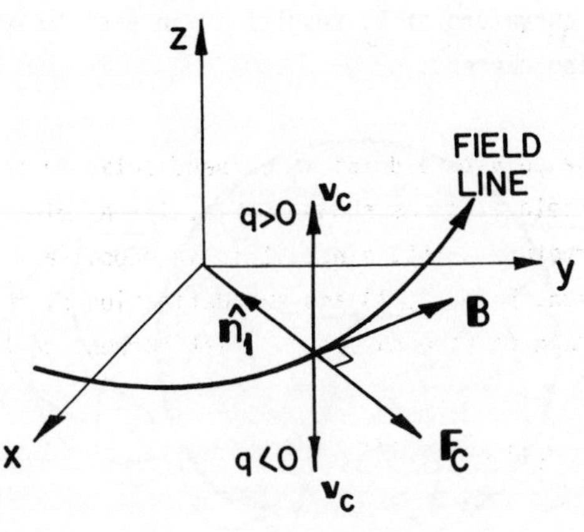

Fig. 17 - Relative direction of the particle guiding center drift velocity $\vec{v}_C$, due to the curvature of the magnetic field line.

illustrates schematically the motion of a charged particle trapped in the Earth's magnetic field. The particle bounces back and forth along the field line between the mirror points M_1 and M_2, and drifts in longitude as a result of the gradient and curvature of the field lines. The trajectory described by the particle is therefore contained in a tire-shaped shell encircling the Earth (Fig. 19). This tire-shaped shell encircling the

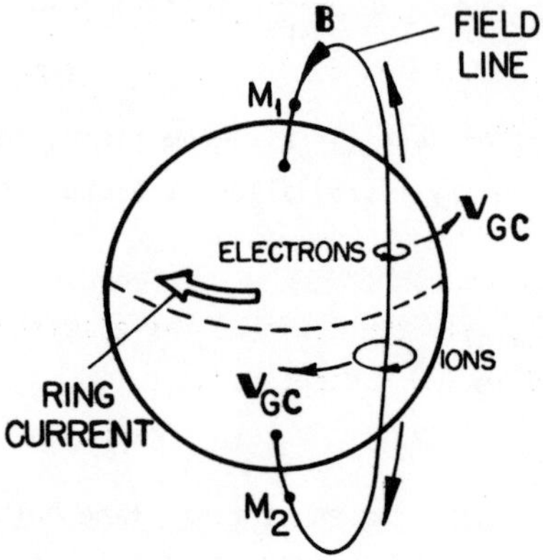

Fig. 18 - Sketch indicating the motion of a charged particle in the Earth's magnetic field. The longitudinal drift velocity $\vec{v}_{GC}$, due to the gradient and curvature of $\vec{B}$, results in an east to west current called the ring current.

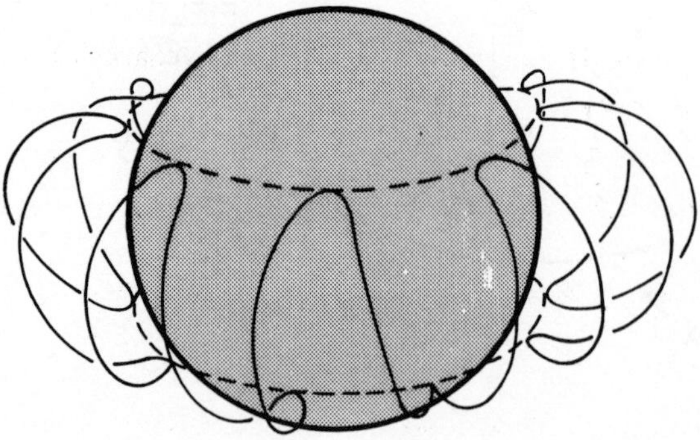

Fig. 19 - Schematic representation of the longitude drift of charged particles around the Earth.

Earth defines a surface on which the particle guiding center drifts slowly around the Earth. Connected with the periodic motion of the particle on this drift surface there is an adiabatic invariant, called the *third adiabatic invariant*, which is the total magnetic flux Φ_m enclosed by the drift surface. Clearly, in a static situation this flux is obviously constant. The significant fact is that the total magnetic flux Φ_m, enclosed by the drift surface, remains invariant when the field varies slowly in time, that is, when the period of motion of the particle on the drift surface is small compared with the time scale for the magnetic field to change significantly. This invariant has few applications because most fluctuations of $\vec{B}$ occur on a time scale short compared with the drift period.

PROBLEMS

3.1 - Describe semiquantitatively the motion of an electron under the presence of a constant electric field in the x direction,

$$\vec{E} = E_0\hat{x}$$

and a space varying magnetic field given by

$$\vec{B} = B_0 \{ \alpha(x+z)\ \hat{x} + [1 + \alpha(x-z)]\ \hat{z} \}$$

where E_0, B_0 and α are positive constants, $|\alpha x| << 1$ and $|\alpha z| << 1$. Assume that initially the electron moves with constant velocity in the z direction, $\vec{v}(t=0) = v_0\hat{z}$. Verify if this magnetic field satisfies Maxwell equations.

3.2 - Verify if there is any drift velocity for a charged particle in a magnetic field given by

$$\vec{B} = B_y(x)\ \hat{y} + B_0\hat{z}$$

where $B_y(x)$ and $\partial B_y/\partial x$ are very small quantities. Does this field satisfy Maxwell equations?

3.3 - Consider a system of two coaxial magnetic mirrors whose axis coincides with the z axis, being symmetrical about the plane z=0, as shown schematically in Fig. P3.1. Describe semiquantitatively the motion of a charged particle in this magnetic mirror system considering that at $z = 0$ the particle has $v_\parallel = v_\parallel^0$ and $v_\perp = v_\perp^0$. What relation must exist between $\vec{B}_0 = \hat{z}\, B(z=0)$, $\vec{B}_m = \hat{z}\, B(z = \pm z_m)$ and α_0 (particle pitch angle at $z = 0$) so that the particle be reflected at z_m?

3.4 - For the magnetic mirror system of Problem 3.3 suppose that the axial magnetic field changes in time, that is $\vec{B}_{axial} = \hat{z}\, B(z,t)$. Considering that the magnetic moment $|\vec{m}| = m v_\perp^2(z,t)/2B(z,t)$ is an adiabatic invariant (note that its value is the same at $z = 0$ and $z = \pm z_m$, and that $v^2 = v_\parallel^2 + v_\perp^2$), show that the *longitudinal adiabatic invariant* can be written in the form

$$\int_{-z_m}^{z_m} [B(z_m,t) - B(z,t)]^{1/2}\, dz = \text{constant}$$

3.5 - Consider the magnetic mirror system shown in Fig. P3.1. Suppose that the axial magnetic field is given by

$$B(z) = B_0\, [1 + (z/a_0)^2]$$

where B_0 and a_0 are positive constants, and that the mirroring planes are given by $z = -z_m$ and $z = z_m$.

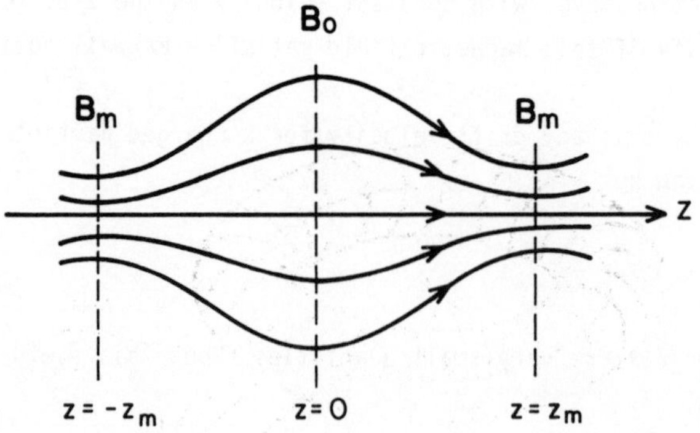

Fig. P3.1

(a) For a charged particle trapped in this mirror system, show that the z component of the particle velocity is given by

$$v_{\parallel}(z) = \{(2|\vec{m}|\ B_0/m)\ [(z_m/a_0)^2 - (z/a_0)^2]\}^{1/2}$$

(b) The average force acting on the particle guiding center, along the z axis, is given by

$$< \vec{F}_{\parallel} > = - |\vec{m}| (\partial B/\partial z)\ \hat{z}$$

Show that the particle performs a simple harmonic motion between the mirroring planes, with a period given by

$$T = 2\pi\ a_0\ [m/(2|\vec{m}|\ B_0)]^{1/2}$$

(c) If the motion of the particle is to be limited to the region $|z| < z_m$, what restriction must be imposed on the total energy and the magnetic moment?

3.6 - Consider a toroidal magnetic field, as shown in Fig. P3.2.
(a) Show that the magnetic flux density along the axis of the torus is given by

$$\vec{B} = \hat{\phi}\ B_a\ a/r$$

where B_a denotes the magnitude of $\vec{B}$ at the radial distance $r = a$.

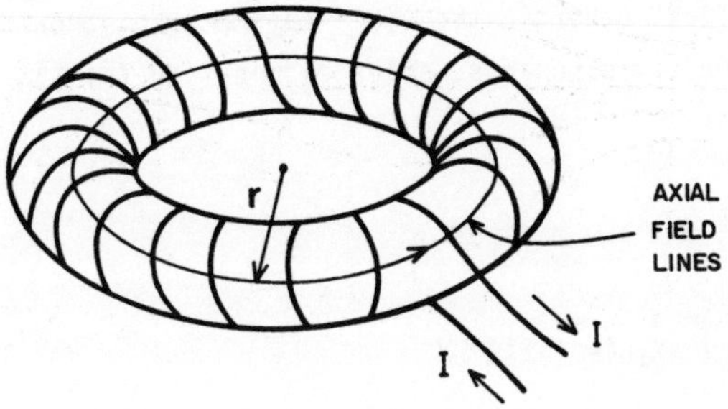

Fig. P3.2

(b) In what direction is the gradient drift associated with the radial variation of B_ϕ? Examine qualitatively the type of charge separation that occurs. Neglect the effect of the magnetic field line curvature.
(c) If $\vec{E}$ denotes the induced electric field due to the charge separation, in what direction is the $\vec{E}\times\vec{B}$ drift?
(d) Show that it is not possible to confine a plasma in a purely toroidal magnetic field, because of the gradient drift and the $\vec{E}\times\vec{B}$ drift.

3.7 - Consider a spatially nonuniform magnetostatic field expressed in terms of a Cartesian coordinate system by

$$\vec{B}(x,z) = B_0 \, [\alpha z \hat{x} + (1 + \alpha x) \, \hat{z}]$$

where B_0 and α are positive constants, $|\alpha x| \ll 1$ and $|\alpha z| \ll 1$.
(a) Show that this magnetic field is consistent with Maxwell equations, so that both gradient and curvature terms are present. Determine the equation of a magnetic flux line.
(b) Write down the Cartesian components of the equation of motion for an electron moving in the region near the origin under the action of this magnetic field.
(c) Consider the following initial conditions for the electron:

$$\vec{r}(0) = (x_0 + v_{\perp 0}/\omega_c) \, \hat{x}$$

$$\dot{\vec{r}}(0) \equiv \vec{v}(0) = v_{\perp 0} \, \hat{y} + v_{z0} \, \hat{z}$$

Solve the equation of motion using a perturbation technique, retaining only terms up to the first order in the small parameter α. Show that the leading terms in the velocity components, after eliminating the time-periodic parts, are given by

$$\vec{v}_x = \alpha \, v_{z0}^2 \, t \, \hat{x}$$

$$\vec{v}_y = - (\alpha/\omega_c) \, (v_{\perp 0}^2/2 + v_{z0}^2) \, \hat{y}$$

$$\vec{v}_z = v_{z0} \, \hat{z}$$

(d) Show that the average position of the electron in the x-z plane follows the magnetic flux line that passes through its initial position.
(e) Show that the gradient and curvature drift velocities are given, respectively, by

$$\vec{v}_G = -(\alpha/\omega_c)(v_\perp^2/2)\,\hat{y}$$

$$\vec{v}_C = -(\alpha/\omega_c)\,v_{zo}^2\,\hat{y}$$

so that the total drift velocity is precisely the nonperiodic part of $\vec{v}_y$.

3.8 - The Earth's magnetic field can be represented, in a first approximation, by a magnetic dipole placed in the Earth's center, at least up to distances of a few Earth radii (R_E).
(a) Using the fact that, at one of the magnetic poles, the field has a magnitude of approximately 0.5 Gauss, calculate the dipole magnetic moment.
(b) Consider the motion of an electron of energy E_0 at a radial distance r_0 ($r_0 > R_E$). Calculate its cyclotron frequency and gyroradius.
(c) Assuming that the electron is confined to move in the equatorial plane, calculate its gradient and curvature drift velocities, and determine the time it takes to drift once around the Earth at the radial distance r_0.
(d) Calculate the period of the bounce motion of the electron, as it gets reflected back and forth between the magnetic mirrors near the poles. What is the altitude of the reflection points? Assume that $W_\parallel = W_\perp$ at the magnetic equatorial plane.
(e) Recalculate (b), (c) and (d), considering E_0 = 1 MeV and r_0 = 4 R_E. Examine these results in terms of typical values for charged particles in the outer Van Allen radiation belt.
(f) Assuming that there is an isotropic population of 1 MeV protons and 100 KeV electrons at about 4 R_E, each having a density $n_e = n_i = 10^7\ m^{-3}$ in the equatorial plane, calculate the *ring current density* in Ampère/m².

3.9 - Imagine an infinite straight wire carrying a current I and electrically charged to a negative potential ϕ. Analyze the motion of an electron in the vicinity of this wire using first-order orbit theory. Sketch the path described by the electron, indicating the relative directions of the electromagnetic, gradient and curvature drift velocities.

3.10 - The field of a *magnetic monopole* can be represented by

$$\vec{B}(\vec{r}) = \lambda\, \vec{r}/r^3$$

where λ is a constant. Solve the equation of motion to determine the trajectory of a charged particle in this field. (You may refer to B. Rossi and S. Olbert, *Introduction to the Physics of Space*, Chapter 2, section 2.5, pg. 29, Mc Graw-Hill, 1970).

3.11 - Analyze the motion of a charged particle in the field of a *magnetic dipole*. Determine the two constants of the motion and analyze their physical meaning. (For this problem, you may refer to S.Störmer, *The Polar Aurora*, University Oxford Press, 1955, or to B. Rossi and S.Olbert, *Introduction to the Physics of Space*, Chapter 3, pg. 45, Mc Graw-Hill, 1970).

CHAPTER 4

Charged Particle Motion in Time-varying Electromagnetic Fields

1. INTRODUCTION

In this chapter we analyze the motion of charged particles in the presence of time-varying fields. Initially, in the two following sections, we consider a time-varying electric field and a constant magnetic field, both fields being spatially uniform. The assumption of a constant and spatially uniform $\vec{B}$ field is well justified if the externally applied magnetostatic field is much larger than the magnetic field associated with the time-varying $\vec{E}$ field. Also, the assumption of an electric field which is spatially uniform is valid if the charged particle cyclotron radius is much smaller than the scale length of the spatial variation of $\vec{E}$. Both these requirements are assumed to hold in the analysis presented in sections 2 and 3. In section 4 we consider a time-varying magnetic field and the corresponding space-varying electric field.

2. SLOWLY TIME-VARYING ELECTRIC FIELD

2.1 Equation of motion and polarization drift

For the moment we shall assume that the characteristic time scale of variation of the electric field is much larger than the particle cyclotron period. The component of the motion of the charged particle along the magnetic field lines is given by (2.5.4), from which we can write in general

$$\vec{v}_{\parallel}(t) - \vec{v}_{\parallel}(0) = \frac{q}{m} \int_0^t \vec{E}_{\parallel}(t')\, dt' \tag{2.1}$$

This result however does not lead to any new interesting information.

Since the $\vec{E}$ field is varying slowly in time, the component of the motion across the magnetic field lines is expected to be not very different from that for a constant $\vec{E}$ field. Therefore, it is reasonable to seek the solution for $\vec{v}_\perp$ in a form similar to (2.5.8). Hence, we take

$$\vec{v}_\perp = \vec{v}_\perp' + \vec{v}_E + \vec{v}_p \tag{2.2}$$

where $\vec{v}_E = \vec{E}_\perp \times \vec{B}/B^2$ is the electromagnetic plasma drift velocity (2.5.11). Note that $\vec{v}_E$ varies slowly with time, since $\vec{E}$ is slowly time-varying. The substitution of (2.2) into the perpendicular component equation of motion (2.5.5) yields

$$m \frac{d}{dt} (\vec{v}_\perp' + \vec{v}_E + \vec{v}_p) = q[\vec{E}_\perp + (\vec{v}_\perp' + \vec{v}_E + \vec{v}_p) \times \vec{B}] \tag{2.3}$$

Using (2.5.9) we can write (2.3) as

$$m \frac{d\vec{v}_\perp'}{dt} + m \frac{d}{dt} (\frac{\vec{E}_\perp \times \vec{B}}{B^2}) + m \frac{d\vec{v}_p}{dt} = q\, \vec{v}_\perp' \times \vec{B} + q\, \vec{v}_p \times \vec{B} \tag{2.4}$$

Thus, if we set

$$\vec{v}_p = (m/q)(\partial \vec{E}_\perp / \partial t)/B^2 \tag{2.5}$$

(2.4) reduces to

$$m\, d\vec{v}_\perp'/dt + m\, d\vec{v}_p/dt = q\, \vec{v}_\perp' \times \vec{B} \tag{2.6}$$

If the second term on the left-hand side can be neglected, then this equation becomes identical to (2.5.12), which describes a circular motion about the magnetic field lines. Comparing the relative magnitudes of the second term on the left, with that on the right-hand side of (2.6), we find

$$\frac{|m\, d\vec{v}_p/dt|}{|q\, \vec{v}_\perp' \times \vec{B}|} = \frac{|m^2(\partial^2 E_\perp/\partial t^2)/qB^2|}{|q\, B\, v_\perp'|} = \left| \frac{(E_\perp/B)}{v_\perp'} \left(\frac{\omega^2 m^2}{q^2 B^2}\right) \right| =$$

$$= |v_E/v_\perp'| (\omega/\omega_c)^2 \tag{2.7}$$

where we have assumed that $\vec{E}_\perp$ has a harmonic time dependence with a characteristic angular frequency ω. Considering that this characteristic frequency is much smaller than the cyclotron frequency,

$$\omega \ll \omega_c \tag{2.8}$$

and, further, if $|\vec{v}_E/\vec{v}'_\perp|$ is also small, then the term m $d\vec{v}_p/dt$ can be neglected in comparison with the other terms of (2.6), and we obtain

$$m\, d\vec{v}'_\perp/dt = q\, \vec{v}'_\perp \times \vec{B} \tag{2.9}$$

which is identical to (2.5.12). Therefore, $\vec{v}'_\perp$ corresponds to the usual circular motion of the charged particle about the magnetic field, and is independent of the variations of the electric field. Superposed upon this circular motion are the drift velocities

$$\vec{v}_E = (\vec{E}_\perp \times \vec{B})/B^2 \tag{2.10}$$

and

$$\vec{v}_p = m(\partial\vec{E}_\perp/\partial t)/(q\, B^2) \tag{2.11}$$

Thus, the effect of a slowly varying electric field is the addition of the velocity $\vec{v}_p$ called the *polarization drift velocity*.

Since $\vec{v}_p$ is in opposite directions for charges of opposite sign, the time dependent electric field produces a net polarization current in a neutral plasma, so that the plasma medium behaves like a dielectric. The *polarization current density* $\vec{J}_p$ is the rate of flow of positive and negative charges across unit area, and is given by

$$\vec{J}_p = (1/\Delta V) \sum_i q_i\, \vec{v}_{pi} = [(1/\Delta V) \sum_i m_i]\, (\partial\vec{E}_\perp/\partial t)/B^2 = \rho_m(\partial\vec{E}_\perp/\partial t)/B^2 \tag{2.12}$$

where the summation is over all positive and negative charges contained in the small volume element ΔV, and ρ_m is the mass density of the plasma.

2.2 Plasma dielectric constant

The polarization effect in a plasma is due to the time variation of the electric field. The application of a steady $\vec{E}$ field does not result in a polarization field, since the ions and electrons can move around to preserve quasineutrality. Since the plasma behaves like a dielectric, the polarization current density $\vec{J}_p$ can be taken into account through the introduction of the dielectric constant of the plasma. For this purpose, we can separate the *total* current density $\vec{J}$ into the polarization current density $\vec{J}_p$ and the current density $\vec{J}_0$ due to other sources,

$$\vec{J} = \vec{J}_p + \vec{J}_0 \tag{2.13}$$

Thus, combining $\vec{J}_p$ with the term $\varepsilon_0 \, \partial\vec{E}_\perp/\partial t$ which appears on the right-hand side of Maxwell $\vec{\nabla} \times \vec{B}$ equation, we obtain

$$\varepsilon_0(\partial\vec{E}_\perp/\partial t) + (\rho_m/B^2)(\partial\vec{E}_\perp/\partial t) = \varepsilon_0(1+\rho_m/\varepsilon_0 B^2)\, \partial\vec{E}/\partial t = \varepsilon(\partial\vec{E}/\partial t) \tag{2.14}$$

where

$$\varepsilon = \varepsilon_0 \, \varepsilon_r = \varepsilon_0(1+\rho_m/\varepsilon_0 B^2) \tag{2.15}$$

is the *effective electric permittivity* perpendicular to the magnetic field. For low frequencies, the *relative permittivity* ε_r of a plasma can be very high; for a number density of 10^{20} particles/m^3 and B = 1 Tesla we have $\varepsilon_r \cong 10^4$.

The resulting charge density ρ_p, which accumulates as a result of the polarization current density $\vec{J}_p$, satisfies the charge continuity equation

$$\partial\rho_p/\partial t + \vec{\nabla}\cdot\vec{J}_p = 0 \tag{2.16}$$

From (2.16) and (2.12) we obtain

$$\rho_p = -(\rho_m/B^2)\, \vec{\nabla}\cdot\vec{E}_\perp \tag{2.17}$$

The total charge density ρ can be separated as

$$\rho = \rho_0 + \rho_p \tag{2.18}$$

where ρ_0 corresponds to $\vec{J}_0$. Assuming that the parallel component of the electric field vanishes, we see that

$$\vec{\nabla}\cdot\vec{E} = (\rho_0 + \rho_p)/\varepsilon_0 = \rho_0/\varepsilon_0 - (\rho_m/\varepsilon_0 B^2)\, \vec{\nabla}\cdot\vec{E} \tag{2.19}$$

from which we find, using (2.15),

$$\vec{\nabla} \cdot \vec{E} = \rho_0 / \varepsilon \qquad (2.20)$$

Thus, the resulting charge density ρ_p can also be correctly taken into account by the introduction of an effective electric permittivity ε.

We can further verify the correctness of introducing the effective electric permittivity of the plasma, by calculating the *total energy density* associated with the $\vec{E}$ field, which for an ordinary dielectric medium of effective permittivity ε is given by $\varepsilon\, E^2/2$. The *energy density* in the electric field is given by

$$W_E = \varepsilon_0\, E^2/2 \qquad (2.21)$$

To calculate the additional drift kinetic energy acquired by the particle as a result of the polarization drift, we note that for a change $\Delta\vec{E}_\perp$ in the electric field, in a time interval Δt, the displacement $\Delta\vec{r}$ of the guiding center due to this change, is

$$\Delta\vec{r} = \vec{v}_p\, \Delta t = (m/qB^2)(\partial\vec{E}_\perp/\partial t)\, \Delta t = (m/qB^2)\, \Delta\vec{E}_\perp \qquad (2.22)$$

The corresponding work done by the electric field is,using (2.22),

$$\Delta W = q\, \vec{E}_\perp \cdot \Delta\vec{r} = (m/B^2)\, \vec{E}_\perp \cdot \Delta\vec{E}_\perp = \Delta(mE_\perp^2/2B^2) \qquad (2.23)$$

Hence, using (2.10), the change in the kinetic energy of the particle associated with the polarization drift is given by

$$\Delta W = \Delta(mv_E^2/2) \qquad (2.24)$$

This result shows that the work done by the electric field, during polarization, is equal to the change in the kinetic energy associated with the particle motion at the electromagnetic drift velocity $\vec{v}_E$. Note that $\vec{v}_E$ does not lead to any energy exchange between field and particle, since the displacement associated with $\vec{v}_E$ is perpendicular to the electric field. Summing (2.24) over all particles in a unit volume gives the change in the total kinetic energy density of the system

$$\Delta W_V = \Delta(\rho_m v_E^2/2) = \Delta(\rho_m E_\perp^2/2B^2) \qquad (2.25)$$

The kinetic energy density associated with the circular motion of the particles is not affected by changes in the electric field. Thus, the total energy density associated with the electric field is

$$W_E + W_V = \varepsilon_0 E^2/2 + \rho_m v_E^2/2$$

$$= (\varepsilon_0 E^2/2)(1 + \rho_m/\varepsilon_0 B^2) = \varepsilon E^2/2 \tag{2.26}$$

assuming that there is no parallel component of the electric field. This result completes our discussion about the legitimacy of the introduction of an effective electric permittivity for the plasma.

3. ELECTRIC FIELD WITH ARBITRARY TIME VARIATION

3.1 Solution of the equation of motion

We consider now an arbitrary time variation of the electric field but, again, the field is uniform in space. The applied magnetic field is static and uniform, as before. Without loss of generality, the time variation of $\vec{E}$ can be assumed to be harmonic with angular frequency ω,

$$\vec{E}(t) = \vec{E}_0 \, e^{-i\omega t} \tag{3.1}$$

where the complex amplitude $\vec{E}_0$ is independent of time. According to the usual convention, only the real part of this expression is to be taken. An arbitrary time variation of $\vec{E}$ can be written as a superposition of terms similar to (3.1), corresponding to all possible values of ω, since the equation of motion (2.5.5) is linear. Using (3.1), the equation of motion becomes

$$m \, d\vec{v}/dt = q \, (\vec{E}_0 \, e^{-i\omega t} + \vec{v} \times \vec{B}) \tag{3.2}$$

It is natural to expect the forced oscillations of the charged particles to have the same frequency as that of the forcing electric field. Thus, the particle velocity vector may be conveniently decomposed into two parts

$$\vec{v} = \vec{v}_m + \vec{v}_e \, e^{-i\omega t} \tag{3.3}$$

where $\vec{v}_m$ is the velocity associated with the magnetic field alone and, thus, contains no time variation at the angular frequency ω, while $\vec{v}_e$ is due to the oscillating electric field. The substitution of (3.3) into (3.2) gives

$$m(d\vec{v}_m/dt - i\omega\vec{v}_e\, e^{-i\omega t}) = q\,[\vec{E}_0\, e^{-i\omega t} + \vec{v}_m \times \vec{B} + (\vec{v}_e \times \vec{B})\, e^{-i\omega t}] \tag{3.4}$$

The terms containing the periodicity with the angular frequency ω are decoupled from those that do not, so that (3.4) separates into two equations,

$$m\, d\vec{v}_m/dt = q\, \vec{v}_m \times \vec{B} \tag{3.5}$$

$$- m\, i\omega\vec{v}_e = q\,(\vec{E}_0 + \vec{v}_e \times \vec{B}) \tag{3.6}$$

Equation (3.5) corresponds to the usual circular motion of the particle about the magnetic field lines at the cyclotron frequency ω_c.

To solve (3.6) for $\vec{v}_e$ it is appropriate to separate it into components parallel and perpendicular to $\vec{B}$. The *parallel* velocity component is obtained immediately as

$$\vec{v}_{e\parallel} = (i/\omega)(q/m)\,\vec{E}_{0\parallel} \tag{3.7}$$

while the *perpendicular* velocity component satisfies the equation

$$[-i\omega + (q/m)\,\vec{B}\,\times]\,\vec{v}_{e\perp} = (q/m)\,\vec{E}_{0\perp} \tag{3.8}$$

Introducing the cyclotron frequency vector

$$\vec{\omega}_c = - q\vec{B}/m \tag{3.9}$$

we can rewrite (3.8) in the form

$$(i\omega + \vec{\omega}_c\,\times)\,\vec{v}_{e\perp} = - (q/m)\,\vec{E}_{0\perp} \tag{3.10}$$

In order to solve this equation for $\vec{v}_{e\perp}$, we multiply both sides by the conjugate operator $-(i\omega - \vec{\omega}_c\,\times)$. First, we note that

$$(i\omega - \vec{\omega}_c \times)(i\omega + \vec{\omega}_c \times)\, \vec{v}_{e\perp} = (i\omega)^2 \vec{v}_{e\perp} - \vec{\omega}_c \times (\vec{\omega}_c \times \vec{v}_{e\perp})$$

$$= (\omega_c^2 - \omega^2)\, \vec{v}_{e\perp} \tag{3.11}$$

Therefore, (3.10) becomes

$$\vec{v}_{e\perp} = (q/m)\,(i\omega - \vec{\omega}_c \times)\, \vec{E}_{o\perp}/(\omega^2 - \omega_c^2) \tag{3.12}$$

Combining the results contained in (3.12), (3.7) and (3.5), we obtain the following expression for the total velocity vector (3.3)

$$\vec{v} = \vec{v}_m + (q/m)[(i/\omega)\, \vec{E}_{o\parallel} + (i\omega - \vec{\omega}_c \times)\, \vec{E}_{o\perp}/(\omega^2 - \omega_c^2)]\, e^{-i\omega t} \tag{3.13}$$

3.2 Physical interpretation

The result contained in (3.13) shows that *along the magnetic field lines* the particle oscillates with frequency ω and amplitude $\vec{v}_{e\parallel}$, as given by (3.7), so that the velocity oscillation lags 90 degrees behind the oscillation of the applied electric field. This result is easily seen by taking the real part of the vectors $\vec{v}_{e\parallel}\, e^{-i\omega t}$ and $\vec{E}_{o\parallel}\, e^{-i\omega t}$

$$\text{Re}\,\{\vec{v}_{e\parallel}\, e^{-i\omega t}\} = (q/\omega m)\, \vec{E}_{o\parallel} \sin(\omega t) \tag{3.14}$$

$$\text{Re}\,\{\vec{E}_{o\parallel}\, e^{-i\omega t}\} = \vec{E}_{o\parallel} \cos(\omega t) \tag{3.15}$$

which clearly are 90 degrees out of phase. In the *plane perpendicular to* $\vec{B}$ the particle motion is the superposition of the circular motion at the cyclotron frequency ω_c, with an oscillation at the frequency ω and amplitude given by (3.12).

In order to analyze the physical meaning of the motion in the *plane perpendicular to* $\vec{B}$, it is convenient to decompose the oscillating electric field vector into two circularly polarized components, with opposite directions of rotation. The advantage of using the two circularly polarized components is that (3.12), for the perpendicular velocity component $\vec{v}_{e\perp}$, uncouples into two separate equations pertaining to the two circular polarizations rotating in opposite directions. Thus, we take

$$\vec{E}_\perp = \vec{E}_R + \vec{E}_L = (\vec{E}_{oR} + \vec{E}_{oL})\, e^{-i\omega t} \tag{3.16}$$

with

$$\vec{E}_{oR} = (\vec{E}_{o\perp} + i\,\hat{B} \times \vec{E}_{o\perp})/2 \tag{3.17}$$

$$\vec{E}_{oL} = (\vec{E}_{o\perp} - i\,\hat{B} \times \vec{E}_{o\perp})/2 \tag{3.18}$$

where $\hat{B} = \vec{B}/B$ is a unit vector pointing in the direction of the magnetic field. The component $\vec{E}_R$ represents a circularly polarized field with its electric vector rotating to the *right* (*clockwise* direction), and the component $\vec{E}_L$ represents a circularly polarized field with its electric vector rotating to the *left* (*counterclockwise* direction), as seen by an observer looking in the direction of the $\vec{B}$ field.

To understand the physical meaning of this decomposition, let us consider a Cartesian coordinate system with the z axis pointing along $\vec{B}$ and the x axis pointing along $\vec{E}_\perp$. Then, we have

$$\hat{B} \times \vec{E}_{o\perp} = \hat{z} \times (\hat{x}\, E_{o\perp}) = E_{o\perp}\, \hat{y} \tag{3.19}$$

and from (3.17) and (3.18),

$$\vec{E}_R = \vec{E}_{oR}\, e^{-i\omega t} = (E_{o\perp}/2)(\hat{x} + i\hat{y})\, e^{-i\omega t} \tag{3.20}$$

$$\vec{E}_L = \vec{E}_{oL}\, e^{-i\omega t} = (E_{o\perp}/2)(\hat{x} - i\hat{y})\, e^{-i\omega t} \tag{3.21}$$

The components of the actual electric field are obtained by taking the real part of these two equations,

$$\text{Re}\,\{E_R\} = (E_{o\perp}/2)(\hat{x}\cos\omega t + \hat{y}\sin\omega t) \tag{3.22}$$

$$\text{Re}\,\{E_L\} = (E_{o\perp}/2)(\hat{x}\cos\omega t - \hat{y}\sin\omega t) \tag{3.23}$$

Thus, the fields $\vec{E}_R$ and $\vec{E}_L$ are constant in magnitude, but sweep around in a circle at frequency ω. For $\vec{E}_R$ the rotation is clockwise looking in the direction of $\vec{B}$, and it is called a *right circularly polarized* component, while for $\vec{E}_L$ the rotation is counterclockwise and it is called a *left circularly polarized* component. This is illustrated in Fig. 1.

Proceeding in the analysis of the particle motion in the perpendicular plane, we must now substitute the decomposed form (3.16) into (3.12). The operator $(i\omega - \vec{\omega}_c \times)$ when applied on the right circularly polarized component $\vec{E}_{oR}$ gives

$$\begin{aligned}(i\omega - \vec{\omega}_c \times)\, \vec{E}_{oR} &= (i\omega - \vec{\omega}_c \times)(\vec{E}_{o\perp} + i\, \hat{B} \times \vec{E}_{o\perp})/2 \\ &= [i\omega\, (\vec{E}_{o\perp} + i\, \hat{B} \times \vec{E}_{o\perp}) + (qB/m)\, \hat{B} \times \vec{E}_{o\perp} - i(qB/m)\, \vec{E}_{o\perp}]/2 \\ &= [i\omega\, (\vec{E}_{o\perp} + i\, \hat{B} \times \vec{E}_{o\perp}) - i(qB/m)\, (\vec{E}_{o\perp} + i\, \hat{B} \times \vec{E}_{o\perp})]/2 \\ &= i(\omega - qB/m)\, \vec{E}_{oR} \end{aligned} \tag{3.24}$$

In a similar fashion we find for the left circularly polarized component,

$$(i\omega - \vec{\omega}_c \times)\, \vec{E}_{oL} = i(\omega + qB/m)\, \vec{E}_{oL} \tag{3.25}$$

Therefore, both $\vec{E}_{oR}$ and $\vec{E}_{oL}$ are eigenvectors of the complex operator appearing in (3.12). Using (3.24) and (3.25), (3.12) becomes

$$\vec{v}_{e\perp} = i(q/m) \cdot \left[\frac{\vec{E}_{oR}}{(\pm\, \omega_c + \omega)} - \frac{\vec{E}_{oL}}{(\pm\, \omega_c - \omega)} \right] \tag{3.26}$$

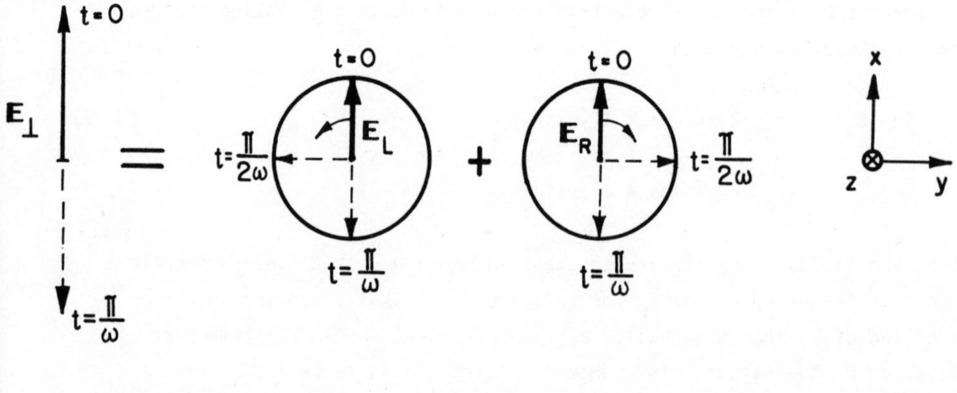

Fig. 1 - A plane polarized wave, indicated by the double-headed arrow $\vec{E}_{\perp}$, is equivalent to the sum of left and right circularly polarized waves $\vec{E}_L$ and $\vec{E}_R$.

The upper sign applies to a positively charged particle and the lower sign to a negatively charged particle, since $\omega_c = |q| B/m$ is always positive. Thus, $\vec{v}_{e\perp}$ also separates into two vectors rotating in opposite directions,

$$\vec{v}_{e\perp} = \vec{v}_R + \vec{v}_L \tag{3.27}$$

where

$$\vec{v}_R = i(q/m) \frac{\vec{E}_{oR}}{(\pm \omega_c + \omega)} \tag{3.28}$$

$$\vec{v}_L = -i(q/m) \frac{\vec{E}_{oL}}{(\pm \omega_c - \omega)} \tag{3.29}$$

For a *positive ion* we see that, as ω approaches the ion cyclotron frequency ω_{ci}, there is resonance between the ion and the *left* circularly polarized component of the $\vec{E}$ field. For an *electron*, as ω approaches the electron cyclotron frequency ω_{ce}, resonance occurs with the *right* circularly polarized component.

3.3 Mobility dyad

The expression for $\vec{v}_e = \vec{v}_{e\parallel} + \vec{v}_R + \vec{v}_L$ can be written in a compact form through the introduction of the *mobility dyad* $\bar{\bar{\mu}}$, defined by

$$\vec{v}_e = \bar{\bar{\mu}} \cdot \vec{E}_0 \tag{3.30}$$

Using (3.7), (3.28) and (3.29) we see that the mobility dyad $\bar{\bar{\mu}}$ is diagonal in the rotating system, so that (3.30) becomes in matrix form,

$$\begin{pmatrix} v_R \\ v_L \\ v_{e\parallel} \end{pmatrix} = \frac{i\,q}{m\,\omega} \begin{pmatrix} \dfrac{\omega}{(\pm \omega_c + \omega)} & 0 & 0 \\ 0 & \dfrac{-\omega}{(\pm \omega_c - \omega)} & 0 \\ 0 & 0 & 1 \end{pmatrix} \begin{pmatrix} E_{oR} \\ E_{oL} \\ E_{o\parallel} \end{pmatrix} \tag{3.31}$$

If, instead of the rotating system, we use a stationary Cartesian coordinate system with the z axis pointing along the magnetostatic field, we can write (3.7) and (3.12) in matrix form as

$$\begin{pmatrix} v_{ex} \\ \\ v_{ey} \\ \\ v_{ez} \end{pmatrix} = \frac{i\,q}{m\,\omega} \begin{pmatrix} \dfrac{\omega^2}{(\omega^2 - \omega_c^2)} & \dfrac{\pm i\omega_c\omega}{(\omega^2 - \omega_c^2)} & 0 \\ -\dfrac{\pm i\omega_c\omega}{(\omega^2 - \omega_c^2)} & \dfrac{\omega^2}{(\omega^2 - \omega_c^2)} & 0 \\ 0 & 0 & 1 \end{pmatrix} \begin{pmatrix} E_{ox} \\ \\ E_{oy} \\ \\ E_{oz} \end{pmatrix} \qquad (3.32)$$

3.4 Plasma conductivity dyad

Denoting by n_0 the number density of the electrons (charge - e) and ions (charge + e) in a plasma, the electric current density is given by

$$\vec{J} = - n_0\, e\, \vec{v}_e + n_0\, e\, \vec{v}_i = -n_0\, e\, \bar{\bar{\mu}}_e \cdot \vec{E} + n_0\, e\, \bar{\bar{\mu}}_i \cdot \vec{E} \qquad (3.33)$$

where $\bar{\bar{\mu}}_e$ and $\bar{\bar{\mu}}_i$ are the mobility dyads for the electrons and ions, respectively. Introducing the conductivity dyad $\bar{\bar{\sigma}}$ by

$$\vec{J} = \bar{\bar{\sigma}} \cdot \vec{E} = (\bar{\bar{\sigma}}_e + \bar{\bar{\sigma}}_i) \cdot \vec{E} \qquad (3.34)$$

we obtain from (3.33), for the electron and ion conductivities, respectively,

$$\bar{\bar{\sigma}}_e = -n_0\, e\, \bar{\bar{\mu}}_e \qquad (3.35)$$

$$\bar{\bar{\sigma}}_i = n_0\, e\, \bar{\bar{\mu}}_i \qquad (3.36)$$

With the help of (3.32), these can be expressed in matrix form as

$$\bar{\bar{\sigma}}_e = \frac{i\, n_0\, e^2}{m_e\, \omega} \begin{pmatrix} \dfrac{\omega^2}{(\omega^2 - \omega_{ce}^2)} & \dfrac{-i\omega_{ce}\omega}{(\omega^2 - \omega_{ce}^2)} & 0 \\ \dfrac{i\omega_{ce}\omega}{(\omega^2 - \omega_{ce}^2)} & \dfrac{\omega^2}{(\omega^2 - \omega_{ce}^2)} & 0 \\ 0 & 0 & 1 \end{pmatrix} \qquad (3.37)$$

$$\bar{\bar{\sigma}}_i = \frac{i\, n_0\, e^2}{m_i\, \omega} \begin{pmatrix} \frac{\omega^2}{(\omega^2 - \omega_{ci}^2)} & \frac{i\omega_{ci}\omega}{(\omega^2 - \omega_{ci}^2)} & 0 \\ \frac{-i\omega_{ci}\omega}{(\omega^2 - \omega_{ci}^2)} & \frac{\omega^2}{(\omega^2 - \omega_{ci}^2)} & 0 \\ 0 & 0 & 1 \end{pmatrix} \qquad (3.38)$$

The fact that the conductivity $\bar{\bar{\sigma}}$ is imaginary implies that $\vec{J}$ and $\vec{E}$ are 90^0 out of phase, since the actual physical expressions for $\vec{J}$ and $\vec{E}$ are obtained by taking the real part of (3.34) and (3.1), respectively.

3.5 Cyclotron resonance

The particle velocity, given in (3.7), (3.28) and (3.29), does not represent correctly the motion of the particle when the frequency ω of the electric field is equal to the particle cyclotron frequency ω_c. A positive particle, for example, when the forcing electric field rotates in the counterclockwise direction looking in the direction of $\vec{B}$ (left circularly polarized component), is able to absorb energy from the electric field, so that its speed increases continually and indefinitely with time. The same is true for an electron, say, and the right circularly polarized component of the forcing electric field when $\omega = \omega_{ce}$. This phenomenon is called *cyclotron resonance*. To investigate the particle motion under resonance conditions, it is necessary to go back to the original equations of motion and solve the problem for the case when $\omega = \omega_c$.

For simplicity, let us assume that the component of the $\vec{E}$ field along $\vec{B}$ vanishes, that is, $\vec{E}_{\parallel} = 0$. Hence, we take

$$\vec{E}(t) = \vec{E}_{0\perp}\, e^{-i\omega_c t} \qquad (3.39)$$

with $\omega_c = |q|\, B/m$. In view of this assumption, the particle velocity *along* $\vec{B}$ is constant and is equal to the initial parallel velocity. The component of the equation of motion in the plane *normal* to the magnetic field becomes

$$d\vec{v}_{\perp}(t)/dt = (q/m)\, [\vec{E}_{0\perp}\, e^{-i\omega_c t} + \vec{v}_{\perp}(t) \times \vec{B}] \qquad (3.40)$$

Differentiating this equation with respect to time, yields

$$d^2\vec{v}_{\perp}/dt^2 = (q/m)\, [-\, i\omega_c\, \vec{E}_{0\perp}\, e^{-i\omega_c t} + (d\vec{v}_{\perp}/dt) \times \vec{B}] \qquad (3.41)$$

and using (3.40) to eliminate $d\vec{v}_\perp/dt$ in (3.41), we obtain, after rearranging the terms,

$$\frac{d^2\vec{v}_\perp}{dt^2} + \omega_c^2\vec{v}_\perp = -\, i\omega_c(q/m)\,\vec{E}_{0\perp}\,e^{-i\omega_c t} - (qB/m)(q/m)(\hat{B}\times\vec{E}_{0\perp})\,e^{-i\omega_c t} \quad (3.42)$$

The solution of this inhomogeneous differential equation is given by the sum of the solution of the homogeneous equation, plus a particular integral of the inhomogeneous equation. The solution of the homogeneous equation

$$d^2\vec{v}_\perp/dt^2 + \omega_c^2\vec{v}_\perp = 0 \quad (3.43)$$

is just the cyclotron motion described previously by the velocity $\vec{v}_m$. A particular integral of (3.42) is provided by the function

$$\vec{v}_\perp = \vec{A}\,t\,e^{-i\omega_c t} \quad (3.44)$$

To determine the vector $\vec{A}$ we differentiate (3.44) twice with respect to time

$$d\vec{v}_\perp/dt = \vec{A}\,(1 - i\omega_c t)\,e^{-i\omega_c t} \quad (3.45)$$

$$d^2\vec{v}_\perp/dt^2 = -\,2\vec{A}\,i\omega_c\,e^{-i\omega_c t} - \omega_c^2\vec{v}_\perp \quad (3.46)$$

Comparing (3.46) with (3.42) we see that (3.44) satisfies (3.42) provided we take

$$\vec{A} = (q/2m)\,\vec{E}_{0\perp} - (i/2\omega_c)(qB/m)(q/m)(\hat{B}\times\vec{E}_{0\perp}) \quad (3.47)$$

Therefore, the complete solution of (3.42) is

$$\vec{v}_\perp(t) = \vec{v}_m(t) + (q/2m)(\vec{E}_{0\perp} \mp i\,\hat{B}\times\vec{E}_{0\perp})t\,e^{-i\omega_c t} \quad (3.48)$$

where we have replaced qB/m by $\pm\omega_c$, the upper and lower signs corresponding, respectively, to positive and negative charges. Using (3.17) and (3.18), which define the right and left circularly polarized components of the electric field, respectively, we can write (3.48) for a *positively* charged particle $(q > 0)$ as

$$\vec{v}_\perp(t) = \vec{v}_m(t) + (q/m)\,\vec{E}_{0L}\;t\,e^{-i\omega_c t} \quad (3.49)$$

and for a *negatively* charged particle (q < 0) as

$$\vec{v}_\perp(t) = \vec{v}_m(t) + (q/m)\, \vec{E}_{oR}\, t\, e^{-i\omega_c t} \tag{3.50}$$

Hence, the particle velocity increases indefinitely with time. Note that the expression $(q/m)\,\vec{E}_{oR,L}$ represents a constant acceleration. A positive charge resonates with $\vec{E}_L$, and a negative one with $\vec{E}_R$. The particle moves in circles of ever increasing radii, with its velocity increasing continually during this spiral motion at the expense of the energy of the electric field. A typical resonant spiral for an electron is shown in Fig. 2. This phenomenon can be used as a method of increasing the particle speed and hence the kinetic temperature of the plasma through collisions. This method is known as *radio frequency heating* of the plasma by *cyclotron resonance*.

4. TIME-VARYING MAGNETIC FIELD AND SPACE VARYING ELECTRIC FIELD

From Maxwell equations it is seen that a time-varying magnetic field is also accompanied by a space-varying electric field. The $\vec{E}$ field associated with the time-dependent $\vec{B}$ field is given by

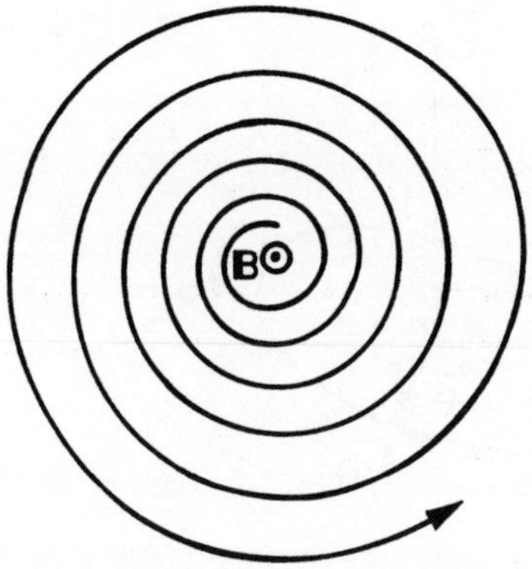

Fig. 2 - Outward spiral motion of an electron, in the plane normal to the magnetic field, under cyclotron resonance.

$$\vec{\nabla} \times \vec{E} = - \partial\vec{B}/\partial t \tag{4.1}$$

We shall assume that the fractional change in the magnetic field, in a time interval of the order of the cyclotron period, is very small.

4.1 Equation of motion and adiabatic invariants

Consider the magnetic field in the z direction and suppose that it is spatially uniform and increases with time within the circular orbit of the particle. From Faraday's law, an electric field is induced along the path of the particle orbit (Fig. 3) which accelerates the particle, with the result that the orbit is no longer a circle. However, since the time variation of the $\vec{B}$ field is small, the azimuthal component of the electric field $\vec{E}_\theta$ is also small, and the orbit will be nearly a circle.

Expressing (4.1) in cylindrical coordinates, taking $\vec{B} = B\hat{z}$, and considering $\vec{E} = \hat{\theta}\, E_\theta(r)$, we find

$$\frac{1}{r}\,\frac{\partial}{\partial r}\,(rE_\theta) = -\,\frac{\partial B}{\partial t} \tag{4.2}$$

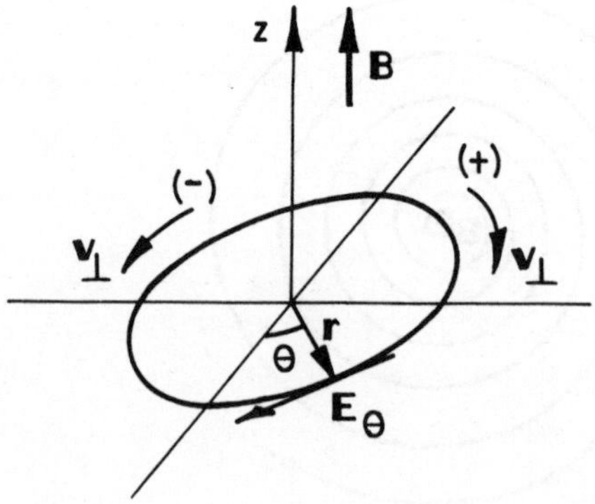

Fig. 3 - Azimuthal electric field $\vec{E}_\theta$ induced by a time-varying magnetic field. The magnetic field inside the orbit is uniform, parallel to the z axis, and increases slowly with time.

Integrating with respect to r, noting that $\partial B/\partial t$ can be taken outside the integral sign since B is a slowly varying function, yields

$$\int_0^r \partial/\partial r' \, (r' E_\theta) \, dr' = - \, \partial B/\partial t \int_0^r r' \, dr' \tag{4.3}$$

from which we obtain the induced electric field as

$$E_\theta = - \, (r/2) \, \partial B/\partial t \tag{4.4}$$

Further, since $\hat{\theta} = -\hat{r} \times \hat{z}$, (4.4) can be writen in vector form as

$$\vec{E}_\theta = \vec{r} \times (\partial \vec{B}/\partial t)/2 \tag{4.5}$$

Using this result in the Lorentz force equation, we obtain for the equation of motion, after some rearrangement,

$$d\vec{v}/dt = (\partial \vec{\omega}_c/\partial t) \times \vec{r}/2 + \vec{\omega}_c \times \vec{v} \tag{4.6}$$

Instead of solving (4.6) directly, we shall determine a relation between the radius vector r and the time rate of change of $\vec{B}$, by calculating the change in the particle transverse kinetic energy over one gyration period, which results from the action of the induced electric field. Since the force acting on the particle due to the electric field is $q \, \vec{E}_\theta$, the increase in the transverse kinetic energy over one gyration period is given by

$$\delta(mv_\perp^2/2) = q \oint \vec{E}_\theta \cdot d\vec{r} \tag{4.7}$$

where $d\vec{r}$ denotes an element of path along the particle trajectory, so that $\vec{v}_\perp = d\vec{r}/dt$. Since the field changes very slowly, we can calculate the line integral in (4.7) as if the orbit were closed. Using Stoke's theorem, we replace the line integral by a surface integral over the unperturbed orbit,

$$\begin{aligned} \delta(mv_\perp^2/2) &= q \int_S (\vec{\nabla} \times \vec{E}_\theta) \cdot d\vec{S} \\ &= - \, q \int_S (\partial \vec{B}/\partial t) \cdot d\vec{S} \end{aligned} \tag{4.8}$$

where use was made of (4.1). Here $\vec{S}$ denotes the surface enclosed by the cyclotron orbit, its direction being such that $\vec{B} \cdot d\vec{S} < 0$ for ions, and

$\vec{B} \cdot d\vec{S} > 0$ for electrons, in view of the diamagnetic character of a plasma. Thus (4.8) becomes

$$\delta(mv_\perp^2/2) = |q|\ (\partial B/\partial t)\,\pi\, r_c^2 \tag{4.9}$$

Now, the change in the magnetic field over one gyration period, $(2\pi/\omega_c)$, is

$$\delta B = (\partial B/\partial t)(2\pi/\omega_c) \tag{4.10}$$

and using the relations $r_c^2 = v_\perp^2/\omega_c^2$ and $\omega_c = |q|\ B/m$, (4.9) can be rewritten as

$$\delta(mv_\perp^2/2) = (mv_\perp^2/2B)\ \delta B = |\vec{m}|\ \delta B \tag{4.11}$$

where the quantity $|\vec{m}| = mv_\perp^2/2B$ is the orbital magnetic moment of the charged particle. Now, since the left-hand side of (4.11) is $\delta(|\vec{m}|B)$, we obtain

$$\delta|\vec{m}| = 0 \tag{4.12}$$

This result shows that the *magnetic moment* is *invariant* in slowly varying magnetic fields for which $(\partial B/\partial t)(2\pi/\omega_c) << B$.

From the constancy of the magnetic moment, we can easily verify that

$$B\ r_c^2 = \text{constant} \tag{4.13}$$

Therefore, as the magnetic field increases, the radius of gyration decreases, as shown in Fig. 4. Further, since the magnetic flux Φ_m through a Larmor orbit is given by

$$\Phi_m = BS = B\pi r_c^2 \tag{4.14}$$

it is clear that the *magnetic flux* is also an *adiabatic invariant*. Hence, as the magnetic field strength increases, the radius of the orbit decreases in such a way that the particle always encircles the same number of magnetic flux lines.

When the time variation of the magnetic field is not spatially uniform within the circular orbit of the gyrating particle, but if it occurs in an unsymmetrical way, then the induced electric field acting on the charged particle can considerably modify its orbit from the one shown in Fig. 4.

In the most general cases the particle orbit can be extremely complicated. In order to obtain a general idea of what may happen, let us consider the simple case of a magnetic field varying in time with cylindrical symmetry over a region of radius R, which is much larger that the cyclotron radius r_c, as shown in Fig. 5. The azimuthal component of the induced electric field, $\vec{E}_\theta$, at the point P (see Fig. 5) is given, from (4.5), by

$$\vec{E}_\theta = \vec{R} \times (\partial\vec{B}/\partial t)/2 \tag{4.15}$$

A charged particle located at the point P is now acted upon by crossed electric and magnetic fields,resulting in a drift velocity given by

$$\vec{v}_E = (\vec{E}_\theta \times \vec{B})/B^2 = (\vec{R} \times \partial\vec{B}/\partial t) \times \vec{B}/2B^2 \tag{4.16}$$

Since $\vec{B}$ is in the z direction (normal to the vector $\vec{R}$) (4.16) gives

$$\vec{v}_E = - \vec{R}\ (\partial B/\partial t)/2B \tag{4.17}$$

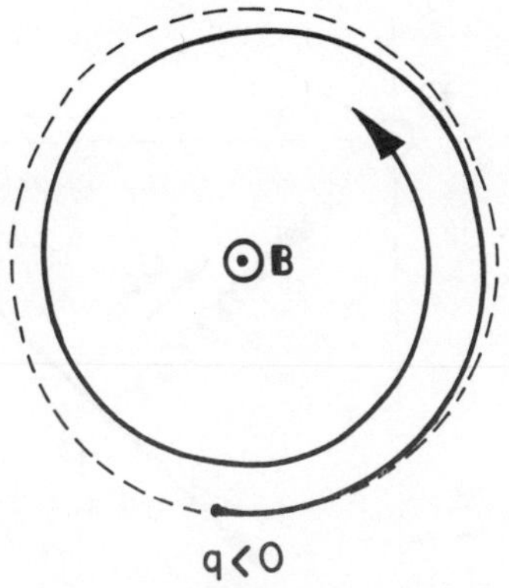

Fig. 4 - Motion of an electron in a time-varying magnetic field. The field is uniform and increases with time inside the particle orbit.

Therefore, the guiding center of the particle drifts radially inward with the drift velocity $\vec{v}_E$ given in (4.17). As the particle drifts radially inward, its radius of gyration decreases in such a way that the flux encircled by the gyrating particle remains constant (Fig. 5). Since the density of the magnetic flux lines increases as the magnetic field strength increases, this radial particle drift can be pictured as a radially inward motion of the magnetic flux lines with the velocity $\vec{v}_E$ given in (4.17), with the guiding center attached to a given flux line.

4.2 Magnetic heating of a plasma

The adiabatic invariance of the orbital magnetic moment of the charged particle ($|\vec{m}| = mv_\perp^2/2B$ = constant) implies that when the magnetic field is increased, the

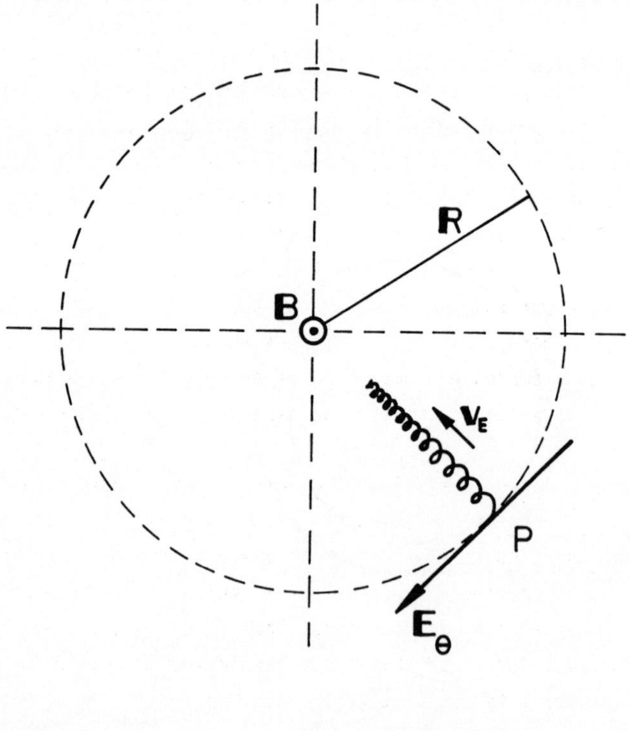

Fig. 5 - Motion of a negatively charged particle in an increasing magnetic field (directed out of the page) with cylindrical symmetry over a region of radius R much larger that the Larmor radius r_c.

particle transverse kinetic energy ($W_{\perp} = mv_{\perp}^2/2$) increases linearly with B. Further, since the magnetic flux encircled by the gyrating particle is also constant, as the magnetic flux density increases, the magnetic flux tube contracts and the particle guiding center moves radially inward, accompanying the radial displacement of the magnetic field lines, as if it was frozen in the field lines. As a consequence, the increase in the magnetic flux density causes the charged particles in a plasma to approach each other, resulting in a *magnetic compression*. In the present case of two-dimensional compression, since the increase in the number density (n) of the particles is proportional to the cross-sectional area πr_c^2, it follows that under magnetic compression in two dimensions n increases linearly with B. Similarly, when B decreases, n also decreases, resulting in a *magnetic decompression*. Thus, for two-dimensional compression we have $W_{\perp} \propto B \propto n$.

This property is used as a method of plasma heating, known as *magnetic pumping*, which consists of periodic magnetic compressions and decompressions of the plasma. The compression and the decompression take place in a time interval that is very large compared to the Larmor period but, at the same time, is very small compared to the relaxation time necessary for the achievement of thermal equilibrium. For the present case of two-dimensional compression-decompression, obtained by varying the axial magnetic field, only the velocity and the energy in the two directions normal to the magnetic field are changed. In order to be able to heat the plasma in a compression--decompression cycle, it is necessary to transfer part of the energy increase in $W_{\perp}$, obtained by compression, to the energy $W_{\parallel}$ which is unaffected by decompression. This energy transfer is brought about by collisional interactions, which promotes equipartition of energy, therefore increasing the particle thermal kinetic energy at the expense of the electric field energy.

Thus, in a complete cycle consisting of a compression, a relaxation time, and a decompression, part of the energy increase in $W_{\perp}$ achieved during compression is transferred to the degree of freedom parallel to B, as a result of collisions, causing an increase in $W_{\parallel}$ which is not affected during the decompression, whereas $W_{\perp}$ is decreased. Therefore, by a periodic repetition of these cycles of adiabatic compression-decompression, the energy of the plasma and therefore its temperature is increased.

5. SUMMARY OF GUIDING CENTER DRIFTS AND CURRENT DENSITIES

5.1 Guiding center drifts

Electric field:	$\vec{v}_E = (\vec{E} \times \vec{B})/B^2$	(2.5.15)
Gravitational field:	$\vec{v}_g = m(\vec{g} \times \vec{B})/(qB^2)$	(2.6.3)
General force:	$\vec{v}_F = (\vec{F} \times \vec{B})/(qB^2)$	(2.6.2)
Gradient of $\vec{B}$:	$\vec{v}_G = -\lvert\vec{m}\rvert(\vec{\nabla}B) \times \vec{B}/(qB^2)$	(3.5.1)
Curvature of $\vec{B}$:	$\vec{v}_C = -mv_{\parallel}^2[(\vec{B}\cdot\vec{\nabla})\vec{B}] \times \vec{B}/(qB^4)$	(3.7.9)
Gradient-curvature of $\vec{B}$: (in vacuum field)	$\vec{v}_{GC} = -m(v_{\parallel}^2 + v_{\perp}^2/2)(\vec{\nabla}B^2/2) \times \vec{B}/(qB^4)$	(3.8.2)
Polarization:	$\vec{v}_p = m(\partial\vec{E}_{\perp}/\partial t)/(qB^2)$	(2.11)

5.2 Current densities

Magnetization:	$\vec{J}_M = \vec{\nabla} \times \vec{M}$	(2.4.40)
Gravitational:	$\vec{J}_g = \rho_m\vec{g} \times \vec{B}/B^2$	(2.6.5)
Gradient of $\vec{B}$:	$\vec{J}_G = -[(1/\Delta V) \sum_i \lvert\vec{m}_i\rvert] (\vec{\nabla}B) \times \vec{B}/B^2$	(3.5.3)
Curvature of $\vec{B}$:	$\vec{J}_C = -2\ [(1/\Delta V) \sum_i W_{i\parallel}][\hat{B}\cdot\vec{\nabla})\hat{B}] \times \vec{B}/B^2$	(3.7.12)
Polarization:	$\vec{J}_p = \rho_m(\partial\vec{E}_{\perp}/\partial t)/B^2$	(2.12)

PROBLEMS

4.1 - With reference to a magnetic field pointing along the z axis ($\vec{B} = B_0\hat{z}$), describe the type of polarization of the following electric field:

$$\vec{E} = E_0\ (\hat{x}\cos\omega t - \hat{y}\sin\omega t)$$

Make a drawing which shows the orientation of $\vec{E}$ for the instants $t = 0$, $t = \pi/2\omega$ and $t = \pi/\omega$. How can you represent this electric field in complex notation?

4.2 - Describe, in a semiquantitative way, the motion of an electron in the presence of a constant magnetic field

$$\vec{B} = B_0 \hat{z}$$

and a time-varying electric field given by

$$\vec{E} = (E_0/2)\ (\hat{x} + i\ \hat{y})\ \exp(-i\omega_c t)$$

where E_0 and B_0 are positive constants and $\omega_c = e\ B_0/m_e$. What type of polarization has this electric field?

4.3 - Solve the equation of motion to determine the transient response of a charged particle in the presence of a spatially uniform a.c. electric field $\vec{E}(t) = \hat{x}\ E_0 \sin(\omega t)$, that is switched on at $t = 0$. Assume that initially, at $t = 0$, the particle is at rest at the origin. Make a plot of the particle trajectory and velocity as a function of time.

4.4 - Consider an electron acted by a constant and uniform magnetic field $\vec{B} = B_0\ \hat{z}$, and a uniform but time-varying electric field $\vec{E} = \hat{y}\ E_{yo} \sin(\omega t)$. Assume that the initial conditions are such that the motion takes place in the x-y plane and that at $t = 0$ the electron is at rest ($\vec{v}_0 = 0$) at the origin.

(a) Show that the orbit of the electron is given by

$$x(t) = -(eE_{yo}/m)[(\omega/\omega_c)(\cos\omega_c t - 1) - (\omega_c/\omega)(\cos\omega t - 1)]/(\omega^2 - \omega_c^2)$$

$$y(t) = -(eE_{yo}/m)[(\omega/\omega_c)\sin(\omega_c t) - \sin(\omega t)]/(\omega^2 - \omega_c^2)$$

(b) In the low-frequency limit, $\omega \ll \omega_c$, show that the electron orbits at the angular frequency ω around an ellipse which has its major axis perpendicular to the electric field. Determine the ratio of the minor to the major axis of the ellipse.

(c) In the high-frequency limit, $\omega \gg \omega_c$, show that the electron moves in a circle at the cyclotron frequency ω_c.

4.5 - Integrate (3.49) and (3.50) to determine the particle trajectory in the plane normal to B and sketch the path of the particle for $q > 0$ and $q < 0$.

4.6 - Consider the motion of an electron in the presence of a uniform magnetostatic field $\vec{B} = B_0\,\hat{z}$, and an electric field which oscillates in time at the electron cyclotron frequency ω_c, according to

$$\vec{E}(t) = E_0\,[\hat{x}\cos(\omega_c t) + \hat{y}\sin(\omega_c t)]$$

(a) What type of polarization has this electric field?

(b) Obtain the following uncoupled differential equations satisfied by the velocity components $v_x(t)$ and $v_y(t)$:

$$d^2v_x/dt^2 + \omega_c^2 v_x = 2(eE_0/m)\,\omega_c \sin(\omega_c t)$$

$$d^2v_y/dt^2 + \omega_c^2 v_y = -2(eE_0/m)\,\omega_c \cos(\omega_c t)$$

(c) Assume that, at $t = 0$, the electron is located at the origin of the coordinate system, with zero velocity. Neglect the time-varying part of $\vec{B}$. Show that the electron velocity is given by

$$v_x(t) = -(eE_0/m)t\cos(\omega_c t)$$

$$v_y(t) = -(eE_0/m)t\sin(\omega_c t)$$

(d) Show that the electron trajectory is given by

$$x(t) = -(eE_0/m)\,[(1/\omega_c^2)\cos(\omega_c t) + (t/\omega_c)\sin(\omega_c t) - 1/\omega_c^2]$$

$$y(t) = -(eE_0/m)\,[(1/\omega_c^2)\sin(\omega_c t) - (t/\omega_c)\cos(\omega_c t)]$$

4.7 - Solve the equation of motion to determine the velocity and the trajectory of an electron in the presence of a uniform magnetostatic field $\vec{B} = B_0\,\hat{z}$, and an oscillating electric field given by

$$\vec{E}(t) = \hat{x}\,E_x \sin(\omega t) + \hat{z}\,E_z \cos(\omega t)$$

Consider the same assumptions and initial conditions as in Problem 4.6.

4.8 - Consider the motion of an electron in a spatially uniform magnetic field $\vec{B} = B_z \hat{z}$, such that B_z has a slow time variation given by

$$B_z(t) = B_0(1 - \alpha t)$$

where B_0 and α are positive constants, and $|\alpha t| << 1$. Assume the following initial conditions: $\vec{r}(0) = (r_c,0,0)$ and $\dot{\vec{r}}(0) \equiv \vec{v}(0) = (0,v_{\perp 0},0)$, where r_c is the Larmor radius, $v_{\perp 0} = \omega_c r_c$, and $\omega_c = |q| B_0/m$.
(a) Write the equation of motion, considering the Lorentz force, and solve it by a perturbation technique including only terms up to the first order in the small parameter α. Show that the particle velocity is given by

$$v_x(t) = - \omega_c r_c \sin(\omega_c t) + \frac{1}{2} \alpha r_c \omega_c t[\sin(\omega_c t) + \omega_c t \cos(\omega_c t)]$$

$$v_y(t) = \omega_c r_c \cos(\omega_c t) + \frac{1}{2} \alpha r_c \omega_c t[-\cos(\omega_c t) + \omega_c t \sin(\omega_c t)]$$

(b) Show that the particle orbit is given by

$$x(t) = r_c[(1 + \alpha t/2) \cos(\omega_c t) + (\alpha/2\omega_c)(\omega_c^2 t^2 - 1) \sin(\omega_c t)]$$

$$y(t) = r_c[(1 + \alpha t/2) \sin(\omega_c t) - (\alpha/2\omega_c)(\omega_c^2 t^2 - 1) \cos(\omega_c t) - \alpha/2\omega_c]$$

$$z(t) = v_{z0} t$$

(c) Determine the orbital magnetic moment and verify its adiabatic invariance, retaining only terms up to the first order in α.

4.9 - Consider the motion of a charged particle in a spatially uniform magnetic field which varies slowly in time as compared to the cyclotron period of the particle.
(a) Show that the equation of motion can be written in vector form as

$$d\vec{v}(t)/dt = \vec{\omega}_c(t) \times \vec{v}(t) + [\partial\vec{\omega}_c(t)/\partial t] \times \vec{r}(t)/2$$

where $\vec{\omega}_c(t) = - q \vec{B}(t)/m$.

(b) Considering that $\vec{B}(t) = \hat{z}\, B_0\, f(t)$, where B_0 is constant, obtain the following equations for the motion of the particle in the plane normal to $\vec{B}$,

$$\frac{d^2x(t)}{dt^2} = \omega_c \left[\frac{dy(t)}{dt} f(t) + \frac{1}{2} y(t) \frac{df(t)}{dt} \right]$$

$$\frac{d^2y(t)}{dt^2} = -\omega_c \left[\frac{dx(t)}{dt} f(t) + \frac{1}{2} x(t) \frac{df(t)}{dt} \right]$$

where $\omega_c = |q|\, B_0/m$.

(c) Define a complex variable $u(t) = x(t) + iy(t)$, and a function $\xi(t)$ by

$$\xi(t) = u(t) \exp\left[(i\omega_c/2) \int_0^t f(t')\, dt'\right]$$

and show that the equation satisfied by $\xi(t)$ is

$$\frac{d^2\xi(t)}{dt^2} + (\omega_c^2/4)\, f^2(t)\, \xi(t) = 0$$

(d) If $\xi_1(t)$ and $\xi_2(t)$ are two linearly independent solutions of this equation, subjected to the initial conditions

$$\xi_1(0) = 0 \qquad d\xi_1/dt\big|_{t=0} = 1$$

$$\xi_2(0) = 1 \qquad d\xi_2/dt\big|_{t=0} = 0$$

show that the solution for $u(t)$ can be written as

$$u(t) = \exp\left[-(i\omega_c/2) \int_0^t f(t')\, dt'\right] \{u_0\, \xi_2(t) + \xi_1(t)\, [\dot{u}_0 + (i\omega_c/2)\, f(0)\, u_0]\}$$

where u_0 and $\dot{u}_0$ represent the initial position and velocity respectively.

(e) Considering now that the particle is initially ($t = 0$) at the origin and moving with velocity v_0 along the negative y axis, that is, $u_0 = 0$ and $\dot{u}_0 = -iv_0$, show that

$$u(t) = \exp\left[-(i\omega_c/2) \int_0^t f(t')\, dt'\right][-iv_0\, \xi_1(t)]$$

and, consequently,

$$x(t) = v_0\ \xi_1(t)\ \sin[(\omega_c/2) \int_0^t f(t')\ dt']$$

$$y(t) = -\ v_0\ \xi_1(t)\ \cos[(\omega_c/2) \int_0^t f(t')\ dt']$$

4.10 - (a) Assume that f(t), in Problem 4.9, is given by $\exp(-\alpha t)$. Show that, in this case, $\xi(t)$ satisfies the Bessel equation of zero order,

$$\frac{d^2\xi(\tau)}{d\tau^2} + (1/\tau)\frac{d\xi(\tau)}{d\tau} + \xi(\tau) = 0$$

where $\tau \equiv (\omega_c/2\alpha)\exp(-\alpha t)$. Determine the two solutions of this equation which satisfy the initial conditions stated in Problem 4.9, and interpret them physically.

(b) Considering now that $f(t) = (1 - \alpha t)$, solve the equation for $\xi(t)$ in Problem 4.9, in a power series in α, and determine the particle trajectory to order α. Show that the ratio $(v_x^2 + v_y^2)/B(t)$ has no terms of order α, thus verifying the adiabatic invariance of the magnetic moment. Compare these results with those of Problem 4.8.

4.11 - For an electron, with initial velocity $v_0\hat{x}$ and initial position $x_0\hat{x}$, acted upon by an electric field $\vec{E} = \hat{x}\ E\cos(kx - \omega t)$, show that its velocity is given by

$$v(t) = v_0 - (eE/m_e) \int_0^t \cos(kx - \omega t')\ dt'$$

Using a perturbation approach, in which to lowest order $\vec{E} = 0$, show that

$$v(t) = v_0 - [(eE/m_e)/(kv_0 - \omega)]\{\sin[kx_0 + (kv_0 - \omega)t] - \sin(kx_0)\}$$

Notice that the velocity perturbation will be large only when v_0 is close to the phase velocity ω/k.

4.12 - Using the Maxwell equation (1.5.3), and (3.34) which defines the plasma conductivity dyad $\bar{\bar{\sigma}}$, and considering the time variation indicated in (3.1), show that

$$\vec{\nabla} \times \vec{B} = -\ i\omega\mu_0\ \bar{\bar{\varepsilon}}\cdot\vec{E}$$

where $\bar{\bar{\varepsilon}}$ is the plasma electric permitivity dyad given by

$$\bar{\bar{\varepsilon}} = \varepsilon_0[\bar{\bar{1}} + (i/\omega\varepsilon_0)\,\bar{\bar{\sigma}}]$$

where $\bar{\bar{1}}$ denotes the unit dyad which in Cartesian coordinates can be written as

$$\bar{\bar{1}} = \hat{x}\hat{x} + \hat{y}\hat{y} + \hat{z}\hat{z}$$

CHAPTER 5

Elements of Plasma Kinetic Theory

1. INTRODUCTION

A plasma is a system containing a very large number of interacting charged particles, so that for its analysis it is appropriate to use a *statistical approach*. In this chapter we present the basic elements of kinetic theory, introducing the concepts of *phase space* and *distribution function* necessary for a statistical description. All physically interesting information about the system is contained in the distribution function. From a knowledge of the distribution function, the average values of the various physical quantities of interest, which can be considered as functions of the particle velocities, can be deduced systematically and used to describe the macroscopic behavior of the plasma.

The differential kinetic equation satisfied by the distribution function, known as the *Boltzmann equation*, is deduced in section 5. At this stage, the effects of collisions are incorporated into this kinetic equation only through a general, unspecified collision term. In Chapter 21 we shall deduce explicit expressions for the collision term, in particular for the Boltzmann collision integral and for the Fokker-Planck collision term. Only a simple approximate expression for the collision term is presented at this point, the so-called *relaxation model* or *Krook collision term*. The *Vlasov equation* for a plasma is introduced in the last section.

2. PHASE SPACE

At any instant of time each particle in the plasma can be localized by a position vector $\vec{r}$ drawn from the origin of a coordinate system to the center of mass of the particle. In the Cartesian frame of reference we have (Fig. 1)

$$\vec{r} = \hat{x}x + \hat{y}y + \hat{z}z \tag{2.1}$$

where $\hat{x}$, $\hat{y}$ and $\hat{z}$ denote unit vectors along the axes x, y and z, respectively. The linear velocity of the center of mass of the particle can be represented by

$$\vec{v} = d\vec{r}/dt = \hat{x}v_x + \hat{y}v_y + \hat{z}v_z \tag{2.2}$$

with $v_x = dx/dt$, $v_y = dy/dt$ and $v_z = dz/dt$.

In analogy with *configuration space* defined by the position coordinates (x,y,z), it is convenient to introduce the *velocity space* defined by the coordinates (v_x,v_y,v_z). In this space the velocity vector $\vec{v}$ can be considered as a position vector drawn from the origin of the coordinate system (v_x,v_y,v_z), as indicated in Fig. 1.

2.1 Single-particle phase space

From the point of view of classical mechanics the instantaneous dynamic state of each particle can be specified by its position and velocity. It is convenient, therefore, to consider the *phase space* defined by the six coordinates x, y, z, v_x, v_y, v_z.

In this six-dimensional space the dynamical state of each particle is appropriately represented by a single point. The coordinates $(\vec{r}, \vec{v})$ of the representative point give the position and velocity of the particle. When the particle moves, its representative point describes a trajectory in phase

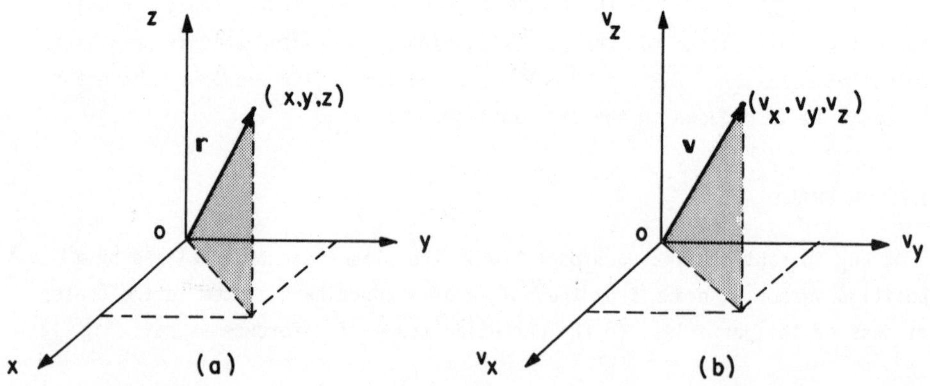

Fig. 1 - Position vectors (a) in configuration space and (b) in velocity space.

space. At each instant of time the dynamical state of a system of N particles is represented by N points in phase space.

2.2 Many-particle phase space

The phase space just defined, often called μ-space, is the phase space for a single particle, in contrast with the many-particle phase space or Γ-*space* for the whole system of particles. In the latter, a system consisting of N particles, with no internal degrees of freedom, is represented by a single point in a 6N-dimensional space defined by the 3N position coordinates $(\vec{r}_1, \vec{r}_2, \ldots, \vec{r}_N)$ and the 3N velocity coordinates $(\vec{v}_1, \vec{v}_2, \ldots, \vec{v}_N)$. Thus, a point in Γ-space corresponds to a single microscopic state for the whole system of particles. This many-particle phase space is ofen used in statistical mechanics and advanced kinetic theory. The single-particle phase space is the one normally used in elementary kinetic theory and basic plasma physics, and is the space to be considered in what follows.

2.3 Volume elements

A small element of volume in configuration space is represented by $d^3r = dx\,dy\,dz$. This differential element of volume should not be taken literally as a mathematically infinitesimal quantity but as a finite element of volume, sufficiently large to contain a very large number of particles, yet sufficiently small in comparison with the characteristic lengths associated with the spatial variation of physical parameters of interest such as, for example, density and temperature. In a gas containing 10^{20} molecules/m^3, for example, if we take $d^3r \sim 10^{-12}$ m^3, which in a macroscopic scale can be considered as a point, there are still 10^8 molecules inside d^3r. Plasmas that do not allow a choice of differential elements of volume as indicated, cannot be analyzed statistically.

When we refer to a particle as being situated inside d^3r, at $\vec{r}$, it is meant that the x coordinate of the particle lies between x and $x + dx$, the y coordinate between y and $y + dy$, and the z coordinate between z and $z + dz$, that is, inside the volume element $d^3r = dx\,dy\,dz$ situated around the terminal point of the position vector $\vec{r} = \hat{x}x + \hat{y}y + \hat{z}z$. It is important to note that the particles localized inside d^3r, at $\vec{r}$, may have completely arbitrary velocities which would be represented by scattered points in velocity space.

A small element of volume in velocity space is represented by $d^3v = dv_x\, dv_y\, dv_z$. For a particle to be included in d^3v, around the terminal point of the velocity vector $\vec{v}$, its v_x component of the velocity must lie between v_x and $v_x + dv_x$, the v_y component between v_y and $v_y + dv_y$, and the v_z component between v_z and $v_z + dv_z$. The differential elements of volume d^3r and d^3v are schematically represented in Fig. 2.

In phase space (μ-space) a differential element of volume may be imagined as a six-dimensional cube, represented by

$$d^3r\, d^3v = dx\, dy\, dz\, dv_x\, dv_y\, dv_z \tag{2.3}$$

as shown schematically in Fig. 3. Note that inside $d^3r\, d^3v$, at the position $(\vec{r}, \vec{v})$ in phase space, there are only the particles inside d^3r around $\vec{r}$ whose velocities lie inside d^3v about $\vec{v}$. The number of representative points inside the volume element $d^3r\, d^3v$ is, in general, a function of time and of the position of this element in phase space. It is important to note that the coordinates $\vec{r}$ and $\vec{v}$ of phase space are considered to be independent variables, since they represent the position of individual volume elements (containing many particles) in phase space.

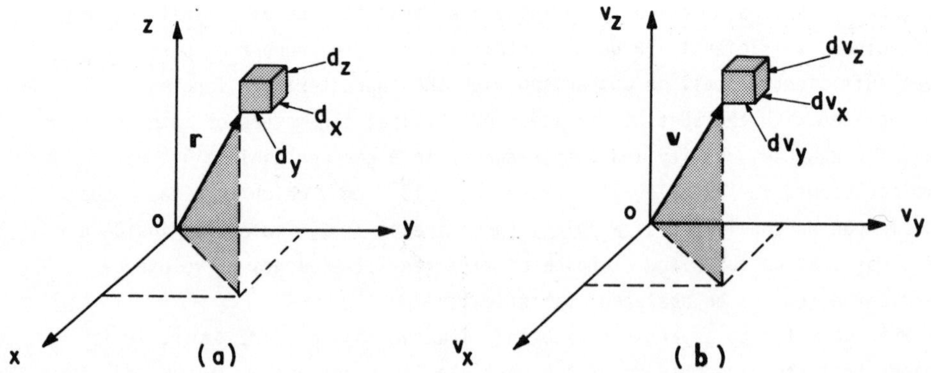

Fig. 2 - (a) The element of volume $d^3r = dx\, dy\, dz$ around the terminal point of $\vec{r}$, in configuration space, and (b) the element of volume $d^3v = dv_x\, dv_y\, dv_z$, in velocity space, around the terminal point of $\vec{v}$.

3. DISTRIBUTION FUNCTION

Let $d^6n_\alpha(\vec{r},\vec{v},t)$ denote the number of particles of type α inside the element of volume $d^3r\,d^3v$ around the coordinates $(\vec{r},\vec{v})$ of phase space, at the instant t. The *distribution function* in phase space, $f_\alpha(\vec{r},\vec{v},t)$, is defined as the density of representative points of the type α particles in phase space, that is,

$$f_\alpha(\vec{r},\vec{v},t) = d^6n_\alpha(\vec{r},\vec{v},t)/(d^3r\,d^3v) \tag{3.1}$$

It is assumed that the density of representative points in phase space does not vary rapidly from one element of volume to the neighboring element, so that $f_\alpha(\vec{r},\vec{v},t)$ can be considered as a continuous function of its arguments. According to its definition $f_\alpha(\vec{r},\vec{v},t)$ is also a positive and finite function at any instant of time. In an element of volume $d^3r\,d^3v$, whose velocity coordinates (v_x,v_y,v_z) are very large, the number of representative points is relatively small since, in any macroscopic system, there must be relatively few particles with very large velocities. Physical considerations require therefore that $f_\alpha(\vec{r},\vec{v},t)$ must tend to zero as the velocity becomes infinitely large.

The distribution function is, in general, a function of the position vector $\vec{r}$. When this is the case the corresponding plasma is said to be *inhomogeneous*. In the absence of external forces, however, a plasma initially

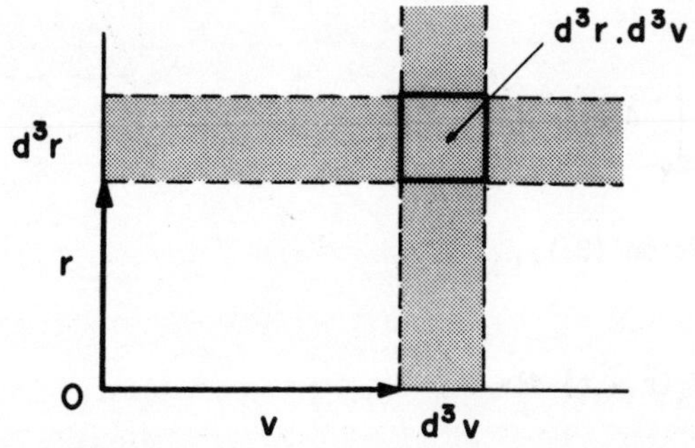

Fig. 3 - Schematic representation of the element of volume $d^3r\,d^3v$ in the six-dimensional phase space, around the point $(\vec{r}, \vec{v})$.

inhomogeneous reaches, in the course of time, an equilibrium state as a result of the mutual particle interactions. In this *homogeneous* state the distribution function does not depend on $\vec{r}$.

In velocity space the distribution function can be *anisotropic*, when it depends on the orientation of the velocity vector $\vec{v}$, or *isotropic*, when it does not depend on the orientation of $\vec{v}$ but only on its magnitude, i.e., on the speed $v = |\vec{v}|$.

The description of different types of plasmas requires the use of inhomogeneous or homogeneous, as well as anisotropic or isotropic distribution functions. A plasma in *thermal equilibrium*, for example, is characterized by a homogeneous, isotropic and time-independent distribution function.

In a statistical sense the distribution function provides a complete description of the system under consideration. Knowing $f_\alpha(\vec{r},\vec{v},t)$ we can deduce all the macroscopic variables of physical interest for the species α. One of the primary problems of kinetic theory consists in determining the form of the distribution function for a given system. The differential equation that governs the temporal and spatial variation of the distribution function under given conditions, known as the Boltzmann equation, will be derived in section 5.

4. NUMBER DENSITY AND AVERAGE VELOCITY

The *number density*, $n_\alpha(\vec{r},t)$, is defined as the number of particles of type α per unit volume, irrespective of velocity. It can be obtained by integrating $d^6n_\alpha(\vec{r},\vec{v},t)$ over all velocity space and dividing the result by the element of volume d^3r of configuration space,

$$n_\alpha(\vec{r},t) = \frac{1}{d^3r} \int_v d^6n_\alpha(\vec{r},\vec{v},t) \tag{4.1}$$

or, using the definition (3.1),

$$n_\alpha(\vec{r},t) = \int_v f_\alpha(\vec{r},\vec{v},t)\, d^3v \tag{4.2}$$

The single integral sign indicated here represents in fact a *triple* integral extending over all velocity space, that is, over each one of the variables

v_x, v_y, v_z from $-\infty$ to $+\infty$. For convenience and simplification of notation only a single integral sign will be used, being implicit the fact that the integral extends over all of velocity space.

The *average velocity* $\vec{u}_\alpha(\vec{r},t)$ is defined as the *macroscopic* flow velocity of the particles of type α in the neighborhood of the position vector $\vec{r}$ at the instant t. In order to relate $\vec{u}_\alpha(\vec{r},t)$ with the distribution function, consider the particles of type α contained in the volume element $d^3r\, d^3v$ about $(\vec{r},\vec{v})$ at the instant t, which we have denoted by $d^6n_\alpha(\vec{r},\vec{v},t)$. The average velocity of the particles of type α can be obtained as follows. First we multiply $d^6n_\alpha(\vec{r},\vec{v},t)$ by the particle velocity $\vec{v}$, next we integrate over all possible velocities, and finally we divide the result by the total number of type α particles contained in d^3r irrespective of velocity. Therefore,

$$\vec{u}_\alpha(\vec{r},t) = \frac{1}{n_\alpha(\vec{r},t)\, d^3r} \int_v \vec{v}\, d^6n_\alpha(\vec{r},\vec{v},t) \tag{4.3}$$

The procedure just described is the usual statistical definition of average values. Using the definition of $f_\alpha(\vec{r},\vec{v},t)$ given in (3.1), we obtain

$$\vec{u}_\alpha(\vec{r},t) = \frac{1}{n_\alpha(\vec{r},t)} \int_v \vec{v}\, f_\alpha(\vec{r},\vec{v},t)\, d^3v \tag{4.4}$$

Note that both $n_\alpha(\vec{r},t)$ and $\vec{u}_\alpha(\vec{r},t)$ are *macroscopic* variables which depend only upon the coordinates $\vec{r}$ and t.

A systematic method for deducing the *macroscopic* variables (such as momentum flux, pressure, temperature, heat flux and so on) in terms of the distribution function, is formally presented in Chapter 6.

5. THE BOLTZMANN EQUATION

In order to calculate the average values of the particle physical properties (the macroscopic variables of interest), it is necessary to know the distribution function for the system under consideration. The dependence of the distribution function on the independent variables $\vec{r}$, $\vec{v}$ and t is governed by an equation known as the *Boltzmann equation*. We present in this section a derivation of the collisionless Boltzmann equation and the general form it takes when the effects of the particle interactions are taken into account, without explicitly deriving any particular expression for the collision term.

5.1 Collisionless Boltzmann equation

Recall that

$$d^6 n_\alpha(\vec{r},\vec{v},t) = f_\alpha(\vec{r},\vec{v},t)\ d^3r\ \ d^3v \tag{5.1}$$

represents the number of particles of type α which, at the instant t, are situated within the volume element $d^3r\ \ d^3v$ of phase space, about the coordinates $(\vec{r},\vec{v})$. Suppose that each particle is subjected to an *external force* $\vec{F}$. In the absence of particle interactions, a particle of type α with coordinates $(\vec{r},\vec{v})$ in phase space, at the instant t, will be found after a time interval dt in the new coordinates $(\vec{r}\,', \vec{v}\,')$, such that

$$\vec{r}\,'(t + dt) = \vec{r}(t) + \vec{v}\ dt \tag{5.2}$$

$$\vec{v}\,'(t + dt) = \vec{v}(t) + \vec{a}\ dt \tag{5.3}$$

where $\vec{a} = \vec{F}/m_\alpha$ is the particle acceleration, and m_α its mass.

Thus, all particles of type α inside the volume element $d^3r\ \ d^3v$ of phase space, about $(\vec{r},\vec{v})$ at the instant t, will occupy a new volume element $d^3r'\ d^3v'$, about $(\vec{r}\,', \vec{v}\,')$ after the interval dt (see Fig. 4). Since we are considering the same particles at t and at $t + dt$, we must have in the absence of collisions,

$$f_\alpha(\vec{r}\,',\vec{v}\,',t + dt)\ d^3r'\ d^3v' = f_\alpha(\vec{r},\vec{v},t)\ d^3r\ d^3v \tag{5.4}$$

The element of volume $d^3r\ d^3v$ may become distorted in shape as a result of the particle motion. The relation between the new element of volume, $d^3r'\ d^3v'$, and the initial one, $d^3r\ d^3v$, is given by

$$d^3r'\ \ d^3v' = |J|\ d^3r\ \ d^3v \tag{5.5}$$

where J stands for the Jacobian of the transformation from the initial coordinates $(\vec{r}, \vec{v})$ to the final ones $(\vec{r}', \vec{v}')$. It will be shown in the next subsection that for the transformation defined by (5.2) and (5.3) we have $|J| = 1$, so that

$$d^3r' \; d^3v' = d^3r \; d^3v \tag{5.6}$$

and (5.4) becomes

$$[f_\alpha(\vec{r}',\vec{v}',t + dt) - f_\alpha(\vec{r},\vec{v},t)] \, d^3r \; d^3v = 0 \tag{5.7}$$

The first term on the left of (5.7) can be expanded in a Taylor series about $f_\alpha(\vec{r},\vec{v},t)$ as follows

$$f_\alpha(\vec{r} + \vec{v}\,dt, \vec{v} + \vec{a}\,dt, t + dt) = f_\alpha(\vec{r},\vec{v},t) +$$

$$+ \left[\frac{\partial f_\alpha}{\partial t} + \left(v_x \frac{\partial f_\alpha}{\partial x} + v_y \frac{\partial f_\alpha}{\partial y} + v_z \frac{\partial f_\alpha}{\partial z} \right) + \right.$$

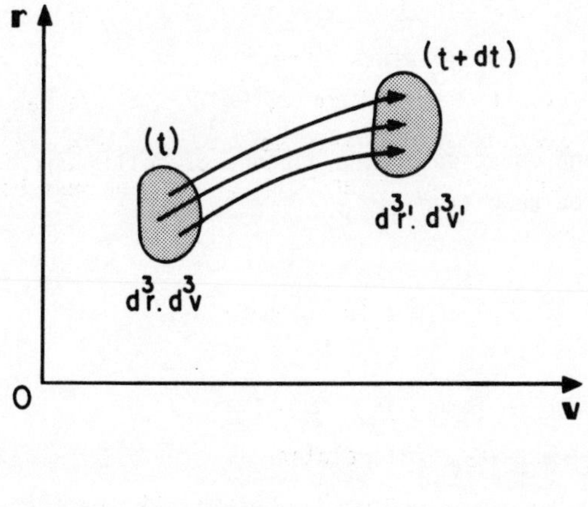

Fig. 4 - In the absence of collisions the particles within the volume element $d^3r\, d^3v$ about $(\vec{r},\vec{v})$, at an instant t, will occupy after a time interval dt a new volume element $d^3r' \; d^3v'$, about $(\vec{r}', \vec{v}')$.

$$+ \left(a_x \frac{\partial f_\alpha}{\partial v_x} + a_y \frac{\partial f_\alpha}{\partial v_y} + a_z \frac{\partial f_\alpha}{\partial v_z} \right) \Bigg] dt \tag{5.8}$$

neglecting terms of order $(dt)^2$ and higher. Using the del operator notation

$$\vec{\nabla} = \hat{x} \frac{\partial}{\partial x} + \hat{y} \frac{\partial}{\partial y} + \hat{z} \frac{\partial}{\partial z} \tag{5.9}$$

and, in a similar way, defining a del operator in velocity space by

$$\vec{\nabla}_v = \hat{x} \frac{\partial}{\partial v_x} + \hat{y} \frac{\partial}{\partial v_y} + \hat{z} \frac{\partial}{\partial v_z} \tag{5.10}$$

we obtain from (5.8)

$$f_\alpha(\vec{r} + \vec{v}\, dt, \vec{v} + \vec{a}\, dt, t + dt) = f_\alpha(\vec{r}, \vec{v}, t) + (\partial f_\alpha/\partial t +$$

$$+ \vec{v} \cdot \vec{\nabla} f_\alpha + \vec{a} \cdot \vec{\nabla}_v f_\alpha)\, dt \tag{5.11}$$

Substituting this result into (5.7) gives

$$\partial f_\alpha/\partial t + \vec{v} \cdot \vec{\nabla} f_\alpha + \vec{a} \cdot \vec{\nabla}_v f_\alpha = 0 \tag{5.12}$$

which is the Boltzmann equation in the absence of collisions.

This equation can be rewritten as

$$\mathcal{D} f_\alpha/\mathcal{D}t = 0 \tag{5.13}$$

where the operator

$$\mathcal{D}/\mathcal{D}t = \partial/\partial t + \vec{v} \cdot \vec{\nabla} + \vec{a} \cdot \vec{\nabla}_v \tag{5.14}$$

represents the *total* derivative with respect to time, in phase space. Eq. (5.13) is a statement of the conservation of the density of representative points in phase space. If we move along with a representative point in phase space and observe the density of representative points f_α in its neighborhood, we find that this density remains constant in time. This result is known as *Liouville's theorem.* Note that it applies only to the special case in which collisions, as well as radiation losses and processes of production and loss of particles are unimportant.

5.2 Jacobian of the transformation in phase space

To determine the Jacobian of the transformation defined in (5.2) and (5.3), recall that, from its definition, we have

$$J = \frac{\partial(\vec{r}\,', \vec{v}\,')}{\partial(\vec{r}, \vec{v})} = \frac{\partial(x', y', z', v_x', v_y', v_z')}{\partial(x, y, z, v_x, v_y, v_z)} \tag{5.15}$$

which corresponds to the determinant of the 6 x 6 matrix

$$(J) = \begin{pmatrix} \partial x'/\partial x & \partial y'/\partial x & \ldots & \partial v_z'/\partial x \\ \partial x'/\partial y & \partial y'/\partial y & \ldots & \partial v_z'/\partial y \\ \cdots & \cdots & \cdots & \cdots \\ \partial x'/\partial v_z & \partial y'/\partial v_z & \ldots & \partial v_z'/\partial v_z \end{pmatrix} \tag{5.16}$$

We can separate the external force $\vec{F}$ into two parts,

$$\vec{F} = \vec{F}\,' + q_\alpha (\vec{v} \times \vec{B}) \tag{5.17}$$

where $\vec{F}\,'$ is a velocity-independent force and the second term is the velocity-dependent force due to an externally applied magnetic field $\vec{B}$, the only velocity-dependent force that may concern us in this treatment. The partial derivatives appearing in (J) are

$$\frac{\partial x_i'}{\partial x_j} = \delta_{ij}, \qquad \frac{\partial v_i'}{\partial x_j} = \frac{1}{m_\alpha} \frac{\partial F_i'}{\partial x_j} dt$$

$$\frac{\partial x_i'}{\partial v_j} = \delta_{ij}\, dt, \quad \frac{\partial v_i'}{\partial v_j} = \delta_{ij} + \frac{q_\alpha}{m_\alpha} \frac{\partial}{\partial v_j} (\vec{v} \times \vec{B})_i dt \tag{5.18}$$

where (5.2), (5.3) and (5.17) have been used, and where $x_{i,j} = x, y, z$ and $v_{i,j} = v_x, v_y, v_z$. The symbol δ_{ij} is the Kronecker delta. The matrix (5.16) can be written in the form

$$(\mathcal{J}) = \begin{pmatrix} (\mathcal{J})_1 & (\mathcal{J})_2 \\ (\mathcal{J})_3 & (\mathcal{J})_4 \end{pmatrix} \tag{5.19}$$

where the $(\mathcal{J})_i$'s, with $i = 1, 2, 3, 4$, represent the following 3×3 submatrices

$$(\mathcal{J})_1 = \begin{pmatrix} 1 & 0 & 0 \\ 0 & 1 & 0 \\ 0 & 0 & 1 \end{pmatrix} \tag{5.20}$$

$$(\mathcal{J})_2 = \frac{dt}{m_\alpha} \begin{pmatrix} \partial F_x'/\partial x & \partial F_y'/\partial x & \partial F_z'/\partial x \\ \partial F_x'/\partial y & \partial F_y'/\partial y & \partial F_z'/\partial y \\ \partial F_x'/\partial z & \partial F_y'/\partial z & \partial F_z'/\partial z \end{pmatrix} \tag{5.21}$$

$$(\mathcal{J})_3 = \begin{pmatrix} dt & 0 & 0 \\ 0 & dt & 0 \\ 0 & 0 & dt \end{pmatrix} \tag{5.22}$$

$$(\mathcal{J})_4 = \begin{pmatrix} 1 & -(q_\alpha/m_\alpha) B_z\, dt & (q_\alpha/m_\alpha) B_y\, dt \\ (q_\alpha/m_\alpha) B_z\, dt & 1 & -(q_\alpha/m_\alpha) B_x\, dt \\ -(q_\alpha/m_\alpha) B_y\, dt & (q_\alpha/m_\alpha) B_x\, dt & 1 \end{pmatrix} \tag{5.23}$$

Neglecting terms of order $(dt)^2$ it can be verified that $|J| = 1$. Thus, up to and including the terms of first order in the infinitesimal dt, we have,

$$d^3r' \quad d^3v' = d^3r \quad d^3v \tag{5.24}$$

which is the result (5.6) used in the previous subsection.

5.3 Effects of particle interactions

When the effects due to the particle interactions are taken into account, (5.12) needs to be modified. As a result of collisions during the time interval dt, some of the particles of type α which were initially within the volume element $d^3r\, d^3v$ of phase space may be removed from it, and particles of type α initially outside this volume element may end up inside it. This is indicated schematically in Fig. 5. Generally, the number of particles of type α inside $d^3r\, d^3v$ about the coordinates $(\vec{r}, \vec{v})$ at an instant t, will be different from the number of particles of type α inside this same volume element about the coordinates $(\vec{r}', \vec{v}')$ at the instant t + dt. We shall denote this net gain or loss of particles of type α, as a result of collisions during the interval dt, in the volume element $d^3r\, d^3v$, by

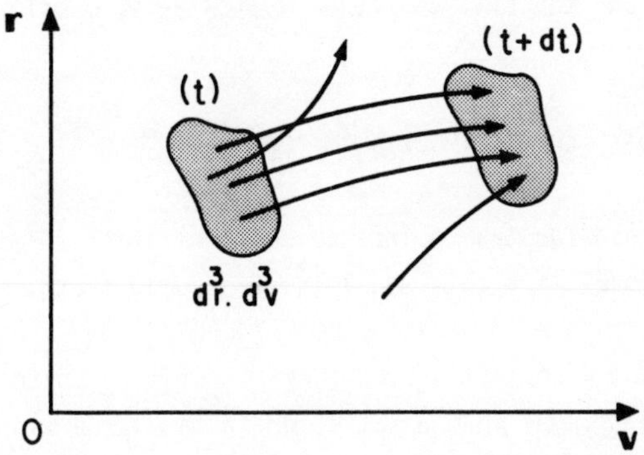

Fig. 5 - Schematic representation of the motion of the volume element $d^3r\, d^3v$ in phase space, showing particles entering and leaving this volume element, as a result of collisions during the time interval dt.

$$(\delta f_\alpha/\delta t)_{coll} \quad d^3r \quad d^3v \quad dt \tag{5.25}$$

where $(\delta f_\alpha/\delta t)_{coll}$ represents the rate of change of $f_\alpha(\vec{r},\vec{v},t)$ due to collisions. Thus, when collisions are considered, (5.7) becomes

$$[f_\alpha(\vec{r}\,', \vec{v}\,', t + dt) - f_\alpha(\vec{r},\vec{v},t)]\; d^3r\; d^3v = (\delta f_\alpha/\delta t)_{coll}\; d^3r\; d^3v\; dt \tag{5.26}$$

and the following modified form of Eq. (5.12) results

$$\partial f_\alpha/\partial t + \vec{v}\cdot \vec{\nabla} f_\alpha + \vec{a}\cdot \vec{\nabla}_v\, f_\alpha = (\delta f_\alpha/\delta t)_{coll} \tag{5.27}$$

Using the total time derivative operator, defined in (5.14), we can rewrite this equation in compact form as

$$\mathcal{D} f_\alpha/\mathcal{D}t = (\delta f_\alpha/\delta t)_{coll} \tag{5.28}$$

This equation is obviously incomplete, since the precise form of the collision term is not known. In the following section we will consider a very simple expression for the collision term, known as the *Krook model* or *relaxation model*. More elaborated expressions, such as the *Boltzmann collision integral* and the *Fokker-Planck collision term* will be considered in chapter 21.

6. RELAXATION MODEL FOR THE COLLISION TERM

A very simple method for taking into account collision effects is provided by the relaxation model. In this model it is assumed that the effect of collisions is to restore a situation of *local equilibrium*, characterized by the local equilibrium distribution function $f_{o\alpha}(\vec{r},\vec{v})$. In the absence of external forces, it assumes that a situation initially not in equilibrium described by a distribution function $f_\alpha(\vec{r},\vec{v},t)$ different from $f_{o\alpha}(\vec{r},\vec{v})$, reaches a local equilibrium condition exponentially with time, as a result of collisions, with a *relaxation time* τ. This relaxation time is of the order of the time between collisions and may also be written as ν^{-1} where ν represents a *relaxation collision frequency*. This model was originally developed by *Krook* and can be written as

$$(\delta f_\alpha/\delta t)_{coll} = - (f_\alpha - f_{o\alpha})/\tau \tag{6.1}$$

According to this expression for the collision term, when $f_\alpha = f_{o\alpha}$ we have $(\delta f_\alpha/\delta t)_{coll} = 0$, so that in a state of local equilibrium the distribution function is not altered as a result of collisions.

In order to put in evidence the physical meaning of the relaxation model, let us consider the Boltzmann equation with this collision term, in the absence of external forces and spatial gradients, and when $f_{o\alpha}$ and τ are time-independent,

$$\partial f_\alpha/\partial t = - (f_\alpha - f_{o\alpha})/\tau \tag{6.2}$$

which may also be written as

$$\partial f_\alpha/\partial t + f_\alpha/\tau = f_{o\alpha}/\tau \tag{6.3}$$

This simple inhomogeneous differential equation has $C\ e^{-t/\tau}$ as the homogeneous solution, where C is a constant, and $f_{o\alpha}$ as a particular solution. Therefore, the complete solution is

$$f_\alpha(\vec{v},t) = f_{o\alpha} + [f_\alpha(\vec{v},0) - f_{o\alpha}]\ \exp\ (-t/\tau) \tag{6.4}$$

Thus, the difference between f_α and $f_{o\alpha}$ decreases exponentially with time at a rate governed by the relaxation collission frequency $\nu = 1/\tau$.

This collisional model has proved to be useful and, in some cases, leads to results almost identical to the ones obtained using the Boltzmann collision integral. It is particularly applicable to a weakly ionized plasma in which only charge-neutral collisions are important. However, it oversimplifies the entire relaxation phenomena and does not predict correctly the different relaxation collision frequencies for the various physical quantities of interest, such as macroscopic velocity, momentum and energy. According to the relaxation model, the macroscopic physical variables approach equilibrium at the same rate ν. A detailed analysis of the collision process, however, shows that this is not the case and that the relaxation time for various macroscopic variables varies to some extent. For nonrelativistic velocities, while the relaxation time for the average velocity and the momentum are found to be the same, approximately τ, that of the average thermal energy is approximately $(m_\beta/2m_\alpha)\tau$. Hence, for collisions between electrons and neutral particles, the relaxation time for the kinetic energy of the electrons is *longer* than that for the average velocity by a factor which is of the order of the ratio of the mass of the neutral particle to the electron mass. The relaxation model is therefore strictly applicable only to the cases of collisions between particles of the same mass. In spite of this limitation, the Krook model is still useful partly because of its simplicity and partly because it usually gives a first approximation to the problem under consideration.

7. THE VLASOV EQUATION

A very useful approximate way to describe the dynamics of a plasma is to consider that the motions of the plasma particles are governed by the applied external fields *plus* the macroscopic average internal fields, smoothed in space and time, due to the presence and motion of all plasma particles. The problem of obtaining the macroscopically smoothed internal electromagnetic fields, however, is still a complex one and requires that a *self-consistent* solution be obtained.

The Vlasov equation is a partial differential equation that describes the evolution of the distribution function in time, and which directly incorporates the macroscopically smoothed internal electromagnetic fields. It may be obtained from the Boltzmann equation with the collision term $(\delta f_\alpha/\delta t)_{coll}$ equal to zero, but including the *internal smoothed fields* in the force term,

$$\partial f_\alpha/\partial t + \vec{v} \cdot \vec{\nabla} f_\alpha + (1/m_\alpha)\ [\vec{F}_{ext} + q_\alpha(\vec{E}_i + \vec{v} \times \vec{B}_i)] \cdot \vec{\nabla}_v f_\alpha = 0 \qquad (7.1)$$

Here $\vec{F}_{ext}$ represents the external forces, including the Lorentz force associated with any *externally* applied electric and magnetic fields, and $\vec{E}_i$ and $\vec{B}_i$ are *internal* smoothed electric and magnetic fields due to the presence and motion of all charged particles inside the plasma. In order that the internal electromagnetic fields $\vec{E}_i$ and $\vec{B}_i$ be consistent with the charge and current densities existing in the plasma itself, they must satisfy Maxwell equations

$$\vec{\nabla} \cdot \vec{E}_i = \rho/\varepsilon_0 \qquad (7.2)$$

$$\vec{\nabla} \cdot \vec{B}_i = 0 \tag{7.3}$$

$$\vec{\nabla} \times \vec{E}_i = - \partial \vec{B}_i / \partial t \tag{7.4}$$

$$\vec{\nabla} \times \vec{B}_i = \mu_0 \, (\vec{J} + \varepsilon_0 \, \partial \, \vec{E}_i / \partial t) \tag{7.5}$$

with the plasma charge density ρ and the plasma current density $\vec{J}$ given by the formulas

$$\rho(\vec{r},t) = \sum_\alpha n_\alpha(\vec{r},t) \, q_\alpha = \sum_\alpha q_\alpha \int_V f_\alpha(\vec{r},\vec{v},t) \, d^3v \tag{7.6}$$

$$\vec{J}(\vec{r},t) = \sum_\alpha n_\alpha(\vec{r},t) \, q_\alpha \, \vec{u}_\alpha(\vec{r},t) = \sum_\alpha q_\alpha \int_V \vec{v} \, f_\alpha(\vec{r},\vec{v},t) \, d^3v \tag{7.7}$$

the summations being over the different charged particle species in the plasma. Here $\vec{u}_\alpha(\vec{r},t)$ denotes the macroscopic *average velocity* for the particles of type α, given in (4.4).

Eqs. (7.1) to (7.7) constitute a complete set of *self-consistent* equations to be solved simultaneously. Assuming values for $\vec{E}_i(\vec{r},t)$ and $\vec{B}_i(\vec{r},t)$, Eq. (7.1) can be solved to yield $f_\alpha(\vec{r},\vec{v},t)$ for the various different species; using the calculated f_α's in (7.6) and (7.7), leads to values for the charge and current densities in the plasma, which can be substituted into Maxwell equations and solved for $\vec{E}_i(\vec{r},t)$ and $\vec{B}_i(\vec{r},t)$; these values are then plugged back into the Vlasov equation, and so on. Hence, a self-consistent solution can be obtained for the single particle distribution function.

Although the Vlasov equation does not explicitly include a collision term in its right-hand side and, hence, does not take into account short-range collisions, it is not so restrictive as it may appear, since a significant part of the effects of the particle interactions has already been included in the Lorentz force, through the internal, self-consistent smoothed fields.

PROBLEMS

5.1 - Consider a system of particles uniformly distributed in space, with a constant particle number density n_0, and characterized by a velocity distribution function $f(v)$ such that

$$f(v) = K_0 \neq 0 \quad \text{for} \quad |v_i| \leq v_0 \quad (i = x,y,z)$$

$$f(v) = 0 \quad \text{otherwise,}$$

where K_0 is a positive constant. Determine the value of K_0 in terms of n_0 and v_0.

5.2 - Consider the following *two-dimensional* Maxwellian distribution function

$$f(v_x,v_y) = n_0\left(\frac{m}{2\pi kT}\right) \exp[-m(v_x^2 + v_y^2)/2kT]$$

(a) Verify that n_0 represents correctly the particle number density, that is, the number of particles per unit area.
(b) Draw, in three dimensions, the surface for this distribution function, plotting f in terms of v_x and v_y. Sketch, on this surface, curves of constant v_x, curves of constant v_y, and curves of constant f.

5.3 - The electrons, inside a system of two coaxial magnetic mirrors, can be described by the so-called *loss-cone distribution function*

$$f(\vec{v}) = \frac{n_0}{(\pi^{3/2}\alpha_\perp^2\, \alpha_\parallel)} \left(\frac{v_\perp}{\alpha_\perp}\right)^2 \exp\left[-\left(\frac{v_\perp}{\alpha_\perp}\right)^2 - \left(\frac{v_\parallel}{\alpha_\parallel}\right)^2\right]$$

where $v_\parallel$ and $v_\perp$ denote the electron velocities in the directions parallel and perpendicular to the magnetic bottle axis, respectively, and where $\alpha_\perp^2 = 2k\,T_\perp/m$ and $\alpha_\parallel^2 = 2k\,T_\parallel/m$.
(a) Verify that the number density of the electrons in the magnetic bottle is given by n_0.
(b) Justify the applicability of the loss-cone distribution function to a magnetic mirror bottle, by analyzing its dependence on $v_\perp$ and $v_\parallel$. Sketch, in three dimensions, the surface for $f(\vec{v})$ as a function of $v_\perp$ and $v_\parallel$.

5.4 - Consider the motion of charged particles, in one dimension only, in an electric potential $V(x)$. Show, by direct substitution, that a function of the form

$$f = f(mv^2/2 + qV)$$

is a solution of the Boltzmann equation under steady state conditions.

5.5 - (a) Show that the Boltzmann equation, in cylindrical coordinates, becomes

$$\frac{\partial f}{\partial t} + \dot{r}\frac{\partial f}{\partial r} + \dot{\phi}\frac{\partial f}{\partial \phi} - r\,\dot{\phi}^2\frac{\partial f}{\partial \dot{r}} + \frac{2\dot{r}\dot{\phi}}{r}\frac{\partial f}{\partial \dot{\phi}} + \dot{z}\frac{\partial f}{\partial z} +$$

$$+ \frac{1}{m}\left(F_r\frac{\partial f}{\partial \dot{r}} + \frac{F_\phi}{r}\frac{\partial f}{\partial \dot{\phi}} + F_z\frac{\partial f}{\partial \dot{z}}\right) = \left(\frac{\delta f}{\delta t}\right)_{coll}$$

where $F_r = m\ddot{r}$, $F_\phi = mr\ddot{\phi}$ and $F_z = m\ddot{z}$.

(b) Show, by direct substitution, that, in the presence of an azimuthally symmetric magnetic field (in the z direction), a function of the form

$$f = f(mv^2/2,\ mr^2\,\dot{\phi} + qr\,A_\phi)$$

is a solution of the Boltzmann equation under steady conditions, where $p_\phi = mr^2\dot{\phi} + qr\,A_\phi$ is the constant canonical momentum, and where A_ϕ denotes the ϕ component of the magnetic potential, defined such that $\vec{B} = \vec{\nabla} \times \vec{A}$.

5.6 - Show that the Vlasov equation for a homogeneous plasma under the influence of a uniform external magnetostatic field, $\vec{B}_0$, in the equilibrium state, is satisfied by any homogeneous distribution function, $f(v_\parallel, v_\perp)$, which is cylindrically symmetric with respect to the magnetostatic field.

5.7 - The entropy of a system can be expressed, in terms of the distribution function, as

$$S = -k\int_r\int_v f\,\ln(f)\,d^3v\,d^3r$$

Show that, for a system which obeys the collisionless Boltzmann equation, the total time derivative of the entropy vanishes.

5.8 - Consider a one-dimensional harmonic oscillator whose energy is given by

$$E = p^2/(2m) + kx^2/2$$

where p denotes its linear momentum and x its displacement coordinate. Show that the trajectory described by the representative point of the oscillator, in phase space, is an ellipse.

CHAPTER 6

Average Values and Macroscopic Variables

1. AVERAGE VALUE OF A PHYSICAL QUANTITY

A systematic method for obtaining the average values of functions of particle velocities is presented in this chapter. The macroscopic variables, such as number density, average velocity, kinetic pressure, thermal energy flux, and so on, can be considered as average values of physical quantities, involving the collective behavior of a large number of particles. These macroscopic variables are related to the *moments* of the distribution function.

To each particle in the plasma we can associate some molecular property, $\chi(\vec{r},\vec{v},t)$, which in general may be a function of the position $\vec{r}$ of the particle, of its velocity $\vec{v}$ and of the time t. This property may be, for example, the mass, the velocity, the momentum, or the energy of the particle.

In order to calculate the average value of $\chi(\vec{r},\vec{v},t)$, recall that $d^6n_\alpha(\vec{r},\vec{v},t)$ represents the number of particles of type α inside the phase space volume element $d^3r\,d^3v$ about $(\vec{r},\vec{v})$ at the instant t. Thus, the total value of $\chi(\vec{r},\vec{v},t)$ for all the particles of type α inside $d^3r\,d^3v$, is given by

$$\chi(\vec{r},\vec{v},t)\; d^6n_\alpha(\vec{r},\vec{v},t) = \chi(\vec{r},\vec{v},t)\; f_\alpha(\vec{r},\vec{v},t)\; d^3r\; d^3v \tag{1.1}$$

The total value of $\chi(\vec{r},\vec{v},t)$ for all the particles of type α inside the volume element d^3r of configuration space, irrespective of velocity, is obtained by integrating (1.1) over all possible velocities,

$$d^3r \int_V \chi(\vec{r},\vec{v},t)\; f_\alpha(\vec{r},\vec{v},t)\; d^3v \tag{1.2}$$

The *average value* of $\chi(\vec{r},\vec{v},t)$ can now be obtained by dividing (1.2) by the number of particles of type α inside d^3r about $\vec{r}$ at the instant t, $n_\alpha(\vec{r},t)\,d^3r$. We define, therefore, the average value of the property $\chi(\vec{r},\vec{v},t)$, for the particles of type α, by

$$\langle\chi(\vec{r},\vec{v},t)\rangle_\alpha = \frac{1}{n_\alpha(\vec{r},t)} \int_V \chi(\vec{r},\vec{v},t)\; f_\alpha(\vec{r},\vec{v},t)\; d^3v \tag{1.3}$$

The symbol $< >_\alpha$ denotes the average value with respect to velocity space for the particles of type α. Note that the average value is independent of $\vec{v}$, being a function of only $\vec{r}$ and t.

If we take $\chi = 1$ in (1.4), the expression for the *number density* given in (5.4.2), is obtained.

2. AVERAGE VELOCITY AND PECULIAR VELOCITY

Consider now $\chi(\vec{r},\vec{v},t)$ as being the velocity $\vec{v}$ of the type α particles in the vicinity of the position $\vec{r}$ at the instant of time t. Eq. (1.3) then gives the macroscopic *average velocity* $\vec{u}_\alpha(\vec{r},t)$ for the particles of type α,

$$\vec{u}_\alpha(\vec{r},t) = <\vec{v}>_\alpha = \frac{1}{n_\alpha(\vec{r},t)} \int_v \vec{v}\, f_\alpha(\vec{r},\vec{v},t)\, d^3v \tag{2.1}$$

which is the same expression given earlier in (5.4.4).

Note that $\vec{r}$, $\vec{v}$ and t are taken as independent variables, whereas the average velocity $\vec{u}_\alpha$ depends on the position $\vec{r}$ and the time t. For the cases in which χ is independent of velocity, we have

$$<\chi(\vec{r},t)>_\alpha = \chi_\alpha(\vec{r},t) \tag{2.2}$$

so that, for example, $<\vec{u}_\alpha> = \vec{u}_\alpha$. In what follows, the index α after the average value symbol will be omitted whenever it is redundant, that is, $<\vec{u}_\alpha>_\alpha = <\vec{u}_\alpha>$.

The *peculiar* or *random velocity* $\vec{c}_\alpha$ is defined as the velocity of a type α particle relative to the average velocity $\vec{u}_\alpha$,

$$\vec{c}_\alpha = \vec{v} - \vec{u}_\alpha \tag{2.3}$$

Consequently we have $<\vec{c}_\alpha> = 0$, since $<\vec{v}>_\alpha = \vec{u}_\alpha$. The peculiar velocity $\vec{c}_\alpha$ is the one associated with the random thermal kinetic energy of the particles of type α. When $\vec{u}_\alpha$ vanishes we have $\vec{c}_\alpha = \vec{v}$.

3. FLUX

From the concept of distribution function many other macroscopic variables can be defined in terms of average values. Macroscopic variables such as the particle current density (particle flux), the pressure dyad or tensor, and the heat-flow vector (thermal energy flux), involve the flux of some molecular property $\chi(\vec{r},\vec{v},t)$. The *flux* of $\chi(\vec{r},\vec{v},t)$ is defined as the amount of the quantity $\chi(\vec{r},\vec{v},t)$ transported across some given surface, per unit area and per unit time.

Consider a surface element $d\vec{S}$ inside the plasma. If the distribution of velocities is isotropic, the flux will be independent of the relative orientation in space of the surface element $d\vec{S}$. However, more generally, when the velocity distribution is anisotropic the flux will depend on the relative spatial orientation of $d\vec{S}$. Suppose, therefore, that the surface element of magnitude dS is oriented along some direction specified by the unit vector $\hat{n}$,

$$d\vec{S} = \hat{n}\, dS \tag{3.1}$$

$\hat{n}$ being normal to the surface element. In the case of an open surface there are two possible directions for the normal $\hat{n}$, one opposite to the other. The direction which is taken as positive is related to the positive sense of traversing the perimeter (bounding curve) of the open surface, according to the following convention: if the positive sense of traversal of the perimeter of a horizontal open surface is taken as counterclockwise, then the positive normal to the open surface is up; if the positive sense of traversal of the perimeter is clockwise, then the positive normal to the open surface is down, as shown in Fig. 1. For a closed surface the normal unit vector is conventionally chosen to point outward.

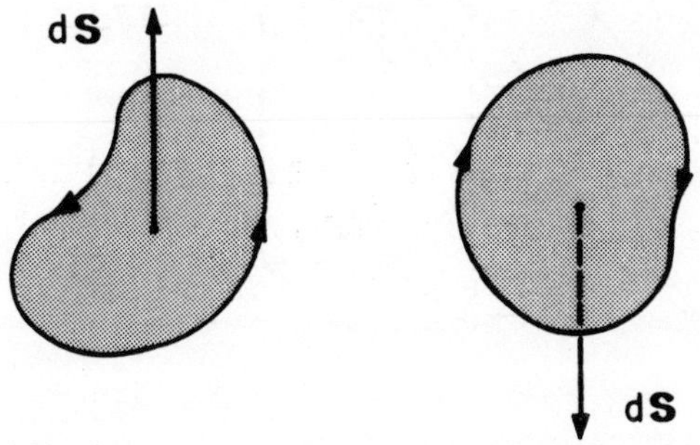

Fig. 1 - Direction of positive normal to the surface element $d\vec{S}$ as related to the sense of traversing the perimeter of $d\vec{S}$.

The particles inside the plasma, due to their velocities, will move across the surface element $d\vec{S}$ carrying the property $\chi(\vec{r},\vec{v},t)$ with them. We want to calculate the number of particles of type α that move across $d\vec{S}$ during the interval dt.

The particles with velocity between $\vec{v}$ and $\vec{v}+d\vec{v}$ that will cross $d\vec{S}$ in the interval between t and t + dt, must lie initially in the volume of the prism of base dS and side v dt, as indicated in Fig. 2. The volume of this prism is

$$d^3r = d\vec{S} \cdot \vec{v}\, dt = \hat{n} \cdot \vec{v}\, dS\, dt \tag{3.2}$$

From the definition of $f_\alpha(\vec{r},\vec{v},t)$, the number of particles of type α in the volume of this prism, that have velocities between $\vec{v}$ and $\vec{v} + d\vec{v}$ is

$$f_\alpha(\vec{r},\vec{v},t)\, d^3r\, d^3v = f_\alpha(\vec{r},\vec{v},t)\, \hat{n} \cdot \vec{v}\, dS\, dt\, d^3v \tag{3.3}$$

so that the total amount of $\chi(\vec{r},\vec{v},t)$ transported across $d\vec{S}$, in the interval dt, is obtained by multiplying this number of particles by $\chi(\vec{r},\vec{v},t)$ and integrating the result over all possible velocities,

$$\int_v \chi(\vec{r},\vec{v},t)\, f_\alpha(\vec{r},\vec{v},t)\, \hat{n} \cdot \vec{v}\, d^3v\, dS\, dt \tag{3.4}$$

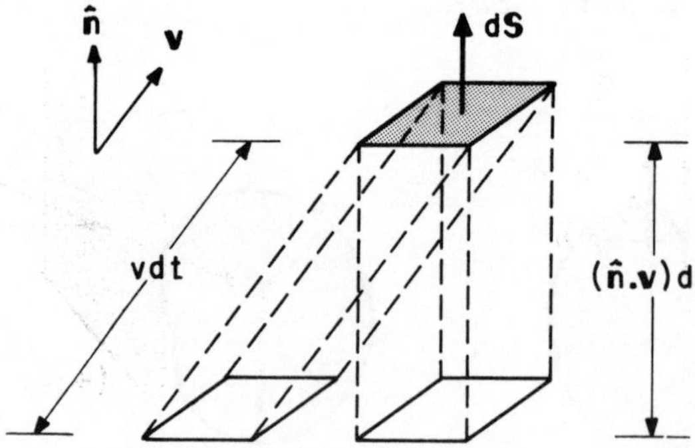

Fig. 2 - Prism of volume $d^3r = d\vec{S} \cdot \vec{v}\, dt = \hat{n} \cdot \vec{v}\, dS\, dt$ containing the particles of type α with velocities between $\vec{v}$ and $\vec{v} + d\vec{v}$, and which will cross $d\vec{S}$ in the time interval dt.

Note that the contributions corresponding to a rotation of the segment v dt over all possible directions about $d\vec{S}$ are taken into account in the integration over velocity space. Particles that cross $d\vec{S}$ in a direction such that $\hat{n}\cdot\vec{v}$ is positive give a positive contribution to the flux in the direction of $\hat{n}$, while particles that cross $d\vec{S}$ in a direction such that $\hat{n}\cdot\vec{v}$ is negative give a negative contribution to the flux in the direction of $\hat{n}$. This is illustrated in Fig. 3.

The net amount of the quantity $\chi(\vec{r},\vec{v},t)$ transported by the particles of type α per unit area and per unit time is obtained by dividing expression (3.4) by dS dt. The *flux* in the direction $\hat{n}$, $\Phi_{\alpha n}(\chi)$, is therefore given by

$$\Phi_{\alpha n}(\chi) = \int_v \chi(\vec{r},\vec{v},t)\ \hat{n}\cdot\vec{v}\ f_\alpha(\vec{r},\vec{v},t)\ d^3v \tag{3.5}$$

or, using the average value symbol,

$$\Phi_{\alpha n} = n_\alpha(\vec{r},t) \quad \langle\chi(\vec{r},\vec{v},t)\ \hat{n}\cdot\vec{v}\rangle_\alpha = n_\alpha \langle\chi v_n\rangle_\alpha \tag{3.6}$$

where $v_n = \hat{n}\cdot\vec{v}$ denotes the component of $\vec{v}$ along the direction specified by the unit vector $\hat{n}$.

When $\chi(\vec{r},\vec{v},t)$ is a scalar quantity, $\Phi_{\alpha n}(\chi)$ can be considered as the component along $\hat{n}$, of a *vector flux* $\vec{\Phi}_\alpha(\chi)$, that is,

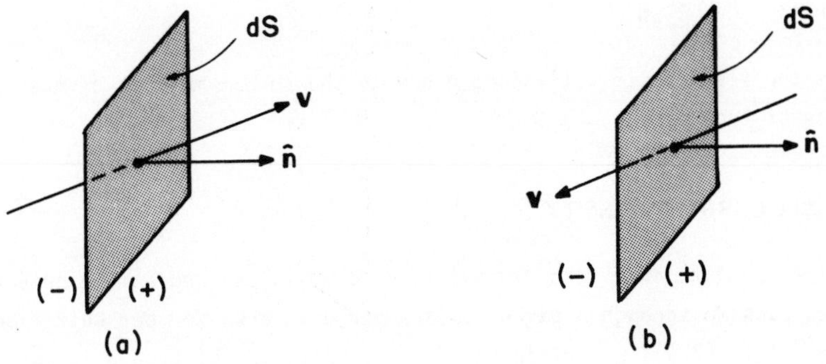

Fig. 3 - (a) Particles that cross $d\vec{S}$ from the (-) region to the (+) region contribute positively to the flux in direction $\hat{n}$, while (b) particles that cross $d\vec{S}$ from the (+) region to the (-) region contribute negatively to the flux in direction $\hat{n}$.

$$\Phi_{\alpha n}(\chi) = \hat{n} \cdot \vec{\Phi}_{\alpha}(\chi) \tag{3.7}$$

with

$$\vec{\Phi}_{\alpha}(\chi) = n_{\alpha} \langle \chi \vec{v} \rangle_{\alpha} \tag{3.8}$$

If $\chi(\vec{r},\vec{v},t)$ represents a *vector* quantity, then we will have a *flux dyad* (or *tensor*),

$$\bar{\bar{\Phi}}_{\alpha}(\chi) = n_{\alpha} \langle \vec{\chi} \vec{v} \rangle_{\alpha} \tag{3.9}$$

and if it represents a *dyad* quantity we will have a *flux triad*, and so on.

In many situations of practical interest it is important to separately consider the contribution to the flux due to the average velocity $\vec{u}_{\alpha}$, and that due to the random velocity $\vec{c}_{\alpha}$ of the particles of type α. Substituting $\vec{v} = \vec{c}_{\alpha} + \vec{u}_{\alpha}$ in (3.6) gives

$$\Phi_{\alpha n}(\chi) = n_{\alpha} \langle \chi c_{\alpha n} \rangle + n_{\alpha} \langle \chi u_{\alpha n} \rangle \tag{3.10}$$

where $c_{\alpha n} = \hat{n} \cdot \vec{c}_{\alpha}$ and $u_{\alpha n} = \hat{n} \cdot \vec{u}_{\alpha}$

For the cases in which the flow velocity $\vec{u}_{\alpha}$ is zero or, equivalently, if we take $d\vec{S}$ to be in a frame of reference moving with the average velocity $\vec{u}_{\alpha}$, (3.10) becomes

$$\Phi_{\alpha n}(\chi) = n_{\alpha} \langle \chi c_{\alpha n} \rangle \tag{3.11}$$

which is the flux of $\chi(\vec{r},\vec{v},t)$ along $\hat{n}$ due to the *random* motions of the particles of type α.

4. PARTICLE CURRENT DENSITY

The *particle current density* (or *particle flux*) is defined as the number of particles passing through a given surface per unit area and per unit time. Taking $\chi = 1$ in (3.6), we obtain the flux of particles of type α in the direction $\hat{n}$,

$$\Gamma_{\alpha n}(\vec{r},t) = n_{\alpha} \langle v_n \rangle_{\alpha} = n_{\alpha} u_{\alpha n} \tag{4.1}$$

since $\langle c_{\alpha n} \rangle = 0$. When $\vec{u}_\alpha$ vanishes, it is of interest to consider only the flux in the positive direction instead of the resulting net flux. The number of particles of type α that cross a given surface along the direction $\hat{n}$ from the same side, per unit area and time, due to their random motions, is given by

$$\Gamma_{\alpha n}^{(+)}(\vec{r},t) = \int_{\substack{v \\ (\hat{n}\cdot\vec{c} > 0)}} \hat{n}\cdot\vec{c}_\alpha \, f_\alpha(\vec{r},\vec{v},t) \, d^3v \tag{4.2}$$

where the integral in velocity space is over only the velocities for which $\hat{n}\cdot\vec{c} > 0$.

The *random mass flux* in the positive direction of $\hat{n}$ is, consequently, $m_\alpha \, \Gamma_{\alpha n}^{(+)}(\vec{r},t)$, where m_α is the mass of the type α particle.

5. MOMENTUM FLOW DYAD OR TENSOR

This quantity is defined as the net momentum transported per unit area and time, through some surface element $\hat{n}$ dS. If we take, in (3.6), $\chi(\vec{r},\vec{v},t)$ as the component of momentum of the type α particles along some direction specified by the unit vector $\hat{j}$, that is,

$$\chi = m_\alpha \, \vec{v} \cdot \hat{j} = m_\alpha \, v_j \tag{5.1}$$

we obtain the element $P_{\alpha jn}(\vec{r},t)$ of the momentum flow tensor

$$P_{\alpha jn}(\vec{r},t) = n_\alpha \langle (m_\alpha \, \vec{v} \cdot \hat{j})(\vec{v} \cdot \hat{n}) \rangle_\alpha = \rho_{m\alpha} \langle v_j v_n \rangle_\alpha \tag{5.2}$$

where $\rho_{m\alpha} = n_\alpha m_\alpha$ denotes the density of the type α particles. Thus, the momentum flow element $P_{\alpha jn}(\vec{r},t)$ represents the flux of the $j^{\underline{th}}$ component of the momentum of the type α particles through a surface element whose normal is oriented along $\hat{n}$. Since $\vec{v} = \vec{c}_\alpha + \vec{u}_\alpha$, we obtain

$$P_{\alpha jn}(\vec{r},t) = \rho_{m\alpha} \langle c_{\alpha j} \, c_{\alpha n} \rangle + \rho_{m\alpha} \, u_{\alpha j} \, u_{\alpha n} \tag{5.3}$$

or, in dyadic form,

$$\bar{\bar{P}}_\alpha(\vec{r},t) = \rho_{m\alpha} \langle \vec{c}_\alpha \, \vec{c}_\alpha \rangle + \rho_{m\alpha} \, \vec{u}_\alpha \, \vec{u}_\alpha \tag{5.4}$$

where we have used the fact that $\langle \vec{u}_\alpha \vec{c}_\alpha \rangle = \vec{u}_\alpha \langle \vec{c}_\alpha \rangle = 0$.

In a Cartesian coordinate system (x,y,z) the momentum flow dyad has the following form, in terms of its components,

$$\begin{aligned} \bar{\bar{P}}_\alpha &= \hat{x}\hat{x}\, P_{\alpha xx} + \hat{x}\hat{y}\, P_{\alpha xy} + \hat{x}\hat{z}\, P_{\alpha xz} \\ &+ \hat{y}\hat{x}\, P_{\alpha yx} + \hat{y}\hat{y}\, P_{\alpha yy} + \hat{y}\hat{z}\, P_{\alpha yz} \\ &+ \hat{z}\hat{x}\, P_{\alpha zx} + \hat{z}\hat{y}\, P_{\alpha zy} + \hat{z}\hat{z}\, P_{\alpha zz} \end{aligned} \tag{5.5}$$

From the rules of matrix multiplication $\bar{\bar{P}}_\alpha$ can be expressed as

$$\bar{\bar{P}}_\alpha = (\hat{x}\ \hat{y}\ \hat{z}) \begin{pmatrix} P_{\alpha xx} & P_{\alpha xy} & P_{\alpha xz} \\ P_{\alpha yx} & P_{\alpha yy} & P_{\alpha yz} \\ P_{\alpha zx} & P_{\alpha zy} & P_{\alpha zz} \end{pmatrix} \begin{pmatrix} \hat{x} \\ \hat{y} \\ \hat{z} \end{pmatrix} \tag{5.6}$$

It is usual, however, to omit the pre and post multiplicative dyadic signs, such as $\hat{x}\ \hat{x}$, etc., and denote the dyad only by the 3 x 3 matrix containing the elements $P_{\alpha ij}$. Thus, $P_{\alpha ij}$ corresponds to the element of the $i^{\underline{th}}$ row and the $j^{\underline{th}}$ column. From (5.3) it is clear that $P_{\alpha ij} = P_{\alpha ji}$ and, consequently, the 3 x 3 matrix in (5.6) is *symmetric*. Therefore, only six of the components of the momentum flow dyad are independent.

6. PRESSURE DYAD

6.1 Concept of pressure

The pressure of a gas is usually defined as the force per unit area exerted by the gas molecules through collisions with the walls of the containing vessel. This force is equal to the rate of transfer of molecular momentum to the walls of the container. This definition applies also to any surface immersed in the gas as, for example, the surface of a material body.

We may generalize this definition of pressure so that it can be applied to any point inside the gas. To this end, we will define pressure in terms of an imaginary surface element $d\vec{S} = \vec{n}\ dS$, inside the gas, moving with its average velocity $\vec{u}(\vec{r},t)$. The pressure on $d\vec{S}$ is then defined as the rate of transport of molecular momentum per unit area, that is, the *flux of momentum* across $d\vec{S}$

due to the random particle motions. When different species of particles are present, as in a plasma, it is useful to define a (partial) pressure due to the particles of type α, as the flux of momentum transported by the type α particles as they move back and forth across the surface element $\hat{n}$ dS, moving with the average velocity $\vec{u}_\alpha(\vec{r},t)$.

In the frame of reference of $d\vec{S}$ (3.11) applies, and taking $\chi(\vec{r},\vec{v},t)$ as the $j^{\underline{th}}$ component of momentum of the type α particles, $m_\alpha c_{\alpha j}$, we obtain the element $p_{\alpha jn}$ of the pressure tensor,

$$p_{\alpha jn} = \rho_{m\alpha} \langle c_{\alpha j}\, c_{\alpha n} \rangle \tag{6.1}$$

The *pressure dyad* is therefore given by

$$\overline{\overline{p}}_\alpha = \rho_{m\alpha} \langle \vec{c}_\alpha\, \vec{c}_\alpha \rangle \tag{6.2}$$

From (5.4) we find the following relation between the pressure dyad $\overline{\overline{p}}_\alpha$ and the momentum flow dyad $\overline{\overline{P}}_\alpha$,

$$\overline{\overline{p}}_\alpha = \overline{\overline{P}}_\alpha - \rho_{m\alpha} \vec{u}_\alpha\, \vec{u}_\alpha \tag{6.3}$$

They are equal only when the flow velocity $\vec{u}_\alpha(\vec{r},t)$ vanishes.

6.2 Force per unit area

Consider now a small element of volume inside the plasma, bounded by the closed surface S, and let $d\vec{S} = \hat{n}\, dS$ be an element of area belonging to S, with the unit vector $\hat{n}$ normal to the surface element and pointing *outward* (Fig. 4). The *force per unit area* acting on the area element $\hat{n}$ dS, as the result of the random particle motions, is given by

$$-\overline{\overline{p}}_\alpha \cdot \hat{n} = -\rho_{m\alpha} \langle \vec{c}_\alpha (\vec{c}_\alpha \cdot \hat{n}) \rangle \tag{6.4}$$

The reason for the minus sign can be seen as follows. Suppose, for the moment, that all type α particles have the same velocity $\vec{c}_\alpha$. If $\vec{c}_\alpha$ forms an angle of less than 90° with $\hat{n}$, then the quantity $n_\alpha(\vec{c}_\alpha \cdot \hat{n})$ dS is the number of type α particles *leaving* per unit time the volume enclosed by the closed surface S, through $d\vec{S}$. The corresponding change (*decrease*) in momentum of the plasma enclosed by the surface S is $-\, n_\alpha\, m_\alpha\, \vec{c}_\alpha\, (\vec{c}_\alpha \cdot \hat{n})$ dS, since $(\vec{c}_\alpha \cdot \hat{n})$ is *positive*. On the other hand, if $\vec{c}_\alpha$ forms an angle greater than 90^0 with $\hat{n}$, then $-n_\alpha (\vec{c}_\alpha \cdot \hat{n})$ dS represents the number of particles *entering* per unit time the bounded volume through $d\vec{S}$, and the corresponding change (*increase*) in momentum of the plasma within the closed surface S is again $-n_\alpha m_\alpha \vec{c}_\alpha (\vec{c}_\alpha \cdot \hat{n})$ dS, since now $(\vec{c}_\alpha \cdot \hat{n})$ is *negative*.

We conclude, by generalizing this result, that for any arbitrary distribution of individual velocities, the vector quantity

$$-n_\alpha m_\alpha \langle \vec{c}_\alpha (\vec{c}_\alpha \cdot \hat{n}) \rangle \, dS = - \bar{\bar{p}}_\alpha \cdot \hat{n} \, dS \tag{6.5}$$

represents the rate of change of plasma momentum within the closed surface S, due to the exchange of type α particles through the surface element $\hat{n}$ dS. Therefore, the force per unit area exerted on an element of area oriented along the unit vector $\hat{n}$ is $-\bar{\bar{p}}_\alpha \cdot \hat{n}$. If we take, for example, an element of area along the $\hat{x}$ direction, that is, $\hat{n} = \hat{x}$, we have

$$-\bar{\bar{p}}_\alpha \cdot \hat{x} = -\hat{x} p_{\alpha xx} - \hat{y} p_{\alpha yx} - \hat{z} p_{\alpha zx} \tag{6.6}$$

where $p_{\alpha xx}$ is normal to the surface and towards it, just like a hydrostatic pressure, whereas the components $p_{\alpha yx}$ and $p_{\alpha zx}$ are pressures due to shear forces which are tangential to the surface, as indicated in Fig. 5. All other components of $\bar{\bar{p}}_\alpha$ are interpreted in an analogous way. Generally, the force per unit area $p_{\alpha jn}$ acts along the negative direction of the axis denoted by the first subscript (j) on a surface whose outward normal is parallel to the axis indicated by the second subscript (n). Alternatively, if the outward normal to the surface is in the negative direction of the axis indicated by the second subscript (n), then the force acts in the same direction as the axis denoted by the first subscript (j).

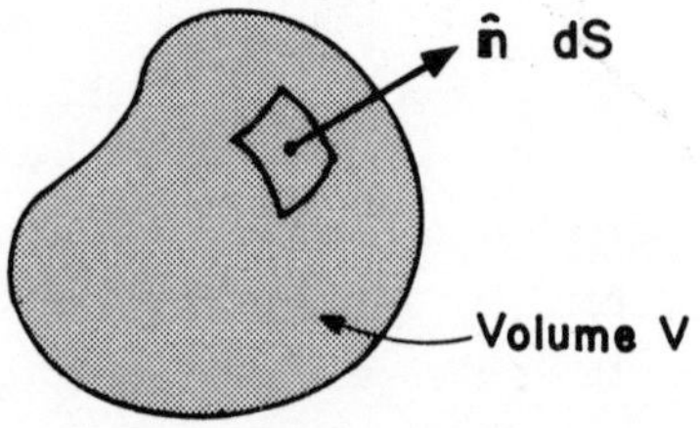

Fig. 4 - Element of volume V bounded by a closed surface S, with the surface element $\hat{n}$ dS pointing outward.

6.3 Force per unit volume

The force per unit volume inside the plasma, due to the random particle motion, can be obtained by integrating (6.5) over the closed surface S bounding the volume element V, dividing the result by V, and then taking the limit as V tends to zero. This procedure is just the definition of the divergence,

$$-\vec{\nabla}\cdot\bar{\bar{p}}_\alpha = -\lim_{V\to 0}\left[\frac{1}{V}\oint \bar{\bar{p}}_\alpha\cdot\hat{n}\, dS\right] \tag{6.7}$$

and, from Gauss' divergence theorem,

$$-\oint \bar{\bar{p}}_\alpha\cdot\hat{n}\, dS = -\int \vec{\nabla}\cdot\bar{\bar{p}}_\alpha\, d^3r \tag{6.8}$$

We conclude, therefore, that the negative divergence of the kinetic pressure dyad ($-\vec{\nabla}\cdot\bar{\bar{p}}_\alpha$) is the force exerted on a unit volume of the plasma due to the random particle motion, and $-\bar{\bar{p}}_\alpha\cdot\hat{n}$ is the force acting on a unit area of a surface normal to the unit vector $\hat{n}$.

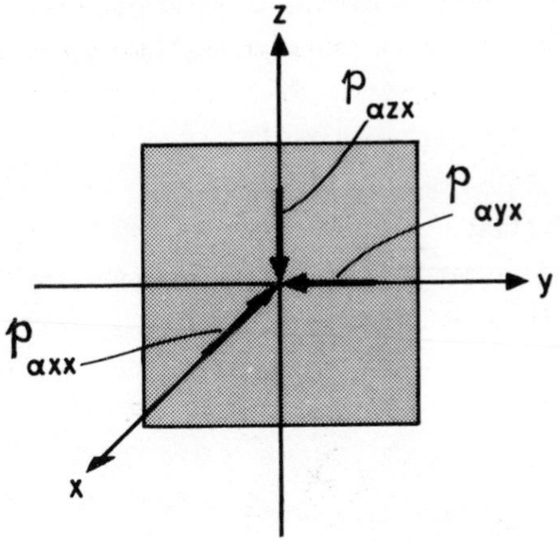

Fig. 5 - Components of the pressure dyad corresponding to the tangential shear stresses, $p_{\alpha yx}$ and $p_{\alpha zx}$, and to the normal stress, $p_{\alpha xx}$, acting on a surface element whose normal is oriented along $\hat{x}$.

6.4 Scalar pressure and absolute temperature

An important macroscopic variable is the *scalar pressure*, or *mean hydrostatic pressure*, p_α. It is defined as one third the trace of the pressure tensor,

$$p_\alpha = \frac{1}{3} \sum_{i,j} p_{\alpha ij}\delta_{ij} = \frac{1}{3} \sum_i p_{\alpha ii} = \frac{1}{3} (p_{\alpha xx} + p_{\alpha yy} + p_{\alpha zz}) \tag{6.9}$$

where δ_{ij} is the Kronecker delta, such that $\delta_{ij} = 1$ for $i = j$ and $\delta_{ij} = 0$ for $i \neq j$. The pressure elements $p_{\alpha ii}$, with $i = x,y,z$, are just the hydrostatic pressures *normal* to the surfaces described by i = constant. Using (6.1),

$$p_\alpha = \rho_{m\alpha} \langle c_{\alpha x}^2 + c_{\alpha y}^2 + c_{\alpha z}^2 \rangle /3 \tag{6.10}$$

Since $c^2 = c_{\alpha x}^2 + c_{\alpha y}^2 + c_{\alpha z}^2$, we have

$$p_\alpha = \rho_{m\alpha} \langle c_\alpha^2 \rangle /3 \tag{6.11}$$

Another important parameter for a macroscopic description of a gas is its temperature. The *absolute temperature* T_α, for the type α particles, is a measure of the *mean kinetic energy* of the random particle motion. According to the thermodynamic definition of absolute temperature, there is a thermal energy of $k\,T_\alpha/2$ associated with each translational degree of freedom, so that

$$kT_\alpha/2 = m_\alpha \langle c_{\alpha i}^2 \rangle /2 \tag{6.12}$$

where k is Boltzmann's constant.

When the distribution of random velocities is isotropic, as is the case of the Maxwell-Boltzmann distribution function (to be considered in the next chapter) which characterizes the state of *thermal equilibrium* of a gas, we have $c_{\alpha x}^2 = c_{\alpha y}^2 = c_{\alpha z}^2 = c_\alpha^2/3$, and therefore,

$$p_\alpha = p_{\alpha xx} = p_{\alpha yy} = p_{\alpha zz} = \rho_{m\alpha} \langle c_{\alpha i}^2 \rangle \tag{6.13}$$

Combining (6.13) and (6.12), gives

$$p_\alpha = n_\alpha\, k\, T_\alpha \tag{6.14}$$

which is the *equation of state* of an *ideal gas*. For the Maxwell-Boltzmann distribution function the nondiagonal elements of the kinetic pressure dyad are all zero and it reduces to

$$\bar{\bar{p}}_\alpha = (\hat{x}\hat{x} + \hat{y}\hat{y} + \hat{z}\hat{z})\, p_\alpha = \bar{\bar{1}}\, p_\alpha \tag{6.15}$$

where $\bar{\bar{1}}$ stands for the *unit dyad*, which in matrix form can be written as

$$\bar{\bar{1}} = \begin{pmatrix} 1 & 0 & 0 \\ 0 & 1 & 0 \\ 0 & 0 & 1 \end{pmatrix} \tag{6.16}$$

In this case the negative divergence of the pressure dyad becomes

$$-\vec{\nabla}\cdot\bar{\bar{p}}_\alpha = -\left(\hat{x}\frac{\partial}{\partial x}p_\alpha + \hat{y}\frac{\partial}{\partial y}p_\alpha + \hat{z}\frac{\partial}{\partial z}p_\alpha\right) = -\vec{\nabla}p_\alpha \tag{6.17}$$

Thus, for an isotropic velocity distribution, the force per unit volume due to the random variations of the peculiar velocities is given by the negative gradient of the scalar pressure.

In some problems a simplification of practical interest for the general form of the kinetic pressure dyad consists in taking

$$\bar{\bar{p}}_\alpha = \hat{x}\hat{x}\, p_{\alpha xx} + \hat{y}\hat{y}\, p_{\alpha yy} + \hat{z}\hat{z}\, p_{\alpha zz} \tag{6.18}$$

or, in matrix form,

$$\bar{\bar{p}}_\alpha = \begin{pmatrix} p_{\alpha xx} & 0 & 0 \\ 0 & p_{\alpha yy} & 0 \\ 0 & 0 & p_{\alpha zz} \end{pmatrix} \tag{6.19}$$

where the diagonal elements are different from one another but all nondiagonal elements vanish. This corresponds to an anisotropy of the peculiar velocities and the absence of shear forces and viscous drag. The effects of viscosity and shear stresses are incorporated in the nondiagonal elements of the pressure dyad. Usually, the effects of viscosity are relatively unimportant for most plasmas and the nondiagonal elements of $\bar{\bar{p}}_\alpha$ can, in many cases, be neglected. In this anisotropic case,a different absolute temperature can be defined for each direction in space, according to (6.12).

7. HEAT FLOW VECTOR

The component of the heat flow vector, $q_{\alpha n}$, is defined as the flux of *random* or *thermal energy* across a surface whose normal points in the direction of the unit vector $\hat{n}$. Taking $\chi(\vec{r},\vec{v},t)$, in (3.11), as the kinetic energy of random motion of the particles of type α, that is, $\chi = m_\alpha c_\alpha^2/2$, we obtain for the component of the heat flow vector along $\hat{n}$,

$$q_{\alpha n} = \vec{q}_\alpha \cdot \hat{n} = \rho_{m\alpha} \langle c_\alpha^2 \, \vec{c}_\alpha \cdot \hat{n} \rangle /2 \tag{7.1}$$

The *heat flow vector* is therefore given by

$$\vec{q}_\alpha = \rho_{m\alpha} \langle c_\alpha^2 \, \vec{c}_\alpha \rangle /2 \tag{7.2}$$

8. HEAT FLOW TRIAD

It is convenient, at this point, to introduce a *triad* of *thermal energy flux* defined by

$$\bar{\bar{\bar{Q}}}_\alpha = \rho_{m\alpha} \langle \vec{c}_\alpha \, \vec{c}_\alpha \, \vec{c}_\alpha \rangle \tag{8.1}$$

Its components are, explicitly,

$$Q_{\alpha ijk} = \rho_{m\alpha} \langle c_{\alpha i} \, c_{\alpha j} \, c_{\alpha k} \rangle \tag{8.2}$$

Using Cartesian coordinates, the thermal energy flux triad can be written in the form

$$\bar{\bar{\bar{Q}}}_\alpha = \bar{\bar{Q}}_{\alpha x} \, \hat{x} + \bar{\bar{Q}}_{\alpha y} \, \hat{y} + \bar{\bar{Q}}_{\alpha z} \, \hat{z} \tag{8.3}$$

where each of the *dyads* $\bar{\bar{Q}}_{\alpha n}$, with n = x,y,z, can be expressed in matrix form as

$$\bar{\bar{Q}}_{\alpha n} = \begin{pmatrix} Q_{\alpha xxn} & Q_{\alpha xyn} & Q_{\alpha xzn} \\ Q_{\alpha yxn} & Q_{\alpha yyn} & Q_{\alpha yzn} \\ Q_{\alpha zxn} & Q_{\alpha zyn} & Q_{\alpha zzn} \end{pmatrix} \tag{8.4}$$

To obtain a relation between $\vec{q}_\alpha$ and $\bar{\bar{\bar{Q}}}_\alpha$, note that (7.1) can be written as

$$q_{\alpha n} = \rho_{m\alpha}(<c^2_{\alpha x}\, c_{\alpha n}> + <c^2_{\alpha y}\, c_{\alpha n}> + <c^2_{\alpha z}\, c_{\alpha n}>)/2 \tag{8.5}$$

and comparing this equation with (8.2), we see that $q_{\alpha n}$ becomes

$$q_{\alpha n} = (Q_{\alpha xxn} + Q_{\alpha yyn} + Q_{\alpha zzn})/2 \tag{8.6}$$

9. TOTAL ENERGY FLUX TRIAD

In analogy with the definition of the heat flow triad $\bar{\bar{\bar{Q}}}$, consider now the quantity

$$E_{\alpha ijk}(\vec{r},t) = \rho_{m\alpha}<v_i\; v_j\; v_k>_\alpha \tag{9.1}$$

which may be called the *total energy flux triad*. This quantity can be considered as the sum of three parts. Substituting $v_i = u_{\alpha i} + c_{\alpha i}$ and expanding,

$$\begin{aligned} <v_i\; v_j\; v_k>_\alpha = <c_{\alpha i}\, c_{\alpha j}\, c_{\alpha k} &+ u_{\alpha i}\, c_{\alpha j}\, c_{\alpha k} + u_{\alpha j}\, c_{\alpha k}\, c_{\alpha i} + \\ &+ u_{\alpha k}\, c_{\alpha i}\, c_{\alpha j} + u_{\alpha i}\, u_{\alpha j}\, c_{\alpha k} + u_{\alpha j}\, u_{\alpha k}\, c_{\alpha i} + \\ &+ u_{\alpha k}\, u_{\alpha i}\, c_{\alpha j} + u_{\alpha i}\, u_{\alpha j}\, u_{\alpha k}> \end{aligned} \tag{9.2}$$

Noting that $<u_{\alpha i}> = u_{\alpha i}$ and $<c_{\alpha i}> = 0$, and using (8.2) and (6.1), we obtain

$$\rho_{m\alpha} < v_i\; v_j\; v_k>_\alpha = \rho_{m\alpha}\, u_{\alpha i}\, u_{\alpha j}\, u_{\alpha k} + (\vec{u}_\alpha, \bar{\bar{p}}_\alpha)_{ijk} + Q_{\alpha ijk} \tag{9.3}$$

where the following notation was employed

$$(\vec{u}_\alpha, \bar{\bar{p}}_\alpha)_{ijk} = u_{\alpha i}\, p_{\alpha jk} + u_{\alpha j}\, p_{\alpha ki} + u_{\alpha k}\, p_{\alpha ij} \tag{9.4}$$

Therefore, we can write (9.3) in triadic form as

$$\rho_{m\alpha}<\vec{v}\;\vec{v}\;\vec{v}>_\alpha = \rho_{m\alpha}\, \vec{u}_\alpha\, \vec{u}_\alpha\, \vec{u}_\alpha + (\vec{u}_\alpha, \bar{\bar{p}}_\alpha) + \bar{\bar{\bar{Q}}}_\alpha \tag{9.5}$$

The total energy flux triad can therefore be considered as the sum of the energy flux transported by the *convective* particle motions, represented by the first two terms in the right-hand side of (9.5), and the *thermal* energy flux $\bar{\bar{\bar{Q}}}_\alpha$ due to the random thermal motions of the particles of type α.

The physical interpretation of the heat flow triad $\bar{\bar{\bar{Q}}}_\alpha$ is, in some sense, analogous to the physical interpretation of the heat flow vector $\vec{q}_\alpha$. For this purpose, consider the quantity

$$\rho_{m\alpha} < v^2\vec{v} >_\alpha /2 \tag{9.6}$$

which represents the *average energy flux* transported by the particles of type α. This quantity can be written as the sum of three terms. Substituting $\vec{v} = \vec{c}_\alpha + \vec{u}_\alpha$ in expression (9.6) and expanding,

$$\rho_{m\alpha} <v^2\vec{v}>_\alpha /2 = \rho_{m\alpha}<u_\alpha^2 \vec{u}_\alpha + 2(\vec{u}_\alpha \cdot \vec{c}_\alpha)\vec{u}_\alpha + c_\alpha^2 \vec{u}_\alpha + u_\alpha^2 \vec{c}_\alpha + \\ + 2(\vec{u}_\alpha \cdot \vec{c}_\alpha)\vec{c}_\alpha + c_\alpha^2 \vec{c}_\alpha> /2 \tag{9.7}$$

and since $<\vec{c}_\alpha> = 0$ and $<\vec{u}_\alpha> = \vec{u}_\alpha$, we obtain

$$\rho_{m\alpha} <v^2\vec{v}>_\alpha /2 = \rho_{m\alpha} (u_\alpha^2 + <c_\alpha^2>)\vec{u}_\alpha/2 + \rho_\alpha \vec{u}_\alpha \cdot <\vec{c}_\alpha \vec{c}_\alpha> + \\ + \rho_{m\alpha}<c_\alpha^2 \vec{c}_\alpha> /2 \tag{9.8}$$

If we now use (6.2) and (7.2), which define $\bar{\bar{p}}_\alpha$ and $\vec{q}_\alpha$, respectively, we obtain the identity

$$\rho_{m\alpha} <v^2\vec{v}>_\alpha /2 = W_\alpha \vec{u}_\alpha + \vec{u}_\alpha \cdot \bar{\bar{p}}_\alpha + \vec{q}_\alpha \tag{9.9}$$

where W_α is the *mean kinetic energy density* of the type α particles,

$$W_\alpha = \rho_{m\alpha} u_\alpha^2/2 + \rho_{m\alpha}<c_\alpha^2>/2 \tag{9.10}$$

Eq. (9.9) is written in a form analogous to (9.5). It shows that the rate of transport per unit area (flux) of the average energy of the type α particles, $\rho_{m\alpha}<v^2 \vec{v}>_\alpha /2$, can be separated into three parts: the first term in the right-hand side of (9.9) represents the flux of the mean kinetic energy transported *convectively*, the second term is the rate of work per unit area done by the kinetic pressure dyad, and the third one is the random thermal energy flux transported by the particles of type α due to their random *thermal* motions. It is instructive to note that in a frame of reference moving with the average velocity $\vec{u}_\alpha$, the particle velocities become identical to their random velocities, that is $\vec{v} = \vec{c}_\alpha$, so that (9.9) reduces to (7.2) which defines the thermal energy flux vector $\vec{q}_\alpha$. When the thermal velocities $\vec{c}_\alpha$ are distributed uniformly in all directions, that is isotropically,it turns out that $\vec{q}_\alpha = 0$ (since the integrand in $<c_\alpha^2 \vec{c}_\alpha>$ is an odd function of $\vec{c}_\alpha$). Consequently, $\vec{q}_\alpha$ can be considered at least as a partial measure of the anisotropies in the distribution of the thermal velocities. The thermal energy flux triad $\bar{\bar{\bar{Q}}}_\alpha$ considerably extends the concept of the heat flux vector and in this sense can be considered as a complete measure of the anisotropies in the distribution of the thermal velocities of the particles.

10. HIGHER MOMENTS OF THE DISTRIBUTION FUNCTION

The first four moments of the distribution function $f_\alpha(\vec{r},\vec{v},t)$ are related to the number density n_α, the average velocity $\vec{u}_\alpha$, the momentum flow dyad $\bar{\bar{P}}_\alpha$ and the energy flow triad $\bar{\bar{\bar{E}}}_\alpha$. For reference, it is convenient to collect them here,

$$n_\alpha(\vec{r},t) = \int_V f_\alpha(\vec{r},\vec{v},t)\, d^3v \tag{10.1}$$

$$u_{\alpha i}(\vec{r},t) = <v_i>_\alpha = \frac{1}{n_\alpha(\vec{r},t)} \int_V v_i\, f_\alpha(\vec{r},\vec{v},t)\, d^3v \tag{10.2}$$

$$P_{\alpha ij}(\vec{r},t) = \rho_{m\alpha} <v_i\, v_j>_\alpha = m_\alpha \int_V v_i\, v_j\, f_\alpha(\vec{r},\vec{v},t)\, d^3v \tag{10.3}$$

$$E_{\alpha ijk}(\vec{r},t) = \rho_{m\alpha} <v_i\, v_j\, v_k>_\alpha = m_\alpha \int_V v_i\, v_j\, v_k\, f_\alpha(\vec{r},\vec{v},t)\, d^3v \tag{10.4}$$

When the average velocity $\vec{u}_\alpha$ vanishes, we have $\vec{v} = \vec{c}_\alpha$, $\bar{\bar{P}}_\alpha$ becomes the same as the pressure dyad $\bar{\bar{p}}_\alpha$, and $\bar{\bar{\bar{E}}}_\alpha$ becomes the same as the thermal energy flux triad $\bar{\bar{\bar{Q}}}_\alpha$.

As a formal extension of these definitions we may, whenever necessary, consider higher moments of the distribution function. The moment of order N can be defined by the expression

$$M^{(N)}_{ij...k}(\vec{r},t) = \int_V \underbrace{v_i\, v_j ... v_k}_{N \text{ times}}\; f_\alpha(\vec{r},\vec{v},t)\, d^3v \tag{10.5}$$

PROBLEMS

6.1 - Consider a system of particles characterized by the distribution function of Problem 5.1.

(a) Show that the absolute temperature of the system is given by $T = m\, v_0^2/3k$, where m is the mass of each particle and k is Boltzmann's constant.

(b) Obtain the following expression for the pressure dyad

$$\bar{\bar{p}} = \rho_m\, v_0^2\, \bar{\bar{1}}/3$$

where $\rho_m = n\, m$ and $\bar{\bar{1}}$ is the unit dyad.

(c) verify that the heat flow vector $\vec{q} = 0$.

6.2 - Suppose that the peculiar (random) velocities of the electrons in a given plasma, satisfy the following modified Maxwell-Boltzmann distribution function (consider $\vec{u} = 0$),

$$f(\vec{c}) = n_0 \left(\frac{m}{2\pi k T_\perp}\right) \left(\frac{m}{2\pi k T_\parallel}\right)^{1/2} \exp\left[-\frac{m}{2k}\left(\frac{c_x^2 + c_y^2}{T_\perp} + \frac{c_z^2}{T_\parallel}\right)\right]$$

(a) Verify that the electron number density is given by n_0.

(b) Show that the kinetic pressure dyad is given by

$$\bar{\bar{p}} = n_0\, k\, [T_\perp(\hat{x}\,\hat{x} + \hat{y}\,\hat{y}) + T_\parallel\, \hat{z}\,\hat{z}]$$

which indicates the presence of an anisotropy in the z direction.

(c) Calculate the heat flow vector $\vec{q}$.

(d) Show that $m\langle v_\parallel^2\rangle/2 = kT_\parallel/2$ and $m\langle v_\perp^2\rangle/2 = kT_\perp$.

6.3 - For the loss-cone distribution function of Problem 5.3, show that $m\langle v_\parallel^2\rangle/2 = m\alpha_\parallel^2/4$ and $m\langle v_\perp^2\rangle/2 = m\alpha_\perp^2$. Compare these results with those of Problem 6.2(d), and provide physical arguments to justify the difference in the perpendicular part of the thermal energy.

6.4 - Convince yourself that there are only *ten* independent elements in the thermal energy flux triad $\bar{\bar{\bar{Q}}}$. Note that $Q_{ijk} = nm\,\langle c_i\, c_j\, c_k\rangle$ is symmetric under the interchange of any two of its three indices.

6.5 - A plasma is made up of a mixture of various particle species, the type α species having mass m_α, number density n_α, average velocity $\vec{u}_\alpha = \langle\vec{v}\rangle_\alpha$, random velocity $\vec{c}_\alpha = \vec{v} - \vec{u}_\alpha$, temperature $T_\alpha = (m_\alpha/3k)\langle c_\alpha^2\rangle$, pressure dyad $\bar{\bar{p}}_\alpha = n_\alpha m_\alpha \langle\vec{c}_\alpha \vec{c}_\alpha\rangle$, and heat flow vector $\vec{q}_\alpha = (n_\alpha m_\alpha/2)\langle c_\alpha^2 \vec{c}_\alpha\rangle$. Similar quantities can be defined for the plasma *as a whole*, for example:

total number density $n_0 = \sum_\alpha n_\alpha$

average mass $m_0 = \sum_\alpha n_\alpha m_\alpha/n_0$

average velocity $\vec{u}_0 = \sum_\alpha n_\alpha m_\alpha \vec{u}_\alpha/(n_0 m_0)$

We can also define an *alternative* random velocity for the type α species as $\vec{c}_{\alpha 0} = \vec{v} - \vec{u}_0$, as well as an alternative temperature $T_{\alpha 0} = (m_\alpha/3k)\langle c_{\alpha o}^2\rangle$, pressure dyad $\bar{\bar{p}}_{\alpha 0} = n_\alpha m_\alpha \langle\vec{c}_{\alpha 0}\, \vec{c}_{\alpha 0}\rangle$, and heat flow vector $\vec{q}_{\alpha 0} = (n_\alpha m_\alpha/2)\langle c_{\alpha o}^2\, \vec{c}_{\alpha o}\rangle$

(a) Show that, for the plasma as a whole, the *total* pressure dyad is given by

$$\bar{\bar{p}}_0 = \sum_\alpha (\bar{\bar{p}}_\alpha + n_\alpha m_\alpha\, \vec{w}_\alpha \vec{w}_\alpha)$$

and the *total* scalar pressure by

$$p_0 = \sum_\alpha (p_\alpha + n_\alpha m_\alpha\, w_\alpha^2/3)$$

where $\vec{w}_\alpha = \vec{u}_\alpha - \vec{u}_0$ is the diffusion velocity.

(b) Assuming that $\vec{c}_\alpha$ is isotropic, that is, $\langle c_{\alpha i}^2\rangle = \langle c_\alpha^2\rangle/3$, for i = x,y,z, show that the *total* heat flow vector is given by

$$\vec{q}_0 = \sum_\alpha (\vec{q}_\alpha + 5p_\alpha \vec{w}_\alpha/2 + n_\alpha m_\alpha\, w_\alpha^2 \vec{w}_\alpha/2)$$

(c) If an *average* temperature T_0, for the plasma as a whole, is defined by requiring that $p_0 = n_0\, k\, T_0$, show that

$$T_0 = (1/n_0) \sum_\alpha n_\alpha\, (T_\alpha + m_\alpha\, w_\alpha^2/3k)$$

(d) Verify that

$$3n_0\, k\, T_0/2 = \sum_\alpha n_\alpha m_\alpha \langle c_{\alpha o}^2\rangle/2$$

6.6 - Consider an infinitesimal element of volume $d^3r = dx\, dy\, dz$ in a gas of number density n.

(a) Show that the time rate of increase of momentum in d^3r, as a result of particles of mass m entering d^3r with average velocity $\vec{u}$, is given by $-\vec{\nabla}\cdot(n\, m\, \vec{u}\, \vec{u})\, d^3r$.

(b) If the infinitesimal volume element d^3r moves with the average particle velocity $\vec{u}$, show that, because of the work done by the kinetic pressure dyad $\bar{\bar{p}}$, the energy of the particle inside d^3r increases at a time rate given by $-\vec{\nabla}\cdot(\vec{u}\cdot\bar{\bar{p}})\, d^3r$.

(c) Verify, by expansion, that:

$$(\bar{\bar{p}}\cdot\hat{n})\cdot\vec{u} = (\vec{u}\cdot\bar{\bar{p}})\cdot\hat{n}$$

where $\hat{n}$ denotes an outward unit vector, normal to the surface bounding the volume element.

6.7 - Consider (5.6.4), which is the solution of the Boltzmann equation with the relaxation model for the collision term, in the absence of external forces and spatial gradients, and when $f_{0\alpha}$ and τ are time-independent. Show that, according to this result, we have

$$G_\alpha(t) = G_{0\alpha} + [G_\alpha(0) - G_{0\alpha}]\exp(-t/\tau)$$

where

$$G_\alpha(t) = \int_v f_\alpha\, \chi\, d^3v = n_\alpha \langle\chi\rangle_\alpha$$

$$G_{0\alpha}(t) = \int_v f_{0\alpha}\, \chi\, d^3v = n_\alpha \langle\chi\rangle_{0\alpha}$$

Thus, according to the relaxation model for the collision term, every average value $\langle\chi\rangle_\alpha$ approaches equilibrium with the same relaxation time τ.

CHAPTER 7

The Equilibrium State

1. THE EQUILIBRIUM STATE DISTRIBUTION FUNCTION

The equilibrium distribution function is the time-independent solution of the Boltzmann equation in the absence of external forces. In the equilibrium state the particle interactions do not cause any change in the distribution function with time, and there are no spatial gradients in the particle number density. We deduce in this section an expression for the equilibrium distribution function, which is known as the *Maxwell-Boltzmann* or *Maxwellian velocity distribution function.*

For simplicity we will consider a gas consisting of only one particle species. The extension to a mixture will be indicated in a subsequent section of this chapter. We assume that there are no external forces acting on the system ($\vec{F}_{ext} = 0$) and that the particles are uniformly distributed in space. Under these conditions, the distribution function is homogeneous ($\vec{\nabla} f = 0$) and, since we are looking for a steady state solution of the Boltzmann equation, it is also time-independent ($\partial f/\partial t = 0$). Therefore, it can be denoted by $f(\vec{v})$. According to the Boltzmann equation (5.5.27), the equilibrium distribution function satisfies the following condition

$$(\delta f/\delta t)_{coll} = 0 \tag{1.1}$$

Hence, under equilibrium conditions, there are no changes in the distribution function as a result of collisions between the particles. In Chapter 21 we shall derive the expression for the equilibrium distribution function using the Boltzmann collision integral. For the moment, however, in order to simplify matters, it is appropriate to consider a simple derivation based on the *general principle of detailed balance* of statistical mechanics.

1.1 The general principle of detailed balance and binary collisions

In general this principle asserts that, under equilibrium conditions, the probability of occurrence of any process is equal to the probability of occurrence of the inverse process. Hence, for the case of a system of interacting particles in the state of *equilibrium*, the principle of detailed balance implies that the effect of each type of collision is exactly compensated by the effect of the corresponding inverse collision.

Consider an elastic collision between two particles having velocities $\vec{v}_1$ and $\vec{v}$ *before* collision, and $\vec{v}_1'$ and $\vec{v}'$ *after* collision. The corresponding inverse collision refers to an elastic collision in which a particle with initial velocity $\vec{v}_1'$ collides with another particle with velocity $\vec{v}'$, the velocities *after* collision being $\vec{v}_1$ and $\vec{v}$, respectively (see Fig. 1).

Assuming that the velocities of the particles before collision are uncorrelated, the number of binary collisions occurring per unit time in a given volume d^3r, about the position $\vec{r}$ in configuration space, between particles having velocities within the velocity space element d^3v, about $\vec{v}$, and particles with velocities within d^3v_1, about $\vec{v}_1$, in the same configuration space element d^3r (see Fig. 2), is proportional to the product of the respective number of particles, that is, to $(f\, d^3r\, d^3v)(f_1\, d^3r\, d^3v_1)$, where f_1 represents $f(\vec{v}_1)$.

In a similar way, assuming the particles to be uncorrelated, the number of corresponding *inverse* binary collisions occurring per unit time in the same volume element d^3r, about $\vec{r}$, in configuration space, between particles having velocities d^3v', about $\vec{v}'$, and particles with velocities within d^3v_1', about $\vec{v}_1'$, is proportional to the product $(f'\, d^3r\, d^3v')\,(f_1'\, d^3r\, d^3v_1')$, where $f' = f(\vec{v}')$ and $f_1' = f(\vec{v}_1')$. According to the principle of detailed balance, in the equilibrium state the effect of each direct collision is compensated by the effect of the corresponding inverse collision, so that

$$f'\, f_1'\, d^3v' d^3v_1' = f\, f_1 d^3v\, d^3v_1 \tag{1.2}$$

Since $d^3v' d^3v_1' = d^3v\, d^3v_1$ (see section 2, Chapter 21), (1.2) yields

$$f(\vec{v})\, f(\vec{v}_1) = f(\vec{v}')\, f(\vec{v}_1') \tag{1.3}$$

The assumption that the velocities of the particles are uncorrelated is known as the *molecular chaos* assumption. It is well justified when the density of the gas is sufficiently small so that the mean free path is larger than the characteristic range of the interparticle forces. Although this is certainly not a general situation for a plasma, the validity of the Maxwell-Boltzmann distribution function is well justified experimentally.

1.2 Summation invariants

It is convenient to introduce at this moment the concept of summation invariants. Consider a collisional interaction between two particles and let $\chi(\vec{v})$ be a physical quantity (scalar or vector) associated with each particle, which in general may be a function of the particle velocity. If the sum of the quantity $\chi(\vec{v})$ for the two particles is conserved during the collision process,

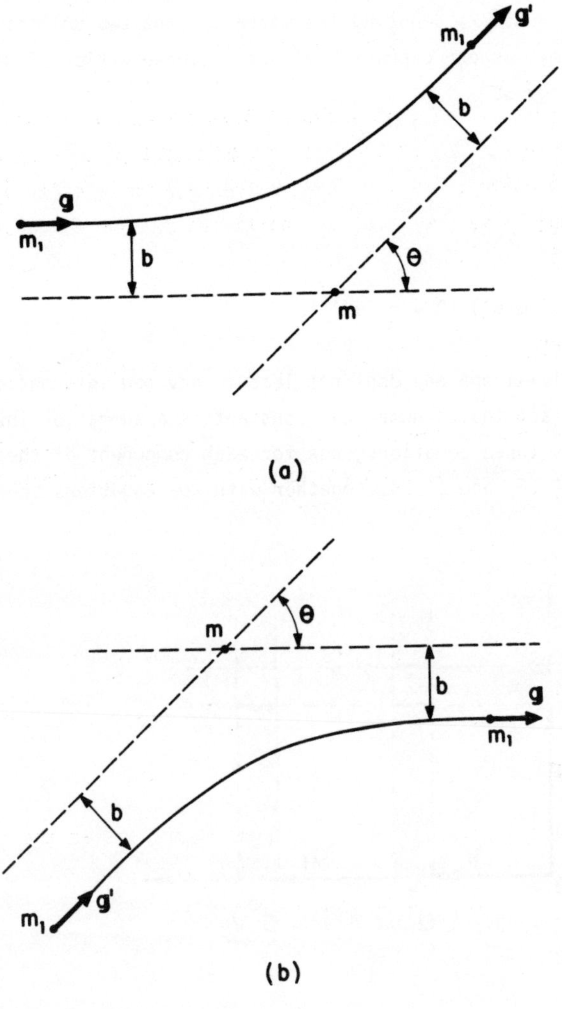

Fig. 1 - (a) Direct collision and (b) inverse collision. Here $\vec{g} = \vec{v}_1 - \vec{v}$ and $\vec{g}' = \vec{v}_1' - \vec{v}'$.

then $\chi(\vec{v})$ is called a summation invariant. For a collision between two particles having initial velocities $\vec{v}$ and $\vec{v}_1$, and velocities after collision $\vec{v}'$ and $\vec{v}_1'$, respectively, we have for a summation invariant quantity $\chi(\vec{v})$,

$$\chi(\vec{v}) + \chi(\vec{v}_1) = \chi(\vec{v}') + \chi(\vec{v}_1') \tag{1.4}$$

From the laws of conservation of mass, of momentum and of energy, the following quantities are summation invariants in binary elastic collisions: $\chi = m$, $\vec{\chi} = m\vec{v}$ and $\chi = mv^2/2$. Denoting the masses of the two colliding particles by m and m_1, we can express the laws of conservation of mass, of momentum and of energy as

$$m + m_1 = m + m_1 \tag{1.5}$$

$$m\vec{v} + m_1\vec{v}_1 = m\vec{v}' + m_1\vec{v}_1' \tag{1.6}$$

$$mv^2/2 + m_1v_1^2/2 = m(v')^2/2 + m_1(v_1')^2/2 \tag{1.7}$$

Eq. (1.5) is a trivial one and does not lead to any new information; it corresponds to the fact that a numerical constant is a summation invariant. Eq. (1.6) represents three equations, one for each component of the momentum. The four equations (1.6) and (1.7), together with the equations specifying the

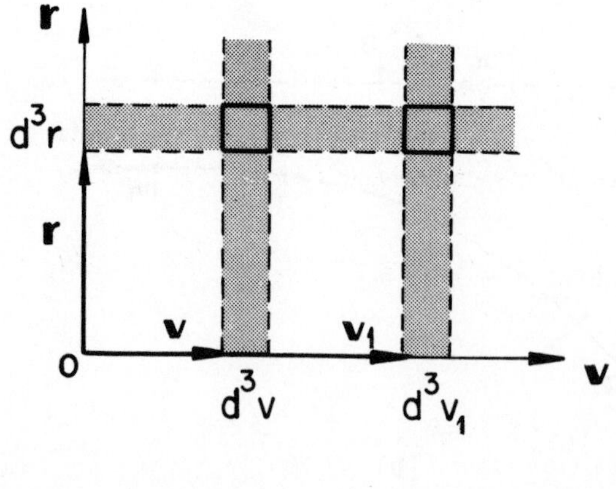

Fig. 2 - Schematic representation of the volume elements $d^3r\, d^3v$ and $d^3r\, d^3v_1$ in phase space.

impact parameter b and the angle ε of the collision plane (obtained through the analysis of the dynamics of binary collisions; see Chapter 20), constitute six equations to be solved for the determination of the six unknown quantities which are the components of the after-collision velocities $\vec{v}'$ and $\vec{v}_1'$, in terms of the initial velocities $\vec{v}$ and $\vec{v}_1$. The binary collision problem is, therefore, uniquely determined by these summation invariants. Any other summation invariant in the collision process gives no additional information and cannot, therefore, be an independent one and may be expressed as a linear combination of the summation invariants m, $m\vec{v}$ and $mv^2/2$.

1.3 Maxwell-Boltzmann distribution function

We proceed now to derive the equilibrium velocity distribution function starting from (1.3) and the concept of summation invariants. Taking the natural logarithm of both sides of (1.3) gives

$$\ln f + \ln f_1 = \ln f' + \ln f_1' \tag{1.8}$$

which shows that $\ln f$ is a summation invariant in the collision process. Therefore, it can be written as a linear combination of the summation invariants m, $m\vec{v}$, and $mv^2/2$,

$$\ln f = m(a_0 + \vec{a}_1 \cdot \vec{v} - a_2 v^2/2) = m[a_0 + a_{1x} v_x + a_{1y} v_y + a_{1z} v_z - - a_2 (v_x^2 + v_y^2 + v_z^2)/2] \tag{1.9}$$

where a_0, $\vec{a}_1 = a_{1x}\hat{x} + a_{1y}\hat{y} + a_{1z}\hat{z}$ and a_2 are constants. The negative sign in front of a_2 is chosen for convenience in the equations that follows. Completing the squares in the right-hand side of (1.9) gives

$$\ln f = m\left[a_0 + (1/2a_2)(a_{1x}^2 + a_{1y}^2 + a_{1z}^2)\right] - (ma_2/2)\left[(v_x - a_{1x}/a_2)^2 + + (v_y - a_{1y}/a_2)^2 + (v_z - a_{1z}/a_2)^2\right] = m(a_0 + a_1^2/2a_2 - - (ma_2/2)(\vec{v} - \vec{a}_1/a_2)^2 \tag{1.10}$$

Defining new constants by

$$\ln C = m(a_0 + a_1^2/2a_2) \tag{1.11}$$

$$\vec{v}_0 = \vec{a}_1/a_2 \tag{1.12}$$

we can write (1.10) in the form

$$f = C \exp[-(ma_2/2)\ (\vec{v} - \vec{v}_0)^2] \tag{1.13}$$

This expression is known as the Maxwell-Boltzmann, or Maxwellian equilibrium distribution function.

1.4 Determination of the constant coefficients

The Maxwellian distribution function (1.13) contains five constant coefficients to be determined, namely C, a_2, v_{ox}, v_{oy}, and v_{oz}. Note that this is exactly the same number of coefficients in the original equation (1.9). These constants can be expressed in terms of observable physical properties of the system, such as the number density n, the average velocity $\vec{u}$ and the kinetic temperature T.

To relate the observables n, $\vec{u}$ and T, with the constant coefficients C, a_2 and $\vec{v}_0$, we proceed as follows. From the definition of the number density we must have

$$n = \int_V f\ d^3v \tag{1.14}$$

Substituting the Maxwellian distribution function (1.13), results in

$$n = C \int_V \exp[-(ma_2/2)\ (\vec{v} - \vec{v}_0)^2]\ d^3v \tag{1.15}$$

If we use the notation $A = ma_2/2$ and $\xi_i = (v_i - v_{oi})$, with i = x, y, z, (1.15) becomes

$$n = C \int\int_{-\infty}^{+\infty}\int \exp[-A\ (\xi_x^2 + \xi_y^2 + \xi_z^2)]\ d\xi_x\ d\xi_y\ d\xi_z \tag{1.16}$$

Performing the integrals over all possible values of ξ_x, ξ_y and ξ_z, yields

$$n = C\,(\pi/A)^{3/2} = C\,(2\pi/ma_2)^{3/2} \tag{1.17}$$

From the definition of the average velocity we have

$$\vec{u} = \langle \vec{v} \rangle = \frac{1}{n} \int_V f\,\vec{v}\,d^3v \tag{1.18}$$

and substituting the Maxwellian distribution function (1.13),

$$\vec{u} = (C/n) \int_V \vec{v}\,\exp\,[-(ma_2/2)(\vec{v} - \vec{v}_0)^2]\,d^3v \tag{1.19}$$

Using the same notation as in (1.16), we can write

$$u = (C/n) \int\int_{-\infty}^{+\infty}\int (\xi_x\hat{x} + \xi_y\hat{y} + \xi_z\hat{z})\,\exp\,[-A\,(\xi_x^2 + \xi_y^2 + \xi_z^2)]\,d\xi_x\,d\xi_y\,d\xi_z\,+$$

$$+\,(C/n)\,\vec{v}_0 \int\int_{-\infty}^{+\infty}\int \exp\,[-A\,(\xi_x^2 + \xi_y^2 + \xi_z^2)]\,d\xi_x\,d\xi_y\,d\xi_z \tag{1.20}$$

The first triple integral in the right-hand side of (1.20) over all possible values of ξ_x, ξ_y and ξ_z vanishes, since the integrand is an odd function of ξ_i. According to (1.16) the second triple integral is equal to n/C. Thus,

$$\vec{u} = \vec{v}_0 \tag{1.21}$$

which shows that the constant $\vec{v}_0$ represents the average (flow) velocity of the particles. Recall that the particle velocity $\vec{v}$ can be written as the sum of the peculiar (random) velocity $\vec{c}$ and the average velocity $\vec{u}$, that is $\vec{v} = \vec{c} + \vec{u}$. If the system has no translational motion as a whole, then $\vec{v}_0 = \vec{u} = 0$.

Consider now the thermodynamic definition of the kinetic temperature T,

$$3kT/2 = m\,\langle c^2 \rangle/2 = (m/2n) \int_V f\,c^2\,d^3v \tag{1.22}$$

where k is Boltzmann's constant. Substituting the Maxwellian distribution function (1.13), noting that $\vec{c} = \vec{v} - \vec{u}$ and $d^3v = d^3c$, we obtain

$$3kT/2 = (mC/2n) \int_c c^2 \exp(-Ac^2)\, d^3c \tag{1.23}$$

Performing the triple integral over all possible valves of c_x, c_y and c_z, gives

$$kT = (C/na_2)(2\pi/ma_2)^{3/2} \tag{1.24}$$

We can now solve (1.17) and (1.24) for C and a_2, which gives

$$C = n \left(\frac{m}{2\pi kT}\right)^{3/2} \tag{1.25}$$

$$a_2 = 1/kT \tag{1.26}$$

Substituting the results obtained for C, a_2 and $\vec{v}_0$, into (1.3), and taking $\vec{c} = \vec{v} - \vec{u}$, the Maxwellian distribution of random velocities becomes

$$f(c) = n \left(\frac{m}{2\pi kT}\right)^{3/2} \exp(-mc^2/2kT) \tag{1.27}$$

This is the equilibrium distribution function for a system of particles distributed in space and free form the action of external forces. Note that the number density n and the temperature T are independent of $\vec{r}$ and t. Once the number density and temperature of the gas have been established, expression (1.27) represents the only permanent mode for the distribution of the particle velocities in the gas. Whatever may be the velocity distribution function of a gas initially not in equillibrium, it tends to the distribution function (1.27) in the course of time, if the gas is maintained isolated from the action of external forces.

When the system has no translational motion as a whole (if it is maintained inside a container, for example), the average (flow) velocity $\vec{u}$ is zero and, consequently, we have $\vec{c} = \vec{v}$ in (1.27). The equilibrium distribution function depends only on the *magnitude* of the random velocity $\vec{c}$ so that, when a perfectly reflecting surface is immersed in the gas, f(c) remains unchanged since the magnitude of the random velocity does not change when the particles are reflected by the surface.

For a plasma in equilibrium, with the various species of particles such as electrons, ions and neutrals at the same temperature, the random velocities of each species are separately described by a Maxwell-Boltzmann distribution function.

1.5 Local Maxwell-Boltzmann distribution function

In many situations of interest we are dealing with a gas that, although not in equilibrium, is not very far from it. It is then a good approximation to consider that, in the neighborhood of any point in the gas, there is an equilibrium situation described by a *local* Maxwell-Boltzmann distribution function of the form

$$f(\vec{r}, v, t) = n(\vec{r}, t) \left[\frac{m}{2\pi k\, T(\vec{r}, t)}\right]^{3/2} \exp\left\{-\frac{m[\vec{v} - \vec{u}(\vec{r}, t)]^2}{2\, k\, T(\vec{r}, t)}\right\} \tag{1.28}$$

where the number density n, the temperature T and the average velocity $\vec{u}$ are slowly varying functions of $\vec{r}$ and t.

2. THE MOST PROBABLE DISTRIBUTION

We have seen that the Maxwell-Boltzmann distribution function is the solution of Boltzmann equation representing the equilibrium state of a gas, in the absence of external forces. One of the important conclusions obtained from the derivation of this distribution function is that it independs of the cross section for the particle collisions, as long as they exist. This means that the Maxwell-Boltzmann distribution function is, in a certain way, universal in the description of the equilibrium state, and it should be possible to derive it without explicitly considering the particle interactions. A derivation in these terms is in fact presented in statistical mechanics, where it is shown that the Maxwell-Boltzmann distribution function represents the *most probable* distribution function satisfying the macroscopic conditions (constraints) imposed on the system.

In statistical mechanics, to a given macroscopic system there corresponds a very large number of possible microscopic states that lead to the same macroscopic parameters specifying the system, such as number density n, average velocity $\vec{u}$ and absolute temperature T. Each microscopic state is considered to be equiprobable. If we choose, at random, any particular microscopic state for the system, amongst all the possible microscopic states consistent with the macroscopic parameters (such as n, $\vec{u}$, T), the probability of choosing a Maxwellian distribution is overwhelmingly larger than any other distribution. It is also shown that the entropy is proportional to the probability of having a given distribution. Consequently, the state having maximum entropy is the most probable state consistent with the macroscopic constraints imposed on the system.

The meaning of the Maxwell-Boltzmann distribution function can be further illustrated by the following example. If a dilute gas is prepared in an arbitrary nonequilibrium initial state, and if there are interactions between the particles so as to allow the gas to pass from the initial state to other states, as time passes the gas will certainly reach the Maxwellian state, since essentially almost all possible microscopic states, consistent with the macroscopic constraints, have a Maxwellian distribution.

The statistical mechanics derivation of the most probable distribution function provides information only about the equilibrium state, and cannot possibly tell, for example, how long (which depends on the collision cross section) a given distribution function, initially not the equilibrium one, takes to become Maxwellian. The Boltzmann equation, on the other hand, is much more general and provides information also for nonequilibrium situations.

3. MIXTURE OF VARIOUS SPECIES OF PARTICLES

For the case of a mixture containing different species of particles, each species having their own number density n_α, average velocity $\vec{u}_\alpha$, and temperature T_α, we can still perform a calculation to determine the *most probable* distribution subjected to the constraints provided by the set of macroscopic parameters n_α, $\vec{u}_\alpha$, T_α. This requires only that we set $f_\alpha f_{\alpha 1} = f'_\alpha f'_{\alpha 1}$ for each particle species, but not necessarily $f_\alpha f_{\beta 1} = f'_\alpha f'_{\beta 1}$ $(\alpha \neq \beta)$. This condition, therefore, is *not* an equilibrium situation, unless the temperatures and mean velocities of all species are equal. To determine the most probable distribution function for this nonequilibrium gas mixture (*each* especies having their *own* number density, mean velocity and temperature), we independently apply (1.3) for each species in order to maximize the entropy for each species. This also maximizes the entropy for the gas mixture with the specified macroscopic constraints. The problem is completely analogous to the one just solved for a one-component gas and leads, in identical fashion, to

$$f_\alpha(v) = n_\alpha \left(\frac{m_\alpha}{2\pi k T_\alpha}\right)^{3/2} \exp\,[-m_\alpha(\vec{v} - \vec{u}_\alpha)^2/2kT_\alpha] \tag{3.1}$$

Thus, each species have a Maxwellian distribution of velocities, but with their *own* density, average velocity and temperature. Although this is *not* an equilibrium distribution for the system, since the equilibrium condition $f'_\alpha f'_{\beta 1} = f_\alpha f_{\beta 1}$ for all α and β is not satisfied, it is, nevertheless, the most probable distribution with the specified constraints. Only if the temperatures and average velocities of all species are equal will this be an equilibrium situation. Indeed, if two systems with different species and at different temperatures are brought together, then, as time passes, there will be a transfer of energy through collisions between the two different species, untill equilibrium is reached with both species at the same temperature.

4. PROPERTIES OF THE MAXWELL–BOLTZMANN DISTRIBUTION FUNCTION

Due to the importance of the equilibrium distribution function, we present in this section some of its basic properties. We consider a gas in thermal equilibrium having no average (flow) velocity, $\vec{u} = 0$. If, however, this average velocity is not zero, we suppose that the observer is moving with the average velocity $\vec{u}$ of the gas; thus, in either case, $\vec{v} = \vec{c}$. According to the definition of the distribution function, the number of particles per unit volume having velocities between $\vec{v}$ and $\vec{v} + d\vec{v}$ is given by

$$f(v)\, d^3v = n \left(\frac{m}{2\pi kT}\right)^{3/2} \exp\,(-mv^2/2kT)\, d^3v \tag{4.1}$$

4.1 Distribution of a velocity component

The distribution function for one component of the velocity, $g(v_i)$, is defined such that $g(v_i)\, dv_i$ represents the number of particles per unit volume, with the i component of the velocity between v_i and $v_i + dv_i$, irrespective of the values of the other two velocity components.

For the x component, for example, $g(v_x)\, dv_x$ is obtained by integrating $f(v)$ over all possible values of the velocity components v_y and v_z,

$$g(v_x)\, dv_x = \int_{v_y}\int_{v_z} f(v)\, dv_x\, dv_y\, dv_z \tag{4.2}$$

Substituting the Maxwell-Boltzmann distribution function,

$$g(v_x)\, dv_x = n\left(\frac{m}{2\pi kT}\right)^{3/2} \exp(-mv_x^2/2kT)\, dv_x \int_{-\infty}^{+\infty} \exp(-mv_y^2/2kT)\, dv_y \;\cdot$$

$$\cdot \int_{-\infty}^{+\infty} \exp(-mv_z^2/2kT)\, dv_z \tag{4.3}$$

Each integral in (4.3) is equal to $(2\pi kT/m)^{1/2}$. Therefore,

$$g(v_x)\, dv_x = n\left(\frac{m}{2\pi kT}\right)^{1/2} \exp(-mv_x^2/2kT)\, dv_x \tag{4.4}$$

Obviously, this expression applies to any of the velocity components. It shows that each of the components, v_x, v_y, v_z, has a *Gaussian* distribution, symmetric about the average value $\langle v_i \rangle = 0$, $i = x, y, z$. The distribution function $g(v_x)$, given by (4.4), is plotted in Fig. 3. Note that it is properly normalized so that

$$\int_{-\infty}^{+\infty} g(v_x)\, dv_x = n \tag{4.5}$$

The fact that the average value $\langle v_i \rangle$ is zero, is physically evident by symmetry, since each component of the velocity can be equally positive or negative. Mathematically, we have

$$< v_i > = (1/n) \int_{-\infty}^{+\infty} g(v_i)\, v_i\, dv_i$$

$$= \left(\frac{m}{2\pi kT}\right)^{1/2} \int_{-\infty}^{+\infty} \exp\,(-mv_i^2/2kT)\, v_i\, dv_i = 0 \tag{4.6}$$

since the integrand is an odd function of v_i. Consequently, if ℓ represents any odd integer number,

$$< v_i^{\ell} > = 0 \; ; \qquad \ell = 1, 3, 5, \ldots \tag{4.7}$$

On the other hand, $< v_i^2 >$ is intrinsically positive and represents the *dispersion*, or *variance* of v_i,

$$< v_i^2 > = \frac{1}{n} \int_{-\infty}^{+\infty} g(v_i)\, v_i^2\, dv_i = kT/m \tag{4.8}$$

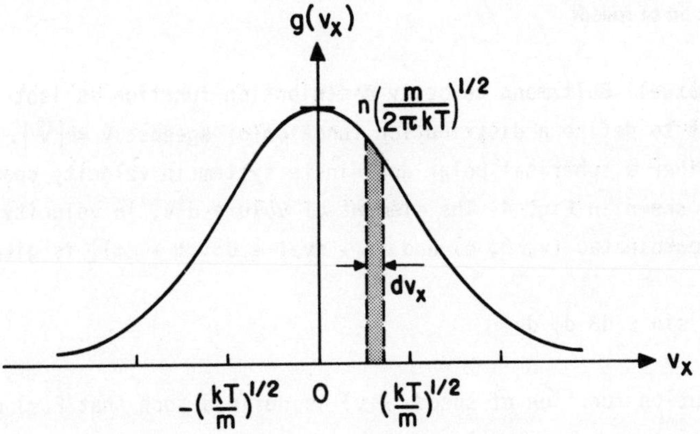

Fig. 3 - The Maxwellian distribution function for a component of the velocity is a Gaussian distribution having zero expectation ($< v_x > = 0$) and root-mean-square width $< v_x^2 >^{1/2} = (kT/m)^{1/2}$.

This result is consistent with the theorem of equipartition of energy, according to which

$$m \langle v_i^2 \rangle /2 = kT/2 \tag{4.9}$$

for i = x, y, z. The *root-mean-square width* of the Gaussian distribution $g(v_i)$ is therefore given by

$$\langle v_i^2 \rangle^{1/2} = (kT/m)^{1/2} \tag{4.10}$$

showing that the higher the temperature, the larger will be the width of the distribution function $g(v_i)$.

The velocity components behave, individually, like statistically independent quantities. Since $v^2 = v_x^2 + v_y^2 + v_z^2$, the probability that the particle lies between $\vec{v}$ and $\vec{v} + d\vec{v}$ is equal to the product of the probabilities that the velocity components lie between v_i and $v_i + dv_i$, for i = x, y, z, that is

$$\frac{f(v)\ d^3v}{n} = \frac{g(v_x)\ dv_x}{n} \cdot \frac{g(v_y)\ dv_y}{n} \cdot \frac{g(v_z)\ dv_z}{n} \tag{4.11}$$

4.2 Distribution of speeds

Since the Maxwell-Boltzmann velocity distribution function is isotropic, it is of interest to define a distribution function of speeds $v = |\vec{v}|$. For this purpose, consider a spherical polar coordinate system in velocity space (v, θ, φ), as shown in Fig. 4. The element of volume d^3v, in velocity space, between the coordinates (v, θ, φ) and (v + dv, θ + dθ, φ + dφ), is given by

$$d^3v = v^2 \sin\theta\ d\theta\ d\phi\ dv \tag{4.12}$$

The distribution function of speeds F(v) is defined such that F(v) dv is the number of particles per unit volume having speed between v and v + dv, irrespective of the direction in space of the velocity vector $\vec{v}$. Hence, to determine F(v) we integrate f(v) over all velocities whose *magnitude* lies between v and v + dv, irrespective of θ and φ, that is, whose velocity vector ends in a spherical shell in velocity space of internal radius v and external radius v + dv, as shown in Fig. 5. Therefore,

$$F(v)\,dv = \int_\theta \int_\phi f(v)\, v^2 \sin\theta \, d\theta \, d\phi \, dv \tag{4.13}$$

Since f(v) depends only on the magnitude of $\vec{v}$, but not on its direction,

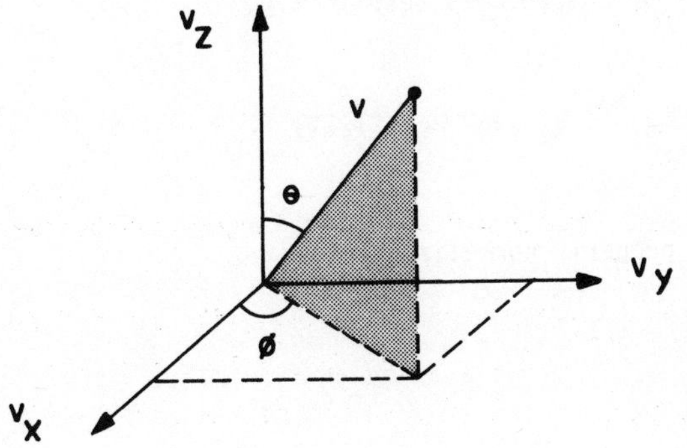

Fig. 4 Spherical coordinate system (v, θ, ϕ) in velocity space.

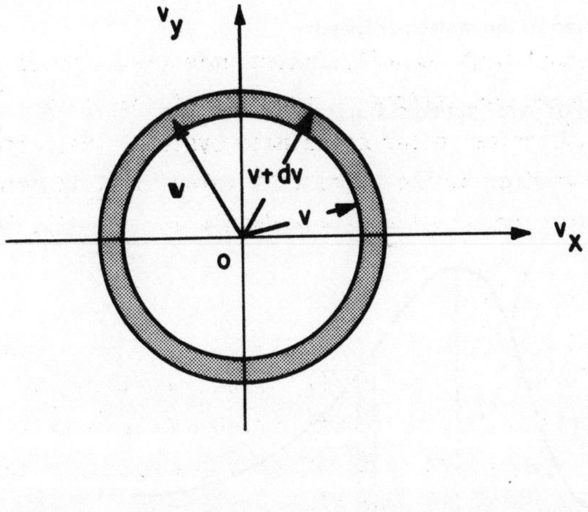

Fig. 5 - Schematic representation, in two dimensions, of a spherical shell in velocity space containing all particles with velocity having magnitude between v and v + dv.

$$F(v)\, dv = f(v)\, v^2\, dv \int_0^{\pi} \sin\theta\, d\theta \int_0^{2\pi} d\phi = 4\,\pi\, v^2\, f(v)\, dv \tag{4.14}$$

Note that $4\pi v^2\, dv$ is the volume of the spherical shell in velocity space shown in Fig. 5. Substituting the Maxwell-Boltzmann distribution function for f(v), the distribution of speeds becomes, explicitly,

$$F(v) = 4\pi n \left(\frac{m}{2\pi kT}\right)^{3/2} v^2 \exp\,(-mv^2/2kT) \tag{4.15}$$

This expression is properly normalized,

$$\int_0^{\infty} F(v)\, dv = n \tag{4.16}$$

From the expression for F(v) we see that as v increases, the exponential factor decreases faster than v^2 increases, resulting in a maximum in F(v) for a given value of v which is called the *most probable speed*. The curve for F(v) is shown in Fig. 6.

4.3 Mean values related to the molecular speeds

The *average value* of the speed is given by

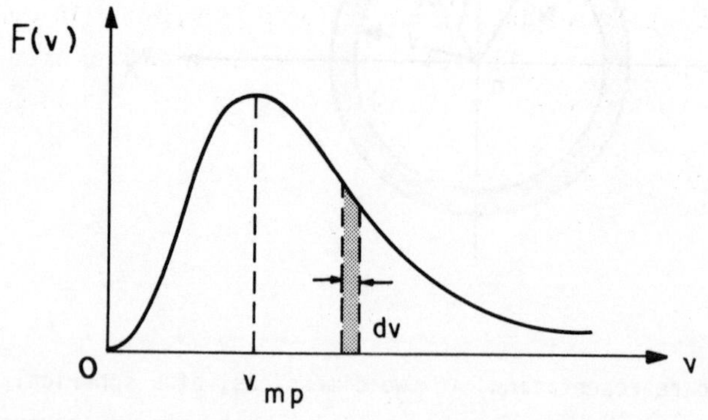

Fig. 6 - Maxwellian distribution of speeds, showing the most probable speed v_{mp}.

$$\langle v \rangle = \frac{1}{n} \int_v f\, v\, d^3v = \frac{1}{n} \int\int_{-\infty}^{+\infty}\int f\, v\, dv_x\, dv_y\, dv_z \tag{4.17}$$

or, equivalently, by

$$\langle v \rangle = \frac{1}{n} \int_0^{\infty} F(v)\, v\, dv \tag{4.18}$$

It is intrinsically a positive quantity, since $v = |\vec{v}|$ is always positive. Using expression (4.15) for F(v), we get

$$\langle v \rangle = 4\pi \left(\frac{m}{2\pi kT}\right)^{3/2} \int_0^{\infty} v^3 \exp\left(-mv^2/2kT\right) dv = 4\pi \left(\frac{m}{2\pi kT}\right)^{3/2} \frac{1}{2}\left(\frac{2kT}{m}\right)^2 \tag{4.19}$$

Therefore,

$$\langle v \rangle = (8/\pi)^{1/2}\, (kT/m)^{1/2} \tag{4.20}$$

Integrals of the type

$$I(j) = \int_0^{\infty} x^j \exp(-\alpha x^2)\, dx \tag{4.21}$$

where j represents a positive integer number, occur frequently in the computation of average values using the Maxwellian distribution of speeds. For this reason, we present here for reference the results for some integrals of the type (4.21),

$$I(0) = (\sqrt{\pi}/2)\, \alpha^{-1/2} \qquad I(1) = (1/2)\, \alpha^{-1}$$

$$I(2) = (\sqrt{\pi}/4)\, \alpha^{-3/2} \qquad I(3) = (1/2)\, \alpha^{-2}$$

$$I(4) = (3\sqrt{\pi}/8)\, \alpha^{-5/2} \qquad I(5) = \alpha^{-3} \tag{4.22}$$

The average of the square of the speed is given by

$$\langle v^2 \rangle = \frac{1}{n} \int \int_{-\infty}^{+\infty} \int f\, v^2\, dv_x\, dv_y\, dv_z = \frac{4\pi}{n} \int_0^{\infty} f(v)\, v^4\, dv \tag{4.23}$$

Substituting the Maxwellian distribution function for f(v),

$$\langle v^2 \rangle = 4\pi \left(\frac{m}{2\pi kT}\right)^{3/2} \int_0^{\infty} v^4 \exp(-mv^2/2kT)\, dv \tag{4.24}$$

which gives

$$\langle v^2 \rangle = 3kT/m \tag{4.25}$$

This result can also be obtained from (4.8), noting that $v^2 = v_x^2 + v_y^2 + v_z^2$ and that $\langle v_x^2 \rangle = \langle v_y^2 \rangle = \langle v_z^2 \rangle$. The *root-mean-square speed* is given by

$$v_{rms} = \langle v^2 \rangle^{1/2} = (3kT/m)^{1/2} \tag{4.26}$$

The *most probable speed* v_{mp} corresponds to the velocity for which F(v) is maximum, and can be obtained by the condition

$$\left.\frac{dF(v)}{dv}\right|_{v_{mp}} = 0 \tag{4.27}$$

Differentiating (4.15) with respect to v, yields

$$\frac{dF(v)}{dv} = 2v \exp(-mv^2/2kT) + v^2\left(-\frac{mv}{kT}\right) \exp(-mv^2/2kT) \tag{4.28}$$

which, for the condition of maximum expressed in (4.27), gives

$$v_{mp} = (2kT/m)^{1/2} \tag{4.29}$$

Note that the mean speeds $<v>$, v_{rms} and v_{mp} are all proportional to $(kT/m)^{1/2}$ and are such that $v_{mp} < (<v>) < v_{rms}$. Therefore, they increase with the temperature and, for a given temperature, particles having a larger mass will move with a smaller speed. We have also seen that the average kinetic energy of the random particle motions satisfies the relation

$$m <v^2>/2 = 3kT/2 \tag{4.30}$$

4.4 Distribution of thermal kinetic energy

The distribution of thermal kinetic energy $G(E)$, where $E = mv^2/2$, is defined such that $G(E)\, dE$ is the number of particles per unit volume having random kinetic energy between E and $E + dE$. It can be obtained from (4.15) substituting v by $(2E/m)^{1/2}$ and dv by $dE/(2mE)^{1/2}$. Hence,

$$G(E)\, dE = 4\pi\, n \left(\frac{m}{2\pi kT}\right)^{3/2} \left(\frac{2E}{m}\right) \exp\,(-E/kT)\, \frac{dE}{(2mE)^{1/2}} \tag{4.31}$$

Simplifying this expression, gives

$$G(E)\, dE = (2n/\sqrt{\pi})\, [E/(kT)^3]^{1/2} \exp\,(-E/kT)\, dE \tag{4.32}$$

The function $G(E)$ is displayed in Fig. 7.

4.5 Random particle flux

We have seen in Chapter 6 that the particle flux, in a given direction $\hat{n}$, is given by

$$\Gamma_n = n <v_n> = \int_v f\, \vec{v} \cdot \hat{n}\, d^3v \tag{4.33}$$

Let us consider a surface element inside the gas. We are interested in determining the number of particles that reach this surface element per unit area and time due to the random particle motions. Eq. (4.33) takes into account particles that reach the surface element, oriented along the direction specified by the unit vector $\hat{n}$, coming from all possible directions. Since we are assuming that the average velocity $\vec{u}$ is zero, the flux given by (4.33) is obviously zero, since $< c > = 0$. In this case, it is of interest to consider only the flux of particles that cross the surface element form the *same side* (such that $\vec{v} \cdot \hat{n}$ is positive, say), due to their random motions.

Let $d\vec{S}$ be a surface element situated at the origin of a Cartesian coordinate system (x,y,z), and oriented along the z azis, that is, $d\vec{S} = \hat{z}dS$, as shown in Fig. 8. Consider the particles that cross $d\vec{S}$ coming from the region $z < 0$, having velocities between $\vec{v}$ and $\vec{v} + d\vec{v}$, making an angle θ with the z axis so that $\vec{v} \cdot \hat{z} = v \cos \theta$. Expressing d^3v in terms of spherical coordinates (v,θ,ϕ)

$$d^3v = v^2 \sin \theta \, d\theta \, d\phi \, dv \tag{4.34}$$

the random particle current density, crossing $d\vec{S}$ from the region $z < 0$, is given by

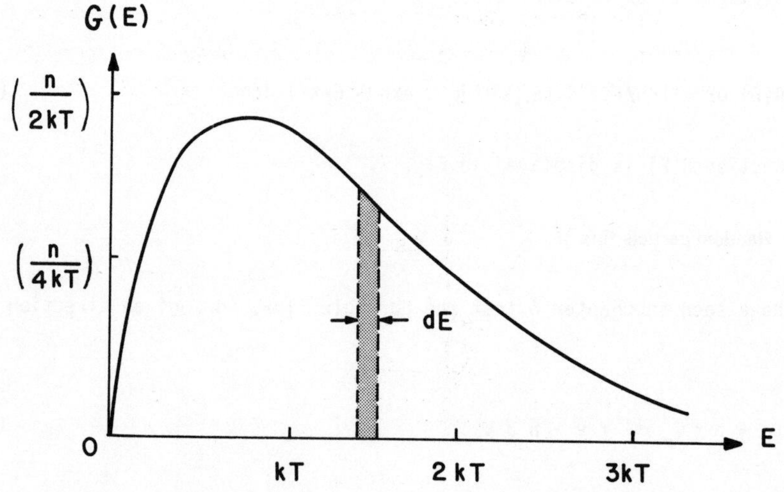

Fig. 7 - Maxwellian distribution of thermal kinetic energies. The shadowed area represents the number of particles having random kinetic energy between E and E + dE.

$$\Gamma_z = \int_0^\infty f\, v^3\, dv \int_0^{\pi/2} \sin\theta \cos\theta\, d\theta \int_0^{2\pi} d\phi = \pi \int_0^\infty f\, v^3\, dv \tag{4.35}$$

Substituting the Maxwellian distribution for f, we find

$$\Gamma_z = \pi\, n \left(\frac{m}{2\pi kT}\right)^{3/2} \int_0^\infty \exp(-mv^2/2kT)\, v^3\, dv \tag{4.36}$$

and solving the integral, we obtain

$$\Gamma = n\, (kT/2\pi m)^{1/2} = n \langle v \rangle /4 \tag{4.37}$$

In this result we have eliminated the index z from Γ, since the Maxwellian distribution function is isotropic; (4.37) applies to any direction inside the gas.

It is important to note that the random particle flux is inversely proportional to the square root of the particle mass. In a plasma the current density for the electrons is therefore much larger than that for the ions (the ratio of the electron mass to the proton mass, for example, is 1/1836). This difference in the particle flux between electrons and ions plays a very

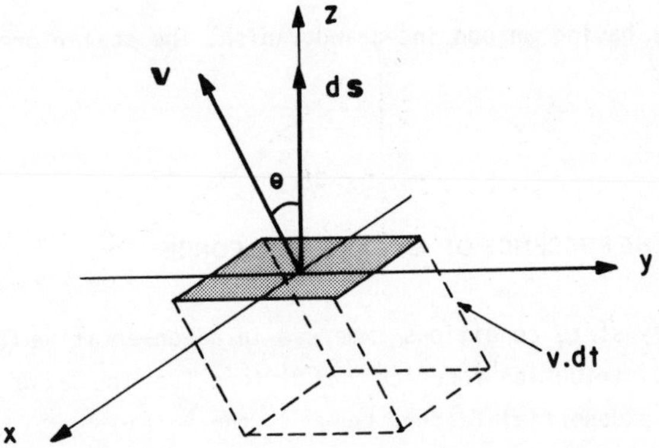

Fig. 8 - Prism of base $d\vec{S} = \hat{z}dS$ containing the particles having velocity between $\vec{v}$ and $\vec{v} + d\vec{v}$ that cross dS during the time interval dt.

important role in the interaction between a plasma and a material body immersed in it (see Chapter 11).

4.6 Kinetic pressure and heat flux

From the definitions of the kinetic pressure dyad

$$\bar{\bar{p}} = \rho_m < \vec{c}\,\vec{c} > = m \int_V \vec{c}\,\vec{c}\, f\, d^3v \tag{4.38}$$

and the heat flux vector

$$\vec{q} = (\rho_m/2) < c^2\vec{c} > = (m/2) \int_V c^2\vec{c}\, f\, d^3v \tag{4.39}$$

we obtain, using the Maxwellian distribution function,

$$\bar{\bar{p}} = \rho_m (< c_x^2 > \hat{x}\,\hat{x} + < c_y^2 > \hat{y}\,\hat{y} + < c_z^2 > \hat{z}\,\hat{z}) = nkT\,(\hat{x}\,\hat{x} + \hat{y}\,\hat{y} + \hat{z}\,\hat{z}) \tag{4.40}$$

and

$$\vec{q} = 0 \tag{4.41}$$

since the integrals having an odd integrand vanish. The scalar pressure is therefore,

$$p = nkT \tag{4.42}$$

5. EQUILIBRIUM IN THE PRESENCE OF AN EXTERNAL FORCE

A gas under steady state conditions, immersed in a conservative force field, is characterized by a distribution function that differs from the Maxwell-Boltzmann distribution by an exponential factor, known as the *Boltzmann factor*. The conservative force field can be specified in terms of a potential energy $U(\vec{r})$

$$\vec{F}(\vec{r}) = - \vec{\nabla}U(\vec{r}) \tag{5.1}$$

Since the conservative force field is a function only of the position vector $\vec{r}$, we expect the steady state solution of the Boltzmann equation for this case to be of the form

$$f(\vec{r},v) = f_0(v)\ \psi(\vec{r}) \tag{5.2}$$

where $f_0(v)$ denotes the Maxwell-Boltzmann equilibrium distribution function and $\psi(\vec{r})$ is a scallar function of $\vec{r}$ only, still to be determined. The function $\psi(\vec{r})$ can be determined by requiring (5.2) to satisfy the Boltzmann equation under equilibrium conditions in the presence of the conservative field,

$$\vec{v}\ .\ \vec{\nabla}\ [f_0(v)\ \psi(\vec{r})] - [\vec{\nabla} U(\vec{r})/m)]\ .\ \vec{\nabla}_v\ [f_0(v)\ \psi(\vec{r})] = 0 \tag{5.3}$$

From the expression for $f_0(v)$ it can be easily verified that

$$\vec{\nabla}_v\ f_0(v) = -\ (m\vec{v}/kT)\ f_0(v) \tag{5.4}$$

Therefore, (5.3) simplifies to

$$f_0(v)\ \vec{v}\ .\ [\vec{\nabla}\ \psi(\vec{r}) + (1/kT)\ \psi(\vec{r})\ \vec{\nabla} U(\vec{r})] = 0 \tag{5.5}$$

from which we can write

$$\vec{\nabla}\ \psi(\vec{r})/\psi(\vec{r}) = -\ (1/kT)\ \vec{\nabla} U(\vec{r}) \tag{5.6}$$

Since $d\psi = \vec{\nabla}\psi\ .\ d\vec{r}$, (5.6) may also be written as

$$d\ \psi(\vec{r})/\psi(\vec{r}) = -\ (1/kT)\ d\,U(\vec{r}) \tag{5.7}$$

The solution of this differential equation is

$$\psi(\vec{r}) = A_0\ \exp[-\,U(\vec{r})/kT] \tag{5.8}$$

where A_0 is a constant that can be determined by requiring that

$$\int_v f(\vec{r},\ v)\ d^3v = n(\vec{r}) \tag{5.9}$$

from which we get

$$n(\vec{r}) = A_0 \exp[-U(\vec{r})/kT] \int_V f_0(v)\, d^3v \tag{5.10}$$

Denoting by n_0 the number density in a region where $U(\vec{r}) = 0$, under equilibrium condition, that is,

$$n_0 = \int_V f_0(v)\, d^3v \tag{5.11}$$

we must choose $A_0 = 1$, so that the equilibrium distribution function for this case is (with $\vec{u} = 0$)

$$f(\vec{r}, v) = f_0(v) \exp[-U(\vec{r})/kT]$$

$$= n_0 \left(\frac{m}{2\pi kT}\right)^{3/2} \exp\{-(1/kT)\,[mv^2/2 + U(\vec{r})]\} \tag{5.12}$$

The number density for a system described by this distribution function is given by

$$n(\vec{r}) = n_0 \exp[-U(\vec{r})/kT] \tag{5.13}$$

The factor $\exp[-U(\vec{r})/kT]$, responsible for the inhomogeneity of $f(\vec{r},v)$, in (5.12), is known as the *Boltzmann factor*.

An important example is provided by a plasma in the presence of a conservative force due to an an electrostatic field

$$\vec{E} = -\vec{\nabla}\phi(\vec{r}) \tag{5.14}$$

where $\phi(\vec{r})$ is the electrostatic scalar potential. The potential energy, in this case, is

$$U(\vec{r}) = q\,\phi(\vec{r}) \tag{5.15}$$

The number density for particles of charge q in equilibrium under the action of an electrostatic field is therefore

$$n(\vec{r}) = n_0 \exp[-q\phi(\vec{r})/kT] \tag{5.16}$$

This expression will be used in Chapter 11 in the analysis of electrostatic shielding in a plasma.

6. DEGREE OF IONIZATION IN EQUILIBRIUM – THE SAHA EQUATION

From the methods of statistical mechanics we can determine the degree of ionization in a gas in thermal equilibrium at some temperature T, without considering the details of the process of ionization. In order to ionize an atom or molecule, it is necessary to provide a certain amount of energy. This ionization energy is conveniently expressed in *electron volts*, and is normally called the *ionization potential*. Values for the first ionization potential of some atoms are given in Table 1. Note that to provide a mean thermal energy kT of 1 eV requires a temperature of 11 600 K. Hence, it is apparent that only at very high temperatures the mean kinetic energy 3kT/2 of a particle exceeds the ionization energy. However, we will show that a considerable degree of ionization can be achieved even when the mean thermal energy of the particles is far below the ionization energy, since some of the particles, the ones with the largest velocities (in the tail of the Maxwellian distribution function), have enough energy to produce ionization by collisions. The equilibrium degree of ionization is then determined by a balance between the rate of ionization by collisions and the rate of recombination.

To calculate the relative numbers of ionized and neutral atoms in a plasma, at a specified temperature, it is necessary to use a particle distribution function similar to that given in (5.13). However, the physical situation is somewhat different because of the necessary quantum mechanical aspects of the problem. Denoting by n_a and n_b the number density of the particles having energies U_a and U_b, respectively, the ratio n_a/n_b is given, from statistical mechanics, by

$$n_a/n_b = (g_a/g_b) \exp[-(U_a - U_b)/kT] \tag{6.1}$$

where g_a and g_b are the statistical weights associated with the energies U_a and U_b, that is, the degeneracy factors giving the number of states having the energies U_a and U_b, respectively. For the case of a system having only two energy levels, U_a and U_b, the fraction (α) of all the particles that are in the higher energy state U_a is given by

TABLE 1

IONIZATION POTENTIAL ENERGY OF SOME ATOMS

Element		Ionization energy U for first electron (eV)
Helium	(He)	24.59
Argon	(A)	15.76
Nitrogen	(N)	14.53
Oxygen	(O)	13.62
Hydrogen	(H)	13.60
Mercury	(Hg)	10.44
Iron	(Fe)	7.87
Sodium	(Na)	5.14
Potassium	(K)	4.34
Cesium	(Cs)	3.89

$$\alpha = n_a/(n_a + n_b) = (n_a/n_b)/(n_a/n_b + 1) \tag{6.2}$$

or using (6.1), with $U = U_a - U_b$,

$$\alpha = \frac{(g_a/g_b)\exp(-U/kT)}{(g_a/g_b)\exp(-U/kT) + 1} \tag{6.3}$$

For the ionization problem, state a is taken as that of the ion-electron pair, state b is that of the neutral atom and $U = U_a - U_b$ is the ionization energy. The temperature $T_{1/2}$ for which $\alpha = 0.5$, that is, for which fifty per cent of all the atoms are in the ionized state ($n_a = n_b$), can be determined by taking

$$(g_a/g_b)\exp(-U/kT_{1/2}) = 1 \tag{6.4}$$

which gives

$$T_{1/2} = U/[k \ln(g_a/g_b)] \tag{6.5}$$

Fig. 9 shows the curve of α as a function of T, according to (6.3).

The fraction of the particles in the ionized state changes from nearly zero to nearly one over a small temperature range. An *estimate* of this temperature range can be obtained from the temperature difference ΔT that would exist between $\alpha = 0$ and $\alpha = 1$, if the curve of $\alpha(T)$ were a straight line with the slope of the true $\alpha(T)$ curve at $T_{1/2}$. Hence, we take

$$\Delta T^{-1} = \left.\frac{d\alpha(T)}{dT}\right|_{T_{1/2}} \tag{6.6}$$

From (6.3) we obtain, assuming $d(g_a/g_b)/dT = 0$,

$$\left.\frac{d\alpha(T)}{dT}\right|_{T_{1/2}} = \left.\frac{U\,\alpha^2}{T^2(g_a/g_b)\exp(-U/kT)}\right|_{T_{1/2}} = \frac{U}{4\,T_{1/2}^2} \tag{6.7}$$

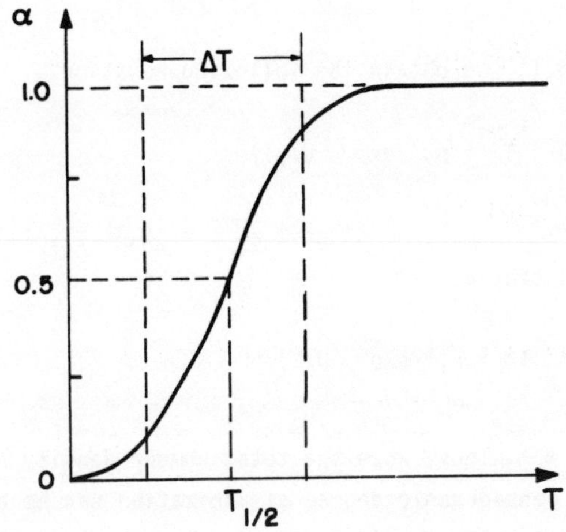

Fig. 9 - The function $\alpha(T)$, which gives the fraction of particles in the ionized state as a function of temperature.

so that

$$\Delta T = 4T_{1/2} / [k \ln(g_a/g_b)] = 4U/[k \ln(g_a/g_b)]^2 \tag{6.8}$$

From this result we can see that the larger is g_a/g_b the smaller is ΔT.

Since the ionized state is much more degenerate than the neutral state ($g_a \gg g_b$), the curve of $\alpha(T)$ presents a very steep inclination near $T_{1/2}$, with most of the transitions from the neutral state to the ionized state occurring near $T_{1/2}$, given by (6.5). Thus, for $g_a \gg g_b$ the curve of $\alpha(T)$ will look approximately like a step function with the ionization occurring near $T_{1/2}$.

The degeneracy factors g_a and g_b can be obtained from a quantum mechanical calculation. If we neglect the small interaction potential between the ion and the free electron, and also the internal degrees of freedom of all the particles, it turns out that

$$g_a/g_b = (2\pi m_e kT/h^2)^{3/2} \; n_i^{-1} \tag{6.9}$$

where h is Planck's constant and n_i the ion number density. For T in degrees Kelvin and n_i in m^{-3},

$$g_a/g_b = 2.405 \times 10^{21} \; T^{3/2} \; n_i^{-1} \tag{6.10}$$

Using this result in (6.1), we obtain the following equation

$$n_i/n_n = 2.405 \times 10^{21} \; T^{3/2} \; n_i^{-1} \exp(-U/kT) \tag{6.11}$$

which is known as the *Saha equation*. Since 1 eV = kT for T = 11 600 K, we can also write the Saha equation as

$$n_i/n_n = 3.00 \times 10^{27} \; T^{3/2} \; n_i^{-1} \exp(-U/T) \tag{6.12}$$

with T in eV and n_i in m^{-3}. Thus, when the total number density $n_t = n_i + n_n$ is sufficiently low, a considerable degree of ionization can be achieved for temperatures that are well below the ionization energy. This is illustrated in Fig. 10, which shows the degree of ionization of hydrogen as a function of temperature, for values of the total number density of 10^{16}, 10^{19}, 10^{22} and 10^{25} m^{-3}. It is clear that, as the number density decreases, the values of

ΔT and $T_{1/2}$ decrease significantly, and a significant degree of ionization can be obtained at temperatures far below the ionization energy of atomic hydrogen (13.60 eV). In a gas like cesium vapor, whose ionization energy is only 3.89 eV, a high degree of ionization can be obtained even at relatively low temperatures of the order of 1000 K.

PROBLEMS

7.1 - A *two-dimensional* gas consisting of only one species and whose particles are restricted to move in a plane (the z = 0 plane), is characterized by a homogeneous, isotropic, two-dimensional Maxwell-Boltzmann distribution (with $\vec{u} = 0$),

$$f(v) = n_0 \left(\frac{m}{2\pi kT}\right) \exp[-m(v_x^2 + v_y^2)/2kT]$$

where n_0 represents the number of particles per unit area.

a) Show that the most probable speed of the particles is $v_{mp} = (kT/m)^{1/2}$.

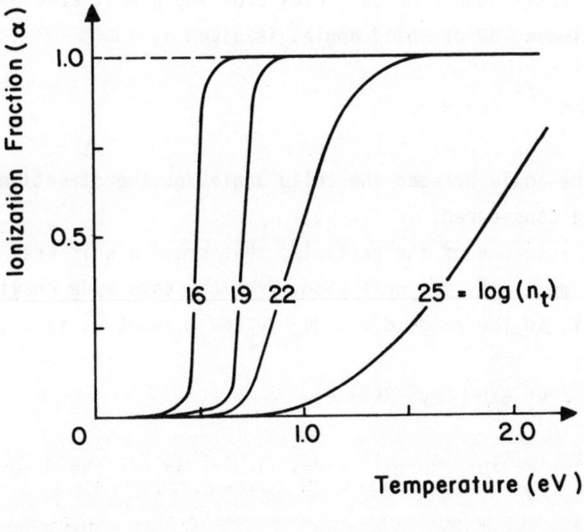

Fig. 10 - Degree of ionization $\alpha = n_i/(n_i + n_n)$ as a function of temperature for atomic hydrogen (U = 13.60 eV). The various curves refer to different number densities $n_t = n_i + n_n$ in m^{-3}.

b) Show that the fraction of the number of particles per unit area, which have speeds greater than the most probable speed is given by $(1/e)^{1/2}$, where e is the base of natural logarithms.

c) Show that the number of particles crossing a unit length per unit time (flux), from one side only, is given by

$$\Gamma = n_0 \langle v \rangle / \pi = n_0 (kT/2\pi m)^{1/2}$$

d) Show that the kinetic pressure dyad is given by

$$\bar{\bar{p}} = n_0 kT (\hat{x}\hat{x} + \hat{y}\hat{y})$$

7.2 - Consider a gas of particles consisting of only one species and characterized by the Maxwell-Boltzmann equilibrium distribution function (with $\vec{u} = 0$)

$$f(v) = n_0 \left(\frac{m}{2\pi kT}\right)^{3/2} \exp(-mv^2/2kT)$$

(a) Show that the total number of particles crossing a unit area per unit time, lying within an element $d\Omega$ of solid angle, is given by

$$(n_0/\pi)\ (kT/2\pi m)^{1/2} \cos\theta\ d\Omega$$

where θ denotes the angle between the solid angle and the direction of the normal to the area considered.

(b) Show that the *fraction* of the particles that cross a unit area perpendicular to the x axis per unit time, from the same side, having the velocity components in the range $d^3v = dv_x\, dv_y\, dv_z$, about $\vec{v}$, is given by

$$(1/2\pi)\ (m/kT)^2\ v_x \exp(-mv^2/2kT)\ d^3v$$

(c) Deduce the value of the thermal energy flux triad for the Maxwellian gas.

7.3 - The distribution of thermal kinetic energies E, for a gas in the Maxwellian state, is given by

$$G(E) = 2n[E/\pi(kT)^3]^{1/2} \exp(-E/kT)$$

Calculate the most probable energy and show that the velocity of the particles, which have this energy, is equal to $(kT/m)^{1/2}$.

7.4 - The entropy of a system can be expressed in terms of the distribution function as

$$S = -k \int_r \int_v f \ln f \, d^3v \, d^3r$$

Prove that, for a Maxwellian distribution function, the entropy satisfies the following thermodynamic relations

$$(\partial S/\partial E)_{V,N} = 1/T$$

$$(\partial S/\partial V)_{E,N} = p/T$$

where N is the total number of particles in the system, V is the total volume, and $E = 3\,NkT/2$ is the total energy.

7.5 - Derive an expression for the Doppler intensity profile (thermal broadening) of a spectral line emitted near the central frequency ν_o, assuming that the emitting atoms have a Maxwellian velocity distribution. Ignore all other factors that contribute to the shape of the line. *Hints:* (1) The change in frequency due to the Doppler effect associated with the relative (nonrelativistic) motion of the emitting atom, with respect to the direction of observation (x direction, e.g.), is given by

$$\nu - \nu_o = -\nu_o \, (v_x/c)$$

where c denotes the velocity of light. (2) The observed intensity in the frequency range between ν and $\nu + d\nu$, that is, $I(\nu)\,d\nu$, is proportional to the number of emitting atoms per unit volume, which have velocities along the direction of observation (x direction) between v_x and $v_x + dv_x$.

7.6 - Consider a gas mixture containing n_e electrons and n_i oxygen ions per unit volume, all in thermal equilibrium at temperature T and having no drift velocity.

(a) Resolve the motion of the particle species into the motion in space of the *center of mass*, plus the *relative motion* of one species with respect to the other, but with the reduced mass. Calculate the Jacobian J of this velocity transformation and show that $|J| = 1$.
(b) Show that the velocities of the center of mass have a Maxwellian distribution, and that the *relative* velocities also have a Maxwellian distribution, but with the reduced mass.
(c) What must be magnitude of T such that 20% of the electrons have a *relative* kinetic energy greater than 2 eV? The following integral will be useful:

$$\int_{x_0}^{\infty} x^2 \exp(-\alpha x^2)dx = \frac{x_0 \exp(-\alpha x_0^2)}{2\alpha} + \frac{\sqrt{\pi}}{4} \alpha^{-3/2} \operatorname{erfc}(\alpha^{1/2} x_0)$$

where erfc $(\alpha^{1/2} x_0)$ denotes the *complementary error function*.

7.7 - A gas of O_2 molecules is in the equilibrium state, with number density n and absolute temperature T. Calculate the average value of the reciprocal of the particle velocity, that is, $\langle 1/v \rangle$.

7.8 - A plasma is in equilibrium under the action of an external electrostatic field $\vec{E}$ and a gravitational field $\vec{g}$. Consider that the plasma as a whole is moving with constant velocity $\vec{u}_0$, with respect to the observer's frame of reference. Write down the distribution function for the species of type α for this plasma.

7.9 - Consider the particles in the Earth's atmosphere under equilibrium conditions in the presence of the Earth's gravitational field. Assume a horizontally stratified atmosphere with constant temperature T and consider a constant value g for the acceleration due to gravity. Derive an expression for the number density n(z) as a function of height z, for the particles of mass m, in terms of the number density $n(z_0)$ at a base level z_0, and of the scale height $H = kT/mg$. How is this expression for n(z) modified, when T and g vary height?

7.10 - The temperature of a plasma, in thermal equilibrium with a neutral gas, can be determined experimentally by measuring the electron density n_e with a microwave transmission experiment, and the neutrals number density in a particular excited state through the rate of transitions to a lower state. Determine the temperature of a plasma, with only one type of ions, whose electron number density is 10^{20} m^{-3}, and which is in equilibrium with a state of ionization potential 2 eV whose population is 10^{15} m^{-3}.

7.11 - Consider two large chambers which communicate with each other only through a small aperture of area A in a very thin wall, as indicated in Fig. P 7.1. The chambers contain an ideal gas at a *very low pressure*, such that the particle mean free path is much larger than the dimensions of A, and are at temperatures T_1 and T_2. Determine the ratio p_1/p_2 of pressures in the two chambers assuming that, under equilibrium conditions, the flux of particles through the aperture A from one chamber must equal that from the other. What would be the result in the case of *normal conditions of pressure?* Give a physical explanation for the two different results.

7.12 - Show that the average thermal energy per particle, for a gas in thermodynamic equilibrium, is equal to 1.292×10^{-4} eV/K.

7.13 - Use the laws of conservation of momentum and energy in a collision, to show that the Maxwell-Boltzmann distribution function

$$f(v) = n \left(\frac{m}{2\pi kT}\right)^{3/2} \exp\,[-\,m(\vec{v} - \vec{u})^2/2kT]$$

satisfies the following equation of detailed balance

$$f' f_1' = f\, f_1$$

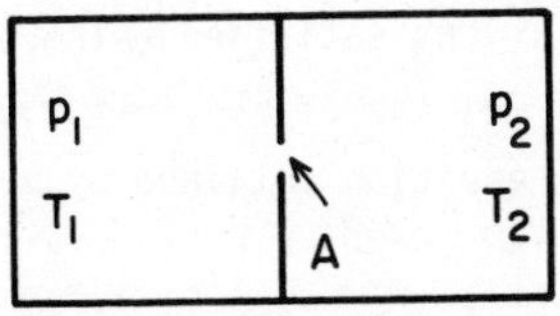

Fig. P7.1

CHAPTER 8

Macroscopic Transport Equations

1. MOMENTS OF THE BOLTZMANN EQUATION

In the previous chapters we have seen that the macroscopic variables of physical interest for a plasma, such as number density n_α, mean velocity $\vec{u}_\alpha$, temperature T_α, and so on, can be calculated if we know the distribution function for the system under consideration. For the case of a system in thermal equilibrium, we have calculated, in Chapter 7, several of these macroscopic parameters using the Maxwell-Boltzmann distribution function. In principle, the distribution function for a system not in equilibrium can be obtained by solving the Bolzmann equation. However, the solution of the Boltzmann equation is generally a matter of great difficulty.

We will see, in this chapter, that it is not necessary to solve the Boltzmann equation for the distribution function in order to determine the macroscopic variables of physical interest. The differential equations governing the temporal and spatial variation of these macroscopic variables can be derived directly from the Boltzmann equation without solving it. These differential equations are known as the *macroscopic transport equations*, and their solutions, under certain assumptions, give us directly the macroscopic variables.

The macroscopic variables are related to the *moments of the distribution function* and the transport equations satisfied by these variables can be obtained by taking the various *moments of the Boltzmann equation*. The first three moments of the Boltzmann equation obtained by multiplying it by m_α, $m_\alpha\vec{v}$ and $m_\alpha v^2/2$, respectively, and integrating over all of velocity space, give us the equation of conservation of mass, the equation of conservation of momentum and the equation of conservation of energy. However, at each stage of the hierarchy of moments of the Boltzmann equation, the resulting set of transport equations is not complete in the sense that the number of equations is not sufficient to determine all the macroscopic variables that appear in them. Each time a higher moment of the Boltzmann equation is calculated in an attempt to obtain a complete set of transport equations, a new macroscopic variable appears. It is necessary, therefore, to truncate the system of transport equations at some stage of the hierarchy and to introduce a simplifying assumption concerning the highest moment of the distribution function that appears in the system. Thus, with such simplifying approximation, we obtain a complete set of transport equations sufficient to determine all the macroscopic variables appearing in the system. Since a plasma is composed of more than one species of particles (electrons, ions, neutral particles) there is, consequently, a system of transport equations for each species. There are several different complete sets of transport equations (or hydrodynamic equations) that can be formed, depending on the assumptions considered. Amongst the possible complete systems of macroscopic equations, there are two which are widely used and which characterize the so-called *cold* and *warm plasma models*. The equations that describe these two simple models, and the corresponding approximations, will be discussed in sections 6 and 7 of this chapter.

2. GENERAL TRANSPORT EQUATION

We derive now a general partial differential equation that describes the temporal and spatial variation of the physically relevant macroscopic parameters. Let $\chi(\vec{v})$ represent some physical property of the particles in the plasma, which may be, in general, a function of the particle velocity. Since the average value of $\chi(\vec{v})$ is obtained by multiplying the distribution function by the property $\chi(\vec{v})$, integrating the product over all of velocity space and dividing the result by the particle number density, the differential equation governing the temporal and spatial variation of the average value of $\chi(\vec{v})$ can be obtained in a similar way by multiplying the Boltzmann equation by the function $\chi(\vec{v})$ and integrating the resulting equation over all of velocity space.

Consider the Boltzmann equation for the type α particles in the general form

$$\partial f_\alpha/\partial t + \vec{v}\cdot\vec{\nabla} f_\alpha + \vec{a}\cdot\vec{\nabla}_v f_\alpha = (\delta f_\alpha/\delta t)_{coll} \tag{2.1}$$

As indicated, we now multiply each term by $\chi(\vec{v})$ and integrate the resulting equation over all of velocity space to obtain

$$\int_V \chi(\partial f_\alpha/\partial t)\, d^3v + \int_V \chi\vec{v}\cdot\vec{\nabla} f_\alpha\, d^3v + \int_V \chi\vec{a}\cdot\vec{\nabla}_v f_\alpha\, d^3v = \int_V \chi(\delta f_\alpha/\delta t)_{coll} \tag{2.2}$$

We proceed next to evaluate separately each of the terms in (2.2).

The *first term* of (2.2) may be rewritten as

$$\int_V \chi(\partial f_\alpha/\partial t)\, d^3v = \left(\frac{\partial}{\partial t}\right)\int_V \chi f_\alpha\, d^3v - \int_V f_\alpha\, (\partial\chi/\partial t)\, d^3v \tag{2.3}$$

since the limits of integration do not depend upon the space and time variables and, therefore, the partial time derivative can be taken inside or outside the integral sign. The last integral in (2.3) vanishes since $\chi(\vec{v})$ does not depend upon t. Using the definition of average values, as given in Chapter 6, we obtain

$$\int_V \chi(\partial f_\alpha/\partial t)\, d^3v = \frac{\partial}{\partial t}(n_\alpha \langle\chi\rangle_\alpha) \tag{2.4}$$

Similarly, for the *second term* of (2.2) we can write

$$\int_V \chi\, \vec{v}\cdot\vec{\nabla} f_\alpha\, d^3v = \vec{\nabla}\cdot\int_V \vec{v}\,\chi\, f_\alpha\, d^3v - \int_V f_\alpha\, \vec{v}\cdot\vec{\nabla}\chi\, d^3v - \int_V f_\alpha \chi\vec{\nabla}\cdot\vec{v}\, d^3v \tag{2.5}$$

The term involving $\vec{\nabla}\cdot\vec{v}$ is zero since $\vec{r}$, $\vec{v}$ and t are independent variables, as well as the term involving $\vec{\nabla}\chi$, since $\chi(\vec{v})$ does not depend upon the space variables. Thus the second term of (2.2) becomes

$$\int_V \chi\, \vec{v}\cdot\vec{\nabla} f_\alpha\, d^3v = \vec{\nabla}\cdot(n_\alpha \langle\chi\vec{v}\rangle_\alpha) \tag{2.6}$$

For the *third term* of (2.2) we have, in a similar way,

$$\int_V \chi\,\vec{a}\cdot\vec{\nabla}_v f_\alpha\, d^3v = \int_V \vec{\nabla}_v\cdot(\vec{a}\,\chi\, f_\alpha)\, d^3v - \int_V f_\alpha\, \vec{a}\cdot\vec{\nabla}_v\chi\, d^3v - \int_V f_\alpha\,\chi\,\vec{\nabla}_v\cdot\vec{a}\, d^3v \tag{2.7}$$

The last integral in (2.7) vanishes if we assume that

$$\vec{\nabla}_v\cdot\vec{a} = (1/m_\alpha)\, \vec{\nabla}_v\cdot\vec{F} = 0 \tag{2.8}$$

that is, if the force component F_i is independent of the corresponding velocity component v_i, for i = x,y,z. Note that this restriction dos not exclude the force due to a magnetic field, $\vec{F} = q_\alpha\, \vec{v}\times\vec{B}$, since in this case F_i is still independent of v_i. For the x component, for example, we have

$$F_x = q_\alpha (v_y B_z - v_z B_y) \tag{2.9}$$

which is **independent** of v_x, and the same holds true for the other two components.

The first integral in the right-hand side of (2.7) consists of a sum of three triple integrals,

$$\int_v \vec{\nabla}_v \cdot (\vec{a} \chi f_\alpha)\, d^3v = \iiint_{-\infty}^{+\infty} \frac{\partial}{\partial v_x} (a_x \chi f_\alpha)\, dv_x\, dv_y\, dv_z + \iiint_{-\infty}^{+\infty} \frac{\partial}{\partial v_y} (a_y \chi f_\alpha)\, dv_x\, dv_y\, dv_z + \iiint_{-\infty}^{+\infty} \frac{\partial}{\partial v_z} (a_z \chi f_\alpha)\, dv_x\, dv_y\, dv_z \tag{2.10}$$

For each one of these triple integrals we have the result

$$\iiint_{-\infty}^{+\infty} \frac{\partial}{\partial v_x} (a_x \chi f_\alpha)\, dv_x\, dv_y\, dv_z = \iint_{-\infty}^{+\infty} dv_y\, dv_z (a \chi f_\alpha \Big|_{-\infty}^{+\infty}) = 0 \tag{2.11}$$

since $f_\alpha(\vec{r},\vec{v},t)$ must be zero when v becomes infinitely large, as there are no particles with infinite velocity. Consequently, the first integral in the right-hand side of (2.7) vanishes. Therefore,

$$\int_v \chi\, \vec{a}\cdot\vec{\nabla}_v f_\alpha\, d^3v = - n_\alpha \langle \vec{a}\cdot\vec{\nabla}_v \chi \rangle_\alpha \tag{2.12}$$

Combining the results contained in Eqs. (2.4), (2.6) and (2.12) we obtain the *general transport equation*,

$$\frac{\partial}{\partial t} (n_\alpha \langle\chi\rangle_\alpha) + \vec{\nabla}\cdot(n_\alpha \langle\chi\vec{v}\rangle_\alpha) - n_\alpha \langle\vec{a}\cdot\vec{\nabla}_v \chi\rangle_\alpha = \left[\frac{\delta}{\delta t} (n_\alpha \langle\chi\rangle_\alpha)\right]_{coll} \tag{2.13}$$

where the term in the right denotes the time rate of change of the quantity χ per unit volume, for the particles of type α, due to collisions,

$$\left[\frac{\delta}{\delta t} (n_\alpha \langle\chi\rangle_\alpha)\right]_{coll} \equiv \int_v \chi (\delta f_\alpha/\delta t)_{coll}\, d^3v \tag{2.14}$$

The equations to be derived in the subsequent sections of this chapter are very general and are not specifically dependent on any particular form of the collision term $(\delta f_\alpha/\delta t)_{coll}$. A derivation of the general transport equation for the case when the property χ is a function of $\vec{r}$, $\vec{v}$ and t is included in Problem 8.6.

3. CONSERVATION OF MASS

3.1 Derivation of the continuity equation

The transport equation (2.13) is a general expression and applies to any arbitrary function $\chi(\vec{v})$. The equation of continuity, or of conservation of mass, can be obtained by taking $\chi = m_\alpha$ in (2.13). Hence,

$$\langle\chi\rangle_\alpha = m_\alpha \tag{3.1a}$$

$$\langle\chi\vec{v}\rangle_\alpha = m_\alpha \langle\vec{v}\rangle_\alpha = m_\alpha\vec{u}_\alpha \tag{3.1b}$$

$$\vec{\nabla}_v\,\chi = \vec{\nabla}_v\, m_\alpha = 0 \tag{3.1c}$$

The substitution of these results into the general transport equation gives the *continuity equation*,

$$\partial\rho_{m\alpha}/\partial t + \vec{\nabla}\cdot(\rho_{m\alpha}\,\vec{u}_\alpha) = S_\alpha \tag{3.2}$$

where $\rho_{m\alpha} = n_\alpha m_\alpha$ represents the mass density and where the collision term S_α, defined by

$$S_\alpha \equiv m_\alpha \int_V (\delta f_\alpha/\delta t)_{coll}\, d^3v = (\delta\rho_{m\alpha}/\delta t)_{coll} \tag{3.3}$$

represents the rate per unit volume at which particles of type α (with mass m_α) are produced or lost as a result of collisions. Contributions to this term are due to processes of particle creation or destruction such as ionization, recombination, attachment, charge transfer, and so on. In the absence of interactions leading to particle creation or destruction, the collision term (3.3) is equal to zero, since in the collision process the mass is conserved. When $S_\alpha = 0$ the continuity equation reduces to

$$\partial\rho_{m\alpha}/\partial t + \vec{\nabla}\cdot(\rho_{m\alpha}\,\vec{u}_\alpha) = 0 \tag{3.4}$$

Dividing each term in (3.4) by m_α, the continuity equation can be written in terms of the number density n_α, as

$$\partial n_\alpha/\partial t + \vec{\nabla}\cdot(n_\alpha \vec{u}_\alpha) = 0 \tag{3.5}$$

The equation of conservation of electric charge follows from (3.5) by multiplying it by the charge q_α,

$$\partial \rho_\alpha/\partial t + \vec{\nabla}\cdot\vec{J}_\alpha = 0 \tag{3.6}$$

where $\rho_\alpha = n_\alpha q_\alpha$ is the charge density and $\vec{J}_\alpha = \rho_\alpha \vec{u}_\alpha$ is the charge current density.

3.2 Derivation by the method of fluid dynamics

The continuity equation can also be derived by the method of fluid dynamics since $n_\alpha(\vec{r},t)$ and $\vec{u}_\alpha(\vec{r},t)$ are macroscopic variables. Consider a volume V in the fluid, limited by the closed surface S and let $d\vec{S} = \hat{n}\, dS$ be an element of area on this surface, such that the unit normal vector $\hat{n}$ points outwards, as shown in Fig. 1. The average number of particles of type α that leave the volume V through the element of area $d\vec{S}$ per unit time is given by

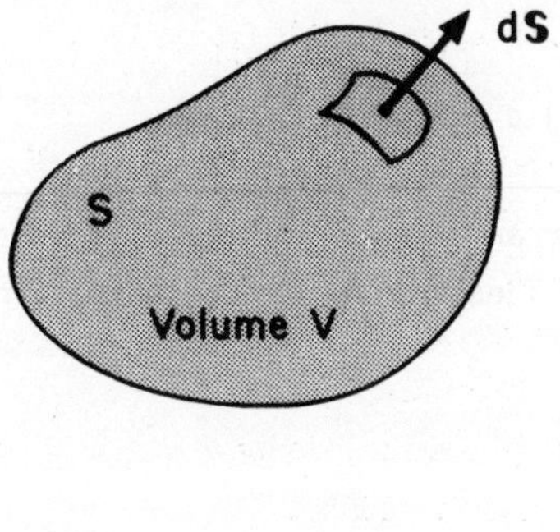

Fig. 1 - Closed surface S surrounding the arbitrary volume V inside the fluid, and the element of area $d\vec{S} = \hat{n}\, dS$ pointing outwards.

$$n_\alpha \vec{u}_\alpha \cdot d\vec{S} \tag{3.7}$$

Therefore, the number of particles of type α that leave the volume V through the whole closed surface S per unit time is obtained by integrating expression (3.7) over the whole surface,

$$\oint_S n_\alpha \vec{u}_\alpha \cdot d\vec{S} \tag{3.8}$$

On the other hand, the total number of particles of type α contained in V, at any time, is given by

$$\int_V n_\alpha \, d^3r \tag{3.9}$$

If we consider that there are no production or loss of particles inside the volume V, then the number of type α particles leaving V, must equal the time rate of decrease of the number of type α particles inside V. We must have therefore

$$\oint_S n_\alpha \vec{u}_\alpha \cdot d\vec{S} = - \frac{\partial}{\partial t} \int_V n_\alpha \, d^3r \tag{3.10}$$

Using Gauss' divergence theorem we can write

$$\oint_S n_\alpha \vec{u}_\alpha \cdot d\vec{S} = \int_V \vec{\nabla}\cdot(n_\alpha \vec{u}_\alpha) \, d^3r \tag{3.11}$$

and (3.10) becomes

$$\int_V [\partial n_\alpha/\partial t + \vec{\nabla} \cdot (n_\alpha \vec{u}_\alpha)] \, d^3r = 0 \tag{3.12}$$

This result must be valid for any arbitrary volume V, which implies that the integrand of (3.12) vanishes identically. Hence, we obtain the continuity equation (3.5).

3.3 The collision term

Let us consider now the form of the collision term S_α for some mechanisms of production and loss of particles in plasmas. The processes leading to production and loss of particles are usually related to inelastic collisions, such as those involving ionization, recombination or electron attachment, for example.

The effect of *ionization* can be included in the continuity equation through a rate coefficient for ionization k_i, defined such that the number of electrons produced per unit time is given by $k_i\, n_e$.

An important process leading to the loss of electrons and ions in a plasma is ion-electron *recombination*. Let k_r denote the recombination coefficient, determined experimentally. The rate of electron recombination is proportional to the product of the electron and ion number densities. Assuming there is only one ion species present, we have $n_i = n_e$, and the electron loss term, due to recombination, can be written as $k_r\, n_e^2$.

Another important mechanism for electron loss is the process of electron *attachment*. In this case, the rate of loss of electrons is proportional to the product of the electron number density and the neutral particle number density. In a weakly ionized plasma the neutral particle number density can be considered to be approximately constant, and the loss term for the electrons by attachment can be written as $k_a n_e$, where k_a is the attachment collision frequency, determined experimentally.

For these inelastic collision mechanisms just described, the collision term S_α is given by

$$S_e = m_e(k_i\, n_e - k_r\, n_e^2 - k_a\, n_e) \tag{3.13}$$

4. CONSERVATION OF MOMENTUM

4.1 Derivation of the equation of motion

In order to derive the momentum transport equation, we replace $\chi(\vec{v})$ by $m_\alpha \vec{v}$ in the general transport equation (2.13). Taking $\vec{v} = \vec{c}_\alpha + \vec{u}_\alpha$ and noting that $\langle \vec{c}_\alpha \rangle = 0$, the terms of the general transport equation become

$$\frac{\partial}{\partial t}(\rho_{m\alpha} \langle \vec{v} \rangle_\alpha) = \rho_{m\alpha} \frac{\partial \vec{u}_\alpha}{\partial t} + \vec{u}_\alpha \frac{\partial \rho_{m\alpha}}{\partial t} \tag{4.1a}$$

$$\vec{\nabla}\cdot(\rho_{m\alpha} \langle \vec{v}\,\vec{v} \rangle_\alpha) = \vec{\nabla}\cdot[\rho_{m\alpha}(\vec{u}_\alpha\,\vec{u}_\alpha + \vec{u}_\alpha \langle \vec{c}_\alpha \rangle + \langle \vec{c}_\alpha \rangle \vec{u}_\alpha + \langle \vec{c}_\alpha\,\vec{c}_\alpha \rangle)]$$

$$= \vec{\nabla}\cdot(\rho_{m\alpha}\,\vec{u}_\alpha\,\vec{u}_\alpha + \rho_{m\alpha} \langle \vec{c}_\alpha\,\vec{c}_\alpha \rangle) \tag{4.1b}$$

$$- n_\alpha \langle \vec{F} \cdot \vec{\nabla}_v \, \vec{v} \rangle_\alpha = - n_\alpha \langle (F_x \frac{\partial}{\partial v_x} + F_y \frac{\partial}{\partial v_y} + F_z \frac{\partial}{\partial v_z}) \, \vec{v} \rangle_\alpha$$

$$= - n_\alpha \langle F_x \, \hat{x} + F_y \, \hat{y} + F_z \, \hat{z} \rangle_\alpha = - n_\alpha \langle \vec{F} \rangle \tag{4.1c}$$

Substituting these expressions into (2.13), results in the momentum equation

$$\rho_{m\alpha} \, \partial \vec{u}_\alpha / \partial t + \vec{u}_\alpha \, \partial \rho_{m\alpha} / \partial t + \vec{\nabla} \cdot (\rho_{m\alpha} \, \vec{u}_\alpha \, \vec{u}_\alpha) + \vec{\nabla} \cdot (\rho_{m\alpha} \, \langle \vec{c}_\alpha \, \vec{c}_\alpha \rangle) - n_\alpha \, \langle \vec{F} \rangle_\alpha = \vec{A}_\alpha \tag{4.2}$$

where $\vec{A}_\alpha$ denotes the collision term

$$\vec{A}_\alpha \equiv m_\alpha \int_v \vec{v} \, (\delta f_\alpha / \delta t)_{coll} \, d^3v = \left[\frac{\delta(\rho_{m\alpha} \, \vec{u}_\alpha)}{\delta t} \right]_{coll} \tag{4.3}$$

The expression $\rho_{m\alpha} \langle \vec{c}_\alpha \, \vec{c}_\alpha \rangle$ is the kinetic pressure dyad $\bar{\bar{p}}_\alpha$ defined in (6.6.2). Therefore,

$$\vec{\nabla} \cdot (\rho_{m\alpha} \, \langle \vec{c}_\alpha \, \vec{c}_\alpha \rangle) = \vec{\nabla} \cdot \bar{\bar{p}}_\alpha \tag{4.4}$$

The third term in the left-hand side of (4.2) can be expanded as follows

$$\vec{\nabla} \cdot (\rho_{m\alpha} \, \vec{u}_\alpha \, \vec{u}_\alpha) = \frac{\partial}{\partial x} (\rho_{m\alpha} \, u_{\alpha x} \, \vec{u}_\alpha) + \frac{\partial}{\partial y} (\rho_{m\alpha} \, u_{\alpha y} \, \vec{u}_\alpha) + \frac{\partial}{\partial z} (\rho_{m\alpha} \, u_{\alpha z} \, \vec{u}_\alpha)$$

$$= \rho_{m\alpha} \, u_{\alpha x} \frac{\partial \vec{u}_\alpha}{\partial x} + \rho_{m\alpha} \, u_{\alpha y} \frac{\partial \vec{u}_\alpha}{\partial y} + \rho_{m\alpha} \, u_{\alpha z} \frac{\partial \vec{u}_\alpha}{\partial z} +$$

$$+ \vec{u}_\alpha \frac{\partial(\rho_{m\alpha} \, u_{\alpha x})}{\partial x} + \vec{u}_\alpha \frac{\partial(\rho_{m\alpha} \, u_{\alpha y})}{\partial y} + \vec{u}_\alpha \frac{\partial(\rho_{m\alpha} \, u_{\alpha z})}{\partial z}$$

$$= \rho_{m\alpha} \, (\vec{u}_\alpha \cdot \vec{\nabla}) \, \vec{u}_\alpha + \vec{u}_\alpha \, [\vec{\nabla} \cdot (\rho_{m\alpha} \, \vec{u}_\alpha)] \tag{4.5}$$

Substituting (4.4) and (4.5) into (4.2), and using the continuity equation (3.2), we obtain

$$\rho_{m\alpha} \, [\partial \vec{u}_\alpha / \partial t + (\vec{u}_\alpha \cdot \vec{\nabla}) \, \vec{u}_\alpha] + \vec{\nabla} \cdot \bar{\bar{p}}_\alpha - n_\alpha \, \langle \vec{F} \rangle_\alpha = \vec{A}_\alpha - \vec{u}_\alpha \, S_\alpha \tag{4.6}$$

For the terms within brackets in this last equation we can use the total (or substantial) time derivative

$$D/Dt = \partial / \partial t + \vec{u}_\alpha \cdot \vec{\nabla} \tag{4.7}$$

which corresponds to the time variation observed in a frame of reference moving with the mean velocity $\vec{u}_\alpha$. If the electromagnetic Lorentz force and the gravitational force are considered, the last term in the left-hand side of (4.6) becomes

$$- n_\alpha \langle \vec{F} \rangle_\alpha = - n_\alpha q_\alpha (\vec{E} + \vec{u}_\alpha \times \vec{B}) - n_\alpha m_\alpha \vec{g} \tag{4.8}$$

where the fields $\vec{E}$ and $\vec{B}$, in this equation, represent smoothed macroscopic fields. The equation of motion can, therefore, be written as

$$\rho_{m\alpha} D\vec{u}_\alpha/Dt = n_\alpha q_\alpha (\vec{E} + \vec{u}_\alpha \times \vec{B}) + \rho_{m_\alpha} \vec{g} - \vec{\nabla} \cdot \overline{\overline{p}}_\alpha + \vec{A}_\alpha - \vec{u}_\alpha S_\alpha \tag{4.9}$$

Physically, this equation expresses the fact that the time rate of change of the mean momentum, in each fluid element, is due to the external forces applied in the fluid, to the shear (viscosity) and pressure forces of the fluid itself, and to the internal forces associated with the collisional interactions. Thus, the equation of motion establishes the condition necessary to guarantee conservation of momentum, just as the continuity equation establishes the condition necessary to guarantee conservation of mass (or number of particles).

In Chapter 6 we have seen that the term $-\vec{\nabla} \cdot \overline{\overline{p}}_\alpha$ represents the force exerted in a unit volume of the plasma, due to the random variation in the particle peculiar velocities. This force per unit volume includes forces associated with the scalar pressure and tangential shear forces (viscous forces). In many cases, the effect of viscosity is relatively unimportant in plasmas and the nondiagonal terms of $\overline{\overline{p}}_\alpha$ can be neglected. Furthermore, in the special case when the distribution of peculiar velocities is isotropic, the diagonal terms of $\overline{\overline{p}}_\alpha$ are all equal and correspond to a scalar kinetic pressure p_α. Thus, neglecting viscosity effects and considering an isotropic velocity distribution, we have $\overline{\overline{p}}_\alpha = \overline{\overline{1}}\, p_\alpha$, and the force per unit volume becomes $- \vec{\nabla} \cdot \overline{\overline{p}}_\alpha = - \vec{\nabla} p_\alpha$, according to (6.6.18).

With these simplifying approximations, and neglecting collisions leading to production or loss of particles ($S_\alpha = 0$), the momentum equation becomes

$$\rho_{m\alpha} D\vec{u}_\alpha/Dt = n_\alpha q_\alpha (\vec{E} + \vec{u}_\alpha \times \vec{B}) + \rho_{m\alpha} \vec{g} - \vec{\nabla} p_\alpha + \vec{A}_\alpha \tag{4.10}$$

The momentum conservation equation can also be derived using a fluid-dynamics approach, in a way similar to the derivation of the mass conservation equation presented in section 3.2, which we shall not discuss here.

4.2 The collision term

The symbol $\vec{A}_\alpha$ denotes the rate of change of mean momentum per unit volume, due to collisions. As a consequence of conservation of the total momentum in an elastic collision, the change in the momentum of one of the particles must be equal and opposite to the change in momentum of the other particle participating in the collision event. This means that, for collisions involving particles of the same species, there is no variation in the total momentum per unit volume and therefore, in this case, $\vec{A}_\alpha = 0$. However, for a fluid composed of particles of different species, as in a plasma, the collision term $\vec{A}_\alpha$ is not zero in general. For collisions between electrons and neutral particles there is a net momentum transfer from the electron gas to the neutral gas. Collisions between electrons and ions also modify the total momentum of the electron gas. Therefore, for the case of collisions between particles of different species, a collision term must be included in the equation of conservation of momentum.

An expression often used for the term of momentum transfer by collisions is

$$\vec{A}_\alpha = - \rho_{m\alpha} \sum_\alpha \nu_{\alpha\beta} (\vec{u}_\alpha - \vec{u}_\beta) \tag{4.11}$$

which assumes that the force per unit volume exerted on the particles of type α due to collisions with particles of some other type β ($\beta \neq \alpha$) is proportional to the difference between the mean velocities $\vec{u}_\alpha$ and $\vec{u}_\beta$. The proportionality constant $\nu_{\alpha\beta}$ (which has dimensions of sec^{-1}) is called the *collision frequency for momentum transfer* between the particles of type α and those of type β. Since the total momentum must be conserved during a collision, we must have

$$\rho_{m\alpha} \nu_{\alpha\beta} (\vec{u}_\alpha - \vec{u}_\beta) + \rho_{m\beta} \nu_{\beta\alpha} (\vec{u}_\beta - \vec{u}_\alpha) = 0 \tag{4.12}$$

The collision frequencies $\nu_{\alpha\beta}$ and $\nu_{\beta\alpha}$ satisfy, therefore, the following important relationship:

$$\rho_{m\alpha} \nu_{\alpha\beta} = \rho_{m\beta} \nu_{\beta\alpha} \tag{4.13}$$

The collision term $\vec{A}_\alpha$, defined in (4.3), will be considered in more detail in Chapter 21. We will see, then, that the expression (4.11) is not generally valid, although this result is obtained when the difference between the mean velocities of the various particle species in the plasma is relatively small and when each species of particles has a Maxwellian velocity distribution.

5. CONSERVATION OF ENERGY

5.1 Derivation of the energy transport equation

To derive the energy transport equation, we substitute $\chi(\vec{v})$ by $m_\alpha v^2/2$ in the general transport equation (2.13). In this case, we have

$$n_\alpha \langle\chi\rangle_\alpha = \rho_{m\alpha}\langle v^2\rangle_\alpha /2 = \rho_{m\alpha}\langle c_\alpha^2\rangle/2 + \rho_{m\alpha} u_\alpha^2/2 =$$

$$= 3p_\alpha/2 + \rho_{m\alpha}\, u_\alpha^2/2 \tag{5.1}$$

$$\vec{\nabla}_v\ \chi = m_\alpha\, \vec{\nabla}_v\, (\vec{v}\cdot\vec{v})/2 = m_\alpha\, (\vec{v}\cdot\vec{\nabla}_v)\, \vec{v} = m_\alpha\, \vec{v} \tag{5.2}$$

Therefore, the terms in the left-hand side of the general transport equation (2.13) become

$$\frac{\partial}{\partial t}(n_\alpha \langle \chi \rangle_\alpha) = \frac{3}{2}\frac{\partial p_\alpha}{\partial t} + \frac{\partial}{\partial t}(\rho_{m\alpha}\, u_\alpha^2/2) \tag{5.3a}$$

$$\vec{\nabla}\cdot(n_\alpha \langle \chi\, \vec{v} \rangle_\alpha) = \vec{\nabla}\cdot\left[\rho_{m\alpha} \langle (\vec{v}\cdot\vec{v})\, \vec{v} \rangle_\alpha /2\right] \tag{5.3b}$$

$$- n_\alpha \langle (\vec{F}/m_\alpha)\cdot\vec{\nabla}_v\, \chi \rangle_\alpha = - n_\alpha \langle \vec{F}\cdot\vec{v} \rangle_\alpha \tag{5.3c}$$

Adding these terms, results in the following *energy conservation equation*

$$\frac{3}{2}\frac{\partial p_\alpha}{\partial t} + \frac{\partial}{\partial t}(\rho_{m\alpha}\, u_\alpha^2/2) + \vec{\nabla}\cdot\left[\rho_{m\alpha} \langle (\vec{v}\cdot\vec{v})\, \vec{v} \rangle_\alpha/2\right]$$

$$- n_\alpha \langle \vec{F}\cdot\vec{v} \rangle_\alpha = M_\alpha \tag{5.4}$$

where M_α represents the rate of energy density change due to collisions.

$$M_\alpha \equiv (m_\alpha/2) \int_v v^2 \left(\frac{\delta f_\alpha}{\delta f}\right)_{coll} d^3v = \left[\frac{\delta(\rho_{m\alpha} < v^2 >_\alpha/2)}{\delta t}\right]_{coll} \tag{5.5}$$

The energy equation (5.4) can be written in an *alternative form* as follows. Consider, initially, the third term in the left-hand side of (5.4). Taking $\vec{v} = \vec{c}_\alpha + \vec{u}_\alpha$, the quantity $< (\vec{v} \cdot \vec{v}) \vec{v} >_\alpha$ can be expanded as

$$< [(\vec{u}_\alpha + \vec{c}_\alpha) \cdot (\vec{u}_\alpha + \vec{c}_\alpha)] \quad (\vec{u}_\alpha + \vec{c}_\alpha) > =$$

$$= < (u_\alpha^2 + 2 \vec{u}_\alpha \cdot \vec{c}_\alpha + c_\alpha^2) (\vec{u}_\alpha + \vec{c}_\alpha) > =$$

$$= u_\alpha^2 \vec{u}_\alpha + < c_\alpha^2 > \vec{u}_\alpha + 2 < \vec{c}_\alpha \vec{c}_\alpha > \cdot \vec{u}_\alpha + < c_\alpha^2 \vec{c}_\alpha > \tag{5.6}$$

The term $\rho_{m\alpha} < \vec{c}_\alpha \vec{c}_\alpha >$ represents the kinetic pressure dyad $\bar{\bar{p}}_\alpha$ and $\rho_{m\alpha} < c_\alpha^2 \vec{c}_\alpha > /2$ is the heat flux vector $\vec{q}_\alpha$, defined in Chapter 6. We have also seen that $\rho_{m\alpha} < c_\alpha^2 >/2 = 3p_\alpha/2$. Therefore,

$$\vec{\nabla} \cdot [\rho_{m\alpha} < (\vec{v}.\vec{v}) \vec{v} >_\alpha/2] = \vec{\nabla} \cdot (\rho_{m\alpha} u_\alpha^2 \vec{u}_\alpha/2 + 3p_\alpha \vec{u}_\alpha/2 + \bar{\bar{p}}_\alpha \cdot \vec{u}_\alpha + \vec{q}_\alpha) =$$

$$= \vec{\nabla} \cdot (\rho_{m\alpha} u_\alpha^2 \vec{u}_\alpha/2) + 3p_\alpha \vec{\nabla} \cdot \vec{u}_\alpha/2 +$$

$$+ 3 (\vec{u}_\alpha \cdot \vec{\nabla}) p_\alpha/2 + \vec{\nabla} \cdot (\bar{\bar{p}}_\alpha \cdot \vec{u}_\alpha) + \vec{\nabla} \cdot \vec{q}_\alpha \tag{5.7}$$

Substituting this result into (5.4) and using the notation D/Dt for the total time derivative (4.7), we obtain

$$\frac{D}{Dt} (3p_\alpha/2) + (3p_\alpha/2) \vec{\nabla} \cdot \vec{u}_\alpha + \frac{\partial}{\partial t} (\rho_{m\alpha} u_\alpha^2/2) + \vec{\nabla} \cdot (\rho_{m\alpha} u_\alpha^2 \vec{u}_\alpha^2/2) +$$

$$+ \vec{\nabla} \cdot (\bar{\bar{p}}_\alpha \cdot \vec{u}_\alpha) + \vec{\nabla} \cdot \vec{q}_\alpha - n_\alpha < \vec{F} \cdot \vec{v} >_\alpha = M_\alpha \tag{5.8}$$

The third and fourth terms in the left-hand side can be written as

$$\frac{\partial}{\partial t}(\rho_{m\alpha}\vec{u}_\alpha \cdot \vec{u}_\alpha/2) + \vec{\nabla}\cdot[\rho_{m\alpha}(\vec{u}_\alpha\cdot\vec{u}_\alpha)\vec{u}_\alpha/2] =$$

$$= (u_\alpha^2/2)\,\partial\rho_{m\alpha}/\partial t + \rho_{m\alpha}\vec{u}_\alpha\cdot(\partial\vec{u}_\alpha/\partial t) +$$

$$+ (u_\alpha^2/2)\,\vec{\nabla}\cdot(\rho_{m\alpha}\vec{u}_\alpha) + \rho_{m\alpha}\vec{u}_\alpha\cdot[(\vec{u}_\alpha\cdot\vec{\nabla})\,\vec{u}_\alpha)] =$$

$$= (u_\alpha^2/2)\left[\frac{\partial\rho_{m\alpha}}{\partial t} + \vec{\nabla}\cdot(\rho_{m\alpha}\vec{u}_\alpha)\right] + \rho_{m\alpha}\vec{u}_\alpha\cdot\left(\frac{D\vec{u}_\alpha}{Dt}\right) \tag{5.9}$$

Using the continuity equation (3.2) and the equation of motion (4.6), this last equation becomes equal to

$$(u_\alpha^2/2)\,S_\alpha + n_\alpha\,\vec{u}_\alpha\cdot\langle\vec{F}\rangle_\alpha - \vec{u}_\alpha\cdot(\vec{\nabla}\cdot\bar{\bar{p}}_\alpha) + \vec{u}_\alpha\cdot\vec{A}_\alpha - u_\alpha^2\,S_\alpha \tag{5.10}$$

Taking this result back into (5.8) we obtain

$$\frac{D}{Dt}(3p_\alpha/2) + (3p_\alpha/2)\,\vec{\nabla}\cdot\vec{u}_\alpha + \vec{\nabla}\cdot(\bar{\bar{p}}_\alpha\cdot\vec{u}_\alpha) - \vec{u}_\alpha\cdot(\vec{\nabla}\cdot\bar{\bar{p}}_\alpha) -$$

$$- n_\alpha\langle\vec{F}\cdot\vec{v}\rangle_\alpha + n_\alpha\vec{u}_\alpha\cdot\langle\vec{F}\rangle_\alpha + \vec{\nabla}\cdot\vec{q}_\alpha =$$

$$= M_\alpha - \vec{u}_\alpha\cdot\vec{A}_\alpha + \frac{1}{2}u_\alpha^2\,S_\alpha \tag{5.11}$$

The third and fourth terms in the left-hand side of this equation can be combined in one single term,

$$\vec{\nabla}\cdot(\bar{\bar{p}}_\alpha\cdot\vec{u}_\alpha) - \vec{u}_\alpha\cdot(\vec{\nabla}\cdot\bar{\bar{p}}_\alpha) = (\bar{\bar{p}}_\alpha\cdot\vec{\nabla})\cdot\vec{u}_\alpha \tag{5.12}$$

as well as the fifth and the sixth terms, which give

$$- n_\alpha\langle\vec{F}\cdot\vec{v}\rangle_\alpha + n_\alpha\vec{u}_\alpha\cdot\langle\vec{F}\rangle_\alpha = - n_\alpha\langle\vec{F}\cdot\vec{c}_\alpha\rangle \tag{5.13}$$

since

$$< \vec{F} \cdot \vec{v} >_\alpha = < \vec{F} \cdot (\vec{u}_\alpha + \vec{c}_\alpha) > = < \vec{F} >_\alpha \cdot \vec{u}_\alpha + < \vec{F} \cdot \vec{c}_\alpha > \tag{5.14}$$

For a velocity-independent force (5.13) vanishes, since, in this case,

$$< \vec{F} \cdot \vec{c}_\alpha > = \vec{F} \cdot < \vec{c}_\alpha > = 0 \tag{5.15}$$

For the force due to a magnetic field $\vec{B}$, the only velocity-dependent force that we are interested here, (5.13) also vanishes,

$$< \vec{F} \cdot \vec{c}_\alpha > = q_\alpha < (\vec{v} \times \vec{B}) \cdot \vec{c}_\alpha >$$

$$= q_\alpha (\vec{u}_\alpha \times \vec{B}) \cdot < \vec{c}_\alpha > + q_\alpha < (\vec{c}_\alpha \times \vec{B}) \cdot \vec{c}_\alpha > = 0 \tag{5.16}$$

where both terms vanish since $< \vec{c}_\alpha > = 0$ and $(\vec{c}_\alpha \times \vec{B})$ is normal to $\vec{c}_\alpha$. We obtain, finally, the following *alternative form* for the equation of *conservation of energy*

$$\frac{D}{Dt} (3p_\alpha/2) + (3p_\alpha/2) (\vec{\nabla} \cdot \vec{u}_\alpha) + (\bar{\bar{p}}_\alpha \cdot \vec{\nabla}) \cdot \vec{u}_\alpha + \vec{\nabla} \cdot \vec{q}_\alpha =$$

$$= M_\alpha - \vec{u}_\alpha \cdot \vec{A}_\alpha + (u_\alpha^2/2) S_\alpha \tag{5.17}$$

5.2 Physical interpretation

The physical interpretation of this equation is as follows. The first term in the left-hand side represents the total rate of change of the particle thermal energy density in a volume element moving with the mean velocity $\vec{u}_\alpha$. Note that the thermal energy density is given by $\rho_{m\alpha} < c_\alpha^2 >/2 = 3p_\alpha/2$.

The other terms of (5.17) contribute to some extent to this total rate of change of the thermal energy density. The second term in the left-hand side of (5.17) can be interpreted as the change in the thermal energy density due to particles entering the volume element with the mean velocity $\vec{u}_\alpha$. The third term is related to the work done on the unit volume by the kinetic pressure dyad acting on its surface, whereas the fourth term represents the change in the thermal energy density due to the heat flux. Finally, the terms in the right-hand side of (5.17) represent the rate of change in the thermal energy density as a consequence of collisions. In the case of a fluid containing only one type of particles, the collision terms vanish, as indicated previously.

The first two terms in the energy equation may also be combined, making use of the continuity equation (3.2). Expanding $\vec{\nabla} \cdot (\rho_{m\alpha} \vec{u}_\alpha)$, (3.2) becomes

$$(\partial/\partial t + \vec{u}_\alpha \cdot \vec{\nabla}) \rho_{m\alpha} + \rho_{m\alpha} \vec{\nabla} \cdot \vec{u}_\alpha = S_\alpha \tag{5.18}$$

which gives

$$\vec{\nabla} \cdot \vec{u}_\alpha = - (1/\rho_{m\alpha}) (D\rho_{m\alpha}/Dt + S_\alpha) \tag{5.19}$$

Substituting this result into (5.17), taking $\rho_{m\alpha} = n_\alpha m_\alpha$ and $p_\alpha = n_\alpha k T_\alpha$, yields the following alternative form for the energy equation in terms of the temperature T_α,

$$\frac{3}{2} n_\alpha k \frac{DT_\alpha}{Dt} + (\bar{\bar{p}}_\alpha \cdot \vec{\nabla}) \cdot \vec{u}_\alpha + \vec{\nabla} \cdot \vec{q}_\alpha = M_\alpha - \vec{u}_\alpha \cdot \vec{A}_\alpha + (\frac{1}{2} u_\alpha^2 - \frac{3}{2} \frac{kT_\alpha}{m_\alpha}) S_\alpha \tag{5.20}$$

5.3 Simplifying approximations

Several simplifying approximations can be considered for the energy equation, depending on the situation of interest.

(a) When the collision terms vanish, or are negligible, and when the mean velocity $\vec{u}_\alpha$ is equal to zero, (5.20) reduces to a *diffusion* equation for T_α, if we take the heat flux vector as

$$\vec{q}_\alpha = - K \vec{\nabla} T_\alpha \tag{5.21}$$

where K denotes the thermal conductivity. Thus, in this case,

$$(3/2) n_\alpha k \, DT_\alpha/Dt = \vec{\nabla} \cdot (K \vec{\nabla} T_\alpha) \tag{5.22}$$

The coefficient of thermal conductivity K is related to the fluid coefficient of viscosity.

b) Consider now a non-viscous fluid, in which the pressure dyad reduces to a scalar pressure without thermal conductivity ($\vec{q}_\alpha = 0$). If we admit also that the collision terms vanish, the energy equation (5.17) becomes

$$\frac{D}{Dt}(3p_\alpha/2) + (3p_\alpha/2)(\vec{\nabla} \cdot \vec{u}_\alpha) + p_\alpha(\vec{\nabla} \cdot \vec{u}_\alpha) = 0 \tag{5.23}$$

Substituting (5.19) for $(\vec{\nabla} \cdot \vec{u}_\alpha)$, with $S_\alpha = 0$, yields

$$\frac{D}{Dt}(3p_\alpha/2) - (5p_\alpha/2\rho_{m\alpha})\frac{D\rho_{m\alpha}}{Dt} = 0 \tag{5.24}$$

from which results

$$(1/p_\alpha)\, Dp_\alpha = (5/3\, \rho_{m\alpha})\, D\rho_{m\alpha} \tag{5.25}$$

Integrating this equation gives

$$p_\alpha/p_0 = (\rho_{m\alpha}/\rho_{mo})^{5/3} \tag{5.26}$$

where p_0 and ρ_{mo} are constants, that is,

$$p_\alpha\, \rho_{m\alpha}^{-5/3} = \text{constant} \tag{5.27}$$

This is the *adiabatic energy equation* for a gas in which the ratio of the specific heats at constant pressure and at constant volume, γ, is equal to 5/3. We emphasize here that the energy equation reduces to this adiabatic equation only when the effects of viscosity, thermal conductivity and energy transfer due to collisions are neglected.

The parameter γ is related to the number of degrees of freedom, N, of a gas by the condition

$$\gamma = (2 + N)/N \tag{5.28}$$

For particles that have no internal degrees of freedom, as for example in a monoatomic gas, where the only degrees of freedom are those associated with the three possible directions of translational motion, we have N = 3 and therefore $\gamma = 5/3$. Other degrees of freedom exist in the case of diatomic or poliatomic gases. The *adiabatic energy equation* often used in thermodynamics is

$$p\,\rho_m^{-\gamma} = \text{constant} \tag{5.29}$$

Differentiating this equation, yields

$$\rho_m^{-\gamma}\,dp - \gamma p \rho_m^{-(\gamma+1)}\,d\rho_m = 0 \tag{5.30}$$

or equivalently,

$$dp = (\gamma p/\rho_m)\,d\rho_m = V_S^2\,d\rho_m \tag{5.31}$$

where we have defined

$$V_s = (\gamma p/\rho_m)^{1/2} = (\gamma kT/m)^{1/2} \tag{5.32}$$

which is the *adiabatic speed of sound* for the fluid.

(c) An equation which is also used in thermodynamics when the temperature is constant inside the fluid, is the *isothermal energy equation*. It can be easily obtained from the equation of state for an ideal gas, p = n k T. For an isothermal process (T = constant), we have

$$dp = k\,T\,dn = (p/\rho_m)\,d\rho_m = V_T^2\,d\rho \tag{5.33}$$

where the *isothermal speed of sound is*

$$V_T = (p/\rho_m)^{1/2} = (k\,T/m)^{1/2} \tag{5.34}$$

6. THE COLD PLASMA MODEL

In the previous sections we have seen that the differential equations governing the temporal and spatial variation of the macroscopic variables can be obtained by taking moments of the Boltzmann equation. The macroscopic parameters are moments of the distribution function $f_\alpha(\vec{r},\vec{v},t)$. The first four moments of the distribution function give us the number density n_α, the mean velocity $\vec{u}_\alpha$, the momentum flow dyad $\bar{\bar{P}}_\alpha$, and the energy flow triad $\bar{\bar{\bar{E}}}_\alpha$. The first moment of the Boltzmann equation gives us the *continuity equation* which relates the number density n_α(or the mass density $\rho_{m\alpha}$) with the mean velocity $\vec{u}_\alpha$ for the particles of type α. In order to determine these two macroscopic variables we need two independent equations. Thus, we evaluate the second moment of the Boltzmann equation, which gives us the *equation of motion,* and which relates the mean velocity $\vec{u}_\alpha$, the number density n_α and the kinetic pressure dyad $\bar{\bar{p}}_\alpha$. We find, therefore, that the set of transport equations derived from the Boltzmann equation always includes more variables than independent equations. This fact is clearly evident in the three transport equations derived in this chapter. The *energy equation,* besides the variables n_α, $\vec{u}_\alpha$ and $\bar{\bar{p}}_\alpha$, also includes the vector of heat flow $\vec{q}_\alpha$. Any finite set of transport equations is, therefore, insufficient to form a *closed* system of equations. Consequently, it is necessary to introduce a scheme of approximation to eliminate some of the independent variables, or to express some of these variables in terms of the others. It is common, therefore, to arbitrarily truncate the system of transport equations at some point in the hierarchy of moments of the Boltzmann equation, and consider some simplifying approximation for the highest moment of the distribution function appearing in the system.

The simplest closed system of transport equations is known as the *cold plasma model*, which encompasses only the equations of conservation of mass and momentum. The highest moment of the distribution function, appearing in the momentum equation, is the kinetic pressure dyad $\bar{\bar{p}}_\alpha$ which, in this model, is taken equal to zero. This means that the effects due to the thermal motion of the particles and the force due to the divergence of the kinetic pressure dyad $\bar{\bar{p}}_\alpha$ are neglected. For convenience, we collect here the two transport equations pertinent to the cold plasma model,

$$\partial\rho_{m\alpha}/\partial t + \vec{\nabla} \cdot (\rho_{m\alpha}\, \vec{u}_\alpha) = S_\alpha \tag{6.1}$$

$$\rho_{m\alpha} \, D\vec{u}_\alpha/Dt = n_\alpha \, q_\alpha \, (\vec{E} + \vec{u}_\alpha \times \vec{B}) + \rho_{m\alpha} \, \vec{g} + \vec{A}_\alpha - \vec{u}_\alpha \, S_\alpha \tag{6.2}$$

In the absence of processes leading to production and loss of particles of type α (such as ionization and recombination), we have $S_\alpha = 0$. The expression normally used for the collision term for momentum transfer $\vec{A}_\alpha$ is the one indicated in (4.11). The cold plasma model assumes, in fact, a zero plasma temperature, so that the corresponding distribution function is a Dirac delta function centered at the macroscopic flow velocity, $f_\alpha(\vec{r},\vec{v},t) = \delta\,[\,\vec{v} - \vec{u}_\alpha\,(\vec{r},t)\,]$.

This model has been successfully applied in the investigation of the properties of small amplitude electromagnetic waves propagating in plasmas, with phase velocities much larger than the thermal velocity of the particles; this theory is known as *magnetoionic theory*.

7. THE WARM PLASMA MODEL

In this closed system of transport equations, the simplifying approximation is introduced in the equation of conservation of energy, in which we neglect the term involving the heat flux vector $\vec{q}_\alpha$. Thus, the approximation consists in taking $\vec{\nabla} \cdot \vec{q}_\alpha = 0$, which means that the processes occurring in the plasma are such that there is no thermal energy flux. This approximation is also called the *adiabatic approximation*. Since the thermal conductivity is zero in this case, it follows that the plasma is non-viscous and, consequently, the non-diagonal terms of the kinetic pressure dyad $\overline{\overline{p}}_\alpha$ are all equal to zero. Further, the diagonal terms of $\overline{\overline{p}}_\alpha$ are assumed to be equal and the kinetic pressure dyad is replaced by a scalar pressure p_α. Thus, the term $\vec{\nabla} \cdot \overline{\overline{p}}_\alpha$ degenerates to $\vec{\nabla} \, p_\alpha$.

The three macroscopic variables appearing in this case are the number density n_α, the mean velocity $\vec{u}_\alpha$ and the scalar pressure p_α. The three transport equations pertaining to the warm plasma model are, therefore,

$$\partial\rho_{m\alpha}/\partial t + \vec{\nabla} \cdot (\rho_{m\alpha} \, \vec{u}_\alpha) = S_\alpha \tag{7.1}$$

$$\rho_{m\alpha} \, D\vec{u}_\alpha/Dt = n_\alpha \, q_\alpha \, (\vec{E} + \vec{u}_\alpha \times \vec{B}) + \rho_{m\alpha} \, \vec{g} - \vec{\nabla} p_\alpha + \vec{A}_\alpha - \vec{u}_\alpha \, S_\alpha \tag{7.2}$$

$$\frac{D}{Dt}(3p_\alpha/2) + (5p_\alpha/2)(\vec{\nabla}\cdot\vec{u}_\alpha) = M_\alpha - \vec{u}_\alpha \cdot \vec{A}_\alpha + u_\alpha^2\, S_\alpha/2 \qquad (7.3)$$

Considering the additional approximation that the change in energy as a result of collisions, is negligible, the energy equation (7.3) reduces to the following *adiabatic equation* (as shown in section 5),

$$p_\alpha\, \rho_{m\alpha}^{-\gamma} = \text{constant} \qquad (7.4)$$

Generally, the warm plasma model gives a more precise description of the behavior of plasma phenomena as compared to the cold plasma model.

In the most general cases, in which the plasma is not in a state of local equilibrium, and when heat flow and viscosity need to be taken into account, it is more convenient and simple to work directly with the phase space distribution function. In this case, the plasma is usually said to be *hot*. After determining the distribution function $f_\alpha(\vec{r},\vec{v},t)$, by solving the differential kinetic equation which governs the evolution of $f_\alpha(\vec{r},\vec{v},t)$ in phase space, for the specific problem under consideration, the macroscopic variables can be obtained according to the systematic method presented in Chapter 6.

PROBLEMS

8.1 - Consider the following simplified steady state equation of motion, for each species in a fluid plasma

$$0 = qn\,(\vec{E} + \vec{u} \times \vec{B}) - \vec{\nabla}p$$

where the electric $(\vec{E})$ and magnetic $(\vec{B})$ fields are uniform, but the number density (n) and the pressure (p) have a gradient. Taking the cross product of this equation with $\vec{B}$ show that, besides the $\vec{E} \times \vec{B}$ drift, there is also a *diamagnetic drift* given by

$$\vec{v}_D = -(1/n)\vec{\nabla}p \times \vec{B}/qB^2$$

Provide physical arguments to justify the physical reason for this fluid drift. Explain if there is any motion of the particle guiding centers, associated with this fluid drift, and why it does not appear in the particle orbit theory.

8.2 - (a) From Maxwell equations,

$$\vec{\nabla} \cdot \vec{E} = \rho/\varepsilon_0$$

$$\vec{\nabla} \cdot \vec{H} = 0$$

$$\vec{\nabla} \times \vec{E} = -\mu_0 \; \partial \vec{H}/\partial t$$

$$\vec{\nabla} \times \vec{H} = \vec{J} + \varepsilon_0 \; \partial \vec{E}/\partial t$$

where $\vec{E}$ and $\vec{H}$ denote the electric and magnetic fields in a plasma, ρ denotes the charge density nq, and $\vec{J}$ the current density $nq\,\vec{u}$, show that

$$\varepsilon_0 \; \mu_0 \; \partial(\vec{E} \times \vec{H})/\partial t = \vec{\nabla} \cdot \overline{\overline{T}} - nq\,(\vec{E} + \vec{u} \times \vec{B}) \qquad \text{(A)}$$

where $\varepsilon_0 \, \mu_0 \, (\vec{E} \times \vec{H})$ is the electromagnetic momentum density, $\overline{\overline{T}}$ is the electromagnetic stress dyad whose components are given by

$$T_{ij} = \varepsilon_0 \, E_i \, E_j + \mu_0 \, H_i \, H_j - \delta_{ij} \, (\varepsilon_0 \, E^2 + \mu_0 \, H^2)/2$$

and δ_{ij} is the Kronecker delta.

(b) From Eq. (A) and the continuity equation

$$\partial n/\partial t + \vec{\nabla} \cdot (n\,\vec{u}) = 0$$

show that the momentum transport equation

$$nm \, D\vec{u}/Dt = nq\,(\vec{E} + \vec{u} \times \vec{B}) - \vec{\nabla} \cdot \overline{\overline{p}}$$

can be written in the form

$$\partial \vec{G}/\partial t + \vec{\nabla} \cdot \overline{\overline{\pi}} = 0$$

where $\vec{G}$ denotes the total momentum density

$$\vec{G} = n\, m\, \vec{u} + \varepsilon_0 \mu_0\ (\vec{E} \times \vec{H})$$

and $\overline{\overline{\pi}}$ is the total momentum flux (rate of transport of momentum through unit area)

$$\overline{\overline{\pi}} = n\, m\, \vec{u}\, \vec{u} + \overline{\overline{p}} - \overline{\overline{T}} = \overline{\overline{P}} - \overline{\overline{T}}$$

(c) Using Maxwell curl equations, show that the energy transport equation

$$\frac{\partial}{\partial t}(n\, m <c^2>/2) + \frac{\partial}{\partial t}(n\, m\, u^2/2) = -\vec{\nabla}\cdot(n\, m <c^2>\vec{u}/2) -$$

$$-\vec{\nabla}\cdot(n\, m\, u^2\,\vec{u}/2) - \vec{\nabla}\cdot\vec{q} - \vec{\nabla}\cdot(\overline{\overline{p}}\cdot\vec{u}) + n\, q\, \vec{u}\cdot\vec{E}$$

[note that $\vec{u}\cdot(\vec{u}\times\vec{B}) = 0$] can be written in the form

$$\partial W/\partial t + \vec{\nabla}\cdot\vec{S} = 0$$

where W denotes the total energy density

$$W = (1/2)\ (\varepsilon_0\, E^2 + \mu_0\, H^2 + n\, m\, u^2 + n\, m\ <c^2>)$$

and $\vec{S}$ is the total energy flux (power per unit area)

$$\vec{S} = \vec{E}\times\vec{H} + \overline{\overline{p}}\cdot\vec{u} + n\, m\, \vec{u}\ (<c^2> + u^2)/2 + \vec{q}$$

8.3 - In order to investigate the effect of the collision term (4.11) in the macroscopic fluid motion, consider a uniform mixture of different fluids (all spatial derivatives vanish), with no external forces, so that the equation of motion for the α species reduces to

$$d\vec{u}_\alpha/dt = -\sum_\beta \nu_{\alpha\beta}\ (\vec{u}_\alpha - \vec{u}_\beta)$$

Solve this equation to determine $\vec{u}_\alpha(t)$, for a two-fluid mixture and a three--fluid mixture. (In the case of the three-fluid mixture it is better to use Laplace transforms). Notice that, at equilibrium (when $d\vec{u}_\alpha/dt = 0$), the velocities of all species must be the same.

8.4 - Consider a uniform mixture of different fluids (all spatial derivatives vanish), with no external forces, such that the equation of motion for the α species becomes

$$d\vec{u}_\alpha/dt = -\sum_\beta \nu_{\alpha\beta} (\vec{u}_\alpha - \vec{u}_\beta)$$

(a) Show that the time rate of change of the total fluid kinetic energy density, W_k, is given by

$$dW_k/dt = -\sum_{\alpha,\beta} (\rho_{m\alpha}/2)\ \nu_{\alpha\beta} (\vec{u}_\alpha - \vec{u}_\beta)^2$$

where

$$W_k = \sum_\alpha \rho_{m\alpha}\ u_\alpha^2/2$$

(b) Consider now the total fluid thermal energy density,

$$W_T = \sum_\alpha 3n_\alpha\, k\, T_\alpha/2$$

If the energy equation for a homogeneous fluid mixture, with no external forces, is

$$\frac{dT_\alpha}{dt} = -\sum_\beta \frac{2m_\alpha \nu_{\alpha\beta}}{(m_\alpha + m_\beta)} \left[(T_\alpha - T_\beta) - \frac{m_\beta}{3k} (\vec{u}_\alpha - \vec{u}_\beta)^2 \right]$$

then, show that the time rate of change of W_T is given by

$$dW_T/dt = \sum_{\alpha,\beta} (\rho_{m\alpha}/2)\ \nu_{\alpha\beta}\ (\vec{u}_\alpha - \vec{u}_\beta)^2$$

Thus, the total thermal energy density W_T increases at exactly the same rate as the total kinetic energy density W_K decreases. *Hint:* for any function which is summed over two indices, the result is unchanged if we interchange the indices, so that

$$\sum_{\alpha,\beta} f_{\alpha\beta} = \sum_{\alpha,\beta} f_{\beta\alpha}$$

or

$$\sum_{\alpha,\beta} f_{\alpha\beta} = \sum_{\alpha,\beta} (f_{\alpha\beta} + f_{\beta\alpha})/2$$

8.5 - Explain the reason why there is no term containing the magnetic flux density $\vec{B}$ in the energy equation (5.17).

8.6 - Derive the following *general transport equation*, similar to (2.13), for the case when the quantity χ depends on $\vec{r}$, $\vec{v}$ and t:

$$\partial(n_\alpha <\chi>_\alpha)/\partial t - n_\alpha <\partial\chi \, \partial t>_\alpha + \vec{\nabla} \cdot (n_\alpha <\chi \, \vec{v}>_\alpha) -$$

$$- n_\alpha <(\vec{v} \cdot \vec{\nabla})\chi>_\alpha - n_\alpha <\vec{a} \cdot \vec{\nabla}_v \chi>_\alpha = [\, \delta(n_\alpha <\chi>_\alpha)/\delta t \,]_{coll}$$

8.7 - Consider the general transport equation of Problem 8.6, and let the property $\chi(\vec{r},\vec{v},t)$ be the random flux of thermal kinetic energy, that is, $m_\alpha c_\alpha^2 \vec{c}_\alpha/2$, where $\vec{c}_\alpha = \vec{v} - \vec{u}_\alpha(\vec{r},t)$. Show that [with $\vec{a} = (q_\alpha/m_\alpha)\,(\vec{E} + \vec{v} \times \vec{B})$],

$$\partial(n_\alpha <\chi>_\alpha)/\partial t = \partial\vec{q}_\alpha/\partial t$$

$$n_\alpha <\partial\chi/\partial t>_\alpha = -\,(\partial\vec{u}_\alpha/\partial t) \cdot (\bar{\bar{p}}_\alpha + 3p_\alpha \bar{\bar{1}}/2)$$

$$\vec{\nabla} \cdot (n_\alpha <\chi\vec{v}>_\alpha) = \vec{\nabla} \cdot (\rho_{m\alpha} <c_\alpha^2 \, \vec{c}_\alpha \, \vec{c}_\alpha>/2 + \vec{u}_\alpha \, \vec{q}_\alpha)$$

$$n_\alpha <(\vec{v} \cdot \vec{\nabla}) \, \chi>_\alpha = -\,(\bar{\bar{\bar{Q}}}_\alpha \cdot \vec{\nabla}) \cdot \vec{u}_\alpha - (\vec{q}_\alpha \cdot \vec{\nabla})\,\vec{u}_\alpha - \left[(\vec{u}_\alpha \cdot \vec{\nabla})\,\vec{u}_\alpha \right] \cdot$$

$$\cdot\, (\bar{\bar{p}}_\alpha + 3p_\alpha \, \bar{\bar{1}}/2)$$

$$n_\alpha \ \langle \vec{a} \cdot \vec{\nabla}_v \ \chi \rangle_\alpha \ = \rho_{m\alpha} \ \langle \vec{a} \cdot (\vec{c}_\alpha \ \vec{c}_\alpha \ + \ c_\alpha^2 \ \bar{\bar{1}}/2) \rangle \ =$$

$$= \ (q_\alpha/m_\alpha) \ \left[(\vec{E} + \vec{u}_\alpha \times \vec{B}) \ \cdot (\bar{\bar{p}}_\alpha + \ 3p_\alpha \ \bar{\bar{1}}/2) + \vec{q}_\alpha \times \vec{B} \right]$$

Using these results in the general transport equation, derive the following equation, known as the *heat flow equation,*

$$\frac{\partial}{\partial t} \vec{q}_\alpha \ - \frac{1}{\rho_{m\alpha}} (\vec{\nabla} \cdot \bar{\bar{p}}_\alpha) \cdot (\bar{\bar{p}}_\alpha + \frac{3}{2} \ p_\alpha \ \bar{\bar{1}}) + \vec{\nabla} \cdot (\frac{1}{2} \ \rho_{m\alpha} \ \langle c_\alpha^2 \ \vec{c}_\alpha \ \vec{c}_\alpha \rangle +$$

$$+ \ \vec{u}_\alpha \ \vec{q}_\alpha) \ + \ (\bar{\bar{\bar{Q}}}_\alpha \cdot \vec{\nabla}) \cdot \vec{u}_\alpha + \ (\vec{q}_\alpha \cdot \vec{\nabla}) \ \vec{u}_\alpha \ - (q_\alpha/m_\alpha) \ (\vec{q}_\alpha \times \ \vec{B}) \ =$$

$$= \ (\delta \vec{q}_\alpha/\delta t)_{coll}$$

8.8 - In the general transport equation of Problem 8.6, consider that the property $\chi(\vec{r},\vec{v},t)$ is the random momentum flux, that is, $m_\alpha c_{\alpha j} c_{\alpha k}$. Show that

$$\partial(n_\alpha \langle \chi \rangle_\alpha)/\partial t \ = \ \partial(p_{\alpha jk})/\partial t$$

$$n_\alpha \ \langle \partial\chi/\partial t \rangle_\alpha \ = \ 0$$

$$\frac{\partial}{\partial x_i} (n_\alpha \ \langle \chi v_i \rangle_\alpha) \ = \ \frac{\partial}{\partial x_i} \ (Q_{\alpha ijk} \ + u_{\alpha i} \ p_{\alpha jk})$$

$$n_\alpha \ \langle v_i (\partial\chi/\partial v_i) \rangle_\alpha \ = \ -p_{\partial ij} \ \frac{\partial u_{\alpha k}}{\partial x_i} \ - \ p_{\alpha ik} \ \frac{\partial u_{\alpha j}}{\partial x_i}$$

$$n_\alpha \ \langle a_i (\partial\chi/\partial v_i) \rangle_\alpha \ = \ \rho_{m\alpha} \ \langle a_j \ c_{\alpha k} \ + a_k \ c_{\alpha j} \rangle$$

where the summation convention on repeated indices is being used. Plug these results in the general transport equation to derive the following equation, known as the *viscous stress equation,*

$$\frac{\partial}{\partial t}(p_{\alpha jk}) + \frac{\partial}{\partial x_i}(Q_{\alpha ijk} + u_{\alpha i}\, p_{\alpha jk}) + p_{\alpha ij}\frac{\partial u_{\alpha k}}{\partial x_i} + p_{\alpha ik}\frac{\partial u_{\alpha j}}{\partial x_i} -$$

$$- \rho_{m\alpha} \langle a_j\, c_{\alpha k} + a_k\, c_{\alpha j} \rangle = (\delta p_{\alpha jk}/\delta t)_{coll}$$

8.9 - Verify that the energy conservation equation (for the random kinetic energy $m_\alpha c_\alpha^2/2$) can be obtained from the viscous stress equation (see Problem 8.8) by letting $j = k$, and summing over k.

8.10 - From the heat flow equation, derived in Problem 8.7, obtain the following simplified equation for heat flow in a stationary ($\vec{u} = 0$) electron gas

$$(5p_e/2)\ \vec{\nabla}(p_e/\rho_{me}) + \omega_{ce}\ (\vec{q}_e \times \hat{B}) = (\delta\vec{q}_e/\delta t)_{coll}$$

State all the assumptions necessary to obtain this result.

8.11 - Using the relaxation model (or Krook collision model) for the collision term, $(\delta f_\alpha/\delta t) = -\nu(f_\alpha - f_{o\alpha})$, and the ideal gas law $p_e = n_e\, k\, T_e$, show that the heat flow equation of Problem 8.10 becomes

$$\vec{q}_e + (\omega_{ce}/\nu)\ (\vec{q}_e \times \hat{B}) = -K_o\ \vec{\nabla}\, T_e$$

where $K_o = 5kp_e/2m_e\nu$ is the thermal conductivity.

CHAPTER 9

Macroscopic Equations for a Conducting Fluid

1. MACROSCOPIC VARIABLES FOR A PLASMA AS A CONDUCTING FLUID

A plasma can also be considered as a conducting fluid, without specifying its various individual species. The macroscopic transport equations, derived in the previous chapter, describe the macroscopic behavior of each individual plasma species (electrons, ions, neutral particles). We will determine now the set of transport equations which describe the behavior of the plasma as a whole. Each macroscopic variable is combined, by adding the contributions of the various particle species in the plasma. This procedure yields the *total* macroscopic parameters of interest, such as the total mass and charge densities, the total mass and charge current densities (or flux), the total kinetic pressure dyad and the total heat flux vector.

The *mass density* is the mass per unit volume of fluid and is given by

$$\rho_m = \sum_\alpha \rho_{m\alpha} = \sum_\alpha n_\alpha m_\alpha \tag{1.1}$$

The *electric charge density* is the electric charge per unit volume of fluid,

$$\rho = \sum_\alpha n_\alpha q_\alpha \tag{1.2}$$

The *mean fluid velocity*, $\vec{u}$, is defined such that the momentum density is the same as if each particle was moving at the mean fluid velocity, according to

$$\rho_m \vec{u} = \sum_\alpha \rho_{m\alpha} \vec{u}_\alpha \tag{1.3}$$

The mean velocity of the plasma, $\vec{u}$, is therefore a weighted mean value, where each species is weighted proportionally to its mass density. The mean velocity of each particle species, when considered in a reference system moving with the global mean velocity $\vec{u}$ of the plasma, is called the *diffusion velocity* $\vec{w}_\alpha$,

$$\vec{w}_\alpha = \vec{u}_\alpha - \vec{u} = \vec{u}_\alpha - (1/\rho_m) \sum_\alpha \rho_{m\alpha} \vec{u}_\alpha \tag{1.4}$$

The *mass current density* or *mass flux* is given by

$$\vec{J}_m = \sum_\alpha n_\alpha m_\alpha \vec{u}_\alpha = \rho_m \vec{u} \tag{1.5}$$

and the *electric current density* or *charge flux* is expressed as

$$\vec{J} = \sum_\alpha n_\alpha q_\alpha \vec{u}_\alpha = \rho \vec{u} + \sum_\alpha n_\alpha q_\alpha \vec{w}_\alpha \tag{1.6}$$

Note that in (1.5) we have $\sum_\alpha \rho_{m\alpha} \vec{w}_\alpha = 0$, in virtue of (1.4), which defines the diffusion velocity $\vec{w}_\alpha$.

The *kinetic pressure dyad* for each species of particles in the plasma is defined in (6.6.2) as

$$\bar{\bar{p}}_\alpha = \rho_{m\alpha} < \vec{c}_\alpha \vec{c}_\alpha > \tag{1.7}$$

where $\vec{c}_\alpha = \vec{v} - \vec{u}_\alpha$ is the peculiar or random velocity of the type α particles. Note that the pressure is defined as the time rate in which momentum is transported by the particles through a surface element moving with the particle mean velocity. For the plasma as a whole it is necessary to define a peculiar velocity $\vec{c}_{\alpha 0}$, for the particles of type α, relative to the global plasma mean velocity $\vec{u}$,

$$\vec{c}_{\alpha 0} = \vec{v} - \vec{u} \tag{1.8}$$

Thus, the total pressure is defined as the rate of momentum transfer due to all particles in the plasma, through a surface element moving with the global mean velocity $\vec{u}$. The *total kinetic pressure dyad* $\bar{\bar{p}}$ is therefore given by

$$\bar{\bar{p}} = \sum_{\alpha} \rho_{m\alpha} < \vec{c}_{\alpha o} \; \vec{c}_{\alpha o} > \tag{1.9}$$

To relate $\bar{\bar{p}}$, given in (1.9), with $\bar{\bar{p}}_{\alpha}$, given in (1.7), we substitute $\vec{u}$ by $\vec{u}_{\alpha} - \vec{w}_{\alpha}$ and $\vec{v}$ by $\vec{c}_{\alpha} + \vec{u}_{\alpha}$ in (1.8), which gives

$$\vec{c}_{\alpha o} = \vec{c}_{\alpha} + \vec{w}_{\alpha} \tag{1.10}$$

Consequently,

$$\bar{\bar{p}} = \sum_{\alpha} \rho_{m\alpha} < (\vec{c}_{\alpha} + \vec{w}_{\alpha})(\vec{c}_{\alpha} + \vec{w}_{\alpha}) > \tag{1.11}$$

and expanding this expression,

$$\bar{\bar{p}} = \sum_{\alpha} \rho_{m_{\alpha}} (< \vec{c}_{\alpha} \; \vec{c}_{\alpha} > + < \vec{c}_{\alpha} \; \vec{w}_{\alpha} > + < \vec{w}_{\alpha} \; \vec{c}_{\alpha} > + < \vec{w}_{\alpha} \; \vec{w}_{\alpha} >) \tag{1.12}$$

From the definition of $\vec{w}_{\alpha}$ we see that $< \vec{w}_{\alpha} > = \vec{w}_{\alpha}$, since it is a macroscopic variable and therefore $< \vec{c}_{\alpha} \; \vec{w}_{\alpha} > = < \vec{c}_{\alpha} > \vec{w}_{\alpha} = 0$. Thus, (1.12) becomes

$$\bar{\bar{p}} = \sum_{\alpha} \bar{\bar{p}}_{\alpha} + \sum_{\alpha} \rho_{m\alpha} \; \vec{w}_{\alpha} \; \vec{w}_{\alpha} \tag{1.13}$$

Note that $\bar{\bar{p}}_{\alpha}$ is a pressure relative to $\vec{u}_{\alpha}$, whereas $\bar{\bar{p}}$ is relative to the global mean velocity $\vec{u}$.

The *total scalar pressure* p is defined as one third the trace of $\bar{\bar{p}}$,

$$p = (1/3) \sum_{i} p_{ii} = \sum_{i} \sum_{\alpha} (\rho_{m\alpha}/3) < c_{\alpha oi} \;\; c_{\alpha oi} > = \sum_{\alpha} (\rho_{m\alpha}/3) < c_{\alpha o}^{2} > \tag{1.14}$$

Using (1.13) we can write

$$p = \sum_{\alpha} p_{\alpha} + \sum_{\alpha} (\rho_{m\alpha}/3) \; w_{\alpha}^{2} \tag{1.15}$$

Finally, we define the *total heat flux vector* $\vec{q}$ as

$$\vec{q} = \sum_\alpha (\rho_{m\alpha}/2) < c_{\alpha 0}^2 \, \vec{c}_{\alpha 0} > \tag{1.16}$$

and the *thermal energy density* of the plasma as a whole as

$$3p/2 = \sum_\alpha (\rho_{m\alpha}/2) < c_{\alpha 0}^2 > \tag{1.17}$$

It is useful to relate $\vec{q}$, defined in (1.16), with the heat flux vector $\vec{q}_\alpha$ for the particles of type α,

$$\vec{q}_\alpha = (\rho_{m\alpha}/2) < c_\alpha^2 \, \vec{c}_\alpha > \tag{1.18}$$

For this purpose, we substitute $\vec{c}_{\alpha 0}$ by $\vec{c}_\alpha + \vec{w}_\alpha$ in (1.16) and expand the resulting expression, obtaining

$$\vec{q} = \sum_\alpha (\rho_{m\alpha}/2) \Big[< c_\alpha^2 \, \vec{c}_\alpha > + \, w_\alpha^2 < \vec{c}_\alpha > + \, 2 < (\vec{w}_\alpha \cdot \vec{c}_\alpha) \, \vec{c}_\alpha > + \\ + < c_\alpha^2 > \vec{w}_\alpha + w_\alpha^2 \, \vec{w}_\alpha + 2 \, (< \vec{c}_\alpha > \cdot \, \vec{w}_\alpha) \, \vec{w}_\alpha \Big] \tag{1.19}$$

The second and sixth terms in the right-hand side of this equation are equal to zero, since $< \vec{c}_\alpha > = 0$. Therefore,

$$\vec{q} = \sum_\alpha (\rho_{m\alpha}/2) \Big[< c_\alpha^2 \, \vec{c}_\alpha > + 2 \, \vec{w}_\alpha \cdot < \vec{c}_\alpha \, \vec{c}_\alpha > + < c_\alpha^2 > \vec{w}_\alpha + w_\alpha^2 \, \vec{w}_\alpha \Big] \tag{1.20}$$

Using (1.18), (1.7) and the relation $p_\alpha = \rho_{m\alpha} < c_\alpha^2 > /3$, we can write (1.20) as

$$\vec{q} = \sum_\alpha (\vec{q}_\alpha + \vec{w}_\alpha \cdot \bar{\bar{p}}_\alpha + 3p_\alpha \vec{w}_\alpha/2 + \rho_{m\alpha} \, w_\alpha^2 \, \vec{w}_\alpha/2) \tag{1.21}$$

In particular, for the isotropic case in which $\bar{\bar{p}}_\alpha = p_\alpha \, \bar{\bar{1}}$, we have $\vec{w}_\alpha \cdot \bar{\bar{p}}_\alpha = \vec{w}_\alpha \, p_\alpha$, so that (1.21) becomes

$$\vec{q} = \sum_\alpha (\vec{q}_\alpha + 5p_\alpha \vec{w}_\alpha/2 + \rho_{m\alpha} \, w_\alpha^2 \, \vec{w}_\alpha/2) \tag{1.22}$$

2. CONTINUITY EQUATION

To obtain the continuity equation for the plasma as a whole, we add (8.3.2) over all particle species in the plasma,

$$\sum_\alpha \partial\rho_{m\alpha}/\partial t + \sum_\alpha \vec{\nabla} \cdot (\rho_{m\alpha} \vec{u}_\alpha) = \sum_\alpha S_\alpha \tag{2.1}$$

which gives

$$\partial\rho_m/\partial t + \vec{\nabla} \cdot (\rho_m \vec{u}) = 0 \tag{2.2}$$

with ρ_m and $\vec{u}$ given by (1.1) and (1.3), respectively. The collision term S_α, when summed over all particle species must certainly vanish, as a consequence of the conservation of the total mass of the system. It is of interest to note that, using the total time derivate $D/Dt = \partial/\partial t + \vec{u} \cdot \vec{\nabla}$, (2.2) can also be written in the form

$$D\rho_m/Dt + \rho_m \vec{\nabla} \cdot \vec{u} = 0 \tag{2.3}$$

3. EQUATION OF MOTION

Similarly, adding the equation of conservation of momentum (8.4.9) over all particle species in the plasma, yields

$$\sum_\alpha \rho_{m\alpha} \left[\partial\vec{u}_\alpha/\partial t + (\vec{u}_\alpha \cdot \vec{\nabla})\vec{u}_\alpha\right] = \sum_\alpha n_\alpha q_\alpha \vec{E} + \sum_\alpha n_\alpha q_\alpha \vec{u}_\alpha \times \vec{B} + \sum_\alpha \rho_{m\alpha} \vec{g} -$$

$$- \sum_\alpha \vec{\nabla} \cdot \bar{\bar{p}}_\alpha + \sum_\alpha \vec{A}_\alpha - \sum_\alpha \vec{u}_\alpha S_\alpha \tag{3.1}$$

Since the total momentum of the particles in the plasma is conserved, the collision term for momentum transfer vanishes when summed over all species. Using the definitions (1.1), (1.2) and (1.6), and the relation (1.13), we can write (3.1) as

$$\sum_\alpha \rho_{m\alpha} \left[\partial\vec{u}_\alpha/\partial t + (\vec{u}_\alpha \cdot \vec{\nabla}) \vec{u}_\alpha\right] = \rho \vec{E} + \vec{J} \times \vec{B} + \rho_m \vec{g} - \vec{\nabla} \cdot \bar{\bar{p}} +$$

$$+ \sum_\alpha \vec{\nabla} \cdot (\rho_{m\alpha} \vec{w}_\alpha \vec{w}_\alpha) - \sum_\alpha \vec{u}_\alpha S_\alpha \tag{3.2}$$

The term involving S_α can be eliminated using the equation of continuity,

$$\sum_\alpha \vec{u}_\alpha S_\alpha = \sum_\alpha \vec{u}_\alpha \left[\partial \rho_{m\alpha}/\partial t + \vec{\nabla} \cdot (\rho_{m\alpha} \vec{u}_\alpha) \right] \tag{3.3}$$

Combining this expression with the terms in the left-hand side of (3.2), results in the expression

$$\sum_\alpha \left[\partial(\rho_{m\alpha} \vec{u}_\alpha)/\partial t + \vec{\nabla} \cdot (\rho_{m\alpha} \vec{u}_\alpha \vec{u}_\alpha) \right] \tag{3.4}$$

We can now substitute the mean velocity $\vec{u}_\alpha$ by $\vec{w}_\alpha + \vec{u}$ and expand the result. Noting that

$$\sum_\alpha \rho_{m\alpha} \vec{w}_\alpha = \sum_\alpha \rho_{m\alpha} (\vec{u}_\alpha - \vec{u}) = \rho_m \vec{u} - \rho_m \vec{u} = 0 \tag{3.5}$$

we can express (3.4) as

$$\sum_\alpha \left[\partial(\rho_{m\alpha} \vec{u}_\alpha)/\partial t + \vec{\nabla} \cdot (\rho_{m\alpha} \vec{u}_\alpha \vec{u}_\alpha) \right] = \partial(\rho_m \vec{u})/\partial t + \vec{\nabla} \cdot (\rho_m \vec{u}\, \vec{u}) +$$

$$+ \sum_\alpha \vec{\nabla} \cdot (\rho_{m\alpha} \vec{w}_\alpha \vec{w}_\alpha) = \rho_m \left[\partial \vec{u}/\partial t + (\vec{u} \cdot \vec{\nabla})\, \vec{u} \right] + \vec{u} \left[\partial \rho_m/\partial t + \vec{\nabla} \cdot (\rho_m \vec{u}) \right] +$$

$$+ \sum_\alpha \vec{\nabla} \cdot (\rho_{m\alpha} \vec{w}_\alpha \vec{w}_\alpha) = \rho_m D\vec{u}/Dt + \sum_\alpha \vec{\nabla} \cdot (\rho_{m\alpha} \vec{w}_\alpha \vec{w}_\alpha) \tag{3.6}$$

where we have used the continuity equation (2.2) and the total time derivative. Taking this result back into the equation of motion (3.2), we obtain the following momentum equation for the plasma as a whole,

$$\rho_m D\vec{u}/Dt = \rho \vec{E} + \vec{J} \times \vec{B} + \rho_m \vec{g} - \vec{\nabla} \cdot \bar{\bar{p}} \tag{3.7}$$

This equation is an expression of Newton's second law of motion.

4. ENERGY EQUATION

To obtain the equation of conservation of energy, for the plasma as a conducting fluid, we start from the energy equation (8.5.4) for the particles of type α, and add this equation over all plasma species,

$$\sum_{\alpha} \frac{\partial}{\partial t} (\rho_{m\alpha} < v^2 >_{\alpha} /2) + \sum_{\alpha} \vec{\nabla} \cdot \left[\rho_{m\alpha} < v^2 \vec{v} >_{\alpha} /2 \right] - \sum_{\alpha} n_{\alpha} < \vec{F} \cdot \vec{v} >_{\alpha} = 0 \qquad (4.1)$$

where the collision term M_{α} vanishes when summed over all species of particles. We substitute now $\vec{v}$ by $\vec{c}_{\alpha o} + \vec{u}$ and expand each term of (4.1). For the *first term* we have

$$\frac{\partial}{\partial t} (\sum_{\alpha} \rho_{m\alpha} < \vec{v} \cdot \vec{v} >_{\alpha} /2) = \frac{\partial}{\partial t} \left[\sum_{\alpha} (\rho_{m\alpha}/2) (< c_{\alpha o}^2 > + u^2 + 2 \vec{w}_{\alpha} \cdot \vec{u}) \right]$$

$$= \frac{\partial}{\partial t} (\sum_{\alpha} (\rho_{m\alpha}/2) < c_{\alpha o}^2 >) + \frac{\partial}{\partial t} (\rho_m u^2/2)$$

$$= \frac{\partial}{\partial t} (3p/2) + \frac{\partial}{\partial t} (\rho_m u^2/2) \qquad (4.2)$$

where we have used the definition (1.17) and the fact that $\sum_{\alpha} \rho_{m\alpha} \vec{w}_{\alpha} = 0$. For the *second term* we note initially that

$$< v^2 \vec{v} >_{\alpha} = < (c_{\alpha o}^2 + u^2 + 2 \vec{c}_{\alpha o} \cdot \vec{u}) (\vec{c}_{\alpha o} + \vec{u}) >$$

$$= < c_{\alpha o}^2 \vec{c}_{\alpha o} > + u^2 \vec{w}_{\alpha} + 2 < \vec{c}_{\alpha o} \vec{c}_{\alpha o} > \cdot \vec{u} +$$

$$+ < c_{\alpha o}^2 > \vec{u} + u^2 \vec{u} + 2 (\vec{w}_{\alpha} \cdot \vec{u}) \vec{u} \qquad (4.3)$$

since $\vec{c}_{\alpha o} = \vec{c}_{\alpha} + \vec{w}_{\alpha}$ and $< \vec{c}_{\alpha} > = 0$. Therefore,

$$\vec{\nabla} \cdot (\sum_{\alpha} \frac{1}{2} \rho_{m\alpha} < v^2 \vec{v} >_{\alpha}) = \vec{\nabla} \cdot (\sum_{\alpha} \frac{1}{2} \rho_{m\alpha} < c_{\alpha o}^2 \vec{c}_{\alpha o} >) + \vec{\nabla} \cdot (\sum_{\alpha} \rho_{m\alpha} < \vec{c}_{\alpha o} \vec{c}_{\alpha o} > \cdot \vec{u}) +$$

$$+ \vec{\nabla} \cdot (\sum_\alpha \frac{1}{2} \rho_{m\alpha} < c^2_{\alpha o} > \vec{u}) + \vec{\nabla} \cdot (\sum_\alpha \frac{1}{2} \rho_{m\alpha} u^2 \vec{u}) \tag{4.4}$$

Using the definitions of the total heat flux vector $\vec{q}$ and of the total kinetic pressure dyad $\bar{\bar{p}}$, we can write (4.4) as

$$\vec{\nabla} \cdot (\sum_\alpha \frac{1}{2} \rho_{m\alpha} < v^2 \vec{v} >_\alpha) = \vec{\nabla} \cdot \vec{q} + \vec{\nabla} \cdot (\bar{\bar{p}} \cdot \vec{u}) + \vec{\nabla} \cdot (\frac{3}{2} p \vec{u}) + \vec{\nabla} \cdot (\rho_m u^2 \vec{u}/2) \tag{4.5}$$

For the *third term* of (4.1) we have

$$\sum_\alpha n_\alpha < \vec{F} \cdot \vec{v} >_\alpha = \sum_\alpha n_\alpha \left[q_\alpha < \vec{E} \cdot \vec{v} >_\alpha + q_\alpha < (\vec{v} \times \vec{B}) \cdot \vec{v} >_\alpha + m_\alpha < \vec{g} \cdot \vec{v} >_\alpha \right] \tag{4.6}$$

where we have considered external forces due to electromagnetic and gravitational fields. Since $< \vec{v} >_\alpha = \vec{u}_\alpha$ and since, for any vector $\vec{v}$, we have $(\vec{v} \times \vec{B}) \cdot \vec{v} = 0$, we obtain

$$\sum_\alpha n_\alpha < \vec{F} \cdot \vec{v} >_\alpha = \vec{J} \cdot \vec{E} + \vec{J}_m \cdot \vec{g} \tag{4.7}$$

where we have used the definitions (1.5) and (1.6), and where $\vec{E}$ and $\vec{g}$ are smoothed macroscopic fields.

Combining the results contained in (4.2), (4.5) and (4.7), the energy equation becomes

$$\frac{\partial}{\partial t} (\frac{3}{2} p) + \vec{\nabla} \cdot (\frac{3}{2} p \vec{u}) + \frac{\partial}{\partial t} (\frac{1}{2} \rho_m u^2) + \vec{\nabla} \cdot (\frac{1}{2} \rho_m u^2 \vec{u}) + \vec{\nabla} \cdot \vec{q} +$$

$$+ \vec{\nabla} \cdot (\bar{\bar{p}} \cdot \vec{u}) - \vec{J} \cdot \vec{E} - \vec{J}_m \cdot \vec{g} = 0 \tag{4.8}$$

This equation can be further simplified as follows. The third and fourth terms of (4.8) can be combined as

$$\frac{\partial}{\partial t} (\frac{1}{2} \rho_m u^2) + \vec{\nabla} \cdot (\frac{1}{2} \rho_m u^2 \vec{u}) \equiv \frac{1}{2} u^2 \left[\frac{\partial \rho_m}{\partial t} + \vec{\nabla} \cdot (\rho_m \vec{u}) \right] + \vec{u} \cdot (\rho_m D\vec{u}/Dt) \tag{4.9}$$

and using the continuity equation (2.2) and the equation of motion (3.7), we can express (4.9) as

$$\rho \vec{u} \cdot \vec{E} + \vec{u} \cdot (\vec{J} \times \vec{B}) + \vec{J}_m \cdot \vec{g} - \vec{u} \cdot (\vec{\nabla} \cdot \bar{\bar{p}}) \tag{4.10}$$

Taking this result back into the energy equation (4.8), yields

$$\frac{D}{Dt}\left(\frac{3}{2}p\right) + \frac{3}{2}p\,\vec{\nabla} \cdot \vec{u} + \vec{\nabla} \cdot \vec{q} + (\bar{\bar{p}} \cdot \vec{\nabla}) \cdot \vec{u} = \vec{J} \cdot \vec{E} - \vec{u} \cdot (\vec{J} \times \vec{B}) - \rho\,\vec{u} \cdot \vec{E} \tag{4.11}$$

The first term in the left-hand side of (4.11) represents the time rate of change of the total thermal energy density of the plasma (3p/2) in a frame of reference moving with the global mean velocity $\vec{u}$. The second term contributes to this rate of change through the thermal energy transfered to this volume element, as a consequence of the particle motions. The third term represents the heat flux and the fourth one the work done on the volume element by the pressure forces (normal and tangential). The terms in the right-hand side of (4.11) represent the work done on the volume element by the electric field existing in the frame of reference moving with the global mean velocity $\vec{u}$. These last terms can be combined as follows. We note, initially, that the charge current density consists of two parts

$$\vec{J} = \sum_\alpha n_\alpha q_\alpha \vec{u}_\alpha = \sum_\alpha n_\alpha q_\alpha \vec{w}_\alpha + \sum_\alpha n_\alpha q_\alpha \vec{u} = \vec{J}' + \rho\vec{u} \tag{4.12}$$

where $\rho\vec{u}$ is the *convection* charge current density, which represents the flux of the space charge with velocity $\vec{u}$, and $\vec{J}'$ is the *conduction* charge current density, which represents the charge current density in the frame of reference moving with the global mean velocity $\vec{u}$. On the other hand, we can write

$$\vec{u} \cdot (\vec{J} \times \vec{B}) = - \vec{J} \cdot (\vec{u} \times \vec{B}) = - \vec{J}' \cdot (\vec{u} \times \vec{B}) \tag{4.13}$$

Substituting (4.13) and (4.12) into the energy equation (4.11), we obtain

$$\frac{D}{Dt}(3p/2) + (3p/2)\,\vec{\nabla} \cdot \vec{u} + \vec{\nabla} \cdot \vec{q} + (\bar{\bar{p}} \cdot \vec{\nabla}) \cdot \vec{u} = \vec{J}' \cdot \vec{E}' \tag{4.14}$$

where $\vec{E}' = \vec{E} + \vec{u} \times \vec{B}$ is the electric field existing in the reference system moving with the global mean velocity $\vec{u}$. The term $\vec{J}' \cdot \vec{E}'$ represents therefore the rate of change in the energy density due to Joule heating.

5. ELECTRODYNAMIC EQUATIONS FOR A CONDUCTING FLUID

In the previous sections we have derived the macroscopic transport equations for conservation of mass, of momentum and of energy in a conducting fluid. As mentioned before, this set of equations does not constitute a complete system, and it is necessary to truncate the hierarchy of macroscopic equations at some stage and to make some simplifying assumptions. The continuity equation relates the mass density ρ_m with the global mean velocity $\vec{u}$; the equation of motion, which specifies the variation of $\vec{u}$, involves also the total kinetic pressure dyad $\bar{\bar{p}}$; the energy equation, which specifies the rate of change of the total thermal energy density (3p/2), includes also the heat flux vector $\vec{q}$ (a more general energy equation would give as the variation of the total kinetic pressure dyad $\bar{\bar{p}}$, which would include the total heat flow triad $\bar{\bar{\bar{Q}}}$). We can continue taking moments of higher order and obtain, for example, the transport equation governing the variation of the heat flow triad $\bar{\bar{\bar{Q}}}$. To obtain a complete system it is essential therefore to truncate the hierarchy of transport equations at some point. However, even after this truncation, the remaining equations include the following electrodynamic variables: electric field $\vec{E}$, magnetic induction $\vec{B}$, charge current density $\vec{J}$, and charge density ρ. Besides the hydrodynamic transport equations, we need therefore ten electrodynamic equations which must relate the variations in $\vec{E}$, $\vec{B}$, $\vec{J}$ and ρ. These equations are considered next.

5.1 Maxwell curl equations

The following Maxwell equations

$$\vec{\nabla} \times \vec{E} = - \partial\vec{B}/\partial t \tag{5.1}$$

$$\vec{\nabla} \times \vec{B} = \mu_0(\vec{J} + \varepsilon_0 \, \partial\vec{E}/\partial t) \tag{5.2}$$

provide six component equations, which can be considered as the equations governing the variations of the electromagnetic fields $\vec{E}$ and $\vec{B}$.

5.2 Conservation of electric charge

The equation of conservation of charge can be obtained by multiplying the equation of conservation of mass (8.3.2) by q_α/m_α, and adding over all species,

$$\frac{\partial}{\partial t}\left(\sum_\alpha n_\alpha q_\alpha\right) + \vec{\nabla}\cdot\left(\sum_\alpha n_\alpha q_\alpha \vec{u}_\alpha\right) = \sum_\alpha (q_\alpha/m_\alpha)\, S_\alpha \tag{5.3}$$

Using the definitions of ρ and $\vec{J}$, and noting that the *total* electric charge does not change as a result of collisions, we obtain

$$\partial\rho/\partial t + \vec{\nabla}\cdot\vec{J} = 0 \tag{5.4}$$

It is worth noting here that (5.4) can also be derived, in an independent way, by considering Maxwell curl equation (5.2) and the Maxwell divergence equation

$$\vec{\nabla}\cdot\vec{E} = \rho/\varepsilon_0 \tag{5.5}$$

Taking the divergence of (5.2), yields

$$\vec{\nabla}\cdot\vec{J} + \varepsilon_0\, \partial(\vec{\nabla}\cdot\vec{E})/\partial t = 0 \tag{5.6}$$

since the divergence of the curl of a vector field vanishes identically. This last equation, combined with (5.5), yields the equation of conservation of charge (5.4). (5.4) and (5.5) cannot therefore be considered as independent. As we have just shown, Maxwell equations (5.2) and (5.5) imply in conservation of electric charge.

Another interesting aspect of Maxwell equations can be seen by taking the divergence of (5.1), which gives

$$\partial(\vec{\nabla}\cdot\vec{B})/\partial t = 0 \tag{5.7}$$

or

$$\vec{\nabla}\cdot\vec{B} = \text{constant} \tag{5.8}$$

Therefore, the Maxwell equation

$$\vec{\nabla} \cdot \vec{B} = 0 \tag{5.9}$$

can be considered as an *initial condition* for (5.1), since if we take $\vec{\nabla} \cdot \vec{B} = 0$ initially, (5.1) implies that this condition will remain satisfied for all subsequent times.

5.3 Generalized Ohm's Law

To obtain a differential equation governing the variation of the charge current density $\vec{J}$, we proceed in a way analogous to the derivation of (5.4). To this end, we multiply the equation of conservation of momentum (8.4.9) by q_α/m_α, and add over all particle species. This procedure leads to

$$\sum_\alpha n_\alpha q_\alpha (\partial \vec{u}_\alpha/\partial t) + \sum_\alpha n_\alpha q_\alpha (\vec{u}_\alpha \cdot \vec{\nabla}) \vec{u}_\alpha = \sum_\alpha n_\alpha (q_\alpha/m_\alpha) < \vec{F} >_\alpha -$$

$$- \vec{\nabla} \cdot \left[\sum_\alpha (q_\alpha/m_\alpha) \bar{\bar{p}}_\alpha \right] + \sum_\alpha (q_\alpha/m_\alpha) \vec{A}_\alpha - \sum_\alpha (q_\alpha/m_\alpha) \vec{u}_\alpha S_\alpha \tag{5.10}$$

We define now the *electrokinetic pressure dyad* $\bar{\bar{p}}_\alpha^E$ for the particles of type α, by

$$\bar{\bar{p}}_\alpha^E = (q_\alpha/m_\alpha) \bar{\bar{p}}_\alpha = n_\alpha q_\alpha < \vec{c}_\alpha \vec{c}_\alpha > \tag{5.11}$$

Consequently, for the plasma as a conducting fluid, we have the following relation analogous to (1.13)

$$\bar{\bar{p}}^E = \sum_\alpha \bar{\bar{p}}_\alpha^E + \sum_\alpha n_\alpha q_\alpha \vec{w}_\alpha \vec{w}_\alpha \tag{5.12}$$

The *second term* in the right-hand side of (5.10) becomes therefore

$$- \vec{\nabla} \cdot \left[\sum_\alpha (q_\alpha/m_\alpha) \bar{\bar{p}}_\alpha \right] = - \vec{\nabla} \cdot \bar{\bar{p}}^E + \vec{\nabla} \cdot (\sum_\alpha n_\alpha q_\alpha \vec{w}_\alpha \vec{w}_\alpha) \tag{5.13}$$

Using the continuity equation (8.3.2) and substituting $\vec{u}_\alpha$ by $\vec{w}_\alpha + \vec{u}$, the *last term* in the right-hand side of (5.10) can be written

$$-\sum_\alpha (q_\alpha/m_\alpha)\, \vec{u}_\alpha\, S_\alpha = -\sum_\alpha \vec{w}_\alpha\, \partial(n_\alpha q_\alpha)/\partial t - \sum_\alpha \vec{w}_\alpha \left[\vec{\nabla}\cdot(n_\alpha\, q_\alpha\, \vec{w}_\alpha)\right] -$$

$$-\sum_\alpha \vec{w}_\alpha \left[\vec{\nabla}\cdot(n_\alpha\, q_\alpha\, \vec{u})\right] - \vec{u}\,(\partial\rho/\partial t) - \vec{u}\,(\vec{\nabla}\cdot\vec{J}) \qquad (5.14)$$

Similarly, the *first* and *second terms* in the left-hand side of (5.10) can be combined in the form

$$\sum_\alpha n_\alpha\, q_\alpha\, (\partial\vec{w}_\alpha/\partial t) + \sum_\alpha (n_\alpha\, q_\alpha\, \vec{w}_\alpha\cdot\vec{\nabla})\, \vec{w}_\alpha + \sum_\alpha (n_\alpha\, q_\alpha\, \vec{u}\cdot\vec{\nabla})\, \vec{w}_\alpha +$$

$$+ \rho\,(\partial\vec{u}/\partial t) + (\vec{J}\cdot\vec{\nabla})\,\vec{u} \qquad (5.15)$$

We can now substitute expressions (5.13), (5.14) and (5.15) into (5.10) and simplify the result. Making use of the following identity for two vectors $\vec{a}$ and $\vec{b}$,

$$\vec{\nabla}\cdot(\vec{a}\,\vec{b}) = \vec{b}\,(\vec{\nabla}\cdot\vec{a}) + (\vec{a}\cdot\vec{\nabla})\,\vec{b} \qquad (5.16)$$

and the relation (4.12), we obtain

$$\partial\vec{J}/\partial t + \vec{\nabla}\cdot(\vec{u}\,\vec{J}' + \vec{J}\,\vec{u}) + \vec{\nabla}\cdot\bar{\bar{p}}^E = \sum_\alpha n_\alpha\,(q_\alpha/m_\alpha) < \vec{F} >_\alpha +$$

$$+ \sum_\alpha (q_\alpha/m_\alpha)\, \vec{A}_\alpha \qquad (5.17)$$

Equations (5.1), (5.2), (5.4) and (5.17) constitute ten component equations which complement the equations of conservation of mass, of momentum and of energy for a conducting fluid. (5.17), however, is still in a very general form of little practical value. A very useful and simple expression exists for the case of a completely ionized plasma consisting of electrons and only one type of ions. In what follows, we simplify (5.17) for this case.

The electric charge current density $\vec{J}$ and the electric charge density ρ

for a completely ionized plasma containing only electrons and one type of ions of charge e are given, respectively, by

$$\vec{J} = \sum_\alpha n_\alpha q_\alpha \vec{u}_\alpha = e\,(n_i \vec{u}_i - n_e \vec{u}_e) \tag{5.18}$$

$$\rho = \sum_\alpha n_\alpha q_\alpha = e\,(n_i - n_e) \tag{5.19}$$

The global mean velocity $\vec{u}$, defined in (1.3), becomes

$$\vec{u} = (1/\rho_m)\,(\rho_{me} \vec{u}_e + \rho_{mi} \vec{u}_i) \tag{5.20}$$

where $\rho_m = \rho_{me} + \rho_{mi}$. Combining this last equation with (5.18) gives

$$\vec{u}_i = (\mu/\rho_{mi})\,(\rho_m \vec{u}/m_e + \vec{J}/e) \tag{5.21}$$

$$\vec{u}_e = (\mu/\rho_{me})\,(\rho_m \vec{u}/m_i - \vec{J}/e) \tag{5.22}$$

where $\mu = m_e m_i/(m_e + m_i)$ denotes the reduced mass.

We assume now that the mean velocity of the electrons and ions, relative to the global mean velocity $\vec{u}$, (that is, the diffusion velocities $\vec{w}_e$ and $\vec{w}_i$) are small compared with the thermal velocities. This condition being satisfied, (5.12) becomes

$$\bar{\bar{p}}^E = \bar{\bar{p}}_i^E + \bar{\bar{p}}_e^E = e\,(\bar{\bar{p}}_i/m_i - \bar{\bar{p}}_e/m_e) \tag{5.23}$$

Considering the conducting fluid immersed in an electromagnetic field, the term containing the external force in (5.17) becomes

$$\sum_\alpha n_\alpha (q_\alpha/m_\alpha) \langle \vec{F} \rangle_\alpha = \sum_\alpha n_\alpha (q_\alpha/m_\alpha)\,[q_\alpha(\vec{E} + \vec{u}_\alpha \times \vec{B})]$$

$$= e^2\,(n_i/m_i + n_e/m_e)\,\vec{E} +$$

$$+ e^2 \, [(n_i/m_i) \, \vec{u}_i + (n_e/m_e) \, \vec{u}_e] \times \vec{B} \tag{5.24}$$

Substituting the relations (5.21) and (5.22) in this last equation and simplifying, yields

$$\sum_\alpha n_\alpha \, (q_\alpha/m_\alpha) < \vec{F} >_\alpha = e^2 \, (n_i/m_i + n_e/m_e) \, \vec{E} + e^2 \, (n_i/m_e + n_e/m_i) \, \vec{u} \times \vec{B} +$$

$$+ e \, (1/m_i - 1/m_e) \, \vec{J} \times \vec{B} \tag{5.25}$$

It is convenient at this moment to simplify this equation by making one additional approximation. Since the ion mass m_i is much larger than the electron mass m_e (for protons and electrons, for example, $m_i/m_e \cong 1836$) and assuming macroscopic charge neutrality, $n_e = n_i = n$, we can take

$$1/m_i - 1/m_e \cong - 1/m_e \tag{5.26}$$

$$n_i/m_i + n_e/m_e \cong n/m_e \tag{5.27}$$

$$n_i/m_e + n_e/m_i \cong n/m_e \tag{5.28}$$

Consequently, from (5.23) we have $\bar{\bar{p}}^E = - \, e \, \bar{\bar{p}}_e/m_e$, and from (5.25)

$$\sum_\alpha n_\alpha \, (q_\alpha/m_\alpha) < \vec{F} >_\alpha = (ne^2/m_e) \, (\vec{E} + \vec{u} \times \vec{B}) - (e/m_e) \, \vec{J} \times \vec{B} \tag{5.29}$$

For the collision term in (5.17) we make use of expression (8.4.11), that is

$$\vec{A}_e = - \, \rho_{me} \, \nu_{ei} \, (\vec{u}_e - \vec{u}_i) \tag{5.30}$$

$$\vec{A}_i = - \, \rho_{mi} \, \nu_{ie} \, (\vec{u}_i - \vec{u}_e) \tag{5.31}$$

From (8.4.13) we have $\rho_{mi} \, \nu_{ie} = \rho_{me} \, \nu_{ei}$, so that

$$\sum_{\alpha} (q_\alpha/m_\alpha)\, \vec{A}_\alpha = e\, \rho_{me}\, \nu_{ei}\, (\vec{u}_e - \vec{u}_i)(1/m_i + 1/m_e) = -\, \nu_{ei}\, \vec{J} \tag{5.32}$$

where we have used (5.18) for $\vec{J}$, and the approximations $m_i \gg m_e$ and $n_e = n_i = n$.

We can now substitute the results contained in (5.23), (5.29) and (5.32) into (5.17), to obtain

$$\partial\vec{J}/\partial t + \vec{\nabla}\cdot(\vec{u}\,\vec{J} + \vec{J}\,\vec{u}) - (e/m_e)\,\vec{\nabla}\cdot\bar{\bar{p}}_e =$$

$$= (ne^2/m_e)\,(\vec{E} + \vec{u}\times\vec{B}) - (e/m_e)\,\vec{J}\times\vec{B} - \nu_{ei}\,\vec{J} \tag{5.33}$$

Note that, since we assumed $n_e = n_i$, we must have $\rho = 0$ and $\vec{J}' = \vec{J}$. In some situations in which $\vec{J}$ and $\vec{u}$ can be considered as small perturbations, the nonlinear terms involving their product may be neglected compared to the other terms. With this simplifying approximation and using the notation

$$\sigma_0 = ne^2/(m_e\, \nu_{ei}) \tag{5.34}$$

which represents the longitudinal electrical conductivity, we get for (5.33)

$$\left(\frac{m_e}{ne^2}\right)\frac{\partial\vec{J}}{\partial t} - \left(\frac{1}{ne}\right)\vec{\nabla}\cdot\bar{\bar{p}}_e = \vec{E} + \vec{u}\times\vec{B} - \left(\frac{1}{ne}\right)\vec{J}\times\vec{B} - \vec{J}/\sigma_0 \tag{5.35}$$

This equation is known as the *generalized Ohm's law.* The terms is the right-hand side are the ones normally retained in magnetohydrodynamics while all the others are neglected. The omission of the terms in the left-hand side of (5.35) is, generally, not always justifiable.

For cases in which $\vec{J}$ does not vary with time, that is, under steady state conditions, we have $\partial\vec{J}/\partial t = 0$. If we consider also that the pressure term in (5.35) is negligible, that is, $\vec{\nabla}\cdot\bar{\bar{p}}_e = 0$, then (5.35) simplifies to

$$\vec{J} = \sigma_0\,(\vec{E} + \vec{u}\times\vec{B}) - (\sigma_0/ne)\,\vec{J}\times\vec{B} \tag{5.36}$$

The last term in this equation is related to a phenomenon called the *Hall effect* in magnetohydrodynamic flow problems and, for this reason, it is normally called the Hall effect term. This term is small when $(\sigma_0|\vec{B}|/ne) \ll 1$, that is, when $\omega_{ce} \ll \nu_{ei}$. Thus, when the collision frequency is much larger than the magnetic gyrofrequency, the Hall effect term can be neglected and (5.36) reduces to

$$\vec{J} = \sigma_0(\vec{E} + \vec{u} \times \vec{B}) \tag{5.37}$$

In the absence of an external magnetic field, (5.37) reduces further to

$$\vec{J} = \sigma_0 \vec{E} \tag{5.38}$$

which is the expression commonly known as *Ohm's law.*

6. SIMPLIFIED MAGNETOHYDRODYNAMIC EQUATIONS

In the last two sections we have shown that the set of macroscopic transport equations for each individual species in the plasma can be substituted by transport equations for the whole plasma as a conducting fluid, complemented by the electrodynamic equations. These total macroscopic equations for a conducting fluid are known as the magnetohydrodynamic (MHD) equations. In their most general form they are essentially equivalent to the set of equations for each individual particle species. In practice, however, the MHD equations are seldom used in their general form. Several simplifying approximations are normally considered, based on physical arguments which permit the elimination of some of the terms in the equations. For steady state situations, or slowly varying problems, the MHD equations are very convenient and, in many cases, lead to results which would not be easily obtained from the individual equations for each particle species.

One of the approximations normally used in MHD consists in neglecting the term $\varepsilon_0\, \partial\vec{E}/\partial t$ in the Maxwell equation (5.2). To analyze the validity of this approximation it is convenient to use dimensional analysis, as follows. We can express, in general, the charge current density as $\vec{J} = \bar{\bar{\sigma}} \cdot \vec{E}$, so that, dimensionally, we have

$$J \cong \sigma E \tag{6.1}$$

$$\varepsilon_0\, |\partial\vec{E}/\partial t| \cong \varepsilon_0\, E/\tau \tag{6.2}$$

where τ represents a characteristic time for changes in the electric field and σ represents a characteristic conductivity. The ratio of the two terms in the right-hand side of (5.2) becomes, therefore,

$$\varepsilon_0 \, |\partial \vec{E}/\partial t| / J \cong \varepsilon_0 / \sigma\tau \tag{6.3}$$

For most of the fluids normally used in MHD problems, σ is typically greater than 1 mho/m, whereas ε_0 is of the order of 10^{-11} Farad/m. Consequently,

$$\varepsilon_0 \, |\partial \vec{E}/\partial t| / J \cong 10^{-11}/\tau \tag{6.4}$$

with τ in seconds, which shows that this approximation is not valid only when we are considering *extremely small* characteristic times.

It is also assumed that macroscopic electric neutrality is maintained with a high degree of accuracy and therefore the electric charge density ρ is set equal to zero.

A questionable approximation in the set of MHD equations is the generalized Ohm's law, in the form given in (5.36). In this form, the terms containing the time derivatives and pressure gradient (or divergence of the pressure dyad) are omitted, even though these terms are considered in other equations of the set. This approximation is not therefore justifiable in a direct manner. It is common to simply assume that all time derivatives are negligibly small and that the plasma is *almost* a cold plasma, so that the generalized Ohm's law reduces to the form given in (5.36).

For convenience, we collect here the following set of simplified magnetohydrodynamic equations

$$\partial \rho_m / \partial t + \vec{\nabla} \cdot (\rho_m \vec{u}) = 0 \tag{6.5}$$

$$\rho_m D\vec{u}/Dt = \vec{J} \times \vec{B} - \vec{\nabla} p \tag{6.6}$$

$$\vec{\nabla} p = V_s^2 \, \vec{\nabla} \rho_m \tag{6.7}$$

$$\vec{\nabla} \times \vec{E} = - \partial \vec{B}/\partial t \tag{6.8}$$

$$\vec{\nabla} \times \vec{B} = \mu_0 \vec{J} \tag{6.9}$$

$$\vec{J} = \sigma_0 (\vec{E} + \vec{u} \times \vec{B}) - (\sigma_0/ne) \vec{J} \times \vec{B} \tag{6.10}$$

In this set of equations, viscosity and thermal conductivity are neglected. The pressure dyad reduces therefore to a scalar pressure. Note that (6.9) implies in

$$\vec{\nabla} \cdot \vec{J} = 0 \tag{6.11}$$

which is the equation of conservation of electric charge in the absence of changes in the total macroscopic charge density ρ. It is for this reason that the equation of conservation of electric charge is not explicitly considered in the set of MHD equations (6.5) to (6.10). Except in some special circunstances, it is also common to neglect the Hall effect term $(\sigma_0/en) \vec{J} \times \vec{B}$ in (6.10).

PROBLEMS

9.1 - Show that the total kinetic energy density of all species in a fluid can be written as the sum of the thermal energy density of the whole fluid plus the kinetic energy of the mass motion, that is

$$\sum_\alpha (\rho_{m\alpha}/2) < v^2 >_\alpha = (3p/2) + \sum_\alpha (\rho_{m\alpha}/2) u_\alpha^2$$

where

$$3p/2 = \sum_\alpha (\rho_{m\alpha}/2) < c_{\alpha 0}^2 > = \sum_\alpha (\rho_{m\alpha}/2) < c_\alpha^2 > + \sum_\alpha (\rho_{m\alpha}/2) w_\alpha^2$$

9.2 - Show that when there is no heat flow ($\vec{q} = 0$) no Joule heating ($\vec{J}' \cdot \vec{E}' = 0$) and when the pressure tensor is isotropic given by $\bar{\bar{p}} = p \bar{\bar{1}}$, the energy equation (4.14) reduces to the following adiabatic equation

$$p \rho_m^{-5/3} = \text{constant}$$

9.3 - From the momentum conservation equation with the MHD approximation [see (6.6)], and the generalized Ohm's law in the simplified form (6.10), but without considering the Hall effect term, derive the following equation:

$$\rho_m \, D\vec{u}/Dt = \sigma_0 \, (\vec{E} \times \vec{B}) + \sigma_0 \, (\vec{u} \times \vec{B}) \times \vec{B} - \vec{\nabla} p$$

Solve this equation, considering that $\vec{E} = 0$ and p = constant, to show that the fluid velocity perpendicular to $\vec{B}$ is given by

$$u_\perp (t) = u_\perp (0) \exp (- t/\tau)$$

where τ is a characteristic time for diffusion of the fluid across the field lines, given by $\tau = \rho_m/\sigma_0 B^2$.

9.4 - In Eqs. (1.5) and (1.6) explain the reason why the mass flux $\vec{J}_m$ is given by $\rho_m \vec{u}$, whereas the electric charge flux J is *not* given by $\rho \, \vec{u}$.

9.5 - Obtain an expression for the heat flux triad $\bar{\bar{\bar{Q}}}$ for the plasma as a whole, defined as

$$\bar{\bar{\bar{Q}}} = \sum_\alpha \rho_{m\alpha} < \vec{c}_{\alpha 0} \, \vec{c}_{\alpha 0} \, \vec{c}_{\alpha 0} >$$

where $\vec{c}_{\alpha 0} = \vec{c}_\alpha + \vec{w}_\alpha$, in terms of a summation over the heat flux triad for each species $\bar{\bar{\bar{Q}}}_\alpha$ and of terms involving the diffusion velocity $\vec{w}_\alpha$. Then, simplify this expression for the isotropic case.

9.6 - Derive an energy equation [of higher order than (4.14)] involving the total time rate of change of the total pressure dyad, that is, $D\bar{\bar{p}}/Dt$.

9.7 - For a perfectly conducting fluid characterized by a scalar pressure, under steady state conditions, use the equation of motion (6.6) and the generalized Ohm's law (6.10) to derive the following equation for the fluid velocity component perpendicular to $\vec{B}$,

$$\vec{u}_\perp = [\vec{E} + (1/ne)\vec{\nabla} p] \times \vec{B}/B^2$$

CHAPTER 10
Plasma Conductivity and Diffusion

1. INTRODUCTION

In the previous chapters we have introduced the fundamental elements of kinetic theory and the macroscopic transport equations necessary for the study of a variety of important phenomena in plasmas. Many plasma phenomena can be analyzed using the macroscopic transport equations, either considering the plasma as a multi-constituent fluid or by treating the whole plasma as a single conducting fluid. In some cases, however, a satisfactory description can only be obtained through the use of kinetic theory.

In this and in the following chapter we investigate a number of basic plasma phenomena which illustrate the use of the cold and warm plasma models, and of the phase space distribution function. Phenomena that can be analyzed treating the whole plasma as a single conducting fluid are usually considered under the general title of magnetohydrodynamics (MHD), and will be studied in Chapters 12, 13 and 15.

2. THE LANGEVIN EQUATION

Before we consider the phenomena of plasma conductivity and diffusion, it is convenient to introduce a very simple form of the equation of motion for a weakly ionized cold plasma, known as the Langevin equation. In a weakly ionized plasma the number density of the charged particles is much smaller than that of the neutral particles. In this case the charge-neutral interactions are dominant. The *macroscopic* equation of motion for an *average* electron, under the action of the Lorentz force and the collisional forces, can be written as

$$m_e(D\vec{u}_e/Dt) = - e\,(\vec{E} + \vec{u}_e \times \vec{B}) + (\vec{F}_{coll})_e \qquad (2.1)$$

where $\vec{u}_e(\vec{r},t)$ is the average electron velocity and $(\vec{F}_{coll})_e$ denotes symbolically the rate of change of the average electron momentum due to collisions with neutral particles. The macroscopic collision term $(\vec{F}_{coll})_e$ can be expressed in a phenomenological way as the product of the average electron momentum with an *effective* constant collision frequency ν_c for momentum transfer between electrons and heavy (neutral) particles,

$$(\vec{F}_{coll})_e = - \nu_c m_e \vec{u}_e \tag{2.2}$$

In doing this we are neglecting the average motion of the neutral particles, as they are much more massive than the electrons. Note that this does not mean that the velocities of the individual neutral particles are zero, but only that they are completely *random* so that their *average* velocity is zero. We obtain therefore the following equation, known as the *Langevin equation*,

$$m_e (D\vec{u}_e/Dt) = - e (\vec{E} + \vec{u}_e \times \vec{B}) - \nu_c m_e \vec{u}_e \tag{2.3}$$

The effect of this collision term can be seen as follows. In the absence of electric and magnetic fields, (2.3) reduces to

$$D\vec{u}_e/Dt = - \nu_c \vec{u}_e \tag{2.4}$$

whose solution is

$$\vec{u}_e(t) = \vec{u}_e(0) \exp (-\nu_c t) \tag{2.5}$$

Thus, the electron-neutral collisions decrease the average electron velocity exponentially in time, at a rate governed by the collision frequency.

An equation analogous to (2.3) can be written for the ions

$$m_i (D\vec{u}_i/Dt) = Ze (\vec{E} + \vec{u}_i \times \vec{B}) + (\vec{F}_{coll})_i \tag{2.6}$$

where $\vec{u}_i$ denotes the average ion velocity and Ze the ion charge. In many cases of interest, as in high-frequency phenomena, we can neglect ion motion and assume $\vec{u}_i = 0$, since the ion mass is typically about 10^3 or 10^4 times greater than the electron mass. The type of plasma in which only the electron motion is important is usually called a *Lorentz gas*. When dealing with very low frequencies, however, the motion of the ions must be considered.

Despite the approximations implicit in the Langevin equation, it has been successfully used to describe a variety of phenomena in plasmas, including the propagation of electromagnetic waves in cold magnetoplasmas. Particularly, the analysis of the characteristics of electromagnetic wave propagation in the Earth's ionosphere has been quite successful. A great advantage of this equation is its simplicity.

3. LINEARIZATION OF THE LANGEVIN EQUATION

In the form presented in (2.3) the Langevin equation contains nonlinear terms which involve the product of two variables. In many situations of interest the difficulty inherent in the nonlinear terms can be eliminated through a linearization approximation, which is valid for small-amplitude variations.

The total time derivative contains the nonlinear convective term $(\vec{u}_e \cdot \vec{\nabla})\, \vec{u}_e$, which is called the inertial term in fluid dynamics. The omission of this inertial term is justified when the average velocity and its space derivatives are small, or when $\vec{u}_e$ is normal to its gradient (such as in the case of transverse waves).

For the nonlinear term $\vec{u}_e \times \vec{B}$, we can separate the magnetic flux density $\vec{B}(\vec{r},t)$ into two parts

$$\vec{B}(\vec{r},t) = \vec{B}_0 + \vec{B}'(\vec{r},t) \tag{3.1}$$

where $\vec{B}_0$ is constant and $\vec{B}'(\vec{r},t)$ is the variable component, so that

$$q\,(\vec{E} + \vec{u}_e \times \vec{B}) = q\,(\vec{E} + \vec{u}_e \times \vec{B}_0 + \vec{u}_e \times \vec{B}') \tag{3.2}$$

For situations in which

$$|\vec{u}_e \times \vec{B}'| << |\,\vec{E}\,| \tag{3.3}$$

the nonlinear term $\vec{u}_e \times \vec{B}'$ in (3.2) can be neglected. With the linearization approximations the Langevin equation becomes

$$m_e\,(\partial \vec{u}_e/\partial t) = -\,e\,(\vec{E} + \vec{u}_e \times \vec{B}_0) - m_e\,\nu_c\,\vec{u}_e \tag{3.4}$$

A case of great practical interest is the one in which the variables $\vec{E}$, $\vec{B}'$ and $\vec{u}_e$ all vary harmonically in space and time. The treatment in terms of plane waves has the advantage of great mathematical simplicity, besides the fact that any complex and physically realizable wave motion can be synthesized in terms of a superposition of plane waves. Let us consider therefore plane wave solutions for $\vec{E}$, $\vec{B}'$ and $\vec{u}_e$ in the form

$$\vec{E}, \vec{B}', \vec{u}_e \propto \exp [i(\vec{k}\cdot\vec{r} - \omega t)] \tag{3.5}$$

where ω denotes the angular frequency, $\vec{k}$ is the wave propagation vector (normal to the wave front) and $\vec{r}$ is a position vector drawn from the origin of a coordinate system to the point considered on the wave front (Fig. 1).

For the space and time dependence given in (3.5), the differential operators $\vec{\nabla}$ and $\partial/\partial t$ are transformed into simple algebraic operators, according to $\vec{\nabla} \to i\vec{k}$ and $\partial/\partial t \to -i\omega$. Substituting (3.1) into Maxwell equation $\vec{\nabla} \times \vec{E} = - \partial\vec{B}/\partial t$, we obtain

$$i\vec{k} \times \vec{E} = i\omega \vec{B}' \tag{3.6}$$

where $\partial\vec{B}_0/\partial t = 0$, since $\vec{B}_0$ is constant. Therefore,

$$\vec{B}' = (\vec{k} \times \vec{E})/\omega \tag{3.7}$$

and plugging this result back into (3.3) yields the condition

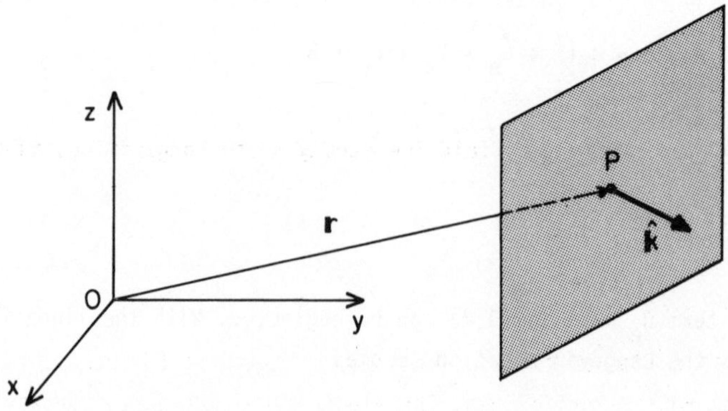

Fig. 1 - Position vector $\vec{r}$ drawn from the origin of a coordinate system (x,y,z) to a point P on the wave front, whose normal is given by the wave propagation vector $\vec{k}$.

$$|\vec{u}_e \times (\vec{k} \times \vec{E})/\omega| << |\vec{E}| \tag{3.8}$$

The magnitude of the nonlinear term $\vec{u}_e \times \vec{B}'$ may be equal to or smaller than $|u_e\ k\ E/\omega|$. Hence, the nonlinear term can be neglected if

$$|u_e\ k\ /\ \omega| << 1 \tag{3.9}$$

or, equivalently, if

$$|u_e| << |\omega/k| \tag{3.10}$$

The term (ω/k) represents the phase velocity of the plane wave. Since this term is usually of the order of the speed of light c, whereas the magnitude of the mean velocity of the electrons u_e is much less than c, the nonlinear term can generally be neglected. However, in cases of resonance ω/k is very small, whereas u_e becomes large. Under these conditions the nonlinear terms are important and a nonlinear analysis must be used.

4. DC CONDUCTIVITY AND ELECTRON MOBILITY

In this section we apply the steady state Langevin equation to derive an expression for the DC (direct current) conductivity of a weakly ionized homogeneous plasma, for which the Lorentz model (electron gas) is applicable. The applied electric field is assumed to be constant and uniform.

4.1 Isotropic Plasma

In the absence of a magnetic field the steady state Langevin equation for the electrons becomes

$$0 = - e\ \vec{E} - m_e\ \nu_c\ \vec{u}_e \tag{4.1}$$

In this case the action of the applied electric field is balanced dynamically by the electron-neutral collisions. The electric current density associated with the electron motion is

$$\vec{J} = - e\ n_e\ \vec{u}_e \tag{4.2}$$

Combining (4.1) and (4.2), gives

$$\vec{J} = (n_e e^2/m_e \nu_c) \vec{E} \tag{4.3}$$

From Ohm's law, $\vec{J} = \sigma_0 \vec{E}$, we identify the following expression for the *DC conductivity* of an isotropic electron gas

$$\sigma_0 = (n_e e^2/m_e \nu_c) \tag{4.4}$$

The *electron mobility* μ_e is defined as the ratio of the mean velocity of the electrons to the applied electric field,

$$\mu_e = u_e/E \tag{4.5}$$

Therefore, from (4.1) we obtain

$$\mu_e = - e/m_e \nu_c = - \sigma_0/n_e e \tag{4.6}$$

4.2 Anisotropic Magnetoplasma

In the presence of a magnetic field the plasma becomes spatially anisotropic. The steady state Langevin equation can be written as

$$0 = - e (\vec{E} + \vec{u}_e \times \vec{B}_0) - m_e \nu_c \vec{u}_e \tag{4.7}$$

where $\vec{B}_0$ is a constant and uniform magnetic field. Using (4.2),

$$(m_e \nu_c/n_e e) \vec{J} = e (\vec{E} + \vec{u}_e \times \vec{B}_0) \tag{4.8}$$

which may be written in the form

$$\vec{J} = \sigma_0 (\vec{E} + \vec{u}_e \times \vec{B}_0) \tag{4.9}$$

where σ_0 is given in (4.4). This last equation is a simplified form of the generalized Ohm's law (see Chapter 9).

At this point it is worth to consider a useful result which arises when the collisional effects are negligible. When $\nu_c \to 0$ the DC conductivity becomes very large ($\sigma_0 \to \infty$) so that we must have, from (4.9),

$$\vec{E} + \vec{u}_e \times \vec{B}_0 = 0 \tag{4.10}$$

This expression represents therefore the simplified form of the generalized Ohm's law for a plasma with a very large conductivity. In this case, taking the cross product of (4.10) with $\vec{B}_0$ and noting that

$$(\vec{u}_e \times \vec{B}_0) \times \vec{B}_0 = - \vec{u}_{e_\perp} B_0^2 \tag{4.11}$$

we obtain

$$\vec{u}_{e_\perp} = (\vec{E} \times \vec{B}_0)/B_0^2 \tag{4.12}$$

This result shows that, in the absence of collisions, the electrons have a *drift velocity* $\vec{u}_{e_\perp}$ perpendicular to both the electric and the magnetic fields. Since this result is independent of the particle mass and charge, the same result will be obtained for the ions if their motion is taken into account. This can be easily shown considering the Langevin equation for the ions. Thus, in the absence of collisions, both electrons and ions move together with the drift velocity (4.12), and there is no electric current ($\vec{J} = 0$). When the collisional effects are not negligible, the motion of the ions suffers a larger retardation than that of the electrons as a result of collisions. In this case, there is an electric current (assuming $n_e = n_i$)

$$\vec{J}_\perp = en_e (\vec{u}_{i_\perp} - \vec{u}_{e_\perp}) \tag{4.13}$$

perpendicular to both $\vec{E}$ and $\vec{B}_0$, known as the *Hall current*. Note that, since $u_{e_\perp} > u_{i_\perp}$, this current is in the direction of $-(\vec{E} \times \vec{B}_0)$, that is, opposite to the drift velocity of both types of particles.

Returning now to the generalized Ohm's law in the simplified form (4.9), let us rewrite it in a way which relates the current density directly to the applied electric field. We define therefore a *DC conductivity dyad* (or *tensor*) $\bar{\bar{\sigma}}$ by the equation

$$\vec{J} = \bar{\bar{\sigma}} \cdot \vec{E} \tag{4.14}$$

In order to obtain an expression for $\bar{\bar{\sigma}}$, consider a Cartesian coordinate system with the z axis parallel to the magnetic field, $\vec{B}_0 = B_0 \hat{z}$. Replacing $\vec{u}_e$ in (4.9) by $-\vec{J}/(en_e)$, we get

$$\vec{J} = \sigma_0 \vec{E} - (\sigma_0 B_0/en_e)(\vec{J} \times \hat{z}) \tag{4.15}$$

Noting that

$$\vec{J} \times \hat{z} = J_y \hat{x} - J_x \hat{y} \tag{4.16}$$

we obtain the following set of equations for the $\hat{x}$, $\hat{y}$ and $\hat{z}$ components of (4.15)

$$\hat{x}: \quad J_x = \sigma_0 E_x - (\omega_{ce}/\nu_c) J_y \tag{4.17}$$

$$\hat{y}: \quad J_y = \sigma_0 E_y + (\omega_{ce}/\nu_c) J_x \tag{4.18}$$

$$\hat{z}: \quad J_z = \sigma_0 E_z \tag{4.19}$$

where ω_{ce} denotes the electron cyclotron frequency. We can combine (4.17) and (4.18) to eliminate J_y from the first one and J_x from the second one, obtaining

$$J_x = \frac{\nu_c^2}{(\nu_c^2 + \omega_{ce}^2)} \sigma_0 E_x - \frac{\nu_c \omega_{ce}}{(\nu_c^2 + \omega_{ce}^2)} \sigma_0 E_y \tag{4.20}$$

$$J_y = \frac{\nu_c \omega_{ce}}{(\nu_c^2 + \omega_{ce}^2)} \sigma_0 E_x + \frac{\nu_c^2}{(\nu_c^2 + \omega_{ce}^2)} \sigma_0 E_y \tag{4.21}$$

In matrix form we can write, therefore,

$$\begin{pmatrix} J_x \\ J_y \\ J_z \end{pmatrix} = \sigma_0 \begin{pmatrix} \dfrac{\nu_c^2}{(\nu_c^2 + \omega_{ce}^2)} & \dfrac{-\nu_c\,\omega_{ce}}{(\nu_c^2 + \omega_{ce}^2)} & 0 \\ \dfrac{\nu_c\,\omega_{ce}}{(\nu_c^2 + \omega_{ce}^2)} & \dfrac{\nu_c^2}{(\nu_c^2 + \omega_{ce}^2)} & 0 \\ 0 & 0 & 1 \end{pmatrix} \begin{pmatrix} E_x \\ E_y \\ E_z \end{pmatrix} \tag{4.22}$$

which is now in the form given in (4.14). The DC conductivity dyad is therefore given by

$$\bar{\bar{\sigma}} = \begin{pmatrix} \sigma_\perp & -\sigma_H & 0 \\ \sigma_H & \sigma_\perp & 0 \\ 0 & 0 & \sigma_\parallel \end{pmatrix} \tag{4.23}$$

where we have used the notation

$$\sigma_\perp \equiv \frac{\nu_c^2}{(\nu_c^2 + \omega_{ce}^2)}\,\sigma_0 \tag{4.24}$$

$$\sigma_H \equiv \frac{\nu_c\,\omega_{ce}}{(\nu_c^2 + \omega_{ce}^2)}\,\sigma_0 \tag{4.25}$$

$$\sigma_\parallel \equiv \sigma_0 = n_e\, e^2/(m_e\, \nu_c) \tag{4.26}$$

To illustrate the physical meaning of the components of $\bar{\bar{\sigma}}$ it is convenient to separate the applied electric field in a component parallel to $\vec{B}_0$, $\vec{E}_\parallel$, and a component in the plane normal to $\vec{B}_0$, $\vec{E}_\perp$, as shown in Fig. 2. The element $\sigma_\perp$ is called the *perpendicular* or *transverse conductivity* (also called Pedersen conductivity) since it governs the electric current in the direction of the component of the electric field normal to the magnetic field ($\parallel \vec{E}_\perp$, $\perp \vec{B}_0$), while σ_H (known as the *Hall conductivity*) governs the electric current in the direction perpendicular to both the electric and magnetic fields ($\perp \vec{E}$, $\perp \vec{B}_0$). The element $\sigma_\parallel$ is the *longitudinal conductivity*, since it governs the electric current in the direction of the electric field component along the magnetic field ($\| \vec{E}_{\parallel}$, $\| \vec{B}_0$). Note that the electric current along $\vec{B}_0$ is governed by the same conductivity (σ_0) as in the isotropic plasma.

The dependence of $\sigma_{\perp}$ and σ_H on the ratio of the cyclotron frequency to the collision frequency is shown in Fig. 3. As the ratio (ω_{ce}/ν_c) increases, $\sigma_{\perp}$ and σ_H decrease rapidly, the effect being more pronounced for $\sigma_{\perp}$. Thus, when (ω_{ce}/ν_c) is relatively large, very little current is produced across the

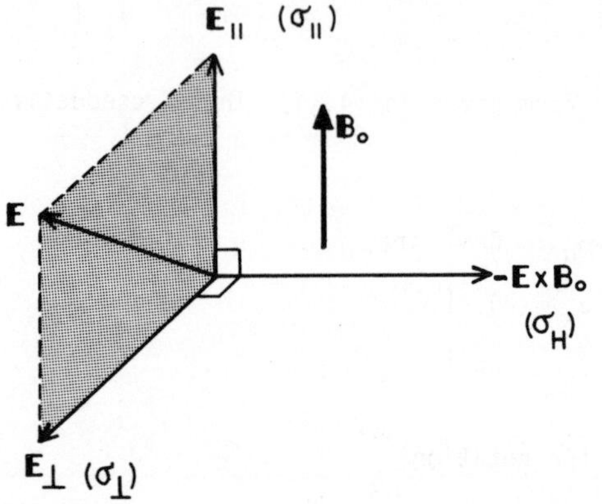

Fig. 2 - Relative orientation of the vector fields $\vec{E}_{\parallel}$, $\vec{E}_{\perp}$ and $-\vec{E} \times \vec{B}_0$; the conductivities $\sigma_{\parallel}$, $\sigma_{\perp}$, and σ_H govern the magnitude of the electric currents along these directions, respectively.

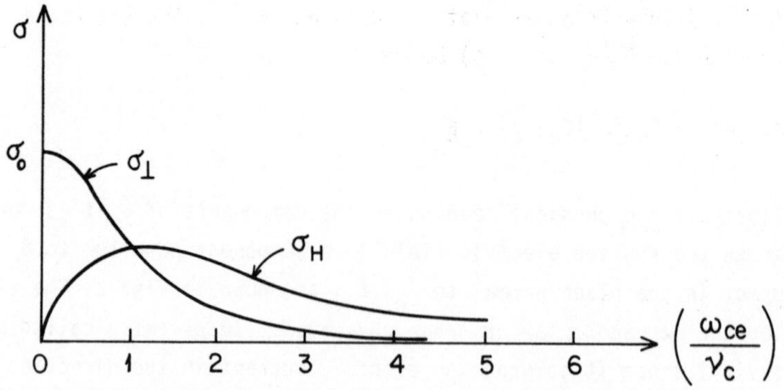

Fig. 3 - Dependence of the Hall conductivity σ_H and the perpendicular conductivity $\sigma_{\perp}$ on the ratio of the cyclotron frequency ω_{ce} to the collision frequency ν_c.

magnetic field lines, as compared with the current produced along the field lines, for the same applied electric field. Note that $\sigma_{\parallel}$ increases as ν_c diminishes and is independent of the magnitude of $\vec{B}$ and therefore of ω_{ce}. Thus, in a rarefied plasma immersed in a relatively strong magnetic field, the electric current flows essentially along the magnetic field lines.

Note that in the absence of a magnetic field ($\omega_{ce} = 0$), (4.24), (4.25) and (4.26) give $\sigma_{\perp} = \sigma_{\parallel} = \sigma_0$ and $\sigma_H = 0$, so that the plasma becomes isotropic.

We deduce next an expression for the electron mobility. Due to the anisotropy introduced by the magnetic field we have in this case a *mobility dyad* $\bar{\bar{\mu}}_e$. We define the electron mobility dyad by the equation

$$\vec{u}_e = \bar{\bar{\mu}}_e \cdot \vec{E} \tag{4.27}$$

Since $\vec{J} = - en_e \vec{u}_e = \bar{\bar{\sigma}} \cdot \vec{E}$, we find that

$$\bar{\bar{\mu}}_e = - (1/n_e e) \bar{\bar{\sigma}} \tag{4.28}$$

Explicit expressions for the components of $\bar{\bar{\mu}}_e$ can be easily written down considering (4.23), (4.24), (4.25) and (4.26).

5. AC CONDUCTIVITY AND ELECTRON MOBILITY

Consider now the case when the electric field $\vec{E}(\vec{r},t)$ and the mean electron velocity $\vec{u}_e(\vec{r},t)$ vary harmonically in time, as exp $(-i\omega t)$. We have seen that for time harmonic disturbances $\partial/\partial t$ is replaced by $-i\omega$. Therefore, the linearized Langevin equation (3.4) becomes

$$- i \omega m_e \vec{u}_e = - e (\vec{E} + \vec{u}_e \times \vec{B}_0) - m_e \nu_c \vec{u}_e \tag{5.1}$$

which can be written as

$$0 = - e (\vec{E} + \vec{u}_e \times \vec{B}_0) - m_e (\nu_c - i\omega) \vec{u}_e \tag{5.2}$$

This equation is identical to (4.7), except for the change in the collision frequency ν_c to $(\nu_c - i\omega)$. We obtain therefore solutions similar to the ones obtained for the DC conductivity dyad in the previous section, except that now we must replace ν_c by $(\nu_c - i\omega)$ in each element of the dyad. Therefore, the expressions for the *frequency-dependent* perpendicular conductivity, Hall conductivity and longitudinal conductivity are, respectively,

$$\sigma_{\perp} = \frac{(\nu_c - i\omega)^2}{(\nu_c - i\omega)^2 + \omega_{ce}^2}\,\sigma_0 \tag{5.3}$$

$$\sigma_H = \frac{(\nu_c - i\omega)\,\omega_{ce}}{(\nu_c - i\omega)^2 + \omega_{ce}^2}\,\sigma_0 \tag{5.4}$$

$$\sigma_{\parallel} = \sigma_0 = \frac{n_e\, e^2}{m_e\,(\nu_c - i\omega)} = \frac{n_e\, e^2\,(\nu_c + i\omega)}{m_e\,(\nu_c^2 + \omega^2)} \tag{5.5}$$

When the electron-neutral collisions can be neglected ($\nu_c = 0$), the expressions for the components of the AC (alternating current) conductivity dyad become

$$\sigma_{\perp} = \frac{\omega^2}{(\omega^2 - \omega_{ce}^2)}\,\sigma_0 \tag{5.6}$$

$$\sigma_H = \frac{i\omega\,\omega_{ce}}{(\omega^2 - \omega_{ce}^2)}\,\sigma_0 \tag{5.7}$$

$$\sigma_{\parallel} = \sigma_0 = i\,(n_e\, e^2/m_e\,\omega) \tag{5.8}$$

A complex conductivity means that there is a phase difference between the current density and the applied electric field.

The electron mobility, in any of the cases considered in this section, can be easily written down considering the relation (4.28).

6. CONDUCTIVITY WITH ION MOTION

The evaluation of the conductivity dyad, when the contribution due to the motion of the ions is included, can be performed in a straightforward way. For this purpose, consider the linearized Langevin equation for the type α species,

$$m_\alpha \, (\partial \vec{u}_\alpha / \partial t) = q_\alpha \, (\vec{E} + \vec{u}_\alpha \times \vec{B}_0) - m_\alpha \, \nu_{c\alpha} \, \vec{u}_\alpha \tag{6.1}$$

where $\nu_{c\alpha}$ is an effective collision frequency or damping term for the type α species resulting from collisions with *neutral* particles. Note that the equations (6.1), for each charged particle species, are uncoupled. Therefore, the net current density is given by

$$\vec{J} = \sum_\alpha n_\alpha \, q_\alpha \, \vec{u}_\alpha = \sum_\alpha \vec{J}_\alpha = \left(\sum_\alpha \bar{\bar{\sigma}}_\alpha \right) \cdot \vec{E} \tag{6.2}$$

and the total conductivity is simply

$$\bar{\bar{\sigma}} = \sum_\alpha \bar{\bar{\sigma}}_\alpha \tag{6.3}$$

For a plasma with electrons and several types of ions (index j) we obtain, using (5.3), (5.4) and (5.5), in terms of the plasma frequency $\omega_{p\alpha}$,

$$\sigma_\perp = \varepsilon_0 \left[\frac{\omega_{pe}^2 \, (\nu_{ce} - i\omega)}{(\nu_{ce} - i\omega)^2 + \omega_{ce}^2} + \sum_j \frac{\omega_{pj}^2 \, (\nu_{cj} - i\omega)}{(\nu_{cj} - i\omega)^2 + \omega_{cj}^2} \right] \tag{6.4}$$

$$\sigma_H = \varepsilon_0 \left[\frac{\omega_{pe}^2 \, \omega_{ce}}{(\nu_{ce} - i\omega)^2 + \omega_{ce}^2} - \sum_j \frac{\omega_{pj}^2 \, \omega_{cj}}{(\nu_{cj} - i\omega)^2 + \omega_{cj}^2} \right] \tag{6.5}$$

$$\sigma_\parallel = \varepsilon_0 \left[\frac{\omega_{pe}^2}{(\nu_{ce} - i\omega)} + \sum_j \frac{\omega_{pj}^2}{(\nu_{cj} - i\omega)} \right] \tag{6.6}$$

7. THE PLASMA AS A DIELECTRIC MEDIUM

The plasma can also be treated as a dielectric medium characterized by a dielectric dyad, in which the internal particle behavior is not considered. So far, we have treated the plasma as a collection of charged and neutral particles moving about in their own internal fields. Thus, as far as the constitutive relations are concerned, we have taken

$$\vec{D} = \varepsilon_0 \, \vec{E} \tag{7.1}$$

$$\vec{B} = \mu_0 \vec{H} \tag{7.2}$$

which is the case for vacuum, and the plasma effects show up through the motion and interaction of the charged particles inside the plasma. A different approach is provided by the use of a dielectric dyad, in which we are concerned only with the gross macroscopic properties of the plasma and not with the particle motions.

Thus, instead of the Langevin equation, let us consider the following Maxwell equation

$$\vec{\nabla} \times \vec{B} = \mu_0 (\vec{J} + \varepsilon_0 \partial\vec{E}/\partial t) \tag{7.3}$$

and incorporate the effects of the plasma in the conductivity dyad $\bar{\bar{\sigma}}$, defined by the equation

$$\vec{J} = \bar{\bar{\sigma}} \cdot \vec{E} \tag{7.4}$$

Substituting (7.4) into (7.3), and assuming time-harmonic variations of the form exp $(- i\omega t)$, we obtain

$$\vec{\nabla} \times \vec{B} = \mu_0 \bar{\bar{\sigma}} \cdot \vec{E} - i\omega \mu_0 \varepsilon_0 \vec{E} \tag{7.5}$$

If we let $\bar{\bar{1}}$ be the unit dyad, we can write

$$\vec{\nabla} \times \vec{B} = - i\omega \mu_0 \varepsilon_0 (\bar{\bar{1}} + i\bar{\bar{\sigma}}/\omega\varepsilon_0) \cdot \vec{E} \tag{7.6}$$

or, equivalently, as

$$\vec{\nabla} \times \vec{B} = - i\omega \mu_0 \bar{\bar{\varepsilon}} \cdot E \tag{7.7}$$

where

$$\bar{\bar{\varepsilon}} = \varepsilon_0 (\bar{\bar{1}} + i\bar{\bar{\sigma}}/\omega\varepsilon_0) \tag{7.8}$$

is called the *dielectric dyad for the plasma*. The use of the dielectric dyad represents therefore a different approach for the treatment of a plasma, as compared to the one we have used so far. Adopting this approach, (7.1) must be replaced by

$$\vec{D} = \bar{\bar{\varepsilon}} \cdot \vec{E} \tag{7.9}$$

and the plasma is considered as a dielectric medium, without bringing into the picture its internal particle behavior. Note that $\bar{\bar{\varepsilon}}$ depends on the frequency ω.

The dielectric dyad can be written in matrix form as

$$\bar{\bar{\varepsilon}} = \varepsilon_0 \begin{pmatrix} \varepsilon_1 & -\varepsilon_2 & 0 \\ \varepsilon_2 & \varepsilon_1 & 0 \\ 0 & 0 & \varepsilon_3 \end{pmatrix} \tag{7.10}$$

where the following notation was introduced

$$\varepsilon_1 = 1 + (i/\omega\varepsilon_0)\, \sigma_\perp \tag{7.11}$$

$$\varepsilon_2 = (i/\omega\varepsilon_0)\, \sigma_H \tag{7.12}$$

$$\varepsilon_3 = 1 + (i/\omega\varepsilon_0)\, \sigma_0 \tag{7.13}$$

For the case of a multi-species plasma the total conductivity must be used in (7.8), so that the expressions to be substituted for $\sigma_\perp$, σ_H and σ_0 are those given in (6.4), (6.5) and (6.6).

8. FREE ELECTRON DIFFUSION

The presence of a pressure gradient term in the momentum transport equation provides a force which tends to smooth out any inhomogeneities in the plasma density. The diffusion of particles in a plasma results from this pressure gradient force. To deduce the electron diffusion coefficient for a warm weakly ionized plasma we will use the momentum transport equation for the electrons with a constant electron-neutral collision frequency. We assume that the deviations from the equilibrium state caused by inhomogeneities in the density are very small, so that they may be considered as first order quantities. This means that the mean velocity of the electrons $\vec{u}_e$ is also a first order quantity, and since the velocity distribution will be approximately isotropic, we can replace the pressure dyad $\overline{\overline{p}}_e$ by a scalar pressure p_e.

Consider the case in which $\vec{E}$ and $\vec{B}$ are zero and the electron temperature T_e is constant. For a *slightly nonuniform* electron number density, we can write

$$n_e(\vec{r},t) = n_0 + n_e'(\vec{r},t) \tag{8.1}$$

$$p_e(\vec{r},t) = n_e(\vec{r},t)\, k\, T_e = (n_0 + n_e')\, k\, T_e \tag{8.2}$$

where $|n_e'| << n_0$ is a first order quantity and n_0 is constant. Since $\vec{u}_e$ is also a first order quantity, the continuity equation for the electron gas becomes

$$\partial n_e'/\partial t + n_0\, \vec{\nabla} \cdot \vec{u}_e = 0 \tag{8.3}$$

where the second order term $n_e' \vec{u}_e$ has been neglected. Similarly, for the momentum transport equation,

$$n_e\, m_e\, [\partial \vec{u}_e/\partial t + (\vec{u}_e \cdot \vec{\nabla})\, \vec{u}_e] = - \vec{\nabla} p_e - n_e\, m_e\, \nu_c\, \vec{u}_e \tag{8.4}$$

we obtain, after linearization,

$$n_0\, (\partial \vec{u}_e/\partial t) = - (kT_e/m_e)\, \vec{\nabla} n_e' - n_0\, \nu_c\, \vec{u}_e \tag{8.5}$$

Taking the divergence of this equation, we obtain

$$n_0\, \partial(\vec{\nabla} \cdot \vec{u}_e)/\partial t = - (kT_e/m_e)\, \nabla^2 n_e' - n_0\, \nu_c\, \vec{\nabla} \cdot \vec{u}_e \tag{8.6}$$

Using (8.3) to substitute for $n_0\, \vec{\nabla} \cdot \vec{u}_e$, yields

$$\partial^2 n_e'/\partial t^2 = (kT_e/m_e)\, \nabla^2 n_e' - \nu_c\, (\partial n_e'/\partial t) \tag{8.7}$$

This equation may also be written in the form

$$\partial n_e'/\partial t = D_e\, \nabla^2 n_e' - (1/\nu_c)\, (\partial^2 n_e'/\partial t^2) \tag{8.8}$$

where we have defined

$$D_e = kT_e/(m_e \nu_c) \tag{8.9}$$

which is called the *electron free-diffusion coefficient.*

To obtain a rough estimate of the order of magnitude of the various terms in (8.8), let τ and L represent, respectively, a characteristic time and a characteristic length over which n_e' varies significantly. Thus, any spatial derivative is of the order of L^{-1} and any time derivative is of the order τ^{-1}, so that the order of magnitude of the terms in (8.8) are

$$\partial n_e'/\partial t \simeq n_e'/\tau \tag{8.10}$$

$$D_e \nabla^2 n_e' \simeq D_e n_e'/L^2 \tag{8.11}$$

$$(1/\nu_c)(\partial^2 n_e'/\partial t^2) \simeq n_e'/(\nu_c \tau^2) \tag{8.12}$$

Comparing (8.10) and (8.12) we see that if $\nu_c \tau >> 1$, that is, if the average number of electron-neutral collisions is large during the time interval τ, then the last term in (8.8) can be neglected and it reduces to the following *diffusion equation*

$$\partial n_e'/\partial t = D_e \nabla^2 n_e' \tag{8.13}$$

Therefore, when the rate of change in the number density is slow compared to the collision frequency, the number density is governed by a diffusion equation with a free electron diffusion coefficient as given by (8.9).

The condition $\nu_c \tau >> 1$ implies in the omission of the acceleration term in the momentum transport equation, that is, $\partial \vec{u}_e/\partial t$ is neglected. From the linearized Eq. (8.5), when there are no time variations in $\vec{u}_e$, we obtain,

$$n_0 \nu_c \vec{u}_e = -(kT_e/m_e) \vec{\nabla} n_e' \tag{8.14}$$

which can be written as

$$\vec{\Gamma}_e = -D_e \vec{\nabla} n_e' \tag{8.15}$$

where $\vec{\Gamma}_e = n_0 \vec{u}_e$ denotes the linearized electron flux. (8.15) is analogous to the simple Ohm's law $\vec{J} = \sigma_0 \vec{E}$, replacing $\vec{J}$ by $\vec{\Gamma}_e$, σ_0 by D_e, and $\vec{E}$ by $-\vec{\nabla} n_e'$. Thus, we see that the electron flux $\vec{\Gamma}_e$ is caused by a density gradient, in a way analogous to the electric current caused by an electric field, under steady state conditions for $\vec{u}_e$.

9. ELECTRON DIFFUSION IN A MAGNETIC FIELD

Consider now the problem of electron diffusion in the presence of a constant and uniform magnetic field B_0. We make the same assumptions as in the previous section, and neglect the acceleration term $\partial \vec{u}_e/\partial t$ in the equation of motion.

In the linearized momentum transport equation (8.5), with the time derivative set equal to zero, we include now a magnetic force term, which results in

$$\vec{\Gamma}_e = - D_e \vec{\nabla} n_e' - (e/m_e \nu_c) (\vec{\Gamma}_e \times \vec{B}_0) \tag{9.1}$$

Choosing a Cartesian coordinate system with the z axis pointing in the direction of the constant $\vec{B}_0$ field, $\vec{B}_0 = B_0 \hat{z}$, we have

$$\vec{\Gamma}_e = - D_e \vec{\nabla} n_e' - (\omega_{ce}/\nu_c) (\vec{\Gamma}_e \times \hat{z}) \tag{9.2}$$

This equation is analogous to (4.15) with $\vec{\Gamma}_e$ replacing $\vec{J}$, D_e replacing σ_0, and $-\vec{\nabla} n_e'$ replacing $\vec{E}$ (note that $\omega_{ce}/\nu_c = \sigma_0 B_0/en_e$). Therefore, in analogy with the expression $\vec{J} = \bar{\bar{\sigma}} \cdot \vec{E}$, we can write

$$\vec{\Gamma}_e = - \bar{\bar{D}} \cdot \vec{\nabla} n_e' \tag{9.3}$$

where $\bar{\bar{D}}$ is the *dyad coefficient for free electron diffusion* given by

$$\bar{\bar{D}} = \begin{pmatrix} D_\perp & D_H & 0 \\ -D_H & D_\perp & 0 \\ 0 & 0 & D_\parallel \end{pmatrix} \tag{9.4}$$

where the following notation is used

$$D_\perp = \frac{\nu_c^2}{(\nu_c^2 + \omega_{ce}^2)} D_e \tag{9.5}$$

$$D_H = \frac{\nu_c \, \omega_{ce}}{(\nu_c^2 + \omega_{ce}^2)} D_e \tag{9.6}$$

$$D_{\parallel} = D_e = kT_e/(m_e \, \nu_c) \tag{9.7}$$

A diffusion equation for n_e', when there is a constant and uniform magnetic field present, can also be derived in the same way as in the previous section. First, we write the continuity equation (8.3) in the form

$$\partial n_e'/\partial t + \vec{\nabla} \cdot \vec{\Gamma}_e = 0 \tag{9.8}$$

Substituting (9.3) for $\vec{\Gamma}_e$, yields

$$\partial n_e'/\partial t = \vec{\nabla} \cdot (\bar{\bar{D}} \cdot \vec{\nabla} n_e') \tag{9.9}$$

Using (9.4) we find, by direct calculation in Cartesian coordinates,

$$\bar{\bar{D}} \cdot \vec{\nabla} n_e' = \hat{x} \left(D_\perp \frac{\partial n_e'}{\partial x} + D_H \frac{\partial n_e'}{\partial y} \right) + \hat{y} \left(-D_H \frac{\partial n_e'}{\partial x} + D_\perp \frac{\partial n_e'}{\partial y} \right) + \hat{z} \, D_e \frac{\partial n_e'}{\partial z} \tag{9.10}$$

Substituting this result into (9.9), yields

$$\frac{\partial n_e'}{\partial t} = D_\perp \left(\frac{\partial^2 n_e'}{\partial x^2} + \frac{\partial^2 n_e'}{\partial y^2} \right) + D_e \frac{\partial^2 n_e'}{\partial z^2} \tag{9.11}$$

Since $D_\perp < D_e$ and since $D_\perp$ decreases with increasing values of ω_{ce}/ν_c (similary to $\sigma_\perp$, as shown in Fig. 3) the diffusion of particles in a direction perpendicular to $\vec{B}$ is always less than that in the direction parallel to $\vec{B}$. For values of ω_{ce} much larger than ν_c the diffusion of particles across the magnetic field lines is greatly reduced (from (9.5) and (9.6) it can be seen that for $\omega_{ce} >> \nu_c$ we have, approximately, $D_\perp \propto 1/B^2$ and $D_H \propto 1/B$).

As a final point in this section we note that the momentum transport equation for an electron gas, neglecting the acceleration term but including the electromagnetic force, and when the temperature is constant, can be written in the general form

$$\vec{\Gamma}_e = \mu_e (n_e \vec{E} + \vec{\Gamma}_e \times \vec{B}) - D_e \vec{\nabla} n_e \tag{9.12}$$

From this equation we can see that the electron flux is produced by either, or both, electromagnetic fields and density gradients. The ratio of the scalar mobility μ_e to the diffusion coefficient is known as the *Einstein relation* and is given by

$$\mu_e/D_e = - e/(kT_e) \tag{9.13}$$

10. AMBIPOLAR DIFFUSION

We have seen in section 8 that the steady state momentum equation, in the absence of electromagnetic forces and when the temperature is constant, gives the following diffusion equation for the electrons

$$\vec{\Gamma}_e = - D_e \vec{\nabla} n_e' \tag{10.1}$$

where the *electron free-diffusion coefficient* is given by

$$D_e = kT_e/(m_e \nu_{ce}) \tag{10.2}$$

The subscript e has been added here to ν_c to indicate that the constant collision frequency ν_{ce} refers to electron-neutral collisions.

If we consider similar equations for the ions in a weakly ionized plasma, under the same assumptions, we obtain the following ion diffusion equation

$$\vec{\Gamma}_i = - D_i \vec{\nabla} n_i' \tag{10.3}$$

where

$$D_i = kT_i/(m_i \nu_{ci}) \tag{10.4}$$

denotes the *ion free-diffusion coefficient* and ν_{ci} is the constant ion-neutral collision frequency.

In deriving the results given by (10.1) and (10.3), the interaction between the electrons and the ions were not taken into account. Since the diffusion coefficient is inversely proportional to the particle mass, the electrons diffuse faster than the ions leaving an excess of positive charge behind them. This gives rise to a space charge electric field in the same direction as the particle diffusion, and which accelerates the diffusion of the ions and slow down that of the electrons. The diffusion in which the effect of the space charge electric field is *not* included is known as *free diffusion*.

For most problems of plasma diffusion, however, the space charge electric field *cannot* be neglected. According to Maxwell equation

$$\vec{\nabla} \cdot \vec{E} = \rho / \varepsilon_0 = e(n_i - n_e)/\varepsilon_0 \tag{10.5}$$

it is clear that an electric field is present whenever the electron density differs from the ion density. To estimate the importance of the space charge electric field in diffusion problems, we may use dimensional analysis and let L represent a characteristic length over which the number density changes significantly. Thus, from (10.5) we may write

$$E \simeq e\,n\,L/\varepsilon_0 \tag{10.6}$$

so that the electric force per unit mass f_E is of the order

$$f_E = e\,E/m \simeq e^2\,n\,L/(m\,\varepsilon_0) \tag{10.7}$$

The diffusion force per unit mass f_D obtained from (10.1), is of the order

$$f_D = (kT/m\,n_0)\,|\vec{\nabla}\,n| \simeq kTn/(m\,n_0\,L) \tag{10.8}$$

Therefore, the space charge electric field can be neglected only if $f_E << f_D$, or equivalently, if

$$L^2 << \varepsilon_0\,k\,T/(n_0\,e^2) = \lambda_D^2 \tag{10.9}$$

where λ_D is the *Debye length*. Since the Debye length is generally very small (see Fig. 2, Chapter 1), the condition $L << \lambda_D$ is rarely satisfied and for most plasma diffusion problems we cannot neglect the space charge electric field. In what follows we will reexamine therefore the problem of plasma diffusion taking into account the motion of both ions and electrons and including the space charge electric field $\vec{E}$. The combined diffusion of the electrons and the ions, forced by the space charge $\vec{E}$ field, is known as *ambipolar diffusion*. Since the electric field retards the electrons and accelerates the ions, the two kinds of charged particles diffuse at a rate which is intermediate in value to their *free* diffusion rates.

To investigate the characteristics of *ambipolar diffusion* we assume that the disturbance for both electrons and ions are small first order quantities, so that (for α = e, i)

$$n_\alpha(\vec{r},t) = n_0 + n'_\alpha(\vec{r},t) \tag{10.10}$$

with $|n'_\alpha| << n_0$, and that the mean velocities $\vec{u}_e$ and $\vec{u}_i$ are of very small amplitude. We obtain, under these assumptions, the following linearized mass conservation equations (α = e, i)

$$\partial n'_\alpha/\partial t + n_0 \vec{\nabla} \cdot \vec{u}_\alpha = 0 \tag{10.11}$$

The linearized momentum equations, assuming that the temperatures are constant, and without a magnetic field, become (α = e, i)

$$\partial \vec{u}_\alpha/\partial t = (q_\alpha/m_\alpha) \vec{E} - (kT_\alpha/m_\alpha n_0) \vec{\nabla} n'_\alpha - \nu_{c\alpha} \vec{u}_\alpha \tag{10.12}$$

where the space charge $\vec{E}$ field satisfies Maxwell equation (10.5). We are assuming that the neutral mean velocity $\vec{u}_n$ is zero and we are neglecting electron-ion collisions, since the plasma is weakly ionized. Taking the divergence of (10.12) and using (10.11), we obtain

$$\frac{\partial^2 n'_\alpha}{\partial t^2} = - \frac{q_\alpha n_0}{m_\alpha} (\vec{\nabla} \cdot \vec{E}) + \frac{k T_\alpha}{m_\alpha} \nabla^2 n'_\alpha - \nu_{c\alpha} \frac{\partial n'_\alpha}{\partial t} \tag{10.13}$$

If we replace $\vec{\nabla} \cdot \vec{E}$ from (10.5), we obtain the following set of coupled equations for the two variables n'_e and n'_i,

$$\frac{\partial^2 n'_e}{\partial t^2} = \omega_{pe}^2 (n'_i - n'_e) + \frac{k T_e}{m_e} \nabla^2 n'_e - \nu_{ce} \frac{\partial n'_e}{\partial t} \tag{10.14}$$

$$\frac{\partial^2 n_i'}{\partial t^2} = - \omega_{pi}^2 \, (n_i' - n_e') + \frac{k\,T_i}{m_i} \nabla^2 n_i' - \nu_{ci} \frac{\partial n_i'}{\partial t} \tag{10.15}$$

These equations, however, are still too complicated for a detailed analytical treatment and to go further we will make some additional simplifying assumptions. Recall that, if $\nu_c \tau >> 1$, that is, if the average electron or ion has many collisions with neutral particles during the characteristic time for diffusion τ, the term $\partial^2 n_\alpha'/\partial t^2$ (originated from the acceleration term in the momentum equation) can be neglected. With this assumption we neglect the term in the left-hand side of (10.14) and (10.15). Combining these equations we obtain

$$0 = k\,T_e \nabla^2 n_e' + k\,T_i \nabla^2 n_i' - m_e \nu_{ce} \frac{\partial n_e'}{\partial t} - m_i \nu_{ci} \frac{\partial n_i'}{\partial t} \tag{10.16}$$

As a second approximation we will set $n_i' = n_e' = n'$ in (10.16) to obtain the following diffusion equation

$$0 = k\,(T_e + T_i) \nabla^2 n' - (m_e \nu_{ce} + m_i \nu_{ci}) \, \partial n'/\partial t \tag{10.17}$$

which can be written in the form

$$\partial n'/\partial t = D_a \nabla^2 n' \tag{10.18}$$

where

$$D_a = k\,(T_e + T_i) / (m_e \nu_{ce} + m_i \nu_{ci}) \tag{10.19}$$

is the *ambipolar diffusion coefficient*. Note that the coupling of the two Eqs. (10.14) and (10.15) is a consequence of the electric field term, and that the simplifying approximation $n_i' = n_e'$ was introduced only after the two equations were combined into (10.16). This approximation implies that the space charge electric field becomes a negligible perturbation with the result that both ions and electrons diffuse together. This situation is known as *perfect ambipolar diffusion*, since the coupling between the two types of charged particles is complete.

Instead of taking $n_i' = n_e'$, a less restrictive simplifying approximation would be to assume

$$n_i' = C\, n_e' \tag{10.20}$$

where C is a constant. Using this approximation in (10.16) we obtain

$$0 = k\,(T_e + CT_i)\, \nabla^2 n_e' = (m_e\, \nu_{ce} + Cm_i\, \nu_{ci})\, \partial n_e'/\partial t \tag{10.21}$$

or

$$\partial n_e'/\partial t = D_a\, \nabla^2 n_e' \tag{10.22}$$

where the ambipolar diffusion coefficient is now given by

$$D_a = k\,(T_e + CT_i) / (m_e\, \nu_{ce} + Cm_i\, \nu_{ci}) \tag{10.23}$$

The space charge density is now

$$\rho = e\,(n_i' - n_e') = e\, n_e'\,(C - 1) \tag{10.24}$$

and the electric field can be obtained from Maxwell equation

$$\vec{\nabla} \cdot \vec{E} = e\, n_e'\,(C - 1)/\varepsilon_0 \tag{10.25}$$

The effect of the electric field is to accelerate the diffusion of the ions and to retard the diffusion of the electrons, as compared to their individual free diffusion rates, so that to a good approximation both species diffuse together.

Whenever there is a significant deviation from charge neutrality ($C \neq 1$), the electric field force becomes very strong as can be seen from the following dimensional argument. A comparison of the magnitude of the electric force per unit mass, $\vec{f}_E = q_\alpha \vec{E}/m_\alpha$, and the diffusion force per unit mass, $\vec{f}_D = -\,(k\, T_\alpha/m_\alpha\, n_0)\, \vec{\nabla}\, n_\alpha'$, which are of the order

$$f_E \simeq Le^2\, n_e'\,(C - 1) / (m\, \varepsilon_0) \tag{10.26}$$

$$f_D \simeq k\,T\, n_e'/(m\, n_0\, L) \tag{10.27}$$

shows that

$$f_E/f_D \simeq L^2 \ (C - 1) / \lambda_D^2 \tag{10.28}$$

Since in most cases L^2 is much larger than λ_D^2, we see that if n_i' differs significantly from n_e' the electric field force (which tends to equalize n_i' and n_e') becomes very strong.

11. DIFFUSION IN A FULLY IONIZED PLASMA

Consider now the problem of diffusion in a fully ionized plasma. For simplicity, we shall describe the plasma as a single conducting fluid for which the equation of motion under steady state conditions, in the presence of magnetic and pressure-gradient forces, is

$$\vec{J} \times \vec{B} = \vec{\nabla} p \tag{11.1}$$

where $\vec{J}$ denotes the total electric current density, $\vec{B}$ is the magnetic induction, and p represents the total pressure of the conducting fluid. Note that the electric force is zero since the plasma, as a whole, is macroscopically neutral ($\rho = 0$). This equation is complemented by the generalized Ohm's law in the following simplified form,

$$\vec{J} = \sigma_0 \ (\vec{E} + \vec{u} \times \vec{B}) \tag{11.2}$$

where σ_0 is the longitudinal electric conductivity and $\vec{u}$ is the total fluid velocity.

Taking the cross-product of (11.2) with $\vec{B}$, yields

$$\vec{J} \times \vec{B} = \sigma_0 \ (\vec{E} \times \vec{B} - B^2 \ \vec{u}_\perp) \tag{11.3}$$

where $\vec{u}_\perp$ is the component of $\vec{u}$ in a direction normal to the external field $\vec{B}$. Using (11.1) and rearranging, (11.3) gives

$$\vec{u}_\perp = (\vec{E} \times \vec{B}) / B^2 - \vec{\nabla} p / (\sigma_0 \ B^2) \tag{11.4}$$

This result shows that the total fluid velocity across the magnetic field is given by the $\vec{E} \times \vec{B}$ drift of the whole plasma plus a diffusion velocity in the direction of $-\vec{\nabla} p$.

The flux associated with the diffusion velocity only, is given by

$$\vec{\Gamma}_\perp = n\, \vec{u}_\perp = -\, n\, \vec{\nabla} p / (\sigma_0\, B^2) \tag{11.5}$$

where n denotes the electron (or total ion) number density. Considering a two--fluid plasma (electrons and one type of ions), we have

$$p = p_e + p_i = n\, k\, (T_e + T_i) \tag{11.6}$$

so that (11.5) becomes, assuming the temperatures to be constant,

$$\vec{\Gamma}_\perp = -\, [n\, k\, (T_e + T_i) / (\sigma_0\, B^2)]\, \vec{\nabla} n = -\, D_\perp\, \vec{\nabla} n \tag{11.7}$$

The quantity

$$D_\perp = n\, k\, (T_e + T_i) / (\sigma_0\, B^2) \tag{11.8}$$

is known as the *classical diffusion coefficient* for a fully ionized plasma.

This diffusion coefficient is proportional to $1/B^2$, just as in the case of a weakly ionized plasma. Nevertheless, there are some fundamental differences between $D_\perp$, as given by (11.8), and the corresponding coefficient for a partially ionized plasma. Initially note that in a fully ionized plasma $D_\perp$ is not constant, but depends on the number density n. Further, since it can be shown that σ_0 is proportional to $T^{3/2}$ for a Maxwellian distribution of velocities, $D_\perp$ decreases with increasing temperature in a fully ionized plasma, while the opposite is true for a weakly ionized plasma. Finally, the diffusion coefficient $D_\perp$ in (11.8) was derived for the whole plasma as a conducting fluid and, since both ions and electrons diffuse together, there is no ambipolar electric field.

In some experiments it has been observed a dependence of $D_\perp$ on the magnetic field as B^{-1}, rather than B^{-2}, and the decay of the plasma was found to be exponential, rather than reciprocal with time. Furthermore, the absolute value of $D_\perp$ was found to be much larger than that given in (11.8). This anomalously poor magnetic confinement was first noted in laboratory by Bohm, in 1946, who obtained the following semiempirical formula

$$D_\perp \equiv D_B = k\,T_e / (16\,e\,B) \tag{11.9}$$

Since this diffusion coefficient does not depend on the density, the decay of the plasma density is exponential with time. This type of diffusion in plasmas is known as *Bohm diffusion*.

PROBLEMS

10.1 - Consider a solid state plasma with the same number of electrons (e) and holes (h). Using the linearized Langevin equation (with α = e, h)

$$m_\alpha\,(\partial \vec{u}_\alpha/\partial t) = q_\alpha\,(\vec{E} + \vec{u}_\alpha \times \vec{B}_0) - \nu_{c\alpha}\, m_\alpha\, \vec{u}_\alpha$$

taking $m_e = m_h$, $\nu_{ce} = \nu_{ch}$, assuming a time dependence for both $\vec{E}$ and $\vec{u}_\alpha$ of the form exp $(-i\omega t)$, and choosing a Cartesian coordinate system with the z axis pointing along the constant and uniform magnetic field $\vec{B}_0$, show that the conductivity dyad is given by

$$\overline{\overline{\sigma}} = 2 \begin{pmatrix} \sigma_\perp & 0 & 0 \\ 0 & \sigma_\perp & 0 \\ 0 & 0 & \sigma_0 \end{pmatrix}$$

with $\sigma_\perp$ and σ_0 given by (5.3) and (5.5), respectively. Explain, in physical terms, why $\sigma_H = 0$ in this case.

10.2 - Assume that the average velocities of the electrons and ions in a completely ionized plasma, in the presence of constant and uniform electric ($\vec{E}$) and magnetic ($\vec{B}_0$) fields, satisfy, respectively, the following equations of motion

$$m_e\,(\partial \vec{u}_e/\partial t) = -\,e\,(\vec{E} + \vec{u}_e \times \vec{B}_0) - m_e\,\nu\,(\vec{u}_e - \vec{u}_i) \qquad \text{(electrons)}$$

$$m_i\,(\partial \vec{u}_i/\partial t) = e\,(\vec{E} + \vec{u}_i \times \vec{B}_0) - m_e\,\nu\,(\vec{u}_i - \vec{u}_e) \qquad \text{(ions)}$$

(a) Determine expressions for the steady state DC conductivities σ_H, $\sigma_\perp$, σ_0.
(b) For $\vec{u}_e$, $\vec{u}_i$ and $\vec{E}$ all proportional to exp $(-i\,\omega\,t)$ and $\vec{B}_0$ constant, calculate the AC conductivity dyad for the plasma.

10.3 - Consider the equation $\vec{J} = \bar{\bar{\sigma}} \cdot \vec{E}$, with $\bar{\bar{\sigma}}$ as given in (4.23). If we choose a Cartesian coordinate system such that $E_x = E_\perp$, $E_y = 0$, $E_z = E_\parallel$ and $\vec{B}_0 = B_0 \hat{z}$ (refer to Fig. 2), verify that in this coordinate system we have

$$J_x = \sigma_\perp E_\perp$$

$$J_y = \sigma_H E_\perp$$

$$J_z = \sigma_\parallel E_\parallel$$

Interpret physically this result with reference to Fig. 2.

10.4 - What is the physical meaning of a complex conductivity, as given in (5.7) and (5.8)? Consider, for example, that $\vec{E}(\vec{r},t) = \vec{E}(\vec{r}) \exp(-i\omega t)$, and calculate the real parts of $\vec{E}(\vec{r},t)$ and of $\vec{J}(\vec{r},t) = \bar{\bar{\sigma}} \cdot \vec{E}(\vec{r},t)$. Interpret physically the results considering the phase differences between $\vec{J}$ and $\vec{E}$.

10.5 - Write expressions for the components of the dielectric dyad $\bar{\bar{\varepsilon}}$ of a multi-constituent magnetized plasma.

10.6 - Consider the electrons in a plasma acted upon by a small, constant and uniform external electric field $\vec{E}$. Under steady state conditions with no spatial gradients, obtain an expression for the nonequilibrium distribution function f for the electrons, by applying a perturbation technique to the Boltzmann equation (take $f = f_0 + f_1$ with $|f_1| \ll f_0$ and neglect all second order terms) using the relaxation model for the collision term

$$(\delta f/\delta t)_{coll} = -\nu\,(f - f_0)$$

where ν is a relaxation collision frequency and f_0 is the equilibrium Maxwellian distribution function. Assuming that ν is independent of velocity, obtain an expression for the electric conductivity σ_0 of the plasma, by taking $\vec{J} = \sigma_0 \vec{E}$.

10.7 - Same as Problem 10.6, but including also a constant and uniform magnetic field $\vec{B}_0$.

10.8 - Imagine a horizontally stratified ionosphere in the absence of a magnetic field, constituted only of electrons (density n, temperature T, charge -e, mass m_e) and one type of ions (density n, temperature T, charge +e, mass m_i), subjected to the gravitational field ($\vec{g}$), vertical pressure gradient ($\vec{\nabla}p$), and the internal electric field ($\vec{E}$) due to the charge separation associated with ambipolar diffusion. Neglect the gravitational force for the electrons and consider the system in equilibrium. Using the collisionless equations of motion for the electrons and the ions, show that the internal electric field acts downward on the electrons with a force $m_i g/2$, and upward on the ions with the same force. Consequently, the net effect is the same as if both ions and electrons had mass $m_i/2$.

10.9 - (a) In order to solve the *diffusion equation*

$$\partial n(\vec{r},t) / \partial t = D \nabla^2 n(\vec{r},t)$$

by the method of separation of variables, let

$$n(\vec{r},t) = S(\vec{r})\, T(t)$$

and show that

$$T_k(t) = (\text{constant}) \cdot \exp(-D k^2 t)$$

$$(\nabla^2 + k^2)\, S(\vec{r}) = 0$$

where k^2 is the separation constant.

(b) Assuming that S depends only on the x coordinate show that

$$S(x) = c(k) \exp(ikx)$$

where k can be either positive or negative, and that

$$n(x,t) = \int_{-\infty}^{+\infty} c(k) \exp(ikx - D k^2 t)\, dk$$

$$n_0(x) = \int_{-\infty}^{+\infty} c(k) \exp(ikx)\, dk$$

where $n_o(x) = n(x,o)$ is the known *initial* density distribution.

(c) Using Fourier transform theory, show that

$$c(k) = \frac{1}{2\pi} \int_{-\infty}^{+\infty} n_o(x) \exp(-ikx)\, dx$$

and, consequently, that

$$n(x,t) = \frac{1}{2(\pi D t)^{1/2}} \int_{-\infty}^{+\infty} n_o(x') \exp[-(x - x')^2 / 4Dt]\, dx'$$

(d) Taking as initial condition

$$n_o(x) = \exp(-x^2/x_0^2)$$

prove that

$$n(x,t) = \left(\frac{\tau_D}{\tau_D + 4t} \right)^{1/2} \exp \left[- \frac{x^2}{x_0^2} \left(\frac{\tau_D}{\tau_D + 4t} \right) \right]$$

where $\tau_D = x_0^2/D$ is a characteristic time for diffusion to smooth out the density n.

(e) Generalize the problem for the three-dimensional case in Cartesian coordinates, when $S = S(\vec{r})$.

10.10 - Consider the solution of the diffusion equation by separation of variables in the linear geometry of the plasma slab indicated in Fig. P 10.1. Show that the solution of the equation

$$d^2 S(x) / dx^2 + k^2 S(x) = 0$$

which satisfies the boundary condition $S = 0$ at $x = \pm L$, is

$$S(x) = \sum_m a_m \cos \left[\frac{(m + 1/2)\pi x}{L} \right]$$

and

$$S(x) = \sum_m b_m \sin\left(\frac{m\pi x}{L}\right)$$

Explain why the solution as a sine series is not a physically acceptable solution for this case. Consequently, from $n(x,t) = S(x)\,T(t)$, show that the number density can be expressed as

$$n(x,t) = \sum_m a_m \exp\left\{- D\left[\frac{\pi(m + 1/2)}{L}\right]^2 t\right\} \cos\left[\frac{\pi(m + 1/2)x}{L}\right]$$

Therefore, the decay time constant for the $m^{\underline{th}}$ mode is

$$\tau_m = \left[\frac{L}{\pi(m + 1/2)}\right]^2 \frac{1}{D}$$

This result shows that the higher modes decay faster than the lower ones. How are the coefficients a_m determined in terms of $n_0(x)$?

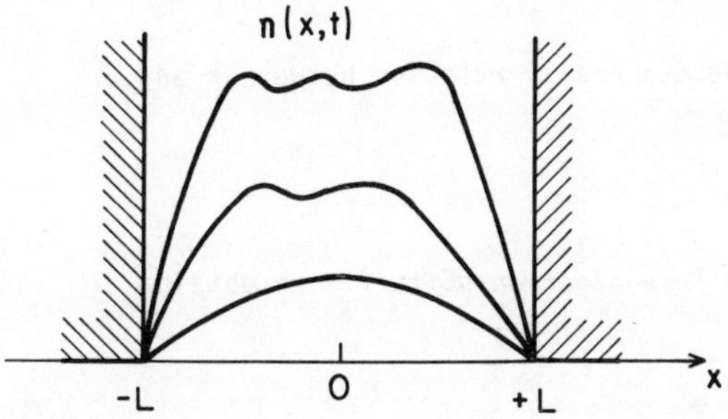

Fig. P 10.1

10.11 - Show that the solution of the diffusion equation in the case of cylindrical geometry (see Fig. P 10.2),

$$\frac{d^2S(r)}{dr^2} + \frac{1}{r}\frac{dS(r)}{dr} + k^2 S(r) = 0$$

is given in terms of Bessel's functions $J_m(kr)$. Explain how k must be determined in order that $n(\vec{r},t)$ satisfies the boundary condition $n = 0$ at $r = R_o$.

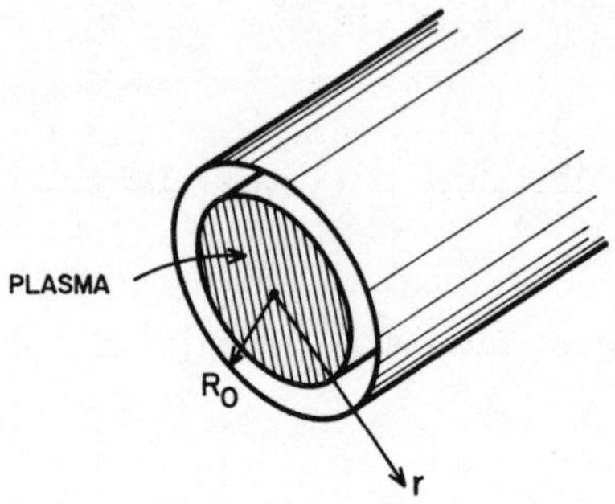

Fig. P 10.2

10.12 - Verify that plane wave solutions to the diffusion equation

$$\partial n(\vec{r},t)/\partial t = D \nabla^2 n(\vec{r},t)$$

yields the following *dispersion relation* between k and ω

$$k^2 D = i\omega$$

Then, show that for *free electron diffusion* we obtain

$$k^2 V_{se}^2 = i\omega \nu_{ce}$$

where $V_{se} = (k_B T_e/m_e)^{1/2} = (p_e/\rho_{me})^{1/2}$ is the isothermal speed of sound in the electron gas and k_B is Boltzmann's constant. Next, show that for *ambipolar diffusion* we obtain

$$k^2 V_{sp}^2 = i\omega \nu_{ci}$$

where

$$V_{sp} = [k_B (T_e + T_i)/m_i]^{1/2} = [(p_e + p_i)/(\rho_{me} + \rho_{mi})]^{1/2}$$

is the isothermal plasma sound speed. Calculate the phase velocity and the damping factor for these waves and verify if they are longitudinal or transverse.

10.13 - Consider a weakly ionized plasma immersed in a uniform magnetostatic field $\vec{B}_0$ oriented along the z axis of a Cartesian coordinate system.
(a) Show that the diffusion equation for the electrons (with $D\vec{u}_e/Dt \simeq 0$) in the presence of the space charge electric field is given by

$$\vec{\Gamma}_e = - \vec{\nabla} \cdot (\bar{\bar{D}}_e n_e) + n_e \bar{\bar{\mu}}_e \cdot \vec{E}$$

where

$$\bar{\bar{D}}_e = \begin{pmatrix} D_{e\perp} & D_{eH} & 0 \\ -D_{eH} & D_{e\perp} & 0 \\ 0 & 0 & D_{e\parallel} \end{pmatrix}$$

with $D_{e\perp}$, D_{eH} and $D_{e\parallel}$ given by (9.5), (9.6) and (9.7), respectively, and where

$$\bar{\bar{\mu}}_e = - (e/kT_e)\, \bar{\bar{D}}_e$$

(b) Deduce the corresponding equation for the ions in the presence of the space charge electric field $\vec{E}$. Combine the equations for the electrons and for the ions to eliminate the space charge electric field. Then, assuming that the electron and ion fluxes are equal, $\vec{\Gamma}_e = \vec{\Gamma}_i$, and that their number densities are also equal, $n_e = n_i$, determine the ambipolar diffusion coefficient, and notice that it is not affected by the presence of the magnetostatic field.

10.14 - Consider the following heat flow equation, derived in Problem 8.11, for a stationary electron gas immersed in a magnetic field,

$$\vec{q}_e + (\omega_{ce}/\nu)\, (\vec{q}_e \times \hat{B}) = - K_o \vec{\nabla} T_e$$

Show that this equation can be written in the form

$$\vec{q}_e = - \bar{\bar{K}} \cdot \vec{\nabla} T_e$$

where $\bar{\bar{K}}$ is the dyadic thermal conductivity coefficient, given by

$$\bar{\bar{K}} = \begin{pmatrix} K_\perp & -K_H & 0 \\ K_H & K_\perp & 0 \\ 0 & 0 & K_0 \end{pmatrix}$$

where

$$K_\perp = \frac{\nu^2}{(\nu^2 + \omega_{ce}^2)} K_0$$

$$K_H = \frac{\nu\, \omega_{ce}}{(\nu^2 + \omega_{ce}^2)} K_0$$

$$K_0 = 5\, k\, p_e/(2\, m_e\, \nu)$$

CHAPTER 11

Some Basic Plasma Phenomena

1. ELECTRON PLASMA OSCILLATIONS

One of the fundamental properties of a plasma is its tendency to maintain electric charge neutrality on a macroscopic scale under equilibrium conditions. When this macroscopic charge neutrality is disturbed, such as to temporarily produce a significant imbalance of charge, large Coulomb forces come into play which tend to restore the macroscopic charge neutrality. Since these Coulomb forces cannot be naturally sustained in the plasma, it breaks into high frequency electron plasma oscillations, which enable the plasma to maintain on the average its electric neutrality.

As a simple example, consider a small spherical region inside a plasma and suppose that a perturbation in the form of an excess of negative charge is introduced in this small region. Because of spherical symmetry, the corresponding electric field is radial and points towards the center (Fig. 1),

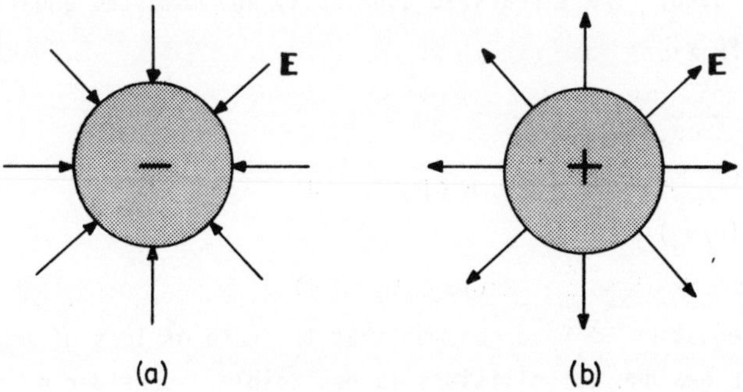

Fig. 1 - The radial electric field $\vec{E}$ produced by a spherical distribution of negative charge (a) forces the electrons to move radially outward, whereas the field produced by a spherical distribution of positive charge (b) forces the electrons to move radially inward.

FPP-J*

forcing the electrons to move radially outward. After a small time interval, since the electrons gain kinetic energy in the course of their motion, more electrons leave the spherical plasma region (due to their inertia) than is necessary to resume the state of electrical neutrality. An excess of positive charge results therefore inside this region and the reversed electric field causes the electrons to move inward now. This movement of electrons inward and outward of the spherical plasma region continues periodically, resulting in electron plasma oscillations. In this way the plasma maintains its macroscopic neutrality on the average, since the total charge inside the spherical region averaged over one period of these oscillations is zero. The frequency of these oscillations is usually very high and since the ions (in view of their mass) are unable to follow the rapidity of these oscillations their motion is often neglected.

To study the characteristics of the electron plasma oscillations we can use the cold plasma model, in which the particle thermal motion and the pressure gradient force are not taken into account. We neglect ion motion and assume a very small electron density perturbation such that

$$n_e(\vec{r},t) = n_0 + n_e'(\vec{r},t) \tag{1.1}$$

where n_0 is a constant number density and $|n_e'| \ll n_0$. Similarly, we assume that the electric field produced, $\vec{E}(\vec{r},t)$, and the average electron velocity, $\vec{u}_e(\vec{r},t)$, are first order perturbations, so that the linearized equations can be used. The linearized continuity and momentum equations become, respectively,

$$\partial n_e'/\partial t + n_0 \ \vec{\nabla} \cdot \vec{u}_e = 0 \tag{1.2}$$

$$\partial \vec{u}_e/\partial t = - (e/m_e) \ \vec{E} \tag{1.3}$$

In the momentum equation we have assumed that the rate of loss of momentum from the electron gas due to collisions is negligible. Considering singly charged ions, the charge density is

$$\rho \ (\vec{r}, t) = - e \ [n_0 + n_e'(\vec{r},t)] + e \ n_0 = - e \ n_e'(\vec{r},t) \tag{1.4}$$

where the ion density was considered to be constant and uniform, and equal to

n_o (neglecting ion motion). Therefore,

$$\vec{\nabla} \cdot \vec{E} = \rho/\varepsilon_o = - (e/\varepsilon_o)\, n_e' \tag{1.5}$$

Eqs. (1.2), (1.3) and (1.5) constitute a complete set of equations to be solved for the variables $n_e'(\vec{r},t)$, $\vec{u}_e(\vec{r},t)$ and $\vec{E}(\vec{r},t)$. Taking the divergence of (1.3) and using (1.2) to substitute for $\vec{\nabla} \cdot \vec{u}_e$, we obtain

$$\partial^2 n_e'/\partial t^2 - (e n_o/m_e)\, \vec{\nabla} \cdot \vec{E} = 0 \tag{1.6}$$

Combining (1.5) and (1.6) to eliminate $\vec{\nabla} \cdot \vec{E}$, yields

$$\partial^2 n_e'/\partial t^2 + \omega_{pe}^2\, n_e' = 0 \tag{1.7}$$

where

$$\omega_{pe} = (n_o\, e^2/m_e\, \varepsilon_o)^{1/2} \tag{1.8}$$

is called the *electron plasma frequency*. (1.7) shows that $n_e'(\vec{r},t)$ varies harmonically in time at the electron plasma frequency,

$$n_e'(\vec{r},t) = n_e'(\vec{r}) \exp\,(-\, i\, \omega_{pe}\, t) \tag{1.9}$$

In fact, all first-order perturbations have a harmonic time variation at the plasma frequency ω_{pe}. To justify this statement it is convenient to start with the assumption that all first order quantities vary harmonically in time, as exp $(-\, i\, \omega\, t)$. Eqs. (1.2) and (1.3) become, in this case,

$$n_e' = -\, (i/\omega)\, n_o\, \vec{\nabla} \cdot \vec{u}_e \tag{1.10}$$

$$\vec{u}_e = -\, (ie/\omega m_e)\, \vec{E} \tag{1.11}$$

which can be combined into

$$n_e' = -\, (n_o e/\omega^2 m_e)\, \vec{\nabla} \cdot \vec{E} \tag{1.12}$$

Substituting this expression for n_e' into (1.5), yields

$$(1 - \omega_{pe}^2/\omega^2)\, \vec{\nabla} \cdot \vec{E} = 0 \tag{1.13}$$

which shows that a nontrivial solution requires $\omega = \omega_{pe}$. Therefore, all the perturbations vary harmonically in time at the electron plasma frequency. Further, for all variables there is no change in phase from point to point, implying in the absence of wave propagation. The oscillations are therefore *stationary*. Also, (1.11) shows that the electron velocity is in the same direction as the electric field, so that these oscillations are *longitudinal*.

The electron plasma oscillations are also *electrostatic* in character. In order to show this aspect of the oscillations consider Maxwell curl equations with a harmonic time variation,

$$\vec{\nabla} \times \vec{E} = i\, \omega\, \vec{B} \tag{1.14}$$

$$\vec{\nabla} \times \vec{B} = \mu_0\, (\vec{J} - i\, \omega\, \varepsilon_0\, \vec{E}) \tag{1.15}$$

The electric current density is given by

$$\vec{J} = - e n_0\, \vec{u}_e = (i\, n_0\, e^2/\omega m_e)\, \vec{E} \tag{1.16}$$

where we have used (1.11) for $\vec{u}_e$. Therefore,

$$\vec{\nabla} \times \vec{B} = - i\, \omega\, \mu_0\, \varepsilon_0\, \varepsilon_r\, \vec{E} \tag{1.17}$$

where we have defined a relative permittivity by

$$\varepsilon_r = 1 - \omega_{pe}^2/\omega^2 \tag{1.18}$$

For the electron plasma oscillations we have $\omega = \omega_{pe}$, so that $\varepsilon_r = 0$, and (1.17) reduces to

$$\vec{\nabla} \times \vec{B} = 0 \tag{1.19}$$

Since the curl of the gradient of any scalar function vanishes identically, we may write

$$\vec{B} = \vec{\nabla} \psi \tag{1.20}$$

where ψ is a magnetic scalar potential. Substituting (1.20) into (1.14) and taking the divergence of both sides we obtain Laplace equation

$$\vec{\nabla} \cdot (\vec{\nabla}\psi) = \nabla^2 \psi = 0 \tag{1.21}$$

since the divergence of the curl of any vector function vanishes identically. The only solution of this equation, which is not singular and finite at infinity, is ψ = constant, so that $\vec{B} = 0$. Hence, there is no magnetic field associated with these space charge oscillations. In summary, the electron plasma oscillations are *stationary*, *longitudinal* and *electrostatic*. They are also referred to as *Langmuir oscillations*. When the effect of the pressure gradient force is included in the equation of motion (1.3), complemented by an adiabatic energy equation,these oscillations become propagating disturbances commonly known as *space charge waves* or *Langmuir waves*. Characteristic values of the electron plasma frequency for various laboratory and cosmic plasmas are given in Fig. 2 of Chapter 1.

2. THE DEBYE SHIELDING PROBLEM

To examine the mechanism by which the plasma strives to shield its interior from a disturbing electric field, consider a plasma whose equilibrium state is perturbed by an electric field due to an external charged particle. For that matter, this electric field may also be considered to be due to one of the charged particles inside the plasma, isolated for observation. For definiteness, we assume this *test particle* to have a positive charge + Q, and choose a spherical coordinate system whose origin coincides with the position of the test particle. We are interested in determining the electrostatic potential $\phi(\vec{r})$ that is established near the test charge Q, due to the *combined* effects of the test charge and the distribution of charged particles surrounding it. Since the positive test charge Q attracts the negatively charged particles and repels the positively charged ones, the number densities of the electrons $n_e(\vec{r})$ and of the ions $n_i(\vec{r})$ will be slightly different near the origin (test particle), whereas at large distances from the origin the electrostatic potential vanhishes so that $n_e = n_i = n_0$. Since this is a steady state problem under the action of a conservative electric field, we have

$$\vec{E}(\vec{r}) = - \vec{\nabla} \phi(\vec{r}) \tag{2.1}$$

and from (7.5.16) it follows that

$$n_e(\vec{r}) = n_0 \exp [e \phi(\vec{r})/kT] \tag{2.2}$$

$$n_i(\vec{r}) = n_0 \exp [- e \phi(\vec{r})/kT] \tag{2.3}$$

where we have assumed that the electrons and ions (of charge e) have the same temperature T.

The total electric charge density $\rho(\vec{r})$, including the test charge Q, can be expressed as

$$\rho(\vec{r}) = - e [n_e(\vec{r}) - n_i(\vec{r})] + Q \delta(\vec{r}) \tag{2.4}$$

where $\delta(\vec{r})$ denotes the Dirac delta function. Using (2.2) and (2.3),

$$\rho(\vec{r}) = - en_0 \{ \exp [e\phi(\vec{r})/kT] - \exp [- e \phi(\vec{r})/kT] \} + Q \delta(\vec{r}) \tag{2.5}$$

Substituting (2.1) and (2.5) into the following Maxwell equation

$$\vec{\nabla} \cdot \vec{E}(\vec{r}) = \rho(\vec{r})/\varepsilon_0 \tag{2.6}$$

gives the differential equation

$$\nabla^2 \phi(\vec{r}) - \frac{en_0}{\varepsilon_0} \left\{ \exp [e \phi(\vec{r})/kT] - \exp [- e \phi(\vec{r})/kT] \right\} = - \frac{Q}{\varepsilon_0} \delta(\vec{r}) \tag{2.7}$$

which permits the evaluation of the electrostatic potential $\phi(\vec{r})$.

We assume now that the perturbing electrostatic potential is weak so that the electrostatic potential energy is much less than the mean thermal energy, that is,

$$e \phi(\vec{r}) \ll kT \tag{2.8}$$

Under this condition we can use the approximation

$$\exp[\pm e\,\phi(\vec{r})/kT] \cong 1 \pm e\,\phi(\vec{r})/kT \tag{2.9}$$

Therefore, (2.7) simplifies to

$$\nabla^2 \phi(\vec{r}) - (2/\lambda_D^2)\,\phi(\vec{r}) = -(Q/\varepsilon_0)\,\delta(\vec{r}) \tag{2.10}$$

where λ_D denotes the Debye length

$$\lambda_D = (\varepsilon_0 kT/n_0 e^2)^{1/2} = (1/\omega_{pe})(kT_e/m_e)^{1/2} \tag{2.11}$$

Since the problem has spherical symmetry, the electrostatic potential depends only on the radial distance r measured from the position of the test particle, being independent of the spatial orientation of $\vec{r}$. Thus, using spherical coordinates, (2.10) can be written (for $r \neq 0$) as

$$\left(\frac{1}{r^2}\right)\frac{d}{dr}\left[r^2\frac{d}{dr}\phi(r)\right] - \frac{2}{\lambda_D^2}\phi(r) = 0; \qquad (r \neq 0) \tag{2.12}$$

In order to solve this equation we note initially that for an isolated particle of charge $+Q$, in free space, the electric field is directed radially outward and is given by

$$\vec{E}(r) = \frac{1}{4\pi\varepsilon_0}\,\frac{Q}{r^2}\,\hat{r} \tag{2.13}$$

so that the electrostatic Coulomb potential $\phi_c(r)$ due to this isolated charged particle in free space is

$$\phi_c(r) = \frac{1}{4\pi\varepsilon_0}\,\frac{Q}{r} \tag{2.14}$$

In the very close proximity of the test particle the electrostatic potential should be the same as that for an isolated particle in free space. Hence, it is appropriate to seek the solution of (2.12) in the form

$$\phi(r) = \phi_c(r)\,F(r) = \frac{Q}{4\pi\varepsilon_0}\,\frac{F(r)}{r} \tag{2.15}$$

where the function $F(r)$ must be such that $F(r) \rightarrow 1$ when $r \rightarrow 0$. Furthermore, the electrostatic potential $\phi(r)$ is required to vanish at infinity, that is, $\phi(r) \rightarrow 0$ when $r \rightarrow \infty$. Substituting (2.15) into (2.12) yields the following differential equation for $F(r)$

$$d^2 F(r)/dr^2 = (2/\lambda_D^2)\, F(r) \tag{2.16}$$

This simple differential equation for $F(r)$ has the solution

$$F(r) = A \exp\,(\sqrt{2}\, r/\lambda_D) + B \exp\,(-\sqrt{2}\, r/\lambda_D) \tag{2.17}$$

The condition that $\phi(r)$ vanishes for large values of r requires $A = 0$. Also, the condition that $F(r)$ tends to one when r tends to zero requires $B = 1$. Therefore, the solution of (2.12) is

$$\phi(r) = \phi_c(r) \exp\,(-\sqrt{2}\, r/\lambda_D) = \frac{1}{4\pi\varepsilon_0}\,\frac{Q}{r} \exp\,(-\sqrt{2}\, r/\lambda_D) \tag{2.18}$$

This result is commonly known as the *Debye potential*, since this nonrigorous derivation was first presented by Debye and Hückel in their theory of electrolytes. It shows that $\phi(r)$ becomes much less than the ordinary Coulomb potential once r exceeds the distance λ_D, called the *Debye length* (see Fig. 2). Hence, we can say in a crude way that a charged particle in a plasma interacts effectively only with particles situated at distances less than one Debye length away, and it has a negligible influence on particles lying at distances greater than one Debye length.

The charge Q of the test particle is neutralized by the charge distribution surrounding the test particle. From (2.5) and (2.9) we obtain for the charge density

$$\rho(\vec{r}) = -\,2 n_0 e^2\, \phi(\vec{r})/kT + Q\, \delta(\vec{r}) \tag{2.19}$$

Substituting $\phi(\vec{r})$ by the Debye potential (2.18), we obtain

$$\rho(\vec{r}) = -\,(Q/2\pi r \lambda_D^2) \exp\,(-\sqrt{2}\, r/\lambda_D) + Q\, \delta(\vec{r}) \tag{2.20}$$

To obtain the total charge we integrate (2.20) over all space,

$$q_t = \iiint \rho(\vec{r})\, d^3r = - \frac{Q}{2\pi\, \lambda_D^2} \int_0^\infty \frac{1}{r} \exp(-\sqrt{2}\, r/\lambda_D)\, 4\pi r^2\, dr +$$

$$+ Q \iiint \delta(\vec{r})\, d^3r = 0 \qquad (2.21)$$

since the first integral gives - Q, whereas the second one is equal to + Q. The principal contribution to the first integral in (2.21) comes from the plasma particles lying in the very close neighborhood of the test particle, since the integrand falls off exponentially with increasing values of r. Thus, the neutralization of the test particle takes place effectively on account of the charged particles inside the Debye sphere. From (2.2) and (2.3) we see that in the neighborhood of the test particle the electron number density is larger than the ion number density, on account of the fact that the positive test particle attracts the electrons and repels the ions. Therefore, in the close proximity of the test particle there is an imbalance of charge and, consequently, an electric field. We have seen that the

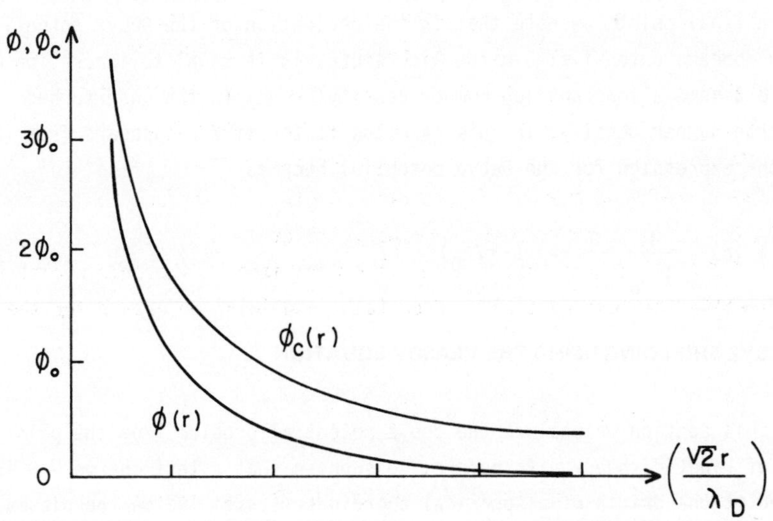

Fig. 2 - Electrostatic Coulomb potential $\phi_C(r)$ and Debye potential $\phi(r)$ as a function of distance r from the test charge Q. ($\phi_0 = \sqrt{2}\, Q/4\pi\varepsilon_0\lambda_D$).

shielding of this electric field is effectively completed over a distance of the order of λ_D. Thus, for a plasma to be considered macroscopically neutral, it is necessary that its typical characteristic dimension L be much greater than λ_D. This quantitative *criterion* for the definition of a plasma has been previously discussed in Chapter 1.

An important point to be noted in the result (2.18) is that, for $r \to 0$, the Debye potential becomes very large and the assumption $e\phi(r) << kT$ is unlikely to be fulfilled. To verify the validity of this approximation, and consequently of (2.18), note that using (2.18) with $Q = e$ we have

$$\frac{e\phi}{kT} = \frac{e^2 \exp(-\sqrt{2}\, r/\lambda_D)}{4\pi\varepsilon_0 rkT} = \frac{\lambda_D}{3N_D} \frac{\exp(-\sqrt{2}\, r/\lambda_D)}{r} \tag{2.22}$$

where N_D is the number of electrons inside a Debye sphere. Since N_D is very large for virtually all plasmas, it is evident that the ratio given in (2.23) is much less than one, except when r is less than λ_D/N_D. Therefore, the Debye potential (2.18) is consistent with the approximation $e\phi << kT$ used to derive it, if we restrict attention to distances from the test particle greater than λ_D/N_D.

As a final point, we note that in the derivation of the Debye potential which appears extensively in the literature, it is usual to ignore ion motion and to assume a constant ion number density equal to the unperturbed electron number density. In this case the factor of 2 disappears from (2.10), and the expression for the Debye potential becomes

$$\phi(r) = \frac{1}{4\pi\varepsilon_0} \frac{Q}{r} \exp(-r/\lambda_D) \tag{2.23}$$

3. DEBYE SHIELDING USING THE VLASOV EQUATION

In this section we analyze the Debye potential problem from the point of view of kinetic theory. As before, we suppose that a test charge + Q is placed at the origin of a spherical coordinate system inside the plasma. In order to determine the steady state electron and ion distribution functions, f_e and f_i, and the electrostatic potential $\phi(\vec{r})$ near the test charge, let us consider the steady state Vlasov equations for the electrons and the ions, with only the electric field $\vec{E} = -\vec{\nabla}\phi$ in the Lorentz force term,

$$\vec{v} \cdot \vec{\nabla} f_e + (e/m_e)\,(\vec{\nabla}\phi) \cdot \vec{\nabla}_v\, f_e = 0 \tag{3.1}$$

$$\vec{v} \cdot \vec{\nabla} f_i - (e/m_i)\,(\vec{\nabla}\phi) \cdot \vec{\nabla}_v\, f_i = 0 \tag{3.2}$$

Since

$$n_\alpha(\vec{r}) = \int_V f_\alpha(\vec{r},v)\, d^3v \tag{3.3}$$

the total charge density (including the test particle) can be expressed as

$$\rho(\vec{r}) = -e \int_V (f_e - f_i)\, d^3v + Q\,\delta(\vec{r}) \tag{3.4}$$

and Poisson equation for this case becomes

$$\nabla^2\phi - \frac{e}{\varepsilon_o} \int_V (f_e - f_i)\, d^3v = -\frac{Q}{\varepsilon_o}\,\delta(\vec{r}) \tag{3.5}$$

Eqs. (3.1), (3.2) and (3.5) constitute three equations to be solved simultaneously to determine f_e, f_i and ϕ.

The solution of the Vlasov equations (3.1) and (3.2) can be expressed in terms of the Maxwellian distribution function and the Boltzmann factor (see section 5, Chapter 7), as

$$f_\alpha(\vec{r},v) = f_{o\alpha}(v)\, \exp\left[-q_\alpha\, \phi(\vec{r})/kT\right] \tag{3.6}$$

When the electrostatic potential vanishes, this distribution function goes into the Maxwellian distribution $f_{o\alpha}(v)$, with zero drift velocity. Substituting (3.6), with $\alpha = e, i$, into (3.5), yields

$$\nabla^2\phi - \frac{e}{\varepsilon_o}\left[\exp(e\,\phi/kT) \int_V f_{oe}(v)\, d^3v - \exp(-e\,\phi/kT) \int_V f_{oi}(v)\, d^3v\right]$$

$$= -\frac{Q}{\varepsilon_o}\,\delta(\vec{r}) \tag{3.7}$$

Denoting the electron and ion number densities under equilibrium conditions (when ϕ vanishes) by n_o,

$$n_o = \int_v f_{o\alpha}(v)\, d^3v \qquad \alpha = e,\ i \tag{3.8}$$

(3.7) becomes

$$\nabla^2\phi - \frac{en_o}{\varepsilon_o}\left[\exp(e\phi/kT) - \exp(-e\phi/kT)\right] = -\frac{Q}{\varepsilon_o}\,\delta(\vec{r}) \tag{3.9}$$

This equation is identical to (2.7), yielding the same result as before for the Debye potential.

4. PLASMA SHEATH

When a material body is immersed in a plasma, the body acquires a net negative charge and therefore a negative potential with respect to the plasma potential. In the region near the wall of the body there is a boundary layer, known as the *plasma sheath*, in which the electron and ion number densities are different. Inside the plasma sheath the potential increases monotonically from a negative value on the wall to the value corresponding to the unperturbed plasma. The thickness of the plasma sheath, where departures from macroscopic electric neutrality occur, is found to be of the order of a Debye length.

A satisfactory mathematical treatment of this phenomenon is quite involved and cannot be presented here. However, the basic physical factors responsible for the formation of the plasma sheath are fairly simple to understand. In this section we will set up the problem mathematically and obtain some approximate results. The problem is strongly dependent on the particular geometry under consideration. For simplicity, we shall assume that the wall bounding the plasma is an infinite plane surface at $x = 0$, with the plasma in the region $x > 0$, and that there are no variations with respect to the coordinates y and z.

4.1 Physical mechanism

We begin with a descriptive account of the physical mechanism responsible for the formation of the plasma sheath. The charged particles in the plasma that strike the wall in virtue of their random thermal motions are for the most part lost to the plasma. The ions generally recombine at the wall and return to the plasma as neutral particles, whereas the electrons may either recombine there or they may enter the conduction band if the surface is a metal. We have seen in section 4, of Chapter 7, that the *random* particle flux, that is, the number of particles that hit the surface per unit area and per unit time, from one side only, for the case of an isotropic velocity distribution function, is given by [see Eqs. (7.4.35) and (7.4.18)]

$$\Gamma_\alpha = n_\alpha \langle v \rangle_\alpha / 4 \tag{4.1}$$

where $\langle v \rangle_\alpha$ is the average particle speed for the α species. For the Maxwell-Boltzmann velocity distribution function we have found that [see Eq. (7.4.20)]

$$\langle v \rangle_\alpha = (8/\pi)^{1/2} \, (kT_\alpha/m_\alpha)^{1/2} \tag{4.2}$$

so that a typical value for the random particle flux in this case is

$$\Gamma_\alpha = n_\alpha \, (kT_\alpha/2\pi m_\alpha)^{1/2} \tag{4.3}$$

It is evident from this result that if initially the electron and ion number densities are equal, then the random particle flux for the electrons, Γ_e, greatly exceeds that for the ions, Γ_i, since in general $(T_e/m_e)^{1/2}$ is much larger than $(T_i/m_i)^{1/2}$. For the least heavy ion, hydrogen, for example, $m_i/m_e = 1836$. Therefore, the wall in contact with the plasma rapidly accumulates a negative charge, since initially more electrons reach the wall than positive ions. This negative potential repels the electrons and attracts the ions so that the electron flux diminishes and the ion flux increases. Eventually, the negative potential at the wall becomes large enough in magnitude to equalize the rate at which electrons and ions hit the surface. At this floating negative potential the wall and the plasma reach a dynamical equilibrium such that the net current at the wall is zero.

4.2 Electric potential on the wall

To estimate the value of the potential on the wall after the plasma sheath has been established, consider a steady state situation and let the electric potential $\phi(x)$ at the wall $(x = 0)$ be given by

$$\phi(0) = \phi_W \tag{4.4}$$

Let us choose the reference potential inside the plasma, at a very large distance from the wall, equal to zero,

$$\phi(\infty) = 0 \tag{4.5}$$

The electrons and the ions are assumed to be in thermodynamic equilibrium at the same temperature T, under the action of the conservative electric field associated with the negative potential on the wall. At $x \rightarrow \infty$ the plasma is unperturbed and the electron and ion number densities are each equal to n_0. According to the results of section 5, Chapter 7, the electron and ion number densities can be expressed as

$$n_e(\vec{r}) = n_0 \exp\left[e\,\phi(\vec{r})/kT\right] \tag{4.6}$$

$$n_i(\vec{r}) = n_0 \exp\left[-e\,\phi(\vec{r})/kT\right] \tag{4.7}$$

It is important to note at this point that (4.6) and (4.7) do not take into account the particle drift velocity towards the wall. Since the electrons and the ions impinging on the wall surface are for the most part lost to the plasma, there must be a steady flux of both species towards the wall to replenish this charge particle loss. Despite this inadequacy, (4.6) and (4.7) will still be used to obtain an approximate expression for the potential on the wall and afterwards, in order to study the inner structure of the plasma sheath, we will take into account the particle drift in an approximate manner using the hydrodynamic equations.

One of the boundary conditions of the problem is that under equilibrium conditions there must be no charge build up at the wall $(x = 0)$, so that

$$J_e(0) = J_i(0) \tag{4.8}$$

Using (4.3), (4.6) and (4.7), considering singly charged ions,

$$(1/m_e)^{1/2} \exp(e\,\phi_W/kT) = (1/m_i)^{1/2} \exp(-\,e\,\phi_W/kT) \tag{4.9}$$

which may be written as

$$\exp(-\,2e\,\phi_W/kT) = (m_i/m_e)^{1/2} \tag{4.10}$$

Taking the natural logarithm of both sides, and solving for the wall potential, we obtain

$$\phi_W = -\,(kT/4e)\,\ell n(m_i/m_e) \tag{4.11}$$

Other more accurate methods of calculating the wall potential yield results which, for $T_e = T_i$, agree qualitatively with the one given in (4.11), despite the inadequacy of (4.6) and (4.7) which neglect the particle drift velocity towards the wall.

Note from (4.11) that the magnitude of the potential energy near the wall $|e\,\phi_W|$ is of the same order as the average thermal energy kT of the particles in the plasma, since

$$|e\,\phi_W|/kT = (1/4)\,\ell n(m_i/m_e) \tag{4.12}$$

For a hydrogen ion, for example, $|e\,\phi_W|/kT$ is approximately equal to 2, whereas for heavier ions it may be close to 3.

4.3 Inner structure of the plasma sheath

To investigate the inner structure of the plasma sheath consider the equations of conservation of particles and momentum for the electrons and ions under steady state conditions, with spatial dependence only on the x direction. The equation of conservation of particles becomes

$$d(n_\alpha u_\alpha)/dx = n_\alpha(du_\alpha/dx) + u_\alpha(dn_\alpha/dx) = 0 \tag{4.13}$$

In the momentum conservation equation we neglect viscosity effects and approximate the kinetic pressure dyad by a scalar pressure. The ideal gas equation of state, $p_\alpha = n_\alpha k T_\alpha$, can be used to introduce the temperature, which is assumed to be constant. Collisions are neglected, since the thickness of the plasma sheath is much less than the mean free path for the plasma particles. With this assumptions and in the absence of a magnetic field, the equation of motion becomes (with $\vec{E} = -\vec{\nabla}\phi$, and $D/DT = \partial/\partial t + \vec{u}_\alpha \cdot \vec{\nabla} = u_\alpha \, d/dx$)

$$m_\alpha u_\alpha (du_\alpha/dx) = -(kT_\alpha/n_\alpha)(dn_\alpha/dx) - q_\alpha \, d\phi/dx \tag{4.14}$$

In order to simplify the analysis we shall make two approximations. From (4.13), we can write

$$dn_\alpha/dx = -(n_\alpha/u_\alpha)(du_\alpha/dx) \tag{4.15}$$

and the ratio of the magnitude of the term in the left-hand side of (4.14) to the first term in the right-hand side can be expressed as

$$\frac{|(m_\alpha u_\alpha)(du_\alpha/dx)|}{|(kT/n_\alpha)(dn_\alpha/dx)|} = \frac{m_\alpha u_\alpha^2}{kT} \tag{4.16}$$

The two approximations consist in neglecting the left-hand side term of (4.14) for the electrons, whereas for the ions we neglect the first term in the right-hand side of (4.14). Explicitly, we take for the electrons (neglecting electron inertia)

$$(kT_e/n_e)\, dn_e/dx - e \, d\phi/dx = 0 \tag{4.17}$$

and for the ions (assuming cold ions)

$$m_i u_i \, du_i/dx + e \, d\phi/dx = 0 \tag{4.18}$$

These two approximations are justified only if the thermal energy of the *electrons* is much *larger* than their kinetic energy, and if the thermal energy of the *ions* is much *smaller* than their kinetic energy. Thus, we require from (4.15) that

$$m_e u_e^2 \ll kT \ll m_i u_i^2 \qquad (4.19)$$

in order to justify the approximations in (4.17) and (4.18). We shall assume that the condition (4.19) is satisfied in the plasma sheath but it remains to be justified later in this section.

If we integrate (4.17), we obtain

$$e\,\phi(x) = kT\,\ln n_e(x) + (\text{constant}) \qquad (4.20)$$

and using the condition that $n_e = n_o$ when $\phi = 0$, we find

$$n_e(x) = n_o \exp\,[e\,\phi(x)/kT] \qquad (4.21)$$

This result is identical to (4.6), which is not surprising, since the condition $m_e u_e^2 \ll kT$ implies in neglecting the electron inertia $(m_e = 0)$ and consequently their kinetic energy.

For the ions, we first integrate (4.13) to find

$$n_i(x)\,u_i(x) = C_1 \qquad (4.22)$$

and then integrate (4.18) to obtain

$$e\,\phi(x) + (m_i/2)\,u_i^2(x) = C_2 \qquad (4.23)$$

where C_1 and C_2 are constants. The boundary conditions require that at $x = \infty$ we must have $\phi(\infty) = 0$, $n_i(\infty) = n_o$ and $u_i(\infty) = u_{oi}$. Thus,

$$C_1 = n_o\,u_{oi} \quad ; \quad C_2 = m_i\,u_{oi}^2/2 \qquad (4.24)$$

and using these results in (4.22) and (4.23), we find

$$n_i(x)\,u_i(x) = n_o\,u_{oi} \qquad (4.25)$$

$$e\,\phi(x) + (m_i/2)\,u_i^2(x) = m_i\,u_{oi}^2/2 \qquad (4.26)$$

These two equations can be combined to eliminate $u_i(x)$ and solved for $n_i(x)$, giving

$$n_i(x) = n_o \, [1 - 2e\,\phi(x)/(m_i \, u_{oi}^2)]^{-1/2} \tag{4.27}$$

This expression for $n_i(x)$ is substantially different from the one given in (4.7) and this difference is due to the importance of the ion drift velocity. We now find that, since $\phi(x) < 0$ in the sheath, $n_i(x)$ decreases slowly towards the wall rather than increasing as predicted by (4.7). Physically, this behavior is due to the fact that the negative potential on the wall causes $u_i(x)$ to increase as the ions approach the wall and since the flux $n_i(x)\ u_i(x)$ must stay constant, in virtue of (4.25), it turns out that $n_i(x)$ must decrease according to (4.27). This behavior is illustrated schematically in Fig. 3.

To obtain the differential equation satisfied by the electrostatic potential $\phi(x)$, we substitute (4.21) and (4.27) into Poisson equation

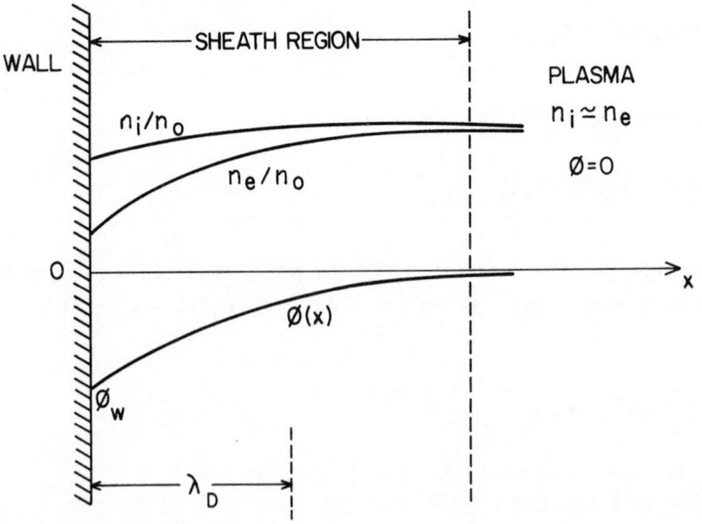

Fig. 3 - Diagram showing the variation of the electrostatic potential $\phi(x)$ and the number densities n_e and n_i inside the plasma sheath near an infinite plane wall.

$$\nabla^2 \phi = e\,(n_e - n_i)/\varepsilon_o \tag{4.28}$$

to obtain

$$\frac{d^2\phi}{dx^2} = \frac{n_o e}{\varepsilon_o}\left[\exp\left(\frac{e\phi}{kT}\right) - \left(1 - \frac{2e\phi}{m_i u_{oi}^2}\right)^{-1/2}\right] \tag{4.29}$$

In this equation, the drift velocity u_{oi}, far away from the wall, still needs to be determined. This equation is nonlinear and in order to facilitate its analytical solution we need to make one more approximation. Since we have seen that $|e\phi|$ ranges from zero in the plasma to a value of the order of kT on the wall and since we have also assumed that $m_i u_{oi}^2$ is larger than kT, we will restrict our attention to the region near the plasma edge of the sheath and assume further that $|e\phi|$ is small compared to both kT and $m_i u_{oi}^2$. Thus, in the region near the edge of the sheath adjacent to the plasma, we can expand the terms in the right-hand side of (4.29), for $(e\phi/kT) \ll 1$ and $(e\phi/m_i u_{oi}^2) \ll 1$, as

$$\exp(e\phi/kT) \cong 1 + (e\phi/kT) \tag{4.30}$$

$$[1 - 2e\phi/(m_i u_{oi}^2)]^{-1/2} \cong 1 + e\phi/(m_i u_{oi}^2) \tag{4.31}$$

with the result that (4.29) reduces to

$$d^2\phi/dx^2 \cong \phi/X^2 \tag{4.32}$$

where

$$X^2 = \lambda_D^2\,[1 - kT/(m_i u_{oi}^2)]^{-1} \tag{4.33}$$

The solution of (4.32), with the boundary condition $\phi(\infty) = 0$, is

$$\phi(x) = A \exp(-x/X) \tag{4.34}$$

where A is a constant. Since we have assumed that $kT \ll m_i u_{oi}^2$, it follows from (4.33) that X is real and approximately equal to λ_D.

Therefore, we find that $\phi(x)$ increases exponentially (since A must be negative) as we move from inside the sheath into the plasma and goes to zero at very large distances from the wall. Since $X \cong \lambda_D$, this increase effectively takes place within a distance of the order of a Debye length. This solution for $\phi(x)$ is strictly valid only near the plasma edge of the sheath, but if it is continued to apply throughout the plasma sheath, we can impose the boundary condition on the wall, that is, $\phi(0) = \phi_w$, which requires that $A = \phi_w$.

If kT were greater than $m_i\, u_{oi}^2$, than X would be imaginary and the electric potential would be an oscillating function of distance near the wall. The condition

$$kT < m_i\, u_{oi}^2 \tag{4.35}$$

must therefore be satisfied for the formation of a plasma sheath. It is known as the *Bohm criterion.*

It is not a trivial matter to determine the potential on the wall using the hydrodynamic equations. All the approximate methods that have been suggested give results which, for $T_e = T_i$, agree reasonably well with the approximate value for ϕ_w given in (4.11). Furthermore, there is no consistent way to determine the ion drift velocity u_{oi} at $x = \infty$, but an approximate estimate can be obtained as follows. Since the ion flux must be constant, from (4.25) we can equate the ion flux $n_o u_{oi}$ at $x = \infty$ to its value $\Gamma_i(0)$ at the wall. Using (4.3) with $n_i(\vec{r})$ as given by (4.7) evaluated at the wall, we find

$$u_{oi} = (kT/2\pi m_i)^{1/2} \exp\,(-e\,\phi_w/kT) \tag{4.36}$$

Similarly, for the electrons we can equate the electron flux $n_o u_{oe}$ at $x = \infty$ to its value $\Gamma_e(0)$ at the wall. From (4.3), with $n_e(\vec{r})$ given by (4.6), evaluated at the wall, we find

$$u_{oe} = (kT/2\pi m_e)^{1/2} \exp\,(e\,\phi_w/kT) \tag{4.37}$$

and using (4.9) we see that

$$u_{oe} = u_{oi} \tag{4.38}$$

To verify the validity of the approximations indicated in (4.19), we note first that from the mass conservation equation the particle flux $n_\alpha(x)\, u_\alpha(x)$ must be constant for all x and equal to the value $n_0 u_{o\alpha}$. From (4.21) we see that the *minimum* value of $n_e(x)$ is $n_0 \exp(e\phi_W/kT)$ since ϕ_W is negative. Therefore,

$$u_e = (n_0/n_e)\, u_{oe} \leq u_{oe} \exp(-e\,\phi_W/kT) \tag{4.39}$$

and using (4.37), we get

$$u_e \leq (kT/2\pi m_e)^{1/2} \tag{4.40}$$

or

$$kT/m_e\, u_e^2 \geq 2\pi \tag{4.41}$$

in agreement with the assumption (4.19). Similarly, from (4.27) the *maximum* value of $n_i(x)$ is n_0, so that

$$u_i = (n_0/n_i)\, u_{oi} \geq u_{oi} \tag{4.42}$$

and using (4.36), we obtain

$$u_i \geq (kT/2\pi m_i)^{1/2} \exp(-e\,\phi_W/kT) \tag{4.43}$$

or

$$kT/m_i\, u_i^2 \leq 2\pi \exp(2e\,\phi_W/kT) \cong 0.1 \tag{4.44}$$

where the result in the right has been obtained substituting ϕ_W by the value given in (4.11). Therefore, in view of (4.41) and (4.44), the thermal energy of the electrons is seen to be greater than their kinetic energy, whereas for the ions the opposite situation is verified. From (4.44) we can verify that the Bohm criterion, for the formation of the plasma sheath, is also satisfied.

Although the quantitative aspects of the discussion presented here are very approximate, it provides, however, a satisfactory qualitative picture of the sheath.

5. THE PLASMA PROBE

The plasma probe is a device that has been widely used to measure the temperature and density of the plasma particles, both in the laboratory and in space vehicles. The electrostatic probe was developed by Langmuir and Mott-Smith, and the physical mechanism of its operation can be well explained using the theory of plasma sheaths presented in the previous section. A conducting probe, or electrode, is immersed in a plasma and the current that flows through it is measured for various potentials applied to the probe. The temperature and number density of the electrons can then be obtained from the characteristics of the resulting current-potential curve. When the surface of the probe is plane, the current-potential curve has a shape like that illustrated in Fig. 4.

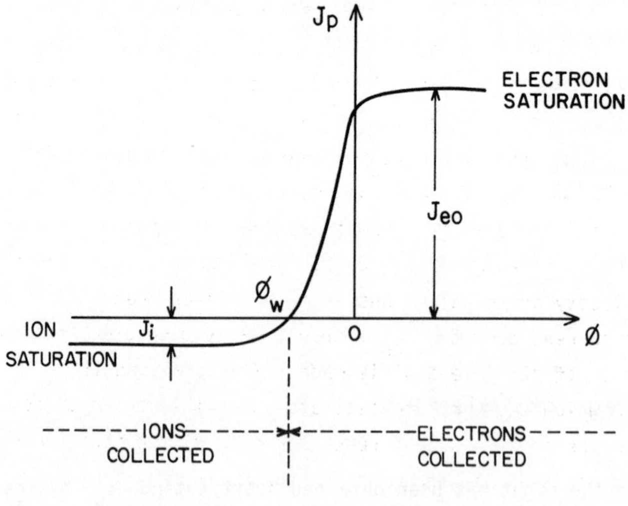

Fig. 4 - Characteristics of the current-potential curve of an electrostatic probe immersed in a plasma. The floating potential of the probe with reference to the plasma potential is denoted by ϕ_W.

The probe, when inserted in the plasma, is surrounded by a plasma sheath which shields the major portion of the plasma from the disturbing probe field. The thickness of the sheath is of the order of a Debye length. When no current flows through the electrode, it is at the negative floating potential ϕ_w, which is the wall potential discussed in the previous section. Under these conditions, the number of electrons reaching the probe per unit time is equal to the number of positive ions reaching the probe per unit time. We assume the current to be positive when it flows in the direction away from the probe. The current associated with the electrons is directed away from the probe and therefore is considered positive. Consequently, the electric current associated with the flow of ions is negative. Under equilibrium conditions, there is no net current flowing through the probe and its potential is the floating potential ϕ_w. When the potential is made more negative than ϕ_w, the electron current is reduced due to the increased repulsive force imposed by the probe electric field on the electrons. As the potential is made more negative, the contribution to the electric current arising from the electrons will eventually become neglibible and the total electric current asymptotically approaches a constant negative value, corresponding to the electric current density J_i associated with only the flow of ions. The ions that reach the edge of the plasma sheath fall into the potential well and their current is practically unaffected if the potential is made even more negative. On the other hand, when the probe potential is increased from the negative value ϕ_w, more electrons reach the probe than ions per unit time due to the decrease in the repulsion force on the electrons and the net electric current becomes positive. When the electric potential is zero, that is, when the probe is at the same potential as the plasma, there is no electric field near the electrode and, since the average thermal velocity of the electrons is much greater than that of the ions, the electron current density J_{eo} (for $\phi = 0$) is much greater than the ion current density. If the potential is made sufficiently positive, a situation arises in which the current associated with the ions becomes negligible, but all the electrons that reach the edge of the sheath are collected by the probe. The electron current density reaches a fairly constant value for sufficiently high positive values of ϕ. This plateau region in the probe current-potential curve is called the region of *saturation* of the electron current. For higher positive values of ϕ there are complications in the current-potential characteristic of the electrode due to the occurrence of another phenomena.

An approximate expression for the magnitude of the electron current density, away from the region of saturation, can be obtained from (4.6) as

$$J_e = J_{eo} \exp (e\,\phi / kT_e) \tag{5.1}$$

where J_{eo} is the electron current density when the electric potential is zero. Since for $\phi = 0$ we have $\Gamma_e = n_e < v >_e/4$, and using (4.2) for the average electron speed, we obtain,

$$J_{eo} = en_e \, (kT_e / 2\pi m_e)^{1/2} \tag{5.2}$$

where n_e is the electron number density in the unperturbed plasma region. Note that when ϕ is negative the ions reaching the edge of the sheath continue to fall into the negative potential of the probe and hence the ion current density is a constant (J_i) in the negative potential region. Thus, we can express the probe current density, in the region where $\phi < 0$, as

$$J_p = J_{eo} \exp (e\,\phi / kT_e) - J_i \qquad (\text{for } \phi < 0) \tag{5.3}$$

From this equation we can deduce the result

$$T_e = \frac{e}{k} \left\{ \frac{d}{d\phi} \left[\ln (J_p + J_i) \right] \right\}^{-1} \tag{5.4}$$

This expression can be used to determine the electron temperature as follows. First, the electrode is made sufficiently negative with reference to the plasma potential, so that the current that flows through the probe is due to the ions only. The measurement of this current gives directly the value J_i. Then, the current-potential characteristic of the probe is measured and a plot of $\ln (J_p + J_i)$ as a function of ϕ is made. This curve has a straight-line section corresponding to the probe potential less than the plasma potential, and the slope of this straight line gives the value of

$$\frac{d}{d\phi} \left[\ln (J_p + J_i) \right] \tag{5.5}$$

which, when substituted in (5.4), gives the electron temperature in the plasma.

After the electron temperature T_e has been determined, we can evaluate the electron number density from (5.2), which can be written as,

$$n_e = (J_{eo}/e)(2\pi m_e/kT_e)^{1/2} \tag{5.6}$$

The value of J_{eo} is determined by measuring the probe current corresponding to the plateau (electron saturation) region of the current-potential characteristic of the probe.

PROBLEMS

11.1 - Consider a stationary plasma (electrons and one type of ions) under steady state conditions at a uniform temperature T_0, when perturbed by a point charge + Q placed at the origin of a coordinate system. Using the collisionless hydrodynamic equation for the electrons and ions

$$\rho_{m\alpha}\, D\vec{u}_\alpha/Dt = n_\alpha\, q_\alpha\, \vec{E} - \vec{\nabla}p_\alpha \quad ; \quad (\alpha = e,i)$$

with the ideal gas law $p_\alpha = n_\alpha k\, T_0$, and Poisson equation

$$\nabla^2\phi(\vec{r}) = -\,\rho(\vec{r})/\varepsilon_0$$

obtain the following differential equation

$$\nabla^2\phi(\vec{r}) - (2/\lambda_D^2)\,\phi(\vec{r}) = -\,(Q/\varepsilon_0)\,\delta(\vec{r})$$

for the Debye potential $\phi(\vec{r})$. Assume that the number density of each species can be expressed as $n_\alpha = n_0 + n'_\alpha$, where n_0 is constant and $|n'_\alpha| << n_0$. What are the approximations necessary to obtain this result?

11.2 - Analyze the Debye potential problem considering only the motion of the electrons (ions stay immobile), and show that in this case the differential equation for the electric potential $\phi(\vec{r})$ is

$$\nabla^2\phi(\vec{r}) - (1/\lambda_D^2)\ \phi(\vec{r}) = -\ (Q/\varepsilon_o)\ \delta(\vec{r})$$

11.3 - When the macroscopic neutrality of a plasma is instantaneously perturbed by external means, the electrons react in a such a way as to give rise to oscillations at the electron plasma frequency $\omega_{pe} = (n_o e^2/m_e \varepsilon_o)^{1/2}$. Consider these oscillations, but including the motion of the ions. Show that in this case the natural frequency of oscillation of the net charge density is given by

$$\omega = (\omega_{pe}^2 + \omega_{pi}^2)^{1/2}.$$

where $\omega_{pi} = (n_o e^2/m_i \varepsilon_o)^{1/2}$. Use the linearized equations of continuity and of momentum for each species, and Poisson equation, considering only the electric force due to the internal charge separation.

11.4 - Evaluate the negative electrostatic potential ϕ_w which appears on an infinite plane wall immersed in a plasma consisting of electrons of charge $-\ e$ and ions of charge Ze, under steady state conditions. Denote the electron and ion temperatures by T_e and T_i, respectively.

11.5 - Deduce an expression for the Debye potential for a test particle of charge $+\ Q$ immersed in a plasma consisting of electrons (charge $-e$) and ions of charge Ze, the temperature of the electrons and the ions being T_e and T_i, respectively. Show that, if $T_e \gg T_i$, then the Debye length is governed by the ion temperature T_i.

11.6 - Using the following expressions for the electron and ion number densities

$$n_e(\vec{r}) = n_o \exp\ [e\ \phi(\vec{r})/kT\ ]$$

$$n_i(\vec{r}) = n_o \exp\ [-e\ \phi(\vec{r})/kT)]$$

in the plasma sheath region formed between an infinite plane and a semi-infinite plasma, deduce the differential equation satisfied by the electric potencial $\phi(x)$ in the plasma sheath. Show that this differential equation can be written in the form

$$d^2F/d\zeta^2 = \sinh (F)$$

where $F = e\phi/kT$ and $\zeta = \sqrt{2}\, x/\lambda_D$. Assuming that at $x = \infty$ we have $n_e = n_i = n_o$, $F = 0$ and $dF/d\zeta = 0$, show that

$$F(\zeta) = 4 \tanh^{-1} \{ \exp [- (\zeta - \zeta_o)] \}$$

where ζ_o is a constant. Denoting the potential at the wall by ϕ_w and assuming that $e\phi/kT << 1$, show that

$$\phi(x) = \phi_w \exp (\sqrt{2}\, x/\lambda_D)$$

with

$$\phi_w = (4kT/e) \exp (\zeta_o)$$

11.7 - For the plasma sheath region formed in the vicinity of a plane wall immersed in a plasma, assume that the ions at the plasma edge of the sheath can be described by a shifted Maxwellian distribution function

$$f_i(v) = n_o \left(\frac{m_i}{2\pi kT} \right)^{3/2} \exp \left[- \frac{m_i (\vec{v} - \vec{u}_o)^2}{2\, kT} \right]$$

with drift velocity $\vec{u}_o = u_o\, \hat{x}$. Prove that the ion flux Γ_{ix}, at the edge of the sheath, is given by

$$\Gamma_{ix} = n_o\, (kT/2\pi m_i)^{1/2} \; \{ \exp(-y^2) + y\sqrt{\pi}\, [1 + \mathrm{erf}(y)] \}$$

where $y = u_o(m_i/2kT)^{1/2}$ and $\mathrm{erf}(y)$ is the error function, defined by

$$\mathrm{erf}(y) = \frac{2}{\sqrt{\pi}} \int_0^y \exp(-s^2)\, ds$$

Calculate $d\Gamma_{ix}/dy$. Note that the error function vanishes for $y = 0$, increases monotonically as y increases and tends asymptotically to unity as

$y \to \infty$. Also note that

$$\frac{d}{dy}\left[\operatorname{erf}(y)\right] = \frac{2}{\sqrt{\pi}} \exp(-y^2)$$

11.8 - From an experimental current-potential curve of a Langmuir probe of area A immersed in a plasma, such as shown in Fig. P11.1, where the electric potentials are measured with respect to a fixed reference potential, explain how you can determine J_i, J_{eo}, the space potential ϕ_s, the floating potential ϕ_f with respect to ϕ_s (note that $\phi_f - \phi_s = \phi_w$), T_e and n_e.

11.9 - The Langmuir plasma probe has been widely used in satellites to measure space plasma properties. In one valuable technique, circuits are arranged that measure directly $dI_p/d\phi$ and $d^2I_p/d\phi^2$, where $I_p = J_p A$ and A is the probe's area. Use (5.3) to show that

$$\frac{(dI_p/d\phi)}{(d^2I_p/d\phi^2)} = \frac{kT_e}{e}$$

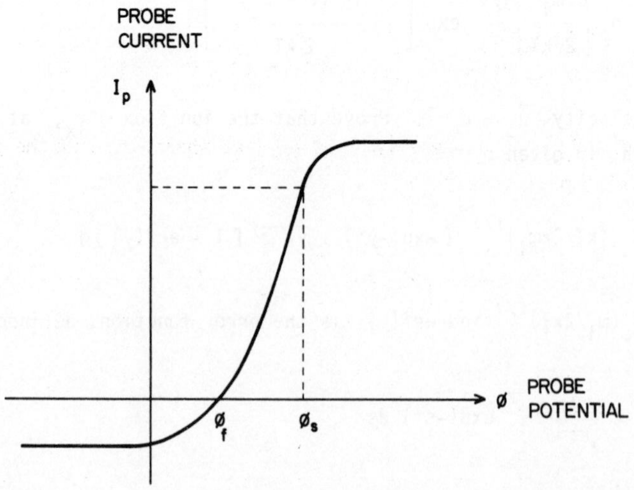

Fig. P11.1

which gives directly the electron temperature T_e. Next show that J_{eo} can be calculated from T_e and $dI_p/d\phi$ at a known value of ϕ, according to

$$J_{eo} = (kT_e/Ae) \exp(-e\phi/kT_e)(dI_p/d\phi)$$

The electron density n can then be calculated from (5.6).

11.10 - An electron gas (Lorentz gas) in a background of stationary ions, is acted upon by a weak, externally applied electric field $\vec{E}$, under steady state conditions. Using the Boltzmann equation for the electrons, with the relaxation model for the collision term (with a constant collision frequency ν),

$$(\delta f/\delta t)_{coll} = -\nu(f - f_0)$$

and considering the adiabatic case for which

$$n(\vec{r})\,[T(\vec{r})]^{-3/2} = \text{constant}$$

show that the electron distribution function is given by

$$f = f_0\left\{1 - \frac{1}{\nu}\left[\left(\frac{mv^2}{2kT}\right)\left(\vec{v}\cdot\frac{\vec{\nabla}T}{T}\right) + \left(\frac{e}{kT}\right)(\vec{v}\cdot\vec{E})\right]\right\}$$

Assume that $f = f_0 + f_1$, where $|f_1| \ll f_0$ and where f_0 is the following modified Maxwellian distribution

$$f_0(\vec{r},v) = n(\vec{r})\left[\frac{m}{2\pi kT(\vec{r})}\right]^{3/2} \exp\left[-\frac{mv^2}{2kT(\vec{r})}\right]$$

and neglect all second-order terms in the Boltzmann equation. Consider the term involving $\vec{\nabla}f_1$ as a second-order quantity.

11.11 - Using the distribution function of the previous problem, evaluate the electric current density $\vec{J}$ to show that the presence of a temperature gradient gives rise to an electric current associated with thermoelectric effects.

11.12 - Consider problem 11.10, but taking $\vec{E} = 0$ and, instead of the adiabatic case, consider a constant kinetic pressure

$$p = n(\vec{r})\ k\ T(\vec{r}) = \text{constant}$$

(a) Show that the electron distribution function is given by

$$f = f_0 \left\{ 1 - \frac{1}{\nu} \left[-\frac{5}{2} + \left(\frac{mv^2}{2kT} \right) \right] \vec{v} \cdot \frac{\vec{\nabla} T}{T} \right\}$$

(b) Evaluate the heat flux vector $\vec{q}$ and show that it can be written as

$$\vec{q} = -K \vec{\nabla} T(\vec{r})$$

where the thermal conductivity K is $5kp/2m\nu$.

(c) What is the value of the electric current density $\vec{J}$ in this case?

11.13 - In the previous problem, consider that n = constant and that f_0 is the following modified Maxwell-Boltzmann distribution function

$$f_0(\vec{r},v) = n \left[\frac{m}{2\pi kT(\vec{r})} \right]^{3/2} \exp\left[-\frac{mv^2}{2kT(\vec{r})} \right]$$

Calculate the electron distribution function $f(\vec{r},\vec{v})$ and show that the heat flux vector is given by

$$\vec{q} = -K\ \vec{\nabla}\ T(\vec{r})$$

Determine the expression for the thermal conductivity K.

11.14 - In Problem 11.12, inculde the presence of an external magnetostatic field $\vec{B}$ in the z direction and deduce the following expression for the nonequilibrium distribution function:

$$f = f_0 \left\{ 1 - \frac{1}{\nu} \left(-\frac{5}{2} + \frac{mv^2}{2kT} \right) \left[\frac{\nu\omega_{ce}}{(\nu^2 + \omega_{ce}^2)} v_x \left(\hat{x} \frac{\nu}{\omega_{ce}} - \hat{y} \right) + \right. \right.$$

$$+ \frac{\nu\omega_{ce}}{(\nu^2 + \omega_{ce}^2)} v_y \left(\hat{x} + \hat{y}\frac{\nu}{\omega_{ce}}\right) + v_z \hat{z}\Bigg] \cdot \frac{\vec{\nabla}T}{T} \Bigg\}$$

Show that the heat flux vector $\vec{q}$ can be expressed as

$$\vec{q} = - \bar{\bar{K}} \cdot \vec{\nabla}T(\vec{r})$$

where $\bar{\bar{K}}$ is the dyadic thermal conductivity, which in matrix form is given by

$$\bar{\bar{K}} = \begin{pmatrix} K_\perp & -K_H & 0 \\ K_H & K_\perp & 0 \\ 0 & 0 & K_\parallel \end{pmatrix}$$

with

$$K_\perp = \frac{\nu^2}{(\nu^2 + \omega_{ce}^2)} K_o$$

$$K_H = \frac{\nu\omega_{ce}}{(\nu^2 + \omega_{ce}^2)} K_o$$

$$K_\parallel = (5kp/2m\nu) \equiv K_o$$

Note: The solution of the differential equation

$$df_1/d\phi + (\nu/\omega_{ce}) f_1 = - (1/\omega_{ce}) \vec{v} \cdot \vec{\nabla} f_o$$

is given by

$$f_1 = - \exp\left(-\frac{\nu}{\omega_{ce}}\phi\right) \frac{1}{\omega_{ce}} \int_{-\infty}^{\phi} d\phi' \exp\left(\frac{\nu}{\omega_{ce}}\phi'\right) \vec{v} \cdot \vec{\nabla} f_o$$

11.15 - The *coefficient of viscosity* η is defined as the shear stress produced by unit velocity gradient. For the p_{xz} component of the kinetic pressure dyad we have

$$p_{xz} = - \eta \frac{d}{dz} u_x(z)$$

Assume the following form for the equilibrium velocity distribution function of the electrons

$$f_0(\vec{r},\vec{v}) = n \left(\frac{m}{2\pi kT}\right)^{3/2} \exp\left[- \frac{m}{2kT} \left\{ [v_x - u_x(z)]^2 + v_y^2 + v_z^2 \right\} \right]$$

which indicates the presence of an average velocity $u_x(z)$ in the x direction with a gradient in the z direction. In the absence of external forces and using the relaxation model for the collision term with a constant collision frequency ν, let $f = f_0 + f_1$ with $|f_1| << f_0$ in the Boltzmann equation, under steady state conditions, to show that

$$f_1(\vec{r},\vec{v}) = - \frac{2}{\nu} \left(\frac{m}{2kT}\right) f_0(\vec{r},\vec{v})\, v_z [v_x - u_x(z)] \frac{\partial}{\partial z} u_x(z)$$

Next, calculate p_{xz} and show that the coefficient of viscosity is given by $\eta = nkT/\nu$.

CHAPTER 12

Simple Applications of Magnetohydrodynamics

1. FUNDAMENTAL EQUATIONS OF MAGNETOHYDRODYNAMICS

The basic equations governing the behavior of a conducting fluid have been presented and discussed in Chapter 9. For convenience, we reproduce here the simplified form of the magnetohydrodynamic equations. They include the equation of continuity for the whole conducting fluid

$$\partial \rho_m/\partial t + \vec{\nabla} \cdot (\rho_m \vec{u}) = 0 \tag{1.1}$$

the equation of motion in the form

$$\rho_m \, (D\vec{u}/Dt) = \vec{J} \times \vec{B} - \vec{\nabla} p \tag{1.2}$$

and the adiabatic equation of conservation of energy

$$\vec{\nabla} p = V_S^2 \, \vec{\nabla} \rho_m \tag{1.3}$$

where ρ_m denotes the total mass density, $\vec{u}$ is the average fluid velocity, $\vec{J}$ is the electric current density, $\vec{B}$ is the magnetic flux density, p is the total scalar pressure, and V_S is the adiabatic sound speed, given by $(\gamma p/\rho_m)^{1/2}$, where γ is the ratio of the specific heats at constant pressure and at constant volume. To these equations we must add Maxwell curl equations, in the following reduced form,

$$\vec{\nabla} \times \vec{B} = \mu_o \vec{J} \tag{1.4}$$

$$\vec{\nabla} \times \vec{B} = - \partial \vec{B}/\partial t \tag{1.5}$$

and the generalized Ohm's law, in the simplified form,

$$\vec{J} = \sigma_0 (\vec{E} + \vec{u} \times \vec{B}) \tag{1.6}$$

where σ_0 denotes the electric conductivity of the fluid, and $\vec{E}$ is the electric field.

In this closed set of simplified MHD equations, it has been assumed that macroscopic electrical neutrality is maintained to a high degree of approximation, so that the electric charge density ρ vanishes. The neglect of the term $\partial\vec{E}/\partial t$, in Maxwell equation (1.4), is justified for *very low frequency* phenomena and *highly conducting fluids*, as discussed in section 6, Chapter 9. As far as the generalized Ohm's law (1.6) is concerned, it is assumed that the time derivatives and pressure gradients are negligible, even though these terms are retained in the other MHD equations. Also, viscosity and thermal conductivity are neglected and the pressure dyad degrades to a scalar pressure.

The advantage of this approximate set of equations is that they reduce substantially the mathematical complexity of the more general equations for a conducting fluid and therefore facilitate the understanding of the physical processes that take place in a highly conducting fluid at very low frequencies.

1.1 Parker modified momentum equation

In the presence of a strong magnetic field the pressure tensor of an inviscid conducting fluid is *anisotropic*. When the cyclotron frequency is much larger than the collision frequency, a charged particle gyrates many times around a line of magnetic force during the time between collisions, so that there is equipartition between the particle kinetic energies in the two independent directions normal to $\vec{B}$ but not, in general, in the direction along $\vec{B}$. If we denote by $p_\perp$ and $p_\parallel$ the scalar pressures in the plane normal to $\vec{B}$ and along $\vec{B}$, respectively, and consider a *local* coordinate system in which the third axis is in the direction of $\vec{B}$, we can write the pressure tensor of an inviscid fluid as

$$\bar{\bar{p}} = \begin{pmatrix} p_\perp & 0 & 0 \\ 0 & p_\perp & 0 \\ 0 & 0 & p_\parallel \end{pmatrix} \tag{1.7}$$

When the magnetic field is not constant, the orientation of the axes of the local coordinate system changes from point to point and this change in direction must be taken into account in evaluating the divergence of the pressure tensor. We can express $\bar{\bar{p}}$, in (1.7), as the sum of a hydrostatic scalar pressure $p_\perp$ and another tensor referred to the local coordinate system, as

$$\bar{\bar{p}} = p_\perp \bar{\bar{1}} + (p_\parallel - p_\perp) \hat{B} \hat{B} \tag{1.8}$$

where $\bar{\bar{1}}$ is the unit dyad

$$\bar{\bar{1}} = \begin{pmatrix} 1 & 0 & 0 \\ 0 & 1 & 0 \\ 0 & 0 & 1 \end{pmatrix} \tag{1.9}$$

and $\hat{B} \hat{B} = \vec{B} \vec{B}/B^2$ is the dyad formed from the unit vector $\hat{B}$,

$$\hat{B} \hat{B} = \begin{pmatrix} 0 & 0 & 0 \\ 0 & 0 & 0 \\ 0 & 0 & 1 \end{pmatrix} \tag{1.10}$$

The momentum equation (1.2) must be modified to include the anisotropy of the pressure dyad. Thus, we write

$$\rho_m (D\vec{u}/Dt) = \vec{J} \times \vec{B} - \vec{\nabla} \cdot \bar{\bar{p}} \tag{1.11}$$

To evaluate $\vec{\nabla} \cdot \bar{\bar{p}}$, with $\bar{\bar{p}}$ as given by (1.8), we note that

$$\vec{\nabla} \cdot (p_\perp \bar{\bar{1}}) = \vec{\nabla} p_\perp \tag{1.12}$$

and using the following identity

$$\vec{\nabla} \cdot [(p_\parallel - p_\perp) \vec{B} \vec{B}/B^2] = (\vec{B} \cdot \vec{\nabla}) [(p_\parallel - p_\perp) \vec{B}/B^2] + [(p_\parallel - p_\perp) \vec{B}/B^2] (\vec{\nabla} \cdot \vec{B}) \tag{1.13}$$

where the second term in the right-hand side vanishes in virtue of $\vec{\nabla} \cdot \vec{B} = 0$, we obtain

$$\vec{\nabla} \cdot \bar{\bar{p}} = \vec{\nabla} p_\perp + (\vec{B} \cdot \vec{\nabla}) [(p_\parallel - p_\perp) \vec{B}/B^2] \tag{1.14}$$

Furthermore, using Maxwell equation (1.4) we can write the magnetic force per unit volume as

$$\vec{J} \times \vec{B} = (\vec{\nabla} \times \vec{B}) \times \vec{B}/\mu_o \tag{1.15}$$

The term in the right-hand side can be expanded, using a vector identity, with the result that

$$\vec{J} \times \vec{B} = (1/\mu_o)(\vec{B} \cdot \vec{\nabla})\vec{B} - \vec{\nabla}(B^2/2\mu_o) \tag{1.16}$$

Substituting expressions (1.14) and (1.16) into the momentum equation (1.11), we obtain, finally,

$$\rho_m \frac{D\vec{u}}{Dt} = -\vec{\nabla}(p_\perp + \frac{B^2}{2\mu_o}) + (\vec{B} \cdot \vec{\nabla}) \left\{ \left[\frac{1}{\mu_o} - \frac{(p_\parallel - p_\perp)}{B^2} \right] \vec{B} \right\} \tag{1.17}$$

This equation differs from the usual momentum equation (1.2), for a highly conducting inviscid fluid, only through the term $(p_\parallel - p_\perp)/B^2$. It was derived, although in a quite different way, by E.N. Parker in 1957 and, for this reason, it is usually referred to as the Parker modified momentum equation.

1.2 The double adiabatic equations of Chew, Goldberger and Low (CGL)

To complement the momentum equation(1.17), we need equations for the time rate of change of $p_\parallel$ and $p_\perp$. These equations will take the place of the familiar adiabatic energy equation (1.3) which applies for the isotropic case. From the general energy equation (9.4.14) for a conducting fluid, if we do not take into account heat conduction and Joule heating, we have

$$\frac{D}{Dt}(3p/2) + (3p/2)\,\vec{\nabla}\cdot\vec{u} + (\bar{\bar{p}} \cdot \vec{\nabla}) \cdot \vec{u} = 0 \tag{1.18}$$

with the pressure dyad $\bar{\bar{p}}$ as given by (1.8), and where the scalar pressure p is one third the trace of $\bar{\bar{p}}$, that is,

$$p = (2p_\perp + p_\parallel)/3 \tag{1.19}$$

Note that 3p/2 is the total thermal energy density. By direct expansion, using (1.8) for $\bar{\bar{p}}$, we find that

$$(\bar{\bar{p}} \cdot \vec{\nabla}) \cdot \vec{u} = [p_\perp \vec{\nabla} + (p_\parallel - p_\perp) (\hat{B}\,\hat{B} \cdot \vec{\nabla})] \cdot \vec{u} \tag{1.20}$$

and taking this expression, together with (1.19), into (1.18), we obtain

$$\frac{D}{Dt} (2p_\perp + p_\parallel) + (p_\parallel + 4p_\perp) \vec{\nabla} \cdot \vec{u} + 2(p_\parallel - p_\perp) (\hat{B}\,\hat{B}\cdot\vec{\nabla}) \cdot \vec{u} = 0 \tag{1.21}$$

A strong magnetic field constrains the charged particle motion only in the direction transverse to $\vec{B}$, but they are still free to move large distances along $\vec{B}$. Thus, it is reasonable to suppose that the contribution to the total thermal energy, arising from the particle motion parallel to $\vec{B}$, also satisfy an equation of conservation of energy similar to (1.18). This leads to the following equation for the part of the total thermal energy due to the random particle motions along $\vec{B}$,

$$Dp_\parallel/Dt + p_\parallel \vec{\nabla} \cdot \vec{u} + 2\, p_\parallel(\hat{B}\,\hat{B} \cdot \vec{\nabla}) \cdot \vec{u} = 0 \tag{1.22}$$

Therefore, decoupling the parallel and perpendicular motions the equation for $p_\perp$ becomes

$$Dp_\perp/Dt + 2\, p_\perp \vec{\nabla} \cdot \vec{u} - p_\perp (\hat{B}\,\hat{B} \cdot \vec{\nabla}) \cdot \vec{u} = 0 \tag{1.23}$$

Eqs. (1.21) and (1.22) can also be obtained from an energy equation of higher order than (9.4.14), involving the total time rate of change of the pressure dyad $\bar{\bar{p}}$. When this equation, involving $D\bar{\bar{p}}/Dt$, is contracted with the unit dyad $\bar{\bar{1}}$ we obtain (1.21), and when contracted with the dyad $\hat{B}\,\hat{B}$ results in (1.22).

Eqs. (1.22) and (1.23) enable $p_\parallel$ and $p_\perp$ to be calculated. They can be written in a more succinct form, as follows. First we note that using Maxwell curl equation

$$\vec{\nabla} \times \vec{E} = - \partial\vec{B}/\partial t \tag{1.24}$$

and considering a perfectly conducting fluid for which

$$\vec{E} + \vec{u} \times \vec{B} = 0 \tag{1.25}$$

we have,

$$\partial\vec{B}/\partial t = \vec{\nabla} \times (\vec{u} \times \vec{B}) \tag{1.26}$$

Expanding the right-hand side using the vector identity $\vec{\nabla} \times (\vec{u} \times \vec{B}) = (\vec{B} \cdot \vec{\nabla})\,\vec{u} - \vec{B}(\vec{\nabla} \cdot \vec{u}) - (\vec{u} \cdot \vec{\nabla})\vec{B} + \vec{u}(\vec{\nabla} \cdot \vec{B})$, and noting that $\vec{\nabla} \cdot \vec{B} = 0$, we obtain

$$D\vec{B}/Dt = (\vec{B} \cdot \vec{\nabla})\,\vec{u} - \vec{B}\,(\vec{\nabla} \cdot \vec{u}) \tag{1.27}$$

If we now take the scalar product of (1.27) with $\vec{B}/B^2$, we obtain

$$(1/2B^2)\, D(B^2)/Dt = \hat{B} \cdot (\hat{B} \cdot \vec{\nabla})\,\vec{u} - \vec{\nabla} \cdot \vec{u} \tag{1.28}$$

which may be written as

$$(1/B)\, DB/Dt = (\hat{B}\,\hat{B} \cdot \vec{\nabla}) \cdot \vec{u} - \vec{\nabla} \cdot \vec{u} \tag{1.29}$$

Furthermore, from the equation of continuity (1.1), we get

$$\vec{\nabla} \cdot \vec{u} = -\,(1/\rho_m)\, D\rho_m/Dt \tag{1.30}$$

and using (1.29) and (1.30), to eliminate $(\hat{B}\,\hat{B} \cdot \vec{\nabla}) \cdot \vec{u}$ and $\vec{\nabla} \cdot \vec{u}$ in (1.22) and (1.23), we obtain

$$(1/p_\parallel)\, Dp_\parallel/Dt - (3/\rho_m)\, D\rho_m/Dt + (2/B)\, DB/Dt = 0 \tag{1.31}$$

$$(1/p_\perp)\, Dp_\perp/Dt - (1/\rho_m)\, D\rho_m/Dt - (1/B)\, DB/Dt = 0 \tag{1.32}$$

These two equations can be written in compact form as

$$\frac{D}{Dt}\,(p_\parallel\, B^2/\rho_m^3) = 0 \tag{1.33}$$

$$\frac{D}{Dt}\,(p_\perp/\rho_m B) = 0 \tag{1.34}$$

They are known as the *double adiabatic equations* for a conducting fluid in a strong magnetic field, and are due to G.F. Chew, M.L. Goldberger and F.E. Low (1956). They are also known as the CGL equations. They take the place of the adiabatic energy equation

$$\frac{D}{Dt}\,(p\,\rho_m^{-\gamma}) = 0 \tag{1.35}$$

1.3 Special cases of the double adiabatic equations

As a simple application of the double adiabatic equations, consider the case in which the only variations are *parallel* to the magnetic field as, for example, in sound waves travelling along the field lines. This situation is usually referred to as *linear compression* parallel to the $\vec{B}$ field or one-dimensional compression. The magnetic field is assumed to be straight and uniform, and directed along the z axis. Thus, $B_x = B_y = 0$ and $\vec{B} = B_z\,\hat{z}$, as well as $\partial/\partial x = \partial/\partial y = 0$. In this case, we find

$$(\hat{B}\,\hat{B}\cdot\vec{\nabla})\cdot\vec{u} = \partial u_z/\partial z = \vec{\nabla}\cdot\vec{u} \tag{1.36}$$

and from (1.29), we see that B stays constant. (1.31) and (1.32), with $DB/Dt = 0$, then yields

$$\frac{D}{Dt}\,(p_\parallel/\rho_m^3) = 0 \tag{1.37}$$

$$\frac{D}{Dt}\,(p_\perp/\rho_m) = 0 \tag{1.38}$$

If we compare these results with (1.35), we find that γ may be assigned the value 3 along the field lines (one-dimensional compression), and the value 1 across the field lines.

It is useful to introduce a parallel and a perpendicular temperature through the relations

$$p_\parallel = n\,k\,T_\parallel \tag{1.39}$$

$$p_\perp = n\,k\,T_\perp \tag{1.40}$$

Therefore, for the case of *one-dimensional compression parallel* to $\vec{B}$,

$$T_\parallel \propto n^2 \tag{1.41}$$

$$T_\perp = \text{constant} \tag{1.42}$$

which shows that this type of compression is isothermal with respect to the perpendicular temperature $T_\perp$. The changes in $p_\perp$ are therefore entirely due to the changes in the number density n, whereas those of $p_\parallel$ are due to changes in both n and $T_\parallel$.

Another special case of interest is the *two-dimensional compression perpendicular* to the $\vec{B}$ field, in which all motion is transverse to the field lines. This situation can be pictured as the motion of magnetic flux tubes, identified by the particles contained in them. Assuming straight field lines along the z axis ($B_x = B_y = 0$, $\vec{B} = B_z\hat{z}$) and variations only in the transverse direction ($\partial/\partial z=0$) we find that

$$(\hat{B}\,\hat{B}\cdot\vec{\nabla})\cdot\vec{u} = (\hat{z}\,\partial/\partial z)\cdot\vec{u} = 0 \tag{1.43}$$

and (1.22) and (1.23) yield

$$Dp_\parallel/Dt - (p_\parallel/\rho_m)\,D\rho_m/Dt = 0 \tag{1.44}$$

$$Dp_\perp/Dt - (2p_\perp/\rho_m)\,D\rho_m/Dt = 0 \tag{1.45}$$

Therefore, in the case of cylindrical compression perpendicular to $\vec{B}$ the adiabatic equations reduce to

$$\frac{D}{Dt}\,(p_\parallel/\rho_m) = 0 \tag{1.46}$$

$$\frac{D}{Dt}\,(p_\perp/\rho_m^2) = 0 \tag{1.47}$$

Comparing with (1.35), γ takes the value 1 parallel to the magnetic field and 2 transverse to it. Using (1.39) and (1.40) it is seen that for a two-dimensional (cylindrically symmetric) compression perpendicular to $\vec{B}$,

$$T_{\parallel} = \text{constant} \tag{1.48}$$

$$T_{\perp} \propto n \tag{1.49}$$

so that this type of compression is isothermal with respect to the parallel temperature. The changes in $p_{\parallel}$ are due entirely to variations in the number density n, whereas those of $p_{\perp}$ result from variations in n as well as in $T_{\perp}$.

In the case of *three-dimensional spherically symmetric* compression, we have

$$p_{\perp} = p_{\parallel} = p \tag{1.50}$$

and (1.21) reduces to

$$3\, Dp/Dt - (5p/\rho_m)\, D\rho_m/Dt = 0 \tag{1.51}$$

Thus, we obtain

$$\frac{D}{Dt}\,(p/\rho_m^{5/3}) = 0 \tag{1.52}$$

which is the familiar adiabatic equation (1.35) of gas dynamics, with $\gamma = 5/3$. In any of the cases of adiabatic compression, the fluid has to be subjected to a certain system of forces in order to achieve the desired type of adiabatic compression. The required system of forces has to be determined from the momentum equation in conjunction with the conditions appropriate to the particular problem under analysis.

1.4 Energy integral

As a final consideration in this section, we will show that the system of hydromagnetic equations (1.1) to (1.6) possesses an energy integral. Using Maxwell equation (1.4), to substitute $\vec{J}$ in the equation of motion (1.2), yields

$$\rho_m\, D\vec{u}/Dt = (\vec{\nabla} \times \vec{B}) \times \vec{B}/\mu_o - \vec{\nabla}\, p \tag{1.53}$$

Now, take the dot product of this equation with $\vec{u}$,

$$\rho_m \vec{u} \cdot (D\vec{u}/Dt) = \vec{u} \cdot (\vec{\nabla} \times \vec{B}) \times \vec{B}/\mu_o - \vec{u} \cdot \vec{\nabla} p \tag{1.54}$$

The term on the left-hand side can be expanded as

$$\rho_m \vec{u} \cdot \frac{D\vec{u}}{Dt} = \rho_m \vec{u} \cdot \left[\frac{\partial \vec{u}}{\partial t} + (\vec{u} \cdot \vec{\nabla}) \vec{u}\right] = \frac{1}{2} \rho_m \left[\frac{\partial u^2}{\partial t} + (\vec{u} \cdot \vec{\nabla}) u^2\right] =$$

$$= \frac{\partial}{\partial t} (\frac{1}{2} \rho_m u^2) - \frac{u^2}{2} \frac{\partial \rho_m}{\partial t} + \frac{1}{2} \rho_m (\vec{u} \cdot \vec{\nabla}) u^2 \tag{1.55}$$

Using the continuity equation (1.1) to eliminate $\partial \rho_m / \partial t$, yields

$$\rho_m \vec{u} \cdot \frac{D\vec{u}}{Dt} = \frac{\partial}{\partial t} (\frac{1}{2} \rho_m u^2) + \frac{u^2}{2} \vec{\nabla} \cdot (\rho_m \vec{u}) + \frac{1}{2} \rho_m (\vec{u} \cdot \vec{\nabla}) u^2 =$$

$$= \frac{\partial}{\partial t} (\frac{1}{2} \rho_m u^2) + \vec{\nabla} \cdot (\frac{1}{2} \rho_m u^2 \vec{u}) \tag{1.56}$$

In order to transform the term $\vec{u} \cdot \vec{\nabla} p$, we write the adiabatic energy equation (1.35) as

$$\rho_m^{-\gamma} (Dp/Dt) - \gamma p \rho_m^{-(\gamma+1)} (D\rho_m/Dt) = 0 \tag{1.57}$$

and use the continuity equation in the form

$$D\rho_m/Dt = - \rho_m (\vec{\nabla} \cdot \vec{u}) \tag{1.58}$$

Combining these two equations, we obtain

$$\partial p/\partial t + \vec{u} \cdot \vec{\nabla} p + \gamma p \vec{\nabla} \cdot \vec{u} = 0 \tag{1.59}$$

which may also be written as

$$\partial p/\partial t + (1-\gamma) \vec{u} \cdot \vec{\nabla} p + \gamma \vec{\nabla} \cdot (p \vec{u}) = 0 \tag{1.60}$$

from which we get

$$\vec{u} \cdot \vec{\nabla} p = \frac{1}{(\gamma - 1)} \frac{\partial p}{\partial t} + \frac{\gamma}{(\gamma - 1)} \vec{\nabla} \cdot (p \vec{u}) \tag{1.61}$$

Finally, for the $\vec{u}\cdot(\vec{\nabla}\times\vec{B})\times\vec{B}$ term in (1.54), considering a perfectly conducting fluid for which $\vec{E} = -\vec{u}\times\vec{B}$, and using a vector identity, we can write

$$\vec{u}\cdot(\vec{\nabla}\times\vec{B})\times\vec{B} = -(\vec{u}\times\vec{B})\cdot(\vec{\nabla}\times\vec{B}) = \vec{E}\cdot(\vec{\nabla}\times\vec{B})$$

$$= \vec{B}\cdot(\vec{\nabla}\times\vec{E}) - \vec{\nabla}\cdot(\vec{E}\times\vec{B}) \tag{1.62}$$

Using Maxwell equation (1.5) we arrive at

$$\frac{1}{\mu_0}\vec{u}\cdot(\vec{\nabla}\times\vec{B})\times\vec{B} = -\frac{\partial}{\partial t}\left(\frac{B^2}{2\mu_0}\right) - \frac{1}{\mu_0}\vec{\nabla}\cdot(\vec{E}\times\vec{B}) \tag{1.63}$$

Substituting (1.56), (1.61) and (1.63), into (1.54), yields the following energy conservation equation

$$\frac{\partial}{\partial t}\left[\frac{1}{2}\rho_m u^2 + \frac{p}{(\gamma-1)} + \frac{B^2}{2\mu_0}\right] + \nabla\cdot\left[\frac{1}{2}\rho_m u^2\vec{u} + \frac{\gamma}{(\gamma-1)}p\,\vec{u} + \vec{E}\times\vec{H}\right] = 0 \tag{1.64}$$

The first three terms of this equation represent the kinetic energy density associated with the macroscopic motion of the fluid, the thermal energy density, and the energy density stored in the magnetic field, respectively, whereas the last three terms denote the flux of macroscopic kinetic energy, the flux of thermal energy transported at the macroscopic mean velocity $\vec{u}$, and the flux of electromagnetic energy (Poynting vector $\vec{E}\times\vec{H}$), respectively.

If we integrate (1.64) over the entire fluid-plus-vacuum volume and use Gauss' divergence theorem to transform the divergence term into a surface integral, we find that the first two terms in the surface integral vanish, since ρ_m, p and $\vec{u}$ are zero outside the fluid. The remaining surface term is the surface integral of the Poynting vector which, for an isolated system, also vanishes. Therefore, we obtain the following energy conservation integral

$$\int_V \left[\frac{1}{2}\rho_m u^2 + \frac{p}{(\gamma-1)} + \frac{B^2}{2\mu_0}\right] d^3r = \text{constant} \tag{1.65}$$

The first integral represents the macroscopic kinetic energy of the fluid, the second term is the thermal free energy, and the last one represents the total energy of the magnetic field. It is usually useful to separate (1.65) into a kinetic energy part

$$K = \int_V \frac{1}{2} \rho_m u^2 \, d^3r \tag{1.66}$$

and a potential energy part

$$U = \int_V \left[\frac{p}{(\gamma - 1)} + \frac{B^2}{2\mu_0} \right] d^3r \tag{1.67}$$

with the energy conservation law K + U = constant. In these equations the integration extends over the entire fluid-plus-vacuum volume.

2. MAGNETIC VISCOSITY AND REYNOLDS NUMBER

The behavior of the magnetic field is of great importance in many MHD problems. To obtain a simple equation for $\vec{B}$, we start by taking the curl of the generalized Ohm's law (1.6),

$$\vec{\nabla} \times \vec{J} = \sigma_0 [\vec{\nabla} \times \vec{E} + \vec{\nabla} \times (\vec{u} \times \vec{B})] \tag{2.1}$$

Replacing $\vec{J}$ and $\vec{\nabla} \times \vec{E}$, using Maxwell curl equations (1.4) and (1.5),

$$\vec{\nabla} \times (\vec{\nabla} \times \vec{B}) = \mu_0 \sigma_0 [-\partial\vec{B}/\partial t + \vec{\nabla} \times (\vec{u} \times \vec{B})] \tag{2.2}$$

Making use of the following identity (with $\vec{\nabla} \cdot \vec{B} = 0$)

$$\vec{\nabla} \times (\vec{\nabla} \times \vec{B}) = - \nabla^2 \vec{B} \tag{2.3}$$

equation (2.2) reduces to

$$\partial\vec{B}/\partial t = \vec{\nabla} \times (\vec{u} \times \vec{B}) + \eta_m \nabla^2 \vec{B} \tag{2.4}$$

where η_m is called the *magnetic viscosity*,

$$\eta_m = 1/(\mu_0 \sigma_0) \tag{2.5}$$

The first term in the right-hand side of (2.4) is called the *flow term*, while the second term is called the *diffusion term*. To compare the relative magnitude of these two terms, we can use dimensional analysis and take, approximately,

$$| \vec{\nabla} \times (\vec{u} \times \vec{B}) | \cong uB/L \tag{2.6}$$

$$\eta_m | \nabla^2 \vec{B} | \cong \eta_m B/L^2 \tag{2.7}$$

where L denotes some characteristic length for variation of the parameters. The ratio of the flow term to the diffusion term is called the *magnetic Reynolds number* and is given by

$$R_m = uL/\eta_m \tag{2.8}$$

In most MHD problems one or the other of these two terms is of predominant importance and R_m is either very large or very small compared to unity.

It is instructive to compare the magnetic viscosity (η_m) and the magnetic Reynolds number (R_m) with the ordinary hydrodynamic viscosity (η_k) and Reynolds number (R). For this purpose, consider the Navier-Stokes equation of hydrodynamics

$$\frac{D\vec{u}}{Dt} = \vec{f} - \frac{1}{\rho_m} \vec{\nabla} p + \eta_k [\nabla^2 \vec{u} + \frac{1}{3} \vec{\nabla} (\vec{\nabla} \cdot \vec{u})] \tag{2.9}$$

where $\vec{f}$ denotes the average force per unit mass of the fluid, and η_k is the kinematic viscosity (viscosity divided by density). Comparing this equation with (2.4) we see that the role played by η_m, in the rate of change of $\vec{B}$, is completely analogous to the role played by η_k, in the rate of change of the mean fluid velocity $\vec{u}$. The ordinary Reynolds number is defined as the ratio of the inertia term $(\vec{u} \cdot \vec{\nabla}) \vec{u}$ to the viscosity term $\eta_k \nabla^2 \vec{u}$. Using dimensional analysis, we have

$$|(\vec{u} \cdot \vec{\nabla})\, \vec{u}| \cong u^2/L \tag{2.10}$$

$$\eta_k\, |\nabla^2 \vec{u}| \cong \eta_k\, u/L^2 \tag{2.11}$$

which gives the following expression, completely analogous to R_m, for the ordinary Reynolds number

$$R = uL/\eta_k \tag{2.12}$$

3. DIFFUSION OF MAGNETIC FIELD LINES

When $R_m << 1$, that is when the diffusion term dominates, (2.4) becomes approximately,

$$\partial \vec{B}/\partial t = \eta_m\, \nabla^2 \vec{B} \qquad (R_m << 1) \tag{3.1}$$

This is the equation of diffusion of a magnetic field in a stationary conductor, resulting in the decay of the magnetic field. It is analogous to the particle diffusion equation studied in Chapter 10. The characteristic decay time of the magnetic field can be obtained by dimensional analysis, taking

$$|\partial \vec{B}/\partial t| \cong B/\tau_D \tag{3.2}$$

$$|\eta_m\, \nabla^2 \vec{B}| \cong \eta_m B/L^2 \tag{3.3}$$

where τ_D represents a characteristic time for variation of the plasma parameters. Thus, according to (3.1), the magnetic field diffuses away with a characteristic decay time of the order of

$$\tau_D = L^2/\eta_m = L^2\, \mu_0\, \sigma_0 \tag{3.4}$$

For ordinary conductors the time of decay is very small. If we consider, for example, a copper sphere of radius 1 meter, we find that τ_D is less than 10 seconds. For a celestial body, however, because of the large dimensions, the time of decay can be very large. For the Earth's core, considering it to be molten iron, the time of free decay is approximately 10^4 years, while for the general magnetic field of the Sun it is found to be of the order of 10^{10} years.

4. FREEZING OF THE MAGNETIC FIELD LINES TO THE PLASMA

A completely different type of behavior appears when $R_m >> 1$. In this case the flow term dominates over the diffusion term and (2.4) reduces to

$$\partial\vec{B}/\partial t = \vec{\nabla} \times (\vec{u} \times \vec{B}) \qquad (R_m >> 1) \tag{4.1}$$

This equation implies that in a highly conducting fluid the magnetic field lines move along exactly with the fluid, rather than simply diffusing out. Alfvēn has expressed this type of behavior by saying that the magnetic field lines are "frozen" in the conducting fluid. In effect, the fluid can flow freely along the magnetic field lines, but any motion of the conducting fluid, perpendicular to the field lines, carries them with the fluid.

In order to show this implication of (4.1), it is convenient to consider initially the concept of magnetic tubes of force which are used to visually describe the direction and magnitude of $\vec{B}$ at various points in space. One can think of the space pervaded by a magnetic field as divided into a large number of elementary magnetic tubes of force, all of them enclosing the same magnetic flux $\Delta\Phi_B$. If S is the local cross sectional area of an elementary magnetic tube of force (Fig. 1), then the magnitude of $\vec{B}$, at the local point P, is equal to $\Delta\Phi_B/\Delta S$. According to this definition, the magnitude of $\vec{B}$ is everywhere inversely propotional to the cross sectional area of the elementary tube of force.

Let us now consider a closed line whose points move with velocity $\vec{u}$ in a space pervaded by a magnetic field. Assume, for the moment, that $\vec{u}$ is an

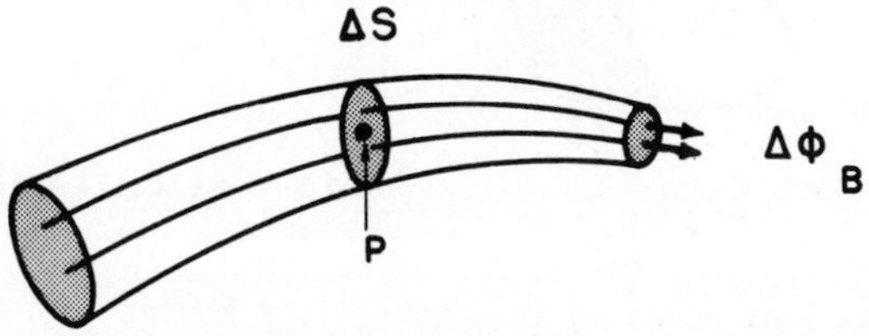

Fig. 1 - Elementary magnetic tube of force. The magnitude of $\vec{B}$, at the point P, is equal to $\Delta\Phi_B/\Delta S$.

arbitrary function of position and time (not necessarily equal to the fluid velocity), with the result that the closed curve may change in shape, as well as undergo translational and rotational motion. Let C_1 denote the closed curve at time t, bounding the open surface $\vec{S}(t) = \vec{S}_1$. At a time Δt later, let C_2 and $\vec{S}(t + \Delta t) = \vec{S}_2$ denote the corresponding closed curve and open surface (Fig. 2). The flux of the magnetic field through an open surface $\vec{S}$, at time t, is given by

$$\Phi_B(t) = \int_S \vec{B}(\vec{r},t) \cdot d\vec{S} \tag{4.2}$$

The rate of change of the magnetic flux through an open surface $\vec{S}$ can be written as

$$\frac{d}{dt}\left[\int_S \vec{B}(\vec{r},t) \cdot d\vec{S}\right] = \lim_{\Delta t \to 0} \frac{1}{\Delta t}\left[\int_{S_2} \vec{B}(\vec{r},t + \Delta t) \cdot d\vec{S} - \int_{S_1} \vec{B}(\vec{r},t) \cdot d\vec{S}\right] \tag{4.3}$$

If we expand $\vec{B}(\vec{r},t + \Delta t)$ in a Taylor series about $\vec{B}(\vec{r},t)$, we have

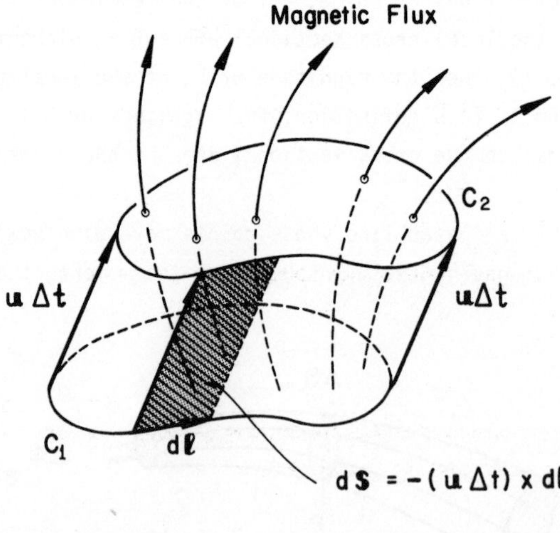

Fig. 2 - A closed line bounding an open surface moving in a magnetic field with velocity $\vec{u}(\vec{r},t)$, viewed at the instants of time t and t + Δt. The shaded area is the part of the cylindrical surface described by an element $d\vec{\ell}$ of the contour curve.

$$\vec{B}(\vec{r},t+\Delta t) = \vec{B}(\vec{r},t) + \frac{\partial \vec{B}(\vec{r},t)}{\partial t}\,\Delta t + \ldots \tag{4.4}$$

so that, in the limit as $\Delta t \rightarrow 0$, (4.3) reduces to

$$\frac{d}{dt}\left[\int_S \vec{B}(\vec{r},t)\cdot d\vec{S}\right] = \lim_{\Delta t \rightarrow 0}\left\{\int_{S_2} \frac{\partial \vec{B}(\vec{r},t)}{\partial t}\cdot d\vec{S} + \frac{1}{\Delta t}\left[\int_{S_2} \vec{B}(\vec{r},t)\cdot d\vec{S} - \int_{S_1} \vec{B}(\vec{r},t)\cdot d\vec{S}\right]\right\} \tag{4.5}$$

To evaluate the term within brackets in the right-hand side of this equation, we can use the fact that for any *closed surface* at time t we have, from Gauss' divergence theorem,

$$\oint \vec{B}\cdot d\vec{S} = \int_V \vec{\nabla}\cdot\vec{B}\, d^3r = 0 \tag{4.6}$$

since $\vec{\nabla}\cdot\vec{B} = 0$. Thus, if we apply this result to the closed surface consisting of $\vec{S}_1$, $\vec{S}_2$ and the sides of the cylindrical surface of length $\vec{u}\,\Delta t$ shown in Fig. 2, we obtain

$$-\int_{S_1} \vec{B}(\vec{r},t)\cdot d\vec{S} + \int_{S_2} \vec{B}(\vec{r},t)\cdot d\vec{S} - \oint_{C_1} \vec{B}(\vec{r},t)\cdot[(\vec{u}\Delta t)\times d\vec{\ell}] = 0 \tag{4.7}$$

where the minus sign in the first term on the left-hand side is due to the fact that the *outwardly drawn* unit normal to the surface S_1 is in a direction opposite to that of the surface S_2, and $-(\vec{u}\Delta t)\times d\vec{\ell}$ is the element of area (pointing outwards) covered by the vector element $d\vec{\ell}$ of the closed curve in the time interval Δt. If (4.7) is substituted into (4.5) and the limit $\Delta t \rightarrow 0$ is evaluated, noting that in this limit $S_2 = S_1 = S(t)$, we obtain

$$\frac{d}{dt}\left[\int_S \vec{B}(\vec{r},t)\cdot d\vec{S}\right] = \int_S \frac{\partial \vec{B}(\vec{r},t)}{\partial t}\cdot d\vec{S} + \oint_C \vec{B}(\vec{r},t)\cdot(\vec{u}\times d\vec{\ell}) \tag{4.8}$$

The last term in the right-hand side of this equation can be transformed using the vector identity

$$\vec{B}(\vec{r},t) \cdot (\vec{u} \times d\vec{\ell}) = - [\vec{u} \times \vec{B}(\vec{r},t)] \cdot d\vec{\ell} \tag{4.9}$$

and from Stokes' theorem we can write

$$\oint_C [\vec{u} \times \vec{B}(\vec{r},t)] \cdot d\vec{\ell} = \int_S \vec{\nabla} \times [\vec{u} \times \vec{B}(\vec{r},t)] \cdot d\vec{S} \tag{4.10}$$

Thus, using this expression in (4.8), we obtain the general result

$$\frac{d}{dt}\left[\int_S \vec{B}(\vec{r},t) \cdot d\vec{S}\right] = \int_S \left\{ \frac{\partial \vec{B}(\vec{r},t)}{\partial t} - \vec{\nabla} \times [\vec{u} \times \vec{B}(\vec{r},t)] \right\} \cdot d\vec{S} \tag{4.11}$$

Suppose now that the space is filled with a highly conducting fluid, so that (4.1), valid for $R_m >> 1$, applies. If the velocity $\vec{u}$ in (4.11) is taken as the fluid velocity, we conclude, from (4.1) and (4.11), that

$$\frac{d}{dt}\left[\int_S \vec{B}(\vec{r},t) \cdot d\vec{S}\right] = 0 \tag{4.12}$$

which is a mathematical statement of the fact that the magnetic flux linked by a closed line (bounding the open surface S) moving with the fluid velocity $\vec{u}$ is constant. Note that this conclusion requires that only the velocity component of the closed line perpendicular to $\vec{B}$ be the same as the fluid velocity component perpendicular to $\vec{B}$, since the velocity component parallel to $\vec{B}$ gives no contribution to the term $\vec{u} \times \vec{B}$. Thus (4.1) implies that the lines of magnetic flux are frozen into the highly conducting fluid and are carried by any motion of the fluid *perpendicular* to the field lines. There is no restriction, however, on the motion along the field lines and the conducting fluid can flow freely in the direction *parallel* to $\vec{B}$.

This result is expected on physical grounds since, as the conducting fluid moves across the magnetic field, it induces an electric field which is proportional to the fluid velocity component perpendicular to $\vec{B}$. However, if the fluid conductivity is infinite, this perpendicular velocity component must be infinitesimally small if the flow of electric current is to remain finite.

In a fluid of *finite* conductivity the result (4.12) is no longer true. Using (2.4) in the general result (4.11), yields

$$\frac{d\Phi_B}{dt} = \frac{1}{\mu_o \sigma_o} \int_S \nabla^2 \vec{B} \cdot d\vec{S} \tag{4.13}$$

where the right-hand side of this equation gives rise to a slipping of magnetic flux through a closed material line.

5. MAGNETIC PRESSURE

5.1 Concept of magnetic pressure

The concept of *magnetic pressure* is very useful in the study of high temperature plasma confinement. Under steady state conditions the MHD equations reduce to the following closed set of *magnetohydrostatic* equations

$$\vec{\nabla} p = \vec{J} \times \vec{B} \tag{5.1}$$

$$\vec{\nabla} \times \vec{B} = \mu_o \vec{J} \tag{5.2}$$

$$\vec{\nabla} \cdot \vec{B} = 0 \tag{5.3}$$

If we eliminate $\vec{J}$ from these equations, we obtain the equivalent set of magnetohydrostatic equations involving only p and $\vec{B}$,

$$\vec{\nabla} p = (\vec{\nabla} \times \vec{B}) \times \vec{B}/\mu_o \tag{5.4}$$

$$\vec{\nabla} \cdot \vec{B} = 0 \tag{5.5}$$

The term in the right-hand side of (5.4) can be written as the divergence of the magnetic part of the electromagnetic stress dyad. Using the vector identity

$$(\vec{\nabla} \times \vec{B}) \times \vec{B} = (\vec{B} \cdot \vec{\nabla})\, \vec{B} - \vec{\nabla}\, (B^2/2) = \vec{\nabla}\cdot(\vec{B}\, \vec{B}) - \vec{\nabla} \cdot (\bar{\bar{1}}\, B^2/2) \tag{5.6}$$

where $\bar{\bar{1}}$ is the unit dyad, and using the following definition of the magnetic stress dyad

$$\bar{\bar{T}}^{(m)} = (\vec{B}\,\vec{B}/\mu_0 - \bar{\bar{1}}\,B^2/2\mu_0) \tag{5.7}$$

which written out in matrix form (in a Cartesian coordinate system) is

$$\bar{\bar{T}}^{(m)} = \frac{1}{\mu_o} \begin{pmatrix} (B_x^2 - B^2/2) & B_x B_y & B_x B_z \\ B_y B_x & (B_y^2 - B^2/2) & B_y B_z \\ B_z B_x & B_z B_y & (B_z^2 - B^2/2) \end{pmatrix} \tag{5.8}$$

we can write (5.4) as

$$\vec{\nabla}\,p = \vec{\nabla} \cdot \bar{\bar{T}}^{(m)} \tag{5.9}$$

or, equivalently,

$$\vec{\nabla} \cdot [\bar{\bar{1}}\,p - \bar{\bar{T}}^{(m)}] = 0 \tag{5.10}$$

The stress is considered to be positive if it is tensile, and negative if it is compressive. Thus, we see that $-\bar{\bar{T}}^{(m)}$ may be defined as the *magnetic pressure* dyad, playing the same role as the fluid pressure dyad.

It is instructive to consider a *local* magnetic coordinate system in which the third axis points along the local direction of $\vec{B}$, as shown in Fig. 3. For this

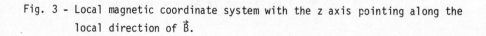

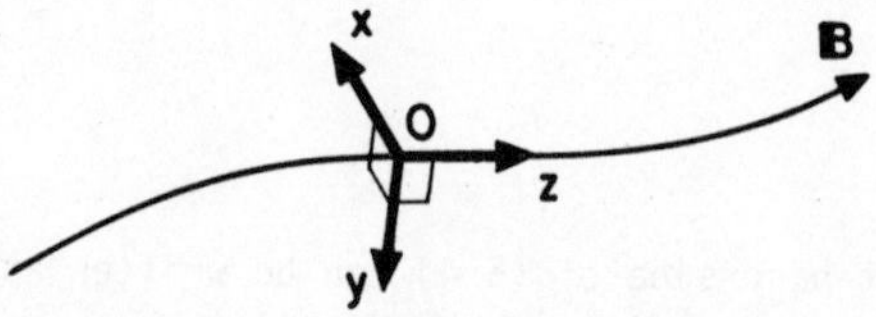

Fig. 3 - Local magnetic coordinate system with the z axis pointing along the local direction of $\vec{B}$.

local coordinate system, the off-diagonal elements of the magnetic stress dyad vanish, since $\vec{B} = (0,0,B)$, so that

$$\bar{\bar{T}}^{(m)} = \begin{pmatrix} -B^2/2\mu_0 & 0 & 0 \\ 0 & -B^2/2\mu_0 & 0 \\ 0 & 0 & B^2/2\mu_0 \end{pmatrix} \tag{5.11}$$

Therefore, the principal stresses are equivalent to a *tension* $B^2/2\mu_0$ along the magnetic field lines, and a *pressure* $B^2/2\mu_0$ perpendicular to the magnetic field lines, similar to a mutual repulsion of the field lines. Alternatively, we can express (5.11) in the form

$$\bar{\bar{T}}^{(m)} = \begin{pmatrix} 0 & 0 & 0 \\ 0 & 0 & 0 \\ 0 & 0 & B^2/\mu_0 \end{pmatrix} + \begin{pmatrix} -B^2/2\mu_0 & 0 & 0 \\ 0 & -B^2/2\mu_0 & 0 \\ 0 & 0 & -B^2/2\mu_0 \end{pmatrix} \tag{5.12}$$

so that the stress caused by the magnetic flux can also be thought of as an *isotropic magnetic pressure* $B^2/2\mu_0$ and a *tension* B^2/μ_0 along the magnetic flux lines as if they were elastic cords (Fig.4). This latter representation is very useful, since the isotropic pressure $B^2/2\mu_0$ can always be superposed on the fluid pressure, resulting in a decrease in the pressure exerted by the fluid.

5.2 Isobaric surfaces

It is convenient to consider in the plasma hypothetical surfaces over which the kinetic pressure is constant, called *isobaric surfaces*. At any point, the vector $\vec{\nabla}p$ is normal to the isobaric surface passing through the point considered. From (5.1) we see that $\vec{\nabla}p$ is normal to the plane containing $\vec{J}$ and $\vec{B}$, that is

$$\vec{J} \cdot \vec{\nabla}p = 0 \tag{5.13}$$

$$\vec{B} \cdot \vec{\nabla}p = 0 \tag{5.14}$$

Therefore, both $\vec{J}$ and $\vec{B}$ lie on isobaric surfaces. To illustrate this fact, consider the particular case in which the isobaric surfaces are closed concentric cylindrical surfaces, with the kinetic pressure increasing in the direction towards the central axis of the concentric cylindrical surfaces. Thus, $\vec{\nabla}p$ is along a radial line directed towards the axis. From (5.13) and (5.14) we see that neither $\vec{B}$, nor $\vec{J}$, pass through the isobaric surfaces and therefore it follows that the cylindrical isobaric surfaces are formed by a network of magnetic field lines and electric currents. Further, in view of (5.1), the magnetic field lines and electric currents, lying on the isobaric surfaces, must cross each other in such a way that $\vec{J} \times \vec{B}$ is equal to $\vec{\nabla}p$. This situation is shown in Fig. 5. The maximum kinetic pressure occurs along the central axis, which also coincides with a magnetic field line. For this reason, this axis is usually called the *magnetic axis* of the magnetoplasma configuration.

6. PLASMA CONFINEMENT IN A MAGNETIC FIELD

The subject of plasma confinement by magnetic fields is of considerable interest in the theory of controlled thermonuclear fusion. Consider, for simplicity, the special case in which the magnetic field is along the z axis,

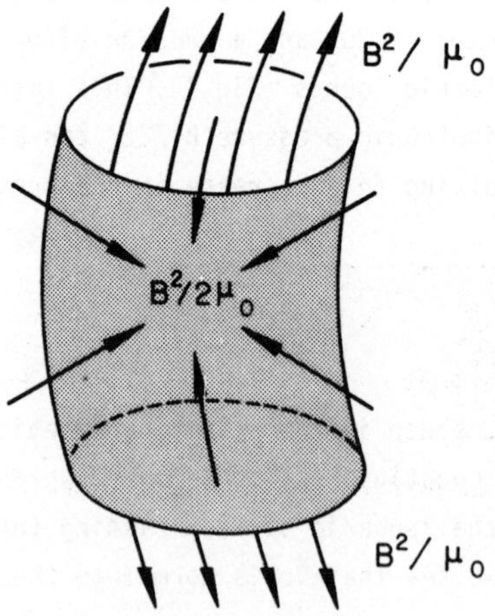

Fig. 4 - The stress caused by the magnetic flux can be decomposed into an isotropic magnetic pressure $B^2/2\mu_0$ and a magnetic tension B^2/μ_0 along the field lines.

that is $\vec{B} = B\,\hat{z}$, so that (5.10) simplifies to

$$\vec{\nabla} \cdot \begin{pmatrix} (p + B^2/2\mu_o) & 0 & 0 \\ 0 & (p + B^2/2\mu_o) & 0 \\ 0 & 0 & (p - B^2/2\mu_o) \end{pmatrix} = 0 \qquad (6.1)$$

from which we obtain

$$\frac{\partial}{\partial x}(p + B^2/2\mu_o) = 0 \qquad (6.2)$$

$$\frac{\partial}{\partial y}(p + B^2/2\mu_o) = 0 \qquad (6.3)$$

$$\frac{\partial}{\partial z}(p - B^2/2\mu_o) = 0 \qquad (6.4)$$

Also, from $\vec{\nabla} \cdot \vec{B} = 0$, we have

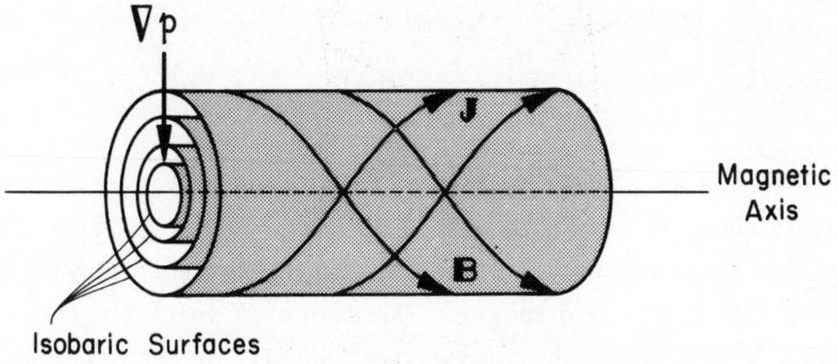

Fig. 5 - Isobaric concentric cylindrical surfaces, with $\vec{\nabla}p$ along a radial directed towards the magnetic axis. The lines of $\vec{B}$ and $\vec{J}$ lie on the isobaric surfaces and cross each other in a such a manner that $\vec{J} \times \vec{B}$ is equal to $\vec{\nabla}p$.

$$\partial B/\partial z = 0 \tag{6.5}$$

since, in the local coordinate system, $\vec{B} = (0,0,B)$. This last equation, together with (6.4), implies that both p and B do not vary in the direction of $\vec{B}$. The solutions of (6.2) and (6.3), combined with this result, give

$$(p + B^2/2\mu_0) = \text{constant} \tag{6.6}$$

Therefore, in the presence of an externally applied magnetic field, if the plasma is bounded, the plasma kinetic pressure decreases from the axis radially outwards, whereas the magnetic pressure increases in the same direction in such a manner that their sum is constant, according to (6.6). The plasma kinetic pressure can be forced to vanish on an outer surface if the applied magnetic field is sufficiently strong, with the result that the plasma is confined within this outer surface by the magnetic field.

Let $\vec{B}_0$ be the value of the magnetic induction external to the plasma (which is the value at the plasma boundary). Since the kinetic pressure at the plasma boundary is zero (ideally), we can evaluate the constant in (6.6) by calculating it at the plasma boundary. Therefore,

$$p + B^2/2\mu_0 = B_0^2/2\mu_0 \tag{6.7}$$

The maximum fluid pressure that can be confined for a given applied field $\vec{B}_0$ is, consequently,

$$p_{max} = B_0^2/2\mu_0 \tag{6.8}$$

A device that can be used to confine a magnetoplasma by straight parallel field lines is shown in Fig. 6, called a *theta* (θ) - *pinch*, since the effect responsible for the confinement is due to electric currents flowing in the plasma in the azimuthal (θ) direction. The plasma is initially confined inside a hollow cylindrical metal tube, whose side is split in the longitudinal direction in such a way as to form a capacitor. When a high voltage is discharged through the capacitor, the large azimuthal current produced in the metal tube produces a magnetic field in the longitudinal direction inside the plasma. The electric current, *induced* in the plasma, is also in the azimuthal direction, but in a sense opposite to that on the metal tube.

The resulting $\vec{J} \times \vec{B}$ force acting on the plasma pushes it inwards, towards the axis, until a balance is reached between the kinetic pressure due to the random particle thermal motions and the magnetic pressure which acts to constrict or *pinch* the plasma.

A parameter β, defined as the ratio of the kinetic pressure at a point inside the plasma, to the confining magnetic pressure at the plasma boundary, is usually introduced as a measure of the relative magnitudes of the kinetic and the magnetic pressures. It is given by

$$\beta = \frac{p}{B_0^2/2\mu_0} \tag{6.9}$$

Note that β ranges between 0 and 1, since the field inside the plasma is less than B_0. From (6.7) we can also express the parameter β as

$$\beta = 1 - (B/B_0)^2 \tag{6.10}$$

Two special cases of plasma confinement schemes are the so called *low* β and *high* β devices. In the low β devices, the kinetic pressure is small in comparison with the magnetic pressure at the plasma boundary, whereas in the high β devices they are of an equal order of magnitude ($\beta \simeq 1$).

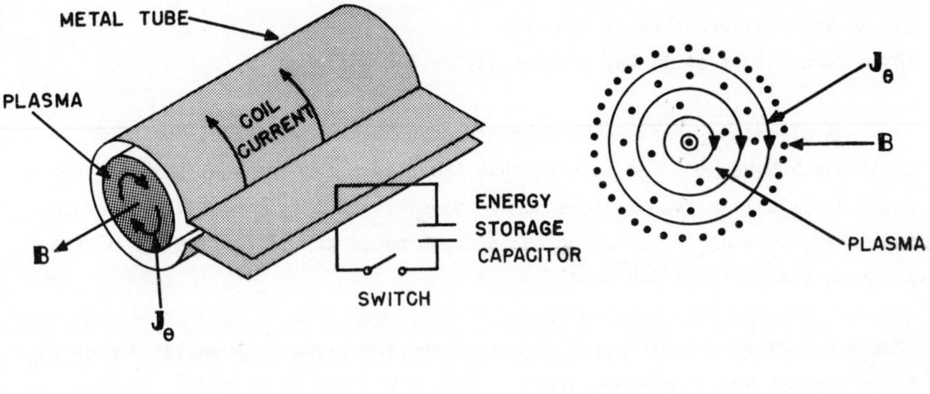

Fig. 6 - Magnetoplasma confined by straight parallel field lines in a theta-pinch device.

An important property of a plasma is its *diamagnetic* character. (6.7) implies that the magnetic field inside the plasma is less than its value at the plasma boundary. As the kinetic pressure increases inside the plasma, the magnetic field decreases. Under the action of the externally aplied $\vec{B}$ field, the particle motions give rise to internal electric currents which induce a magnetic field opposite to the externally applied field. Consequently, the *resultant* magnetic field inside the plasma is reduced to a value less than the plasma boundary value. The electric current, induced in the plasma, depends on the number density of the charged particles and their velocity. Therefore, as the plasma kinetic pressure increases, the induced electric current and the induced magnetic field also increase, thus enchancing the diamagnetic effect.

PROBLEMS

12.1 - Consider the energy equation involving the time rate of change of the total pressure dyad $\bar{\bar{p}}$, derived in Problem 9.6. Show that when this equation is contracted with the unit dyad $\bar{\bar{1}}$ results in (1.21), whereas when contracted with the dyad $\hat{B}\,\hat{B}$ yields (1.22).

12.2 - Derive an energy conservation equation, similar to (1.64), but considering the Parker modified momentum equation and the CGL energy equations, instead of (1.2) and (1.3).

12.3 - Calculate the *minimum* intensity of the magnetic induction ($\vec{B}_o$) necessary to confine a plasma at:

(a) an internal pressure of 100 atm.

(b) a temperature of 10 keV and density of $8 \times 10^{21} m^{-3}$.

12.4 - A plasma is confined by a unidirectional magnetic induction $\vec{B}$ of magnitude 5 Weber/m^2. Considering that the plasma temperature is 10 keV and $\beta = 0.4$, calculate the particle number density. If the temperature increases to 50 keV, what is the value of the $\vec{B}$ field necessary to confine the plasma, assuming that β stays the same?

12.5 - Calculate the diffusion time (τ_D) and the magnetic Reynolds number (R_m) for a typical MHD generator, with L = 0.1 m, u = 10^3 m/sec and σ_o = 100 mho/m. Verify that in this case τ_D is very short, so that inhomogeneities in the magnetic field are smoothed out rapidly.

12.6 - Consider a plasma in the form of a straight circular cylinder with a helical magnetic field given by

$$\vec{B} = B_\theta(r)\,\hat{\theta} + B_z(r)\,\hat{z}$$

Show that the force per unit volume, associated with the inward magnetic pressure for this configuration, is

$$-\vec{\nabla}_\perp (B^2/2\mu_o) = -\hat{r}\frac{\partial}{\partial r}[B^2(r)/2\mu_o]$$

and the force per unit volume, associated with the magnetic tension due to the curvature of magnetic field lines, is

$$(B^2/\mu_o)\,\hat{B}\cdot\vec{\nabla}\,\hat{B} = -\hat{r}\,B_\theta^2(r)/\mu_o r$$

12.7 - Use (4.1), for a perfectly conducting fluid, and the nonlinear equation of continuity (1.1), to show that the change of $\vec{B}$ with time in a fluid element is related to changes of density according to

$$\frac{D}{Dt}(\vec{B}/\rho_m) = [(\vec{B}/\rho_m)\cdot\vec{\nabla}]\,\vec{u}$$

Use this relation to establish that, in a perfectly conducting fluid, the fluid elements which lie initially on a magnetic flux line continue to lie on a flux line.

12.8 - The boundary of the Earth's magnetosphere, in the direction of the Earth-Sun line, occurs at a distance where the kinetic pressure of the solar wind particles is equal to the (modified) Earth's magnetic field pressure. Show that the distance of the magnetopause from the center of the Earth, along the Earth-Sun line, is given approximately by

$$R_M = [2B_o^2/(\mu_o\rho_m u_s^2)]^{1/6}\,R_E$$

where R_E is the Earth's radius, ρ_m is the mass density of the solar wind, u_s is its undisturbed speed, and B_o is the surface value of the undisturbed Earth's magnetic field.

12.9 - Consider a cylindrically symmetric plasma column($\partial/\partial z = 0$, $\partial/\partial\theta = 0$) under equilibrium conditions, confined by a magnetic field. Verify that in cylindrical coordinates the radial component of (5.1) becomes

$$dp(r)/dr = J_\theta(r)\, B_z(r) - J_z(r)\, B_\theta(r)$$

Using Maxwell equation (5.2), show that

$$J_\theta = -\ (1/\mu_o)\ dB_z/dr$$

$$J_z = (1/\mu_o r)\ d(rB_\theta)/dr$$

Therefore, obtain the following basic equation for the equilibrium of a plasma column with cylindrical symmetry

$$\frac{d}{dr}\left(p + \frac{B_z^2}{2\mu_o} + \frac{B_\theta^2}{2\mu_o} \right) = -\ \frac{1}{\mu_o}\ \frac{B_\theta^2}{r}$$

Give a physical interpretation for the various terms in this equation.

CHAPTER 13

The Pinch Effect

1. INTRODUCTION

In view of the importance of plasma confinement by a magnetic field in controlled thermonuclear research, as well as in other applications, we present in this chapter a detailed treatment of plasma confinement for the special case in which the confinement is produced by an azimuthal (Θ) self-magnetic field, due to an axial current in the plasma generated by an appropriately applied electric field.

Consider an infinite cylindrical column of conducting fluid with an axial current density $\vec{J} = J_z(r)\,\hat{z}$ and a resulting azimuthal induction $\vec{B} = B_\Theta(r)\,\hat{\Theta}$, as depicted in Fig. 1. The $\vec{J} \times \vec{B}$ force, acting on the plasma, forces the column to contract laterally. This lateral constriction of the plasma column is known as the *pinch effect*. The isobaric surfaces, for which p = constant, are, in this case, concentric cylinders.

As the plasma is compressed laterally, the plasma number density and the temperature increase. The plasma kinetic pressure counteracts to hinder the constriction of the plasma column, whereas the magnetic force acts to confine the plasma. When these counteracting forces are balanced, a steady state condition results in which the material is mainly confined within a certain radius R, which remains constant in time. This situation is commonly referred to as the *equilibrium pinch*. When the self-magnetic pressure exceeds the plasma kinetic pressure, the column radius changes whith time resulting in a situation

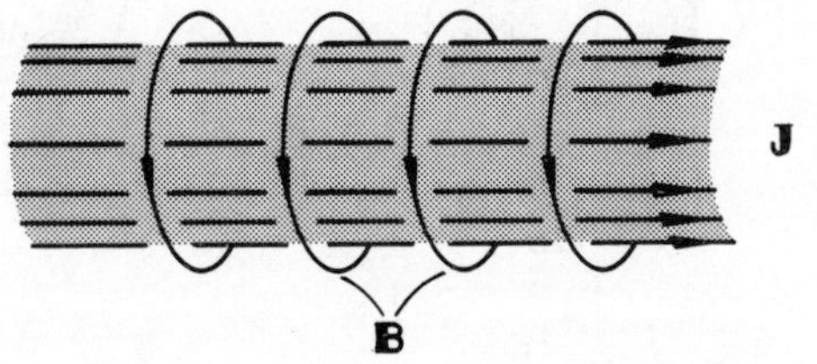

Fig. 1 - Pinch configuration in which a magnetoplasma is confined by azimuthal magnetic fields generated by axial currents flowing along the plasma column.

known as the *dynamic pinch*. In what follows we investigate first the equilibrium pinch and afterwards the dynamic pinch.

2. THE EQUILIBRIUM PINCH

For simplicity, the current density, the magnetic field and the plasma kinetic pressure are assumed to depend only on the distance from the cylinder axis. For steady state conditions, none of the variables change with time. The various parameters of the equilibrium pinch are schematically shown in Fig.2. Since the system is cylindrically symmetric, only the radial component of (12.5.1) must be considered,

$$\frac{dp(r)}{dr} = - J_z(r)\, B_\Theta(r) \tag{2.1}$$

Inside a cylinder of general radius r, the total enclosed current $I_z(r)$ is

$$I_z(r) = \int_0^r J_z(r)\, 2\pi r\, dr \tag{2.2}$$

Note that the variable r inside the integrand is a dummy variable. From (2.2) we obtain

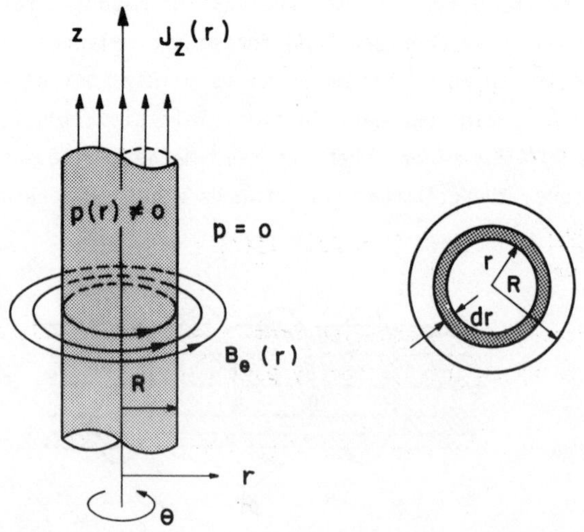

Fig. 2 - Schematic diagram illustrating the various parameters relevant to the study of the equilibrium longitudinal pinch configuration.

$$\frac{dI_z(r)}{dr} = 2\pi r\, J_z(r) \tag{2.3}$$

Ampēre's law in integral form relates $B_\Theta(r)$ to the total enclosed current, giving for the magnetic induction

$$B_\Theta(r) = \frac{\mu_0}{2\pi r} I_z(r) = \frac{\mu_0}{r} \int_0^r J_z(r)\, r\, dr \tag{2.4}$$

A number of results can be obtained even without specifying the precise form of $J_z(r)$. If the conducting fluid lies almost entirely inside $r = R$, then the magnetic induction $B_\Theta(r)$ outside the plasma is

$$B_\Theta(r) = \mu_0 I_0 / 2\pi r \qquad (r \geqq R) \tag{2.5}$$

where

$$I_0 = \int_0^R J_z(r)\, 2\pi r\, dr = I_z(R) \tag{2.6}$$

which is the total current flowing inside the cylindrical plasma column. The substitution of $B_\Theta(r)$ and $J_z(r)$, from (2.4) and (2.3), respectively, into (2.1), gives

$$\frac{dp(r)}{dr} = - \frac{\mu_0}{4\pi^2 r^2} I_z(r) \frac{dI_z(r)}{dr} \tag{2.7}$$

which can be written as

$$4\pi^2 r^2 \frac{dp(r)}{dr} = - \frac{d}{dr} [(\mu_0/2)\; I_z^2(r)] \tag{2.8}$$

If we now integrate this equation from $r = 0$ to $r = R$, and simplify the left-hand side by an integration by parts, we obtain

$$(4\pi^2 r^2\, p(r) \Big|_0^R) - 4\pi \int_0^R 2\pi r\, p(r)\, dr = - \mu_0 I_0^2/2 \tag{2.9}$$

where $I_0 = I_z(R)$ is the total current flowing through the entire cross section of the plasma column and, obviously, $I_z(0) = 0$. Considering the plasma column to be confined to the range $0 \leqq r < R$, it follows that $p(r)$ is zero for $r \geqq R$ and finite for $0 \leq r < R$, so that the first term in the left-hand side of (2.9) vanishes. Therefore, we find that

$$I_0^2 = (8\pi/\mu_0) \int_0^R 2\pi r\, p(r)\, dr \tag{2.10}$$

If the partial pressures of the electrons and the ions are governed by the ideal gas law,

$$p_e(r) = n(r)\, k\, T_e \tag{2.11}$$

$$p_i(r) = n(r)\, k\, T_i \tag{2.12}$$

assuming that the electron and ion temperatures, T_e and T_i, respectively, are constants throughout the column, we have

$$p(r) = p_e(r) + p_i(r) = n(r)\, k\, (T_e + T_i) \tag{2.13}$$

Therefore, (2.10) becomes

$$I_0^2 = (8\pi/\mu_0)\, k\, (T_e + T_i) \int_0^R 2\pi r\, n(r)\, dr \tag{2.14}$$

or

$$I_0^2 = (8\pi/\mu_0)\, k\, (T_e + T_i)\, N_\ell \tag{2.15}$$

where

$$N_\ell = \int_0^R 2\pi r\, n(r)\, dr \tag{2.16}$$

is the number of particles per unit length of the plasma column.

Eq. (2.15) is known as the *Bennett relation*. It gives the total current that must be discharged through the plasma column in order to confine a plasma at a specified temperature and a given number of particles (N_ℓ) per unit length. The current required for the confinement of hot plasmas is usually very large. As an example, suppose that $N_\ell = 10^{19}\ m^{-1}$ and that the plasma temperature is such that $T_e + T_i = 10^8$K. Since $\mu_0 = 4\pi \times 10^{-7}$ H/m and $k = 1.38 \times 10^{-23}$ J/K, it follows that the required current I_0 is of the order of one million Ampéres.

To obtain the radial distribution of $p(r)$ in terms of $B_\Theta(r)$, it is convenient to start from (2.1) and proceed in a different way. First, we note that from Maxwell equation $\vec{\nabla} \times \vec{B} = \mu_0 \vec{J}$ we have, in cylindrical coordinates, with only radial dependence,

$$\frac{1}{r}\frac{d}{dr}\left[r\, B_\Theta(r) \right] = \mu_0\, J_z(r) \tag{2.17}$$

from which we get

$$J_z(r) = (1/\mu_0)\, dB_\Theta(r)/dr + (1/\mu_0)\, B_\Theta(r)/r \tag{2.18}$$

Substitution of this result for $J_z(r)$ into (2.1), yields

$$\frac{dp(r)}{dr} = -\frac{1}{2\mu_0 r^2}\frac{d}{dr}\left[r^2 B_\Theta^2(r) \right] \tag{2.19}$$

We now integrate this equation from $r = 0$ to a general radius r,

$$p(r) = p(0) - \frac{1}{2\mu_0}\int_0^r \frac{1}{r^2}\frac{d}{dr}\left[r^2\, B_\Theta^2(r) \right] dr \tag{2.20}$$

In particular, since for $r = R$ we have $p(R) = 0$,

$$p(0) = \frac{1}{2\mu_0}\int_0^R \frac{1}{r^2}\frac{d}{dr}\left[r^2 B_\Theta^2(r) \right] dr \tag{2.21}$$

and substituting this result into (2.20),

$$p(r) = \frac{1}{2\mu_0}\int_r^R \frac{1}{r^2}\frac{d}{dr}\left[r^2 B_\Theta^2(r) \right] dr \tag{2.22}$$

The *average* pressure inside the cylinder can be related to the total current I_0 and the column radius R without knowing the detailed radial dependence. The average value of the kinetic pressure inside the column is defined by

$$\bar{p} = \frac{1}{\pi R^2} \int_0^R 2\pi r\, p(r)\, dr \tag{2.23}$$

Simplifying this expression by an integration by parts, yields

$$\bar{p} = -\frac{1}{R^2} \int_0^R r^2 \frac{d}{dr} p(r)\, dr \tag{2.24}$$

since the integrated term is zero, because $p(R) = 0$. Replacing $dp(r)/dr$, using (2.19), we get

$$\bar{p} = B_\Theta^2(R)/2\mu_0 = \mu_0 I_0^2/(8\pi^2 R^2) \tag{2.25}$$

This result shows that the average kinetic pressure in the equilibrium plasma column is balanced by the magnetic pressure at the boundary.

From (2.2), (2.4) and (2.22), we can deduce the radial distribution of $I_z(r)$, $B_\Theta(r)$ and $p(r)$ if we know the radial dependence of $J_z(r)$. So far, the radial dependence of $J_z(r)$ has not been discussed. In what follows, we will consider two simple possibilities, in order to illustrate the use of the above-mentioned equations.

As a simple example consider the case in which the current density $J_z(r)$ is constant for $r < R$. Taking $J_z = I_0/\pi R^2$ in (2.4), we obtain for $r < R$,

$$B_\Theta(r) = \frac{\mu_0 I_0}{\pi R^2 r} \int_0^r r\, dr = \frac{\mu_0 I_0}{2\pi R^2} r \qquad (r < R) \tag{2.26}$$

Substituting this result into (2.22) we obtain a parabolic dependence for the pressure versus radius,

$$p(r) = \frac{1}{2\mu_0} \int_r^R \frac{1}{r^2} \frac{d}{dr} \left(\frac{\mu_0^2\, I_0^2\, r^4}{4\pi^2\, R^4}\right) dr = \frac{\mu_0\, I_0^2}{4\pi^2 R^2} (1 - r^2/R^2) \tag{2.27}$$

Note that, in this case, the axial pressure $p(0)$ is twice the average pressure $\bar{p}$ given in (2.25). The radial dependence of the various quantities is shown in Fig. 3.

Another radial distribution of $J_z(r)$ which is also of interest in the investigation of the dynamic pinch is the one in which the current density is confined to a very thin layer on the surface of the column. This model is appropriate for a highly conducting fluid. In a perfectly conducting plasma, the current cannot penetrate the plasma and exists only on the column surface. This surface current density can be conveniently represented by a Dirac delta function at $r = R$. In this case there is no magnetic field inside the plasma and $B_\Theta(r)$ exists only for $r > R$. From (2.5) the magnetic induction is given by

$$B_\Theta(r) = \mu_o I_o / 2\pi r \qquad (r > R) \qquad (2.28)$$

where I_o is the total axial current. Therefore, from (2.20), we have

$$p(r) = p(0) \qquad (0 < r < R) \qquad (2.29)$$

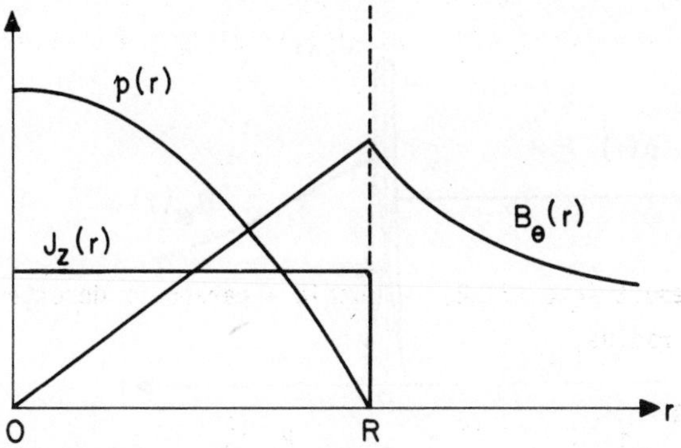

Fig. 3 - Radial dependence of the azimuthal magnetic induction $B_\Theta(r)$ and plasma pressure $p(r)$ in a cylindrical plasma column with a constant current density $J_z(r)$. The radius of the column is R.

so that the plasma kinetic pressure is constant inside the cylindrical column and equal to the average value given in (2.25). The radial dependence of the various quantities for this model is sketched in Fig. 4. Thus, for a perfectly conducting plasma column, the magnetic induction vanishes inside the column and falls off as $1/r$ outside the column. The plasma kinetic pressure is constant inside the column and vanishes outside it. The pinch effect, in this special case, can be thought of as due to an abrupt build up of the magnetic pressure $B_\Theta^2/2\mu_o$ in the region external to the column.

3. THE BENNETT PINCH

W.H. Bennett, the discoverer of the pinch effect, has investigated a special model of the equilibrium longitudinal pinch in which the radial distribution of the various quantities are such that the drift velocity of the plasma particles is constant throughout the column cross section. As an instructive application of the previous equations for the equilibrium pinch configuration, we investigate this particular model in what follows. In view

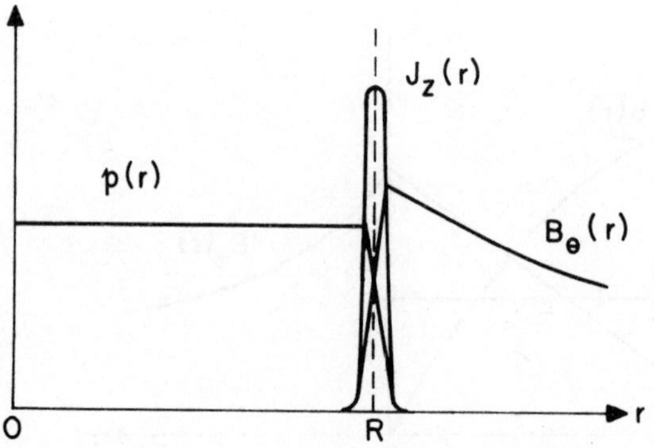

Fig. 4 - Radial dependence of the azimuthal magnetic induction $B_\Theta(r)$ and plasma pressure $p(r)$ in a cylindrical plasma column with a surface current density $J_z(r)$. The radius of the column is R.

of the fact that the ion mass is much larger than the electron mass, the drift velocity of the ions is much smaller than that of the electrons and can therefore be neglected. Thus, we take the current density to be given by

$$\vec{J}(r) = - e\, n(r)\, \vec{u}_e \tag{3.1}$$

Since the applied electric field is in the z direction, we have $\vec{J}(r) = J_z(r)\,\hat{z}$ and $\vec{u}_e = - u_{ez}\,\hat{z}$, where u_{ez} is positive and constant, independent of r. Therefore,

$$J_z(r) = e\, n(r)\, u_{ez} \tag{3.2}$$

Substitution of this equation for $J_z(r)$, and (2.13) for $p(r)$, into the hydrostatic equation of motion (2.1), yields

$$k\,(T_e + T_i)\, dn(r)/dr = - e\, n(r)\, u_{ez}\, B_\Theta(r) \tag{3.3}$$

If we multiply this equation by $r/[n(r)\, k\,(T_e + T_i)]$ and differentiate it with respect to r, we obtain

$$\frac{d}{dr}\left[\frac{r}{n(r)}\frac{dn(r)}{dr}\right] = - \frac{e\,u_{ez}}{k\,(T_e + T_i)}\frac{d}{dr}\left[r\, B_\Theta(r)\right] \tag{3.4}$$

From (2.17) and (3.2), we have

$$\frac{d}{dr}\left[r\, B_\Theta(r)\right] = \mu_o\, e\, u_{ez}\, r\, n(r) \tag{3.5}$$

and using this result in (3.4),

$$\frac{d}{dr}\left[\frac{r}{n(r)}\frac{dn(r)}{dr}\right] + \left[\frac{\mu_o\, e^2\, u_{ez}^2}{k\,(T_e + T_i)}\right] r\, n(r) = 0 \tag{3.6}$$

The solution of this nonlinear differential equation gives the radial dependence of the number density n(r). Bennett has obtained the solution of this nonlinear equation subject to the boundary condition that n(r) is symmetric about the z axis, where r = 0, and is a smoothly varying function of r, so that

$$\left.\frac{dn(r)}{dr}\right|_{r=0} = 0 \tag{3.7}$$

The solution of (3.6), subjected to the boundary condition (3.7), is known as the *Bennett distribution* and is given by

$$n(r) = n_0/(1 + n_0 b r^2)^2 \tag{3.8}$$

where $n_0 = n(r=0)$, which is the number density on the axis, and

$$b = \frac{\mu_0 e^2 u_{ez}^2}{8k(T_e + T_i)} \tag{3.9}$$

which has dimensions of length. This radial dependence of the number density is sketched in Fig. 5. From (3.2) and (2.13) we see that the radial dependence of $J_z(r)$ and $p(r)$ is the same as that of $n(r)$. It can be used to determine $B_\Theta(r)$ according to (2.4).

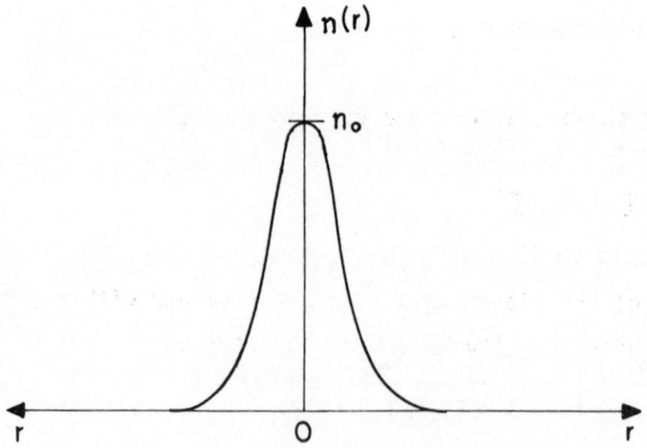

Fig. 5 - The Bennett distribution of the number density n(r) of the particles in an equilibrium pinched plasma column.

The Bennett distribution (3.8) shows that particles are present up to infinity but, since $n(r)$ falls off very rapidly with increasing values of r, we can consider, for all practical purposes, that the plasma is essentially concentrated symmetrically in a small cylindrical region about the z axis. Using (3.8) we can obtain the number of particles $N_\ell(R)$ per unit lenght contained in a cylindrical column of radius R,

$$N_\ell(R) = \int_0^R n(r)\, 2\pi r\, dr = 2\pi n_0 \int_0^R \frac{r}{(1+n_0 b r^2)^2}\, dr \tag{3.10}$$

Evaluating the integral yields

$$N_\ell(R) = n_0 \pi R^2/(1+n_0 b R^2) \tag{3.11}$$

Since particles are present up to infinity, the *total* number of particles per unit length can be obtained from (3.11) by taking the limit as $R \to \infty$, which gives

$$N_\ell(\infty) = \pi/b \tag{3.12}$$

If we let α denote the fraction of the number of particles per unit length that is contained in a cylinder of radius R, that is,

$$\alpha = N_\ell(R)/N_\ell(\infty) = bN_\ell(R)/\pi \tag{3.13}$$

and use (3.11), we obtain, after some rearrangement,

$$(n_0 b)^{1/2} R = [\alpha/(1-\alpha)]^{1/2} \tag{3.14}$$

Therefore, if 90% of the plasma particles are confined within the cylindrical plasma column of radius R, that is $\alpha = 0.9$, we have

$$(n_0 b)^{1/2} R = 3 \tag{3.15}$$

Thus, even though the particles extend up to infinity, the major portion of them lies in a small neighborhood around the z axis. Note that, since

$(n_0\ b)^{1/2}$ has dimensions of an inverse length, we can think of $(n_0\ b)^{1/2}\ R$ as a normalized radius of the cylindrical plasma column. If we assume arbitrarily that a plasma is confined within a cylindrical surface of radius R if 90% of the particles are within this cylindrical column, then the radius R of the cylindrical surface is given by (3.15).

4. DYNAMIC MODEL OF THE PINCH

The simple theory of the equilibrium pinch, considered in section 2, is valid when the plasma column radius is constant in time or when it is varying very slowly compared with the time required for the plasma to attain a constant temperature. In actual practice, however, static or quasi-static situations do not arise and it is necessary to consider the dynamical behavior of the pinch effect. Initially, when the current starts flowing down the plasma column, the kinetic pressure is generally too small to resist the force due to the external magnetic pressure, so that the radius of the plasma cylinder is forced inwards and the plasma column is pinched. The essential dynamic features of the time-varying pinch are illustrated by the following simple model.

Suppose that a fully ionized plasma fills the interior region $(0 < r < R_0)$ of a hollow dielectric cylinder of radius R_0 and length L. A voltage difference V is applied between the ends of the cylinder, so that a current I flows in the plasma. This current produces an azimuthal magnetic induction $B_\Theta(r)$ which causes the plasma to pinch inwards. The plasma is assumed to be perfectly conducting, so that all the current flows on the surface and there is no magnetic flux inside the plasma. Also, the plasma kinetic pressure is neglected. Let $R(t)$ be the plasma column radius at time t (Fig. 6). The magnitude of the azimuthal magnetic induction just adjacent to the current sheath at radius $R(t)$ is given by

$$B_\Theta(R) = \mu_0 I(t)/2\pi R \tag{4.1}$$

where $I(t)$ is the total axial current at the instant t. In particular, for $t = 0$ we have $R = R_0$, and this equation gives the inital value $B_\Theta(R_0)$ of the magnetic induction. The magnetic pressure $p_m(R)$ produced by this magnetic induction, acting on the current sheath radially inwards, is given by

$$p_m(R) = B_\Theta^2(R)/2\mu_0 = \mu_0 I^2(t)/(8\pi^2 R^2) \quad (4.2)$$

The force per unit length of the current sheath, acting radially inwards, is obtained from (4.2) as

$$\vec{F}(R) = \hat{r}\ F(R) = -\ 2\pi R\ p_m(R)\ \hat{r} = -\hat{r}\,\mu_0 I^2(t)/(4\pi R) \quad (4.3)$$

To set up the equation of motion, relating $I(t)$ to the instantaneous radius $R(t)$ of the pinch discharge, we must make some assumption about the plasma. We shall consider the so-called *snowplow model*, in which the current sheath is imagined to carry along with it all the material which it hits as it moves inward. If ρ_m is the original mass density of the plasma, then the mass per unit length carried by the interface as it moves in, at time t, when the radius of the current sheath is R, is given by

$$M(R) = \pi(R_0^2 - R^2)\rho_m \quad (4.4)$$

Fig. 7 illustrates the cross sectional area swept by the current sheath as it moves inward. From Newton's second law, the magnetic pressure force and the rate of change of momentum are related by

$$\frac{d}{dt}\left[M(R)\ \frac{dR}{dt}\right] = F(R)$$

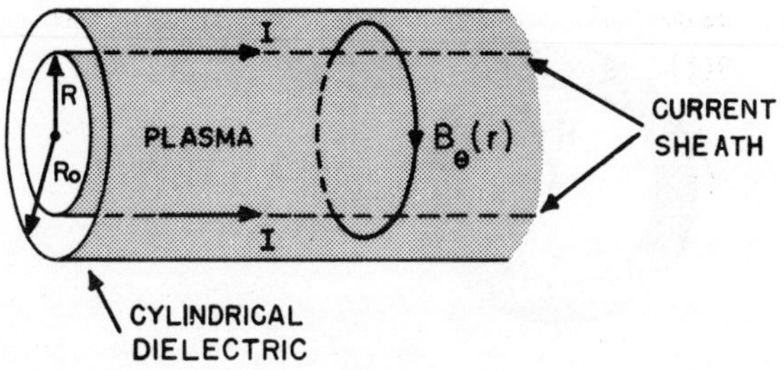

Fig. 6 - Plasma column of infinite conductivity, inside a hollow cylindrical dielectric, with a current sheath on its surface.

or, using (4.3) and (4.4),

$$\frac{d}{dt}\left[\pi\,\rho_m(R_0^2 - R^2)\,\frac{dR}{dt}\right] = -\,\frac{\mu_0\, I^2(t)}{4\pi R} \tag{4.5}$$

If the functional dependence of the pinch current $I(t)$ is known, (4.5) permits the evaluation of the pinch discharge radius as a function of time.

A standard inductive relation between the applied voltage, the current and the dimensions (inductance) of the plasma column can be obtained using Faraday's law of induction. For this purpose consider the closed loop shown in Fig. 8, in which the inner arm lies on the interface and moves inward with it. Applying Faraday's law to this dotted loop,

$$\oint \vec{E}\cdot d\vec{\ell} = -\,\frac{d}{dt}\left(\int_S \vec{B}\cdot d\vec{S}\right) \tag{4.6}$$

and noting that the only contribution to the line integral of $\vec{E}$ comes from the side of the loop lying in the conducting wall, we obtain

$$-\,\frac{V}{L} = -\,\frac{d}{dt}\int_{R(t)}^{R_0} B_\Theta(r)\,dr \tag{4.7}$$

Using (4.1), and performing the integral, yields

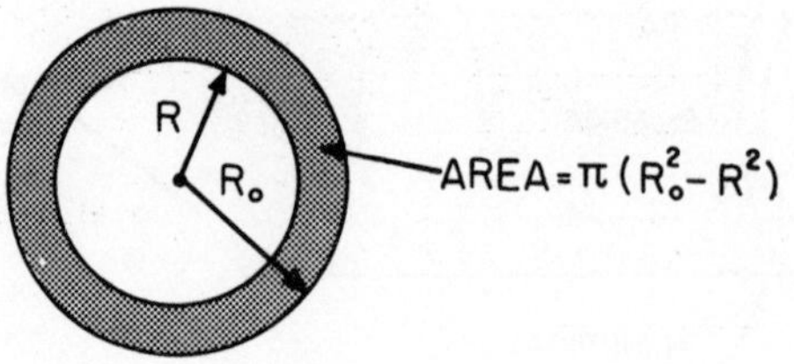

Fig. 7 - Area swept by the current sheath as it moves inward from the radius R_0 to $R(t)$.

$$\frac{V}{L} = \frac{\mu_0}{2\pi}\frac{d}{dt}\left\{I(t)\,\ell n\,[R_0/R(t)]\right\} \tag{4.8}$$

If we denote the applied electric field V/L by $E_0 f(t)$, where the function f(t) is assumed known and is normalized so that the peak value of the applied electric field is E_0, (4.8) becomes

$$I(t)\,\ell n\,[R_0/R(t)] = \frac{2\pi}{\mu_0}E_0\int_0^t f(t')\,dt' \tag{4.9}$$

This equation can be used to eliminate I(t) from the equation of motion (4.5), resulting in the following equation for the rate of change of R(t)

$$\frac{d}{dt}\left[(R_0^2 - R^2)\frac{dR}{dt}\right] = -\frac{E_0^2\left[\int_0^t f(t')\,dt'\right]^2}{\mu_0\,\rho_m\,R\left[\ell n\,(R_0/R)\right]^2} \tag{4.10}$$

It is convenient to introduce the following dimensionless variables

$$x = R/R_0 \tag{4.11}$$

$$\tau = [E_0^2/(\mu_0\rho_m R_0^4)]^{1/4}\,t \tag{4.12}$$

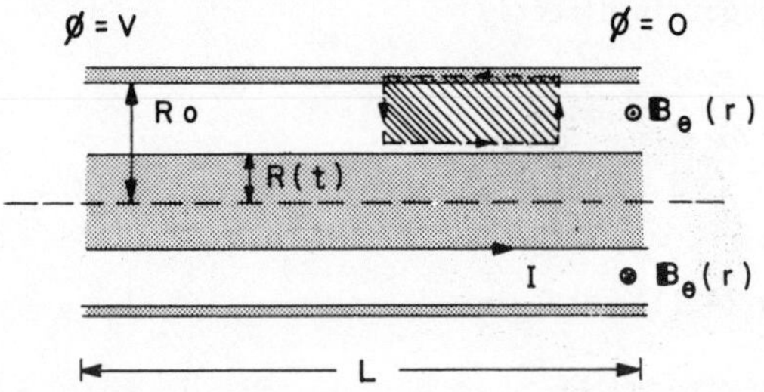

Fig. 8 - Schematic representation of a closed loop for application of Faraday's law, with the inner side lying on the interface and moving inwards with it.

and recast (4.10) in normalized form as

$$\frac{d}{d\tau}\left[(1-x^2)\frac{dx}{d\tau}\right] = -\frac{\left[\int_0^{\tau} f(\tau')\, d\tau'\right]^2}{x\,(\ell n\, x)^2} \tag{4.13}$$

This equation cannot be solved without knowing the function $f(t)$. However, some idea of the results can be obtained, without solving this equation, by noting that x changes significantly for time periods such that $\tau = 1$. Thus, from (4.12), the scaling law for the radial velocity of the pinch is, approximately,

$$|dR/dt| \cong v_0 = (E_0^2/\mu_0\rho_m)^{1/4} \tag{4.14}$$

The typical experimental conditions involved in a small scale pinch column of hydrogen or deuterium plasma are initial mass densities of the order of 10^{-8} gm/cm³ and applied electric fields of the order of 10^3 Volts/cm, which give a velocity v_0 of the order of 10^7 cm/sec. For these conditions, in a tube of 10cm radius, the current measured is of the order of 10^5 or 10^6 Ampēres.

It is instructive to consider a particular case in which the pinch current varies in time according to

$$I(t) = I_0 \sin(\omega t) \cong I_0\, \omega t \tag{4.15}$$

Then, from (4.5) we obtain directly

$$\frac{d}{d\tau}\left[(1-x^2)\frac{dx}{d\tau}\right] = -\frac{\tau^2}{x} \tag{4.16}$$

with x as given by (4.11), and

$$\tau = [\mu_0 I_0^2 \omega^2/(4\pi^2 \rho_m R^4)]^{1/4}\; t \tag{4.17}$$

Equation (4.16) has to be solved numerically to determine $x(\tau)$. The resulting relation between the normalized radius of the dynamic pinch and the normalized time is sketched in Fig. 9. This simplified model indicates that the plasma column radius goes to zero in a time slightly greater than τ. This is a consequence of neglecting the kinetic pressure of the plasma. The above discussion is therefore valid only for very short time periods after the onset of the current flow.

An important phenomenon that usually occurs in the dynamic pinch has not been considered in this analysis. As the current sheath moves radially inwards, compressing the plasma, the behavior just discussed is modified. A radial wave motion is usually set up by the pinch and this wave travels faster than the current sheath. These waves, travelling inwards, get reflected off the axis and move outwards striking the interface and retarding the inward motion of the current sheath or even reversing it. This phenomenon is known as *bouncing*. This sequence of events takes place periodically and the amplitude of each succeeding bounce becomes smaller. The plasma column radius presumably reaches an equilibrium state at some radius less than R_0. Fig. 10 illustrates the general behavior expected for the column radius R as a function of time.

5. INSTABILITIES IN A PINCHED PLASMA COLUMN

Although it is possible to achieve an equilibrium state for plasma confinement with the pinch effect, this equilibrium state is not stable. A small departure from the cylindrical geometry of the equilibrium state, results in the growth of the original perturbations with time and the desintegration of the plasma column. The growth of instabilities is the

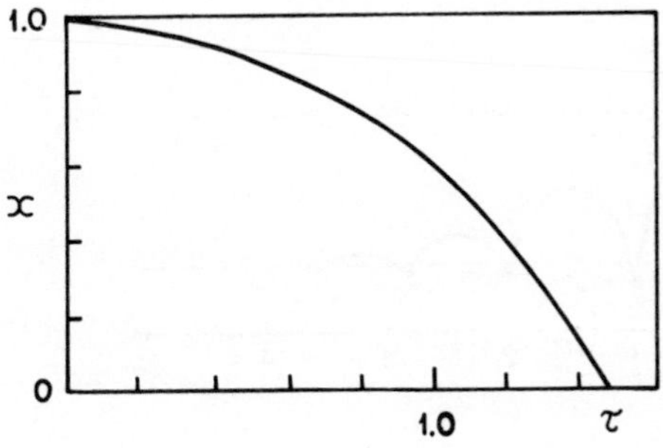

Fig. 9 - Normalized radius $x = R/R_0$ of the dynamic pinch column as a function of the normalized time τ, according to (4.16).

reason why it is diffucult to sustain reasonably long-lived pinched plasmas in the laboratory.

A detailed mathematical treatment of these instabilities is out of the scope of this text. For simplicity, in the following discussion of instabilities we shall consider a perfectly diamagnetic plasma column confined by a static magnetic field. Since the plasma is perfectly diamagnetic, there is no magnetic field and consequently no magnetic pressure inside the plasma column. The plasma kinetic pressure is assumed to be uniform inside the plasma and vanishes outside it. In the *equilibrium state*, the magnetic pressure at the plasma surface p_{mo} must be equal to the kinetic pressure p of the plasma,

$$p = p_{mo} = B_0^2/2\mu_0 \tag{5.1}$$

where B_0 is the magnitude of the magnetic flux density at the plasma surface. This situation of a sharp plasma boundary is an idealized one and is difficult to create in laboratory, since the plasma particles diffuse through the magnetic field lines in a diffussion time of the order of $\mu_0\, \sigma_0\, L^2$, in view of the finite plasma conductivity σ_0, as discussed in section 3, Chapter 12.

In the cylindrical pinch column, the confining magnetic field lines have a curvature such that they are concave towards the plasma and the field strength decreases whith increasing distance from the center of curvature of the field lines (Fig. 11). According to Ampēre's law, this azimuthal magnetic field is inversellyproportional to the radial distance r from the column axis.

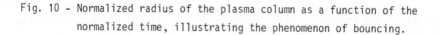

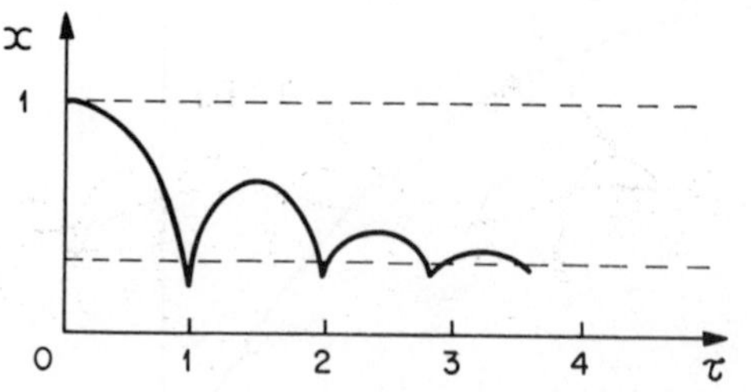

Fig. 10 - Normalized radius of the plasma column as a function of the normalized time, illustrating the phenomenon of bouncing.

6. THE SAUSAGE INSTABILITY

Suppose that the equilibrium state of the pinched plasma column, shown in Fig. 11, is disturbed by a wave-like pertubation, with the crests and troughs on the surface of the plasma column and cylindrically symmetric about the column axis, as indicated in Fig. 12. We shall consider that the plasma is constricted in some locations and expanded at others, in such a way that its volume does not change. Consequently, the uniform kinetic pressure of the plasma is left unchanged. However, in view of the 1/r radial dependence of the azimuthal magnetic field, the magnitude of this field at the surface of the disturbed plasma column will vary from place to place on the surface. At the locations where the radius has decreased, in relation to the equilibrium value, the magnetic pressure at the constricted plasma surface will be larger than the plasma kinetic pressure, and will force the plasma surface radially inwards, thus enhancing the constriction. At the locations where the radius has become larger than the equilibrium value, the plasma kinetic pressure

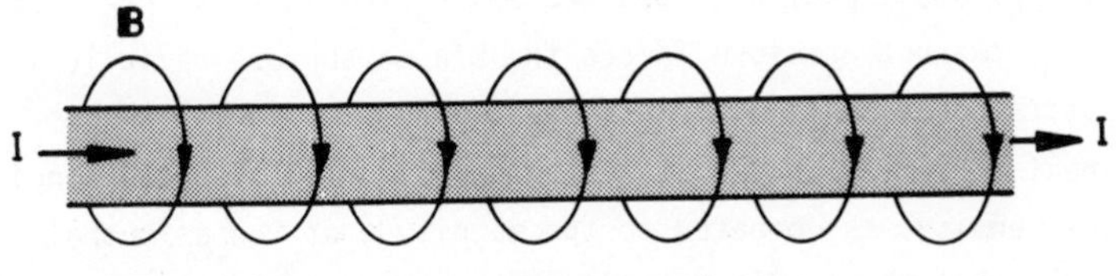

Fig. 11 - Unstable equilibrium configuration of a cylindrical plasma column. The azimuthal $\vec{B}$ field decreases radially outwards.

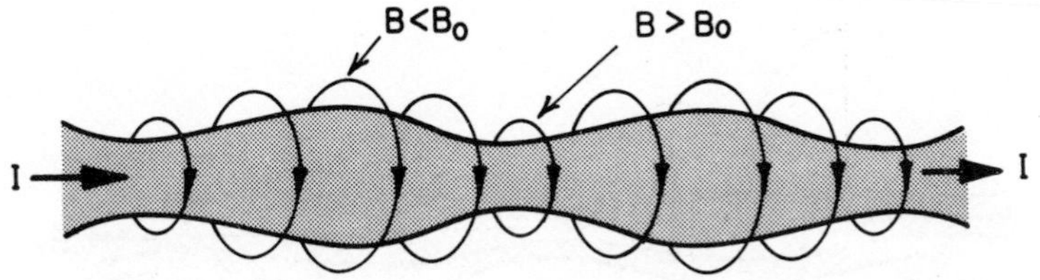

Fig. 12 - The sausage instability.

will be larger than the magnetic pressure at the expanded plasma surface and will force the surface radially outward, increasing the local expansion of the plasma. Therefore, the troughs will become deeper and the crests higher. The initial perturbation gives rise to forces that tend to further increase the initial disturbance, so that the initial equilibrium state is unstable. When the constrictions reach the axis, the column appears like a string of sausages and, for this reason, this type of instability has been called a *sausage* instability.

The sausage instability can be inhibited by a longitudinal magnetic field applied inside the plasma column. This longitudinal magnetic field is produced by passing a current through a solenoidal coil wound around the column. Because of the plasma high electric conducticity, the longitudinal field lines are frozen in the plasma. When the sausage distortion starts to grow, the longitudinal magnetic field lines are compressed at the constrictions, causing an increase in the pressure inside the plasma that opposes the increased magnetic pressure of the azimuthal field at the constricted surface, and forces the constriction to expand. At the locations where the column radius has increased, the longitudinal field lines move apart with the plasma expansion, thus decreasing the internal pressure, with the result that the net pressure forces the plasma surface radially inwards. This is illustrated schematically in Fig. 13.

We shall next determine what must be the magnitude B_z of the longitudinal magnetic flux density, as compared to the magnitude of the azimuthal B_Θ field, in order that the longitudinal field be able to stabilize the plasma column against the setting of the sausage instability. If the radius r of the column, at the constriction, is decreased by an amount dr, and considering that the magnetic flux $(\Phi_m = B_z \pi r^2)$ through the cross sectional

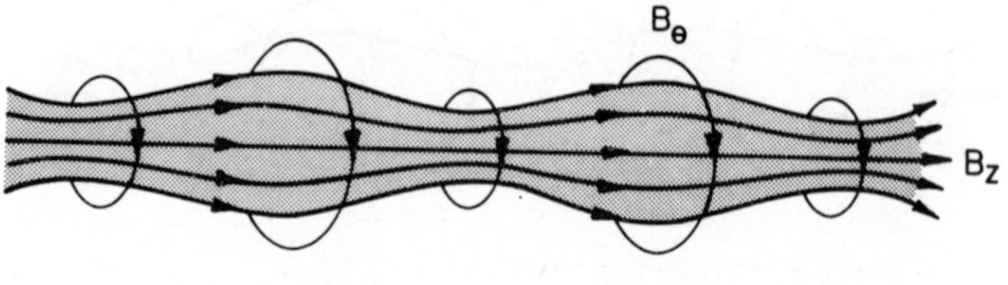

Fig. 13 - A longitudinal magnetic flux density B_z can be used to inhibit the sausage instability.

area of the column remains constant during compression, then we have

$$d\Phi_m = \pi r^2 \, dB_z + B_z \, 2\pi r \, dr = 0 \tag{6.1}$$

Hence, the longitudinal magnetic flux density is increased by the amount

$$dB_z = - 2 \, B_z \, dr/r \tag{6.2}$$

Consequently, the corresponding internal magnetic pressure increases by

$$dp_z = \frac{(B_z + dB_z)^2}{2\mu_0} - \frac{B_z^2}{2\mu_0} = \frac{1}{\mu_0} B_z \, dB_z \tag{6.3}$$

or, using (6.2),

$$dp_z = - (2B_z^2/\mu_0) \, dr/r \tag{6.4}$$

Considering now the azimuthal magnetic flux density B_Θ, it is easily seen, from Ampère's law, that external to the column we have

$$r \, B_\Theta(r) = \text{constant} \tag{6.5}$$

so that the azimuthal magnetic flux density, at the constricted surface, increases by the amount

$$dB_\Theta = - B_\Theta \, dr/r \tag{6.6}$$

Hence, the corresponding increase in the external magnetic pressure is

$$dp_\Theta = (B_\Theta/\mu_0) \, dB_\Theta = - (B_\Theta^2/\mu_0) \, dr/r \tag{6.7}$$

Therefore, in order that the plasma column be stable against the sausage distortion, we must have $dp_z > dp_\Theta$, or, using (6.4) and (6.7),

$$B_z^2 > B_\Theta^2/2 \tag{6.8}$$

7. THE KINK INSTABILITY

Another type of instability of the pinched plasma column is the so-called *kink instability*. The kink distortion consists of a perturbation in the form of a bend or kink in the column, but with the disturbed column maintaining its uniform circular cross section, as shown in Fig. 14. Usually there may be several kinks along the column length. In the neighborhood of the column, where the kink has developed, the magnetic field lines are brought closer together on the concave side, and separated on the convex side, so that the external magnetic pressure is increased on the concave side and decreased on the convex side. Therefore, the changes in the external magnetic pressure are in such a way as to accentuate the distortion still further. This type of distortion is therefore unstable.

The kink instability can be hindered by the application of a longitudinal magnetic field within the plasma column, as in the case of the sausage instability. In the kink distortion, the longitudinal magnetic field lines, frozen inside the plasma column, are stretched and the increased tension acting along the longitudinal magnetic field lines opposes the external forces. The net result is the stabilization of the column (Fig. 15).

In actual pratice, however, the plasma is not perfectly diamagnetic and other fields may also be present. The calculation of the stability of the pinched plasma column is not, in general, a very simple task.

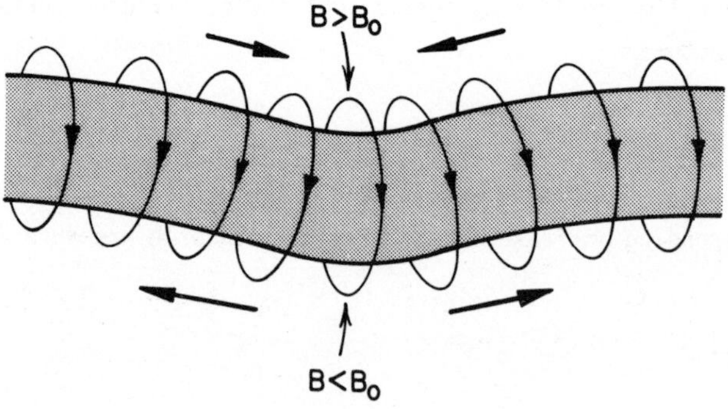

Fig. 14 - The kink instability.

8. CONVEX FIELD CONFIGURATIONS

In the linear pinch configuration, the azimuthal magnetic field confining the plasma column is produced by a longitudinal current flowing along the column. The configuration of this field is such that the magnetic flux lines are *concave* towards the plasma. Configurations of this type are unstable, as we have seen with the sausage and the kink instabilities.

Configurations for which the field lines are *convex* towards the plasma lead to a stable equilibrium, since the magnetic field strength increases in a direction away from the plasma. If the plasma surface is perturbed by a wave-like disturbance, the magnetic pressure at the crests will be larger than the internal kinetic pressure and the plasma is forced to return to its equilibrium configuration (assuming that the kinetic pressure is not affected by the perturbations). At the troughs, the internal kinetic pressure will be larger than the magnetic pressure acting on the plasma surface and will force the plasma to expand. Therefore, for plasma confinement, it is desirable to use a magnetic field configuration in which the magnetic flux lines are everywhere convex towards the plasma. An example of this type of configuration is the *cusp field*, which can be produced by an array of four current-carrying wires, as shown in Fig. 16. The presence of sharp edges and cusps at the plasma boundary, however, can lead to escape of the plasma particles. Although edges and cusps are characteristics of these configurations, modifications of the cusp field geometry are commonly employed for confinement of high temperature plasmas. Higher order cusp fields can be produced by lining up several pairs of current-carrying wires as, for example, in the *picket-fence* field geometry illustrated in Fig. 17.

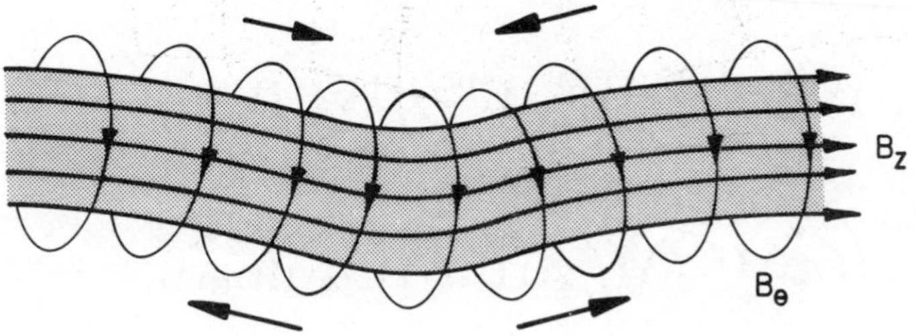

Fig. 15 - The increased tension of a longitudinal magnetic field, applied inside the column, inhibits the kink instability.

PROBLEMS

13.1 - The minimum intensity of the magnetic induction ($\vec{B}_o$) necessary to confine a plasma at an internal pressure of 100 atm is 5 Weber/m² (see Problem 12.3). Assuming that this field is produced by an axial current flowing in a cylindrical plasma column (as in the longitudinal pinch effect) of 10cm radius, show (applying Ampère's law) that the total current necessary to produce this magnetic field ($\vec{B}_o$) at the column surface is 2.5×10^6 Ampères. (1 atm = 10^5 Newton/m^2; $\mu_o = 4\pi \times 10^{-7}$ Henry/m).

13.2 - For the equilibrium Bennett pinch with cylindrical geometry, calculate $B_\theta(r)$ using (2.4) and the expression for $n(r)$ given in (3.8). Make a plot showing the radial distribution of $p(r)$, $J_z(r)$ and $B_\theta(r)$.

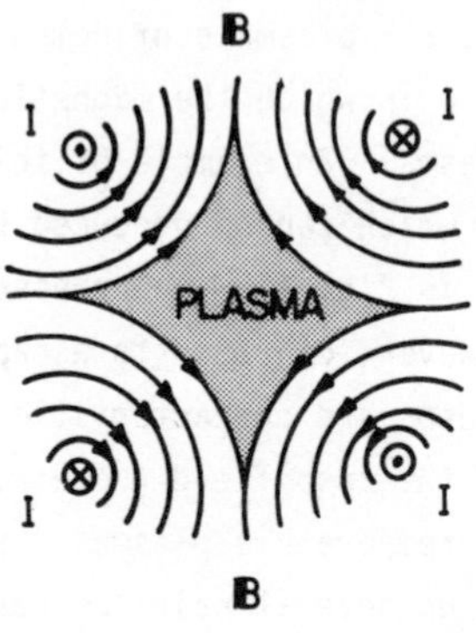

Fig. 16 - Plasma confinement by a cusped magnetic field, produced by four current-carrying wires.

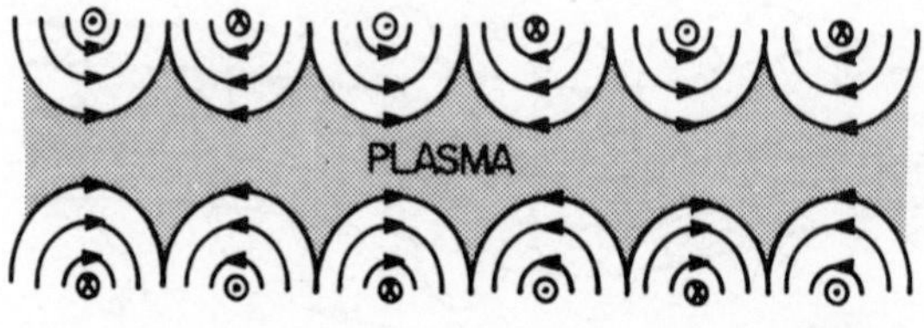

Fig. 17 - Picket-fence field configuration for magnetic confinement of a plasma.

13.3 - For the *equilibrium* theta-pinch produced by an azimuthal current in the theta-direction ($\vec{J}_\theta$), as illustrated in Fig. 6 of Chapter 12, determine expressions for the radial distributions of $J_\theta(r)$ and $p(r)$ in terms of $B_z(r)$. Draw a diagram illustrating these radial distributions for the special case when B_z is constant.

13.4 - Use the equation for the fluid velocity component ($\vec{u}_\perp$) normal to $\vec{B}$, derived in Problem 9.7, to determine the relative orientations of $\vec{u}$, $\vec{B}$, $\vec{E}$, $\vec{J}$ and $\vec{\nabla}p$ in a theta-pinch device.

13.5 - In the longitudinal equilibrium pinch shown schematically in Fig. 1, assume that the radial dependence of the current density $J_z(r)$ is such that

$$J_z = 0 \quad \text{for} \quad 0 < r < a$$

$$J_z = J_o = \text{constant for } a < r < b$$

$$J_z = 0 \quad \text{for} \quad r > b$$

Calculate $p(r)$ and $B_\theta(r)$ and make a plot showing the radial dependence. Show that, as $a \to b$, the magnetic pressure $B_\theta^2/2\mu_o$ at $r = b$ becomes equal to p at $r = 0$, while as $a \to 0$, B_θ^2/μ_o at $r = b$ becomes equal to p at $r = 0$.

13.6 - (a) Show that a *force-free* magnetic field satisfies the relation

$$(\vec{\nabla} \times \vec{B}) \times \vec{B} = 0$$

(b) Let $\vec{\nabla} \times \vec{B} = \alpha(\vec{r})\,\vec{B}$ and show that

$$\vec{B} \cdot \vec{\nabla}\alpha = 0$$

(c) Verify that the surfaces α = constant are made up of magnetic field lines.

(d) Show that $\alpha(\vec{r})$, as defined in part (b), for the force-free field, can be expressed as

$$\alpha = \hat{B} \cdot (\vec{\nabla} \times \hat{B})$$

(e) Prove that for the force-free field $\vec{\nabla} B$ lies on the osculating plane ($\bar{B}$, $\hat{n}$ plane, where $\hat{n}$ is the principal normal to the field line).

13.7 - Consider the following basic equation for the equilibrium of a plasma column with cylindrical geometry (see Problem 12.9)

$$\frac{d}{dr}\left(p + \frac{B_z^2}{2\mu_o} + \frac{B_\theta^2}{2\mu_o} \right) = -\frac{1}{\mu_o}\frac{B_\theta^2}{r}$$

(a) Verify that, for the θ-pinch, this equation reduces to

$$p + B_z^2/2\mu_o = \text{constant}$$

whereas, for the longitudinal pinch, it becomes

$$\frac{d}{dr}(p + B_\theta^2/2\mu_o) = -B_\theta^2/\mu_o r$$

(b) For the cylindrical screw-pinch, in which both B_θ and B_z are nonzero, assume that the longitudinal current density and the kinetic pressure are given, respectively, by

$$J_z(r) = J_o\ (1 - r^2/a^2) \qquad (r \leq a)$$

$$p(r) = p_o = \text{constant} \qquad (r < a)$$

Verify that

$$B_\theta(r) = (\mu_o J_o/2)\ (1 - r^2/2a^2)\ r \qquad (r \leq a)$$

$$B_\theta(r) = (\mu_o J_o a^2/4)\ (1/r) \qquad (r \geq a)$$

Show that $B_z(r)$ is given by

$$\frac{B_z^2}{2\mu_o} = -\frac{B_\theta^2}{2\mu_o} - \frac{1}{\mu_o}\int_r \frac{B_\theta^2(r)}{r}\,dr$$

From this equation determine $B_z(r)$ and make a plot showing $p(r)$, $J_z(r)$, $B_\theta(r)$ and $B_z(r)$ as a function of r.

CHAPTER 14

Electromagnetic Waves in Free Space

1. THE WAVE EQUATION

Plasmas are able to sustain a great variety of *wave phenomena*. Before we get started in the study of wave phenomena in plasmas, we review in this chapter some of the basic features of electromagnetic waves propagating in free space. The starting point for deriving the partial differential equation for electromagnetic waves in free space is Maxwell equations, which for free space ($\rho = 0$ and $\vec{J} = 0$) may be written as

$$\vec{\nabla} \cdot \vec{E} = 0 \tag{1.1}$$

$$\vec{\nabla} \cdot \vec{B} = 0 \tag{1.2}$$

$$\vec{\nabla} \times \vec{E} = - \partial\vec{B}/\partial t \tag{1.3}$$

$$\vec{\nabla} \times \vec{B} = (1/c^2)\, \partial\vec{E}/\partial t \tag{1.4}$$

Taking the time derivative of both sides of (1.4), yields

$$\vec{\nabla} \times (\partial\vec{B}/\partial t) = (1/c^2)\, \partial^2\vec{E}/\partial t^2 \tag{1.5}$$

Substituting $\partial\vec{B}/\partial t$ from (1.3), gives

$$- \vec{\nabla} \times (\vec{\nabla} \times \vec{E}) = (1/c^2)\, \partial^2\vec{E}/\partial t^2 \tag{1.6}$$

Using the vector operator identity

$$\vec{\nabla} \times (\vec{\nabla} \times \vec{E}) = \vec{\nabla}\, (\vec{\nabla} \cdot \vec{E}) - \nabla^2\, \vec{E} \tag{1.7}$$

and noting that $\vec{\nabla} \cdot \vec{E} = 0$ in free space, we obtain

$$\nabla^2 \vec{E} - (1/c^2) \quad \partial^2\vec{E}/\partial t^2 = 0 \tag{1.8}$$

In a similar way we can perform the same operations on (1.3) to obtain

$$\nabla^2 \vec{B} - (1/c^2)\partial^2\vec{B}/\partial t^2 = 0 \tag{1.9}$$

Eqs. (1.8) and (1.9) are the *vector* wave equations satisfied by the electromagnetic field vectors $\vec{E}$ and $\vec{B}$ in free space. The velocity of propagation of such waves is $c = 1/\sqrt{\mu_0\varepsilon_0}$. Since these equations are satisfied by each component of the field vectors $\vec{E}$ and $\vec{B}$, we way as well write a *scalar* wave equation

$$\nabla^2\psi - (1/c^2)\partial^2\psi/\partial t^2 = 0 \tag{1.10}$$

where ψ is used to denote any one of the components of $\vec{E}$ and $\vec{B}$.

2. SOLUTION IN PLANE WAVES

We are interested in *transverse plane wave* solutions of the partial differential wave equation (1.10), since these are the simplest and most fundamental electromagnetic waves. For transverse plane waves, the field vectors $\vec{E}$ and $\vec{B}$ lye on a plane perpendicular to the propagation direction and are functions only of the perpendicular distance from the origin to this plane and, of course, of time also. This plane, normal to the propagation direction, is called the *wave front*. Let ζ denote the perpendicular distance from the origin to the wave front plane and let $\hat{k}$ be a vector normal to this plane (Fig. 1). Any point P, in the wave front, can be represented by a position vector $\vec{r}$ drawn form the origin of the coordinate system. Thus, for any point in a given wave front we have $\hat{k} \cdot \vec{r} = \zeta$ = constant, which is the equation specifying this wave front plane. The direction cosines of the unit vector $\hat{k}$, in a Cartesian coordinate system, are given by $\hat{k} \cdot \hat{x}$, $\hat{k}.\hat{y}$ and $\hat{k}.\hat{z}$.

Since the field vectors $\vec{E}$ and $\vec{B}$ are spatially constant along the wave front (normal to $\hat{k}$), but vary only in the ζ direction (and with time), the del operator can be written as

$$\vec{\nabla} = \hat{x}\ \partial/\partial x + \hat{y}\ \partial/\partial y + \hat{z}\ \partial/\partial z = \hat{k}\ \partial/\partial\zeta \tag{2.1}$$

Hence, the wave equation is *one-dimensional* in form for the variable ζ,

$$\partial^2\psi/\partial\zeta^2 - (1/c^2)\,\partial^2\psi/\partial t^2 = 0 \tag{2.2}$$

The general solution of this one-dimensional wave equation is a combination of arbitrary functions of the variables $(\zeta - ct)$ and $(\zeta + ct)$,

$$\psi(\zeta,t) = f(\zeta - ct) + g(\zeta + ct) \tag{2.3}$$

The function $f(\zeta - ct)$ represents a plane waveform propagating in the positive ζ direction, whereas $g(\zeta + ct)$ corresponds to a waveform propagating in the negative ζ direction, the velocity of propagation being $c = 1/\sqrt{\mu_o \varepsilon_o}$.

3. HARMONIC WAVES

A particularly important type of plane waveform is the harmonic wave, which has the form (for propagation in the positive ζ direction)

$$\psi(\zeta,t) = A \cos[k(\zeta - ct)] = A \cos(k\zeta - \omega t) \tag{3.1}$$

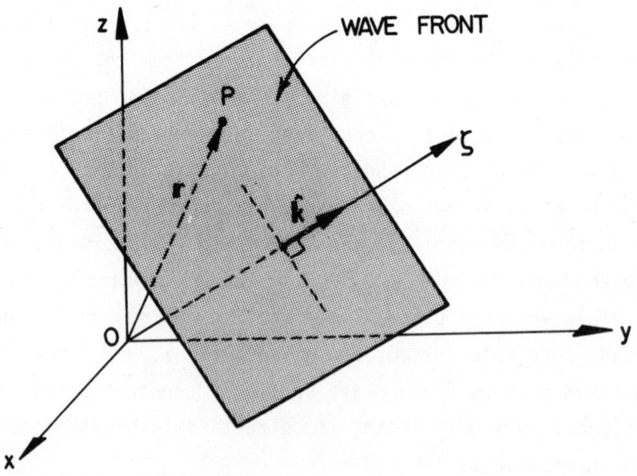

Fig. 1 - In a transverse plane wave the field vectors do not vary along a given wave front plane, normal to the direction of propagation ζ. They vary only along the ζ direction and with time.

where $\omega = kc$ is the *angular frequency* of the oscillation and k is the *wave number* or *propagation constant*. Fig. 2 shows the space and time dependence of the harmonic wave $\psi(\zeta,t)$ given by (3.1). The wavelength λ and the period T of the wave motion are given, respectively, by

$$\lambda = 2\pi/k \; ; \qquad T = 2\pi/\omega = 1/\nu \tag{3.2}$$

where ν is the frequency in cycles per second.

If we define a *propagation vector* $\vec{k}$, whose direction is that of the normal to the wave front ($\hat{k}$) and whose magnitude is the wave number ($|\vec{k}| = k$), then, since ζ is the perpendicular distance from the origin to the wave front, we have

$$\hat{k} \cdot \vec{r} = (\vec{k}/k) \cdot \vec{r} = \zeta \tag{3.3}$$

Hence, for a harmonic plane wave travelling in some arbitrary direction specified by the unit vector $\hat{k}$,

$$\psi(\vec{r},t) = A \cos (\vec{k} \cdot \vec{r} - \omega t) \tag{3.4}$$

In view of the argument of the cosine function (3.4), the planes of *constant phase* are defined by the condition

$$\vec{k} \cdot \vec{r} - \omega t = k\zeta - \omega t = \text{constant} \tag{3.5}$$

The *phase velocity* is defined as the velocity of propagation of these planes of constant phase ($d\zeta/dt$) and is found by differentiating (3.5) with respect

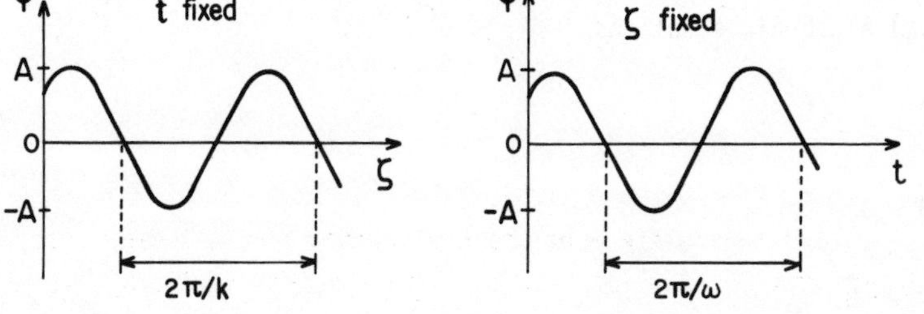

Fig. 2 - Amplitude of a harmonic plane wave as a function of space and time.

to time,

$$v_{ph} = \omega/k \tag{3.6}$$

The phase velocity is positive for a wave moving in the positive ζ direction, that is, ζ increases as t increases in order to keep $k\zeta - \omega t$ constant. If we had taken

$$\psi(\vec{r},t) = A \cos (\vec{k} \cdot \vec{r} + \omega t) \tag{3.7}$$

then the phase velocity would be negative for a wave moving in the negative ζ direction.

It is generally convenient and extremelly useful to write (3.4) in complex form

$$\psi(\vec{r},t) = A \exp [i(\vec{k} \cdot \vec{r} - \omega t)] \tag{3.8}$$

where implicit in this expression is the understanding that the field quantities are obtained by taking the real part of (3.8). The true physical quantity involved (the one we would measure in an experiment) is the real part of the complex form. The use of complex expressions, however, simplifies the mathematical calculations and, at the end, we take the real part of the result. This can be done as long as the equations are linear.

Using the complex notation (3.8) for the harmonic plane wave solutions for the field vectors $\vec{E}$ and $\vec{B}$, the operators $\vec{\nabla}$ and $\partial/\partial t$ become

$$\vec{\nabla} = i \vec{k} \quad ; \quad \partial/\partial t = - i\omega \tag{3.9}$$

so that Maxwell equations (1.1) to (1.4) in free space become simple (linear, homogeneous) algebraic equations

$$\vec{k} \cdot \vec{E} = 0 \tag{3.10}$$

$$\vec{k} \cdot \vec{B} = 0 \tag{3.11}$$

$$\vec{k} \times \vec{E} = \omega\vec{B} \tag{3.12}$$

$$\vec{k} \times \vec{B} = - (\omega/c^2) \vec{E} \tag{3.13}$$

Thus, from (3.10) and (3.11) we see that $\vec{E}$ and $\vec{B}$ are both perpendicular to $\vec{k}$ and, for this reason, these waves are called *transverse waves*. In addition, from (3.12) or (3.13) we see that $\vec{E}$ and $\vec{B}$ are also perpendicular to each other. This situation is illustrated in Fig. 3. The set of vectors ($\vec{E}$, $\vec{B}$, $\vec{k}$), taken in this order, constitute a right-handed orthogonal set. Fig. 4 illustrates the relation between the $\vec{E}$ and $\vec{B}$ vectors of a plane electromagnetic wave propagating in free space.

4. POLARIZATION

From (3.8) it is seen that the field vectors $\vec{E}$ and $\vec{B}$, which are solutions of the wave equation in the form of harmonic plane waves propagating in the positive ζ direction, can be written as

$$\vec{E}(\vec{r},t) = \vec{E}_0 \exp [i(\vec{k} \cdot \vec{r} - \omega t)] = \vec{E}_0 \exp [i(k\zeta - \omega t)] \tag{4.1}$$

$$\vec{B}(\vec{r},t) = \vec{B}_0 \exp [i(\vec{k} \cdot \vec{r} - \omega t)] = \vec{B}_0 \exp [i(k\zeta - \omega t)] \tag{4.2}$$

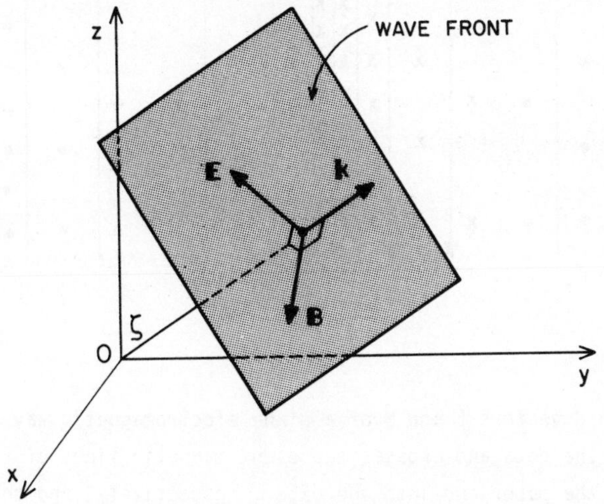

Fig. 3 - The propagation vector $\vec{k}$ and the two field vectors $\vec{E}$ and $\vec{B}$ are orthogonal to one another.

where $\vec{E}_0$ and $\vec{B}_0$ are constant vector amplitudes, which may be complex quantities. The phenomenon of polarization can be discussed entirely from the point of view of the electric field vector, since the magnetic induction vector may always be obtained from $\vec{E}$ using (3.12) for the case of plane waves.

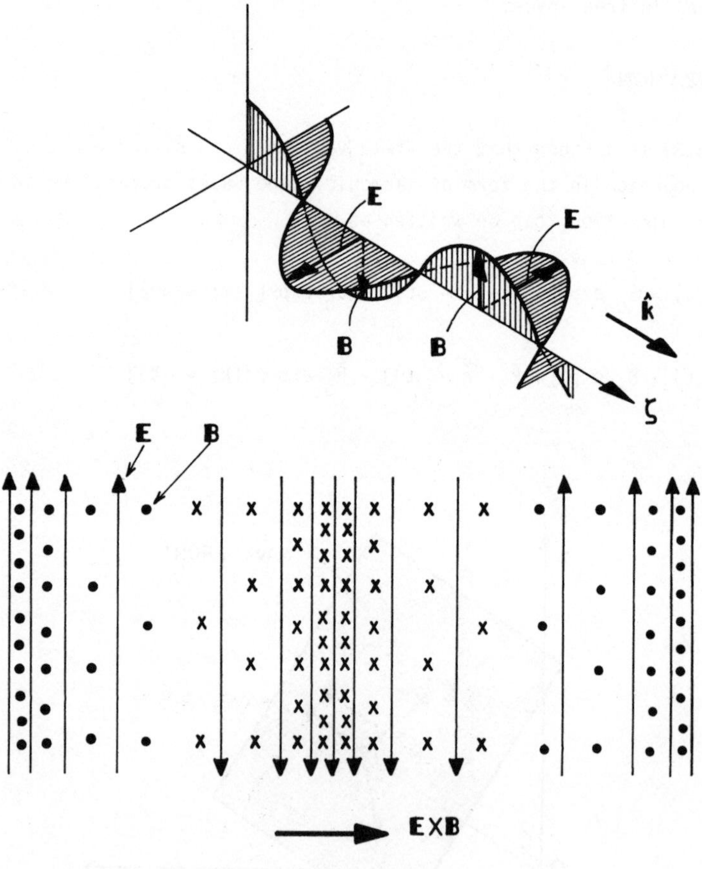

Fig. 4 - The field vectors $\vec{E}$ and $\vec{B}$ of a plane electromagnetic wave in free space. The dots and crosses represent magnetic lines of force coming out of the paper and into the paper, respectively, and the vertical lines represent the electric field. The direction of propagation is given by $\hat{k}$ which, in free space, is along $\vec{E} \times \vec{B}$.

According to (4.1), the direction of the electric field vector $\vec{E}$ is always the same, and the wave is said to be *linearly polarized*. This is the simplest type of polarization state.

In general, in order to describe an arbitrary state of polarization, we must consider the electric field vector in a plane as the superposition of two linearly independent, linearly polarized waves. Two such linearly independent waves can be represented by

$$\vec{E}_1(\vec{r},t) = \hat{e}_1 \ E_1 \ \exp\left[i(\vec{k} \cdot \vec{r} - \omega t)\right] \tag{4.3}$$

$$\vec{E}_2(\vec{r},t) = \hat{e}_2 \ E_2 \ \exp\left[i(\vec{k} \cdot \vec{r} - \omega t)\right] \tag{4.4}$$

with the associated magnetic induction vectors given, respectively, by

$$\vec{B}_1(\vec{r},t) = (1/\omega) \ \vec{k} \times \vec{E}_1(\vec{r},t) \tag{4.5}$$

$$\vec{B}_2(\vec{r},t) = (1/\omega) \ \vec{k} \times \vec{E}_2(\vec{r},t) \tag{4.6}$$

The unit vectors $\hat{e}_1$ and $\hat{e}_2$, called the *polarization vectors*, are perpendicular to each other and lie in the plane normal to the propagation vector $\hat{k}$, and are such that $(\hat{e}_1, \hat{e}_2, \hat{k})$, taken in this order, form a right-handed orthogonal set of unit vectors (Fig. 5). The amplitudes E_1 and E_2 are complex numbers, in

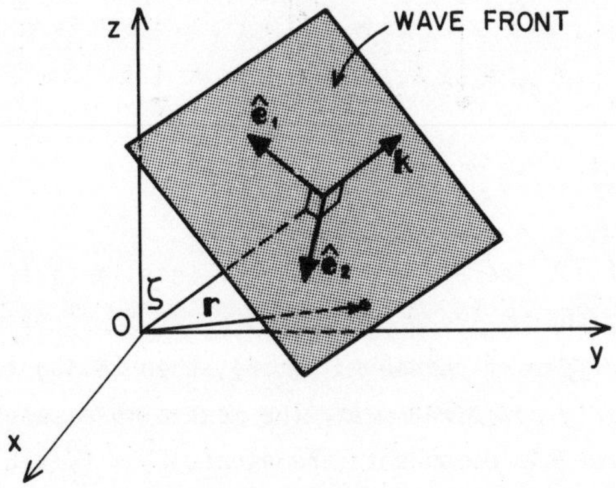

Fig. 5 - The two orthogonal polarization vectors $\hat{e}_1$ and $\hat{e}_2$, and the propagation vector $\hat{k}$.

order to allow for the possibility of phase difference between the two waves $\vec{E}_1(\vec{r},t)$ and $\vec{E}_2(\vec{r},t)$. Therefore, a general solution for a harmonic plane wave propagating in the direction $\hat{k}$ can be written as a linear combination of the two linearly independent solutions $\vec{E}_1$ and $\vec{E}_2$,

$$\vec{E}(\vec{r},t) = (\hat{e}_1\, E_1 + \hat{e}_2\, E_2) \exp [i(\vec{k} \cdot \vec{r} - \omega t)] \tag{4.7}$$

Since any complex quantity can be expressed as the product of a real quantity and a complex phase factor, we way write

$$E_1 = E_1^0 \exp (i\alpha_1) \tag{4.8}$$

$$E_2 = E_2^0 \exp (i\alpha_2) \tag{4.9}$$

where the amplitudes E_1^0 and E_2^0 are real numbers, and α_1 and α_2 represent the phase of the complex amplitudes E_1 and E_2, respectively. Note that it is only the *phase difference* that is physically significant and not the absolute magnitudes of α_1 and α_2. Hence,(4.7) becomes

$$\vec{E}(\vec{r},t) = [\hat{e}_1\, E_1^0 \exp (i\alpha_1) + \hat{e}_2\, E_2^0 \exp (i\alpha_2)] \exp [i(\vec{k} \cdot \vec{r} - \omega t)] \tag{4.10}$$

If E_1 and E_2 have the *same phase* ($\alpha_1 = \alpha_2$), then (4.10) represents a *linearly polarized wave*. The magnitude of $\vec{E}$ is

$$E = (E_1^2 + E_2^2)^{1/2} \tag{4.11}$$

and its polarization vector makes an angle

$$\theta = \tan^{-1} (E_2/E_1) \tag{4.12}$$

with the direction of $\hat{e}_1$. The electric field of this linearly polarized wave is represented in Fig. 6.

If E_1 and E_2 have *different phases* ($\alpha_1 \neq \alpha_2$), then (4.10) represents, in general, an *elliptically polarized wave*. The most simple case is the one in which the amplitudes of the components are equal, $E_1^0 = E_2^0 = E^0$, but the phases differ by $\pi/2$, that is, $\alpha_1 = 0$ and $\alpha_2 = \pm\, \pi/2$. Then, the wave (4.10) becomes [since $\exp (\pm i\pi/2) = \pm i$],

$$\vec{E}(\vec{r},t) = E^0 \ (\hat{e}_1 \pm i\hat{e}_2) \exp [i(\vec{k} \cdot \vec{r} - \omega t)] \tag{4.13}$$

This wave is said to be *circularly polarized*. In order to illustrate this point, consider a Cartesian coordinate system such that the wave is propagating in the z direction, $\hat{k} = \hat{z}$, with $\hat{e}_1 = \hat{x}$ and $\hat{e}_2 = \hat{y}$. Then, taking the *real part* of (4.13) we obtain for the x and y components of the *actual* electric field,

$$\vec{E}_x(z,t) = E^0 \ \hat{x} \cos (kz - \omega t) \tag{4.14}$$

$$\vec{E}_y(z,t) = \mp E^0 \ \hat{y} \sin (kz - \omega t) \tag{4.15}$$

At a fixed plane z = constant, the fields are such that the $\vec{E}$ vector has a constant magnitude E^0, but rotates around in a circle at the frequency ω. For the upper sign $(\hat{e}_1 + i\hat{e}_2)$ the direction of rotation is *clockwise* if the wave is observed by looking at the outgoing wave front (i.e. looking along the positive z direction), as shown in Fig. 7 (a). Such a wave is said to be *right circularly polarized*. For the lower sign $(\hat{e}_1 - i\hat{e}_2)$ the rotation is in the *counterclockwise* direction as shown in Fig. 7 (b), and the wave is said to be *left circularly polarized.*

In the most general case the amplitudes are different ($E_1^0 \neq E_2^0$) and the fields $\vec{E}_1$ and $\vec{E}_2$ have an arbitrary phase difference ($\alpha_1 \neq \alpha_2$). Considering

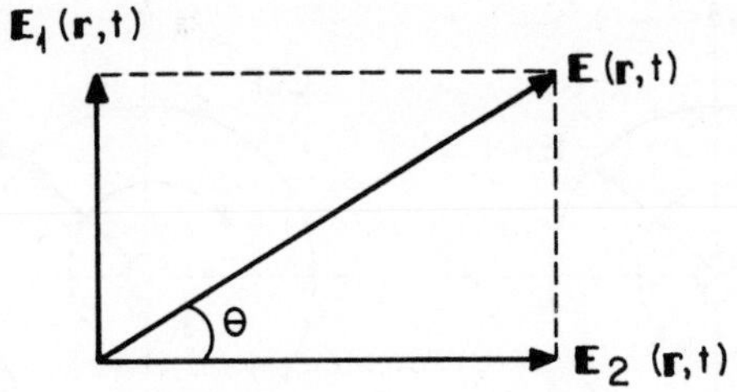

Fig. 6 - Electric field vector of a linearly polarized wave, represented by the superposition of two independent, linearly polarized waves $\vec{E}_1$ and $\vec{E}_2$, having the same phase.

$(\hat{e}_1, \hat{e}_2, \hat{k}) = (\hat{x}, \hat{y}, \hat{z})$, we have, from (4.10),

$$\vec{E}_x(z,t) = \hat{x}\, E_x^0 \exp\,(i\alpha_1) \exp\,[i(kz - \omega t)] \tag{4.16}$$

$$\vec{E}_y(z,t) = \hat{y}\, E_y^0 \exp\,(i\alpha_2) \exp\,[i(kz - \omega t)] \tag{4.17}$$

Taking the *real part* of these expressions we obtain

$$E_x(z,t) = E_x^0 \cos\,(kz - \omega t + \alpha_1) \tag{4.18}$$

$$E_y(z,t) = E_y^0 \cos\,(kz - \omega t + \alpha_2) \tag{4.19}$$

Squaring and adding (4.18) and (4.19), yields

$$\left(\frac{E_x}{E_x^0}\right)^2 + \left(\frac{E_y}{E_x^0}\right)^2 = \cos^2(kz - \omega t + \alpha_1) + \cos^2(kz - \omega t + \alpha_2) \tag{4.20}$$

This result shows that at a fixed plane z = constant the electric field vector

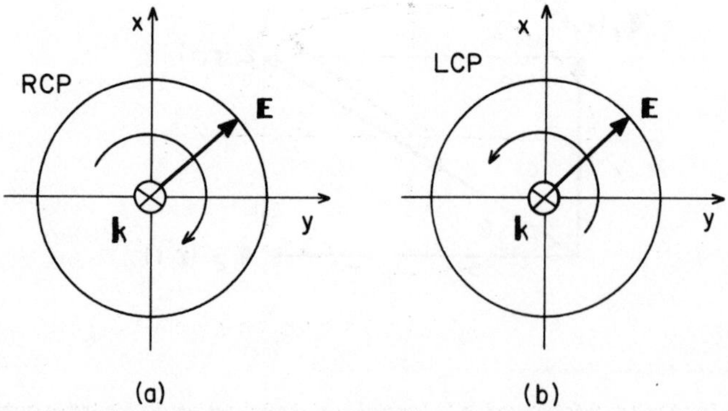

Fig. 7 - Electric field vector of a right circularly polarized wave (a), and a left circularly polarized wave (b). The propagation vector $\vec{k}$ points into the paper.

$\vec{E}$ performs, in general, an elliptical motion as a function of time.

For the case when $\alpha_1 = 0$ and $\alpha_2 = \pi/2$, (4.20) becomes, at the plane $z = 0$,

$$\left(\frac{E_x}{E_x^0}\right)^2 + \left(\frac{E_y}{E_y^0}\right)^2 = \cos^2(-\omega t) + \cos^2(-\omega t + \pi/2) = 1$$

since for any angle ϕ we have $\cos(\phi + \pi/2) = -\sin(\phi)$. In this case, the x and y components of the electric field are given by

$$E_x(0,t) = E_x^0 \cos(\omega t) \tag{4.22}$$

$$E_y(0,t) = E_y^0 \sin(\omega t) \tag{4.23}$$

Therefore, the tip of the electric field vector traces an ellipse in the x - y plane rotating in the *clockwise* direction *(right-hand polarization)* when the wave is propagating away from the observer, as illustrated in Fig. 8.

For the case when $\alpha_1 = 0$ and $\alpha_2 = -\pi/2$, the electric field vector rotates in the *counterclockwise* direction when the wave is propagating away from the

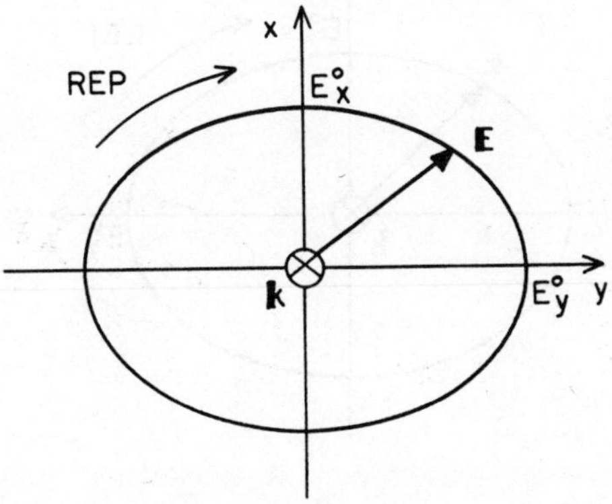

Fig. 8 - Electric field vector of a right-hand elliptically polarized wave ($\alpha_1 = 0$, $\alpha_2 = \pi/2$), rotating in the clockwise direction for an observer looking at the outgoing wave ($\vec{k}$ points into the paper).

observer, and the wave is said to be *left-hand elliptically polarized,* as shown in Fig. 9.

The ellipticity of the ellipse traced and the orientation of the major axis with respect to the x axis, depend on the relative magnitude of the amplitudes E_x^0 and E_y^0, and on the phase difference $(\alpha_2 - \alpha_1)$. Fig. 10 illustrates various wave polarizations, obtained from (4.18) and (4.19) at the plane z = 0, for the case when $E_x^0 = E_y^0$. The angle ϕ indicates the phase difference, $\phi = (\alpha_2 - \alpha_1)$. In the figures shown, the sense of rotation of the $\vec{E}$ vector in the x - y plane depends on whether ϕ lies between 0 and π, or between π and 2π .

5. ENERGY FLOW

Associated with the electric and magnetic field vectors $\vec{E}$ and $\vec{H}$ of an electromagnetic wave, there is a *flow of energy* (energy per unit area and per unit time) in the direction perpendicular to both $\vec{E}$ and $\vec{H}$. This energy flow is given the *Poynting vector*

$$\vec{S} = \vec{E} \times \vec{H} \tag{5.1}$$

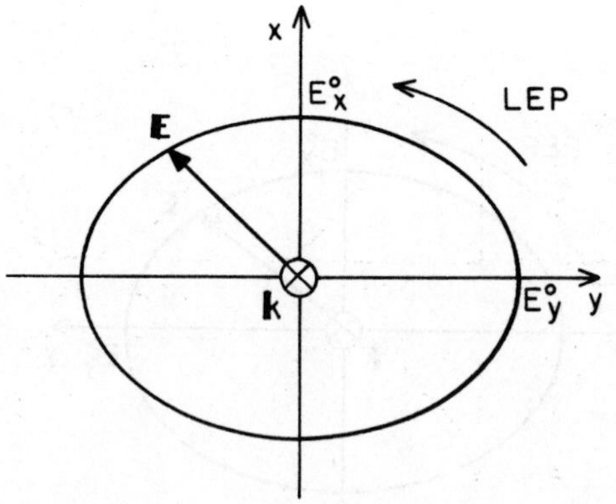

Fig. 9 - Electric field vector of a left-hand elliptically polarized wave ($\alpha_1 = 0, \alpha_2 = -\pi/2$), rotating in the counterclockwise direction for an observer looking at the outgoing wave ($\vec{k}$ points into the paper).

In free space $\vec{H}$ and $\vec{B}$ are related by $\vec{B} = \mu_0 \vec{H}$. Also, in free space the field vectors $\vec{E}$ and $\vec{H}$, and the propagation vector $\vec{k}$, form a mutually orthogonal set of vectors, but this is not the case in a conducting media as, for example, in a plasma. In such a medium, because of the presence of polarization charges, Maxwell equation $\vec{\nabla} \cdot \vec{E} = \rho/\varepsilon_0$ requires that $\vec{k} \cdot \vec{E} \neq 0$. Nevertheless, the direction of energy flow, given by the Poynting vector, is always perpendicular to $\vec{E}$ and $\vec{H}$.

When the field vectors $\vec{E}$ and $\vec{H}$ are expressed in terms of complex quantities, the energy flow vector given by (5.1) is called the *complex Poynting vector*. In this case, however, in order to obtain real physical quantities some caution must be taken when combining complex quantities. The real energy flow in the wave is obtained by using the *real part of* both $\vec{E}$ and $\vec{H}$ in (5.1).

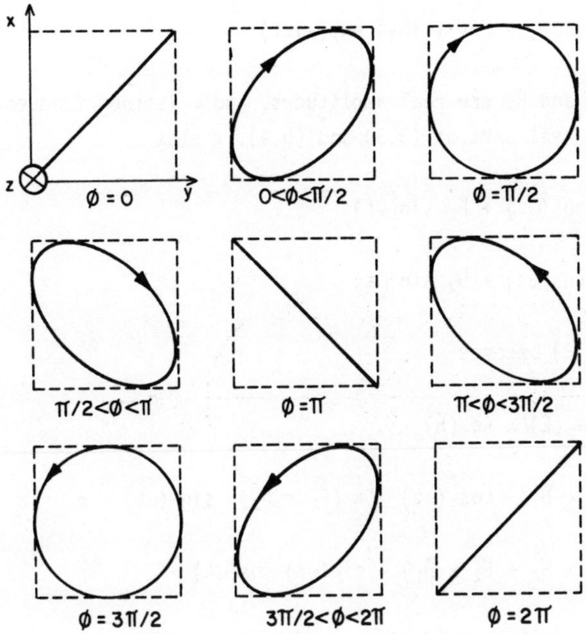

Fig. 10 - Various wave polarizations for the case when $E_x^0 = E_y^0$ and $(\alpha_2 - \alpha_1) = \phi$, obtained from (4.18) and (4.19) at the plane $z = 0$.

For the harmonic waves, the quantity of interest is the *average* energy flow (over one cycle) given by

$$< \vec{S} > = < \text{Re}\,(\vec{E}) \times \text{Re}\,(\vec{H}) > \tag{5.2}$$

where the time average is obtained by integrating the quantity over one period of the harmonic wave and dividing by the period, and where Re means "the real part of". It must be noted that the average energy flow $< \vec{S} >$ is *not* given by $< \text{Re}\,(\vec{E} \times \vec{H}) >$, nor by $\text{Re} < \vec{E} \times \vec{H} >$.

In what follows we show that the time-averaged energy flow $< \vec{S} >$, given by (5.2), can also be computed directly from the complex field vectors without performing an average over one cycle of the oscillation. For this purpose let us write

$$\vec{E} = \vec{E}_0 \exp(-i\omega t) = (\vec{E}_1 + i\vec{E}_2) \exp(-i\omega t) \tag{5.3}$$

$$\vec{H} = \vec{H}_0 \exp(-i\omega t) = (\vec{H}_1 + i\vec{H}_2) \exp(-i\omega t) \tag{5.4}$$

where $\vec{E}_1$, $\vec{E}_2$, $\vec{H}_1$ and $\vec{H}_2$ are real amplitudes, and ω is the frequency of the wave. Taking the real part of (5.3) and (5.4), yields

$$\text{Re}\,(\vec{E}) = \vec{E}_1 \cos(\omega t) + \vec{E}_2 \sin(\omega t) \tag{5.5}$$

$$\text{Re}\,(\vec{H}) = \vec{H}_1 \cos(\omega t) + \vec{H}_2 \sin(\omega t) \tag{5.6}$$

Consequently, (5.2) becomes

$$\begin{aligned} < \vec{S} > &= < \text{Re}\,(\vec{E}) \times \text{Re}\,(\vec{H}) > \\ &= (\vec{E}_1 \times \vec{H}_1) < \cos^2(\omega t) > + (\vec{E}_2 \times \vec{H}_2) < \sin^2(\omega t) > + \\ &\quad + (\vec{E}_1 \times \vec{H}_2 + \vec{E}_2 \times \vec{H}_1) < \sin(\omega t) \cos(\omega t) > \end{aligned} \tag{5.7}$$

The following integrals are easily verified,

$$< \cos^2(\omega t) > = \frac{1}{T} \int_0^T \cos^2(\omega t)\, dt = \frac{1}{2} \tag{5.8}$$

$$< \sin^2(\omega t) > = \frac{1}{T} \int_0^T \sin^2(\omega t)\, dt = \frac{1}{2} \tag{5.9}$$

$$< \sin(\omega t) \cos(\omega t) > = \frac{1}{T} \int_0^T \sin(\omega t) \cos(\omega t)\, dt = 0 \tag{5.10}$$

Therefore,

$$< \vec{S} > = (\vec{E}_1 \times \vec{H}_1 + \vec{E}_2 \times \vec{H}_2)/2 \tag{5.11}$$

Let us consider now the following quantity

$$\text{Re } (\vec{E} \times \vec{H}^*) = \text{Re } \{[(\vec{E}_1 + i\vec{E}_2)(\cos \omega t - i \sin \omega t)] \times$$

$$\times \; [(\vec{H}_1 - i \vec{H}_2)(\cos \omega t - i \sin \omega t)]\}$$

$$= \vec{E}_1 \times \vec{H}_1 + \vec{E}_2 \times \vec{H}_2 \tag{5.12}$$

Comparing this result with (5.11) we obtain the following alternative form for the time-averaged energy flow (5.2)

$$< \vec{S} > = (1/2) \text{ Re } (\vec{E} \times \vec{H}^*) \tag{5.13}$$

In a medium (not free space) the time-averaged energy flow is given by

$$< \vec{S} > = (c/4) \, (\vec{E}.\vec{D}^* + \vec{B}.\vec{H}^*) \, \hat{e}_3 \tag{5.14}$$

where $\hat{e}_3$ is a unit vector in the direction of $\vec{E} \times \vec{H}$.

6. WAVE PACKETS AND GROUP VELOCITY

So far we have considered simple waves having one specific value for k and ω. In practice, however, real disturbances consist of waves having some finite spread in the wave number k and in the frequency ω. A *wave packet* is a superposition of waves with different values of k and ω. This is equivalent to the statement that any periodic motion can be decomposed by Fourier analysis into a superposition of harmonic oscillations with different frequencies ω and wave numbers k.

A wave packet, consisting of a superposition of plane harmonic waves propagating in the ζ direction, can be represented by

$$\psi(\zeta,t) = \int_{-\infty}^{+\infty} A(k) \exp\left[i(k\zeta - \omega t)\right] dk \tag{6.1}$$

The wave packet concept is particularly useful when there is only a small spread in wave numbers about a central wave number k_0. This is equivalent to a small spread in frequencies about a central frequency ω_0, since for any wave there is a functional relationship between k and ω, which depends on the medium, and which is called the *dispersion relation*. Thus, the amplitude function A(k) is usually assumed to be peaked about some central wave number k_0. In (6.1) the wave number k has been taken as the independent variable and ω is considered to be a function of k, determined by the dispersion relation $\omega = \omega(k)$.

The amplitude function A(k) can be determined from the Fourier transform of $\psi(\zeta,0)$. For this purpose, we multiply (6.1) (with t = 0) by $\exp(-ik\zeta)$ and integrate the resultant equation over all possible values of ζ. Thus,

$$\int_{-\infty}^{+\infty} \psi(\zeta,0) \exp(-ik\zeta)\, d\zeta =$$

$$= \int_{-\infty}^{+\infty} A(k')\, dk' \int_{-\infty}^{+\infty} \exp\left[i(k' - k)\zeta\right] d\zeta \tag{6.2}$$

Using the following representation for the Dirac delta function

$$\delta(k' - k) = \frac{1}{2\pi} \int_{-\infty}^{+\infty} \exp\left[i(k' - k)\zeta\right] d\zeta \tag{6.3}$$

and the property

$$\int_{-\infty}^{+\infty} A(k')\ \delta(k' - k)\ dk' = A(k) \tag{6.4}$$

we obtain, from (6.2),

$$A(k) = \frac{1}{2\pi} \int_{-\infty}^{+\infty} \psi(\zeta,0) \exp\ (-i\ k\ \zeta)\ d\zeta \tag{6.5}$$

The field quantity $\psi(\zeta,t)$ can also be synthesized in terms of time periodic functions of all possible frequencies, using the relation

$$\psi(\zeta,t) = \int_{-\infty}^{+\infty} A(\omega) \exp\ [i(k\zeta - \omega t)]\ d\omega \tag{6.6}$$

where the amplitude function $A(\omega)$ gives the frequency spectrum. In this case, ω is taken as the independent variable and the dispersion relation gives $k = k(\omega)$. The amplitude function $A(\omega)$ is usually peaked about some central frequency ω_0 and it can be determined from the Fourier transform of $\psi(0,t)$,

$$A(\omega) = \frac{1}{2\pi} \int_{-\infty}^{+\infty} \psi(0,t) \exp\ (i\omega t)\ dt \tag{6.7}$$

Consider now that we have a wave packet represented by (6.1), in which the range of values of k is small and is centered about some specific wave number k_0,

$$k_0 - \delta k \leq k \leq k_0 + \delta k \tag{6.8}$$

If $\omega(k)$ is a slowly varying function of k, then $\omega(k)$ deviates only very slightly from its value $\omega_0 = \omega(k_0)$, and we can expand $\omega(k)$ in a Taylor series about k_0 retaining only the first two terms,

$$\omega(k) = \omega_0 + (k - k_0)\ (\partial\omega/\partial k)_{k_0} \tag{6.9}$$

The phase factor in (6.1) can thus be written as

$$[k\zeta - \omega(k)t] = k_0\zeta - \omega_0 t + (k - k_0)\ [\zeta - (\partial\omega/\partial k)_{k_0}\ t\ ] \tag{6.10}$$

This allows the wave packet $\psi(\zeta,t)$ of (6.1) to be recast in the form

$$\psi(\zeta,t) = \psi_m(\zeta,t) \exp\ [i(k_0\zeta - \omega_0 t)] \tag{6.11}$$

where

$$\psi_m(\zeta,t) = \int_{k_0 - \delta k}^{k_0 + \delta k} A(k) \exp\{i\ (k - k_0)\ [\zeta - (\partial\omega/\partial k)_{k_0}\ t]\ \}\ dk \tag{6.12}$$

Therefore, the wave packet $\psi(\zeta,t)$ corresponds to a carrier wave at the frequency ω_0, modulated by the amplitude function $\psi_m(\zeta,t)$. A typical form for a wave packet is shown in Fig. 11, at a fixed instant of time.

The phases of constant packet amplitude are given by

$$\zeta - (\partial\omega/\partial k)_{k_0}\ t = 0 \tag{6.13}$$

and the velocity of propagation of these planes of constant phase, $d\zeta/dt$, called the *group velocity,* is given by differentiating (6.13) with respect to time,

$$v_g = (\partial\omega/\partial k)_{k_0} \tag{6.14}$$

The group velocity can be written in terms of the phase velocity ω/k as

$$v_g = \left(\frac{\partial\omega}{\partial k}\right)_{k_0} = \left[\frac{\partial(v_{ph}k)}{\partial k}\right]_{k_0} = v_{ph_0} + k_0\left(\frac{\partial v_{ph}}{\partial k}\right)_{k_0} \tag{6.15}$$

If the medium is such that the phase velocity is not a function of k (i.e. $\omega(k)$ = constant x k), then the group velocity is equal to the phase velocity and the medium is *nondispersive.* This is the case of free space, for which $\omega = ck$. If the phase velocity decreases with increasing k (i.e. $\partial v_{ph}/\partial k < 0$), then v_g is less than v_{ph} and the medium is said to be normally *dispersive.* If the phase velocity increases with increasing k (i.e $\partial v_{ph}/\partial k > 0$), then v_g is larger than v_{ph} and the medium shows what is called *anomalous dispersion.*

It should be noted that, in many cases, the phase velocity of a wave in a plasma exceeds the velocity of light c. This fact is not in conflict with the relativity theory, because an infinitely long wave train of constant amplitude does not carry information. This information, however, is contained in the modulation of the carrier wave, and this modulation travels at the group velocity, which is always less than the velocity of light.

In obtaining (6.11) and (6.12), the Taylor expansion of $\omega(k)$ was carried out only up to a term linear in $(k - k_0)$. In this case the spatial variation of the amplitude function $\psi_m(\zeta,t)$, which gives the shape of the wave packet, remains the same at any time. If the higher order terms in $(k - k_0)$ are included in (6.10), then the wave packet shape will change with time and the packet will spread out. Hence, the group velocity concept is useful only for a wave packet with a very small spread in wave numbers and frequencies.

PROBLEMS

14.1 - Derive the vector wave equations, analogous to (1.8) and (1.9) for the electric ($\vec{E}$) and magnetic ($\vec{B}$) fields, for a medium in which there is a space charge density distribution $\rho(\vec{r},t)$ and a charge current density distribution $\vec{J}(\vec{r},t)$.

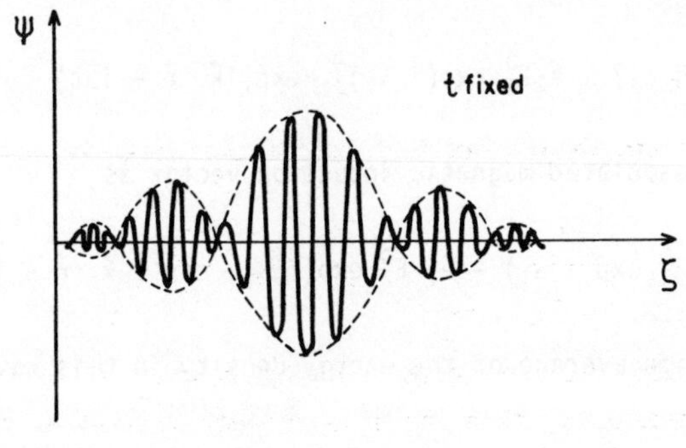

Fig. 11 - Amplitude function $\psi(\zeta,t)$ of a typical wave packet. The dotted curve is the amplitude modulation function $\psi_m(\zeta,t)$.

14.2 - Consider a plane electromagnetic harmonic wave travelling toward the positive x direction in free space, having the frequency $\nu = 5 \times 10^{15}$ Hertz.

(a) What is the wavelength λ in Angstroms?

(b) Calculate the corresponding values of ω and k.

(c) Considering that

$$\vec{E} = \hat{y}\, E_0 \exp(ikx - i\omega t)$$

calcule $\vec{\nabla} \times \vec{E}$ and $-\partial\vec{B}/\partial t$, and compare the results. Calculate also $\vec{\nabla} \times \vec{B}$ and $\partial\vec{E}/\partial t$, and compare the results.

(d) Calculate the Poynting vector $\vec{S}$ and the average Poynting vector $<\vec{S}>$.

14.3 - Consider a plane electromagnetic harmonic wave propagating along the positive x direction in free space, which can be decomposed in the sum of two waves:

$$\vec{E} = \hat{y}\, E_1 \exp(ikx - i\omega t) + \hat{z}\, E_2 \exp(ikx - i\omega t + i\alpha)$$

Show that

$$\vec{S} = \hat{x}\, (\varepsilon_0/\mu_0)^{1/2} [E_1^2 \cos^2(kx - \omega t) + E_2^2 \cos^2(kx - \omega t + \alpha)]$$

$$< \vec{S} > = \hat{x}\, (\varepsilon_0/\mu_0)^{1/2}\, (E_1^2 + E_2^2)/2$$

14.4 - The electric field vector of an elliptically polarized plane wave, propagating in free space, can be expressed as

$$\vec{E} = [\hat{e}_1 E_1^0 \exp(i\,\alpha_1) + \hat{e}_2 E_2^0 \exp(i\,\alpha_2)]\ \exp(i\vec{k}.\,\vec{r} - i\omega t)$$

(a) Show that the associated magnetic induction vector is

$$\vec{B} = (1/c)\ [\hat{e}_2\, E_1^0 \exp(i\,\alpha_1) - \hat{e}_1\, E_2^0 \exp(i\alpha_2)] \exp(i\,\vec{k}.\,\vec{r} - i\omega t)$$

(b) Show that the time average of the energy density in this wave is

$$< W > = (1/4)\ (\varepsilon_0\ E^2 + B^2/\mu_0)$$

Note that $E^2 = (E_1^0)^2 + (E_2^0)^2 = c^2\, B^2$.

(c) Show that the time-averaged energy flow $<\vec{S}>$ is equal to the phase velocity ($\omega/k = c$) times the average energy density of the wave, that is,

$$< \vec{S} > = c < W > \hat{E} \times \hat{H} = (c/4)\ (\varepsilon_0 E^2 + B^2/\mu_0)\ \hat{E} \times \hat{H}$$

14.5 - Show that in a medium (not free space), the time average of the energy density (over one cicle) for harmonic waves is given by

$$< W > = (1/4)\ (\vec{E}\ .\ \vec{D}^* + \vec{B}\ .\ \vec{H}^*)$$

Consequently, show that the average Poynting vector (energy flow) is given by

$$< \vec{S} > = (c/4)\ (\vec{E}\ .\ \vec{D}^* + \vec{B}\ .\ \vec{H}^*)\ \hat{E} \times \hat{H}$$

14.6 - Consider the superposition ($\vec{E} = \vec{E}_1 + \vec{E}_2$) of the following waves:

$$\vec{E}_1 = \hat{e}_1\ E_1^0 \exp\ (i\alpha_1) \exp\ (i\vec{k}\ .\ \vec{r} - i\omega t)$$

$$\vec{E}_2 = \hat{e}_2\ E_2^0 \exp\ (i\alpha_2) \exp\ (i\vec{k}\ .\ \vec{r} - i\omega t)$$

Analyze the resultant polarization for the following cases:

(a) $\alpha_2 = \alpha_1$ and $E_2^0 \neq E_1^0$

(b) $\alpha_2 = \alpha_1 \pm \pi/2$ and $E_2^0 = E_1^0 = E^0$

(c) $\alpha_2 = \alpha_1 \pm \pi/2$ and $E_2^0 > E_1^0$

(d) $\alpha_2 = \alpha_1 + \phi$ and $E_2^0 = E_1^0$, with $\alpha_1 = 0$ and ϕ varying from 0 to 2π

(e) Same as in (d), but with $E_2^0 > E_1^0$ and $\alpha_1 = \pi/2$.

14.7 - Generalize Eqs. (6.1) through (6.5) for the three-dimensional case in Cartesian coordinates.

14.8 - Show that the time evolution of a wave packet $\Psi(\vec{r},t)$ can be expressed in terms of the initial form of the wave packet $\Psi(\vec{r},t_0)$ as

$$\Psi(\vec{r},t\ ) = \int_{-\infty}^{+\infty} G(\vec{r},t;\ \vec{r}\,',t_0)\ \Psi(\vec{r}\,',t_0)\ d^3r'$$

where $G(\vec{r},t;\ \vec{r}\,',t_0)$ denotes the Green function, or kernel of the integral [which depends on the dispersion relation $\omega(k)$],

$$G(\vec{r},t;\ \vec{r}\,',t_0) = \left(\frac{1}{2\pi}\right)^3 \int_{-\infty}^{+\infty} \exp\ [i\vec{k}.(\vec{r} - \vec{r}\,') - i\omega(t - t_0)]\ d^3k$$

14.9 - Analyze the meaning of the Green function of Problem 14.8 for the case when $\Psi(\vec{r},t_0)$ is given by the Dirac delta function

$$\Psi(\vec{r},t_0) = \delta(\vec{r} - \vec{r}_0)$$

14.10 - Calculate the Green function of Problem 14.8 for the case of free space, in which $\omega = ck$, where c is the speed of light in free space. Show that, in this case, the initial wave packet $\Psi(\vec{r},t_0)$ maintains its original shape and there is no dispersion.

14.11 - Consider a one-dimensional wave packet at the instant t = 0, whose amplitude function A(k) is given by

$$A(k) = 1 \quad \text{for} \quad -a \leq k \leq a$$

$$A(k) = 0 \quad \text{otherwise.}$$

Show that

$$\Psi(x,0) = 2\ [\sin\ (ax)\ ]/x$$

Make a plot of both A(k) and $\Psi(x,0)$, and verify that the uncertainties in x and k satisfy the relation

$$(\Delta x)\ (\Delta k) \geq 4\pi$$

14.12 - Consider a one-dimensional wave packet at the instant t = 0, whose amplitude function A(k) is given by the Gaussian function

$$A(k) = \exp\ [-\ a^2\ (k - k_0)^2]$$

where a and k are constants.

(a) Show that $\Psi(x,0)$ is also a Gaussian function given by

$$\Psi(x,0) = (\sqrt{\pi}/a)\ \exp\ (i\ k_0\ x)\ \exp\ (-\ x^2/4a^2)$$

Make a plot of both A(k) and Re{$\Psi(x,0)$}, considering that $k_0 \gg 1/a$.

(b) The average extension of the wave packet $\overline{\Delta x}$ can be defined in terms of the root mean square deviation, $\overline{\Delta x} = <(\Delta x)^2>^{1/2}$, where $<(\Delta x)^2>$ is the dispersion,

$$<(\Delta x)^2> = <(x - <x>)^2> = \frac{\int_{-\infty}^{+\infty} |\Psi|^2 (x - <x>)^2 \, dx}{\int_{-\infty}^{+\infty} |\Psi|^2 \, dx}$$

Similarly, $\overline{\Delta k} = <(\Delta k)^2>^{1/2}$, where

$$<(\Delta k)^2> = <(k - <k>)^2> = \frac{\int_{-\infty}^{+\infty} |A|^2 (k - <k>)^2 \, dk}{\int_{-\infty}^{+\infty} |A|^2 \, dk}$$

Show that for this Gaussian wave packet we wave

$\overline{\Delta x} = a$

$\overline{\Delta k} = 1/(2a)$

Consequently, $\overline{\Delta x}\ \overline{\Delta k} = 1/2$. It can be shown that the Gaussian wave packet is the minimum uncertainty wave packet and that, in general, $\overline{\Delta x}\ \overline{\Delta k} \geq 1/2$.

14.13 - Calculate $\Psi(x,0)$ for a one-dimensional wave packet, when the amplitude function A(k) is given by

(a) $A(k) = \exp(-i k x_0)$

(b) $A(k) = \delta(k - k_0)$

For both cases, verify the validity of the uncertainty principle for wave packets (which states that, in general, ($\overline{\Delta x}\ \overline{\Delta k} \geq 1/2$).

14.14 - Consider an *evanescent* plane electromagnetic wave (for which $\vec{k} = i\alpha\hat{x}$, with α real), with the field vectors $\vec{E}$ and $\vec{H}$ proportional to $\exp(i\vec{k}.\vec{r} - i\omega t)$. Show that the average value of the Poynting vector $<\vec{S}>$ is zero for an evanescent wave.

CHAPTER 15

Magnetohydrodynamic Waves

1. INTRODUCTION

The most fundamental type of wave motion that propagates in a compressible, nonconducting fluid is that of longitudinal *sound waves*. For these waves the variations in pressure (p) and in density (ρ_m), associated with the fluid compressions and rarefactions, obey the adiabatic energy equation commonly used in thermodynamics,

$$p\rho_m^{-\gamma} = \text{constant} \tag{1.1}$$

where γ denotes the ratio of the specific heats at constant pressure and at constant volume. Differentiating (1.1) gives

$$\vec{\nabla} p = (\gamma p/\rho_m)\, \vec{\nabla}\rho_m = V_s^2\, \vec{\nabla}\rho_m \tag{1.2}$$

where

$$V_s = (\gamma p/\rho_m)^{1/2} = (\gamma kT/m)^{1/2} \tag{1.3}$$

is the wave propagation velocity, known as the *adiabatic sound velocity*. Fig. 1 illustrates the regions of fluid compression and rarefaction associated with the longitudinal motion of sound waves.

1.1 Alfvén waves

In the case of a compressible, conducting fluid immersed in a magnetic field, other types of wave motion are possible.

We have seen that, in a magnetic field of intensity $\vec{B}_o$, the magnetic stresses are equivalent to a tension B_o^2/μ_o along the field lines and an isotropic hydrostatic pressure $B_o^2/2\mu_o$ (see section 5, Chapter 12). Since the latter can always be superposed on the fluid pressure, the magnetic field lines behave effectively as elastic cords under a tension B_o^2/μ_o. Further, in a perfectly conducting fluid the plasma particles behave as if they were tied to the magnetic field lines (see section 4, Chapter 12), so that the lines of force act as if they were mass-loaded strings under tension. Thus, by analogy with the transverse vibrations of elastic strings, we expect that, whenever the conducting fluid is slightly disturbed from the equilibrium conditions, the magnetic field lines will perform transverse vibrations. The velocity of propagation of these transverse vibrations are expected to be given by

$$V_A = (\text{Tension/Density})^{1/2} = (B_0^2/\mu_o\rho_m)^{1/2} \tag{1.4}$$

This velocity is known as the *Alfvén velocity*, since the existence of this type of wave motion was first pointed out by Alfvén, in 1942. An important property of these waves, as will be shown later, is the absence of any fluctuations in density (ρ_m) or fluid pressure (p). Fig. 2 illustrates the

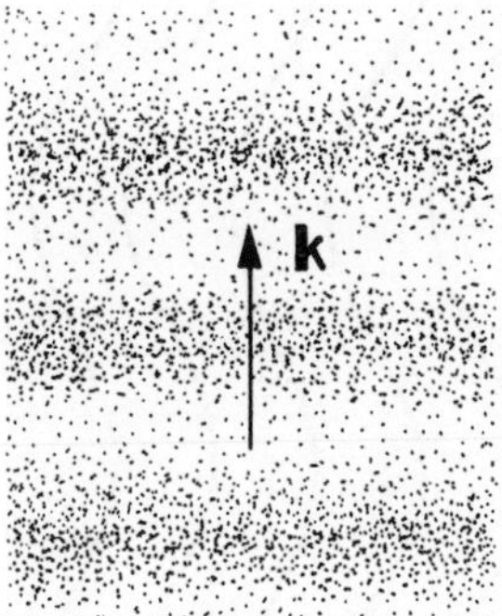

Fig. 1 - Schematic representation of longitudinal *sound waves* that propagate in a compressible, nonconducting fluid, showing the regions of compression and rarefaction associated with the longitudinal wave motion.

transverse motions of the fluid (and of the "frozen in" field lines) for the Alfvēn wave.

1.2 Magnetosonic waves

Longitudinal oscillations are also expected to occur in a compressible, conducting fluid in a magnetic field. For motion of the particles (and propagation of the wave) along the magnetic field there will be no field perturbation, since the particles are free to move in this direction. Thus, in this case, the waves will be ordinary longitudinal sound waves propagating at the the velocity V_S along the field lines (Fig. 3).

On the other hand, for motion of the particles (and propagation of the wave) in the direction perpendicular to the magnetic field, a new type of longitudinal wave motion is possible since now, in addition to the fluid pressure p, there is also the magnetic pressure $B^2/2\mu_o$ in the plane normal to $\vec{B}_o$. Hence, the total pressure is $p + B^2/2\mu_o$ and, consequently, the velocity (V_M) of propagation of these so-called *magnetosonic* or *magnetoacoustic waves* (Fig. 4) satisfies the following relation, analogous to (1.1),

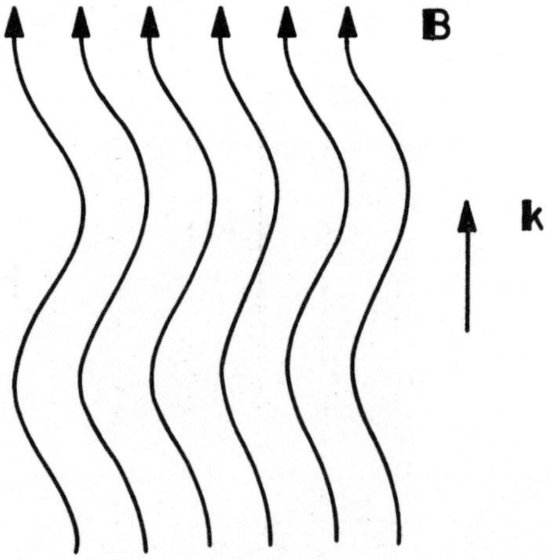

Fig. 2 - Transverse *Alfvén waves* in a compressible, conducting magnetofluid. The velocity of propagation is along the magnetic field lines, and the fluid motion and magnetic field perturbations are perpendicular to the field lines.

$$\vec{\nabla}(p + B^2/2\mu_o) = V_M^2 \, \vec{\nabla}\rho_m \tag{1.5}$$

Therefore, we can write

$$V_M^2 = \frac{d}{d\rho_m}(p + B^2/2\mu_o)_{\rho_{mo}} = V_s^2 + \frac{d}{d\rho_m}(B^2/2\mu_o)_{\rho_{mo}} \tag{1.6}$$

where the suffix zero, in ρ_m, refers to the undisturbed state, and V_s is the adiabatic sound velocity. Since the lines of force are frozen in the conducting fluid, the magnetic flux BdS across an element of surface $d\vec{S}$ (whose normal is oriented along the magnetic field) and the mass ρ_m dS of a unit length of column having $d\vec{S}$ as base, are both conserved during the oscillation, in such a way that $(B/\rho_m) = (B_o/\rho_{mo})$. Consequently, (1.6) becomes

$$V_M^2 = V_s^2 + \frac{d}{d\rho_m}(B_o^2\rho_m^2/2\mu_o\rho_{mo}^2)_{\rho_{mo}} = V_s^2 + V_A^2 \tag{1.7}$$

where V_A is the Alfvẽn velocity defined in (1.4).

For propagation in a direction inclined with respect to the magnetic field the waves are more complex. This subject will be considered in some detail in section 5.

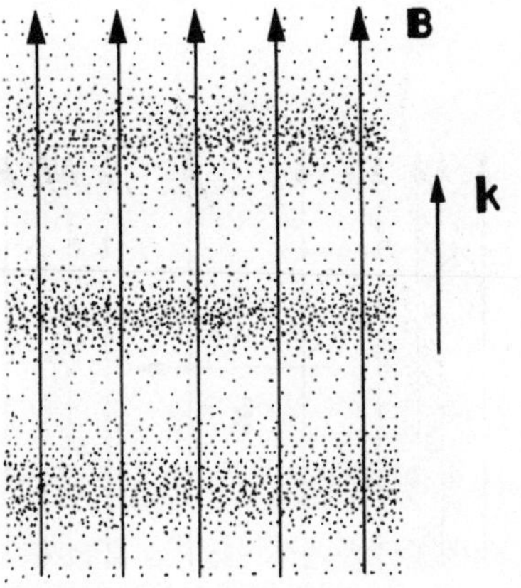

Fig. 3 - Longitudinal *sound waves* propagating along the magnetic field lines in a compressible, conducting magnetofluid.

2. MHD EQUATIONS FOR A COMPRESSIBLE NONVISCOUS CONDUCTING FLUID

2.1 Basic equations

To investigate the propagation of waves in a conducting magnetofluid, let us consider a compressible, nonviscous, perfectly conducting fluid immersed in a magnetic field. The appropriate system of equations governing the behavior of this type of fluid, and the assumptions involved, have been summarized in section 1, of Chapter 12. These equations are

$$\partial \rho_m / \partial t + \vec{\nabla} \cdot (\rho_m \vec{u}) = 0 \tag{2.1}$$

$$\rho_m \partial \vec{u}/\partial t + \rho_m (\vec{u} \cdot \vec{\nabla}) \vec{u} = - \vec{\nabla} p + \vec{J} \times \vec{B} \tag{2.2}$$

$$\vec{\nabla}_p = V_s^2 \vec{\nabla} \rho_m \tag{2.3}$$

$$\vec{\nabla} \times \vec{B} = \mu_o \vec{J} \tag{2.4}$$

$$\vec{\nabla} \times \vec{E} = - \partial \vec{B}/\partial t \tag{2.5}$$

$$\vec{E} + \vec{u} \times \vec{B} = 0 \tag{2.6}$$

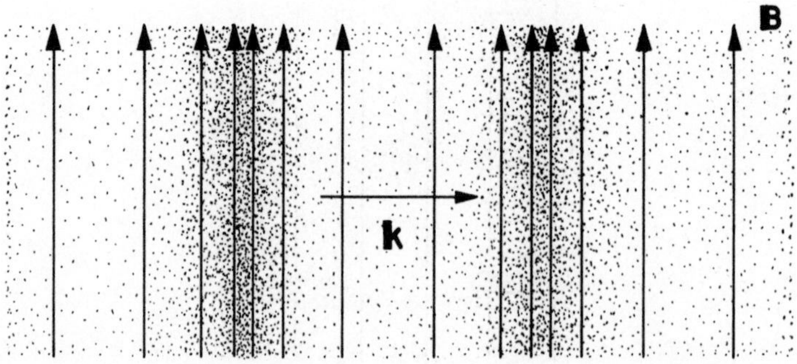

Fig. 4 - The longitudinal *magnetosonic wave* propagates perpendicularly to the magnetic field, causing compressions and rarefactions of both the lines of force and the conducting fluid.

To reduce this system of equations we combine (2.2) to (2.4) in the form

$$\rho_m \, \partial\vec{u}/\partial t + \rho_m(\vec{u} \cdot \vec{\nabla})\, \vec{u} = - V_s^2 \, \vec{\nabla}\rho_m + (\vec{\nabla} \times \vec{B}) \times \vec{B}/\mu_o \tag{2.7}$$

as well as (2.5) and (2.6) in the form

$$\vec{\nabla} \times (\vec{u} \times \vec{B}) = \partial\vec{B}/\partial t \tag{2.8}$$

Under equilibrium conditions, the fluid is assumed to be uniform with constant density ρ_{mo}, the equilibrium velocity is zero, and throughout the fluid the magnetic induction $\vec{B}_o$ is uniform and constant.

In order to develop a dispersion relation for small-amplitude waves, consider small-amplitude departures from the equilibrium values, so that

$$\vec{B}(\vec{r},t) = \vec{B}_o + \vec{B}_1(\vec{r},t) \tag{2.9}$$

$$\rho_m(\vec{r},t) = \rho_{mo} + \rho_{m_1}(\vec{r},t) \tag{2.10}$$

$$\vec{u}(\vec{r},t) = \vec{u}_1(\vec{r},t) \tag{2.11}$$

Substituting (2.9) to (2.11) into (2.1), (2.7) and (2.8), and neglecting second-order terms, we obtain the following linearized equations in the small first-order quantities

$$\partial\rho_{m_1}/\partial t + \rho_{mo}(\vec{\nabla} \cdot \vec{u}_1) = 0 \tag{2.12}$$

$$\rho_{mo} \, \partial\vec{u}_1/\partial t + V_s^2 \, \vec{\nabla}\rho_{m_1} + \vec{B}_o \times (\vec{\nabla} \times \vec{B}_1)/\mu_o = 0 \tag{2.13}$$

$$\partial\vec{B}_1/\partial t - \vec{\nabla} \times (\vec{u}_1 \times \vec{B}_o) = 0 \tag{2.14}$$

2.2 Development of an equation for u_1

Eqs. (2.12) to (2.14) can be combined to yield an equation for $\vec{u}_1$ alone. For this purpose, we first differentiate (2.13) with respect to time, obtaining

$$\rho_{mo} \frac{\partial^2\vec{u}_1}{\partial t^2} + V_s^2 \, \vec{\nabla}\left(\frac{\partial\rho_{m_1}}{\partial t}\right) + \frac{1}{\mu_o} \vec{B}_o \times \left[\vec{\nabla} \times \left(\frac{\partial\vec{B}_1}{\partial t}\right) \right] = 0 \tag{2.15}$$

Next, using (2.12) and (2.14), we can write (2.15) as

$$\partial^2\vec{u}_1/\partial t^2 - V_s^2\, \vec{\nabla}\, (\vec{\nabla} \cdot \vec{u}_1) + \vec{V}_A \times \{\vec{\nabla} \times [\vec{\nabla} \times (\vec{u}_1 \times \vec{V}_A)]\} = 0 \qquad (2.16)$$

where we have introduced the *vector* Alfvēn velocity

$$\vec{V}_A = \vec{B}_0/(\mu_0 \rho_{mo})^{1/2} \qquad (2.17)$$

Without loss of generality we can consider plane wave solutions of the form

$$\vec{u}_1(\vec{r},t) = \vec{u}_1 \exp\, (i\vec{k} \cdot \vec{r} - i\,\omega\, t) \qquad (2.18)$$

In what follows $\vec{u}_1$ can stand for either the amplitude or the entire expression (2.18). Thus, in (2.16) we can replace the operator $\vec{\nabla}$ by $i\vec{k}$ and the time derivative by $- i\omega$, so that

$$-\omega^2\, \vec{u}_1 + V_s^2\, (\vec{k} \cdot \vec{u}_1)\, \vec{k} - \vec{V}_A \times \{\vec{k} \times [\vec{k} \times (u_1 \times \vec{V}_A)]\} = 0 \qquad (2.19)$$

Since for any three vectors $\vec{A}$, $\vec{B}$ and $\vec{C}$ we have the vector identity

$$\vec{A} \times (\vec{B} \times \vec{C}) = (\vec{A} \cdot \vec{C})\, \vec{B} - (\vec{A} \cdot \vec{B})\, \vec{C} \qquad (2.20)$$

we can rearrange (2.19) to read

$$-\omega^2\, \vec{u}_1 + (V_s^2 + V_A^2)\, (\vec{k} \cdot \vec{u}_1)\, \vec{k} + (\vec{k} \cdot \vec{V}_A)\, [(\vec{k} \cdot \vec{V}_A)\, \vec{u}_1 -$$

$$- (\vec{V}_A \cdot \vec{u}_1)\, \vec{k} - (\vec{k} \cdot \vec{u}_1)\, \vec{V}_A] = 0 \qquad (2.21)$$

Although this expression appears to be somewhat involved, it leads to remarkably simple solutions for waves propagating in the directions parallel or perpendicular to the magnetic field.

3. PROPAGATION PERPENDICULAR TO THE MAGNETIC FIELD

When the wave vector $\vec{k}$ is perpendicular to the magnetic induction $\vec{B}_0$, we have $\vec{k} \cdot \vec{V}_A = 0$, and (2.21) simplifies to

$$-\omega^2 \vec{u}_1 + (V_S^2 + V_A^2)(\vec{k} \cdot \vec{u}_1)\vec{k} = 0 \tag{3.1}$$

from which we obtain

$$\vec{u}_1 = (V_S^2 + V_A^2)(\vec{k} \cdot \vec{u}_1)\vec{k}/\omega^2 \tag{3.2}$$

Therefore, $\vec{u}_1$ is parallel to $\vec{k}$, so that $\vec{k} \cdot \vec{u}_1 = ku_1$, and the solution for $\vec{u}_1$ is a *longitudinal* wave with the *phase velocity*

$$(\omega/k) = (V_S^2 + V_A^2)^{1/2} \tag{3.3}$$

The magnetic field associated with this longitudinal wave can be obtained from (2.14). Taking

$$\vec{B}_1(\vec{r},t) = \vec{B}_1 \exp(i\,\vec{k} \cdot \vec{r} - i\,\omega t) \tag{3.4}$$

we obtain

$$-\omega \vec{B}_1 - \vec{k} \times (\vec{u}_1 \times \vec{B}_0) = 0 \tag{3.5}$$

Using the vector identity (2.20), and noting that $\vec{k} \cdot \vec{B}_0 = 0$, we find

$$\vec{B}_1 = \frac{u_1}{(\omega/k)} \vec{B}_0 \tag{3.6}$$

The electric field associated with this wave is seen, from (2.6), to be given by

$$\vec{E} = -\vec{u}_1 \times \vec{B}_0 \tag{3.7}$$

This wave is somewhat similar to an electromagnetic wave, since the time-varying magnetic field is perpendicular to the direction of propagation, but parallel to the magnetostatic field, whereas the time varying electric field is perpendicular to both the direction of propagation and the magnetostatic field. It is a longitudinal wave, however, since the velocity of mass flow and the fluctuating mass density associated with it are both in the direction of wave propagation. For these reasons, this wave is called the *magnetosonic wave.* The phase velocity of this wave is independent of frequency, so that it is a nondispersive wave. As illustrated in Fig. 4, the magnetosonic wave produces compressions and rarefactions in the magnetic field lines without changing their direction. Since the fluid is perfectly conducting, the lines of force and the fluid move together.

The restoring forces, operating in the magnetosonic wave, are the fluid pressure gradient and the gradient of the compressional stresses between the magnetic field lines. If the fluid pressure is much greater than the magnetic pressure, the effect of the magnetic field is negligible, so that $\omega/k \cong V_S$ and the magnetosonic wave becomes essentially an acoustic wave. On the other hand, if the magnetic field is very strong, so that the magnetic pressure is much larger than the fluid pressure, then the phase velocity of the magnetosonic wave becomes equal to the Alfvén wave velocity V_A.

The magnetosonic wave mode is also known variously as the *compressional Alfvén wave* or the *fast Alfvén wave*.

4. PROPAGATION PARALLEL TO THE MAGNETIC FIELD

For waves propagating along the magnetic field $(\vec{k} \parallel \vec{B}_0)$, we have $\vec{k} \cdot \vec{V}_A = k V_A$, and (2.21) simplifies to

$$(k^2 V_A^2 - \omega^2)\, \vec{u}_1 + (V_S^2/V_A^2 - 1)\, k^2\, (\vec{u}_1 \cdot \vec{V}_A)\, \vec{V}_A = 0 \tag{4.1}$$

In this case, there are two types of wave motion possible.

For $\vec{u}_1$ parallel to $\vec{B}_0$ and $\vec{k}$, we find from (4.1) that a longitudinal mode is possible, with the phase velocity

$$(\omega/k) = V_S \tag{4.2}$$

This is an ordinary *longitudinal sound wave*, in which the velocity of mass flow is in the propagation direction (Fig. 3). There is no electric field, electric current density and magnetic field associated with this wave.

A transverse wave, with $\vec{u}_1$ perpendicular to $\vec{B}_0$ and $\vec{k}$, is the other possibility. In this case $\vec{u}_1 \cdot \vec{V}_A = 0$, and (4.1) gives for the phase velocity of this transverse wave, known as the *Alfvén wave*,

$$(\omega/k) = V_A \tag{4.3}$$

Since the phase velocity is independent of frequency, there is no dispersion.

The magnetic field associated with the Alfvén wave is found, from (2.14) and (3.5), to be given by

$$\vec{B}_1 = - B_o \frac{\vec{u}_1}{(\omega/k)} \tag{4.4}$$

Hence, the magnetic field disturbance is normal to the original magnetostatic induction $\vec{B}_o$. The small component $\vec{B}_1$, when added to $\vec{B}_o$, gives the lines of force a sinusoidal ripple, shown in Fig. 5. The associated electric field is given by (3.8).

The Alfvén wave involves no fluctuations in the fluid density or pressure, although both the fluid and the magnetic field lines oscillate back and forth laterally, in the plane normal to $\vec{B}_o$. The magnetic energy density of the wave motion ($\vec{B}_1^2/2\mu_o$) is equal to the kinetic energy density of the fluid motion ($\rho_{mo} \vec{u}_1^2/2$). This equipartition of energy is easily verified from (4.4),

$$\frac{\vec{B}_1^2}{2\mu_o} = \frac{B_o^2 \, \vec{u}_1^2}{2\mu_o \, (\omega/k)^2} = \frac{B_o^2 \, \vec{u}_1^2}{2\mu_o \, V_A^2} = \frac{1}{2} \rho_{mo} \, \vec{u}_1^2 \tag{4.5}$$

where we have used (4.3) and (2.17).

The Alfvén wave mode is also known variously as the *shear Alfvén wave* or the *slow Alfvén wave*.

5. PROPAGATION IN AN ARBITRARY DIRECTION

Proceeding further, let us now investigate the case of wave propagation in an arbitrary direction with respect to the magnetic induction $\vec{B}_o$. With no loss of generality, we introduce a Cartesian coordinate system such that the y axis is normal to the plane defined by the direction of propagation $\hat{k}$ and the magnetic induction $\vec{B}_o$, and choose $\hat{z}$ to be along $\vec{B}_o$, as shown in Fig. 6. Denoting the angle between $\vec{k}$ and $\vec{B}_o$ by θ, we have

$$\vec{k} = k \, (\hat{x} \sin \theta + \hat{z} \cos \theta) \tag{5.1}$$

$$\vec{V}_A = V_A \, \hat{z} \tag{5.2}$$

$$\vec{u}_1 = u_{1x}\,\hat{x} + u_{1y}\,\hat{y} + u_{1z}\,\hat{z} \tag{5.3}$$

$$\vec{k} \cdot \vec{V}_A = k\,V_A \cos\Theta \tag{5.4}$$

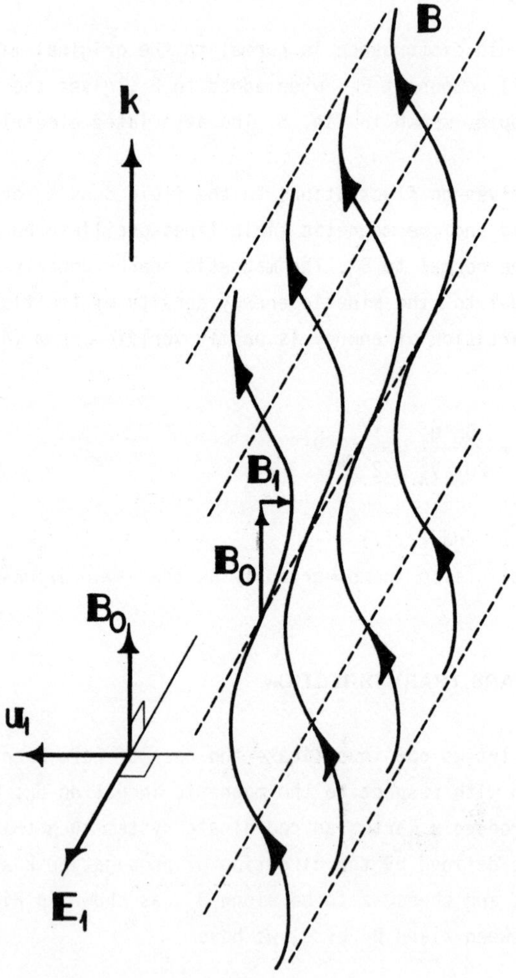

Fig. 5 - Schematic illustration for Alfvén waves propagating along $\vec{B}_0$, showing the relations between the oscillating quantities.

$$\vec{k} \cdot \vec{u}_1 = k\,(u_{1x} \sin\Theta + u_{1z} \cos\Theta) \tag{5.5}$$

$$\vec{V}_A \cdot \vec{u}_1 = V_A\, u_{1z} \tag{5.6}$$

Substituting these expressions into (2.21), performing the required algebra and rearranging the terms, we obtain for the x component equation,

$$u_{1x}\,(-\omega^2 + k^2 V_A^2 + k^2 V_S^2 \sin^2\Theta) + u_{1z}\,(k^2 V_S^2 \sin\Theta \cos\Theta) = 0 \tag{5.7}$$

for the y component equation,

$$u_{1y}\,(-\omega^2 + k^2 V_A^2 \cos^2\Theta) = 0 \tag{5.8}$$

and for the z component equation,

$$u_{1x}\,(k^2 V_S^2 \sin\Theta \cos\Theta) + u_{1z}\,(-\omega^2 + k^2 V_S^2 \cos^2\Theta) = 0 \tag{5.9}$$

5.1 Pure Alfvén wave

From (5.8) we see that there is a linearly polarized wave involving oscillations in the direction perpendicular to both $\vec{k}$ and $\vec{B}_0$ ($u_{1y} \neq 0$), with a phase velocity given by

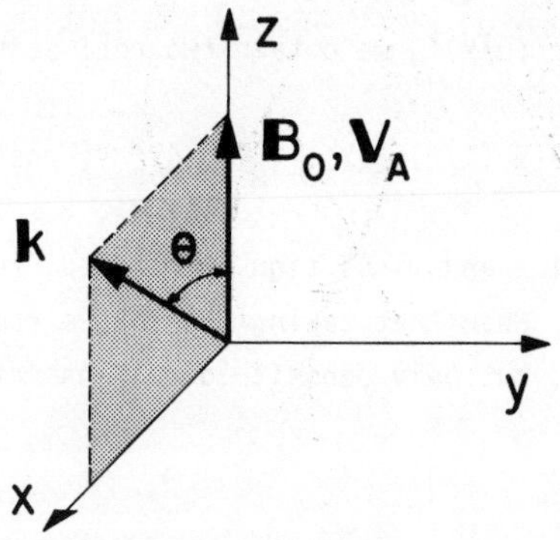

Fig. 6 - Cartesian coordinate system with the relative directions of the vectors $\vec{k}$ and $\vec{B}_0$.

$$(\omega/k) = V_A \cos\Theta \tag{5.10}$$

The field components associated with this wave can be seen to be B_{1y}, u_{1y}, E_{1x}, and J_{1x}, so that it is a transverse Alfvén wave. For this reason, this wave is generally referred to as the *pure Alfvén wave*. Note that for propagation along the magnetostatic field (Θ=0) Eq. (5.10) gives $\omega/k = V_A$, while for propagation across the magnetostatic field ($\Theta = 90^o$) this wave disappears, since $\omega/k = 0$. This mode is also known as the *oblique Alfvén wave*.

5.2 Fast and slow MHD waves

Eqs. (5.7) and (5.9) constitute a system of two simultaneous equations for the amplitudes of u_{1x} and u_{1z}. To have a solution in which u_{1x} and u_{1z} are nonzero, the determinant of the coefficients of this system must vanish. Therefore, setting

$$\begin{vmatrix} (-\omega^2 + k^2 V_A^2 + k^2 V_S^2 \sin^2\Theta) & (k^2 V_S^2 \sin\theta \cos\Theta) \\ (k^2 V_S^2 \sin\Theta\cos\Theta) & (-\omega^2 + k^2 V_S^2 \cos^2\Theta) \end{vmatrix} = 0 \tag{5.11}$$

we obtain the following dispersion relation, expressed in terms of the phase velocity ω/k,

$$(\omega/k)^4 - (V_S^2 + V_A^2)\,(\omega/k)^2 + V_S^2\,V_A^2 \cos^2\Theta = 0 \tag{5.12}$$

Solving this equation for $(\omega/k)^2$, we obtain two real solutions

$$(\omega/k)^2 = (V_S^2 + V_A^2)/2 \pm (1/2)\,[(V_S^2 + V_A^2)^2 - 4V_S^2\,V_A^2 \cos^2\Theta]^{1/2} \tag{5.13}$$

The solutions with the plus-and-minus sign are called, respectively, the *fast* and *slow MHD wave modes*. Note that taking the square root of $(\omega/k)^2$ does not give two different modes, but only opposite directions of propagation.

5.3 Phase velocities

All the three MHD waves have constant phase velocities, given by (5.10) and (5.13), for all frequencies, and hence there is no dispersion. Fig. 7 displays the phase velocity, for each of these waves, as a function of the angle Θ between $\vec{k}$ and $\vec{B}_o$, for both cases when $V_A > V_s$ and when $V_s > V_A$. The phase velocity of the fast MHD wave increases from V_A (or V_s if $V_s > V_A$) when $\Theta = 0^0$, to $(V_s^2 + V_A^2)^{1/2}$ when $\Theta = 90^0$, while that of the slow MHD wave decreases from V_s (or V_A if $V_s > V_A$) when $\Theta = 0^0$, to zero when $\Theta = 90^0$. Therefore, if $V_A > V_s$, the fast MHD wave becomes the Alfvēn wave for $\Theta = 0^0$, and the magnetosonic wave for $\Theta = 90^0$, while the slow MHD wave becomes the sound wave for $\Theta = 0^0$, and disappears for $\Theta = 90^0$. On the other hand, if $V_s > V_A$, the fast MHD wave becomes the sound wave for $\Theta = 0^0$ and the magnetosonic wave for $\Theta = 90^0$, while the slow MHD wave becomes the Alfvēn wave for $\Theta = 0^0$ and disappears for $\Theta = 90^0$.

5.4 Wave normal surfaces

The propagation of these waves are conveniently represented by means of diagrams called *phase velocity* or *wave normal surfaces*, which give the variations of the magnitude of the phase velocity of plane waves with respect to the magnetic field direction. Fig. 8 shows the wave normal diagram for the pure Alfvēn wave, constructed from (5.10). The vector drawn from the origin to a point P on the curve represents the phase velocity of a plane wave and the direction of the wave normal with respect to $\vec{B}_o$. The actual state of affairs, in three dimensions, is obtained by rotating the circles of Fig. 8 about the axis oriented along $\vec{B}_o$. The three-dimensional surface, thus obtained, is called the *wave normal surface*.

Fig. 9 shows the wave normal diagrams for propagation of the pure Alfvēn, the fast and the slow MHD waves, for the two cases $V_A > V_s$ and $V_A < V_s$. The three-dimensional wave normal surfaces are obtained by rotating Fig. 9 about the axis oriented along $\vec{B}_o$. The wave normal surface corresponding to the fast MHD wave is a smooth, closed surface enclosing the two spheres passing through 0 which correspond to the pure Alfvēn wave. Within each of these spheres, there is another smooth, closed wave normal surface corresponding to the slow MHD wave.

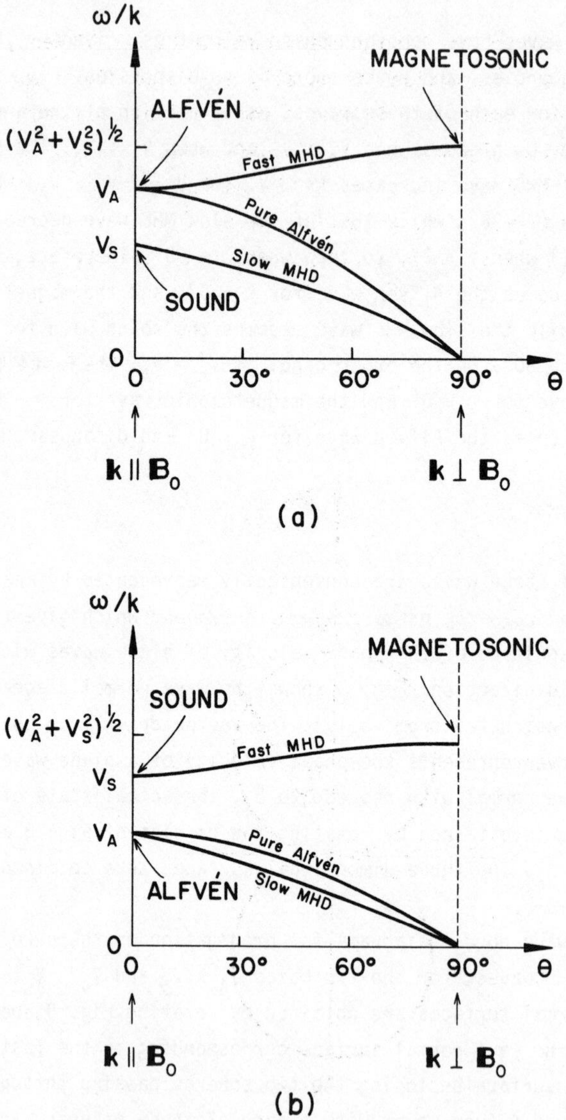

Fig. 7 - Phase velocity curves (independent of frequency) as a function of the angle between $\vec{k}$ and $\vec{B}_0$, for the pure Alfvén, the fast, and the slow MHD waves, for the cases (a) $V_A > V_s$ and (b) $V_s > V_A$.

6. EFFECT OF DISPLACEMENT CURRENT

In magnetohydrodynamics the displacement current ($\varepsilon_0 \partial\vec{E}/\partial t$) term, which appears in Maxwell $\vec{\nabla} \times \vec{B}$ equation, is usually neglected. This approximation is valid only for high conductivity fluids at comparatively low frequencies (well below the ion cyclotron frequency), as discussed in section 6, Chapter 9. The inclusion of the displacement current in the basic equations modify the propagation of the Alfvēn and magnetosonic waves. The results obtained, however, are valid only at frequencies where charge separation effects are unimportant.

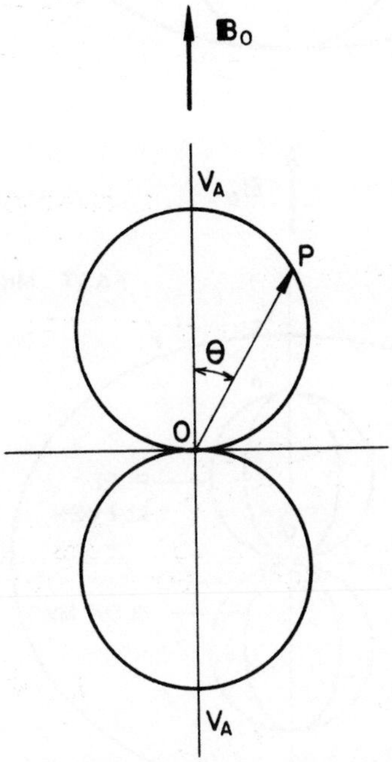

Fig. 8 - Wave normal diagram illustrating the variations of the phase velocity and the direction of the wave normal at any angle Θ with respect to $\vec{B}_0$, for the pure Alfvēn wave.

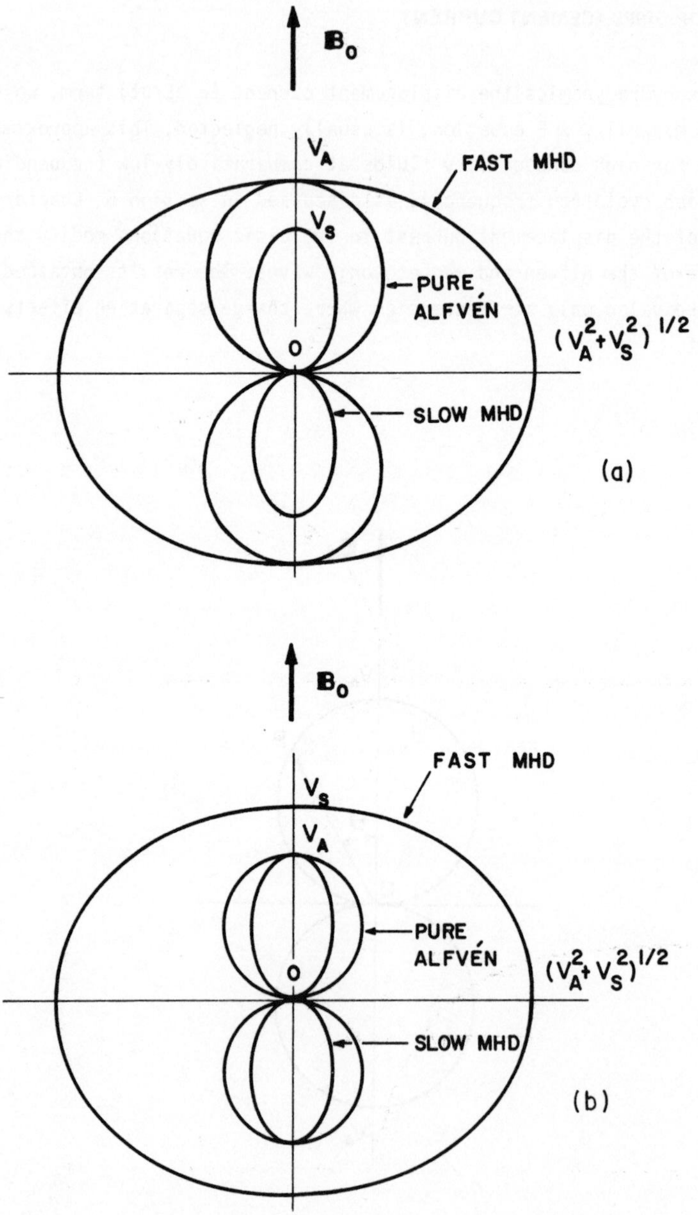

Fig. 9 - Wave normal diagrams for the pure Alfvén, the fast and the slow MHD waves, for (a) $V_A > V_s$ and (b) $V_A < V_s$.

6.1 Basic equations

To investigate the effect of the displacement current on the propagation of MHD waves in a compressible, nonviscous, perfectly conducting fluid, (2.4) must be modified to read

$$\vec{\nabla} \times \vec{B} = \mu_o \vec{J} + (1/c^2)\, \partial\vec{E}/\partial t \tag{6.1}$$

Consequently, the current density to be inserted into the $\vec{J} \times \vec{B}$ term, in the equation of motion (2.2), is now

$$\vec{J} = \frac{1}{\mu_o} [\, \vec{\nabla} \times \vec{B} + \frac{1}{c^2} \frac{\partial}{\partial t} (\vec{u} \times \vec{B})\,] \tag{6.2}$$

where use was made of (2.6). Using expressions (2.9) to (2.11) for small-amplitude waves, the set of linearized equations (2.12) to (2.14) for the small quantities ρ_1, $\vec{u}_1$ and $\vec{B}_1$, become now

$$\partial\rho_{m1}/\partial t + \rho_{mo} (\vec{\nabla} \cdot \vec{u}_1) = 0 \tag{6.3}$$

$$\rho_{mo} \frac{\partial\vec{u}_1}{\partial t} + V_S^2\, \vec{\nabla}\, \rho_{m1} + \frac{1}{\mu_o} \vec{B} \times (\vec{\nabla} \times \vec{B}_1 + \frac{1}{c^2} \frac{\partial\vec{u}_1}{\partial t} \times \vec{B}_o) = 0 \tag{6.4}$$

$$\partial\vec{B}_1/\partial t - \vec{\nabla} \times (\vec{u}_1 \times \vec{B}_o) = 0 \tag{6.5}$$

6.2 Equation for u_1

To obtain an equation for $\vec{u}_1$ alone, we take the time derivative of (6.4), and use (6.3) and (6.5), which gives

$$\frac{\partial^2\vec{u}_1}{\partial t^2} - V_S^2\, \vec{\nabla} (\vec{\nabla} \cdot \vec{u}_1) + \vec{V}_A \times \{\, \vec{\nabla} \times [\, \vec{\nabla} \times (\vec{u}_1 \times \vec{V}_A)]\} + $$

$$+ \frac{1}{c^2} \vec{V}_A \times (\, \frac{\partial^2\vec{u}_1}{\partial t^2} \times \vec{V}_A) = 0 \tag{6.6}$$

where $\vec{V}_A$ is the vector Alfvén velocity, defined in (2.17). From the vector indentity (2.20), we have

$$\vec{V}_A \times \left(\frac{\partial^2 \vec{u}_1}{\partial t^2} \times \vec{V}_A \right) = \frac{\partial^2}{\partial t^2} [\, V_A^2 \, \vec{u}_1 - (\vec{V}_A \cdot \vec{u}_1) \, \vec{V}_A \,] \tag{6.7}$$

so that (6.6) can be rearranged in the form

$$\frac{\partial^2}{\partial t^2} [(1 + \frac{V_A^2}{c^2})\vec{u}_1 - (\vec{V}_A \cdot \vec{u}_1) \frac{\vec{V}_A}{c^2}] - V_S^2 \, \vec{\nabla} \, (\vec{\nabla} \cdot \vec{u}_1) +$$

$$+ \vec{V}_A \times \{\vec{\nabla} \times [\, \vec{\nabla} \times (\vec{u}_1 \times \vec{V}_A)]\} = 0 \tag{6.8}$$

It is evident that this equation reduces to (2.16), if $V_A^2/c^2 << 1$.

Plane wave solutions of (6.8), in the form (2.18), give

$$- \omega^2 \, [(1 + \frac{V_A^2}{c^2})\vec{u}_1 - (\vec{V}_A \cdot \vec{u}_1) \frac{\vec{V}_A}{c^2}] + (V_S^2 + V_A^2) \, (\vec{k} \cdot \vec{u}_1)\vec{k} +$$

$$+ (\vec{k} \cdot \vec{V}_A) \, [(\vec{k} \cdot \vec{V}_A)\vec{u}_1 - (\vec{V}_A \cdot \vec{u}_1)\vec{k} - (\vec{k} \cdot \vec{u}_1) \, \vec{V}_A] = 0 \tag{6.9}$$

6.3 Propagation across the magnetostatic field

For $\vec{k} \perp \vec{B}_o$ we have $\vec{k} \cdot \vec{V}_A = 0$, so that (6.9) gives $(\vec{V}_A \cdot \vec{u}_1) = 0$ and

$$- \omega^2 \, (1 + V_A^2/c^2)\vec{u}_1 + (V_S^2 + V_A^2) \, (\vec{k} \cdot \vec{u}_1)\vec{k} = 0 \tag{6.10}$$

This equation is similar to (3.1), except that the square of the frequency is multiplied by the factor $(1 + V_A^2/c^2)$. Thus, the phase velocity of the *longitudinal magnetosonic wave* propagating across $\vec{B}_o$ becomes now

$$(\omega/k) = (V_S^2 + V_A^2)^{1/2}/(1 + V_A^2/c^2)^{1/2} \tag{6.11}$$

6.4 Propagation along the magnetostatic field

For $\vec{k} \parallel \vec{B}_o$, inspection of (6.9) shows that for $\vec{u}_1$ parallel to $\vec{V}_A$ (i.e. $\vec{B}_o$) it becomes identical to (2.21). Thus, for the *longitudinal sound wave* propagating along $\vec{B}_o$ there is no change from the results obtained before.

However, for the *transverse Alfvén wave* $(\vec{u}_1 \perp \vec{k})$ we have $(\vec{V}_A \cdot \vec{u}_1) = 0$ and (6.9) reduces to

$$- \omega^2 (1 + V_A^2/c^2)\vec{u}_1 + k^2 V_A^2 \vec{u}_1 = 0 \tag{6.12}$$

Consequently, the modification introduced in the Alfvēn wave by the displacement current is that the square of the frequency must be multiplied by the factor $(1 + V_A^2/c^2)$. Thus, the phase velocity of the Alfvēn wave becomes

$$(\omega/k) = V_A/(1 + V_A^2/c^2)^{1/2} \tag{6.13}$$

In the usual limit of $V_A^2/c^2 << 1$, (6.13) reduces to (4.3) and the effect of the displacement current is unimportant. On the other hand, if $V_A^2/c^2 >> 1$, then ω/k becomes equal to the speed of light. In using these results, however, it must be kept in mind that they are valid only at frequencies where charge separation effects are negligible, since the electric force term has been neglected in the equation of motion (2.13).

7. DAMPING OF MHD WAVES

In this section it is shown that when the fluid is not perfectly conducting, but has a finite conductivity, or if viscous effects are present, the MHD oscillations will be damped. Denoting the kinematic viscosity (viscosity divided by mass density) of the fluid by η_k, and the magnetic viscosity by η_m [see Eq. (6.2.5)], the linearized set of equations (2.12) to (2.14) are modified to include additional terms as follows,

$$\partial\rho_{m1}/\partial t + \rho_{mo} (\vec{\nabla} \cdot \vec{u}_1) = 0 \tag{7.1}$$

$$\rho_{mo}\partial\vec{u}_1/\partial t + V_S^2 \vec{\nabla} \rho_{m1} + \vec{B}_o \times (\vec{\nabla} \times \vec{B}_1)/\mu_o - \rho_{mo} \eta_k \nabla^2\vec{u}_1 = 0 \tag{7.2}$$

$$\partial\vec{B}_1/\partial t - \vec{\nabla} \times (\vec{u}_1 \times \vec{B}_o) - \eta_m\nabla^2\vec{B}_1 = 0 \tag{7.3}$$

Although for a compressible fluid the use of the simple viscous force term $\rho_{mo}\eta_k\nabla^2\vec{u}_1$ is not really allowed, it is, nevertheless, expected to give the correct order of magnitude behavior. The displacement current is not included in the treatment presented in this section.

For plane wave solutions, the differential operators $\partial/\partial t$ and $\vec{\nabla}$ are replaced, respectively, by $-i\omega$ and $i\vec{k}$, so that the set of differential equations (7.1) to (7.3) are replaced by a correspondent set of algebraic equations. Thus, we have

$$\rho_{m1} = \rho_{mo}\,(\vec{k}\cdot\vec{u}_1)/\omega \tag{7.4}$$

$$\omega\vec{u}_1 = (\rho_{m1}/\rho_{mo})V_s^2\,\vec{k} + \vec{B}_o \times (\vec{k}\times\vec{B}_1)/\mu_o\rho_{mo} - i\eta_k k^2\vec{u}_1 \tag{7.5}$$

$$\vec{B}_1 = -\,\vec{k}\times(\vec{u}_1\times\vec{B}_o)/(\omega + i\eta_m k^2) \tag{7.6}$$

Substituting (7.4) and (7.6) into (7.5), and rearranging, we obtain

$$-\omega^2\,(1 + i\eta_k k^2/\omega)\,(1 + i\eta_m k^2/\omega)\vec{u}_1 + (1 + i\eta_m k^2/\omega)V_s^2\,(\vec{k}\cdot\vec{u}_1)\vec{k} - \vec{V}_A \times \{\vec{k}\times[\vec{k}\times(\vec{u}\times\vec{V}_A)]\} = 0 \tag{7.7}$$

Comparing this equation for $\vec{u}_1$ with (2.19), we see that we obtain the same results as before, except that ω^2 must be multiplied by the factor $(1 + i\eta_k k^2/\omega)\,(1 + i\eta_m k^2/\omega)$, and V_s^2 must be multiplied by the factor $(1 + i\eta_m k^2/\omega)$.

7.1 Alfvén waves

For the case of the transverse Alfvén waves propagating along $\vec{B}_o$, the relation (4.3) between ω and k becomes

$$\begin{aligned} k^2\,V_A^2 &= \omega^2(1 + i\eta_k k^2/\omega)\,(1 + i\eta_m k^2/\omega) \\ &= \omega^2\,[1 + i(\eta_k + \eta_m)k^2/\omega - \eta_k\eta_m k^4/\omega^2 \end{aligned} \tag{7.8}$$

In order to simplify this result we shall assume that the correction terms corresponding to the kinematic and magnetic viscosity are small, so that the term in the right-hand side of (7.8) can be neglected. Thus,

$$\begin{aligned} k^2\,V_A^2 &\cong \omega^2\,[\,1 + i(\eta_k + \eta_m)k^2/\omega\,] \\ &\cong \omega^2\,[\,1 + i(\eta_k + \eta_m)\omega/V_A^2\,] \end{aligned} \tag{7.9}$$

where we have replaced ω/k, in the right-hand side, by the first order result (V_A). Eq. (7.9) can be further simplified to the form [using the binomial expansion $(1 + x)^{1/2} \cong 1 + x/2$, $x \ll 1$]

$$k \cong \omega/V_A + i(\eta_k + \eta_m)\omega^2/(2V_A^3) \tag{7.10}$$

The positive imaginary part in this expression for $k(\omega)$ implies in wave damping. This is easily seen by writing $k = k_r + i\,k_i$, with k_r and k_i real numbers, and noting that

$$\exp(ikz) = \exp(-k_i z)\,\exp(ik_r z) \tag{7.11}$$

which represents a wave propagating along the z axis with wave number k_r, but with an exponentially decreasing amplitude, the amplitude falling to 1/e of its original intensity in a distance of $1/k_i$.

Expression (7.10) shows that the attenuation of Alfvén waves increases rapidly with frequency (or wave number), but decreases rapidly with increasing magnetic field intensity. Also, the attenuation increases with the fluid viscosity and the magnetic viscosity. The latter increases as the fluid conductivity decreases.

7.2 Sound waves

For the longitudinal sound waves propagating along $\vec{B}_o$, (4.2) is modified to read

$$k^2\,V_S^2 = \omega^2\,(1 + i\eta_k k^2/\omega) \tag{7.12}$$

Considering that the resistive and viscous correction terms are small, we find

$$k \cong \omega/V_S + i\eta_k\omega^2/(2V_S^3) \tag{7.13}$$

This shows that attenuation of sound waves also increases rapidly with frequency, but decreases with increasing sound velocity. It also increases with increasing fluid viscosity, as expected.

7.3 Magnetosonic waves

For longitudinal magnetosonic waves propagating across $\vec{B}_o$, the dispersion relation becomes [see (3.3)]

$$k^2\,V_S^2\,(1 + i\eta_m k^2/\omega) + k^2V_A^2 = \omega^2(1 + i\eta_k k^2/\omega)(1 + i\eta_m k^2/\omega) \tag{7.14}$$

To simplify this expression we consider that the kinematic and magnetic viscosities are small, and neglect the term involving the product $\eta_m \eta_k k^4/\omega^2$. Hence, (7.14) becomes, after some rearrangement,

$$k^2 (V_S^2 + V_A^2) \cong \omega^2 \{ 1 + i \frac{k^2}{\omega} [\eta_k + \eta_m (1 - \frac{k^2 V_S^2}{\omega^2})] \} \tag{7.15}$$

In the terms in the right-hand side of (7.15) we can replace ω^2/k^2 by the approximate result $(V_S^2 + V_A^2)$, so that (7.15) can be further simplified to give the following dispersion relation

$$k = \frac{\omega}{(V_S^2 + V_A^2)^{1/2}} + i \frac{\omega^2}{2(V_S^2 + V_A^2)^{3/2}} [\eta_k + \frac{\eta_m}{(1 + V_S^2/V_A^2)}] \tag{7.16}$$

Thus, the attenuation of magnetosonic waves also increases with frequency, and with kinematic and magnetic viscosity, but decreases with increasing magnetic field strength.

PROBLEMS

15.1 - Calculate the speed of an Alfvén wave for the following cases:
(a) In the Earth's ionosphere, considering that $n_e = 10^5$ cm^{-3}, B = 0.5 Gauss and that the positive charge carriers are atomic oxygen ions.
(b) In the solar corona, assuming $n_e = 10^6$ cm^{-3}, B = 10 Gauss and that the positive charge carriers are protons.
(c) In the interstellar space, considering $n = 10^7$ m^{-3} and $B = 10^{-7}$ Weber/m^2, the positive charge carriers being protons.

15.2 - Show that Alfvén waves, propagating along the magnetic field, are circularly polarized.

15.3 - For the pure Alfvén wave, propagating at an angle θ with respect to the magnetosonic field $\vec{B}_o$, with phase velocity given by (5.10), determine the associated field components B_{1y}, u_{1y}, E_{1x}, and J_{1x}.

15.4 - Include the effect of finite conductivity in the derivation of the equations for the plane Alfvén wave propagating along the magnetic field. Show that the linearized equations are satisfied by solutions of the form $\exp(\alpha z - i\omega t)$ and determine the coefficient α.

15.5 - A plane electromagnetic wave is incident normally on the surface of a conducting fluid of large but finite conductivity (σ), immersed in a uniform magnetic field $\vec{B}_o$ such that $\vec{k} \perp \vec{B}_o$. Assume that the magnetic field ($\vec{B}$) of the incoming wave is parallel to $\vec{B}_o$. Show that there are two wave modes which penetrate the fluid: an unattenuated magnetosonic wave, and another mode which has an effective skin depth $\delta = (V_S/V_M)\ \delta_{rc}$, where V_S and V_M are the sound and magnetosonic velocities, respectively, and δ_{rc} is the skin depth in a rigid conductor.

15.6 - For the fast and slow MHD waves, let u_ℓ and u_t be the components of the velocity of mass flow which are longitudinal and transverse, respectively, to the direction of propagation. Show that u_ℓ and u_t are in phase for the fast wave and 180° out of phase for the slow wave. Also, show that the perturbations of the kinetic and magnetic pressures are in phase for the fast wave and 180° out of phase for the slow wave.

15.7 - Consider the following closed set of MHD equations in the so-called Chew, Goldberger and Low approximation:

$$\partial\rho_m/\partial t + \vec{\nabla}.(\rho_m\vec{u}) = 0$$

$$\rho_m \frac{D}{Dt}\vec{u} = \rho\vec{E} - \vec{\nabla}\left(p_\perp + \frac{B^2}{2\mu_o}\right) + (\vec{B}\,.\,\vec{\nabla})\left\{\left[\frac{1}{\mu_o} - \frac{(p_\parallel - p_\perp)}{B^2}\right]\vec{B}\right\}$$

$$\frac{D}{Dt}(p_\parallel B^2/\rho_m^{\,3}) = 0$$

$$\frac{D}{Dt}(p_\perp/\rho_m B) = 0$$

$$\vec{\nabla} \times \vec{B} = \mu_o\vec{J}$$

$$\vec{\nabla} \times \vec{E} = -\partial\vec{B}/\partial t$$

$$\vec{\nabla} \cdot \vec{E} = \rho/\varepsilon_o$$

$$\vec{E} + \vec{u} \times \vec{B} = 0$$

In the equations of this set, involving the pressure tensor $\bar{\bar{p}}$, it is considered that

$$\bar{\bar{p}} = \begin{pmatrix} p_{\perp} & 0 & 0 \\ 0 & p_{\perp} & 0 \\ 0 & 0 & p_{\parallel} \end{pmatrix}$$

(a) Taking the *equilibrium* mean velocity equal to zero, show that the dispersion relation for the magnetohydrodynamic waves is given by

$$\rho_{mo}\omega^2 + \cos\Theta \left(p_{\parallel} - p_{\perp} - \frac{B_o^2}{\mu_o} \right) - k^2 \sin^2\Theta \left(2p_{\perp} + \frac{B_o^2}{\mu_o} \right) =$$

$$= \frac{p_{\perp}^2 \, k^4 \, \sin^2\Theta \, \cos^2\Theta}{\rho_{mo}\omega^2 - 3p_{\parallel}k^2\cos^2\Theta}$$

where Θ is the angle between $\vec{k}$ and $\vec{B}_o$, and ρ_{mo}, $p_{\parallel}$, $p_{\perp}$ and $\vec{B}_o$ stand for the unperturbed quantities.

(b) Show that these waves are unstable for all values of Θ less than a critical angle Θ_c, which satisfies the equation

$$\frac{B_o^2}{\mu_o} + p_{\perp} \left(1 + \sin^2\Theta_c \right) = \frac{p_{\perp}^2}{3p_{\parallel}} \sin^2\Theta_c + 2p_{\parallel} \cos^2\Theta_c$$

CHAPTER 16

Waves in Cold Plasmas

1. INTRODUCTION

In this chapter we analyze the problem of wave propagation in cold plasmas, in which the thermal kinetic energy of the particles is ignored. The corresponding velocity distribution function is a Dirac delta function centered at the macroscopic fluid velocity. The study of waves in plasmas is very useful for plasma diagnostics, since it provides information on the plasma properties. The theory of wave propagation in a cold homogeneous plasma immersed in a magnetic field is known as *magnetoionic theory*.

There are two main different methods of approach that are normally used in analyzing the problem of waves in plasmas. In one of them, the plasma is characterized as a medium having either a conductivity or a dielectric constant and the wave equation for this medium is derived from Maxwell equations. In the presence of an externally applied magnetostatic field, the plasma is equivalent to an anisotropic dielectric characterized by a dielectric tensor or dyad. In another approach, Maxwell equations are solved simultaneously with the equations describing the particle motions. In this case we do not explicitly derive a wave equation and expressions for the dielectric or conductivity dyad are not obtained directly. Instead, we obtain a *dispersion relation*, which relates the wave number $\vec{k}$ to the wave frequency ω. All the information about the propagation of a given plasma mode is contained in the appropriate dispersion relation. This method of approach is often straightforward and simpler than the other one, and is the method we shall adopt in this treatment.

The neglect of the pressure term is justified if the particle thermal velocity is small when compared with the wave phase velocity. Therefore, the cold plasma model gives a satisfactory description except for waves with extremely small phase velocities. For waves with such small phase velocities, the pressure term becomes important and must be considered for a correct description. The propagation of waves in warm plasmas is the subject of the next chapter.

The study presented here is restricted to *small amplitude waves*, so that the analysis will be based on a *linear perturbation theory* under the assumption that the variations in the plasma parameters, due to the presence of waves, are small (to the first order) as compared to the undisturbed parameters. The plasma is assumed to be homogeneous and infinite (no boundary effects), and the externally applied magnetostatic field is assumed to be uniform. This medium is called a *magnetoionic medium*. Because of mathematical simplicity, the analysis will be made in terms of *plane waves*. This does not imply in loss of generality, since any physically realizable wave motion can always be synthesized in terms of plane waves.

In magnetoionic theory only the electron motion is considered. This is valid for *high frequency* waves i.e. for frequencies large compared to the ion cyclotron frequency. The theory of high-frequency small-amplitude plane waves propagating in an arbitrary direction with respect to the magnetostatic field, in a magnetoionic medium, is known as the Appleton-Hartree theory, in honor to E.V. Appleton and D.R. Hartree, who developed this theory when studying the problem of wave propagation in the Earth's ionosphere. At frequencies of the order of the ion cyclotron frequency, and smaller, the ion motion must be considered. The theory of wave propagation in a cold multicomponent plasma is commonly referred to as the hydromagnetic extension of magnetoionic theory.

2. BASIC EQUATIONS OF MAGNETOIONIC THEORY

In a cold electron gas the two hydrodynamic variables involved are the electron number density $n(\vec{r},t)$ and the average electron velocity $\vec{u}(\vec{r},t)$. They satisfy the continuity equation

$$\partial n/\partial t + \vec{\nabla} \cdot (n\vec{u}) = 0 \tag{2.1}$$

and the Langevin equation of motion

$$m(D\vec{u}/Dt) = q(\vec{E} + \vec{u} \times \vec{B}) - m\nu\vec{u} \tag{2.2}$$

These two equations are complemented by Maxwell equations

$$\vec{\nabla} \cdot \vec{E} = \rho/\varepsilon_0 \tag{2.3}$$

$$\vec{\nabla} \cdot \vec{B} = 0 \tag{2.4}$$

$$\vec{\nabla} \times \vec{E} = - \partial\vec{B}/\partial t \tag{2.5}$$

$$\vec{\nabla} \times \vec{B} = \mu_0(\vec{J} + \varepsilon_0 \partial\vec{E}/\partial t) \tag{2.6}$$

where the electron charge density in given by

$$\rho = - en \tag{2.7}$$

and the electric current density by

$$\vec{J} = \rho\vec{u} = - en\vec{u} \tag{2.8}$$

As previously discussed (2.4) is actually considered as an initial condition for (2.5). Furthermore, (2.3) and (2.6) can be combined to yield the electric charge conservation equation.

3. PLANE WAVE SOLUTIONS AND LINEARIZATION

Let us separate the total magnetic induction and the electron number density into two parts,

$$\vec{B}(\vec{r},t) = \vec{B}_0 + \vec{B}_1(\vec{r},t) \tag{3.1}$$

$$n(\vec{r},t) = n_0 + n_1(\vec{r},t) \tag{3.2}$$

where $\vec{B}_0$ is a constant and uniform field, and n_0 is the undisturbed electron number density in the absence of waves. Denoting by ψ_j any one of the components of the quantities $\vec{E}$, $\vec{B}_1$, $\vec{u}$ and n_1 we can write, for harmonic plane wave solutions,

$$\psi_j(\vec{r},t) = \psi_j \exp\ [\ i(\vec{k} \cdot \vec{r} - \omega t)\ ] \tag{3.3}$$

where $\vec{k}$ is the wave propagation vector and ω is the wave frequency. The use of the same symbol to denote the complex amplitude as well as the entire expression in Eq. (3.3) should lead to no confusion, because in linear wave theory the same exponential factor will occur on both sides of any equation and can be cancelled out.

Eq. (2.2) is not yet quite tractable because of the *nonlinear* terms $(\vec{u} \cdot \vec{\nabla})\vec{u}$ and $\vec{u} \times \vec{B}$. This difficulty can be eliminated considering $\vec{u}$ and $\vec{B}_1$ as first order quantities and neglecting second order terms. As discussed in section 3, Chapter 10, when dealing with wave phenomena in plasmas the second order nonlinear term $\vec{u} \times \vec{B}_1$ can be neglected provided the average electron velocity is much less than the wave phase velocity ($u \ll \omega/k$).

For harmonic plane wave solutions the differential operators $\vec{\nabla}$ and $\partial/\partial t$ are replaced, respectively, by $i\vec{k}$ and $-i\omega$. Consequently, the differential equations become simple algebraic equations. Therefore, (2.2), (2.5) and (2.6) become, respectively, neglecting second order terms,

$$- i\omega m\vec{u} = -e(\vec{E} + \vec{u} \times \vec{B}_0) - m\nu\vec{u} \tag{3.4}$$

$$\vec{k} \times \vec{E} = \omega \vec{B}_1 \tag{3.5}$$

$$i\vec{k} \times \vec{B}_1 = \mu_0 (-e\, n_0\, \vec{u} - i\,\omega\,\varepsilon_0\, \vec{E}) \tag{3.6}$$

where use was made of (2.8) linearized. These three equations involving the dependent variables $\vec{u}$, $\vec{E}$ and $\vec{B}_1$ can be used to derive a *dispersion relation* for wave propagation in a cold electron gas. In order to keep matters as simple as possible, we investigate initially the characteristics of wave propagation in a cold isotropic plasma with $\vec{B}_0 = 0$.

4. WAVE PROPAGATION IN ISOTROPIC ELECTRON PLASMAS

4.1 Derivation of the dispersion relation

In the absence of an externally applied magnetic field ($\vec{B}_0 = 0$) the Langevin equation (3.4) gives

$$\vec{u} = - \frac{e}{m(\nu - i\omega)} \vec{E} \tag{4.1}$$

Combining (3.5), (3.6) and (4.1), we obtain

$$\vec{k} \times (\vec{k} \times \vec{E}) = - \frac{i\omega\mu_0 e^2 n_0}{m(\nu - i\omega)} \vec{E} - \frac{\omega^2}{c^2} \vec{E} \tag{4.2}$$

It is convenient to separate the electric field vector in a *longitudinal component* $\vec{E}_\ell$ (parallel to $\vec{k}$) and a *transverse component* $\vec{E}_t$ (perpendicular to $\vec{k}$),

$$\vec{E} = \vec{E}_\ell + \vec{E}_t \tag{4.3}$$

as shown in Fig. 1. Therefore,

$$(\vec{k} \times \vec{E}_\ell) = 0 \tag{4.4}$$

$$\vec{k} \times (\vec{k} \times \vec{E}_t) = -k^2 \vec{E}_t \tag{4.5}$$

and (4.2) becomes

$$- k^2 \vec{E}_t = - \left[\frac{i\omega\mu_0 e^2 n_0}{m(\nu - i\omega)} + \frac{\omega^2}{c^2} \right] (\vec{E}_\ell + \vec{E}_t) \tag{4.6}$$

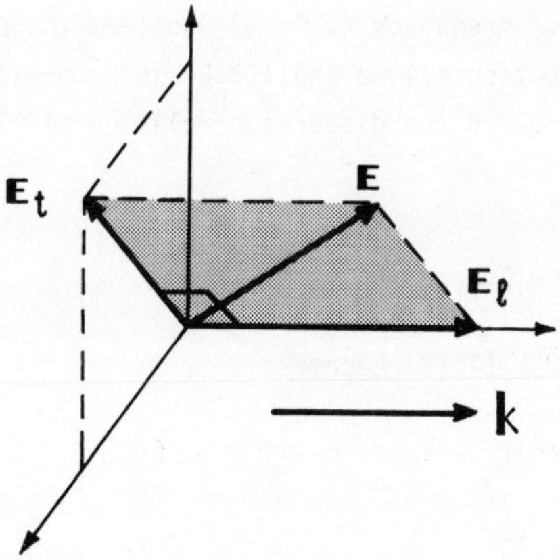

Fig. 1 - Longitudinal and transverse components of the electric field vector with respect to the propagation vector $\vec{k}$.

This equation can be separated in a *longitudinal component*,

$$\left[\frac{\omega_{pe}^2}{c^2(1 + i\nu/\omega)} - \frac{\omega^2}{c^2} \right] \vec{E}_\ell = 0 \tag{4.7}$$

and a *transverse component*,

$$- k^2 \vec{E}_t = \left[\frac{\omega_{pe}^2}{c^2(1 + i\nu/\omega)} - \frac{\omega^2}{c^2} \right] \vec{E}_t \tag{4.8}$$

(4.7) yields the following dispersion relation for a *longitudinal mode*

$$\omega^2(1 + i\nu/\omega) - \omega_{pe}^2 = 0; \; (\vec{E}_\ell \neq 0) \tag{4.9}$$

For a *transverse mode* the dispersion relations is, from (4.8),

$$(\omega^2 - k^2c^2)\,(1 + i\nu/\omega) - \omega_{pe}^2 = 0; \; (\vec{E}_t \neq 0) \tag{4.10}$$

4.2 Collisionless plasma

For simplicity we consider first the case in which the collision frequency is much less than the wave frequency ($\nu \ll \omega$), so that the effect of collisions can be ignored. In sub-section 4.4 we shall take into consideration the effect of collisions. Thus, for $\nu = 0$ the dispersion relation (4.9) for *longitudinal waves* becomes

$$\omega^2 = \omega_{pe}^2 \tag{4.11}$$

while, for *transverse waves* (4.10) becomes

$$k^2c^2 = \omega^2 - \omega_{pe}^2 \tag{4.12}$$

Eq. (4.11) shows that *longitudinal oscillations* ($\vec{E}_\ell \neq 0$) can occur just at the plasma frequency ω_{pe}. These longitudinal oscillations are just the *plasma oscillations* discussed in section 1, Chapter 11. It is seen from (4.1), that the electrons oscillate with a velocity given by

$$\vec{u} = - (ie/m\omega)\, \vec{E}_\ell \tag{4.13}$$

From (4.4) and (3.5) it is clear that $\vec{B}_1 = 0$, so that there is no magnetic field associated with these longitudinal oscillations. Further, there is no wave propagation, since there is no relative phase variation from point to point. These oscillations are therefore longitudinal, electrostatic and stationary. In the next chapter we consider wave propagation in a *warm* plasma, where we show that these electron plasma oscillations correspond to the limit, when the electron temperature goes to zero, of the longitudinal mode of propagation called the *electron plasma wave*.

Considering now the dispersion relation (4.12) for *transverse waves* ($\vec{E}_t \neq 0$), it is seen that k^2 is positive for $\omega > \omega_{pe}$ and negative for $\omega < \omega_{pe}$. Hence, for travelling waves (with ω real) k becomes imaginary for $\omega < \omega_{pe}$. Writing $k = \beta + i\alpha$, where β and α are real quantities, it is seen, from (4.12), that for $\omega > \omega_{pe}$ ($k = \beta$; $\alpha = 0$) the transverse wave propagates with a *phase velocity* (ω divided by the real part of k) given by

$$v_{ph} = \omega/k = c/(1 + \omega_{pe}^2/\omega^2)^{1/2} \qquad (\omega > \omega_{pe}) \qquad (4.14)$$

Also, for $\omega > \omega_{pe}$ the *group velocity* of the transverse wave can be obtained differentiating (4.12) with respect to k,

$$v_g = \partial\omega/\partial k = c^2/v_{ph} \qquad (\omega > \omega_{pe}) \qquad (4.15)$$

For $\omega < \omega_{pe}$, k is imaginary ($k = i\alpha$) and the transverse wave is exponentially damped, since

$$E_t \propto \exp(ik\zeta - i\omega t) = \exp(-\alpha\zeta)\exp(-i\omega t) \qquad (4.16)$$

so that the wave dies out with increasing values of ζ. Such exponentially damped fields are called *evanescent waves* and do not transport any time-averaged power. Since $\beta = 0$, it is easily seen that in this case,

$$v_{ph} = \infty \qquad (\omega < \omega_{pe}) \qquad (4.17)$$

$$v_g = 0 \qquad (\omega < \omega_{pe}) \qquad (4.18)$$

Also, from (4.12) we find, for $\omega < \omega_{pe}$,

$$\alpha = \mathrm{Im}(k) = (\omega_{pe}^2 - \omega^2)^{1/2}/c \qquad (4.19)$$

where Im denotes the "imaginary part of".

A plot of phase velocity and group velocity as a function of frequency is shown in Fig. 2(a), whereas the frequency dependence of the attenuation factor α is shown in Fig. 2(b). Note that the phase velocity is always greater than the velocity of light c, but the group velocity, which is the velocity at which a signal propagates, is always less than c, in agreement with the requirements of the theory of relativity. For $\omega >> \omega_{pe}$ we find, from (4.14) and (4.15),

$$v_{ph} = v_g = c \qquad (\omega >> \omega_{pe}) \tag{4.20}$$

which shows that for very high frequencies the plane wave characteristics of a plasma degenerate to those of free space. This is expected on a physical basis, since in the limiting case of infinite frequency even the electrons are unable to respond to the oscillating electric field.

The dispersion relation (4.12) is plotted in Fig. 3 in terms of ω as a function of the real part of k. In this text we shall follow the usual graphic representation of dispersion relations, which is plotting ω versus k, rather than k versus ω. The frequency region in which the transverse wave is evanescent is the region for which $\omega < \omega_{pe}$.

From the relation $v_g = \partial\omega/\partial k$ we notice that at a given point in the $\omega(k)$ curve the group velocity is equal to the slope of the tangent to the curve at that point, whereas the phase velocity ω/k is equal to the slope of the line drawn from the origin to this point. This geometrical representation is illustrated in Fig. 3.

4.3 Time - averaged Poynting vector

We evaluate next the time-averaged Poynting vector $< \vec{S} >$, which gives the time averaged power flow for the transverse wave. From (3.5), taking $\vec{B}_1 = \mu_0 \vec{H}_1$, we have

$$\vec{H}_1 = \vec{k} \times \vec{E}/(\mu_0 \omega) \tag{4.21}$$

and the expression for $< \vec{S} >$, given in (14.5.13), becomes

$$< \vec{S} > = (1/2) \text{ Re } (\vec{E} \times \vec{H}_1^*) = \text{Re } [\vec{E} \times (\vec{k}^* \times \vec{E}^*)]/2\mu_0\omega$$

$$= \hat{n} \text{ Re } [k^* E(\vec{r},t) E^*(\vec{r},t)]/2\mu_0\omega \tag{4.22}$$

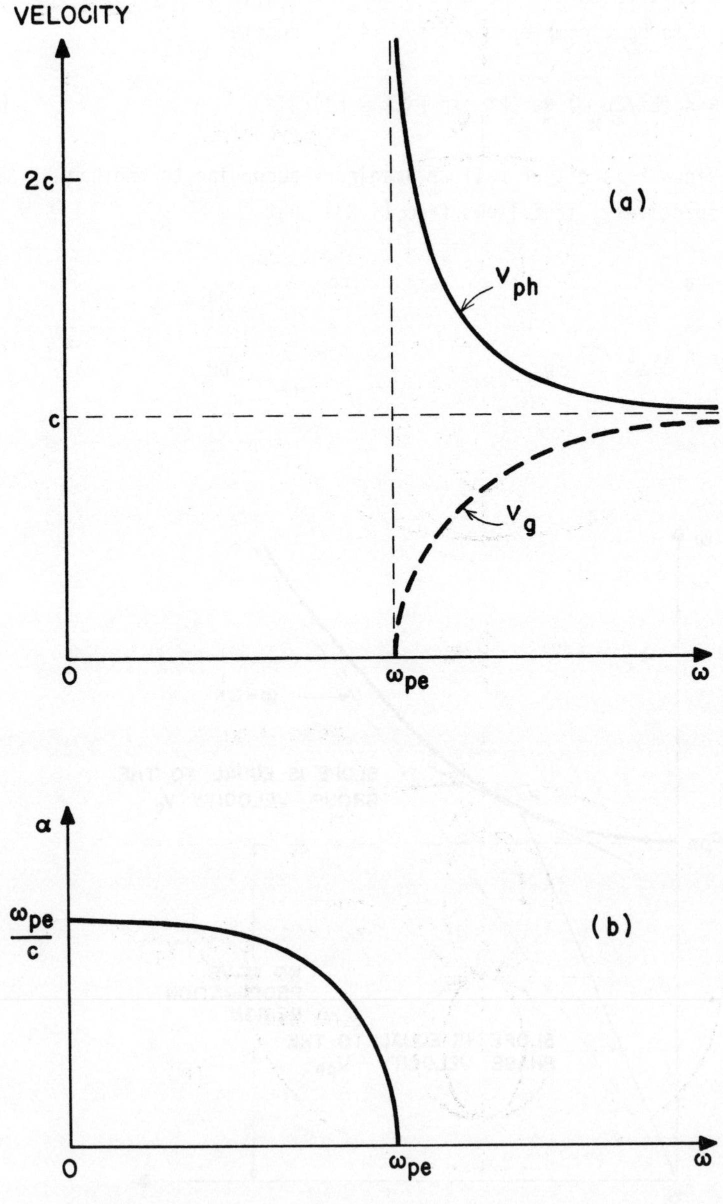

Fig. 2 - Frequency dependence of the phase velocity, group velocity and the attenuation factor α for transverse waves in a collisionless isotropic cold electron gas.

where $\hat{n}$ is a unit vector in the direction of $\vec{E} \times \vec{H}_1$. Using (3.3) and considering k to be a complex quantity, (4.22) becomes

$$< \vec{S} > = \hat{n} \, (E^2/2\mu_0\omega) \; \text{Re} \, \{k^* \exp [i(k - k^*)\zeta]\} \tag{4.23}$$

Therefore, since k is either real or imaginary according to whether $\omega > \omega_{pe}$ or $\omega < \omega_{pe}$, respectively, if follows from (4.23) that

$$< \vec{S} > = 0 \qquad \text{for} \quad \omega < \omega_{pe} \tag{4.24}$$

$$< \vec{S} > = \hat{n} \, (\varepsilon_0 E^2/2) \, v_g \qquad \text{for} \quad \omega > \omega_{pe} \tag{4.25}$$

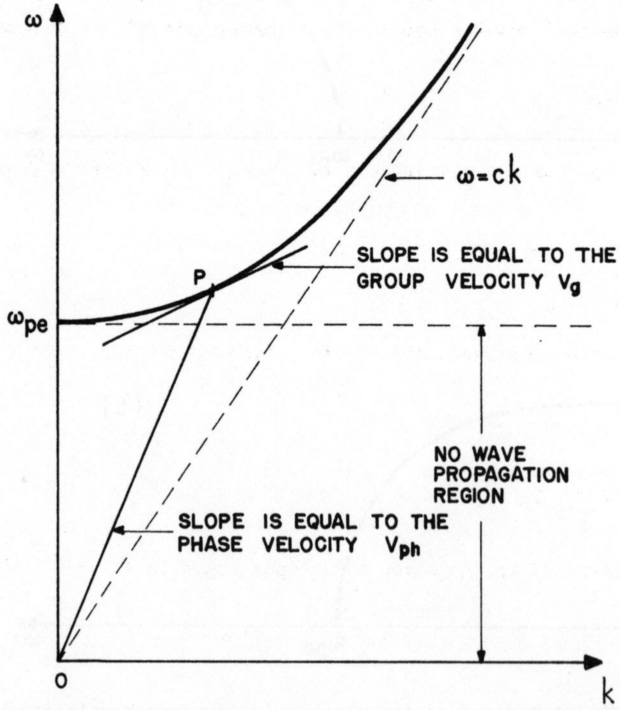

Fig. 3 - Dispersion relation $\omega(k)$ for the tranverse wave propagating in an isotropic cold electron plasma. Note the geometrical representation of the phase and group velocities at the point P.

where, in (4.25), we have used the relation $c^2k/\omega = v_g$ given in (4.15). Thus, for $\omega > \omega_{pe}$ the fields transport power in the direction $\hat{E} \times \hat{H}_1$, whereas for $\omega < \omega_{pe}$ there is no power flow and the wave is evanescent. For this reason, the region $\omega > \omega_{pe}$ is called the *propagation region*. Since the wave is totally reflected for $\omega < \omega_{pe}$, the frequency ω_{pe} is often called a *reflection point* (where v_{ph} is infinite). It can be shown that the power transmitted into a *semi-infinite* slab of plasma is zero if $\beta = \text{Re}(k)$ is zero, so that in a more general sense any frequency for which $\beta = 0$ ($v_{ph} = \infty$) is referred to as a reflection point. However, if the plasma medium is *finite*, some energy can be transmitted through the finite plasma slab even if $\beta = 0$. This effect is known as the *tunneling effect*.

4.4 The effect of collisions

The principal effect of collisions is to produce a *damping* of the waves. Before considering the dispersion relations (4.9) and (4.10), it is useful to discuss some general results concerning dispersion relations of the form

$$k^2 = A + iB \tag{4.26}$$

where A and B are real quantities. If we separate k into its real and imaginary parts,

$$k = \beta + i\alpha \tag{4.27}$$

where β and α are both real, then it is a simple matter to verify that

$$A = \text{Re}(k^2) = \beta^2 - \alpha^2 \tag{4.28}$$

$$B = \text{Im}(k^2) = 2\beta\alpha \tag{4.29}$$

On the other hand, since the waves are proportional to $\exp(ik\zeta - i\omega t)$, we have

$$\exp(ik\zeta - i\omega t) = \exp(-\alpha\zeta)\exp(i\beta\zeta - i\omega t) \tag{4.30}$$

Thus, the sign of β determines the *direction* of wave propagation i.e. $\beta > 0$ implies propagation in the *positive* ζ direction, whereas $\beta < 0$ implies propagation in the *negative* ζ direction. The sign of α is related to *growing* or *damping* of the wave amplitude as the wave propagates. If both α and β are positive then the wave travels in the positive ζ direction and is exponentially damped. If both α and β are negative then the wave travels in the negative ζ direction and is also exponentially damped. On the other hand, if α and β have opposite signs then the wave is exponentially growing (see Fig. 4). In any case, the sign of B determines whether the travelling wave is growing or decaying. For $B > 0$ the wave is *damped* with distance, whereas for $B < 0$ the wave *grows*.

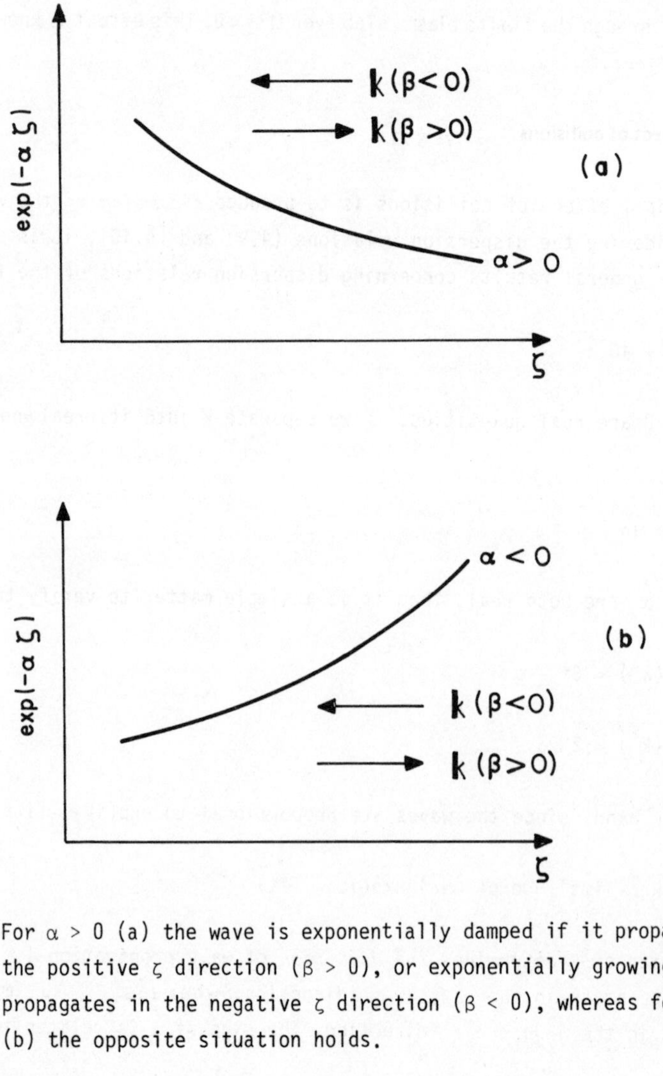

Fig. 4 - For $\alpha > 0$ (a) the wave is exponentially damped if it propagates in the positive ζ direction ($\beta > 0$), or exponentially growing if it propagates in the negative ζ direction ($\beta < 0$), whereas for $\alpha < 0$ (b) the opposite situation holds.

Similarly, for a dispersion relation having the form

$$\omega^2 = C + iD \tag{4.31}$$

it can be easily verified that, for standing waves, if $D > 0$ the wave *grows* in time, whereas if $D < 0$ the wave is *damped*.

Consider now the dispersion relation (4.9) for the *longitudinal oscillation*,

$$\omega^2 + i\nu\omega + \omega_{pe}^2 = 0 \tag{4.32}$$

or

$$\omega = [\,-\,i\nu \pm (4\omega_{pe}^2 - \nu^2)^{1/2}]/2 \tag{4.33}$$

This equation shows that for any value of ν the imaginary part of ω is negative, so that the oscillation is *damped*, since it is proportional to $\exp(-\,i\omega t)$.

For the *transverse mode* the dispersion relation (4.10) gives

$$k^2c^2 = \omega^2 - \frac{\omega_{pe}^2}{1 + (\nu/\omega)^2} + \frac{i\omega_{pe}^2(\nu/\omega)}{1 + (\nu/\omega)^2} \tag{4.34}$$

Consequently, in the propagation band B is negative and the travelling waves are *damped* for all frequencies. A plot of the attenuation factor $\alpha = \mathrm{Im}(k)$, as a function of the collision frequency ν, is shown in Fig. 5, calculated from (4.34) for a given frequency $\omega \gg \omega_{pe}$. Fig. 6 shows the dispersion relation $\omega(k)$ for the transverse mode of propagation, considering several values of ν, where $\nu_3 > \nu_2 > \nu_1 > 0$.

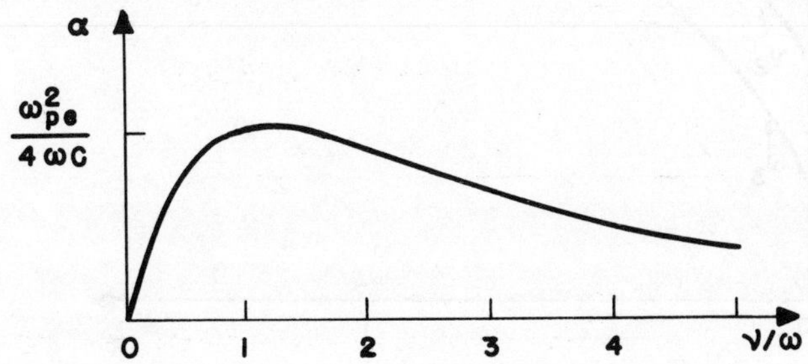

Fig. 5 - Attenuation factor, α, as a function of collision frequency for a given frequency such that $\omega \gg \omega_{pe}$, for the transverse wave.

5. WAVE PROPAGATION IN MAGNETIZED COLD PLASMAS

Consider now the problem of wave propagation in a cold electron plasma when there is a uniform magnetostatic field externally applied. The presence of the magnetostatic field $\vec{B}_0$ introduces an anisotropy in the plasma.

5.1 Derivation of the dispersion relation

To derive the dispersion relation for this case, we start from the coupled set of equations (3.4), (3.5) and (3.6). Combining (3.5) and (3.6), and rearranging, this set reduces to

$$\vec{k} \times (\vec{k} \times \vec{E}) + (\omega^2/c^2)\,\vec{E} = (i\omega e n_0/c^2\varepsilon_0)\,\vec{u} \tag{5.1}$$

$$(1 + i\nu/\omega)\,\vec{u} + (ie/\omega m)\,(\vec{u} \times \vec{B}_0) = -\,(ie/\omega m)\,\vec{E} \tag{5.2}$$

If we denote the angle between $\vec{B}_0$ and $\vec{k}$ by θ and choose a Cartesian

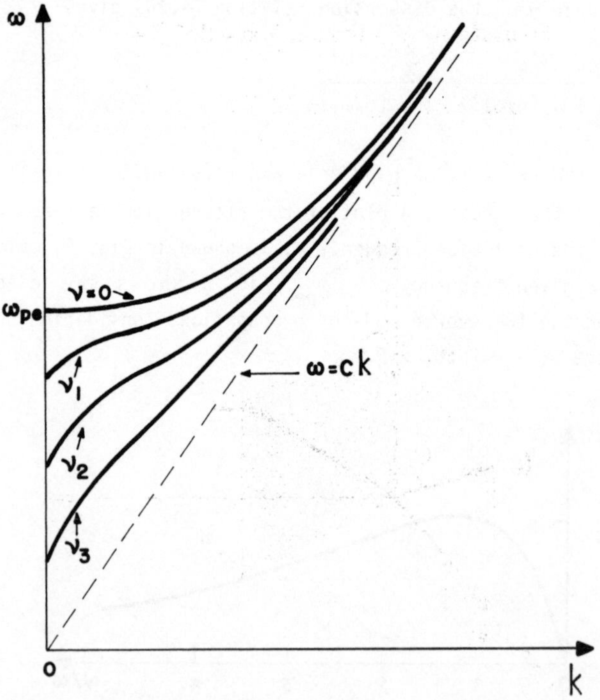

Fig. 6 - Plot of the dispersion relation $\omega(k)$ for transverse waves in an isotropic cold electron plasma, considering the effects of collisions ($\nu_3 > \nu_2 > \nu_1 > 0$).

coordinate system in which $\hat{z}$ is in the direction of $\vec{B}_0$, and $\hat{y}$ is perpendicular to the plane formed by $\vec{B}_0$ and $\vec{k}$ (Fig. 7), we have

$$\vec{B}_0 = B_0\,\hat{z} \tag{5.3}$$

$$\vec{k} = k_\perp\hat{x} + k_\parallel\hat{z} = k\sin\theta\,\hat{x} + k\cos\theta\,\hat{z} \tag{5.4}$$

Note that the index symbols $\perp$ and $\parallel$ are used to denote components perpendicular and parallel to the direction of the magnetostatic field, $\vec{B}_0$, whereas the indices ℓ and t (used in the previous section) refer to components longitudinal and transverse with respect to the wave vector $\vec{k}$, respectively.

With this choice of coordinate system, we have

$$\vec{k}\times(\vec{k}\times\vec{E}) = (k_\perp\hat{x} + k_\parallel\hat{z})\times[(k_\perp\hat{x} + k_\parallel\hat{z})\times(E_x\hat{x} + E_y\hat{y} + E_z\hat{z})]$$

$$= k^2\cos\theta(\sin\theta\,E_z - \cos\theta\,E_x)\,\hat{x} - k^2E_y\hat{y} +$$

$$+ k^2\sin\theta\,(\cos\theta\,E_x - \sin\theta\,E_z)\,\hat{z} \tag{5.5}$$

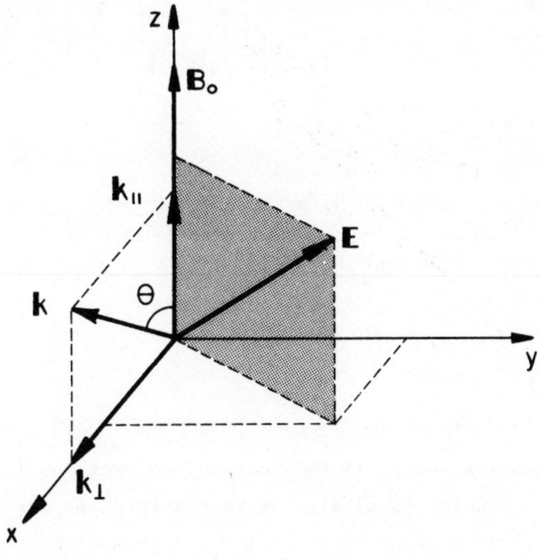

Fig. 7 - Set of retangular coordinates (x,y,z), chosen with $\hat{z}$ along $\vec{B}_0$ and $\hat{y}$ perpendicular to the plane formed by $\vec{B}_0$ and $\vec{k}$.

Using this result in (5.1), we find the following relations for the x, y and z components of this equation,

$$\hat{x}: \quad \left(1 - \frac{k^2c^2}{\omega^2} \cos^2 \theta\right) E_x + \left(\frac{k^2c^2}{\omega^2} \sin\theta \cos \theta\right) E_z = \left(\frac{ien_0}{\omega\varepsilon_0}\right) u_x \tag{5.6}$$

$$\hat{y}: \quad \left(1 - \frac{k^2c^2}{\omega^2}\right) E_y = \left(\frac{ien_0}{\omega\varepsilon_0}\right) u_y \tag{5.7}$$

$$\hat{z}: \quad \left(\frac{k^2c^2}{\omega^2} \sin\theta \cos\theta\right) E_x + \left(1 - \frac{k^2c^2}{\omega^2} \sin^2 \theta\right) E_z = \left(\frac{ien_0}{\omega\varepsilon_0}\right) u_z \tag{5.8}$$

which can be written, in matrix form, as

$$\begin{pmatrix} \left(1 - \frac{k^2c^2}{\omega^2} \cos^2 \theta\right) & 0 & \frac{k^2c^2}{\omega^2} \sin\theta \cos\theta \\ 0 & \left(1 - \frac{k^2c^2}{\omega^2}\right) & 0 \\ \frac{k^2c^2}{\omega^2} \sin\theta \cos\theta & 0 & \left(1 - \frac{k^2c^2}{\omega^2} \sin^2\theta\right) \end{pmatrix} \begin{pmatrix} E_x \\ E_y \\ E_z \end{pmatrix} =$$

$$= \frac{ien_0}{\omega\varepsilon_0} \begin{pmatrix} u_x \\ u_y \\ u_z \end{pmatrix} \tag{5.9}$$

Note that the quantity kc/ω is the *index of refraction* of the medium.

Next, in order to write (5.2) also in matrix form, we first note that (Fig. 7)

$$\vec{u} \times \vec{B}_0 = B_0 \, (u_y\hat{x} - u_x\hat{y}) \tag{5.10}$$

Using this result in (5.2), and after some algebraic manipulations, we obtain for the x, y, and z components of this equation

$$\hat{x}: \quad (1 + i\nu/\omega)\, u_x + (i\omega_{ce}/\omega)\, u_y = - (ie/\omega m)\, E_x \tag{5.11}$$

$$\hat{y}: \quad - (i\omega_{ce}/\omega)\, u_x + (1 + i\nu/\omega)\, u_y = - (ie/\omega m)\, E_y \tag{5.12}$$

$$\hat{z}: \quad (1 + i\nu/\omega)\, u_z = - (ie/\omega m)\, E_z \tag{5.13}$$

Introducing now the notation

$$U = 1 + i\nu/\omega \tag{5.14}$$

$$Y = \omega_{ce}/\omega \tag{5.15}$$

$$X = \omega_{pe}^2/\omega^2 \tag{5.16}$$

we can write Eqs. (5.11) to (5.13) in matrix form as

$$\begin{pmatrix} U & iY & 0 \\ -iY & U & 0 \\ 0 & 0 & U \end{pmatrix} \begin{pmatrix} u_x \\ u_y \\ u_z \end{pmatrix} = - \frac{ie}{m\omega} \begin{pmatrix} E_x \\ E_y \\ E_z \end{pmatrix} \tag{5.17}$$

Inverting the 3 x 3 matrix of (5.17), and multiplying this equation by the inverted matrix, we find

$$- \left(\frac{ie}{m\omega}\right) \frac{1}{U(U^2 - Y^2)} \begin{pmatrix} U^2 & -iUY & 0 \\ iUY & U^2 & 0 \\ 0 & 0 & (U^2 - Y^2) \end{pmatrix} \begin{pmatrix} E_x \\ E_y \\ E_z \end{pmatrix} = \begin{pmatrix} u_x \\ u_y \\ u_z \end{pmatrix} \tag{5.18}$$

Eqs. (5.9) and (5.18) can now be combined to eliminate the velocity components u_x, u_y and u_z, yielding the following component equations involving only the electric field,

$$\hat{x}: \left(-\frac{XU}{U^2-Y^2} + 1 - \frac{k^2c^2}{\omega^2}\cos^2\theta\right) E_x + \left(\frac{iXY}{U^2-Y^2}\right) E_y + $$

$$+ \left(\frac{k^2c^2}{\omega^2}\sin\theta\cos\theta\right) E_z = 0 \qquad (5.19)$$

$$\hat{y}: -\left(\frac{iXY}{U^2-Y^2}\right) E_x + \left(-\frac{XU}{U^2-Y^2} + 1 - \frac{k^2c^2}{\omega^2}\right) E_y = 0 \qquad (5.20)$$

$$\hat{z}: \left(\frac{k^2c^2}{\omega^2}\sin\theta\cos\theta\right) E_x + \left(-\frac{X}{U} + 1 - \frac{k^2c^2}{\omega^2}\sin^2\theta\right) E_z = 0 \qquad (5.21)$$

For reasons to become apparent later, it is appropriate to define the following quantities

$$S = 1 - XU/(U^2-Y^2) \qquad (5.22)$$

$$D = -XY/(U^2-Y^2) \qquad (5.23)$$

$$P = 1 - X/U \qquad (5.24)$$

With this notation, Eqs. (5.19) to (5.21) can be written in matrix form as

$$\begin{pmatrix} \left(S - \frac{k^2c^2}{\omega^2}\cos^2\theta\right) & -iD & \frac{k^2c^2}{\omega^2}\sin\theta\cos\theta \\ iD & \left(S - \frac{k^2c^2}{\omega^2}\right) & 0 \\ \frac{k^2c^2}{\omega^2}\sin\theta\cos\theta & 0 & \left(P - \frac{k^2c^2}{\omega^2}\sin^2\theta\right) \end{pmatrix} \begin{pmatrix} E_x \\ E_y \\ E_z \end{pmatrix} = 0 \qquad (5.25)$$

In order to have a nontrivial solution ($\vec{E} \neq 0$), the determinant of the 3 x 3 matrix in (5.25) must vanish. This condition gives the following *dispersion relation*, by direct calculation of the determinant,

$$(S\sin^2\theta + P\cos^2\theta)(k^2c^2/\omega^2)^2 - [RL\sin^2\theta + SP(1+\cos^2\theta)](k^2c^2/\omega^2) + PRL = 0 \qquad (5.26)$$

where

$$R = S + D \qquad \text{or} \qquad S = (R + L)/2 \tag{5.27}$$

$$L = S - D \qquad\qquad D = (R - L)/2 \tag{5.28}$$

Since (5.26) is a quadratic equation in k^2c^2/ω^2, there will be two solutions. Thus, at each frequency there can be in general two types of waves that propagate or two *modes of propagation*. Note, however, that if we take the square root of k^2c^2/ω^2 we have two values for kc/ω, which correspond to opposite directions of propagation and not to two different modes.

5.2 The Appleton - Hartree equation

This well known equation has been used with considerable success to study radio wave propagation in the ionosphere, taking account of the Earth's magnetic field. It is just the dispersion relation (5.26), but written in a different form. In order to obtain the Appleton-Hartree equation, we first write (5.26) as

$$A\,(k^2c^2/\omega^2)^2 - B\,(k^2c^2/\omega^2) + C = 0 \tag{5.29}$$

where

$$A = (S \sin^2\theta + P \cos^2\theta) \tag{5.30}$$

$$B = RL \sin^2\theta + SP\,(1 + \cos^2\theta) \tag{5.31}$$

$$C = PRL \tag{5.32}$$

Solving (5.29) for k^2c^2/ω^2, we find

$$k^2c^2/\omega^2 = (B \pm \sqrt{B^2 - 4AC})/2A \tag{5.33}$$

Now, we add the quantity $A(k^2c^2/\omega^2)$ to both sides of (5.29) and rearrange, to obtain

$$\frac{k^2c^2}{\omega^2} = \frac{A(k^2c^2/\omega^2) - C}{A(k^2c^2/\omega^2) + A - B} \tag{5.34}$$

Next, we substitute k^2c^2/ω^2 from (5.33), into the right-hand side of (5.34) and manipulate, obtaining

$$\frac{k^2c^2}{\omega^2} = 1 - \frac{2(A - B + C)}{2A - B \pm \sqrt{B^2 - 4AC}} \tag{5.35}$$

Finally, we substitute the appropriate expressions which define the quantities A, B, C and S, D, P, R, L, to obtain

$$\frac{k^2c^2}{\omega^2} = 1 - \frac{X}{U - \dfrac{Y^2\sin^2\theta}{2(U - X)} \pm \left[\dfrac{Y^4\sin^4\theta}{4(U - X)^2} + Y^2 \cos^2\theta\right]^{1/2}} \tag{5.36}$$

This is the *Appleton-Hartree equation*. It is valid for high wave frequencies as compared to the *ion cyclotron* frequency, since ion motion was neglected.

Because of the complexity of either (5.36) or (5.26), in order to simplify matters we shall first analyze the wave propagation problem when $\vec{k}$ is either parallel to, or perpendicular to $\vec{B}_0$. Afterwards, it will be easier to analyze some important aspects of wave propagation at an arbitrary angle θ with respect to $\vec{B}_0$ using the dispersion relation (5.36) or (5.26).

6. PROPAGATION PARALLEL TO B

For wave propagation in the direction of $\vec{B}_0$ ($\vec{k} \parallel \vec{B}_0$) we have $\theta = 0$, and (5.25) simplifies to

$$\begin{pmatrix} (S - k^2c^2/\omega^2) & -iD & 0 \\ iD & (S - k^2c^2/\omega^2) & 0 \\ 0 & 0 & P \end{pmatrix} \begin{pmatrix} E_x \\ E_y \\ E_z \end{pmatrix} = 0 \tag{6.1}$$

For a nontrivial solution ($\vec{E} \neq 0$), we must require the determinant of the 3 x 3 matrix in (6.1) to vanish. Thus, by direct calculation of the determinant we find three independent conditions

$$P = 0 \; ; \; (\vec{E}_{\parallel} \equiv \vec{E}_{\ell} \neq 0) \tag{6.2}$$

$$(k^2c^2/\omega^2) = S + D \equiv R \; ; \; (\vec{E}_\perp \equiv \vec{E}_t \neq 0) \tag{6.3}$$

$$(k^2c^2/\omega^2) = S - D \equiv L \; ; \; (\vec{E}_\perp \equiv \vec{E}_t \neq 0) \tag{6.4}$$

Using Eqs. (5.22) to (5.24), and (5.14) to (5.16), which define S,D, P and U, Y, X, respectively, we obtain from (6.2), neglecting collisions ($\nu = 0$),

$$\omega^2 = \omega_{pe}^2 \tag{6.5}$$

which corresponds to the *longitudinal electron plasma oscillations* discussed previously in section 4. Thus, these oscillations along $\vec{B}_0$ are not affected by the presence of the magnetic field. Since there is no wave propagation in this case, these plasma oscillations do not constitute a mode of propagation.

Eq. (6.3) corresponds to *transverse right-hand circularly polarized waves* (RCP), with the dispersion relation

$$(k^2c^2/\omega^2)_R = 1 - X/(U-Y) = R \tag{6.6}$$

or, neglecting collisions ($\nu = 0$),

$$(k^2c^2/\omega^2)_R = 1 - \omega_{pe}^2/[\omega(\omega-\omega_{ce})] \tag{6.7}$$

Eq. (6.4) corresponds to *transverse left-hand circularly polarized waves* (LCP), with the dispersion relation

$$(k^2c^2/\omega^2)_L = 1 - X/(U+Y) = L \tag{6.8}$$

or, neglecting collisions ($\nu = 0$),

$$(k^2c^2/\omega^2)_L = 1 - \omega_{pe}^2/[\omega(\omega+\omega_{ce})] \tag{6.9}$$

The *polarization* of these two modes of propagation can be obtained from the x component of (6.1), which gives

$$E_x/E_y = iD/(S - k^2c^2/\omega^2) \tag{6.10}$$

Thus, for the *RCP wave*, substituting $k^2c^2/\omega^2 = R$,

$$E_x/E_y = -i \tag{6.11}$$

whereas for the *LCP wave*, substituting $k^2c^2/\omega^2 = L$,

$$E_x/E_y = i \tag{6.12}$$

Since the time dependence of $\vec{E}$ is of the form exp $(-i\omega t)$, if we take $E_x \propto \cos(\omega t)$ then for the RCP wave we have $E_y \propto \sin(\omega t)$, whereas for the LCP wave we have $E_y \propto -\sin(\omega t)$. Therefore, for an observer looking at the outgoing wave, as time passes the transverse electric field vector $\vec{E}_t$ rotates in the clockwise direction for the RCP wave, and in the counterclockwise direction for the LCP wave. This is illustrated in Fig. 8. Note that the RCP wave rotates in the same direction as the electrons about the $\vec{B}_0$ field. This means that, when $\omega = \omega_{ce}$, the RCP wave is in *resonance* with the electron cyclotron motion and therefore energy is transferred from the wave to the electrons. This absorption of energy by the electrons, from the RCP electromagnetic wave, at the electron cyclotron frequency, is used as a means of heating the plasma electrons. When the motion of the ions is taken into account, a resonance exists between the ion cyclotron motion and the LCP wave, at $\omega = \omega_{ci}$, since the ions gyrate in the same direction as the $\vec{E}_t$ vector of the LCP wave.

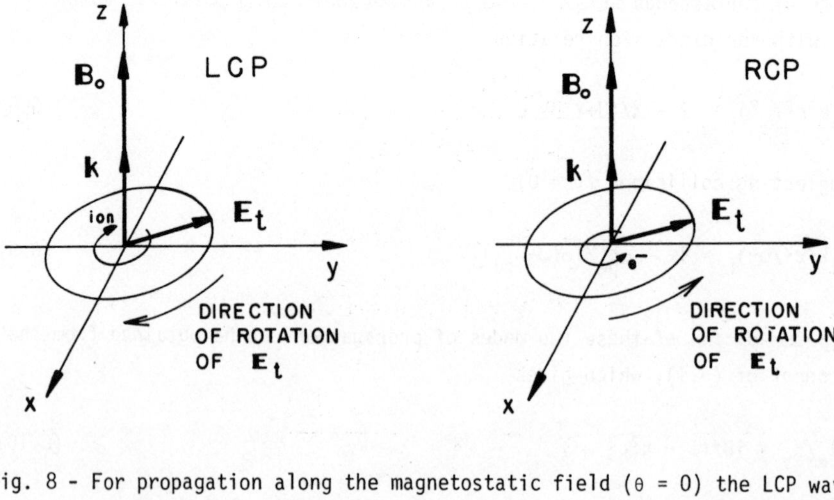

Fig. 8 - For propagation along the magnetostatic field ($\theta = 0$) the LCP wave rotates in the counterclockwise direction, and the RCP wave rotates in the clockwise direction, for an observer looking at the outgoing wave.

The phenomenon of *resonance* occurs when the phase velocity goes to zero, $v_{ph} = 0$ (or $kc/\omega \to \infty$), whereas *reflection* occurs when $v_{ph} \to \infty$ (or $kc/\omega = 0$). Thus, it is clear from (6.7) and from the physical argument just given that the RCP wave has a resonance at $\omega = \omega_{ce}$, whereas (6.9) indicates no resonance for the LCP wave (when ion motion is included the LCP wave has a resonance at ω_{ci}). Also, from (6.9) it is easily verified that the LCP wave has a reflection point (L = 0) at

$$\omega_{01} = [-\omega_{ce} + (\omega_{ce}^2 + 4\omega_{pe}^2)^{1/2}]/2 \tag{6.13}$$

and, from (6.7), the RCP wave has a reflection point (R = 0) at

$$\omega_{02} = [\omega_{ce} + (\omega_{ce}^2 + 4\omega_{pe}^2)^{1/2}]/2 = \omega_{01} + \omega_{ce} \tag{6.14}$$

The *phase velocity* of the *LCP wave* is obtained, from (6.9), as

$$(v_{ph})_L = \left(\frac{\omega}{k}\right)_L = \frac{c(1+\omega_{ce}/\omega)^{1/2}}{(1+\omega_{ce}/\omega-\omega_{pe}^2/\omega^2)^{1/2}} \quad ; \ (\omega > \omega_{01}) \tag{6.15}$$

For $\omega < \omega_{01}$, the wave number k is imaginary and the LCP wave is evanescent. Thus, the LCP wave propagates only for $\omega > \omega_{01}$.

Similarly, the *phase velocity* of the *RCP wave* is obtained, from (6.7), as

$$(v_{ph})_R = \left(\frac{\omega}{k}\right)_R = \frac{c(1-\omega_{ce}/\omega)^{1/2}}{(1-\omega_{ce}/\omega-\omega_{pe}^2/\omega^2)^{1/2}} \quad ; \ (\omega<\omega_{ce};\ \omega>\omega_{02}) \tag{6.16}$$

Thus, the RCP wave propagates in two frequency ranges: $0<\omega<\omega_{ce}$ and $\omega_{02}<\omega<\infty$; it is evanescent for $\omega_{ce}<\omega<\omega_{02}$.

The *group velocity* for the LCP and RCP waves in the propagation bands are given, respectively by

$$(v_g)_L = \left(\frac{\partial\omega}{\partial k}\right)_L = \frac{2c(\omega + \omega_{ce})^{3/2}\ [\omega(\omega^2 + \omega\omega_{ce} - \omega_{pe}^2)]^{1/2}}{2\omega(\omega + \omega_{ce})^2 - \omega_{ce}\ \omega_{pe}^2} \tag{6.17}$$

$$(v_g)_R = \left(\frac{\partial \omega}{\partial k}\right)_R = \frac{2c(\omega - \omega_{ce})^{3/2}[\omega(\omega^2 + \omega\omega_{ce} - \omega_{pe}^2)]^{1/2}}{2\omega\,(\omega - \omega_{ce})^2 + \omega_{ce}\,\omega_{pe}^2} \tag{6.18}$$

A plot of phase velocity and group velocity as a function of frequency for these two transverse modes is shown in Fig. 9. The same dispersion relations (6.7) and (6.9) are plotted, in a different form, in Figs. 10 and 11, respectively, where it is shown the frequency ω as a function of the real part of the wave number k. The frequency bands for which there is no wave propagation are indicated.

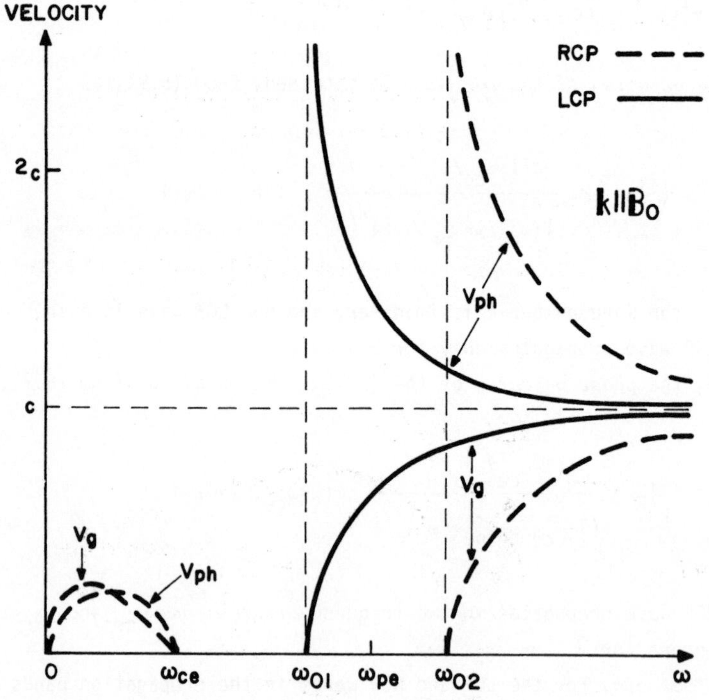

Fig. 9 - Phase velocity and group velocity as a function of frequency for the transverse RCP and LCP waves propagating along the magnetostatic field ($\vec{k} \parallel \vec{B}_0$).

The RCP waves in the lower branch which have $\omega \lesssim \omega_{ce}$ are commonly known as *electron cyclotron waves*. Similarly, when ion motion is taken into account, the LCP mode has also a lower branch of propagation for $0 < \omega < \omega_{ci}$ with a resonance at ω_{ci}. The LCP waves having $\omega \lesssim \omega_{ci}$ are commonly known as *ion cyclotron waves*.

7. PROPAGATION PERPENDICULAR TO B

We consider now wave propagation in the direction perpendicular to $\vec{B}_0 (\vec{k} \perp \vec{B}_0)$. For $\theta = 90^0$, (5.25) simplifies to

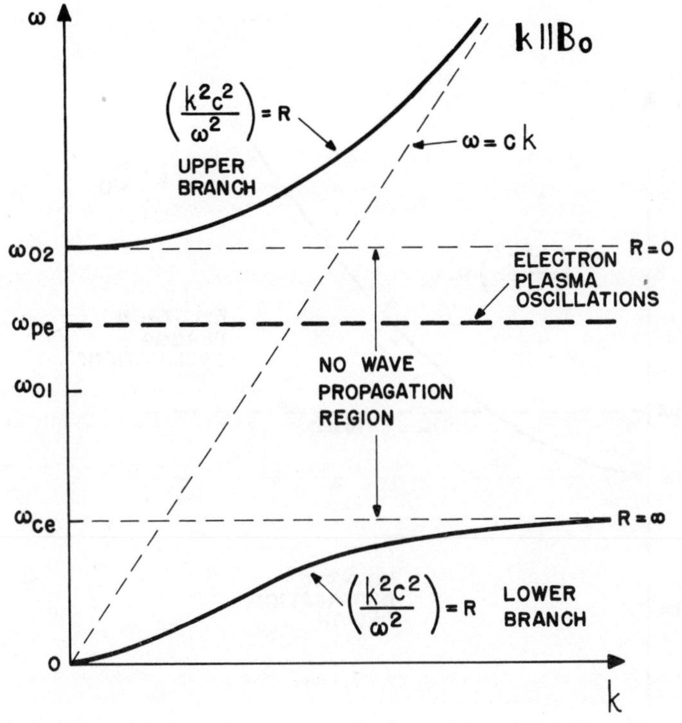

Fig. 10 - Dispersion plot for the RCP wave propagating along the magnetostatic field $(\vec{k} \parallel \vec{B}_0)$.

$$\begin{pmatrix} S & -iD & 0 \\ iD & (S - k^2c^2/\omega^2) & 0 \\ 0 & 0 & (P - k^2c^2/\omega^2) \end{pmatrix} \begin{pmatrix} E_x \\ E_y \\ E_z \end{pmatrix} = 0 \tag{7.1}$$

Again, for a nontrivial solution ($\vec{E} \neq 0$) the determinant of the 3 x 3 matrix in (7.1) must vanish. Direct calculation of this determinant yields the following two independent modes of propagation:

$$(k^2c^2/\omega^2)_O = P \qquad (\vec{E}_\parallel \neq 0) \tag{7.2}$$

$$(k^2c^2/\omega^2)_X = \frac{RL}{S} \qquad (\vec{E}_\perp \neq 0) \tag{7.3}$$

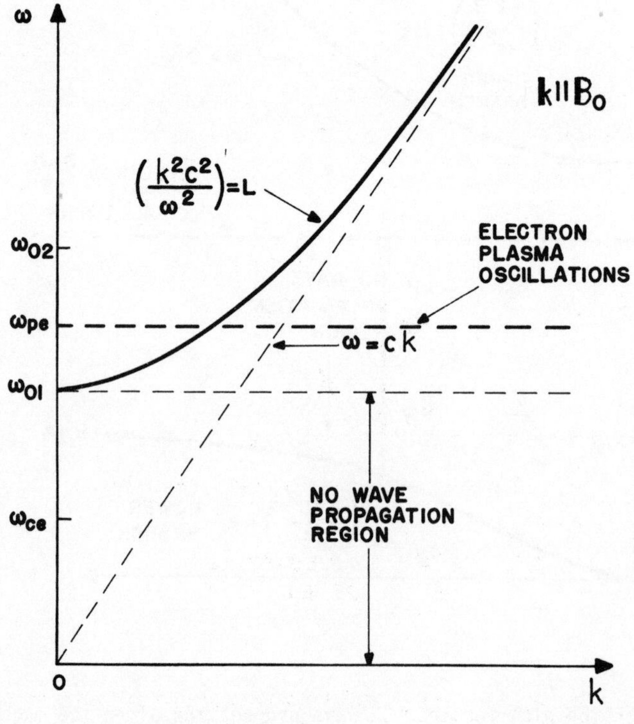

Fig. 11 - Dispersion plot for the LCP wave propagating along the magnetostatic field ($\vec{k} \parallel \vec{B}_0$).

The indices 0 and X refer to the *ordinary* and *extraordinary* modes, respectively, as will be explained shortly.

From (7.2) and using (5.24), we obtain the dispersion relation

$$(k^2c^2/\omega^2)_0 = 1 - X/U \tag{7.4}$$

or using (5.14) and (5.16), neglecting collisions ($\nu = 0$),

$$(k^2c^2/\omega^2)_0 = 1 - \omega_{pe}^2/\omega^2 \tag{7.5}$$

This relation is identical to (4.12) for *transverse waves* in an isotropic plasma. Hence, this mode of propagation is not affected by the presence of the magnetic field $\vec{B}_0$ and, for this reason, it is called an *ordinary wave*. For this mode propagating perpendicular to $\vec{B}_0$, the wave electric field ($\vec{E}_\parallel \neq 0$) is parallel to $\vec{B}_0$, so that it involves electron velocities solely in the direction of $\vec{B}_0$. Consequently, the magnetic force term $\vec{u} \times \vec{B}_0$ is zero and the wave propagates as if $\vec{B}_0$ were zero. The *ordinary mode* is also called a *TEM (Transverse Electric-Magnetic) mode*, since both the electric and the magnetic fields are transverse to the direction of propagation ($\vec{E}_\parallel \perp \vec{k}$, $\vec{B} \perp \vec{k}$; see Fig. 12). The electric field is *linearly polarized* in the direction $\vec{B}_0$.

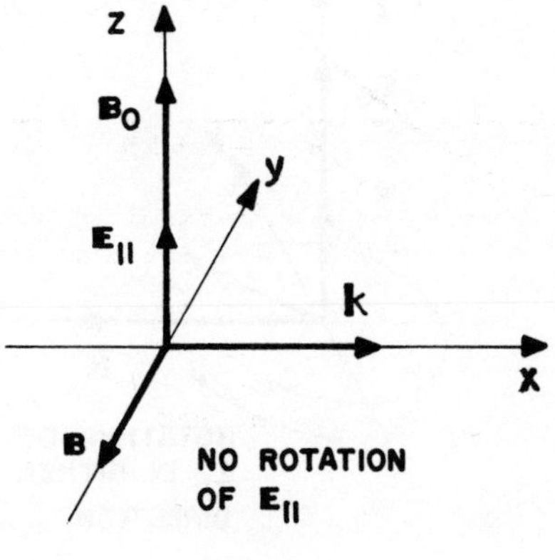

Fig. 12 - Vector diagram for the ordinary wave propagating perpendicular to $\vec{B}_0$ ($\theta = \pi/2$).

The other mode of propagation ($\vec{E}_\perp \neq 0$) is called the *extraordinary mode*, since it depends on the $\vec{B}_0$ field, with the dispersion relation given by (7.3),

$$\left(\frac{k^2c^2}{\omega^2}\right)_X = \frac{RL}{S} = \frac{1}{[1 - XU/(U^2 - Y^2)]}\left[\left(1 - \frac{XU}{U^2 - Y^2}\right)^2 - \left(\frac{XY}{U^2 - Y^2}\right)^2\right] \tag{7.6}$$

or, using (5.14), (5.15) and (5.16), after neglecting collisions ($\nu = 0$),

$$\left(\frac{k^2c^2}{\omega^2}\right)_X = \frac{(\omega^2 + \omega\omega_{ce} - \omega_{pe}^2)(\omega^2 - \omega\omega_{ce} - \omega_{pe}^2)}{\omega^2(\omega^2 - \omega_{ce}^2 - \omega_{pe}^2)} = \frac{(\omega^2 - \omega_{01}^2)(\omega^2 - \omega_{02}^2)}{\omega^2(\omega^2 - \omega_{UH}^2)} \tag{7.7}$$

where ω_{01} and ω_{02} are given by (6.13) and (6.14), respectively, and where ω_{UH} denotes the *upper hybrid frequency*, defined by

$$\omega_{UH} = (\omega_{pe}^2 + \omega_{ce}^2)^{1/2} \tag{7.8}$$

For the *extraordinary mode*, the wave electric field ($\vec{E}_\perp \neq 0$) has in general a longitudinal component (along $\vec{k}$) and a transverse component (normal to $\vec{k}$), as shown in Fig. 13. Hence, these waves are *partially longitudinal* and *partially transverse*. From (7.1), the *polarization* of the extraordinary mode is determined

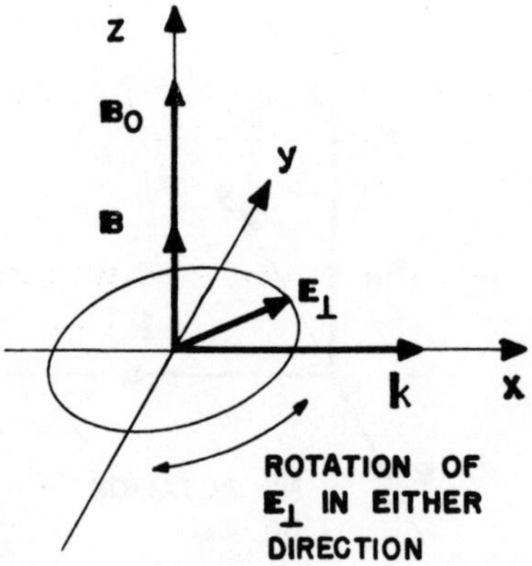

Fig. 13 - Vector diagram for the extraordinary wave propagating perpendicular to $\vec{B}_0$ ($\theta = \pi/2$).

by

$$E_x/E_y = i(D/S) \tag{7.9}$$

so that this mode is in general *elliptically polarized*. The extraordinary mode is also called a *TM (Transverse Magnetic) mode*, since the magnetic field of this wave is transverse to the direction of propagation.

From (7.5) it is clear that the *ordinary wave* has a *reflection point* ($v_{ph} \to \infty$ or $kc/\omega = 0$) at $\omega = \omega_{pe}$, and no *resonances* ($v_{ph} = 0$ or $kc/\omega \to \infty$). For the *extraordinary wave*, (7.7) indicates a *resonance* at the upper hybrid frequency $\omega_{UH} = (\omega_{pe}^2 + \omega_{ce}^2)^{1/2}$ and *reflection points* at ω_{01} and ω_{02} (when ion motion is included it turns out that the extraordinary wave has also a resonance at the *lower hybrid frequency*, given approximately by $\omega_{LH}^2 \simeq \omega_{ce}\,\omega_{ci}$). The dispersion plot for the ordinary wave is the same as that presented in Fig. 3 for the transverse wave in an isotropic plasma. For the extraordinary mode, the dispersion plot shown in Fig. 14 indicates that there is wave propagation only for $\omega > \omega_{02}$ and for ω in a band of frequencies between ω_{01} and ω_{UH}. For other frequencies k is imaginary and the phase velocity is infinite.

The *phase velocities* of the ordinary and extraordinary waves are obtained from (7.5) and (7.7), respectively, as

$$(v_{ph})_O = \left(\frac{\omega}{k}\right)_O = \frac{c}{(1 - \omega_{pe}^2/\omega^2)^{1/2}} \; ; \quad (\omega > \omega_{pe}) \tag{7.10}$$

$$(v_{ph})_X = \left(\frac{\omega}{k}\right)_X = c\left[\frac{\omega^2\,(\omega^2 - \omega_{UH}^2)}{(\omega^2 - \omega_{01}^2)\,(\omega^2 - \omega_{02}^2)}\right]^{1/2} \; ; \quad (\omega>\omega_{02};\ \omega_{01}<\omega<\omega_{UH}) \tag{7.11}$$

Expressions for the *group velocities* of these two modes can be derived with the help of (7.5) and (7.7),

$$(v_g)_O = \left(\frac{\partial\omega}{\partial k}\right)_O = c\,(1 - \omega_{pe}^2/\omega^2)^{1/2} \; ; \quad (\omega>\omega_{pe}) \tag{7.12}$$

$$(v_g)_X = \left(\frac{\partial\omega}{\partial k}\right)_X = \frac{c(\omega^2 - \omega_{UH}^2)^2}{\omega\left[\omega^4 - 2\omega^2(\omega_{ce}^2 + \omega_{pe}^2) + \omega_{ce}^4 + 3\omega_{ce}^2\,\omega_{pe}^2 + \omega_{pe}^4\right]} \cdot$$

$$\cdot\left[\frac{(\omega^2 - \omega_{01}^2)(\omega^2 - \omega_{02}^2)}{(\omega^2 - \omega_{UH}^2)}\right]^{1/2} \; ; \quad (\omega>\omega_{02};\omega_{01}<\omega<\omega_{UH}) \tag{7.13}$$

The plot of phase velocity and group velocity versus frequency for the extraordinary (TM) mode has the form depicted in Fig. 15. A similar plot for the ordinary (TEM) mode is shown in Fig. 2 (the same one for the transverse wave in an isotropic plasma).

8. PROPAGATION AT ARBITRARY DIRECTIONS

8.1 Resonances and reflection points

Going back now to (5.26), we shall first determine the resonances ($v_{ph} = 0$ or $kc/\omega \to \infty$) and the reflection points ($v_{ph} \to \infty$ or $kc/\omega = 0$) for arbitrary angles of propagation with respect to $\vec{B}_0$. From (5.33) and (5.30) it is seen

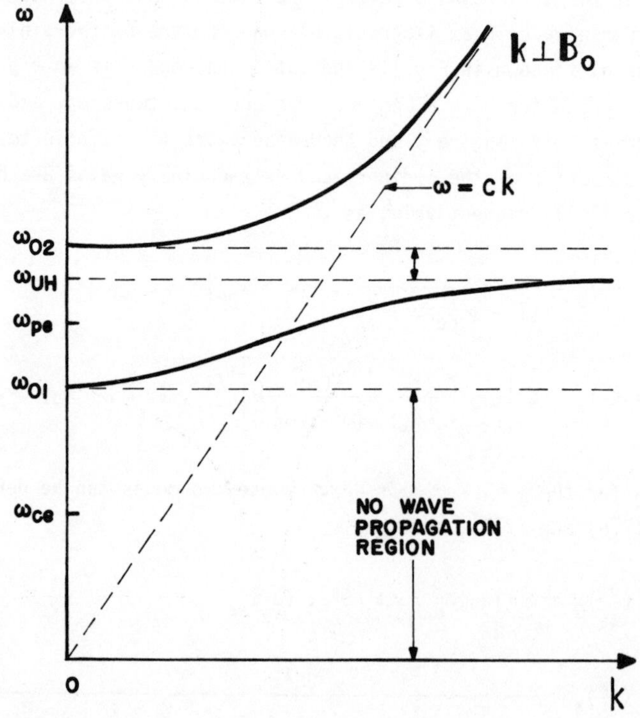

Fig. 14 - Dispersion relation for the extraordinary wave ($k^2c^2/\omega^2 = RL/S$) propagating perpendicular to the magnetostatic field ($\vec{k} \perp \vec{B}_0$).

that *resonance* occurs when

$$S \sin^2\theta + P \cos^2\theta = 0 \tag{8.1}$$

or

$$\tan^2\theta = -P/S \tag{8.2}$$

Using (5.22) and (5.24), and neglecting collisions ($\nu = 0$), (8.2) yields

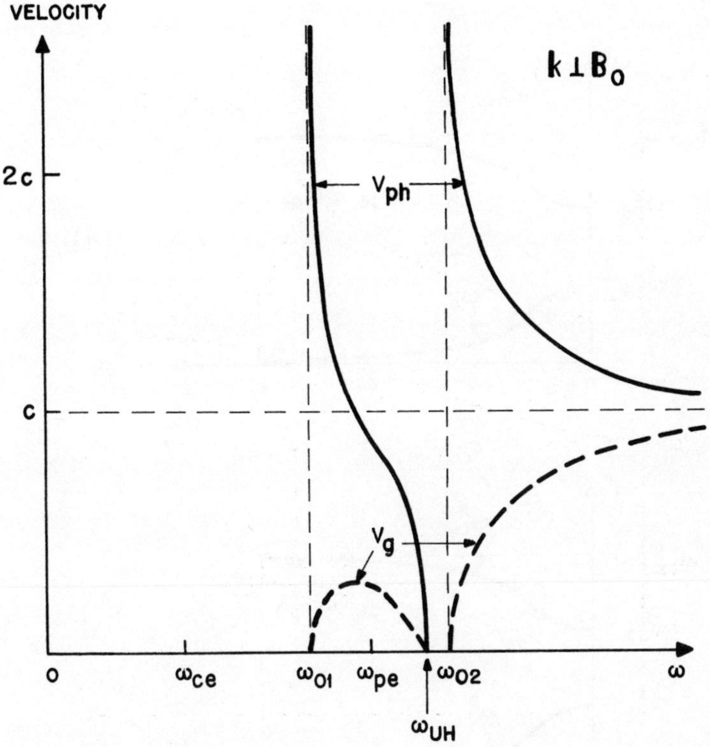

Fig. 15 - Phase velocity and group velocity as a function of frequency for the extraordinary (TM) mode propagating perpendicular to the magnetic field ($\vec{k} \perp \vec{B}_0$).

$$1 - X = Y^2 (1 - X \cos^2\theta) \tag{8.3}$$

or, using (5.15) and (5.16),

$$\omega^4 - \omega^2 (\omega_{pe}^2 + \omega_{ce}^2) + \omega_{pe}^2 \, \omega_{ce}^2 \cos^2\theta = 0 \tag{8.4}$$

Thus, the *resonance frequencies* as functions of θ are given by

$$\omega_{0\pm}^2 = (\omega_{pe}^2 + \omega_{ce}^2)/2 \pm \left[(\omega_{pe}^2 + \omega_{ce}^2)^2/4 - \omega_{pe}^2 \omega_{ce}^2 \cos^2\theta\right]^{1/2} \tag{8.5}$$

These two resonance frequencies are plotted against θ in Fig. 16. From (8.5) it is clear that the sum of the square of these two frequencies $(\omega_{0+}^2 + \omega_{0-}^2)$ is

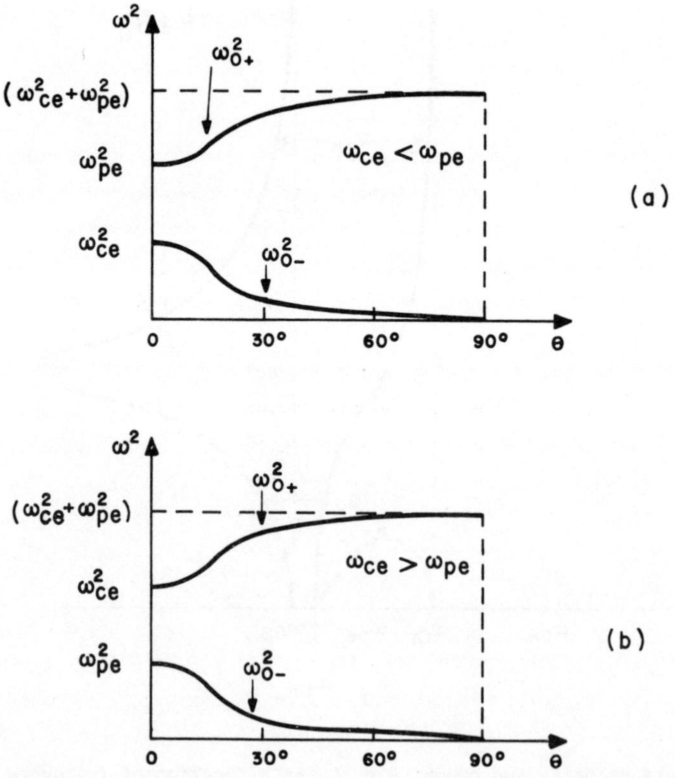

Fig. 16 - Resonance frequencies as functions of the angle θ between $\vec{B}_0$ and the direction of wave propagation in a cold electron plasma, for (a) $\omega_{ce} < \omega_{pe}$ and (b) $\omega_{ce} > \omega_{pe}$.

always equal to $(\omega_{pe}^2 + \omega_{ce}^2)$ for any angle θ. From Fig. 16 we see that the high-frequency resonance increases with increasing θ, from the larger of ω_{pe} and ω_{ce}, at $\theta = 0^0$, to the upper hybrid resonance frequency, $(\omega_{pe}^2 + \omega_{ce}^2)^{1/2}$, at $\theta = 90^0$. The low-frequency resonance decreases correspondingly from the frequency which is the smaller between ω_{pe} and ω_{ce}, at $\theta = 0^0$, to zero, at $\theta = 90^0$. The resonances at $\theta = 0^0$ and $\theta = 90^0$ are called the *principal resonances*. At $\theta = 90^0$ the principal resonances are given by $S = \infty$ and $P = 0$, whereas at $\theta = 90^0$ the principal resonance is given by $S = 0$.

The *reflection points* are seen, from (5.26), to be given by

$$PRL = 0 \tag{8.6}$$

This equation is satisfied whenever $P = 0$, or $R = 0$, or $L = 0$. However, for $\theta = 0^0$ (5.26) simplifies to

$$k^2c^2/\omega^2 - 2S\,(k^2c^2/\omega^2) + RL = 0 \tag{8.7}$$

so that $P = 0$ is no longer a reflection point for $\theta = 0^0$. Thus, for propagation exactly along the $\vec{B}$ field the reflection points are given by $R = 0$ and $L = 0$. But for $\theta \neq 0$, irrespective of how small θ is, $P = 0$ also corresponds to a reflection point. Note that these cut-off frequencies are, otherwise, *independent* of θ. The cut-off frequencies and the principal resonances are summarized in Table 1.

Expressions for the phase velocity and group velocity for arbitrary angles of propagation can be obtained from the dispersion relation (5.26) or (5.36). Since this involves considerable algebra they will not be presented here. For this case the curves of k, v_{ph} and v_g, as functions of ω, must lie somewhere between the corresponding curves for $\theta = 0^0$ (see Figs. 9, 10 and 11) and for $\theta = 90^0$ (see Figs 2, 3, 14 and 15). If the angle θ is continuously changed from 0^0 to 90^0 then the curves for $\theta = 0^0$ must change continuously into those for $\theta = 90^0$.

Fig. 17 shows ω as a function of the real part of k, while Fig. 18 shows v_{ph} and v_g as functions of ω, for the two modes of propagation at an angle $\theta = 45^0$ with respect to $\vec{B}_0$. It is interesting to note that the branch of mode 2 (that propagates for $\omega_{01} < \omega < \omega_{0+}$) and the branch of mode 1 (that propagates for $\omega > \omega_{pe}$) are transformed, as θ goes to zero, into the LCP waves and the electron plasma oscillations at ω_{pe}. This is indicated in Fig. 19.

TABLE 1

CUT-OFFS AND PRINCIPAL RESONANCES FOR WAVES IN A COLD ELECTRON PLASMA

Cut-offs	Principal Resonances	
	$\theta = 0^0$	$\theta = 90^0$
$P = 0$ $(\theta \neq 0)$	$P = 0$	$S = 0$
$R = 0$	$R = \infty$ } $S = \infty$	
$L = 0$	$L = \infty$ } $S = \infty$	

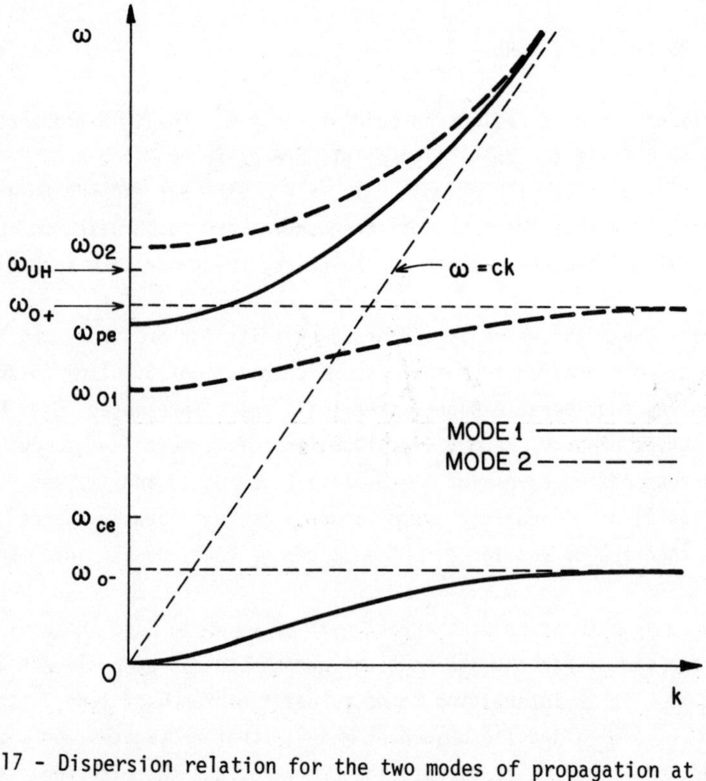

Fig. 17 - Dispersion relation for the two modes of propagation at an angle $\theta = 45^0$ with respect to the magnetostatic field in a cold electron plasma.

Fig. 20 is a plot of the phase velocity versus frequency illustrating how the two modes of wave propagation when $\theta = 0^{0}$ (LCP and RCP waves) evolve into the two modes of wave propagation when $\theta = 90^{0}$ (ordinary and extraordinary waves).

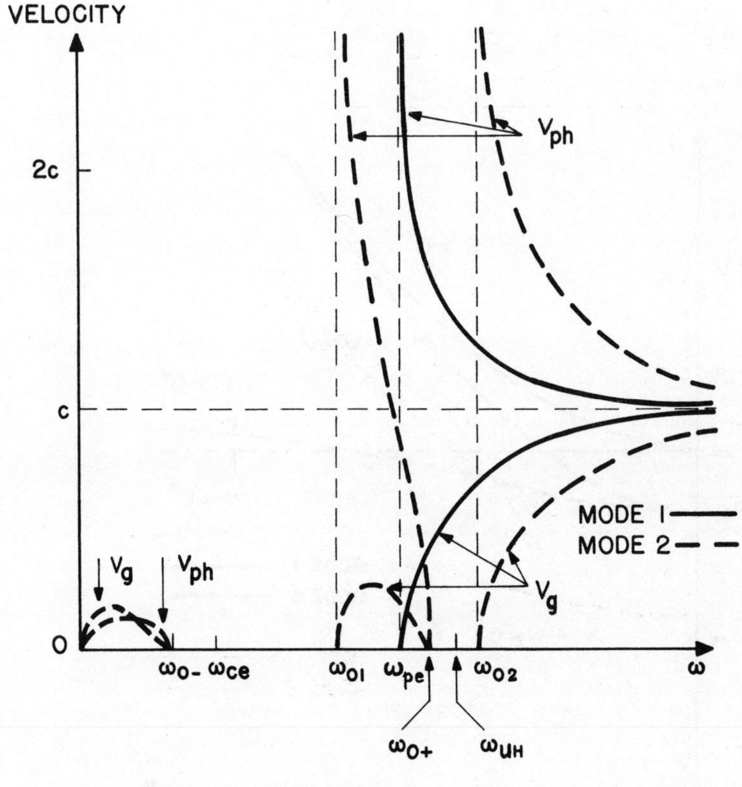

Fig. 18 - Phase velocity (v_{ph}) and group velocity (v_g) as a function of frequency for the two modes of propagation at an angle $\theta = 45^{0}$ with respect to the magnetostatic field in a cold electron plasma.

8.2 Wave normal surfaces

The *wave normal surface*, also known as the normalized *phase velocity surface*, is a polar plot of v_{ph}/c as a function of θ. Because of the symmetry in the azimuthal angle ϕ, it is a surface of revolution about $\vec{B}_0$. For any direction of propagation, the "length" (properly normalized) of the line drawn from the origin to intersect this surface is v_{ph}/c. This surface is therefore the loci of points of constant phase emitted from the origin. The shape of the wave

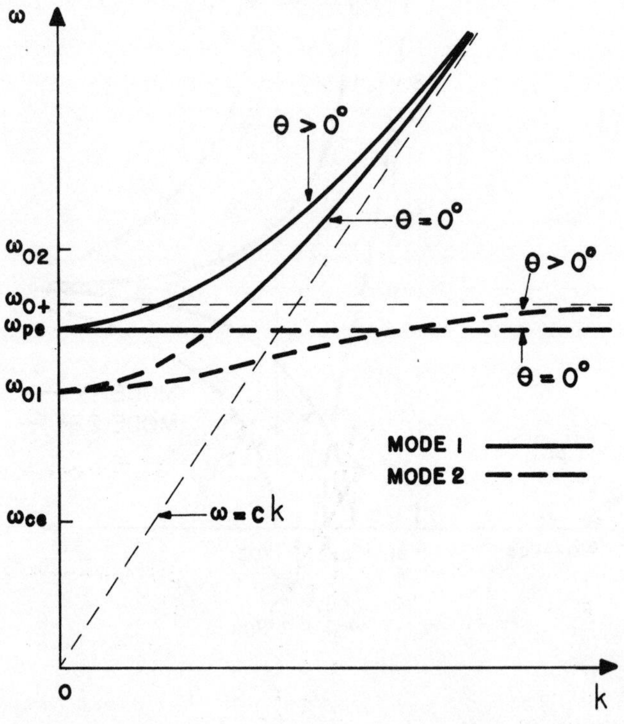

Fig. 19 - Illustrating how the branch $\omega_{01} < \omega < \omega_{0+}$ of mode 2 and the branch $\omega > \omega_{pe}$ of mode 1, for $\theta > 0^0$, are related to the LCP wave and the electron plasma oscillations when $\theta = 0^0$.

normal surface is generally not the same as the shape of a wave front. A typical wave normal surface is presented in Fig. 21, in which the velocity of light is shown as a dashed circle. The two solutions for $(kc/\omega)^2$, from (5.26), are superimposed on the same coordinate axes as slow and fast waves. The denomination slow wave refers to the mode with the largest value of $(kc/\omega)^2$, whereas the fast wave refers to the mode with the smallest value of $(kc/\omega)^2$. With some exceptions, the fast wave has generally a phase velocity greater than c, while the phase velocity of the slow wave is generally less than c.

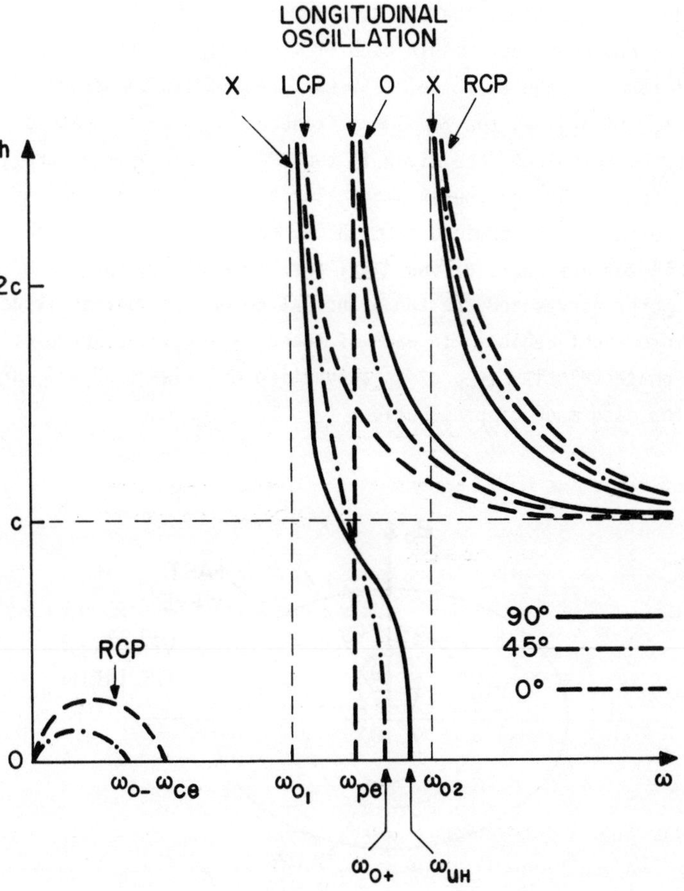

Fig. 20 - Phase velocity versus frequency for waves in a cold electron plasma, illustrating how the two modes of propagation for $\theta = 0^0$ (LCP and RCP) evolute into the two modes for $\theta = 90^0$ (O and X).

8.3 The CMA diagram

The CMA (Clemmow - Mullaly-Allis) diagram is a very compact alternative way for presenting solutions of the dispersion relation. This diagram is constructed in a two-dimensional parameter space having $X = \omega_{pe}^2/\omega^2$ as the *horizontal* axis and $Y^2 = \omega_{ce}^2/\omega^2$ as the *vertical* axis, and displays all the resonances and reflection points as a function of both X and Y^2. Thus, in this diagram the magnetic field increases in the vertical direction, the plasma electron density increases in the horizontal direction, and the electromagnetic wave frequency decreases in the radial direction (in each case, considering all other parameters fixed). Furthermore, the CMA diagram divides the (X, Y^2) plane into a number of regions such that within each of these regions the characteristic topological forms of the phase velocity surfaces remain unchanged.

From (8.3), which gives the *resonance* frequencies, we see that for $\theta = 0^o$ the loci of the resonances are given in the CMA diagram by the straight line $Y^2 = 1$, and for $\theta = 90^o$ by the straight line $Y^2 = 1 - X$. The loci of the *reflection* points, as determined from (8.6), can be shown to be the curves $Y^2 = (1 - X)^2$ for any angle θ, and $X = 1$ for any angle except $\theta = 0^o$. The two reflection point curves and the two principal resonance curves divide the (X, Y^2) plane into eight regions. In each of these regions, a polar plot ot the normalized phase velocity (v_{ph}/c) as a function of θ (wave normal surface) is presented for each mode of propagation.

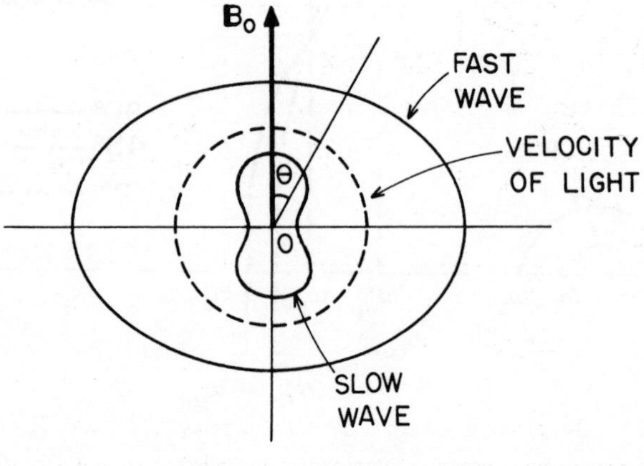

Fig. 21 - Typical wave normal surface, or phase velocity surface.

Fig. 22 shows the CMA diagram for wave propagation in a cold electron plasma. The dashed lines are the loci of the reflection points and the solid lines are the loci of the principal resonances (the dotted line indicates the loci of the resonances when $\theta = 30^0$). The dashed circles represent the wave normal surface corresponding to the velocity of light. The "slow" and "fast" wave notation, used in Fig. 21, becomes now apparent. The labels R (right-hand polarization) and L (left-hand polarization) appear on the phase velocity surface only along the magnetic axis (up in the diagram). The labels O (ordinary) and X (extraordinary) appear only at 90^0 with respect to the magnetic field axis.

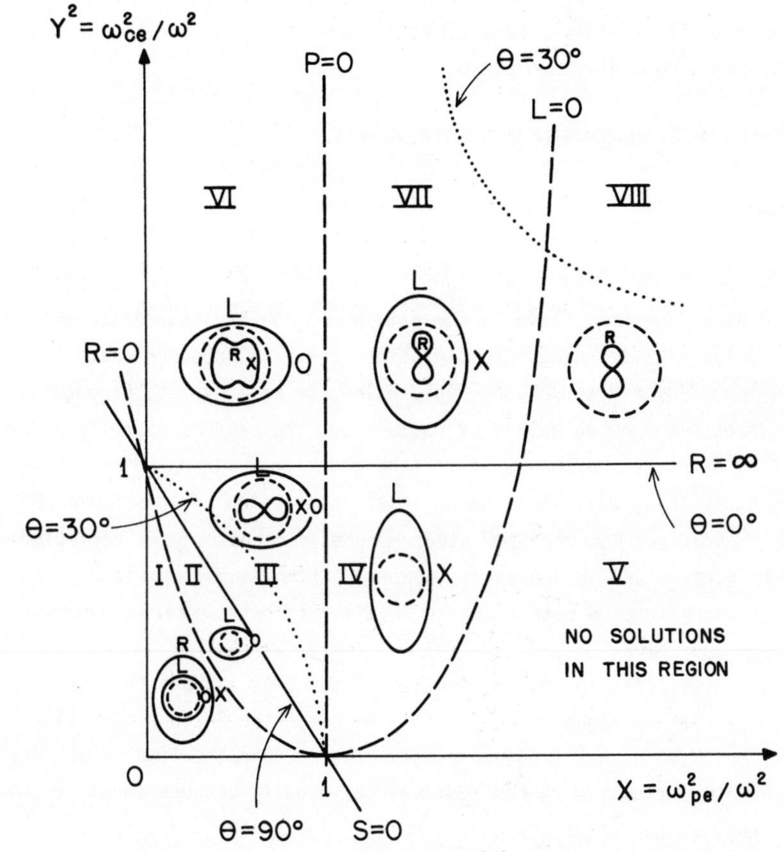

Fig. 22 - The CMA diagram for waves in a cold electron gas. The solid lines represent the principal resonances and the dashed lines the reflection points.

In some regions of the CMA diagram certain modes are present and others are not. As the boundaries of these regions are crossed, the wave normal surfaces for the modes change shape, and a given mode may appear or disappear. For instance, in region I both modes are present, but when we move to region II the fast wave disappears. Similarly, if the parameters are changed along a path that goes from region VIII to VII (decreasing electron density), the fast wave appears as the boundary L = 0 is crossed, and so on. Note that the same frequency may appear in the modes of different regions, depending upon the values of electron density and magnetic field. Note also that, although the characteristic shapes of the wave normal surfaces remain the same inside each bounded region, their magnitudes may change. A detailed examination of the CMA diagram shows that it provides a very broad picture of the nature of the waves that propagate in a cold electron plasma.

9. SOME SPECIAL WAVE PHENOMENA IN COLD PLASMAS

9.1 Atmospheric whistlers

The propagation of *whistlers* is a naturally occurring phenomenon which can be originated by a lightning flash in the atmosphere. During thunderstorms and lightning, a pulse of electromagnetic radiation energy is produced which is rich in very low frequency components. This pulse, or wave packet, propagates through the ionosphere, being guided by ducts along the Earth's magnetic field to a distant point at the Earth's surface (the magnetic conjugate point), where it may be detected (Fig. 23). When the whistler is detected at this point it is called a *short whistler*. However, the electromagnetic signal may be reflected at the Earth's surface and guided back along the Earth's magnetic field to a point close to where it originated; if the whistler is detected at this point it is called a *long whistler*.

As the wave packet, rich in low frequencies, propagates through the ionosphere along the Earth's magnetic field, it gets dispersed in course of time in such a way that the higher frequencies move faster than the lower ones. The frequencies in a whistler are in the audio range, usually between about 100 Hz and 10 kHz. Thus, at the point of detection, the high frequencies arrive at the receiver sooner than the low ones, and if the receiver is attached to a loudspeaker we hear a discending pitch whistle. These frequencies are usually much smaller than the electron cyclotron frequency ($\omega \ll \omega_{ce}$) in the Earth's ionosphere.

At various locations on the Earth there are stations that continuously record *sonograms* of whistler activity. A sonogram is a spectrum of the frequency versus time of arrival, as illustrated in Fig. 24. These sonograms are used as an effective diagnostic tool for studying the ionospheric conditions.

The phenomenon of atmospheric whistler propagation can be explained in terms of the very low frequency ($\omega \ll \omega_{ce}$) region of propagation of the *right circularly polarized wave* (see Fig. 20). For a simplified analysis of this phenomenon, consider the Appleton-Hartree equation (5.36), neglecting collisions ($U = 1$). For propagation nearly along the magnetic field lines, and for $\omega \ll \omega_{ce}$ and $\omega \ll \omega_{pe}$, we have $Y \cos\theta \gg Y^2 \sin^2\theta/[2(1 - X)]$, so that (5.36) simplifies to (using the "minus" sign),

$$k^2c^2/\omega^2 = 1 - X/(1 - Y\cos\theta) \tag{9.1}$$

This equation is often referred to as the dispersion relation for the *quasi-longitudinal mode*. For $Y \cos\theta \gg 1$ (i.e. $\omega \ll \omega_{ce} \cos\theta$), (9.1) becomes

$$k^2c^2/\omega^2 = 1 + X/(Y\cos\theta) \tag{9.2}$$

and considering, further, that $X \gg Y$ (i.e $\omega_{pe}^2 \gg \omega\omega_{ce}$), we obtain

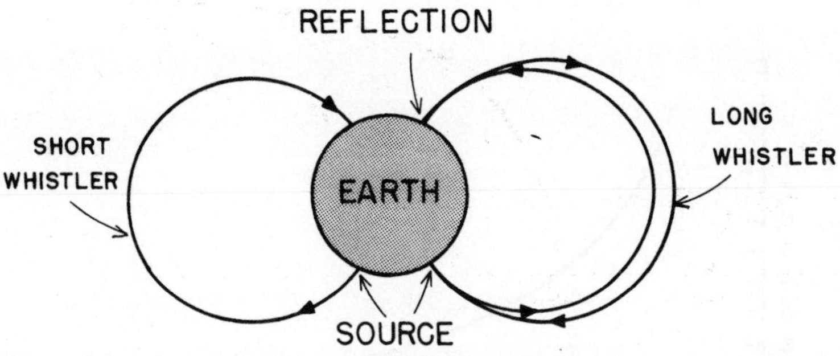

Fig. 23 - Atmospheric whistler propagation, illustrating the detection of a short whistler and a long whistler.

$$kc/\omega = (X/Y\cos\theta)^{1/2} \tag{9.3}$$

The *phase velocity* is found directly from (9.3),

$$v_{ph} = \omega/k = c\,(Y\cos\theta/X)^{1/2} \tag{9.4}$$

or, substituting $Y = \omega_{ce}/\omega$ and $X = \omega_{pe}^2/\omega^2$,

$$v_{ph} = c\,(\omega\omega_{ce}\cos\theta)^{1/2}/\omega_{pe} \tag{9.5}$$

Also, from (9.3) we obtain the *group velocity* as

$$v_g = \partial\omega/\partial k = 2c\,(\omega\omega_{ce}\cos\theta)^{1/2}/\omega_{pe} \tag{9.6}$$

Thus, the group velocity is proportional to the square root of the frequency and consequently the higher frequencies arrive at the receiver slightly ahead of the lower frequencies, producing a descending pitch whistle when received with a simple antenna and loudspeaker.

The characteristics of atmospheric whistler propagation are such that they are situated in region VIII of the CMA diagram. In this region, the wave normal diagram for the RCP wave is a lemniscate, as shown in Fig. 25. This wave normal

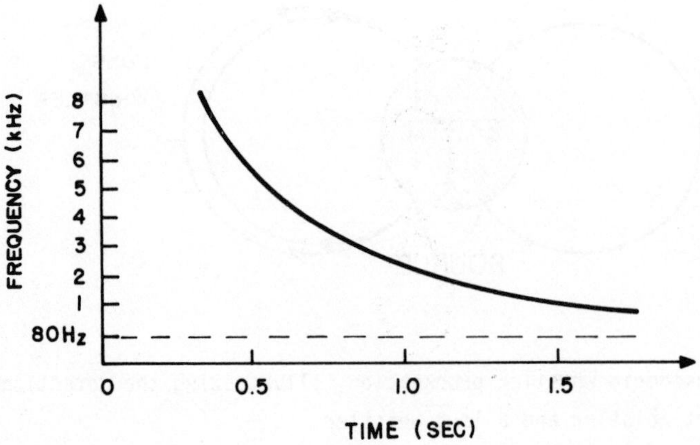

Fig. 24 - Typical sonogram of a whistler.

surface has a resonant cone, which gives the maximum value that the angle θ may have. The angle between the direction of propagation of the wave packet and the magnetic field also has a maximum value, which specifies the maximum angular deviation (from the magnetic field) of the direction in which a wave packet can propagate. It can be shown that the maximum value of this angle is about 19.5^{0}. Therefore, the wave packet is confined to a beam of less than 20^{0} about the magnetic field lines.

Experiments carried out on whistlers have verified the results presented here. In addition, when the frequency is near (but smaller than) the electron cyclotron frequency, it is possible to have the frequency increasing with the time of arrival, and these have been called *ascending frequency whistlers*. The whistlers in the frequency regime where they change from the ascending to the descending tone are known as the *nose whistlers*. These types of whistlers have also been observed experimentally.

9.2 Helicons

The experimentally observed *helicon* waves, in a solid state plasma, is also a phenomenon related with the very low frequency propagation of the *right circularly polarized wave*. The reason for the name helicons comes from the fact the tip of the wave $\vec{B}$ vector traces a helix.

Consider a solid state plasma slab of thickness d, the other two dimensions being very large, oriented perpendicularly to an externally applied $\vec{B}$ field,

Fig. 25 - Wave normal surface for whistlers and helicons.

as indicated in Fig. 26. Suppose that a low frequency ($\omega \ll \omega_{ce}$) right circularly polarized wave is launched in the direction of the $\vec{B}$ field.

From the dispersion relation (6.7) for the RCP wave we obtain, approximately, for $\omega \ll \omega_{ce}$,

$$k = (\omega_{pe}/c)\ (\omega/\omega_{ce})^{1/2}\ ; \qquad (\omega \ll \omega_{ce}) \tag{9.7}$$

Denoting the propagation coefficient of the electromagnetic wave in the medium *external* to the plasma slab by k_v, the magnitude of the reflection coefficient at the plasma boundary is given by $(k_v - k)/(k_v + k) \simeq 1$, since $\omega \ll \omega_{ce}$. Consequently, the reflection of the waves at the plasma boundary is nearly complete. Therefore, the wave will be successively reflected at the boundaries of the plasma slab and will form a standing wave, whose resonances are given approximately by

$$n\lambda/2 = d \tag{9.8}$$

where λ is the wavelength inside the slab of thickness d and n is an integer.

Since $\lambda = 2\pi/k$, we can combine (9.7) and (9.8) to obtain

$$(\pi nc/\omega_{pe})\ (\omega_{ce}/\omega)^{1/2} = d \tag{9.9}$$

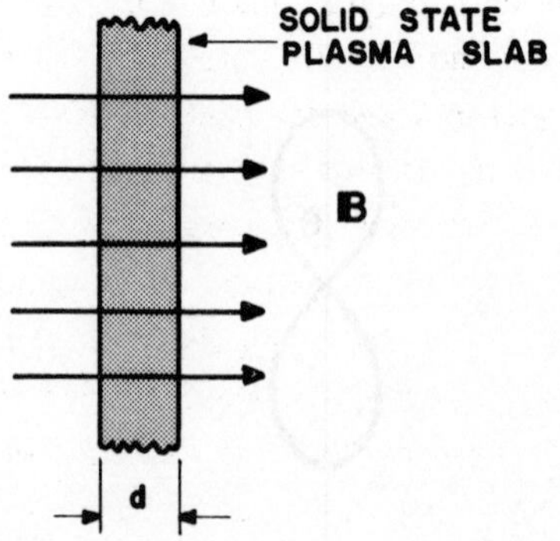

Fig. 26 - Geometrical arrangement for the detection of helicon waves.

This is the condition for standing wave resonance. It is appropriate to add the subscript n to ω, in order to identify the resonance frequency with the corresponding values of n which gives the number of the standing-wave pattern in the slab. Thus, (9.9) can be rearranged in the following convenient form

$$\omega_n = (\pi nc/\omega_{pe}d)^2\omega_{ce} \tag{9.10}$$

In some experiments carried out in helicons, the frequency of the wave excited along $\vec{B}$ is continuously varied, maintaining constant the values of ω_{pe}, ω_{ce} and d. At the frequencies where $\omega = \omega_n$, given by (9.10), there are standing wave resonances inside the plasma slab, resulting in large wave amplitudes, which can be measured. A plot of wave amplitude inside the plasma slab, versus frequency, permits the identification of the resonant frequencies ω_n.

In sodium, which contains 10^{28} electrons/m^3, the first (n = 1) standing-wave resonant frequency is of the order of 100Hz. Note that ω_n is proportional to n^2.

In some other experimental investigations, the parameters d, ω_{pe} and ω are kept fixed, and the $\vec{B}$ field is varied. Then, the standing-wave resonant frequencies occur for those values of the $\vec{B}$ field for which

$$\omega_{ce} = (\omega_{pe})_n = (\omega_{pe}d/\pi nc)^2\omega \tag{9.11}$$

9.3 Faraday rotation

We consider now a phenomenon, known as *Faraday rotation*, which occurs in the range of frequencies where both the RCP and LCP waves propagate. When a plane polarized wave is sent along the magnetic field in a plasma, the plane of polarization of the wave gets rotated as it propagates in the plasma. Since a plane polarized wave can be considered as a superposition of RCP and LCP waves (Fig. 27), which propagate independently, this phenomenon can be understood in terms of the *difference in phase velocity* of the RCP and LCP waves.

If we take a look at Fig. 9, we see that the RCP wave (for frequencies greater than ω_{02}) propagates faster than the LCP wave. After travelling a given distance, in which the RCP wave has undergone N cycles, the LCP wave (which travels more slowly) will have undergone N + ε (with $\varepsilon > 0$) cycles. Obviously, both waves are considered to be at the same frequency. Therefore, the plane of polarization of the plane wave is rotated counterclockwise (looking along $\vec{B}$), as shown in Fig. 28.

In order to obtain an expression for the angle of rotation θ_F, as the plane wave propagates a given distance in the plasma, let us consider a Cartesian coordinate system with propagation along the z axis (also the direction of $\vec{B}_0$) and such that, at z = 0, the electric field has only the x component, as indicated in Fig. 27. Therefore, without loss of generality, we take

$$\vec{E}\,(z = 0,\, t) = E_0 \hat{x} \exp\,(-\, i\omega t) \tag{9.12}$$

This equation can be rewritten as

$$\vec{E}\,(z = 0,\, t) = \left[(E_0/2)\,(\hat{x} + i\hat{y}) + (E_0/2)\,(\hat{x} - i\hat{y})\right] \exp\,(-\, i\omega t) \tag{9.13}$$

where the first and the second terms in the right-hand side are, respectively, the RCP and the LCP components. These two components propagate independently, so that, for any z > 0, the electric field vector is given by

$$\vec{E}(z,\, t) = \left[(E_0/2)\,(\hat{x} + i\hat{y})\exp(ik_R z) + (E_0/2)(\hat{x} - i\hat{y})\exp(ik_L z)\right]\exp(-\, i\omega t) \tag{9.14}$$

where $\vec{k}_R$ and $\vec{k}_L$ denote the wave number vectors for the RCP and LCP waves, respectively. Eq. (9.14) can be rearranged as follows

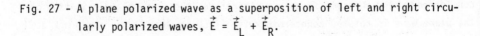

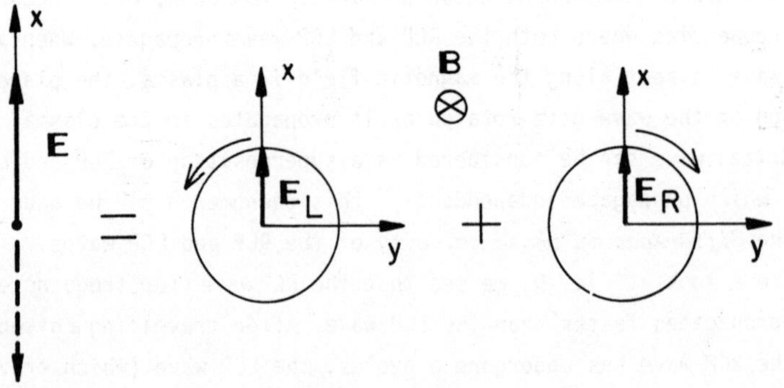

Fig. 27 - A plane polarized wave as a superposition of left and right circularly polarized waves, $\vec{E} = \vec{E}_L + \vec{E}_R$.

$$\vec{E}(z,t) = (E_0/2)\exp\left[i(k_R + k_L)\,z/2\right]\left\{(\hat{x} + i\hat{y})\exp\left[i(k_R - k_L)\,z/2\right] + \right.$$

$$\left. + (\hat{x} - i\hat{y})\exp\left[-i(k_R - k_L)\,z/2\right]\right\}\exp(-i\omega t) =$$

$$= E_0 \exp\left[i(k_R + k_L)\,z/2\right]\left\{\hat{x}\cos\left[(k_R - k_L)\,z/2\right] - \right.$$

$$\left. - \hat{y}\sin\left[(k_R - k_L)\,z/2\right]\right\}\exp(-i\omega t) \qquad (9.15)$$

Eq. (9.12) represents a linearly polarized wave in the $\hat{x}$ direction at $z = 0$, and (9.15) is also a linearly polarized wave, but with the polarization direction rotated in the counterclockwise direction (looking along $\vec{B}$) by the angle

$$\theta_F = (k_R - k_L)\,z/2 \qquad (9.16)$$

Therefore, the angle of rotation per unit distance (θ_F/z) depends on the difference between the propagation coefficients of the RCP and LCP waves. Expressions for k_R and k_L are given in (6.6) and (6.8), respectively.

The measurement of Faraday rotation is a useful tool in plasma diagnostic and it has been widely used in the investigation of ionospheric properties. A linearly polarized wave, emitted by an orbiting satellite, has its plane of polarization rotated as it traverses the ionospheric plasma. A measurement of the rotation angle θ_F, after the wave has traversed the plasma, provides information on the total electron content along the wave path.

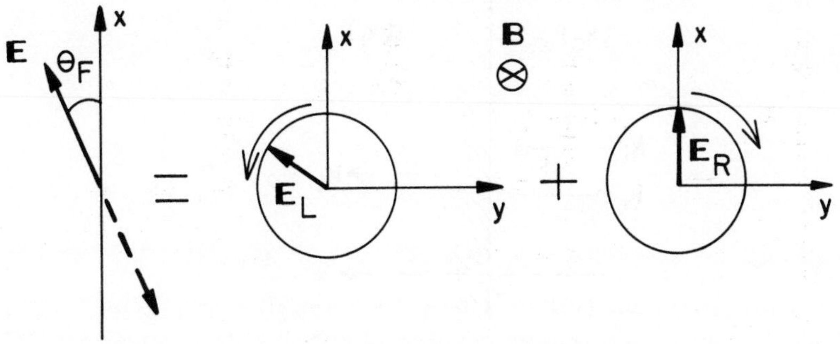

Fig. 28 - After travelling a given distance in the plasma, the plane of polarization of the plane wave is rotated, since the LCP wave moves slower than the RCP wave.

PROBLEMS

16.1 - Consider a plane electromagnetic wave incident normally on a semi-infinite plasma occupying the semi-space $x \geq 0$, with vacuum for $x < 0$, (Fig. P 16.1). Denote the incident, reflected and transmitted waves, respectively, by

$$\vec{E}_i = \hat{y} \exp(ik_0 x - i\omega t)$$

$$\vec{E}_r = \hat{y} E_r \exp(-ik_0 x - i\omega t)$$

$$\vec{E}_t = \hat{y} E_t \exp(ik_1 x - i\omega t)$$

(a) Show that the associated magnetic fields are given by

$$\vec{H}_i = \hat{z} (k_0/\omega\mu_0) \exp(ik_0 x - i\omega t)$$

$$\vec{H}_r = \hat{z} (k_0/\omega\mu_0) E_r \exp(-ik_0 x - i\omega t)$$

$$\vec{H}_t = \hat{z} (k_1/\omega\mu_0) E_t \exp(ik_1 x - i\omega t)$$

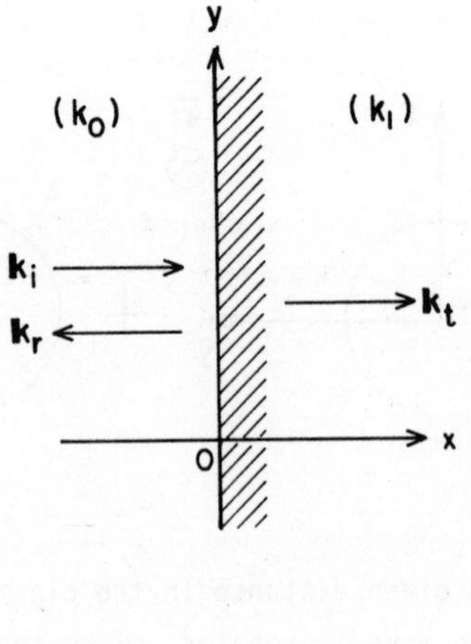

Fig. P 16.1

(b) From the continuity of E_y and H_z at the boundary x = 0, show that

$$E_r = (k_o - k_1)/(k_o + k_1); \; E_t = 2k_o/(k_o + k_1)$$

(c) Prove that the ratio of the transmitted average power to the incident average power, at the boundary x = 0, is

$$T = \left. \frac{\text{Re}\,\{\vec{E}_t \times \vec{H}_t^*\}}{\text{Re}\,\{\vec{E}_i \times \vec{H}_i^*\}} \right|_{x=0} = \left(\frac{E_t\, E_t^*}{k_o} \right) \beta$$

where k_o is real and $\beta = \text{Re}\,\{k_1\}$. Show that T = 0 both at a reflection point and a resonance.

16.2 - Consider a plane electromagnetic wave incident normally on an infinite plane plasma slab occupying the space $0 \leq x \leq L$, with vacuum for x < 0 and x > L (Fig. 16.2). Use the following representation for the wave electric field vector, as indicated in Fig. P 16.2,

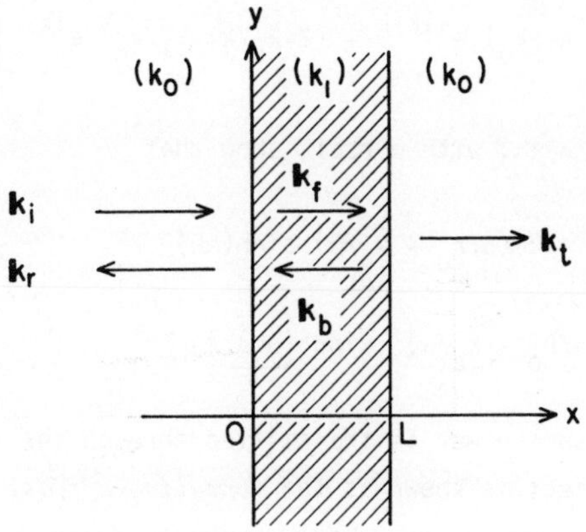

Fig. P 16.2

$\vec{E}_i = \hat{y} \exp(ik_o x - i\omega t)$ (incident wave)

$\vec{E}_r = \hat{y}\, E_r \exp(-ik_o x - i\omega t)$ (reflected wave)

$\vec{E}_f = \hat{y}\, E_f \exp(ik_1 x - i\omega t)$ (forward wave)

$\vec{E}_b = \hat{y}\, E_b \exp(-ik_1 x - i\omega t)$ (backward wave)

$\vec{E}_t = \hat{y}\, E_t \exp[ik_o(x-L) - i\omega t]$ (transmitted wave)

(a) Calculate the corresponding expressions for the associated magnetic fields.
(b) Calculate the amplitudes E_r, E_f, E_b and E_t by applying the condition of continuity of E_y and H_z at the boundaries $x = 0$ and $x = L$.
(c) Show that the ratio of the average power transmitted out of the plasma slab to the incident average power is given by

$$T = \frac{\text{Re}\,\{\vec{E}_t \times \vec{H}_t^*\}_{x=L}}{\text{Re}\,\{\vec{E}_i \times \vec{H}_i^*\}_{x=0}} = E_t\, E_t^*$$

where

$$E_t = 4\left[(2+k_o/k_1 + k_1/k_o)\, e^{-ik_1 L} + (2-k_o/k_1-k_1/k_o)\, e^{ik_1 L}\right]^{-1}$$

(d) For $\omega<\omega_{pe}$, where $k_1 = i\alpha$, with α real, show that

$$E_t = 4\,[4 \cosh(\alpha L) + 2i(\alpha/k_o - k_o/\alpha) \sinh(\alpha L)]^{-1}$$

$$T = [\cosh^2(\alpha L) + (\alpha/k_o - k_o/\alpha)^2 \sinh^2(\alpha L)/4]^{-1}$$

This result shows that some power is transmitted through the slab, even with $\beta = \text{Re}\{k_1\} = 0$. This effect is known as the *tunneling effect*.

16.3 - Derive expressions for the phase velocity v_{ph} and the group velocity v_g from the dispersion relation (5.26), for wave propagation at arbitrary angles in a cold magnetoplasma.

16.4 - Use the dispersion relation (4.10), for the transverse mode of propagation in a cold isotropic electron gas (with $\vec{B}_0 = 0$), to calculate the damping factor $\alpha = \text{Im}\{k\}$. Show that, when $\omega \gg \omega_{pe}$, the damping factor is given approximately by

$$\alpha \cong (\nu/\omega)\omega_{pe}^2/[1 + (\nu/\omega)^2]\ 2\omega c$$

16.5 - Consider the propagation of high-frequency waves in a solid state plasma with equal number of electrons and holes (considering $m_e = m_h$ and $\nu_e = \nu_h$), immersed in a magnetostatic field $\vec{B}_0$. Let $\vec{k} = k\hat{x}$ and $\vec{B}_0 = B_0(\cos\theta\,\hat{x} + \sin\theta\,\hat{y})$. Use the Langevin equations for electrons and holes, and Maxwell equations, to show that

$$(-2UX - Y^2 \sin^2\theta + U^2)\, u_x + (Y^2 \sin\theta \cos\theta)\, u_y = 0$$

$$(Y^2 \sin\theta \cos\theta)\, u_x + (-2U\phi - Y^2 \cos^2\theta + U^2)\, u_y = 0$$

$$(-2U\phi - Y^2 + U^2)\, u_z = 0$$

where $\vec{u} = \vec{u}_e - \vec{u}_h$ and

$$U = 1 + i(\nu/\omega)$$

$$X = \omega_{pe}^2/\omega^2$$

$$Y = \omega_{ce}/\omega$$

$$\phi = X/(1 - k^2c^2/\omega^2)$$

From these component equations derive the following dispersion relations

$$\phi = \frac{U}{2} - \frac{Y^2\cos^2\theta}{2U} - \frac{Y^4 \sin^2\theta \cos^2\theta}{2U(-2UX - Y^2\sin^2\theta + U^2)}$$

$$\phi = \frac{U}{2} - \frac{Y^2}{2U}$$

Obtain expressions for the reflection points and the resonances. In particular, for the collisionless case (ν=0; U=1) show that the conditions for resonance are

$$\omega^2 = \frac{1}{2}\left\{\omega_{ce}^2 + 2\omega_{pe}^2 \pm \left[(\omega_{pe}^2 + 2\omega_{pe}^2)^2 - 8\omega_{pe}^2\,\omega_{ce}^2 \cos^2\theta\right]^{1/2}\right\}$$

$$\omega^2 = \omega_{ce}^2$$

and the reflection points are given by

$$\omega^2 = (\omega_{ce}^2 + 4\omega_{pe}^2 \pm \omega_{ce}^2)/2$$

$$\omega^2 = \omega_{ce}^2 + 2\omega_{pe}^2$$

16.6 - Use (6.17) and (6.18) for the group velocities of the left and the right circularly polarized waves, respectively, to show that the group velocity vanishes at the resonances and reflection points.

16.7 - Consider the problem of wave propagation at an arbitrary direction in a cold magnetoplasma, but including the motion of the ions (one type only).
(a) Show that the dispersion relation is obtained from an equation identical to (5.25), except that now we have (neglecting collisions)

$$S = 1 - X_e/(1 - Y_e^2) - X_i/(1 - Y_i^2)$$

$$D = - X_e Y_e/(1 - Y_e^2) + X_i Y_i/(1 - Y_i^2)$$

$$P = 1 - X_e - X_i$$

where (with $\alpha = e,i$)

$$X_\alpha = \omega_{p\alpha}^2/\omega^2$$

$$Y_\alpha = \omega_{c\alpha}/\omega$$

(b) Obtain the dispersion relation and show that it can be written in the form

$$\tan^2\theta = - \frac{P(k^2c^2/\omega^2 - R)\,(k^2c^2/\omega^2 - L)}{(S\,k^2c^2/\omega^2 - RL)\,(k^2c^2/\omega^2 - P)}$$

where θ is the angle between $\vec{k}$ and $\vec{B}_0$, $R = S + D$ and $L = S - D$.

(c) Determine and plot the resonances and reflection points as a function of θ.

16.8 - Using the results of the previous problem, analyze the various modes of propagation for the particular cases when $\theta = 0$ and $\theta = \pi/2$. Compare the results with those for a cold electron gas. Make a plot analogous to the one presented in Fig. 20.

16.9 - From the dispersion relations obtained in Problem 16.8 show that, in the limit of $\omega \ll \omega_{ci}$, we obtain the dispersion relation for the (shear) Alfvēn wave when $\vec{k} \,//\, \vec{B}_0$, and the dispersion relation for the compressional Alfvēn wave (cold plasma limit of the magnetosonic wave) when $\vec{k} \perp \vec{B}_0$. Furthermore, for $\omega \lesssim \omega_{ci}$ and $\vec{k} \,//\, \vec{B}_0$ determine the following approximate dispersion relation for ion cyclotron waves (assume $1 + c^2/V_A^2 \ll k^2c^2/\omega^2$),

$$\frac{k^2c^2}{\omega^2} = \frac{2c^2\,\omega_{ci}^2}{V_A^2(\omega_{ci}^2 - \omega^2)} = \frac{2\omega_{pi}^2}{(\omega_{ci}^2 - \omega^2)}$$

16.10 - From Eq. (5.25) show that the polarization of the waves propagating at an angle θ with respect to $\vec{B}_0$ (considering the perpendicular electric field vector component) is determined by

$$i(E_x/E_y) = (k^2c^2/\omega^2 - S)/D$$

From this result verify that for $\theta = 0$ the waves are left and right circularly polarized, whereas for $\theta = \pi/2$ the polarization is given by (for the extraordinary mode)

$$i(E_x/E_y)_X = -\,D/S$$

so that this mode is in general elliptically polarized.

16.11 - For a helicon wave, or a circularly polarized wave, show that the tip of the wave magnetic field vector traces out a helix.

16.12 - Make a plot analogous to Fig. 20 for wave propagation in a cold magnetoplasma, but in terms of ω as a function of the real part of k.

16.13 - Consider a plasma slab of thickness L and number density specified by n(x), where the x axis is normal to the slab. A plane-polarized monochromatic electromagnetic wave is normally incident on the slab (assume ω sufficiently large that $k^2c^2/\omega^2 > 0$). Neglecting reflection from the slab surfaces, determine an expression for the Faraday rotation angle as the wave traverses the plasma slab. Then, simplify this expression considering the cases n(x) = constant and n(x) = x, for $0 \leq x \leq L$.

CHAPTER 17

Waves in Warm Plasmas

1. INTRODUCTION

In the previous chapter we have analyzed the wave propagation problem in a cold plasma. Now we want to extend the theory developed in the previous chapter to include the pressure gradient term in the momentum equation. We shall consider wave propagation in a warm electron gas (in which ion motion is ignored) and in a fully ionized warm plasma (considering only one ion species), in the absence as well as in the presence of an externally applied magnetic field.

2. WAVES IN A FULLY IONIZED ISOTROPIC WARM PLASMA

2.1 Derivation of the equations for the electron and ion velocities

Consider now a fully ionized warm plasma having only one ion species, with no externally applied magnetic field ($\vec{B}_0 = 0$). The equations of conservation of mass and of momentum, for the electrons and the ions, can be written as

$$\partial n_\alpha/\partial t + \vec{\nabla}.\ (n_\alpha \vec{u}_\alpha) = 0 \tag{2.1}$$

$$m_\alpha\ (D\vec{u}_\alpha/Dt) = q_\alpha\ (\vec{E} + \vec{u}_\alpha \times \vec{B}) - (1/n_\alpha)\vec{\nabla}\, p_\alpha - m_\alpha \nu_{\alpha\beta}(\vec{u}_\alpha - \vec{u}_\beta) \tag{2.2}$$

where for the electrons $\alpha = e$, $\beta = i$, and for the ions $\alpha = i$, $\beta = e$. These equations are complemented by the following adiabatic energy equation for each species ($\alpha = e,i$),

$$p_\alpha\ n_\alpha^{-\gamma} = \text{constant} \tag{2.3}$$

where $\gamma = 1 + 2/N$ is the adiabatic constant and N denotes the number of degrees of freedom. Applying the $\vec{\nabla}$ operator to (2.3) and using the ideal gas law $p_\alpha = n_\alpha\ k_B\ T_\alpha$, we can rewrite (2.3) in the form

$$\vec{\nabla}p_\alpha = \gamma\, k_B\, T_\alpha\, \vec{\nabla}n_\alpha \tag{2.4}$$

We restrict our attention to *small-amplitude* waves in order to linearize the equations, and assume that

$$n_\alpha(\vec{r},t) = n_0 + n'_\alpha \exp(i\vec{k}.\vec{r} - i\omega t) \qquad |n'_\alpha| << n_0 \tag{2.5}$$

$$\vec{u}_\alpha(\vec{r},t) = \vec{u}_\alpha \exp(i\vec{k}.\vec{r} - i\omega t) \qquad u_\alpha << |\omega/k| \tag{2.6}$$

$$\vec{E}(\vec{r},t) = \vec{E} \exp(i\vec{k}.\vec{r}\ i\omega t) \tag{2.7}$$

$$\vec{B}(\vec{r},t) = \vec{B} \exp(i\vec{k}.\vec{r} - i\omega t) \tag{2.8}$$

Using these expressions in (2.1), and neglecting second order terms, we find

$$n'_\alpha / n_0 = (\vec{k}.\vec{u}_\alpha)/\omega \tag{2.9}$$

Similarly, we obtain for (2.2), after the substitution of $\vec{\nabla}p_\alpha$ from (2.4) and linearizing,

$$- i\, \omega\, \vec{u}_\alpha = (q_\alpha/m_\alpha)\, \vec{E} - V_{S\alpha}^2 i\vec{k}\, (n'_\alpha / n_0) - \nu_{\alpha\beta}(\vec{u}_\alpha - \vec{u}_\beta) \tag{2.10}$$

where $V_{S\alpha} = (\gamma k_B T_\alpha/m_\alpha)^{1/2}$ is the adiabatic sound speed for the type α particles.

Substituting (2.9) into (2.10), and multiplying by $i\omega$, we obtain the following equation involving the variables $\vec{u}_\alpha$, $\vec{u}_\beta$ and $\vec{E}$,

$$\omega^2\, \vec{u}_\alpha = i\omega(q_\alpha/m_\alpha)\, \vec{E} + V_{S\alpha}^2\, \vec{k}\, (\vec{k}.\vec{u}_\alpha) - i\omega\, \nu_{\alpha\beta}\, (\vec{u}_\alpha - \vec{u}_\beta) \tag{2.11}$$

A relationship between the electric field, and the electron and ion velocities, can be obtained from Maxwell curl equations with harmonic variations of $\vec{E}$ and $\vec{B}$, according to (2.7) and (2.8),

$$\vec{k} \times \vec{E} = \omega\, \vec{B} \tag{2.12}$$

$$i\, \vec{k} \times \vec{B} = \mu_0\, \vec{J} - (i\omega/c^2)\, \vec{E} \tag{2.13}$$

and the linearized expression for the plasma current density,

$$\vec{J} = n_0 \sum_\alpha q_\alpha \vec{u}_\alpha = n_0\, e\, (\vec{u}_i - \vec{u}_e). \tag{2.14}$$

Combining (2.12), (2.13) and (2.14) we find

$$\vec{E}_\ell = (ien_0/\omega\varepsilon_0)\,(\vec{u}_{e\ell} - \vec{u}_{i\ell}) \tag{2.15}$$

$$\vec{E}_t = (ien_0/\omega\varepsilon_0)\,(\vec{u}_{et} - \vec{u}_{it})/(1 - k^2c^2/\omega^2) \tag{2.16}$$

where the subscripts ℓ and t indicate components *longitudinal* and *transverse*, respectively, with respect to the wave propagation vector $\vec{k}$ (see Fig. 1 of Chapter 16).

Substituting (2.15) and (2.16) into (2.11), and writing this equation for each type of particles (electrons and ions), we have the following set of coupled equations for the *longitudinal* components of the electron and ion velocities,

$$\vec{u}_{e\ell}\,(\omega^2 - \omega_{pe}^2 - k^2 V_{se}^2 + i\omega\,\nu_{ei}) + \vec{u}_{i\ell}(\omega_{pe}^2 - i\omega\,\nu_{ei}) = 0 \tag{2.17}$$

$$\vec{u}_{e\ell}\,(\omega_{pi}^2 - i\omega\,\nu_{ie}) + \vec{u}_{i\ell}(\omega^2 - \omega_{pi}^2 - k^2 V_{si}^2 + i\omega\,\nu_{ie}) = 0 \tag{2.18}$$

and for the *transverse* components,

$$\vec{u}_{et}\left[\omega^2 - \frac{\omega_{pe}^2}{(1 - k^2c^2/\omega^2)} + i\omega\,\nu_{ei}\right] + \vec{u}_{it}\left[\frac{\omega_{pe}^2}{(1 - k^2c^2/\omega^2)} - i\omega\,\nu_{ei}\right] = 0 \tag{2.19}$$

$$\vec{u}_{et}\left[\frac{\omega_{pi}^2}{(1 - k^2c^2/\omega^2)} - i\omega\,\nu_{ie}\right] + \vec{u}_{it}\left[\omega^2 - \frac{\omega_{pi}^2}{(1 - k^2c^2/\omega^2)} + i\omega\,\nu_{ie}\right] = 0 \tag{2.20}$$

Note that the effect of the pressure gradient term appears only on the longitudinal component of the motion and, consequently, the transverse modes of propagation are the same ones as in the cold plasma model, but with the motion of the ions included.

2.2 Longitudinal waves

In what follows, in order to simplify the algebra, we shall neglect collisions ($\nu_{ei} = \nu_{ie} = 0$). For *longitudinal waves* the determinant of the coefficients in the system of Eqs. (2.17) and (2.18) must vanish. This condition gives

$$(\omega^2 - \omega_{pe}^2 - k^2 V_{se}^2)(\omega^2 - \omega_{pi}^2 - k^2 V_{si}^2) - \omega_{pe}^2 \omega_{pi}^2 = 0 \tag{2.21}$$

Multiplying the terms within parenthesis, this equation can be recast into the form

$$k^4 (V_{se}^2 V_{si}^2) + k^2 \left[\omega_{pe}^2 V_{si}^2 + \omega_{pi}^2 V_{se}^2 - \omega^2(V_{se}^2 + V_{si}^2)\right] + \omega^2(\omega^2 - \omega_{pe}^2 - \omega_{pi}^2) = 0 \tag{2.22}$$

Note that in the special case of the cold plasma model, in which the pressure gradient terms are ignored (i. e. $V_{se} = V_{si} = 0$), (2.22) gives $\omega^2 = \omega_{pe}^2 + \omega_{pi}^2$, which corresponds to the longitudinal plasma oscillations when the motion of both electrons and ions are taken into account. Eq. (2.22) has two roots for k^2, so that there are *two longitudinal modes* of propagation. One of these is termed the longitudinal *electron plasma wave* and the other is the longitudinal *ion plasma wave*. These plasma modes are electrostatic in charater,and contain all the charge accumulation and no magnetic field, whereas the transverse electromagnetic mode contains the entire magnetic field and has no charge accumulation (section 2.3).

Although it is not difficult to obtain the two exact solutions for k^2 from (2.22), it is more convenient to analyze it for some special cases which emphasize the role played by the inclusion of ion motion and the pressure gradient term. For this purpose, let us first rewrite (2.22) for the case when ion motion is not taken into account, which becomes

$$- k^2 V_{se}^2 \omega^2 + \omega^2(\omega^2 - \omega_{pe}^2) = 0 \tag{2.23}$$

or

$$\omega^2 = \omega_{pe}^2 + k^2 V_{se}^2 \tag{2.24}$$

Now, $V_{se}^2 = \gamma k_B T_e/m_e$, and since for a plane wave the compression is one-dimensional we have $\gamma = 3$, so that

$$\omega^2 = \omega_{pe}^2 + (3 k_B T_e/m_e) k^2 \tag{2.25}$$

This equation is known as the *Bohm-Gross dispersion relation* for the longitudinal *electron plasma* wave. This relation shows a reflection point (k = 0) for $\omega = \omega_{pe}$. For very high frequencies ($\omega \gg \omega_{pe}$) the phase velocity is $\omega/k = V_{se}$, which represents an electron acoustic wave.

Next, let us include the motion of the ions but under the assumption that its temperature is such that $V_{si} = 0$. Then, (2.22) simplifies to

$$k^2 V_{se}^2 (\omega_{pi}^2 - \omega^2) + \omega^2(\omega^2 - \omega_{pe}^2 - \omega_{pi}^2) = 0 \tag{2.26}$$

At very high frequencies ($\omega \gg \omega_{pe}$) we still have $\omega/k = V_{se}$, but now (2.26) shows a reflection point ($k = 0$) at $\omega = (\omega_{pe}^2 + \omega_{pi}^2)^{1/2}$.

Finally, let us analyze (2.22) in the limits of high and low frequencies. From the definitions of ω_{pe} and V_{si}, we have

$$\omega_{pe}^2 V_{si}^2 = (T_i/T_e)\, \omega_{pi}^2 V_{se}^2 \tag{2.27}$$

Therefore, (2.22) can be rewritten as

$$k^4 V_{se}^2 V_{si}^2 + k^2 \left[\omega_{pi}^2 V_{se}^2 (1 + T_i/T_e) - \omega^2(V_{se}^2 + V_{si}^2)\right] +$$

$$+ \omega^2 (\omega^2 - \omega_{pe}^2 - \omega_{pi}^2) = 0 \tag{2.28}$$

For *high frequencies,* such that $\omega^2 \gg \omega_{pi}^2 (1 + T_i/T_e)$, (2.28) becomes

$$k^4 V_{se}^2 V_{si}^2 - k^2 \omega^2(V_{se}^2 + V_{si}^2) + \omega^2 (\omega^2 - \omega_{pe}^2 - \omega_{pi}^2) = 0 \tag{2.29}$$

Further, considering $V_{se}^2 \omega^2 \gg V_{si}^2(\omega_{pe}^2 + \omega_{pi}^2)$, or equivalently $\omega^2 \gg \omega_{pi}^2 (T_i/T_e)(1 + m_e/m_i)$, a condition which also satisfies $\omega^2 \gg \omega_{pi}^2 (1 + T_i/T_e)$, we can add the term $k^2 V_{si}^2 (\omega_{pe}^2 + \omega_{pi}^2)$ to the left-hand side of (2.29) and rearrange this equation in the following aproximate form

$$(k^2 V_{si}^2 - \omega^2)(k^2 V_{se}^2 - \omega^2 + \omega_{pe}^2 + \omega_{pi}^2) \cong 0 \tag{2.30}$$

From this equation we see that for high frequencies $[\omega^2 \gg \omega_{pi}^2(1+T_i/T_e)]$ the dispersion relation for the longitudinal *ion plasma wave* is

$$\omega^2 = k^2 V_{si}^2 \tag{2.31}$$

while, for the *electron plasma wave*, the dispersion relation is

$$\omega^2 = \omega_{pe}^2 + \omega_{pi}^2 + k^2 V_{se}^2 \tag{2.32}$$

Next, for *low frequencies*, such that $\omega^2 << \omega_{pi}^2 (1 + T_i/T_e)$, (2.28) becomes

$$k^4 V_{se}^2 V_{si}^2 + k^2 V_{se}^2 \omega_{pi}^2 (1+T_i/T_e) - \omega^2 \omega_{pe}^2 = 0 \tag{2.33}$$

Multiplying this equation by $- \omega^2/(\omega_{pe}^2 k^4)$, assuming $k \neq 0$, it can be rewritten as

$$\left(\frac{\omega}{k}\right)^4 - \left(\frac{\omega}{k}\right)^2 V_{se}^2 \frac{\omega_{pi}^2}{\omega_{pe}^2} \left(1 + \frac{T_i}{T_e}\right) - V_{se}^2 V_{si}^2 \frac{\omega^2}{\omega_{pe}^2} = 0 \tag{2.34}$$

Since we are considering low frequencies, $\omega^2 << \omega_{pi}^2 (1 + T_i/T_e)$, and as long as (ω/k) is not much larger than V_{si}, the last term in the left hand side of (2.34) can be neglected as compared to the second one. Therefore, (2.34) gives, for low frequencies,

$$(\omega/k)^2 = V_{se}^2 (\omega_{pi}^2/\omega_{pe}^2) (1 + T_i/T_e) \tag{2.35}$$

Using the relation (2.27), this equation can be rewritten in the form

$$\omega^2 = k^2 V_{sp}^2 \tag{2.36}$$

where

$$V_{sp}^2 = \gamma k_B (T_e + T_i)/m_i \tag{2.37}$$

which is known as the *plasma sound speed*. It can be verified that the other root of (2.33) gives an evanescent wave at very low frequencies.

A plot of phase velocity versus frequency for the longitudinal wave is shown in Fig. 1. The longitudinal waves with phase velocities equal to V_{se} or V_{si} at high frequencies represent, respectively, acoustic oscillations due to the electrons and the ions. The low frequency wave travelling at the plasma sound speed, V_{sp}, represents an acoustic oscillation of *both* the electrons and the ions, and is usually referred to as the *ion-acoustic wave*.

2.3 Transverse wave

For a *transverse mode* of propagation ($\vec{u}_{et} \neq 0$; $\vec{u}_{it} \neq 0$) the determinant of the coefficients in the system of Eqs. (2.19) and (2.20) must vanish. Neglecting collisions ($\nu_{ei} = \nu_{ie} = 0$), we find

$$(\omega^2 - \frac{\omega_{pe}^2}{1-k^2c^2/\omega^2})(\omega^2 - \frac{\omega_{pi}^2}{1-k^2c^2/\omega^2}) - \frac{\omega_{pe}^2\ \omega_{pi}^2}{(1-k^2c^2/\omega^2)^2} = 0 \qquad (2.38)$$

which simplifies to

$$k^2\ c^2 = \omega^2 - (\omega_{pe}^2 + \omega_{pi}^2) \qquad (2.39)$$

This equation is similar to the dispersion relation (16.4.12) for the

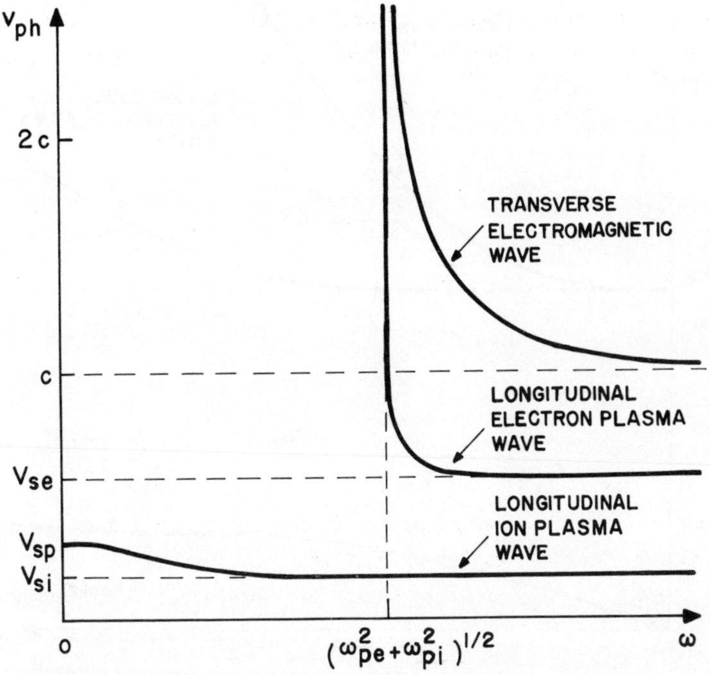

Fig. 1 - Phase velocity as a function of frequency for waves in a fully ionized isotropic ($\vec{B}_0 = 0$) warm plasma. The curves for the longitudinal waves also hold for propagation in the direction of $\vec{B}_0$, when $\vec{B}_o \neq 0$.

propagation of transverse waves in a cold isotropic plasma, except that the reflection point is now $(\omega_{pe}^2+\omega_{pi}^2)^{1/2}$ as a consequence of the inclusion of ion motion. A plot of phase velocity as a function of frequency for the dispersion relation (2.39) is also shown in Fig. 1. A dispersion plot in terms of ω as a function of k is displayed in Fig. 2 for the three modes of propagation.

In summary there are *three modes of wave propagation* in a warm fully ionized isotropic plasma (as compared to only one mode in the case of a cold isotropic plasma). They are the *transverse electromagnetic mode* (also present in the case of a cold plasma), the *longitudinal electron plasma mode* and the *longitudinal ion plasma mode*.

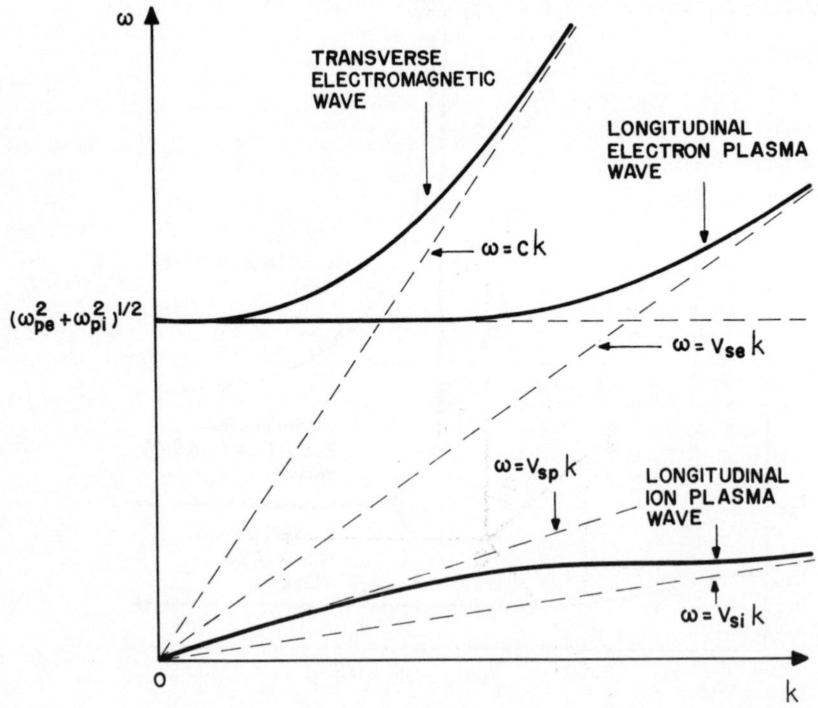

Fig. 2 - Dispersion relation for the three modes of wave propagation in a warm isotropic fully ionized plasma.

3. BASIC EQUATIONS FOR WAVES IN A WARM MAGNETOPLASMA

The basic equations for the study of wave propagation in a warm fully ionized magnetoplasma are (2.1), (2.2) and (2.3). Proceeding in the same manner as in the previous section, but now considering an externally applied uniform magnetostatic field $\vec{B}_0$, we obtain, in place of (2.11),

$$\omega^2\vec{u}_\alpha = i\omega\,(q_\alpha/m_\alpha)(\vec{E} + \vec{u}_\alpha \times \vec{B}_0) + V_{S\alpha}^2\,\vec{k}(\vec{k}.\vec{u}_\alpha) - i\omega\,\nu_{\alpha\beta}\,(\vec{u}_\alpha - \vec{u}_\beta) \qquad (3.1)$$

This equation is complemented by (2.15) and (2.16) or, equivalently, by

$$\vec{k} \times (\vec{k} \times \vec{E}) + (\omega^2/c^2)\,\vec{E} = -\,(i\omega\, en_0/\varepsilon_0\, c^2)(\vec{u}_i - \vec{u}_e) \qquad (3.2)$$

If we choose a Cartesian coordinate system such that the z axis is along $\vec{B}_0$ and $\vec{k}$ is in the x-z plane (Fig. 3), we have

$$\vec{B}_0 = B_0\,\hat{z} \qquad (3.3)$$

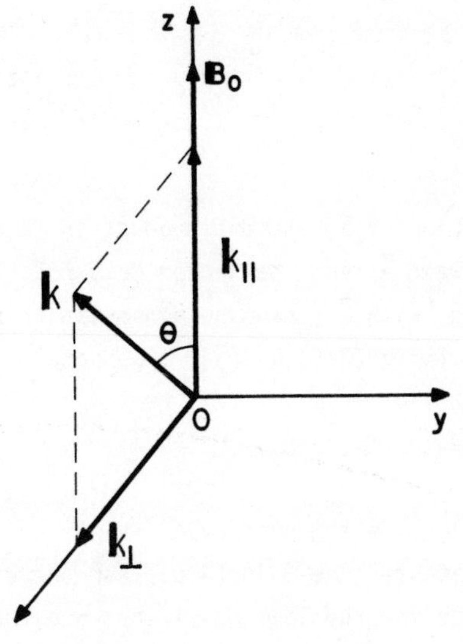

Fig. 3 - Cartesian coordinate system chosen with $\vec{B}_0$ along the z axis and $\vec{k}$ in the x-z plane.

$$\vec{k} = \vec{k}_{\parallel} + \vec{k}_{\perp} = k \sin\Theta \,\hat{x} + k \cos\Theta \,\hat{z} \tag{3.4}$$

and consequently (3.1) and (3.2) become, respectively, [see (16.5.10) and (16.5.5)]

$$\omega^2 \vec{u}_\alpha - i\omega(q_\alpha/m_\alpha)\, B_0\, (u_{\alpha y}\, \hat{x} - u_{\alpha x}\, \hat{y}) - V_{s\alpha}^2\, k^2 (\sin\Theta\, u_{\alpha x} + \cos\Theta\, u_{\alpha z}) \,.$$

$$.\, (\sin\Theta\, \hat{x} + \cos\Theta\, \hat{z}) + i\omega\, \nu_{\alpha\beta}\, (\vec{u}_\alpha - \vec{u}_\beta) = i\omega(q_\alpha/m_\alpha)\, \vec{E} \tag{3.5}$$

and

$$\bar{\bar{a}} \,.\, \vec{E} = -\,(ien_0/\omega\varepsilon_0)\, (\vec{u}_i - \vec{u}_e) \tag{3.6}$$

where the components of the dyad $\bar{\bar{a}}$, which represents the operator $[(c^2/\omega^2)\, \vec{k} \times (\vec{k} \times \ldots) + (\ldots)]$, can be arranged in matrix form as

$$\bar{\bar{a}} = \begin{pmatrix} 1 - \dfrac{k^2c^2}{\omega^2}\cos^2\Theta & 0 & \dfrac{k^2c^2}{\omega^2}\sin\Theta\cos\Theta \\ 0 & 1 - \dfrac{k^2c^2}{\omega^2} & 0 \\ \dfrac{k^2c^2}{\omega^2}\sin\Theta\cos\Theta & 0 & 1 - \dfrac{k^2c^2}{\omega^2}\sin^2\Theta \end{pmatrix} \tag{3.7}$$

With this matrix definition of $\bar{\bar{a}}$, the dot product in (3.6) can be thought of as a matrix product between $\bar{\bar{a}}$ and the vector column $\vec{E}$. Taking the inverse of the matrix associated with $\bar{\bar{a}}$ (assuming a non-vanishing determinant of its elements) and multiplying (3.6) by $(\bar{\bar{a}})^{-1}$, we obtain

$$\vec{E} = -\,(ien_0/\omega\varepsilon_0)\, (\bar{\bar{a}})^{-1} .\, (\vec{u}_i - \vec{u}_e) \tag{3.8}$$

since $(\bar{\bar{a}})^{-1} .\, \bar{\bar{a}} = \bar{\bar{1}}$, where $\bar{\bar{1}}$ is the unit dyad.

Eq. (3.8) can be used to replace $\vec{E}$ in (3.5). For the electrons we take α=e and β=i in (3.5), whereas for the ions α = i and β = e. We obtain therefore a system of six equations with the six unknowns $u_{\alpha j}$(with j = x, y, z, and α = e, i). The requirement that the determinant of its coefficients be equal to zero gives the dispersion relation.

4. WAVES IN A WARM ELECTRON GAS IN A MAGNETIC FIELD

In view of the complexity of the algebra involved, we shall initially consider the simple case of an electron gas immersed in an externally applied magnetic field, neglecting for the moment the macroscopic ion motion ($\vec{u}_i = 0$).

4.1 Derivation of the dispersion relation

From (3.5) we obtain for the electrons (taking $\vec{u}_i = 0$)

$$\vec{u}_e + i\,(\omega_{ce}/\omega)(u_{ey}\,\hat{x} - u_{ex}\,\hat{y}) - (V_{se}^2 k^2/\omega^2)(\sin\Theta\, u_{ex} + \cos\Theta\, u_{ez})(\sin\Theta\,\hat{x} + $$

$$+ \cos\Theta\,\hat{z}) + i\,(\nu_e/\omega)\,\vec{u}_e = (\omega_{pe}^2/\omega^2)\,(\bar{\bar{a}})^{-1} \cdot \vec{u}_e \tag{4.1}$$

Using the notation introduced in (16.5.14), (16.5.15) and (16.5.16), (4.1) can be rewritten in the form

$$U\,\vec{u}_e + \left(-\frac{V_{se}^2 k^2}{\omega^2}\sin^2\Theta\, u_{ex} + i\,Y\,u_{ey} - \frac{V_{se}^2 k^2}{\omega^2}\sin\Theta\cos\Theta\, u_{ez}\right)\hat{x} -$$

$$- i\,Y\,u_{ex}\,\hat{y} + \left(-\frac{V_{se}^2 k^2}{\omega^2}\sin\Theta\cos\Theta\, u_{ex} - \frac{V_{se}^2 k^2}{\omega^2}\cos^2\Theta\, u_{ez}\right)\hat{z} =$$

$$= X\,(\bar{\bar{a}})^{-1} \cdot \vec{u}_e \tag{4.2}$$

Defining a dyad $\bar{\bar{b}}$ through the matrix

$$\bar{\bar{b}} = \begin{pmatrix} \left(U - \dfrac{V_{se}^2 k^2}{\omega^2}\sin^2\Theta\right) & i\,Y & -\dfrac{V_{se}^2 k^2}{\omega^2}\sin\Theta\cos\Theta \\ -\,i\,Y & U & 0 \\ -\dfrac{V_{se}^2 k^2}{\omega^2}\sin\Theta\cos\Theta & 0 & \left(U - \dfrac{V_{se}^2 k^2}{\omega^2}\cos^2\Theta\right) \end{pmatrix} \tag{4.3}$$

equation (4.2) becomes

$$[\bar{\bar{b}} - X\,(\bar{\bar{a}})^{-1}] \cdot \vec{u}_e = 0 \tag{4.4}$$

This equation is of the form $\bar{\bar{c}} \cdot \vec{u}_e = 0$, with $\bar{\bar{c}} \equiv \bar{\bar{b}} - X\,(\bar{\bar{a}})^{-1}$. A nontrivial solution exists only if the determinant of the matrix $\bar{\bar{c}}$ vanishes. Therefore, in order to have nontrivial solutions ($\vec{u}_e \neq 0$) we must have

$$\det\,[\bar{\bar{b}} - X\,(\bar{\bar{a}})^{-1}] = 0 \tag{4.5}$$

This condition gives the dispersion relation for wave propagation in a warm electron gas immersed in a magnetic field.

In order to simplify matters, in the two following subsections we examine the dispersion relation (4.5) for the special cases of propagation parallel and perpendicular to the magnetic field.

4.2 Wave propagation along the magnetic field

For propagation *along* the magnetic field ($\vec{k} \parallel \vec{B}_0$) we have $\vec{k} = k\,\hat{z}$ and $\Theta = 0^0$, so that (3.7) and (4.3) simplify to

$$\bar{\bar{a}} = \begin{pmatrix} (1 - k^2c^2/\omega^2) & 0 & 0 \\ 0 & (1 - k^2c^2/\omega^2) & 0 \\ 0 & 0 & 1 \end{pmatrix} \tag{4.6}$$

$$\bar{\bar{b}} = \begin{pmatrix} U & i\,Y & 0 \\ -i\,Y & U & 0 \\ 0 & 0 & (U - V_{se}^2 k^2/\omega^2) \end{pmatrix} \tag{4.7}$$

Therefore, the determinant (4.5) becomes

$$\begin{vmatrix} U - \dfrac{X}{(1 - k^2c^2/\omega^2)} & iY & 0 \\ -iY & U - \dfrac{X}{(1 - k^2c^2/\omega^2)} & 0 \\ 0 & 0 & U - \dfrac{V_{se}^2 k^2}{\omega^2} - X \end{vmatrix} = 0 \tag{4.8}$$

which gives the following dispersion relations for *transverse waves* ($u_{ex} \neq 0$; $u_{ey} \neq 0$),

$$U - X/(1 - k^2c^2/\omega^2) = \pm Y \tag{4.9}$$

and for a *longitudinal wave* ($u_{ez} \neq 0$),

$$U - V_{se}^2 k^2/\omega^2 - X = 0 \tag{4.10}$$

Note that in this case the z component of (4.4) is uncoupled from the x and y components, so that the longitudinal mode is independent of the transverse modes.

Eq. (4.9) yields the following expressions corresponding, respectively, to the "plus" and "minus" signs,

$$k^2c^2/\omega^2 = 1 - X/(U - Y) \tag{4.11}$$

$$k^2c^2/\omega^2 = 1 - X/(U + Y) \tag{4.12}$$

These dispersion relations correspond, respectively, to the *right* and *left circularly polarized waves* (RCP and LCP) obtained in section 6, Chapter 16 [see (16.6.6) and (16.6.8)] , for *transverse waves* in a cold plasma.

For the longitudinal wave, substituting $U = 1 + i\nu_e/\omega$ and $X = \omega_{pe}^2/\omega^2$ in (4.10), the dispersion relation becomes

$$\omega^2 + i\nu_e\omega = \omega_{pe}^2 + k^2V_{se}^2 \tag{4.13}$$

Hence, as compared to the cold plasma model, instead of the longitudinal oscillation at ω_{pe}^2 (present in the cold plasma) there is in this case an additional mode of propagation, known as the *electron plasma wave*. Neglecting collisions ($\nu_e = 0$), (4.13) becomes the same dispersion relation as obtained in section 2 [Eq. (2.24)] for waves in an isotropic warm plasma. Hence, for propagation along the magnetic field, the longitudinal electron plasma wave is not affected by the presence of the magnetic field.

In summary, there are three modes of propagation in a warm electron gas for $\vec{k}$ parallel to the magnetic field: the transverse *RCP* and *LCP waves*, and the longitudinal *electron plasma wave*. The addition of the pressure gradient term in the equation of motion for the electrons has no effect on the transverse waves. A plot of phase velocity versus frequency for these three modes is displayed in Fig. 4. The corresponding $\omega(k)$ dispersion plot is shown in Fig. 5.

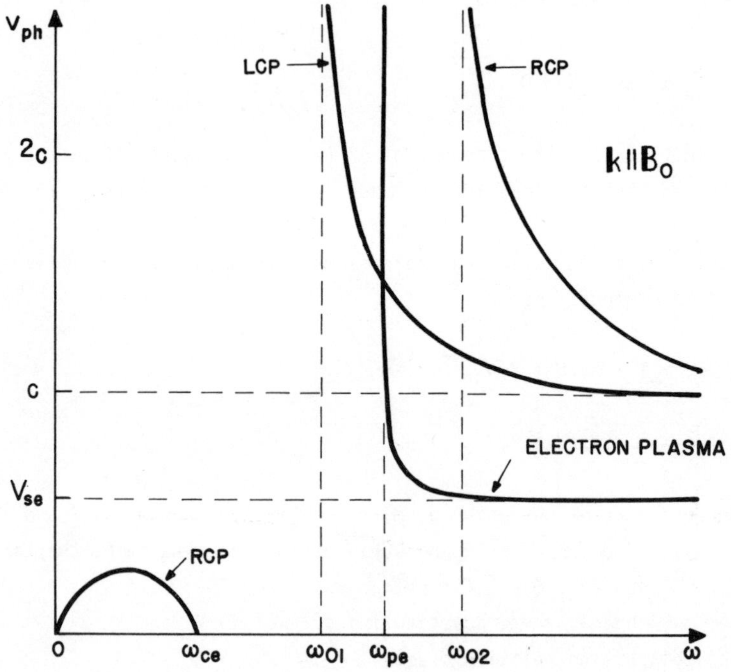

Fig. 4 - Phase velocity as a function of frequency for waves propagating in the direction of $\vec{B}_0$ in a warm electron gas.

4.3 Wave propagation normal to the magnetic field

For the case of propagation *across* the magnetic field ($\vec{k} \perp \vec{B}_0$) we have $\vec{k} = k\,\hat{x}$ and $\Theta = 90^0$, so that (3.7) and (4.3) simplify to

$$\bar{\bar{a}} = \begin{pmatrix} 1 & 0 & 0 \\ 0 & (1 - k^2c^2/\omega^2) & 0 \\ 0 & 0 & (1 - k^2c^2/\omega^2) \end{pmatrix} \qquad (4.11)$$

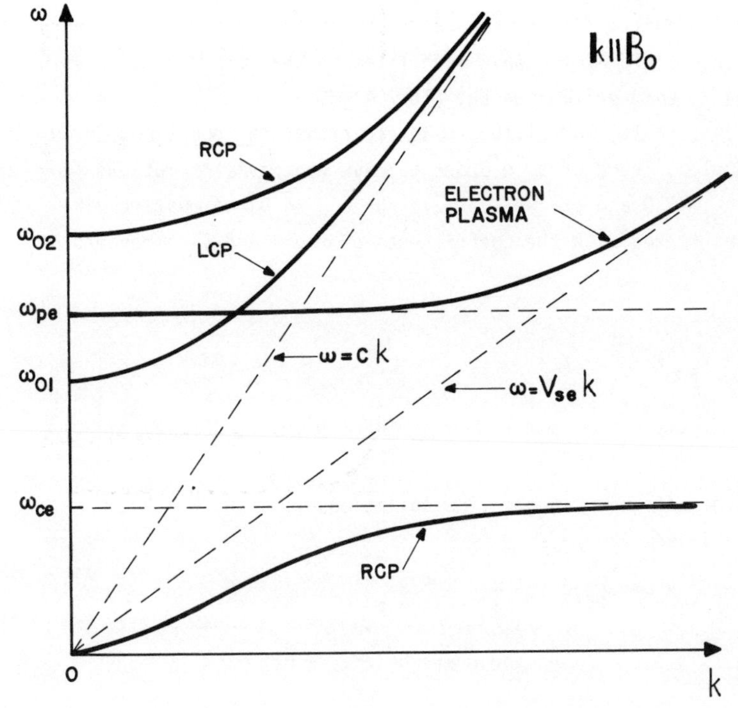

Fig. 5 - Dispersion plot for waves propagating parallel to $\vec{B}_0$ in a warm electron gas.

$$\bar{\bar{b}} = \begin{pmatrix} (U - V_{se}^2\, k^2/\omega^2) & i\,Y & 0 \\ -\,i\,Y & U & 0 \\ 0 & 0 & U \end{pmatrix} \tag{4.15}$$

From these expressions it is clear that the z component of (4.4) is uncoupled from the x and y components. Thus, for a *transverse wave* with electron motion along the z axis ($u_{ez} \neq 0$), we must have from the z component of (4.4),

$$U - X/(1 - k^2c^2/\omega^2) = 0 \tag{4.16}$$

or

$$k^2c^2/\omega^2 = 1 - X/U \tag{4.17}$$

which is the familiar dispersion relation for the transverse *ordinary* *wave* (the electric field of the wave oscillates in the same direction as $\vec{B}_0$) found in section 7, Chapter 16 [see Eq. (16.7.4)] .

From (4.4), (4.14) and (4.15) it is clear that the equations for u_{ex} and u_{ey} are coupled. Therefore, in order to have nontrivial solutions (*longitudinal* wave for $u_{ex} \neq 0$ and *transverse* wave for $u_{ey} \neq 0$) we must require the determinant formed with the coefficients of the x and y components of (4.4) to vanish,

$$\begin{vmatrix} U - V_{se}^2\, k^2/\omega^2 - X & i\,Y \\ -\,i\,Y & U - X/(1 - k^2c^2/\omega^2) \end{vmatrix} = 0 \tag{4.18}$$

This determinant gives, neglecting collisions ($\nu_e = 0$; $U = 1$),

$$(\omega^2 - V_{se}^2 k^2 - \omega_{pe}^2)(\omega^2 - \frac{\omega_{pe}^2}{1 - k^2c^2/\omega^2}) - \omega^2\,\omega_{ce}^2 = 0 \tag{4.19}$$

Expanding this expression, and rearranging, we get

$$k^4\,(c^2\,V_{se}^2) - k^2\left[V_{se}^2(\omega^2 - \omega_{pe}^2) + c^2(\omega^2 - \omega_{pe}^2 - \omega_{ce}^2)\right] + (\omega^2 - \omega_{pe}^2)^2 -$$

$$- \omega^2 \omega_{ce}^2 = 0 \tag{4.20}$$

This dispersion relation is quadratic in k^2, so that there will be two values of k^2 as a function of ω, that is, *two modes of propagation*. Since generally we have $V_{se} << c$, the first term within brackets in the left-hand side of (4.20) can be neglected as compared to the other one. With this approximation, (4.20) becomes

$$k^4 (c^2 V_{se}^2) - k^2c^2(\omega^2 - \omega_{pe}^2 - \omega_{ce}^2) + (\omega^2 - \omega_{pe}^2)^2 - \omega^2 \omega_{ce}^2 = 0 \tag{4.21}$$

Although it is not difficult to obtain the exact solution of this equation , it is more instructive to analyze it for some special limiting cases. First, let us obtain the approximate solution of (4.21) in the region where $\omega^2 >> k^2V_{se}^2$, that is, when the term $k^4c^2 V_{se}^2$ is smaller than any of the others. For k^2 positive this condition implies in phase velocities much larger than V_{se} and, for this reason, it will be referred to as the *high phase velocity* limit. With this condition, (4.21) reduces to

$$- k^2c^2(\omega^2 - \omega_{pe}^2 - \omega_{ce}^2) + (\omega^2 - \omega_{pe}^2)^2 - \omega^2 \omega_{ce}^2 = 0 \tag{4.22}$$

or

$$k^2c^2 = \frac{(\omega^2 + \omega\omega_{ce} + \omega_{pe}^2)(\omega^2 - \omega\omega_{ce} - \omega_{pe}^2)}{(\omega^2 - \omega_{pe}^2 - \omega_{ce}^2)} \quad ; (\omega^2 >> k^2 V_{se}^2) \tag{4.23}$$

This equation is exactly the same dispersion relation found in section 7, Chapter 16 [Eq. (16.7.7)], for the *transverse extraordinary wave* in a cold plasma, except that now the condition $\omega^2 >> k^2 V_{se}^2$ must be satisfied for (4.23) to be applicable.

Next, let us obtain the approximate solution of (4.21) in the region where $\omega^2 << k^2c^2$. For k^2 positive this condition implies in phase velocities much smaller than the the velocity of light and, for this reason, it will be referred to as the *low phase velocity* limit. Thus, for $\omega^2 << k^2c^2$, (4.21) reduces to

$$k^4 (c^2 V_{se}^2) - k^2c^2 (\omega^2 - \omega_{pe}^2 - \omega_{ce}^2) = 0 \tag{4.24}$$

or

$$\omega^2 = \omega_{pe}^2 + \omega_{ce}^2 + k^2 V_{se}^2 \; ; \quad (\omega^2 << k^2c^2) \tag{4.25}$$

When $\vec{B}_0 = 0$ (i.e. $\omega_{ce} = 0$) this equation becomes identical to the dispersion relation for the longitudinal *electron plasma wave* [see Eq. (2.24)]. It is a valid solution for (4.21) only under the condition $\omega^2 << k^2c^2$.

Fig. 6 displays the phase velocityas a function of frequency for the *transverse ordinary mode* (4.17) and for the two modes described by (4.20). Note that, of these last two modes, one is a purely transverse *extraordinary wave*, while the other one is partially transverse (electromagnetic *extraordinary wave* in the high phase velocity limit) and partially longitudinal *(electron plasma wave* in the low phase velocity limit). In this last mode, the transition from a basically transverse electromagnetic wave to a basically longitudinal electron plasma wave occurs in the frequency region where the phase velocity

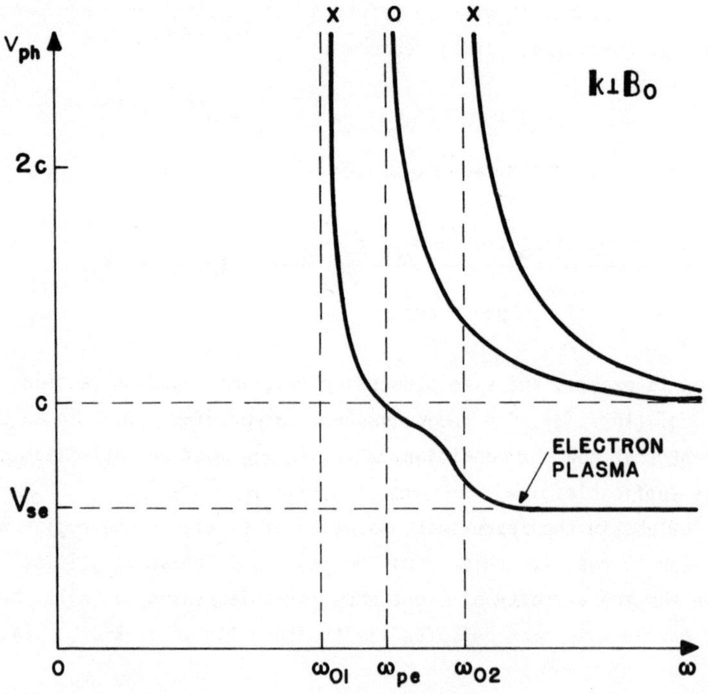

Fig. 6 - Phase velocity as a function of frequency for waves propagating perpendicular to $\vec{B}_0$ in a warm electron gas.

lies between c and V_{se}. The corresponding $\omega(k)$ dispersion plot is shown in Fig. 7.

4.4 Wave propagation in an arbitrary direction

For propagation in an arbitrary direction with respect to the magnetic field, the dispersion relation is obtained from (4.5) with the dyads $\bar{\bar{a}}$ and $\bar{\bar{b}}$ as given by (3.7) and (4.3). For an arbitrary angle between 0^0 and 90^0, we expect the phase velocity versus frequency curves to lie between those of Figs. 4 and 6. Therefore, instead of getting involved in the cumbersome algebra behind (4.5), we present only the dispersion curves of Fig. 8, in which the shaded area illustrates how the transition occurs from $\Theta = 0^0$ to $\Theta = 90^0$. It can be easily verified that the only resonance which exists for an arbitrary angle is at the frequency $\omega = \omega_{ce} \cos\Theta$. The reflection points for any angle of propagation occur at the frequencies ω_{01}, ω_{pe} and ω_{02}.

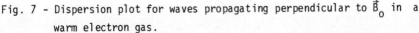

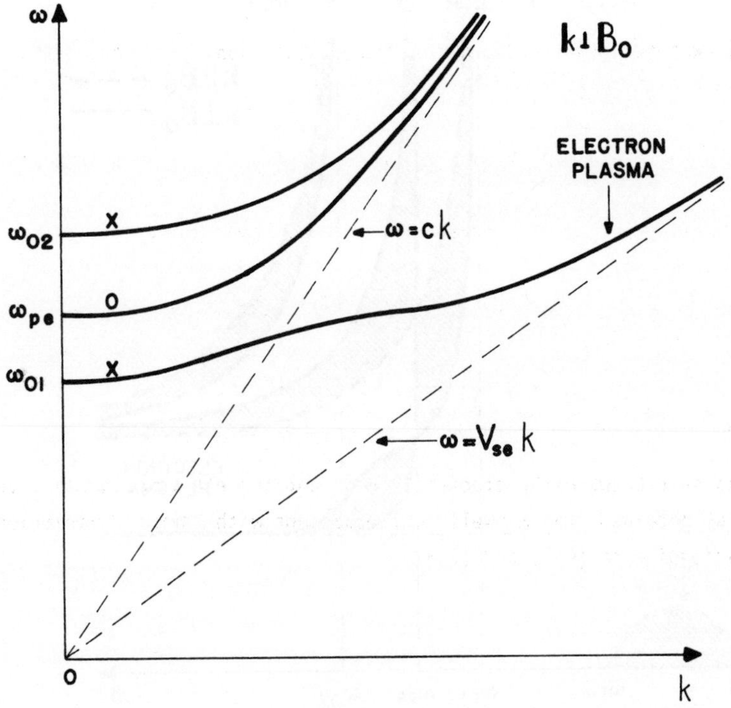

Fig. 7 - Dispersion plot for waves propagating perpendicular to $\vec{B}_0$ in a warm electron gas.

5. WAVES IN A FULLY IONIZED WARM MAGNETOPLASMA

Consider now the propagation of plane waves in a fully ionized warm plasma having only one ion species, immersed in an externally applied uniform magnetostatic field.

5.1 Derivation of the dispersion relation

The equation of motion for the *electrons* is, from (3.5),

$$\omega^2 \vec{u}_e + i\omega\omega_{ce} (u_{ey} \hat{x} - u_{ex} \hat{y}) -$$

$$- V_{se}^2 k^2 (\sin\Theta\, u_{ex} + \cos\Theta\, u_{ez}) (\sin\Theta\, \hat{x} + \cos\Theta\, \hat{z}) +$$

$$+ i\omega\, \nu_{ei} (\vec{u}_e - \vec{u}_i) = - i\omega\, (e/m_e) \vec{E} \tag{5.1}$$

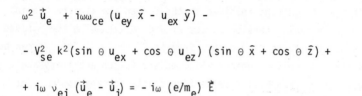

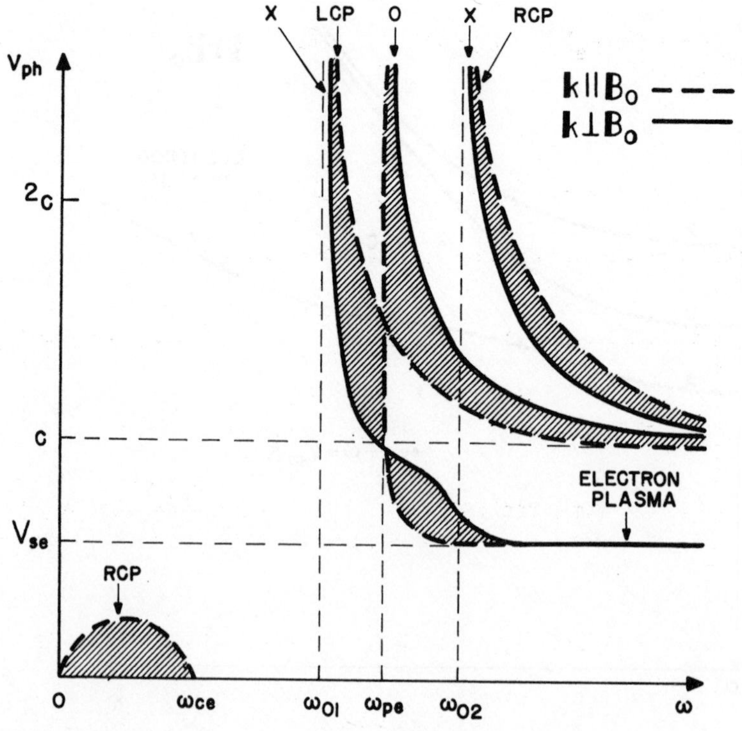

Fig. 8 - Phase velocity versus frequency for wave propagation in a warm electron gas immersed in a magnetic field.

and for the ions,

$$\omega^2 \vec{u}_i - i\omega\omega_{ci} (u_{iy} \hat{x} - u_{ix} \hat{y}) -$$

$$- V_{si}^2 k^2 (\sin \Theta u_{ix} + \cos \Theta u_{iz}) (\sin \Theta \hat{x} + \cos \Theta \hat{z}) +$$

$$+ i\omega \nu_{ie} (\vec{u}_i - \vec{u}_e) = i\omega (e/m_i) \vec{E} \tag{5.2}$$

Eqs. (5.1) and (5.2), involving the variables $\vec{u}_e$, $\vec{u}_i$ and $\vec{E}$, are complemented by (3.6),

$$\bar{\bar{a}} \cdot \vec{E} = (ien_0/\omega\varepsilon_0) (\vec{u}_e - \vec{u}_i) \tag{5.3}$$

where the dyad $\bar{\bar{a}}$ is defined according to (3.7).

Eqs. (5.1) and (5.2) can be written, respectively, in compact form, as

$$\bar{\bar{b}}_e \cdot \vec{u}_e = - i (e/\omega m_e) \vec{E} - i (\nu_{ei}/\omega) (\vec{u}_e - \vec{u}_i) \tag{5.4}$$

and

$$\bar{\bar{b}}_i \cdot \vec{u}_i = i (e/\omega m_i) \vec{E} - i (\nu_{ie}/\omega) (\vec{u}_i - \vec{u}_e) \tag{5.5}$$

where the dyads $\bar{\bar{b}}_e$ and $\bar{\bar{b}}_i$ are appropriately defined by

$$\bar{\bar{b}}_e = \begin{pmatrix} \left(1 - \dfrac{V_{se}^2 k^2}{\omega^2} \sin^2\Theta\right) & i Y_e & - \dfrac{V_{se}^2 k^2}{\omega^2} \sin \Theta \cos\Theta \\ - i Y_e & 1 & 0 \\ - \dfrac{V_{se}^2 k^2}{\omega^2} \sin \Theta \cos \Theta & 0 & \left(1 - \dfrac{V_{se}^2 k^2}{\omega^2} \cos^2 \Theta \right) \end{pmatrix} \tag{5.6}$$

$$\bar{\bar{b}}_i = \begin{pmatrix} \left(1 - \dfrac{V_{si}^2 k^2}{\omega^2} \sin^2 \Theta\right) & - i Y_i & - \dfrac{V_{si}^2 k^2}{\omega^2} \sin \Theta \cos\Theta \\ i Y_i & 1 & 0 \\ - \dfrac{V_{si}^2 k^2}{\omega^2} \sin \Theta \cos \Theta & 0 & \left(1 - \dfrac{V_{si}^2 k^2}{\omega^2} \cos^2 \Theta\right) \end{pmatrix} \tag{5.7}$$

where $Y_e = \omega_{ce}/\omega$ and $Y_i = \omega_{ci}/\omega$. Multiplying (5.4) and (5.5), respectively, by the inverse matrices corresponding to $\bar{\bar{b}}_e$ and $\bar{\bar{b}}_i$, we get

$$\vec{u}_e = - i\,(e/\omega m_e)\,(\bar{\bar{b}}_e)^{-1} \cdot \vec{E} - i\,(\nu_{ei}/\omega)\,(\bar{\bar{b}}_e)^{-1} \cdot (\vec{u}_e - \vec{u}_i) \tag{5.8}$$

$$\vec{u}_i = i(e/\omega m_i)\,(\bar{\bar{b}}_i)^{-1} \cdot \vec{E} + i\,(\nu_{ie}/\omega)\,(\bar{\bar{b}}_i)^{-1} \cdot (\vec{u}_e - \vec{u}_i) \tag{5.9}$$

Subtracting (5.9) from (5.8), and rearranging, yields

$$[\,\bar{\bar{1}} + i\,(\nu_{ei}/\omega)(\bar{\bar{b}}_e)^{-1} + i\,(\nu_{ie}/\omega)(\bar{\bar{b}}_i)^{-1}] \cdot (\vec{u}_e - \vec{u}_i) +$$

$$+\, i\,(e/\omega)\,[\,(1/m_e)(\bar{\bar{b}}_e)^{-1} + (1/m_i)(\bar{\bar{b}}_i)^{-1}] \cdot \vec{E} = 0 \tag{5.10}$$

Combining (5.10) and (5.3) to eliminate the variable $(\vec{u}_e - \vec{u}_i)$, results in the following equation involving only the electric field vector

$$[\,\bar{\bar{1}} + i\,(\nu_{ei}/\omega)\,(\bar{\bar{b}}_e)^{-1} + i\,(\nu_{ie}/\omega)\,(\bar{\bar{b}}_i)^{-1}] \cdot (\bar{\bar{a}} \cdot \vec{E}) -$$

$$-\,[(\omega_{pe}^2/\omega^2)\,(\bar{\bar{b}}_e)^{-1} - (\omega_{pi}^2/\omega^2)\,(\bar{\bar{b}}_i)^{-1}] \cdot \vec{E} = 0 \tag{5.11}$$

or

$$\Big\{[\bar{\bar{1}} + i\,(\nu_{ei}/\omega)\,(\bar{\bar{b}}_e)^{-1} + i\,(\nu_{ie}/\omega)\,(\bar{\bar{b}}_i)^{-1}\,] \cdot \bar{\bar{a}} - X_e\,(\bar{\bar{b}}_e)^{-1} -$$

$$-\,X_i\,(\bar{\bar{b}}_i)^{-1}\Big\} \cdot \vec{E} = 0 \tag{5.12}$$

where $X_e = \omega_{pe}^2/\omega^2$ and $X_i = \omega_{pi}^2/\omega^2$.

As before, the dispersion relation is obtained by setting the determinant of the 3 x 3 matrix in (5.12) equal to zero,

$$\det\Big\{\,[\bar{\bar{1}} + i\,(\nu_{ei}/\omega)\,(\bar{\bar{b}}_e)^{-1} + i\,(\nu_{ie}/\omega)\,(\bar{\bar{b}}_i)^{-1}] \cdot \bar{\bar{a}} - X_e\,(\bar{\bar{b}}_e)^{-1}$$

$$-\,X_i\,(\bar{\bar{b}}_i)^{-1}\Big\} = 0 \tag{5.13}$$

If collisions are neglected ($\nu_{ei} = \nu_{ie} = 0$), (5.13) simplifies to

$$\det \left[\bar{\bar{a}} - X_e (\bar{b}_e)^{-1} - X_i (\bar{b}_i)^{-1} \right] = 0 \qquad (5.14)$$

In the following subsections, in order to simplify the algebra involved, we shall neglect collisions and analyze the problem using (5.14).

5.2 Wave propagation along the magnetic field

For $\Theta = 0^0$ ($\vec{k} \,||\, B_0$) we have from (3.7), (5.6) and (5.7), respectively,

$$\bar{\bar{a}} = \begin{pmatrix} (1-k^2c^2/\omega^2) & 0 & 0 \\ 0 & (1-k^2c^2/\omega^2) & 0 \\ 0 & 0 & 1 \end{pmatrix} \qquad 5.15)$$

$$\bar{b}_e = \begin{pmatrix} 1 & i\,Y_e & 0 \\ -i\,Y_e & 1 & 0 \\ 0 & 0 & (1-V_{se}^2 k^2/\omega^2) \end{pmatrix} \qquad (5.16)$$

$$\bar{b}_i = \begin{pmatrix} 1 & -i\,Y_i & 0 \\ i\,Y_i & 1 & 0 \\ 0 & 0 & (1-V_{si}^2 k^2/\omega^2) \end{pmatrix} \qquad (5.17)$$

The inverse of the matrices (5.16) and (5.17) are, respectively,

$$(\bar{b}_e)^{-1} = \begin{pmatrix} \dfrac{1}{(1-Y_e^2)} & \dfrac{-\,i\,Y_e}{(1-Y_e^2)} & 0 \\ \dfrac{i\,Y_e}{(1-Y_e^2)} & \dfrac{1}{(1-Y_e^2)} & 0 \\ 0 & 0 & \dfrac{1}{(1-V_{se}^2 k^2/\omega^2)} \end{pmatrix} \qquad (5.18)$$

$$(\bar{\bar{b}}_i)^{-1} = \begin{pmatrix} \dfrac{1}{(1 - Y_i^2)} & \dfrac{i\,Y_i}{(1 - Y_i^2)} & 0 \\ \dfrac{-i\,Y_i}{(1 - Y_i^2)} & \dfrac{1}{(1 - Y_i^2)} & 0 \\ 0 & 0 & \dfrac{1}{(1-V_{si}^2\,k^2/\,\omega^2)} \end{pmatrix} \tag{5.19}$$

Substituting the matrices (5.15), (5.18) and (5.19), into (5.12), and setting $\nu_{ei} = \nu_{ie} = 0$, we obtain

$$\begin{pmatrix} A_1 & A_2 & 0 \\ -A_2 & A_1 & 0 \\ 0 & 0 & A_3 \end{pmatrix} \begin{pmatrix} E_x \\ E_y \\ E_z \end{pmatrix} = 0 \tag{5.20}$$

where

$$A_1 = 1 - \frac{k^2c^2}{\omega^2} - \frac{X_i}{(1 - Y_i^2)} - \frac{X_e}{(1 - Y_e^2)} \tag{5.20a}$$

$$A_2 = - \frac{i\,X_i\,Y_i}{(1 - Y_i^2)} + \frac{i\,X_e\,Y_e}{(1 - Y_e^2)} \tag{5.20b}$$

$$A_3 = 1 - \frac{X_i}{(1 - V_{si}^2\,k^2/\omega^2)} - \frac{X_e}{(1 - V_{se}^2\,k^2/\omega^2)} \tag{5.20c}$$

It is clear from this matrix equation that the *longitudinal* component of the electric field (E_z) is uncoupled from the *transverse* component (E_x and E_y). Therefore, for longitudinal waves ($E_z \neq 0$), the coefficient of E_z in (5.20) must be equal to zero, which gives the following dispersion relation for *longitudinal waves*,

$$1 - \frac{X_i}{(1 - V_{si}^2 k^2/\omega^2)} - \frac{X_e}{(1 - V_{se}^2 k^2/\omega^2)} = 0 \tag{5.21}$$

This dispersion relation can be rearranged in the following form

$$k^4 (V_{se}^2 V_{si}^2) + k^2 [\omega_{pe}^2 V_{si}^2 + \omega_{pi}^2 V_{se}^2 - \omega^2 (V_{se}^2 + V_{si}^2)] +$$

$$+ \omega^2 (\omega^2 - \omega_{pe}^2 - \omega_{pi}^2) = 0 \tag{5.22}$$

which is identical to (2.22). Therefore, since it is a quadratic equation in k^2, there are in general *two longitudinal modes* of propagation. Note that these two longitudinal modes, propagating along $\vec{B}_0$, are not affected by the magnetic field strength. This dispersion relation has already been analyzed in section 2, where it was shown that the two longitudinal modes are the *electron plasma wave* and the *ion plasma wave*.

The dispersion relation for transverse waves ($E_x \neq 0$; $E_y \neq 0$) are seen, from (5.20), to be given by

$$\left[1 - \frac{k^2c^2}{\omega^2} - \frac{X_i}{(1 - Y_i^2)} - \frac{X_e}{(1 - Y_e^2)}\right]^2 - \left[\frac{X_i\, Y_i}{(1 - Y_i^2)} - \frac{X_e\, Y_e}{(1 - Y_e^2)}\right]^2 = 0 \tag{5.23}$$

Using the notation

$$S = 1 - X_i/(1 - Y_i^2) - X_e/(1 - Y_e^2) \tag{5.24}$$

$$D = X_i\, Y_i/(1 - Y_i^2) - X_e\, Y_e/(1 - Y_e^2) \tag{5.25}$$

and letting

$$R = S + D \tag{5.26}$$

$$L = S - D \tag{5.27}$$

then (5.23) becomes

$$(k^2c^2/\omega^2 - R)\ (k^2c^2/\omega^2 - L) = 0 \tag{5.28}$$

There are therefore *two transverse modes* that propagate along the magnetic field with dispersion relations given by

$$(k^2c^2/\omega^2)_R = R \tag{5.29}$$

and

$$(k^2c^2/\omega^2)_L = L \tag{5.30}$$

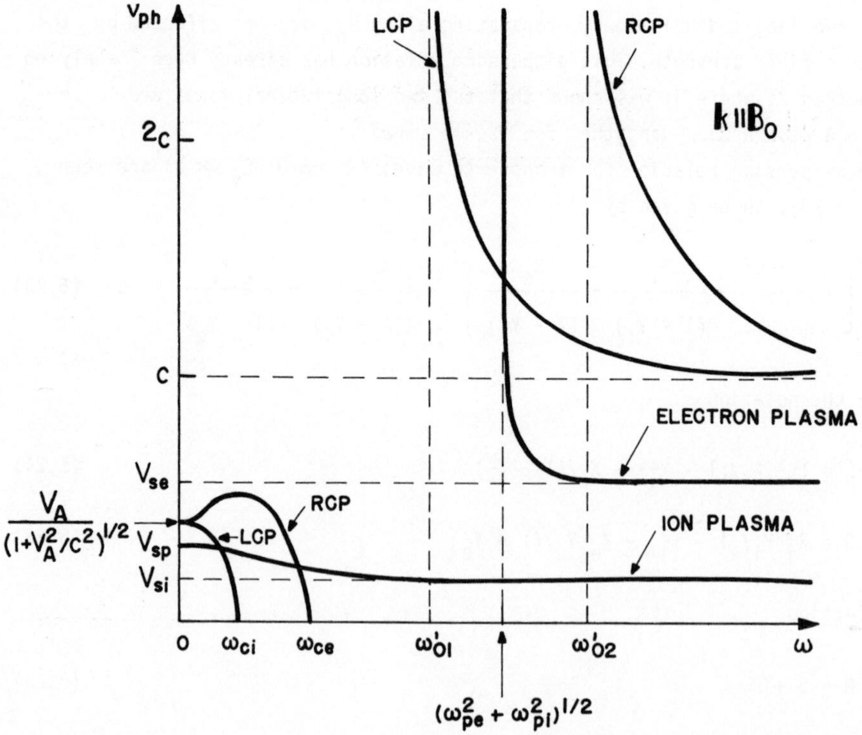

Fig. 9 - Phase velocity as a function of frequency for plane waves travelling along the magnetic field in a warm fully ionized magnetoplasma.

From the x component of (5.20) we have

$$E_y/E_x = (S - k^2c^2/\omega^2)/i\ D \tag{5.31}$$

so that, using (5.29), we obtain

$$(E_y/E_x)_R = i \tag{5.32}$$

whereas, using (5.30),

$$(E_y/E_x)_L = -\ i \tag{5.33}$$

Therefore, the dispersion relation (5.29) corresponds to a *right-hand circularly polarized wave*, and (5.30) to a *left-hand circularly polarized wave*.

The phase velocity as a function of frequency for propagation along $\vec{B}_0$ is shown in Fig. 9. The reflection points at ω'_{01} and ω'_{02} are not exactly the same ones given by equations (16.6.13) and (16.6.14), but are slightly different as a result of the inclusion of ion motion. Also, because ion motion has been taken into account, besides the resonance at $\omega = \omega_{ce}$ for the RCP wave, there is also a resonance at $\omega = \omega_{ci}$ for the LCP wave.

In the very low-frequency limit, the phase velocities of the RCP and LCP waves tend to $V_A/(1 + V_A^2/c^2)^{1/2}$, instead of going to zero as in the case of the cold electron plasma model. This result can be seen as follows. For very low frequency waves such that

$$\omega \ll \omega_{ci} \tag{5.34}$$

we obtain, using (5.24) and (5.25),

$$R = L = 1 + \omega_{pe}^2/\omega_{ci}\ \omega_{ce} \qquad (\omega \ll \omega_{ci}) \tag{5.35}$$

Therefore, using the definitions of ω_{pe}, ω_{ci} and ω_{ce}, the dispersion relation for the RCP and LCP waves becomes

$$k^2c^2/\omega^2 = 1 + n_0\ m_i/\varepsilon_0\ B_0^2 \tag{5.36}$$

The average mass density is $\rho_m = n_0\ (m_e + m_i) \simeq n_0\ m_i$, and since $\varepsilon_0 = 1/(\mu_0 c^2)$,

(5.36) can be rewritten as

$$k^2c^2/\omega^2 = 1 + c^2 \mu_o \rho_m/B_o^2 \tag{5.37}$$

or

$$k^2c^2/\omega^2 = 1 + c^2/V_A^2 \tag{5.38}$$

where $V_A = (B_o^2/\mu_o \rho_m)^{1/2}$ is the Alfvēn velocity, defined in (15.1.4). Thus, from (5.38), in the very low-frequency limit the phase velocity of both transverse waves is given by

$$v_{ph} = \omega/k = V_A/(1 + V_A^2/c^2)^{1/2} \tag{5.39}$$

Note that, for plasmas in which $V_A^2 << c^2$ (weak B_o field or high density), (5.39) reduces to $v_{ph} = V_A$. This very low-frequency limit corresponds to the Alfvēn wave discussed in Chapter 15.

5.3 Wave propagation normal to the magnetic field

Considering now $\vec{k} \perp \vec{B}_o$, we set $\Theta = 90^o$ in (3.7), (5.6) and (5.7), to obtain,

$$\bar{\bar{a}} = \begin{pmatrix} 1 & 0 & 0 \\ 0 & (1 - k^2c^2/\omega^2) & 0 \\ 0 & 0 & (1 - k^2c^2/\omega^2) \end{pmatrix} \tag{5.40}$$

$$\bar{\bar{b}}_e = \begin{pmatrix} (1 - V_{se}^2k^2/\omega^2) & i\,Y_e & 0 \\ -i\,Y_e & 1 & 0 \\ 0 & 0 & 1 \end{pmatrix} \tag{5.41}$$

$$\bar{\bar{b}}_i = \begin{pmatrix} (1 - V_{si}^2k^2/\omega^2) & -i\,Y_i & 0 \\ i\,Y_i & 1 & 0 \\ 0 & 0 & 1 \end{pmatrix} \tag{5.42}$$

Taking the inverse of the matrices in (5.41) and (5.42), we obtain for (5.12) (neglecting collisions),

$$\begin{pmatrix} S_{II} & -i\,D_I & 0 \\ i\,D_I & (S_I - k^2c^2/\omega^2) & 0 \\ 0 & 0 & (P-k^2c^2/\omega^2) \end{pmatrix} \begin{pmatrix} E_x \\ E_y \\ E_z \end{pmatrix} = 0 \tag{5.43}$$

where

$$S_I = 1 - \frac{X_i\,(1 - k^2V_{si}^2/\omega^2)}{(1 - Y_i^2 - k^2V_{si}^2/\omega^2)} - \frac{X_e\,(1 - k^2V_{se}^2/\omega^2)}{(1 - Y_e^2 - k^2V_{se}^2/\omega^2)} \tag{5.44}$$

$$S_{II} = 1 - \frac{X_i}{(1 - Y_i^2 - k^2V_{si}^2/\omega^2)} - \frac{X_e}{(1 - Y_e^2 - k^2V_{se}^2/\omega^2)} \tag{5.45}$$

$$D_I = \frac{X_i\,Y_i}{(1 - Y_i^2 - k^2V_{si}^2/\omega^2)} - \frac{X_e\,Y_e}{(1 - Y_e^2 - k^2V_{se}^2/\omega^2)} \tag{5.46}$$

$$P = 1 - X_i - X_e \tag{5.47}$$

From (5.43) it it clear that E_z is uncoupled from the electric field components E_x and E_y so that the *ordinary mode* (the *transverse* mode which has $E_z \neq 0$ and is not affected by the presence of the magnetostatic field) has the dispersion relation

$$k^2c^2/\omega^2 = P \tag{5.48}$$

or

$$k^2c^2 = \omega^2(\omega_{pe}^2 + \omega_{pi}^2) \tag{5.49}$$

which is the same expression obtained in (2.39).

The modes involving the field components E_x and E_y (longitudinal for $E_x \neq 0$ and transverse for $E_y \neq 0$) are seen, from (5.43), to be coupled and have the following dispersion relation

$$S_{II}\,(S_I - k^2c^2/\omega^2) - D_I^2 = 0 \tag{5.50}$$

Substituting the expressions for S_I, S_{II} and D_I into (5.50), results in a cubic equation in k^2, showing that in general there are three modes of propagation. A detailed analysis of this dispersion relation shows that these three modes of propagation are the *partially transverse extraordinary wave,* the *longitudinal electron plasma wave* and the *longitudinal ion plasma wave.*

Fig. 10 shows the phase velocity as a function of frequency for the four

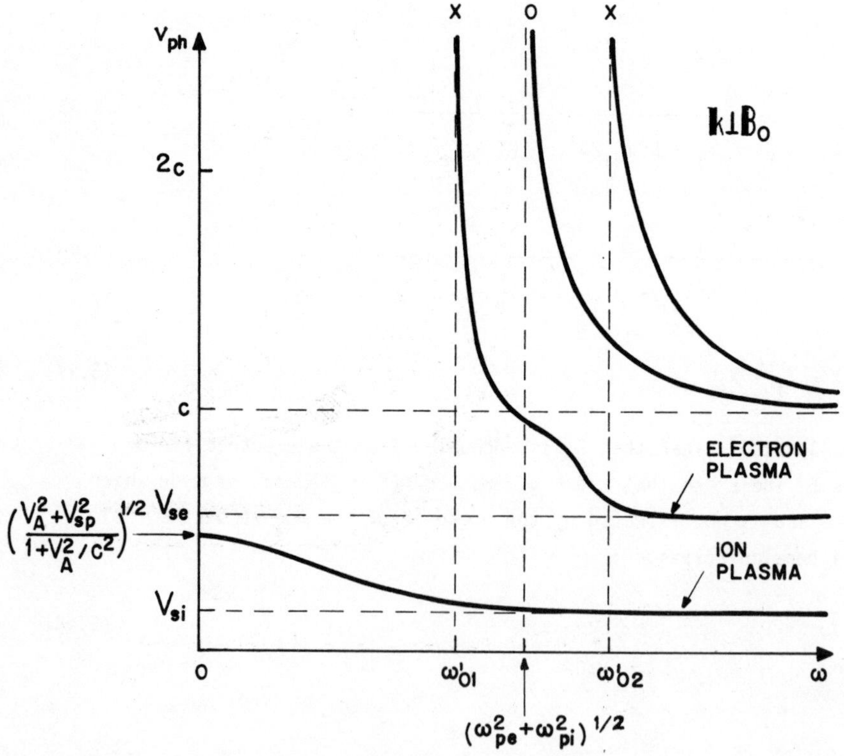

Fig. 10 - Phase velocity as a function of frequency for waves propagating in a direction normal to the magnetic field in a warm fully ionized magnetoplasma.

modes of propagation in a direction normal to the magnetic field. The basic points to be noted in this plot are: (1) the presence of the reflection points at $(\omega_{pe}^2 + \omega_{pi}^2)^{1/2}$, ω'_{01} and ω'_{02}; (2) the transition from a basically longitudinal (electron plasma) wave to a basically transverse electromagnetic (extraordinary) wave, in the frequency region where the phase velocity lies between V_{se} and c; and (3) in the very low-frequency limit the phase velocity of the ion plasma wave tends to $[(V_A^2 + V_{sp}^2)/(1 + V_A^2/c^2)]^{1/2}$.

5.4 Wave propagation in an arbitrary direction

For arbitrary directions of propagation the dispersion relation is given by (5.14). Since a detailed analysis of this dispersion relation is a rather non-

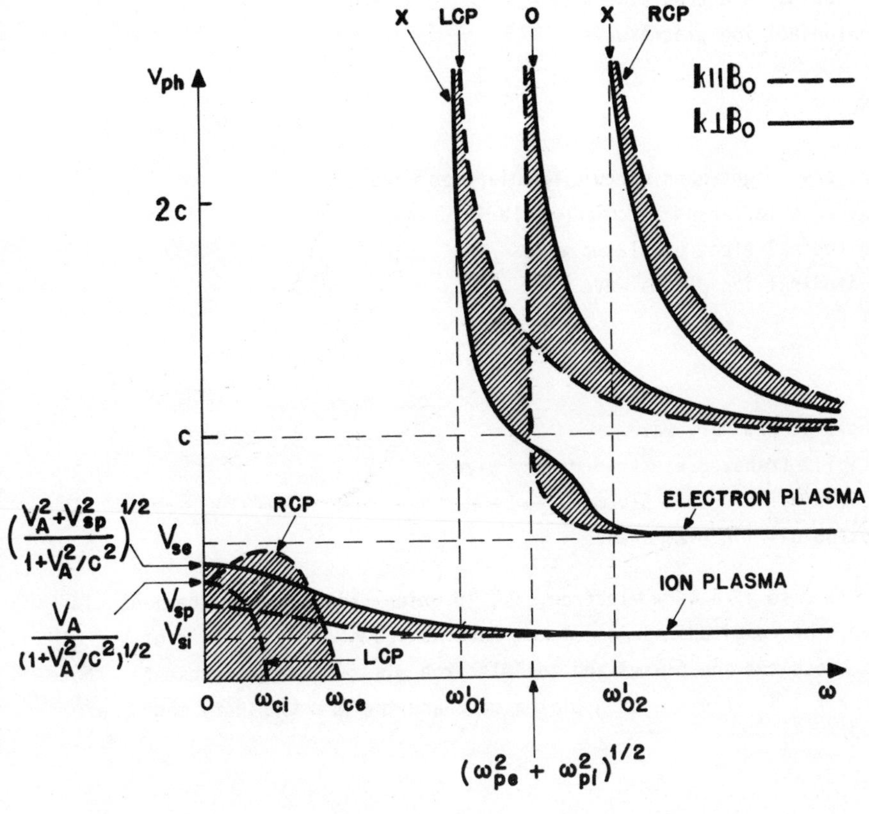

Fig. 11 - Phase velocity as a function of frequency for wave propagation in a warm fully ionized magnetoplasma.

instructive and tedious affair, we shall content ourselves by merely presenting the plot of phase velocity versus frequency in Fig. 11, in which the shaded areas give an indication of how the curves evolve from $\theta = 0^o$ to $\theta = 90^o$.

6. SUMMARY

The modes for wave propagation in a *warm* fully ionized plasma can be summarized as follows:

$\vec{B}_o = 0$:

Transverse electromagnetic wave
Longitudinal electron plasma wave
Longitudinal ion plasma wave

$\vec{k} \parallel \vec{B}_o$:

Transverse right-hand circularly polarized wave
Transverse left-hand circularly polarized wave
Longitudinal electron plasma wave
Longitudinal ion plasma wave

$\vec{k} \perp \vec{B}_o$:

Transverse ordinary wave
Partially transverse extraordinary wave
Longitudinal electron plasma wave
Longitudinal ion plasma wave

For the case of a warm electron gas, in which the motion of the ions is ignored, the longitudinal ion plasma mode is absent. For the case of a cold plasma, both the ion plasma and the electron plasma modes are absent. Note that for $\vec{k} \perp \vec{B}_o$ the electron plasma mode and the extraordinary mode are coupled.

PROBLEMS

17.1 - Show that one of the roots of the dispersion relation (2.33), at very low frequencies, corresponds to an evanescent wave.

17.2 - Make a plot analogous to Fig. 8 for wave propagation in a warm electron gas immersed in a magnetic field, but in terms of ω as a function of the real part of k.

17.3 - Show that the reflection points ω'_{01} and ω'_{02}, for the LCP and RCP waves propagating along $\vec{B}_0$ in a fully ionized warm plasma (see Fig. 9) are given, respectively, by

$$\omega'_{01} = \{-(\omega_{ce} - \omega_{ci}) + [(\omega_{ce} + \omega_{ci})^2 + 4\omega_{pe}^2]^{1/2}\}/2$$

$$\omega'_{02} = \{(\omega_{ce} - \omega_{ci}) + [(\omega_{ce} + \omega_{ci})^2 + 4\omega_{pe}^2]^{1/2}\}/2$$

Compare these expressions with (16.6.13) and (16.6.14).

17.4 - Starting from (5.12), (5.40), (5.41) and (5.42) provide all the necessary steps to obtain (5.43).

17.5 - Obtain a cubic equation in k^2, from (5.50), and analyze the dispersion relations for these three modes of wave propagation across $\vec{B}_0$ in a fully ionized warm plasma.

17.6 - Make plots analogous to Figs. 9, 10 and 11 for wave propagation in a fully ionized warm plasma, but in terms of ω as a function of the real part of k.

17.7 - Show that the resonances in a warm fully ionized magnetoplasma, neglecting collisions, occur at the frequencies $\omega = \omega_{ce} \cos\Theta$ and $\omega = \omega_{ci} \cos\Theta$.

CHAPTER 18

Waves in Hot Isotropic Plasmas

1. INTRODUCTION

We consider in this chapter the propagation of small amplitude waves in unbounded hot plasmas which are close to equilibrium conditions, from the kinetic theory point of view. The problem is examined using the Vlasov equation and only electron motion is considered. The ions, in view of their greater inertia, are assumed to stay immobile. A major point of this chapter will be to emphasize those effects which arise when the Vlasov equation is used, and which were missing when the problem was treated using the cold and warm plasma models (Chapters 16 and 17).

The treatment presented here is restricted to *isotropic* plasmas, in the absence of an externally applied magnetic field. It is shown that the plasma waves can be separated into three groups, the first group being the *longitudinal plasma wave* (also known as space charge wave or Langmuir wave), and the second and third groups being the two different polarizations of the *transverse electromagnetic wave*. The chapter ends with a brief discussion of plasma instabilities which arise from the interaction of the plasma particles with the wave electric field. To illustrate the wave-particle interaction phenomenon we describe one important example, the two-stream instability.

2. BASIC EQUATIONS

The relevant equations for the kinetic theory treatment of small amplitude waves in a electron gas of infinite extent are the Vlasov and Maxwell equations. The Vlasov equation, satisfied by the electron distribution function $f(\vec{r},\vec{v},t)$, can be written as

$$\frac{\partial f(\vec{r},\vec{v},t)}{\partial t} + \vec{v}.\vec{\nabla}f(\vec{r},\vec{v},t) + \left\{ -\frac{e}{m_e}[\vec{E}(\vec{r},t) + \vec{v} \times \vec{B}(\vec{r},t)] + \frac{\vec{F}_{ext}}{m_e} \right\} .$$

$$. \vec{\nabla}_v f(\vec{r},\vec{v},t) = 0 \tag{2.1}$$

where $\vec{F}_{ext}$ denotes any force *externally* applied to the plasma, and $\vec{E}(\vec{r},t)$ and $\vec{B}(\vec{r},t)$ are the *internal* smoothed, self-consistent, macroscopic electric and magnetic induction fields associated with the distributions of charge density and charge current density inside the plasma. The fields $\vec{E}(\vec{r},t)$ and $\vec{B}(\vec{r},t)$ satisfy Maxwell equations

$$\vec{\nabla}.\vec{E}(\vec{r},t) = \rho(\vec{r},t)/\epsilon_0 \tag{2.2}$$

$$\vec{\nabla}.\vec{B}(\vec{r},t) = 0 \tag{2.3}$$

$$\vec{\nabla} \times \vec{E}(\vec{r},t) = - \partial\vec{B}(\vec{r},t)/\partial t \tag{2.4}$$

$$\vec{\nabla} \times \vec{B}(\vec{r},t) = \mu_0\vec{J}(\vec{r},t) + (1/c^2)\partial\vec{E}(\vec{r},t)/\partial t \tag{2.5}$$

where the charge and current densities are given, respectively, by

$$\rho(\vec{r},t) = \sum_\alpha q_\alpha n_\alpha(\vec{r},t) = \sum_\alpha q_\alpha \int_V f_\alpha(\vec{r},\vec{v},t)\, d^3v \tag{2.6}$$

$$\vec{J}(\vec{r},t) = \sum_\alpha q_\alpha n_\alpha(\vec{r},t)\vec{u}_\alpha(\vec{r},t) = \sum_\alpha q_\alpha \int_V \vec{v}\, f_\alpha(\vec{r},\vec{v},t)\, d^3v \tag{2.7}$$

Eqs. (2.1) to (2.7) form a complete self-consistent set of equations, which were first introduced in section 7, Chapter 5. It is worth noting that even though there is no explicit collision term in the Vlasov equation (2.1), an important contribution to the charged particle interactions is included through the internal self-consistent electromagnetic fields.

3. GENERAL RESULTS FOR A PLANE WAVE IN A HOT ISOTROPIC PLASMA

Consider an unbounded uniform electron plasma with a fixed neutralizing ion background and without any external field present. This is obviously an equilibrium arrangement. Suppose that some electrons are slightly displaced from their equilibrium position. As a result of this small space-dependent perturbation in the electron gas, some sort of oscillatory or wave phenomenon can be expected to arise as a consequence of the electric fields produced by charge separation. The ions, because of their much larger mass, can be assumed to remain nearly stationary during the process, since the frequencies of interest are sufficiently high. Since we are dealing with small deviations from equilibrium, the equations can be linearized, that is, the products of two nonequilibrium quantities, which are considered to be of second order, can be neglected.

3.1 Perturbation charge density and current density

To describe small deviations from equilibrium we express the electron distribution function in the form

$$f(\vec{r},\vec{v},t) = f_0(v) + f_1(\vec{r},\vec{v},t) \qquad (|f_1| \ll f) \tag{3.1}$$

where $f_0(v)$ is the equilibrium distribution function, considered to be homogeneous and isotropic, and $f_1(\vec{r},\vec{v},t)$ is a perturbation in the distribution function, always small compared to $f_0(v)$. Before the application of the perturbation the plasma is in equilibrium, so that the macroscopic self-consistent electric and magnetic fields as well as the charge and current densities vanish throughout the plasma. The equilibrium number density of the electrons is everywhere the same as that of the ions, and is given by

$$n_0 = \int_v f_0(v)\, d^3v \tag{3.2}$$

Since $f_1(\vec{r},\vec{v},t)$ is a first order quantity, the internal electric and magnetic fields that arise due to the perturbation are also first order quantities. From (2.6) the perturbation charge density is given by

$$\rho(\vec{r},t) = en_0 - \int_v f(\vec{r},\vec{v},t)\, d^3v \tag{3.3}$$

Using (3.1) and (3.2), we obtain

$$\rho(\vec{r},t) = -e \int_v f_1(\vec{r},\vec{v},t)\, d^3v \tag{3.4}$$

The perturbation current density is obtained from (2.7), noting that the ions are assumed to stay immobile,

$$\vec{J}(\vec{r},t) = -e \int_v \vec{v} f(\vec{r},\vec{v},t)\, d^3v \tag{3.5}$$

Substituting (3.1) into (3.5), and considering that the current density in the equilibrium state vanishes, that is,

$$- e \int_V \vec{v} f_0(v) \, d^3v = 0 \qquad (3.6)$$

we find

$$\vec{J}(\vec{r},t) = - e \int_V \vec{v} f_1(\vec{r},\vec{v},t) \, d^3v \qquad (3.7)$$

3.2 Solution of the linearized Vlasov equation

Substituting (3.1) into the Vlasov equation (2.1), without any external fields present, we obtain

$$\frac{\partial f_1(\vec{r},\vec{v},t)}{\partial t} + \vec{v} \cdot \vec{\nabla} f_1(\vec{r},\vec{v},t) - \frac{e}{m_e} [\vec{E}(\vec{r},t) + \vec{v} \times \vec{B}(\vec{r},t)] \cdot \vec{\nabla}_v f_0(v) -$$

$$- \frac{e}{m_e} [\vec{E}(\vec{r},t) + \vec{v} \times \vec{B}(\vec{r},t)] \cdot \vec{\nabla}_v f_1(\vec{r},\vec{v},t) = 0 \qquad (3.8)$$

Since $\vec{E}(\vec{r},t)$, $\vec{B}(\vec{r},t)$ and $f_1(\vec{r},\vec{v},t)$ are first order quantities, the last term in the left-hand side of (3.8) involves the product of two first order quantities and therefore it is of second order and can be neglected as compared to the remaining terms. Thus, the *linearized* Vlasov equation becomes

$$\frac{\partial f_1(\vec{r},\vec{v},t)}{\partial t} + \vec{v} \cdot \vec{\nabla} f_1(\vec{r},\vec{v},t) - \frac{e}{m_e} [\vec{E}(\vec{r},t) + \vec{v} \times \vec{B}(\vec{r},t)] \cdot \vec{\nabla}_v f_0(v) = 0 \qquad (3.9)$$

A convenient way to solve this equation is to use the method of integral transforms. For an initial-value problem the equation is simplified by taking its Laplace transform in the time domain and the Fourier transform with respect to the space variables. This method reduces the differential equation to an algebraic equation which can then be solved for the desired transform variables. Next, in order to regain the original variables, we have to invert the Laplace and Fourier transforms of the dependent variables. This mathematical treatment, however, involves the calculation of some complicated contour integrals in the complex plane, which is out of the scope of this text. Therefore, in order to simplify the mathematical analysis of the problem, without losing the essentials of the plasma behavior under consideration, we shall look for periodic harmonic solutions of $f_1(\vec{r},\vec{v},t)$ in the space and time variables, according to

$$f_1(\vec{r},\vec{v},t) = f_1(\vec{v}) \exp(i\vec{k}\cdot\vec{r} - i\omega t) \tag{3.10}$$

where the vectors involved are referred to a Cartesian coordinate system. With this special choice for $f_1(\vec{r},\vec{v},t)$, (3.4) and (3.7) become

$$\rho(\vec{r},t) = \rho \exp(i\vec{k}\cdot\vec{r} - i\omega t) \tag{3.11}$$

$$\vec{J}(\vec{r},t) = \vec{J} \exp(i\vec{k}\cdot\vec{r} - i\omega t) \tag{3.12}$$

where

$$\rho = -e \int_v f_1(\vec{v})\, d^3v \tag{3.13}$$

$$\vec{J} = -e \int_v \vec{v} f_1(\vec{v})\, d^3v \tag{3.14}$$

Consequently, the macroscopic self-consistent electric and magnetic fields have the same harmonic time and space dependence,

$$\vec{E}(\vec{r},t) = \vec{E} \exp(i\vec{k}\cdot\vec{r} - i\omega t) \tag{3.15}$$

$$\vec{B}(\vec{r},t) = \vec{B} \exp(i\vec{k}\cdot\vec{r} - i\omega t) \tag{3.16}$$

Furthermore, since we are assuming that the equilibrium distribution function $f_0(v)$ is a function of only the magnitude of $\vec{v}$, we have the very useful identity

$$\vec{\nabla}_v f_0(v) = \frac{\vec{v}}{v} \frac{df_0(v)}{dv} \tag{3.17}$$

so that, for the term involving the magnetic force in (3.9), we have

$$[\vec{v} \times \vec{B}(\vec{r},t)]\cdot\vec{\nabla}_v f_0(v) = [\vec{v} \times \vec{B}(\vec{r},t)] \cdot \frac{\vec{v}\, df_0(v)}{v\, dv} = 0 \tag{3.18}$$

Substituting (3.10), (3.15), (3.16) and (3.18) into the linearized Vlasov equation (3.9), we get

$$-i\omega f_1(\vec{v}) + i\vec{k}\cdot\vec{v} f_1(\vec{v}) - \frac{e}{m_e} \vec{E}\cdot\vec{\nabla}_v f_0(v) = 0 \tag{3.19}$$

whose solution is

$$f_1(\vec{v}) = \frac{ie}{m_e} \frac{\vec{E}.\vec{\nabla}_v f_0(v)}{(\omega-\vec{k} \, . \, \vec{v})} \tag{3.20}$$

For definiteness we shall consider the direction of propagation of the plane waves as being the x direction, that is, $\vec{k} = k\,\hat{x}$.Therefore, $\vec{k} \, . \, \vec{v} = kv_x$ and (3.20) becomes

$$f_1(\vec{v}) = \frac{ie}{m_e} \frac{\vec{E}. \, \vec{\nabla}_v f_0(v)}{(\omega - kv_x)} \tag{3.21}$$

With this orientation chosen for the coordinate system, the longitudinal componentof the wave electric field is $\vec{E}_\ell = E_x\hat{x}$, whereas the transverse component is $\vec{E}_t = E_y\hat{y} + E_z\hat{z}$, as illustrated in Fig. 1.

3.3 Expression for the current density

Next we derive expressions for the Cartesian components of the charge current

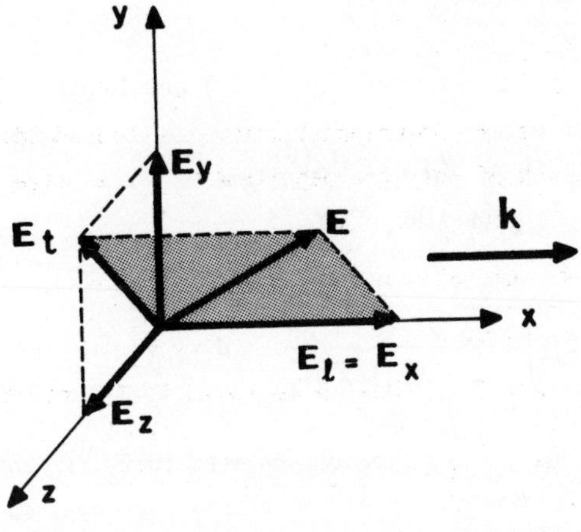

Fig. 1 - Illustrating the relative orientations of the wave propagation vector $\vec{k}$ and the wave electric field $\vec{E}$ in a Cartesian coordinate system.

density $\vec{J}$. Substituting (3.21) into (3.14), we obtain

$$\vec{J} = -\frac{ie^2}{m_e}\int_v \frac{\vec{v}\,[\vec{E}\,.\,\vec{\nabla}_v f_0(v)]}{(\omega - kv_x)}\, d^3v \tag{3.22}$$

Note that the x component of this equation in given by

$$J_x = -\frac{ie^2}{m_e}\int_v \frac{v_x\,[\vec{E}.\,\vec{\nabla}_v f_0(v)]}{(\omega - kv_x)}\, d^3v \tag{3.23}$$

where the triple integral with respect to the three variables v_x, v_y and v_z range from $-\infty$ to $+\infty$. Using the identity (3.17), we note that

$$\int_v \frac{v_x}{(\omega - kv_x)}\,\frac{E_j\, v_j}{v}\,\frac{df_0(v)}{dv}\, d^3v = 0 \quad \text{(for j=y,z)} \tag{3.24}$$

since the integrand is an odd function of v_j, for j = y,z. Consequently, the only contribution from $\vec{E}\,.\,\vec{\nabla}_v f_0(v)$ to the x component of $\vec{J}$ comes from the term $E_x \partial f_0(v)/\partial v_x$, so that (3.23) can be written as

$$J_x = -\frac{ie^2}{m_e} E_x \int_v \frac{v_x}{(\omega - kv_x)}\,\frac{\partial f_0(v)}{\partial v_x}\, d^3v \tag{3.25}$$

Similarly, the y and z components of (3.22) are found to be given by

$$J_y = -\frac{ie^2}{m_e} E_y \int_v \frac{v_y}{(\omega - kv_x)}\,\frac{\partial f_0(v)}{\partial v_y}\, d^3v \tag{3.26}$$

$$J_z = -\frac{ie^2}{m_e} E_z \int_v \frac{v_z}{(\omega - kv_x)}\,\frac{\partial f_0(v)}{\partial v_z}\, d^3v \tag{3.27}$$

Note that J_x, J_y and J_z are linearly related to E_x, E_y and E_z, respectively, a feature which is a consequence of the plasma isotropy, as expected in the absence of an external magnetic field.

3.4 Separation into the various modes

To complete the specification of the problem we use the two Maxwell curl equations (2.4) and (2.5), which for the fields given by (3.15) and (3.16) reduce to

$$i\, k\, \hat{x} \times \vec{E} = i\omega\vec{B} \tag{3.28}$$

$$i\, k\, \hat{x} \times \vec{B} = \mu_0 \vec{J}\ (i\omega/c^2)\ \vec{E} \tag{3.29}$$

In Cartesian coordinates, $\hat{x} \times \vec{E} = E_y\, \hat{z} - E_z \hat{y}$ so that the components of the vector equations (3.28) and (3.29) become , respectively,

$$\omega B_x = 0 \tag{3.30}$$

$$\omega B_y = -\, kE_z \tag{3.31}$$

$$\omega B_z = kE_y \tag{3.32}$$

and

$$0 = \mu_0 J_x - (i\omega/c^2)\ E_x \tag{3.33}$$

$$-\, i\, k\, B_z = \mu_0 J_y - (i\omega/c^2)\ E_y \tag{3.34}$$

$$i\, k\, B_y = \mu_0 J_z - (i\omega/c^2)\ E_z \tag{3.35}$$

where the components of $\vec{J}$ are given by (3.25) to (3.27).

An examination of these equations shows that the electromagnetic fields can be separated into *four independent groups*, each one of them involving the following variables:

(a) J_x, E_x [Eq. (3.33)]

(b) B_x [Eq. (3.30)]

(c) J_y, E_y, B_z [Eqs. (3.32) and (3.34)]

(d) J_z, E_z, B_y [Eqs. (3.31) and (3.35)]

The first group contains an electric field and a current density in the direction of the propagation coefficient $\vec{k}$, that is, parallel to the wave normal of the initial plane wave disturbance produced in the plasma, but contains no magnetic field. This group gives the *longitudinal plasma wave* mode, since the average particle velocity is also in the direction of $\vec{k}$. The second group does not constitute a natural wave mode, since it has no current associated with it and therefore is not influenced by the collective electron motion. It only indicates that there is no magnetic field associated with the longitudinal plasma wave so that these waves are electrostatic in character. The third and fourth groups involve electric and magnetic fields which are perpendicular to $\vec{k}$. The electric current density and therefore the average particle velocity are also perpendicular to the wave normal direction. Note that $\vec{E}$, $\vec{B}$ and $\vec{k}$ form a mutually perpendicular triad. These two groups constitute the two different polarizations of the *transverse electromagnetic wave* mode. In the next section we discuss the characteristics of the longitudinal plasma wave. The characteristics of the transverse electromagnetic wave are discussed in section 5.

4. ELECTROSTATIC LONGITUDINAL WAVE IN A HOT ISOTROPIC PLASMA

4.1 Development of the dispersion relation

The intrinsic behavior of the longitudinal plasma wave is contained in the dispersion relation. This equation, which relates the variables k and ω, determines the natural wave modes of the system. To obtain the dispersion relation for the longitudinal plasma wave we use (3.33) with J_x as given by (3.25),

$$E_x = \frac{\omega_{pe}^2}{n_0\omega} E_x \int_v \frac{v_x}{(kv_x - \omega)} \frac{\partial f_0(v)}{\partial v_x} d^3v \tag{4.1}$$

Dividing this equation by $E_x \neq 0$, yields the *dispersion relation* for the *longitudinal plasma wave*

$$1 = \frac{\omega_{pe}^2}{n_0\omega} \int_v \frac{v_x}{(kv_x - \omega)} \frac{\partial f_0(v)}{\partial v_x} d^3v \tag{4.2}$$

It is convenient to simplify (4.2) by noting that

$$\int_v \frac{v_x}{(kv_x - \omega)} \frac{\partial f_0(v)}{\partial v_x} d^3v = \frac{1}{k} \int_v \frac{\partial f_0(v)}{\partial v_x} \left[\frac{\omega}{(kv_x - \omega)} + 1 \right] d^3v =$$

$$= \frac{\omega}{k} \int_V \frac{\partial f_0(v)/\partial v_x}{(kv_x - \omega)} \, d^3v \tag{4.3}$$

since

$$\int_V \frac{\partial f_0(v)}{\partial v_x} \, d^3v = \int_{-\infty}^{+\infty} dv_y \int_{-\infty}^{+\infty} dv_z \left[f_0(v)\right] \Bigg|_{v_x = -\infty}^{v_x = +\infty} = 0 \tag{4.4}$$

because $f_0(v)$ vanishes at both limits. Therefore, the dispersion relation (4.2) becomes

$$1 = \frac{\omega_{pe}^2}{n_0 k^2} \int_V \frac{\partial f_0(v)/\partial v_x}{(v_x - \omega/k)} \, d^3v \tag{4.5}$$

A useful alternative form of this dispersion relation can be obtained by an integration by parts in the v_x variable. Thus, using the relation

$$\int_a^b U \, dV = UV \Bigg|_a^b - \int_a^b V \, dU \tag{4.6}$$

for the integration with respect to v_x in (4.5), where

$$\begin{aligned} U &= (v_x - \omega/k)^{-1} \qquad & dU &= -(v_x - \omega/k)^{-2} \, dv_x \\ V &= f_0(v) \qquad & dV &= \frac{\partial f_0(v)}{\partial v_x} \, dv_x \end{aligned} \tag{4.7}$$

the triple integral in (4.5) becomes

$$\int\int\int_{-\infty}^{+\infty} \frac{\partial f_0(v)/\partial v_x}{(v_x - \omega/k)} \, dv_x \, dv_y \, dv_z = \int_{-\infty}^{+\infty} dv_y \int_{-\infty}^{+\infty} dv_z \left[\frac{f_0(v)}{(v_x - \omega/k)} \Bigg|_{v_x = -\infty}^{v_x = +\infty} + \int_{-\infty}^{+\infty} \frac{f_0(v)}{(v_x - \omega/k)^2} \, dv_x \right]$$

$$= \int_V \frac{f_0(v)}{(v_x - \omega/k)^2} \, d^3v \tag{4.8}$$

Therefore, the dispersion relation (4.5) can also be written as

$$1 = \frac{\omega_{pe}^2}{n_0 k^2} \int_V \frac{f_0(v)}{(v_x - \omega/k)^2} \, d^3v$$

$$= (\omega_{pe}^2/k^2) < (v_x - \omega/k)^{-2} >_0 \tag{4.9}$$

where the average value with the subscript 0 is calculated using the equilibrium distribution function f_0.

4.2 Limiting case of a cold plasma

Before proceeding further with the analysis of the dispersion relation (4.9), it is instructive to examine the results for the limiting case of a cold plasma, for which the electron velocity distribution, under equilibrium conditions and at rest, is given by

$$f_0(v) = n_0 \, \delta(v_x) \, \delta(v_y) \, \delta(v_z) \tag{4.10}$$

where $\delta(x)$ is the Dirac delta function, defined by

$$\delta(x) = 0 \quad \text{for} \quad x = 0; \qquad \int_{-\infty}^{+\infty} \delta(x) \, dx = 1 \tag{4.11}$$

Substituting (4.10) into the dispersion relation (4. 9) and using the following property of the Dirac delta function

$$\int_{-\infty}^{+\infty} f(x) \, \delta(x - x_0) \, dx = f(x_0) \tag{4.12}$$

we obtain

$$1 = \frac{\omega_{pe}^2}{k^2} \int_V \frac{\delta(v_x) \, \delta(v_y) \, \delta(v_z)}{(v_x - \omega/k)^2} \, d^3v \tag{4.13}$$

or

$$\omega^2 = \omega_{pe}^2 \tag{4.14}$$

in agreement with the cold plasma result (section 4 of Chapter 16).

4.3 High phase velocity limit

Another important result can be immediately obtained from the dispersion relation (4.9), for the limiting case in which the wave phase velocity, ω/k, is very large compared to the velocity of almost all of the electrons. In this high phase velocity limit it is reasonable to expand $(1 - kv_x/\omega)^{-2}$ into a binomial series and retain only the first few terms, since $kv_x/\omega \ll 1$. Thus, recalling that for any $|\varepsilon| < 1$ we have

$$(1 - \varepsilon)^{-2} = 1 + 2\varepsilon + 3\varepsilon^2 + 4\varepsilon^3 + \ldots \tag{4.15}$$

the dispersion relation (4.9) becomes (for $|v_x| \ll |\omega/k|$),

$$(\omega_{pe}^2/\omega^2) < (1 - kv_x/\omega)^{-2} >_0 = 1$$

$$= \frac{\omega_{pe}^2}{\omega^2} \left(1 + 2 \frac{k}{\omega} < v_x >_0 + 3 \frac{k^2}{\omega^2} < v_x^2 >_0 + \ldots \right) \tag{4.16}$$

Since the plasma is considered to be stationary, we have $< v_x >_0 = u_x = 0$, so that the second term in the right-hand side of (4.16) vanishes. To a first degree of approximation we obtain $\omega^2 = \omega_{pe}^2$, which is again the cold plasma result given in (4.14). For a small correction to the cold plasma result, we consider the next non-zero term in the expansion (4.16). Assuming that the equilibrium distribution function is isotropic and using the definition of absolute temperature,

$$<v_x^2>_0 = <c_x^2>_0 = (1/3) <c^2>_0 = k_B T_e/m_e \tag{4.17}$$

where T_e is the temperature of the electron gas at equilibrium and k_B is Boltzmann's constant, the dispersion relation (4.16) becomes

$$\omega^2 = \omega_{pe}^2 (1 + 3 k_B T_e k^2/m_e\omega^2 + \ldots) \tag{4.18}$$

Since the second term in the right-hand side of (4.18) is very small in the high phase velocity limit, we can replace ω, in just this small term, by ω_{pe} (which is the value of ω when this term is zero) and write (4.18) as

$$\omega^2 = \omega_{pe}^2 + 3 (k_B T_e/m_e) k^2 \tag{4.19}$$

This result is known as the *Bohm-Gross dispersion relation.* Note that it is identical to the result obtained using the warm plasma model when collisions are neglected and when the ratio of specific heats γ is taken equal to 3. Since γ is related to the number of degrees of freedom N by the relation

$$\gamma = (2 + N)/N \tag{4.20}$$

we see that $\gamma = 3$ corresponds to the case when the electron gas has one degree of freedom ($N = 1$), so that the electrons move only in the direction of wave propagation.

If additional terms are retained in the binomial series expansion (4.16), additional approximations can be obtained for the dispersion relation $\omega(k)$. In all these approximations we find that ω remains real, so that the longitudinal plasma wave has a constant amplitude in time. There is neither temporal growth nor decay. It is usual to terminate the approximations to $\omega(k)$ at the stage given by (4.19). Using the definition of the Debye length, λ_D, the Bohm-Gross dispersion relation can be rewritten as

$$\omega^2 = \omega_{pe}^2 \; (1 + 3\; k^2\lambda_D^2) \tag{4.21}$$

4.4 Dispersion relation for Maxwellian distribution function

The longitudinal wave dispersion relation (4.5) is now evaluated for the important case when $f_0(v)$ is the Maxwellian distribution function for a stationary equilibrium plasma ($\vec{u} = 0$),

$$f_0(v) = n_0 \left(\frac{m_e}{2\pi k_B T_e}\right)^{3/2} \exp\left(-\frac{m_e v^2}{2\, k_B T_e}\right) \tag{4.22}$$

In this case, a careful analysis of (4.5) shows that ω has a negative imaginary part, causing a temporal damping of the electron plasma wave. This temporal damping, which arises in the *absence* of collisions, is known as *Landau damping* and will be discussed in the next sub-section.

For the moment, we evaluate the dispersion relation for the longitudinal electron wave using the Maxwell-Boltzmann equilibrium distribution function. Substituting (4.22) into (4.5) yields

$$1 = \frac{\omega_{pe}^2}{n_0 k^2} \int_V \frac{\partial f_0(v)/\partial v_x}{(v_x - \omega/k)} \; d^3v$$

$$= - \frac{\omega_{pe}^2}{n_0 k^2} \int_v \frac{(m_e/k_B T_e) v_x f_0(v)}{(v_x - \omega/k)} d^3v$$

$$= - \frac{\omega_{pe}^2}{k^2} \left(\frac{m_e}{k_B T_e}\right) \left(\frac{m_e}{2\pi k_B T_e}\right)^{3/2} \int_{-\infty}^{+\infty} \frac{v_x}{(v_x - \omega/k)} \exp\left(- \frac{m_e v_x^2}{2k_B T_e}\right) dv_x \cdot$$

$$\cdot \int_{-\infty}^{+\infty} \exp\left(- \frac{m_e v_y^2}{2k_B T_e}\right) dv_y \int_{-\infty}^{+\infty} \exp\left(- \frac{m_e v_z^2}{2k_B T_e}\right) dv_z \quad (4.23)$$

The second and third integrals are each equal to $(2\pi k_B T_e/m_e)^{1/2}$. It is convenient to introduce the following dimensionless parameters

$$C = \frac{(\omega/k)}{(2k_B T_e/m_e)^{1/2}} \quad (4.24)$$

$$q = \frac{v_x}{(2k_B T_e/m_e)^{1/2}} \quad (4.25)$$

so that the dispersion relation (4.23) reduces to

$$1 = - \frac{\omega_{pe}^2}{k^2} \left(\frac{m_e}{k_B T_e}\right) \frac{1}{\sqrt{\pi}} \int_{-\infty}^{+\infty} \frac{q \exp(- q^2)}{(q - C)} dq \quad (4.26)$$

Using the notation

$$I(C) = \frac{1}{\sqrt{\pi}} \int_{-\infty}^{+\infty} \frac{q \exp(- q^2)}{(q - C)} dq \quad (4.27)$$

and substituting $(k_B T_e/m_e)/\omega_{pe}^2$ by λ_D^2, (4.26) becomes

$$k^2 \lambda_D^2 + I(C) = 0 \quad (4.28)$$

The evaluation of the integral I(C) is not straightforward because of the singularity at q = C, since for real $\omega(k)$ the denominator vanishes on the real v_x axis. For complex $\omega(k)$, which corresponds to damped $[\mathrm{Im}(\omega) < 0]$ or unstable $[\mathrm{Im}(\omega) > 0]$ waves, the singularity lies off the path of integration along the real v_x axis. However, this simplified derivation of the dispersion relation gives no indication of the proper integration contour to be chosen in the complex v_x plane. Possible contours of integration are shown in Fig. 2 for the cases: (a) unstable wave, with $Im(\omega) > 0$; (b) real $\omega(k)$; and (c) damped wave, with $Im(\omega) < 0$. Landau was the first to treat this problem properly as an *initial-value* problem. If we are interested in the evaluation of the plasma after an initial perturbation, then the causality principle demands that there should be no fields before the starting of the source. According to the well known theorem of residues in complex variables, the value of an integral in the complex domain with a closed contour of integration, such as in Fig. 2, is equal to $2\pi i$ times the sum of the residues within the closed path. The integral vanishes if there are no singularities enclosed by the integration path. Thus, the nature of the singularities of the integrand determines the behavior of the fields after the initial perturbation. The correct contour prescribed by Landau is along the real v_x axis, indented such as to pass below the singularities, and closed by an infinite semicircular path in the upper half of the complex v_x plane, as shown in Fig. 2.

This technique of integration around a contour closed by an infinite semicircle in the upper half plane works if the contribution of the integral from the semicircular path vanishes as its radius goes to infinity. The integral $I(C)$ given in (4.27), the way it stands, cannot be handled by the usual method of residues, since the integrand diverges for $q = \pm i\infty$. To put this integral in a form suitable for evaluation by the method of residues, or by any other method, note first that we can write

$$q/(q - C) = 1 + C/(q - C) \tag{4.29}$$

so that we have

$$I(C) = \frac{1}{\sqrt{\pi}} \int_{-\infty}^{+\infty} \left(1 + \frac{C}{q - C}\right) \exp(-q^2)\, dq \tag{4.30}$$

The first integral in the right-hand side of this equation is equal to unity. Therefore,

$$I(C) = 1 + \frac{C}{\sqrt{\pi}} \int_{-\infty}^{+\infty} \frac{\exp(-q^2)}{(q - C)}\, dq \tag{4.31}$$

For purposes of integration it is convenient to introduce a parameter s in the integral of (4.31), by defining

$$G(C,s) = \frac{1}{\sqrt{\pi}} \int_{-\infty}^{+\infty} \frac{\exp(-s\,q^2)}{(q - C)}\, dq \tag{4.32}$$

Hence, we identify the integral I(C) as

$$I(C) = 1 + C\, G(C,1) \tag{4.33}$$

so that the dispersion relation (4.28) becomes

$$k^2\, \lambda_D^2 + 1 + C\, G(C,1) = 0 \tag{4.34}$$

The purpose of defining G(C,s), as in (4.32), is that this relation allows us to evaluate G(C,1) through a transformation of the integral into a differential equation. Initially, note that the integral in (4.32) can also be written as

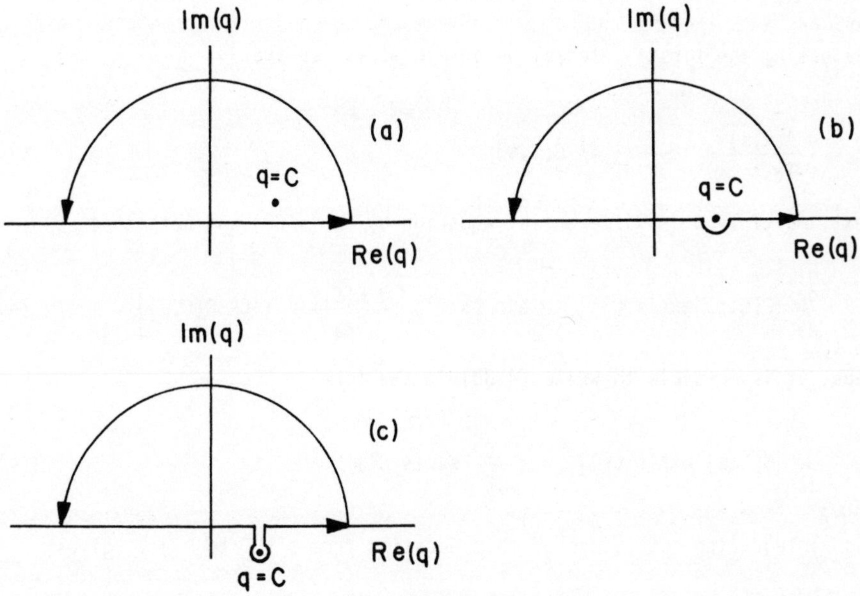

Fig. 2 - Contour of integration in the complex v_x plane for (a) $Im(\omega) > 0$, (b) $Im(\omega) = 0$ and (c) $Im(\omega) < 0$.

$$G(C,s) = \frac{1}{\sqrt{\pi}} \int_{-\infty}^{+\infty} \frac{(q + C)}{(q^2 - C^2)} \exp(-s\, q^2)\, dq \tag{4.35}$$

The first integral in the right-hand side of this equation vanishes, since the integrand is an odd function of q. Therefore, an alternative expression for G(C,s) is

$$G(C,s) = \frac{C}{\sqrt{\pi}} \int_{-\infty}^{+\infty} \frac{\exp(-s\, q^2)}{(q^2 - C^2)}\, dq \tag{4.36}$$

Taking the derivative of (4.36) with respect to s, yields

$$\frac{d\, G(C,s)}{ds} = - \frac{C}{\sqrt{\pi}} \int_{-\infty}^{+\infty} \frac{q^2 \exp(-s\, q^2)}{(q^2 - C^2)}\, dq$$

$$= - \frac{C}{\sqrt{\pi}} \int_{-\infty}^{+\infty} \left[1 + \frac{C^2}{(q^2 - C^2)}\right] \exp(-s\, q^2)\, dq \tag{4.37}$$

Evaluating the first integral we obtain $-C/\sqrt{s}$ so that

$$\frac{d\, G(C,s)}{ds} = - \frac{C}{\sqrt{s}} - C^2\, G(C,s) \tag{4.38}$$

Next, multiply this differential equation by $\exp(s\, C^2)$ and note that

$$\frac{d}{ds} [G(C,s) \exp(s\, C^2)] = \exp(s\, C^2) \left[\frac{d\, G(C,s)}{ds} + C^2\, G(C,s)\right] \tag{4.39}$$

Thus, it is possible to write (4.38) in the form

$$\frac{d}{ds} [G(C,s) \exp(s\, C^2)] = - \frac{C}{\sqrt{s}} \exp(s\, C^2) \tag{4.40}$$

Upon integrating both sides of this equation from s = 0 to s = 1, gives

$$G(C,1) \exp(C^2) - G(C,0) = - C \int_0^1 \frac{\exp(s\, C^2)}{\sqrt{s}}\, ds \tag{4.41}$$

or, rearranging,

$$G(C,1) = G(C,0)\exp(-C^2) - C\exp(-C^2)\int_0^1 \frac{\exp(s\,C^2)}{\sqrt{s}}\,ds \tag{4.42}$$

The integral G(C,0) is easily evaluated for the case of weak damping (large phase velocity). In this case, the pole at $v_x = \omega/k$ lies near the real v_x axis, and G(C,0) can be evaluated as an improper integral as follows:

$$G(C,0) = \frac{1}{\sqrt{\pi}}\int_{-\infty}^{+\infty}\frac{dq}{(q-C)} = \lim_{X\to\infty}\left[\frac{1}{\sqrt{\pi}}\int_{-X}^{+X}\frac{dq}{(q-C)}\right]$$

$$= \lim_{X\to\infty}\left[\frac{1}{\sqrt{\pi}}\ln\left(-\frac{X-C}{X+C}\right)\right]$$

$$= \frac{1}{\sqrt{\pi}}\ln(-1) = \frac{1}{\sqrt{\pi}}\ln(e^{i\pi}) = i\sqrt{\pi} \tag{4.43}$$

The integral G(C,0) can also be evaluated by the method of residues, using an appropriate contour of integration in the complex q-plane [as shown in Fig. 2(b)] , which gives the same result (4.43) for the *Cauchy principal value* of the integral. Therefore, (4.42) becomes

$$G(C,1) = i\sqrt{\pi}\exp(-C^2) - C\exp(-C^2)\int_0^1 \frac{\exp(s\,C^2)}{\sqrt{s}}\,ds \tag{4.44}$$

The remaining integral in the right-hand side of (4.44) can be rewritten in a different form by changing the variable s to W^2/C^2. Consequently, $ds/\sqrt{s} = 2\,dW/C$ and

$$G(C,1) = i\sqrt{\pi}\exp(-C^2) - 2\int_0^C \exp(W^2 - C^2)\,dW \tag{4.45}$$

Although this integral cannot be evaluated explicitly, it is now in a more convenient form for numerical calculation.

Substituting (4.45) into (4.34), results in the following expression for the *dispersion relation*

$$-k^2\lambda_D^2 = 1 + i\sqrt{\pi}\,C\exp(-C^2) - 2C\int_0^C \exp(W^2 - C^2)\,dW \tag{4.46}$$

The integral remaining here can be evaluated numerically and its values have been extensively tabulated*, while the imaginary term is known as the *Landau damping* term. The formal procedure to evaluate k as a function of ω (or vice versa) from this dispersion relation, consists in choosing a given value of C i.e. of $(\omega/k)/(2\, k_B\, T_e/m_e)^{1/2}$ and find the (tabulated) corresponding value of the dispersion function from the tables. Eq. (4.46) can then be used to evaluate the propagation coefficient k.

4.5 Landau damping

In order to show that (4.46) predicts the temporal damping of the longitudinal plasma wave, it is convenient to perform an approximate evaluation of the dispersion ralation. The special case of high phase velocity and weak damping can be obtained in a straightforward way and, at the same time, provides a partial check on the accuracy of the Bohm-Gross dispersion relation obtained earlier. Furthermore, an explicit expression is obtained for the imaginary part of ω. Thus, for the limiting case of $C \gg 1$, lets us find an approximate expression for the dispersion function integral

$$I_1 = 2C \int_0^C \exp(W^2 - C^2)\, dW \tag{4.47}$$

As the first step, Eq. (4.47) can be rewritten by transforming the variable of integration to $\xi = C^2 - W^2$, which gives

$$I_1 = \int_0^{C^2} \left(1 - \frac{\xi}{C^2}\right)^{-1/2} \exp(-\xi)\, d\xi \tag{4.48}$$

Since ξ is less than C^2 over the entire range of integration, we can expand $(1 - \xi/C^2)^{-1/2}$ in a binomial series,

---

* *Remark. See, for example B.D. Fried and S.D. Conte, "The Plasma Dispersion Function", Acad. Press, N.Y., 1961. The function G(C,1), defined here, is the same as the plasma dispersion function defined by Fried & Conte as*

$$Z(\zeta) = \frac{1}{\sqrt{\pi}} \int_{-\infty}^{+\infty} \frac{\exp(-x^2)}{(x - \zeta)}\, dx$$

with $\zeta \equiv C$ *and* $x \equiv q$.

$$\left(1 - \frac{\xi}{C^2}\right)^{-1/2} = 1 + \frac{\xi}{2C^2} + \frac{3\xi^2}{8C^4} + \dots +$$

$$+ \frac{1 \times 3 \times \dots \times (2n - 1)}{2^n \, n!} \left(\frac{\xi}{C^2}\right)^n + \dots \qquad (4.49)$$

If this expansion is substituted into (4.48), and each term is integrated by noting that

$$\int_0^{C^2} \left(\frac{\xi}{C^2}\right)^n \exp(-\xi) \, d\xi =$$

$$= \frac{n!}{(C^2)^n} - \exp(-C^2) \left[1 + \frac{n}{C^2} + \frac{n(n-1)}{C^4} + \dots + \frac{n!}{(C^2)^n}\right] \qquad (4.50)$$

we find

$$I_1 = 1 + \frac{1}{2C^2} + \frac{3}{4C^4} + \dots + \frac{1 \times 3 \times \dots \times (2n - 1)}{(2C^2)^n} + \dots +$$

$$+ O[\exp(-C^2)] \qquad (4.51)$$

where $O[\exp(-C^2)]$ denotes terms of order $\exp(-C^2)$. Although this is an asymptotic expansion, and actually diverges when $n \to \infty$, a good estimate of I_1 can be obtained by retaining only the first few terms, provided C is large. Therefore, on retaining only the first three terms of (4.51), the dispersion relation (4.46) in the high phase velocity limit becomes

$$k^2 \lambda_D^2 = \frac{1}{2C^2} + \frac{3}{4C^4} - i \sqrt{\pi} \, C \exp(-C^2) \qquad (4.52)$$

With the help of Eq. (4.24), which defines C, and the definition of the Debye length λ_D, Eq. (4.52) can be written as

$$\omega^2/\omega_{pe}^2 = 1 + 3k^2 \lambda_D^2 (\omega_{pe}/\omega)^2 -$$

$$- i \frac{\sqrt{\pi/2}}{k^3 \lambda_D^3} \left(\frac{\omega}{\omega_{pe}}\right)^3 \exp\left[- \frac{1}{2k^2 \lambda_D^2} \left(\frac{\omega}{\omega_{pe}}\right)^2\right] \qquad (4.53)$$

In the high phase velocity limit the second term in the right-hand side of (4.53) is small as compared to the first one, and the third term is exponentially small as compared to the first one, so that in this limit the plasma oscillates very close to the plasma frequency ω_{pe}. Note that this limit corresponds also to a long-wavelength limit. Thus, (4.53) can be further approximated as

$$\omega^2 = \omega_{pe}^2 + 3k^2\,(k_B\,T_e/m_e) - \frac{i\,\sqrt{\pi/2}\,\omega_{pe}^5}{k^3(k_B\,T_e/m_e)^{3/2}} \exp\left[-\frac{\omega_{pe}^2}{2k^2\,(k_B\,T_e/m_e)} - \frac{3}{2}\right] \tag{3.54}$$

where in the right-hand side of (4.53) we have replaced ω by ω_{pe}, except in the exponential term where ω^2 has been replaced by the Bohm-Gross result (4.19). Note that the first two terms in (4.54) correspond to the Bohm-Gross result, whereas the imaginary term is new. Separating ω in its real and imaginary parts, according to $\omega = \omega_r + i\omega_i$, and noting that $\omega_i = (\omega^2)_i/(2\omega_r)$, we obtain (taking $\omega_r \simeq \omega_{pe}$)

$$\omega_i = -\frac{\sqrt{\pi/8}\;\omega_{pe}^4}{k^3(k_B\,T_e/m_e)^{3/2}} \exp\left[-\frac{\omega_{pe}^2}{2k^2(k_B\,T_e/m_e)} - \frac{3}{2}\right] \tag{4.55}$$

This *negative* imaginary term in ω leads to *temporal decay*, since for a standing wave problem (where k is real) the waves are proportional to

$$\exp\,(ikx - i\omega t) = \exp\,(ikx - i\omega_r t)\,\exp\,(\omega_i t) \tag{4.56}$$

This damping of the longitudinal plasma wave with time was first pointed out by L.D. Landau and, for this reason, the expression (4.55) is usually called the *Landau damping factor.*

This temporal decay of the longitudinal plasma wave amplitude arises in the absence of dissipative mechanisms, such as collisions of the electrons with heavy particles. The physical mechanism responsible for collisionless Landau damping is the *wave - particle interaction* i.e. the interaction of the electrons with the electric field $E\hat{x}\cos(kx - \omega t)$ of the wave. The electrons that initially have velocities quite close to the phase velocity of the wave are trapped inside the moving potential wells of the wave and this trapping results in a net interchange of energy between the electrons and the wave. For the Maxwell-Boltzmann velocity distribution function we find that, for small k, the phase velocity lies far out on the tail and the damping is negligible, but for values of k close to $1/\lambda_D$ the phase velocity lies within the tail, as shown in Fig. 3, so that there is a velocity band, Δv, around $v = \omega/k$, where there are more electrons in Δv moving initially slower than ω/k, than moving faster than ω/k. Consequently the trapping of the electrons in the potential troughs of the wave will cause a net increase in the electron energy at the expense of the wave energy. This happens in the region where $\partial f_0/\partial v_x$ is negative, like the one shown in Fig. 3. In some cases, the initial velocity distribution of the electrons may be appropriately chosen in such a way that ω_i becomes positive. This would indicate an unstable situation, with the wave amplitude growing with time. This happens when $\partial f_0/\partial v_x$ is positive at $v_x=\omega/k$.

It is important to note that the Landau damping factor, ω_i, is essentially due to the pole of the integrand in (4.31), which occurs at the value of the electron velocity component v_x (parallel to $\vec{k}$) equal to the phase velocity of the wave (ω/k). This property is a mathematical manifestation of the fact

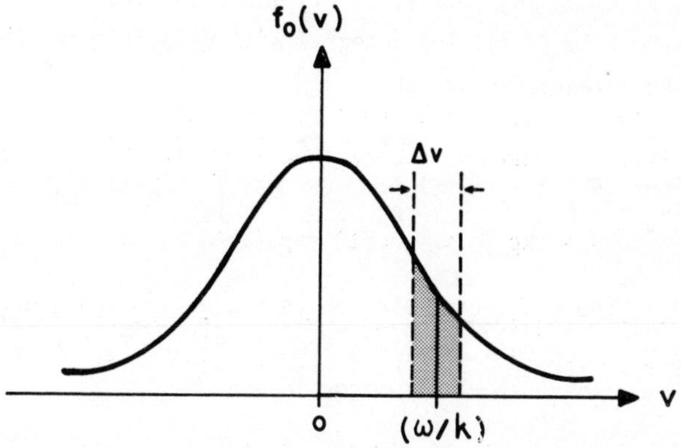

Fig. 3 - Equilibrium distribution function of the electrons showing a velocity band Δv around the phase velocity (ω/k), in which there are more electrons moving slower than (ω/k), than moving faster than (ω/k).

that the wave-particle interaction is effective only when the velocity of the electrons are very close to the phase velocity of the wave.

5. TRANSVERSE WAVE IN A HOT ISOTROPIC PLASMA

5.1 Development of the dispersion relation

The third and fourth independent groups of fields, consisting of J_y, E_y, B_z and J_z, E_z, B_y, respectively, constitute the two different polarizations of the transverse wave mode. In order to deduce the dispersion relation for the transverse electromagnetic wave, let us consider initially Eqs. (3.26), (3.32) and (3.34). Substituting B_z form (3.32) into (3.34), yields

$$E_y = \frac{i\omega}{\varepsilon_0 (k^2c^2 - \omega^2)} J_y \tag{5.1}$$

Combining this equation with (3.26), to eliminate J_y, we obtain

$$E_y = \frac{\omega_{pe}^2 \omega \, E_y}{n_0(\omega^2 - k^2c^2)} \int_v \frac{v_y}{(kv_x - \omega)} \frac{\partial f_0(v)}{\partial v_y} \, d^3v \tag{5.2}$$

In a similar way, combining (3.27), (3.31) and (3.35) we find that the equation for E_z is identical to (5.2). The integral with respect to v_y in (5.2) can be simplified by an integration by parts

$$\int_{-\infty}^{+\infty} v_y \frac{\partial f_0(v)}{\partial v_y} \, dv_y = v_y \, f_0(v) \Bigg|_{v_y = -\infty}^{v_y = +\infty} - \int_{-\infty}^{+\infty} f_0(v) \, dv_y \tag{5.3}$$

The first term in the right-hand side of this equation vanishes, since $f_0(v)$ vanishes at $v_y = \pm \infty$. Thus, we obtain from (5.2) the following dispersion relation for the transverse electromagnetic wave

$$k^2c^2 - \omega^2 = \frac{\omega_{pe}^2 \omega}{n_0 \, k} \int_v \frac{f_0(v)}{(v_x - \omega/k)} \, d^3v \tag{5.4}$$

5.2 Cold plasma result

Again, we examine first the limiting case of a cold plasma characterized by the distribution function (4.10). Substituting (4.10) into (5.4) and using the property (4.12) of the Dirac delta function, we find

$$k^2c^2 = \omega^2 - \omega_{pe}^2 \tag{5.5}$$

This result is identical to the one obtained in Chapter 16 using the cold plasma approximation [see (16.4.12)].

5.3 Dispersion relation for Maxwellian distribution function

Considering $f_0(v)$, in (5.4), as the Maxwell-Boltzmann distribution function, we find, after integrating over v_y and v_z,

$$k^2c^2 - \omega^2 = \frac{\omega_{pe}^2 C}{\sqrt{\pi}} \int_{-\infty}^{+\infty} \frac{\exp(-q^2)\, dq}{(q - C)} \tag{5.6}$$

where, as before, we have introduced the dimensionless parameters C and q, defined in (4.24) and (4.25), respectively. The integral appearing in (5.6) is the same as the integral G(C,s), for s = 1, defined in (4.32), so that we can write the dispersion relation (5.6) as

$$k^2c^2 - \omega^2 = \omega_{pe}^2\, C\, G(C,1) \tag{5.7}$$

For weak damping we can use (4.45), obtaining

$$k^2c^2 - \omega^2 = \omega_{pe}^2 \left[i\sqrt{\pi}\; C \exp(-C^2) - 2C \int_0^C \exp(W^2 - C^2)\, dW\right] \tag{5.8}$$

5.4 Landau damping of the transverse wave

In contrast with the Landau damping of the longitudinal plasma wave, the Landau damping of the transverse electromagnetic wave, which is due to the small negative imaginary part of ω in (5.8), is negligibly small. For the purpose of establishing this result, it is convenient to evaluate approximately the dispersion relation (5.8) in the high phase velocity limit. In the limit when C is very large we can use (4.51). To obtain a first approximation to the real part of ω, it is sufficient to retain only the first term in (4.51), so that in the high phase velocity limit (5.8) reduces to

$$k^2c^2 = \omega^2 - \omega_{pe}^2 + i\sqrt{\pi}\,\omega_{pe}^2\, C\exp(-C^2) \tag{5.9}$$

This result is identical to the dispersion relation obtained using the cold plasma model without collisions, except for the Landau damping term.

In the high phase velocity limit ($C \gg 1$) the Landau damping factor is very small and can be omitted in a first approximation, with the result that (5.12) reduces to the cold plasma result (5.5). From (5.5) we see that for $\omega > \omega_{pe}$ the phase velocity ω/k is greater than c (the velocity of electromagnetic waves in free space). Thus, C is of the order of $c/(2k_B T_e/m_e)^{1/2}$, and is therefore a very large number. Since C is very large, the Landau damping of the transverse electromagnetic wave is negligible. As a matter of fact it can be argued that, for this case, the Landau damping term is really zero, since the integration over v_x should really extend only from -c to +c, while the phase velocity is always greater than c. This implies that the pole at $v_x = (\omega/k)$, or equivalently at $q = C$, lies outside the path of integration along the real axis. Therefore, the conditions for efficient wave-particle interaction are not met for the transverse electromagnetic wave throughout the frequency range of propagation (since ω/k is greater than c), resulting in no wave damping. On the other hand, for the longitudinal plasma wave there are frequencies for which the wave phase velocity is of the order of the electron thermal velocities, so that wave-particle interaction can take place efficiently, with the result that the Landau damping factor becomes important for the lower phase velocity longitudinal waves.

6. THE TWO-STREAM INSTABILITY

As an example of a situation in which wave-particle interaction leads to a *growing wave amplitude*, at the expense of the kinetic energy of the plasma particles, we consider in this section the *two-stream instability*. Although the instability arises under a wide range of beam conditions, we shall consider only the simple case of two contrastreaming uniform beams of electrons with the same number density $n_o/2$. The first stream travels in the x direction with drift velocity $\vec{v}_D = v_D\hat{x}$, and the second stream in the opposite direction with drift velocity $\vec{v}_D = -v_D\hat{x}$. We shall assume that each particle, in each stream, has exactly the stream velocity i.e. the particles are assumed to be cold, so that the electron distribution function can be written in terms of the Dirac delta function as

$$f_0(v) = (n_0/2)\,[\delta(v_x - v_D) + \delta(v_x + v_D)]\,\delta(v_y)\,\delta(v_z) \tag{6.1}$$

This distribution function is illustrated in Fig. 4 for the v_x component only.

For longitudinal plasma waves propagating in the x direction ($\vec{k} = k\hat{x}$) in an electron gas, described by the Vlasov equation, the dispersion relation is, from (4.9),

$$1 = \frac{\omega_{pe}^2}{n_0\,k^2}\int_v \frac{f_0(\vec{v})}{(v_x - \omega/k)^2}\,d^3v \tag{6.2}$$

Substituting (6.1) into (6.2), yields

$$1 = \frac{\omega_{pe}^2}{2}\int_{-\infty}^{+\infty} \frac{[\delta(v_x - v_D) + \delta(v_x + v_D)]}{(kv_x - \omega)^2}\,dv_x \int_{-\infty}^{+\infty} \delta(v_y)dv_y \int_{-\infty}^{+\infty} \delta(v_z)dv_z \tag{6.3}$$

and integrating over each of the δ functions, we obtain

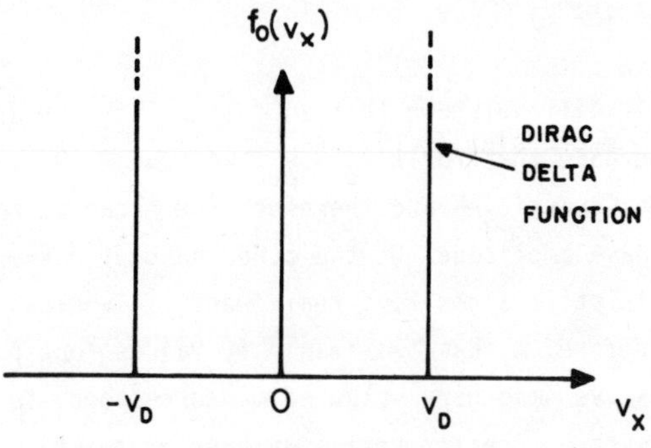

Fig. 4 - Illustrating the v_x component of the distribution function (6.1).

$$1 = \frac{\omega_{pe}^2}{2}\left[\frac{1}{(kv_D - \omega)^2} + \frac{1}{(kv_D + \omega)^2}\right] \tag{6.4}$$

This is the dispersion relation for longitudinal waves (with the wave normal in the direction of the first electron stream) in a constrastreaming electron plasma characterized by the distribution function (6.1). We assume that the propagation coefficient, k, of the longitudinal plasma wave is real (standing waves), and investigate the existence of temporal growth or damping of the wave amplitude.

Eq. (6.4) can be rearranged in the following polynomial form

$$\omega^4 - B\omega^2 + C = 0 \tag{6.5}$$

where

$$B \equiv \omega_{pe}^2 + 2k^2v_D^2 \tag{6.6}$$

$$C \equiv k^2\, v_D^2\, (k^2v_D^2 - \omega_{pe}^2) \tag{6.7}$$

Note that B is always positive, whereas C can be either positive or negative, depending on whether $k^2v_D^2 > \omega_{pe}^2$ or $k^2v_D^2 < \omega_{pe}^2$, respectively. The polynomial equation (6.5) has two roots for ω^2, which are

$$\omega_1^2 = B/2 + (B^2/4 - C)^{1/2} \tag{6.8}$$

$$\omega_2^2 = B/2 - (B^2/4 - C)^{1/2} \tag{6.9}$$

In what follows it will be shown that an instability can arise only when $k^2v_D^2 < \omega_{pe}^2$. First we note that for $k^2v_D^2 > \omega_{pe}^2$ we have $C > 0$, so that both ω_1^2 and ω_2^2 are positive real quantities and therefore there can be no temporal growth or damping of the wave amplitude. On the other hand, for $k^2v_D^2 < \omega_{pe}^2$ we have $C < 0$, so that ω_1^2 is still a positive real quantity, whereas ω_2^2 is a negative real quantity. Therefore, ω_2 has two imaginary values (one positive and one negative). The positive imaginary value of ω_2 corresponds to an unstable mode, since for $\omega_2 = i\omega_{2i}$ (with ω_{2i} real, positive) we have $\exp(-i\omega t) = \exp(\omega_{2i}t)$. Hence the growth rate is given by

$$\omega_{2i} = \left[- B/2 + (B^2/4 - C)^{1/2}\right]^{1/2} \quad ; \qquad C < 0 \tag{6.10}$$

or, using (6.6) and (6.7),

$$\omega_{2i} = \{ - (\omega_{pe}^2/2 + k^2 v_D^2) + [(\omega_{pe}^2/2 + k^2 v_D^2)^2 - k^2 v_D^2 (k^2 v_D^2 - \omega_{pe}^2)]^{1/2}\}^{1/2}; \quad k^2 v_D^2 < \omega_{pe}^2 \tag{6.11}$$

The maximum value of the growth rate (6.11) corresponds to the minimum value of ω_2^2 in (6.9), since $\omega_{2i}^2 = -\omega_2^2$. Examining the derivate of ω_2^2 with respect to k, we find that the minimum value of ω_2^2 occurs when $k^2 v_D^2 = (3/8)\,\omega_{pe}^2$, and the corresponding value of ω_2^2 is $-\omega_{pe}^2/8$. Consequently, the maximum value of the growth rate is

$$(\omega_{2i})_{max} = (1/\sqrt{8})\,\omega_{pe} \tag{6.12}$$

7. SUMMARY

7.1 Longitudinal mode

The *dispersion relation* is (for $\vec{k} = k\hat{x}$)

$$1 = \frac{\omega_{pe}^2}{n_0\,\omega} \int_v \frac{v_x}{(kv_x - \omega)} \frac{\partial f_0(v)}{\partial v_x}\, d^3v \tag{4.2}$$

Alternative forms for this dispersion relation are

$$1 = \frac{\omega_{pe}^2}{n_0\, k^2} \int_v \frac{\partial f_0(v)/\partial v_x}{(v_x - \omega/k)}\, d^3v \tag{4.5}$$

$$1 = \frac{\omega_{pe}^2}{n_0\, k^2} \int_v \frac{f_0(v)}{(v_x - \omega/k)^2}\, d^3v \tag{4.9}$$

When $f_0(v)$ is the Maxwell-Boltzmann distribution function,

$$-k^2\lambda_D^2 = 1 + i\sqrt{\pi}\, C \exp(-C^2) - 2C \int_0^C \exp(W^2 - C^2)\, dW \tag{4.46}$$

The *cold plasma* limit gives stationary electrostatic oscillations at the plasma frequency,

$$\omega^2 = \omega_{pe}^2 \tag{4.14}$$

The high phase velocity limit gives the *warm plasma* model result (Bohm-Gross dispersion relation) for the electron plasma wave,

$$\omega^2 = \omega_{pe}^2 + 3\,(k_B\,T_e/m_e)\,k^2 \tag{4.19}$$

The *Landau* (temporal) *damping factor* is (with $\omega = \omega_r + i\omega_i$)

$$\omega_i = -\frac{\sqrt{\pi/8}\;\omega_{pe}^4}{k^3(k_B\,T_e/m_e)^{3/2}}\exp-\left[\frac{\omega_{pe}^2}{2k^2(k_B\,T_e/m_e)} - \frac{3}{2}\right] \tag{4.55}$$

7.2 Transverse mode

The *dispersion relation* is (for $\vec{k} = k\hat{x}$)

$$k^2c^2 - \omega^2 = \frac{\omega_{pe}^2\,\omega}{n_0\,k}\int_v \frac{f_0(v)}{(v_x - \omega/k)}\,d^3v \tag{5.4}$$

For $f_0(v)$ as the Maxwell-Boltzmann distribution function,

$$k^2c^2 - \omega^2 = \omega_{pe}^2\left[\,i\,\sqrt{\pi}\,C\exp(-C^2) - 2C\int_0^C \exp(W^2 - C^2)\,dW\,\right] \tag{5.8}$$

The *cold* and *warm plasma* limits give

$$k^2c^2 = \omega^2 - \omega_{pe}^2 \tag{5.5}$$

The high phase velocity limit gives

$$k^2c^2 - \omega^2 = \omega_{pe}^2 + i\,\sqrt{\pi}\,\omega_{pe}^2\,C\exp(-C^2) \tag{5.9}$$

The *Landau damping term* is negligible, since $v_{ph} \geqq c$.

PROBLEMS

18.1 - Since the longitudinal plasma wave is an electrostatic oscillation, it is possible to derive its dispersion relation using Poisson equation, satisfied by the electrostatic potential $\phi(\vec{r},t)$, instead of Maxwell equations. Consider the problem of small amplitude longitudinal waves propagating in the x direction in an electron gas (only electrons move in a background of stationary ions), in the absence of a magnetic field. Assume that

$$f(\vec{r},\vec{v},t) = f_0(\vec{v}) + f_1(\vec{v}) \exp(ikx - i\omega t)$$

$$\vec{E}(\vec{r},t) = \hat{x} E \exp(ikx - i\omega t)$$

where $|f_1| << f_0$, with $f_0(\vec{v})$ the nonperturbed equilibrium distribution function, and where $\vec{E}(\vec{r},t)$ is the internal electric field due to the small amplitude perturbation in the electron gas. Using the linearized Vlasov equation (neglecting second-order terms) determine the expression for $f_1(\vec{v})$ in terms of $\vec{E} = -\vec{\nabla}\phi$ and $\vec{\nabla}_v f_0$. Using this result in Poisson equation, obtain the following *dispersion relation* for longitudinal waves propagating in the x direction:

$$1 = \frac{\omega_{pe}^2}{n_0 k^2} \int_v \frac{(\partial f_0/\partial v_x)}{(v_x - \omega/k)} d^3v$$

18.2 - Show that

$$2C \int_0^C \exp(W^2 - C^2)\, dW = 2C^2 \sum_{n=0}^{\infty} (-1)^n \frac{2^n C^{2n}}{1.3.5 \ldots (2n-1)(2n+1)}$$

by making a series expansion of the integrand. For $C << 1$, that is, for $(\omega/k) << (2k_B T_e/m_e)^{1/2}$, show that the dispersion relation for the longitudinal plasma wave reduces to

$$k^2 \lambda_D^2 = -1$$

or

$$k^2 (k_B T_e/m_e)^{1/2} = -\omega_{pe}^2$$

This result is the low-frequency limit of the result obtained from the macroscopic warm plasma model, using the isothermal sound speed of the electron gas $V_{se} = (k_B T_e/m_e)^{1/2}$.

18.3 - (a) Show that the dispersion relation for the longitudinal plasma wave (with $\vec{k} = k\hat{x}$), for the case of an unbounded homogeneous plasma in which the motion of the electrons and the ions is taken into account, can be written as

$$1 = \frac{\omega_{pe}^2}{n_0 k^2} \int_v \frac{\partial f_{oe}(v)/\partial v_x}{(v_x - \omega/k)} d^3v + \frac{\omega_{pi}^2}{n_0 k^2} \int_v \frac{\partial f_{oi}(v)/\partial v_x}{(v_x - \omega/k)} d^3v$$

Show that this dispersion relation can be recast into the form

$$1 = \frac{1}{k^2} \left[\omega_{pe}^2 \langle (v_x - \omega/k)^{-2} \rangle_{oe} + \omega_{pi}^2 \langle (v_x - \omega/k)^{-2} \rangle_{oi} \right]$$

where (with $\alpha = e,i$)

$$\langle (v_x - \omega/k)^{-2} \rangle_{o\alpha} = \frac{1}{n_0} \int_v \frac{f_{o\alpha}(v)}{(v_x - \omega/k)^2} d^3v$$

(b) For the cold plasma model, for which

$$f_{o\alpha}(v) = n_0\, \delta(v_x)\, \delta(v_y)\, \delta(v_z)$$

show that the dispersion relation reduces to

$$\omega^2 = \omega_{pe}^2 + \omega_{pi}^2 = n_0 e^2/\mu\varepsilon_0$$

where $\mu = m_e m_i/(m_e + m_i)$ is the reduced mass of an electron and an ion.
(c) In the high-phase velocity limit, show, by making a binomial expansion, that the dispersion relation becomes

$$(\omega_{pe}^2/\omega^2)\, [1 + 3\,(k^2/\omega^2)\,(k_B T_e/m_e) + \ldots] +$$

$$+(\omega_{pi}^2/\omega^2)\, [1 + 3\,(k^2/\omega^2)\,(k_B T_i/m_i) + \ldots] = 1$$

Show that this equation can be written as

$$1 = [(\omega_{pe}^2 + \omega_{pi}^2)/\omega^2][\,1 + 3\,(k^2/\omega^2)\,(k_B T_h/\mu) + \dots\,]$$

where T_h is a "hybrid" temperature given by

$$T_h = (m_i^2\, T_e + m_e^2\, T_i)/(m_e + m_i)^2$$

Under what conditions does this relation reduce to the Bohm-Gross dispersion relation for a warm electron plasma?

(d) Show that the dispersion relation of part (a) can be expressed as

$$1 = -\frac{1}{k^2\lambda_{De}^2}\left[1 + i\sqrt{\pi}\, C_e \exp(-C_e^2) - 2C_e \int_0^{C_e} \exp(W^2 - C_e^2)\,dW\right] -$$

$$-\frac{1}{k^2\,\lambda_{Di}^2}\left[1 + i\,\sqrt{\pi}\, C_i \exp(-C_i^2) - 2C_i \int_0^{C_i} \exp(W^2 - C_i^2)\,dW\right]$$

where (with $\alpha = e,i$)

$$\lambda_{D\alpha} = \left(\frac{\varepsilon_0\, kT_\alpha}{n_0\, e^2}\right)^{1/2} \quad ; \quad C_\alpha = \frac{(\omega/k)}{(2k_B\, T_\alpha/m_\alpha)^{1/2}}$$

For weakly damped oscillations ($\omega_i \ll \omega_r$) and in the low frequency and low phase velocity range specified by the condition

$$C_i \gg 1 \gg C_e$$

show that the dispersion relation reduces to

$$1 = -\,(1/k^2\,\lambda_{De}^2)(1 + i\,\sqrt{\pi}\, C_e - m_e/2m_i C_e^2)$$

Consequently, verify that the frequency of oscillation and the Landau damping constant are given by

$$\omega_r = (k_B\, T_e/m_i)^{1/2}\, k(1 + k^2\,\lambda_{De}^2)^{-1/2}$$

$$\omega_i = -\,(\pi m_e/8m_i)^{1/2}\,(k_B\, T_e/m_i)^{1/2}\, k(1 + k^2\,\lambda_{De}^2)^{-2}$$

Note that the condition $C_i >> 1 >> C_e$ is fulfilled only if $T_e/T_i >> (1 + k^2\lambda_{De}^2)$, which implies in a strongly nonisothermal plasma, with hot electrons and cold ions. Show that in the long-wave range we find

$$\omega_r = k^2(k_B T_e/m_i)^{1/2}$$

which are essentially the same as the low frequency ion acoustic waves that propagate at a sound speed determined by the ion mass and the electron temperature.

18.4 - A longitudinal plasma wave is set up propagating in the x direction ($\vec{k} = k\,\hat{x}$) in a plasma whose equilibrium state is characterized by the following so-called *resonance distribution* of velocities in the direction of the wave normal of the longitudinal plasma wave

$$f_0(\vec{v}) = n_0 \frac{A}{\pi} \frac{1}{(v_x^2 + A^2)} \delta(v_y)\, \delta(v_z)$$

where A is a constant.

(a) Using this expression for $f_0(\vec{v})$ in the dispersion relation for the longitudinal plasma wave [Eq. (4.9)] , obtain the result

$$1 = \frac{\omega_{pe}^2}{k^2} \frac{A}{\pi} \int_{-\infty}^{+\infty} \frac{dv_x}{(v_x - \omega/k)^2\,(v_x^2 + A^2)}$$

(b) Evaluate the integral of part (a) by closing the contour in the upper half plane (note that there is a double pole at $v_x = \omega/k$ and a simple pole at $v_x = i\,A$), to obtain the dispersion relation

$$1 = \frac{\omega_{pe}^2}{k^2} \frac{1}{(\omega/k + i\,A^2\,)}$$

(c) Analyze this dispersion relation ($\omega = \omega_r + i\omega_i$) to show that the longitudinal wave in this plasma is not unstable and determine the frequency of oscillation (ω_r) and the Landau damping constant (ω_i). Compare this Landau damping constant with the corresponding value for a Maxwellian distribution of velocities, for the cases when $k\,\lambda_D << 1$ and $k\,\lambda_D \geq 1$.

18.5 - Solve the linearized Vlasov equation (3.9) by method of integral transforms, taking its Laplace transform in the time domain and the Fourier transform with respect to the space variables. Then, determine the dispersion relation for the modes of wave propagation in a hot isotropic plasma.

18.6 - Evaluate the integral $G(C,0)$, defined in (4.32) with $s = 0$, by the method of residues using the contours of integration in the complex plane shown in Fig. 2.

18.7 - Consider a longitudinal wave propagating along the x direction in a plasma, whose electric field is given by

$$E_x(x,t) = E_0 \sin (kx - \omega t)$$

(a) Show that, for small displacements, the electrons which are moving with a velocity approximately equal to the phase velocity of the wave will oscillate with a frequency given by

$$\omega' = (e E_0 k/m_e)^{1/2}$$

(b) Establish the necessary conditions for trapping of the electrons by the wave.

18.8 - Consider the two-stream problem using the macroscopic cold plasma equations for two beams of electrons having number densities given by

$$n_{1,2} = (n_0/2) + n_{1,2} \exp [i(kx - \omega t)]$$

and average velocities given by

$$u_{1,2} = \pm u_0 + u_{1,2} \exp [i(kx - \omega t)]$$

Consider that the electric field is given by

$$E_x = E_0 \exp [i(kx - \omega t)]$$

Determine the dispersion relation for this two-stream problem and verify if the oscillations with real k are stable or unstable.

CHAPTER 19

Waves in Hot Magnetized Plasmas

1. INTRODUCTION

The analysis of small amplitude waves propagating in a plasma, presented in the previous chapter, is now extended to *anisotropic* plasmas immersed in an externally applied magnetic field. Emphasis is given to the study of plasma waves having their propagation vector $\vec{k}$ either parallel or perpendicular to the externally applied magnetostatic field.

For propagation along the magnetostatic field the plasma waves separate again into three independent groups. The first group is the *longitudinal plasma wave*, and the second and third groups are the *left* and the *right circularly polarized transverse electromagnetic waves*. For propagation across the magnetostatic field the plasma waves separate into two groups, which are designated as the *TM (Transverse Magnetic)* and the *TEM (Transverse Electric-Magnetic)* modes. The longitudinal plasma wave does not exist independently for any orientation of the magnetostatic field other than parallel to $\vec{k}$.

The mathematical analysis of the problem of wave propagation at an arbitrary direction relative to the magnetostatic field is more complicated insofar as the details are concerned and will not be presented here.

2. WAVE PROPAGATION ALONG THE MAGNETOSTATIC FIELD IN A HOT PLASMA

In this section we study the problem of wave propagation in an unbounded plasma consisting of mobile electrons in a neutralizing background of stationary ions, immersed in a uniform magnetostatic field $\vec{B}_0$. In the equilibrium state, the electron number density (which is the same as that of the ions) is denoted by n_0.

In the absence of perturbations, the homogeneous electron equilibrium distribution function has to satisfy the zero-order Vlasov equation

$$(\vec{v} \times \vec{B}_0) \cdot \vec{\nabla}_v f_0(\vec{v}) = 0 \tag{2.1}$$

The presence of the magnetostatic field introduces an anisotropy in the distribution function, so that the equilibrium distribution function is denoted by $f_0(v_\parallel, v_\perp)$, where $v_\parallel$ and $v_\perp$ represent the electron velocity in directions parallel and perpendicular to $\vec{B}_0$, respectively.

2.1 Linearized Vlasov equation

As before, the perturbed distribution function is assumed to consist of a small perturbation, $f_1(\vec{r},\vec{v},t)$, superimposed on the equilibrium distribution function, $f_0(v_\parallel, v_\perp)$, that is

$$f(\vec{r},\vec{v},t) = f_0(v_\parallel, v_\perp) + f_1(\vec{r},\vec{v},t) \tag{2.2}$$

where $|f_1| \ll f_0$. The electric field $\vec{E}(\vec{r},t)$ and the magnetic field $\vec{B}(\vec{r},t)$, related to the charge density and current density inside the plasma, and which are associated with the first order perturbation $f_1(\vec{r},\vec{v},t)$, are also first order quantities. Note, however, that $\vec{E}(\vec{r},t)$ denotes the total electric field inside the plasma, whereas the total magnetic field $\vec{B}_T(\vec{r},t)$ is given by

$$\vec{B}_T(\vec{r},t) = \vec{B}_0 + \vec{B}(\vec{r},t) \tag{2.3}$$

Substituting (2.2) and (2.3) into the Vlasov equation (18.2.1), neglecting all second order terms, and noting that the equilibrium distribution function is homogeneous, results in the following *linearized* Vlasov equation

$$\frac{\partial}{\partial t} f_1(\vec{r},\vec{v},t) + \vec{v} \,.\, \vec{\nabla} f_1(\vec{r},\vec{v},t) - (e/m_e) \left[\vec{E}(\vec{r},t) + \vec{v} \times \vec{B}(\vec{r},t) \right] .$$

$$. \, \vec{\nabla}_v f_0(v_\parallel, v_\perp) - (e/m_e)(\vec{v} \times \vec{B}_0) . \, \vec{\nabla}_v f_1(\vec{r},\vec{v},t) = 0 \tag{2.4}$$

2.2 Solution of the linearized Vlasov equation

For the purpose of investigating the characteristics of plane waves propagating along the magnetostatic field, we shall assume that the space-time dependence of all physical quantities is a periodic harmonic dependence of the form $\exp(i\vec{k} \,.\, \vec{r} - i\omega t)$, that is,

$$\vec{E}(\vec{r},t) = \vec{E} \exp(i \vec{k} \,.\, \vec{r} - i\omega t) \tag{2.5}$$

$$\vec{B}(\vec{r},t) = \vec{B} \exp(i \vec{k} \,.\, \vec{r} - i\omega t) \tag{2.6}$$

$$f_1(\vec{r},\vec{v},t) = f_1\,(\vec{v})\,\exp\,(i\,\vec{k}\,.\,\vec{r} - i\omega t) \tag{2.7}$$

where $\vec{E}$, $\vec{B}$ and $f_1(\vec{v})$ are phasor amplitudes (which in general may be complex quantities) independent of space and time. With this space-time dependence, the linearized Vlasov equation (2.4) reduces to

$$- i\,(\omega - \vec{k}.\vec{v})\,f_1(\vec{v}) - (e/m_e)(\vec{v} \times \vec{B}_0)\,.\,\vec{\nabla}_v\,f_1(\vec{v}) =$$

$$= (e/m_e)(\vec{E} + \vec{v} \times \vec{B})\,.\,\vec{\nabla}_v\,f_0(v_\parallel, v_\perp) \tag{2.8}$$

To solve this differential equation for $f_1(\vec{v})$ in velocity space, we introduce cylindrical coordinates $(v_\perp, \phi, v_\parallel)$ with the vector component $\vec{v}_\parallel$ along the magnetostatic field, as shown in Fig. 1. Therefore, $\vec{B}_0 = B_0\,\hat{z}$ and

$$v_x = v_\perp \cos\phi\,; \quad v_y = v_\perp \sin\phi\,; \quad v_z = v_\parallel \tag{2.9}$$

Also, using these relations, we have

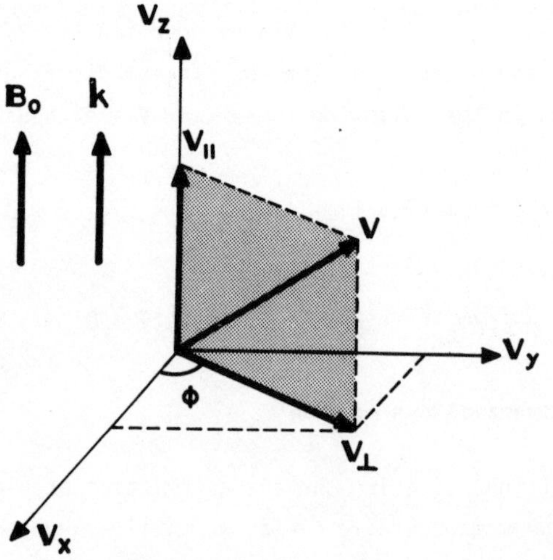

Fig. 1 - Cylindrical coordinate system $(v_\perp, \phi, v_\parallel)$ in velocity space, with the $v_\parallel$ axis along the magnetostatic field $\vec{B}_0$ and $v_\perp$ in the (v_x, v_y) plane normal to $\vec{B}_0$.

$$\frac{df_1(\vec{v})}{d\phi} = \left(\frac{dv_x}{d\phi}\frac{\partial}{\partial v_x} + \frac{dv_y}{d\phi}\frac{\partial}{\partial v_y} + \frac{dv_z}{d\phi}\frac{\partial}{\partial v_z}\right) f_1(\vec{v})$$

$$= \left(- v_y \frac{\partial}{\partial v_x} + v_x \frac{\partial}{\partial v_y}\right) f_1(\vec{v})$$

$$= - (\vec{v} \times \hat{z}) \cdot \vec{\nabla}_v f_1(\vec{v}) \tag{2.10}$$

Substituting this result into (2.8), we obtain

$$- i(\omega - \vec{k} \cdot \vec{v}) f_1(\vec{v}) + \frac{eB_0}{m_e}\frac{df_1(\vec{v})}{d\phi} = \frac{e}{m_e}(\vec{E} + \vec{v} \times \vec{B}) . \vec{\nabla}_v f_0(v_\parallel, v_\perp) \tag{2.11}$$

Using the electron cyclotron frequency $\omega_{ce} = e B_0/m_e$, (2.11) can be rewritten as

$$\frac{df_1(\vec{v})}{d\phi} - \frac{i(\omega - \vec{k} \cdot \vec{v})}{\omega_{ce}} f_1(\vec{v}) = \frac{e}{m_e \omega_{ce}} (\vec{E} + \vec{v} \times \vec{B}) . \vec{\nabla}_v f_0(v_\parallel, v_\perp) \tag{2.12}$$

From Maxwell $\vec{\nabla} \times \vec{E}$ equation we can express the magnetic field as

$$\vec{B} = (\vec{k} \times \vec{E})/\omega \tag{2.13}$$

Substituting (2.13) into (2.12), and making use of the vector identity $\vec{v} \times (\vec{k} \times \vec{E}) = (\vec{v} \cdot \vec{E}) \vec{k} - (\vec{k} \cdot \vec{v}) \vec{E}$, we obtain for the right-hand side of (2.12)

$$\frac{e}{m_e \omega_{ce}} (\vec{E} + \vec{v} \times \vec{B}) . \vec{\nabla}_v f_0 = \frac{e}{m_e \omega_{ce}} \left[\left(1 - \frac{kv_\parallel}{\omega}\right) \vec{E} . \vec{\nabla}_v f_0 + \right.$$

$$\left. + \frac{k (\vec{v} \cdot \vec{E})}{\omega} \frac{\partial f_0}{\partial v_\parallel} \right]$$

$$= \frac{e}{m_e \omega_{ce}} \left\{ \left(1 - \frac{kv_\parallel}{\omega}\right) \left[(E_x \cos\phi + E_y \sin\phi) \frac{\partial f_0}{\partial v_\perp} + E_\parallel \frac{\partial f_0}{\partial v_\parallel}\right] + \right.$$

$$+ \left(\frac{k}{\omega}\right) \left[(E_x \cos\phi + E_y \sin\phi)\, v_\perp + E_\parallel\, v_\parallel\right] \frac{\partial f_0}{\partial v_\parallel} \Bigg\}$$

$$= \frac{e}{m_e\, \omega_{ce}} \Bigg\{ \left[\left(1 - \frac{k v_\parallel}{\omega}\right) \frac{f_0}{\partial v_\perp} + \left(\frac{k v_\perp}{\omega}\right) \frac{f_0}{\partial v_\parallel} \right] (E_x \cos\phi + E_y \sin\phi) +$$

$$+ E_\parallel \frac{\partial f_0}{\partial v_\parallel} \Bigg\} \tag{2.14}$$

where we have taken $\vec{k} \cdot \vec{v} = k\, v_\parallel$ and $\vec{k} \cdot \vec{\nabla}_v = k\, \partial/\partial v_\parallel$, since we are considering wave propagation parallel to the magnetic field ($\vec{k} \parallel \vec{B}_0$).

At this point it is convenient to express the component of the electric field vector in the plane perpendicular to $\vec{B}_0$ as a linear superposition of two, oppositely directed , circularly polarized components. Noting that $(\hat{x} + i\hat{y})/\sqrt{2}$ and $(\hat{x} - i\hat{y})/\sqrt{2}$ are unit complex vectors, the Cartesian components of the electric field vector

$$\vec{E} = \hat{x}\, E_x + \hat{y}\, E_y + \hat{z}\, E_\parallel \tag{2.15}$$

can be appropriately rewritten as

$$\vec{E} = E_+ \frac{(\hat{x} + i\hat{y})}{\sqrt{2}} + E_- \frac{(\hat{x} - i\hat{y})}{\sqrt{2}} + \hat{z}\, E_\parallel \tag{2.16}$$

where the following notation is used

$$E_\pm = (E_x \mp iE_y)/\sqrt{2} \tag{2.17}$$

The first term in the right-hand side of (2.16) represents a circularly polarized field with the electric field vector rotating in the clockwise direction, whereas the second term represents a circularly polarized field with the electric field vector rotating in the counterclockwise direction. For the right (left) circularly polarized field, with the thumb of the right (left) hand pointing in the direction of propagation ($\hat{z}$), the fingers curl in the direction of rotation of the electric field vector. Thus, the two linearly polarized perpendicular components of the electric field in the plane ($\hat{x}$,y), normal to $\vec{B}_0$, can be recast as a superposition of two circularly polarized components with opposite directions of rotation. The advantage of using the two circularly polarized components is that it permits the final equations, involving the transverse modes of propagation, to be separated into two independent sets of transverse fields.

It is a trivial matter to verify the relation

$$E_x \cos\phi + E_y \sin\phi = (E_+ e^{i\phi} + E_- e^{-i\phi})/\sqrt{2} \tag{2.18}$$

so that (2.12) can be rewritten as

$$\frac{df_1(\vec{v})}{d\phi} - \frac{i(\omega - kv_\parallel)}{\omega_{ce}} f_1(\vec{v}) = \frac{e}{m_e \omega_{ce}} \left\{ \left[(1 - \frac{kv_\parallel}{\omega}) \frac{\partial f_0}{v_\perp} + (\frac{k v_\perp}{\omega}) \frac{\partial f_0}{\partial v_\parallel} \right] \frac{1}{\sqrt{2}} (E_+ e^{i\phi} + E_- e^{-i\phi}) + E_\parallel \frac{\partial f_0}{\partial v_\parallel} \right\} \tag{2.19}$$

Introducing the notation

$$F_+(\vec{v}) = F_+(v_\parallel, v_\perp) e^{i\phi} \tag{2.20}$$

$$F_-(\vec{v}) = F_-(v_\parallel, v_\perp) e^{-i\phi} \tag{2.21}$$

$$F_\parallel(\vec{v}) = F_\parallel(v_\parallel, v_\perp) \tag{2.22}$$

where

$$F_+(v_\parallel, v_\perp) = \frac{e}{m_e \omega_{ce}} \left[(1 - \frac{kv_\parallel}{\omega}) \frac{\partial f_0}{\partial v_\perp} + (\frac{kv_\perp}{\omega}) \frac{\partial f_0}{d v_\parallel} \right] \frac{E_+}{\sqrt{2}} \tag{2.23}$$

$$F_-(v_\parallel, v_\perp) = \frac{e}{m_e \omega_{ce}} \left[(1 - \frac{kv_\parallel}{\omega}) \frac{\partial f_0}{\partial v_\perp} + (\frac{kv_\perp}{\omega}) \frac{\partial f_0}{\partial v_\parallel} \right] \frac{E_-}{\sqrt{2}} \tag{2.24}$$

$$F_\parallel(v_\parallel, v_\perp) = \frac{e}{m_e \omega_{ce}} (\frac{\partial f_0}{\partial v_\parallel}) E_\parallel \tag{2.25}$$

Eq. (2.19) becomes

$$\frac{df_1(\vec{v})}{d\phi} - \frac{i(\omega - kv_\parallel)}{\omega_{ce}} f_1(\vec{v}) = F_+(\vec{v}) + F_-(\vec{v}) + F_\parallel(\vec{v}) \tag{2.26}$$

In order to solve this differential equation, let

$$f_1(\vec{v}) = f_{1+}(\vec{v}) + f_{1-}(\vec{v}) + f_{1\parallel}(\vec{v}) \tag{2.27}$$

where $f_{1+}(\vec{v})$, $f_{1-}(\vec{v})$ and $f_{1\parallel}(\vec{v})$ are the solutions of (2.26) corresponding, respectively, to $F_+(\vec{v})$, $F_-(\vec{v})$ and $F_\parallel(\vec{v})$, individually, in the right-hand side of (2.26). Thus, the differential equation for $f_{1+}(v)$, for example, can be written as

$$\frac{d}{d\phi}\left\{ f_{1+}(\vec{v}) \exp\left[\frac{-i\,(\omega - kv_\parallel)}{\omega_{ce}}\,\phi\right]\right\} =$$

$$= F_+(v_\parallel, v_\perp) \exp\left[\frac{-i\,(\omega - kv_\parallel)}{\omega_{ce}}\,\phi + i\phi\right] \tag{2.28}$$

Integrating both sides of this equation with respect to ϕ, from $\phi = -\infty$ to an arbitrary value of ϕ, and noting that the exponential term vanishes at $\phi = -\infty$, since ω has a vanishingly small positive imaginary part, yields

$$f_{1+}(\vec{v}) = \frac{i\omega_{ce}}{(\omega - kv_\parallel - \omega_{ce})} F_+(v_\parallel, v_\perp) e^{i\phi} + C_+ \exp\left[\frac{i\,(\omega - kv_\parallel)}{\omega_{ce}}\,\phi\right] \tag{2.29}$$

The value of $f_{1+}(\vec{v})$ must not change if ϕ is increased or decreased by integral multiples of 2π, since by physical arguments $f_{1+}(\vec{v})$ must be a unique function of $\vec{v}$. This requirement can be satisfied only if $C_+ = 0$. Therefore, we obtain

$$f_{1+}(\vec{v}) = f_{1+}(v_\parallel, v_\perp)\, e^{i\phi} \tag{2.30}$$

where

$$f_{1+}(v_\parallel, v_\perp) = \frac{i\,\omega_{ce}}{(\omega - kv_\parallel - \omega_{ce})} F_+(v_\parallel, v_\perp) \tag{2.31}$$

In a similar way, we find

$$f_{1-}(\vec{v}) = f_{1-}(v_\parallel, v_\perp)\, e^{-i\phi} \tag{2.32}$$

where

$$f_{1-}(v_\parallel, v_\perp) = \frac{i\omega_{ce}}{(\omega - kv_\parallel + \omega_{ce})} F_-(v_\parallel, v_\perp) \tag{2.33}$$

and

$$f_{1\parallel}(\vec{v}) = f_{1\parallel}(v_\parallel, v_\perp) = \frac{i\,\omega_{ce}}{(\omega - kv_\parallel)} F_\parallel(v_\parallel, v_\perp) \tag{2.34}$$

Substituting expressions (2.23), (2.24) and (2.25), respectively, into (2.31), (2.33) and (2.34), yields the following explicit expression for the phasor amplitude $f_1(\vec{v})$ of the perturbation of the velocity distribution function, in terms of the equilibrium distribution function of the electrons,

$$f_1(\vec{v}) = \frac{i\,e}{m_e(\omega - kv_\parallel - \omega_{ce})}\left[\left(1 - \frac{kv_\parallel}{\omega}\right)\frac{\partial f_0}{\partial v_\perp} + \left(\frac{kv_\perp}{\omega}\right)\frac{\partial f_0}{\partial v_\parallel}\right]\frac{E_+}{\sqrt{2}}\,e^{i\phi} +$$

$$+ \frac{i\,e}{m_e(\omega - kv_\parallel + \omega_{ce})}\left[\left(1 - \frac{kv_\parallel}{\omega}\right)\frac{\partial f_0}{\partial v_\perp} + \left(\frac{kv_\perp}{\omega}\right)\frac{\partial f_0}{\partial v_\parallel}\right]\frac{E_-}{\sqrt{2}}\,e^{-i\phi} +$$

$$+ \frac{i\,e}{m_e(\omega - kv_\parallel)}\left(\frac{\partial f_0}{\partial v_\parallel}\right) E_\parallel \tag{2.35}$$

2.3 Perturbation current density

Since the space-time dependence of the electromagnetic fields are of the form $\exp(i\,\vec{k}\cdot\vec{r} - i\,\omega\,t)$, we expect the current density to behave also as

$$\vec{J}(\vec{r},t) = \vec{J}\exp(i\vec{k}.\vec{r} - i\,\omega\,t) \tag{2.36}$$

where the phasor amplitude of the current density is given by

$$\vec{J} = -e\int_V \vec{v}\, f_1(\vec{v})\, d^3v \tag{2.37}$$

where the integration is to be performed over all of velocity space. It is also convenient to separate $\vec{J}$ into two, oppositely directed, circularly polarized components, and a longitudinal component along $\vec{B}_0$. For this purpose, we express the electron velocity in a form analogous to (2.16),

$$\vec{v} = v_+\frac{(x + i\hat{y})}{\sqrt{2}} + v_-\frac{(\hat{x} - i\hat{y})}{\sqrt{2}} + \hat{z}\,v_\parallel \tag{2.38}$$

where

$$v_\pm = (v_x \mp iv_y)/\sqrt{2} \tag{2.39}$$

Thus, with this representation for $\vec{v}$, we obtain the following corresponding components for $\vec{J}$,

$$J_+ = -e\int_v v_+ f_1(\vec{v})\, d^3v \tag{2.40}$$

$$J_- = -e\int_v v_- f_1(\vec{v})\, d^3v \tag{2.41}$$

$$J_\parallel = -e\int_v v_\parallel\, f_1(\vec{v})\, d^3v \tag{2.42}$$

According to (2.27), (2.30), (2.32) and (2.34), we can replace $f_1(\vec{v})$ by

$$f_1(\vec{v}) = f_{1+}(v_\parallel, v_\perp)\, e^{i\phi} + f_{1-}(v_\parallel, v_\perp)\, e^{-i\phi} + f_{1\parallel}(v_\parallel, v_\perp) \tag{2.43}$$

Further, in view of (2.9), we also have

$$v_+ = \frac{1}{\sqrt{2}}\, v_\perp\, e^{-i\phi} \qquad v_- = \frac{1}{\sqrt{2}}\, v_\perp\, e^{+i\phi} \tag{2.44}$$

so that Eqs. (2.40), (2.41) and (2.42) become

$$J_+ = -\frac{e}{\sqrt{2}}\int_v v_\perp\, e^{-i\phi}[f_{1+}(v_\parallel, v_\perp)\, e^{i\phi} + f_{1-}(v_\parallel, v_\perp)\, e^{-i\phi} +$$

$$+ f_{1\parallel}(v_\parallel, v_\perp)]\, d^3v \tag{2.45}$$

$$J_- = -\frac{e}{\sqrt{2}}\int_v v_\perp\, e^{+i\phi}[f_{1+}(v_\parallel, v_\perp)\, e^{i\phi} + f_{1-}(v_\parallel, v_\perp)\, e^{-i\phi} +$$

$$+ f_{1\parallel}(v_\parallel, v_\perp)]\, d^3v \tag{4.46}$$

$$J_\parallel = -e\int_v v_\parallel\, [f_{1+}(v_\parallel, v_\perp)\, e^{i\phi} + f_{1-}(v_\parallel, v_\perp)\, e^{-i\phi} +$$

$$f_{1\parallel}(v_\parallel, v_\perp)]\, d^3v \tag{2.47}$$

In cylindrical coordinates we have $d^3v = v_\perp \, dv_\perp \, dv_\parallel \, d\phi$. Evaluating the integrals with respect to ϕ, from $\phi = 0$ to $\phi = 2\pi$,yields the following simple results

$$J_+ = - e\pi \sqrt{2} \int_0^{\infty} v_\perp^2 \, dv_\perp \int_{-\infty}^{+\infty} f_{1+}(v_\parallel , v_\perp) \, dv_\parallel \tag{2.48}$$

$$J_- = - e\pi \sqrt{2} \int_0^{\infty} v_\perp^2 \, dv_\perp \int_{-\infty}^{+\infty} f_{1-}(v_\parallel , v_\perp) \, dv_\parallel \tag{2.49}$$

$$J_\parallel = - e\pi \, 2 \int_0^{\infty} v_\perp \, dv_\perp \int_{-\infty}^{+\infty} v_\parallel \; f_{1\parallel}(v_\parallel , v_\perp) \, dv_\parallel \tag{2.50}$$

since

$$\int_0^{2\pi} e^{\pm in\phi} \, d\phi = 0; \quad \text{for } n = 1, 2, 3, \ldots \tag{2.51}$$

$$= 2\pi \; ; \text{for } n = 0$$

From (2.31), (2.33) and (2.34), together with (2.23), (2.24) and (2.25), we see that J_+, J_- and $J_\parallel$ depend, respectively, on *only* E_+, E_- and $E_\parallel$. This result justifies the use of the method of decomposition of the vectors into the sum of two, oppositely directed, circularly polarized components in the plane normal to $\vec{B}_0$, and a longitudinal component along $\vec{B}_0$.

2.4 Separation into the various modes

From Maxwell equations, and for the special case in which all field vectors vary as exp $(i \, \vec{k} . \vec{r} - i \omega t)$, with $\vec{k} = k\hat{z}$, we have

$$k\hat{z} \times \vec{E} = \omega \vec{B} \tag{2.52}$$

$$i \, k \, \hat{z} \times \vec{B} = \mu_0 \vec{J} - (i\omega/c^2) \vec{E} \tag{2.53}$$

Noting that $\hat{z} \times \vec{E} = \hat{y} E_x - \hat{x} E_y$, (2.52) and (2.53) can be rewritten in component form as

$$\omega B_x = -k E_y \tag{2.54}$$

$$\omega\, B_y = k\, E_x \tag{2.55}$$

$$\omega\, B_{\parallel} = 0 \tag{2.56}$$

and

$$-\, i\, k\, B_y = \mu_0\, J_x - (i\omega/c^2)\, E_x \tag{2.57}$$

$$i\, k\, B_x = \mu_0\, J_y - (i\omega/c^2)\, E_y \tag{2.58}$$

$$0 = \mu_0\, J_{\parallel} - (i\omega/c^2)\, E_{\parallel} \tag{2.59}$$

Now we define, as in (2.17),

$$B_{\pm} = (B_x \mp i\, B_y)/\sqrt{2} \tag{2.60}$$

Multiplying (2.54) by $1/\sqrt{2}$, and (2.55) by $\mp i/\sqrt{2}$, and adding the resulting expressions, yields

$$B_{\pm} = \mp\, i\, (k/\omega)\, E_{\pm} \tag{2.61}$$

Note that the signs are coupled, that is, *either* upper signs or lower signs are to be used. Similarly, combining (2.57) and (2.58), and noting that

$$J_{\pm} = (J_x \mp i\, J_y)/\sqrt{2} \tag{2.62}$$

we obtain

$$\mp k\, B_{\pm} = -\, \mu_0\, J_{\pm} + (i\omega/c^2)\, E_{\pm} \tag{2.63}$$

From these equations it is clear that the total electromagnetic field can be separated into four *independent* groups, involving the following quantities:

1. $J_{\parallel}$, $E_{\parallel}$ [Eq. (2.59)]

2. $B_{\parallel}$ [Eq. (2.56)]

3. J_-, E_-, B_- [Eqs. (2.61) and (2.63), lower signs]

4. J_+, E_+, B_+ [Eqs. (2.61) and (2.63), upper signs]

Note that J_+, J_- and $J_{\parallel}$ depend, respectively, only on E_+, E_- and $E_{\parallel}$. The first group contains an electric field and an electric current in the direction of $\vec{k}$ which, in this section, is also the direction of $\vec{B}_0$. Further, there is no associated magnetic field. Therefore, it represents the electrostatic *longitudinal plasma wave.* The second group does not constitute a mode of propagation but only shows, through (2.56), that for a wave propagating parallel to $\vec{B}_0$ the time-varying magnetic field in the parallel direction is zero. The third and fourth groups represent, respectively, the *left circularly polarized* and the *right circularly polarized transverse electromagnetic waves.* Thus, we can separately analyze the characteristics of the longitudinal plasma wave and the two polarized transverse electromagnetic waves.

2.5 Longitudinal plasma wave

To deduce the dispersion relation for the longitudinal plasma wave propagating along the magnetostatic field $\vec{B}_0$, we substitute $J_\parallel$, from (2.50), into (2.59), obtaining

$$E_\parallel = + \frac{i\, 2\pi\, e}{\varepsilon_0\, \omega} \int_0^\infty v_\perp\, dv_\perp \int_{-\infty}^{+\infty} v_\parallel f_{1\parallel}(v_\parallel, v_\perp)\, dv_\parallel \tag{2.64}$$

From (2.34) and (2.25) we can replace $f_{1\parallel}(v_\parallel, v_\perp)$ in Eq. (2.64), to obtain the following *dispersion relation*

$$1 = - \frac{\omega_{pe}^2}{n_0\, \omega}\, 2\pi \int_0^\infty v_\perp\, dv_\perp \int_{-\infty}^{+\infty} \frac{v_\parallel (\partial f_0/\partial v_\parallel)}{(\omega - k v_\parallel)}\, dv_\parallel \tag{2.65}$$

This dispersion relation can be conveniently recasted as

$$1 = - \frac{\omega_{pe}^2}{n_0\, \omega} \int_v \frac{v_\parallel (\partial f_0/\partial v_\parallel)}{(\omega - k v_\parallel)}\, d^3v \tag{2.66}$$

since in cylindrical coordinates $d^3v = v_\perp\, dv_\perp\, dv_\parallel\, d\phi$ and $\int_0^{2\pi} d\phi = 2\pi$.

This equation is identical to the dispersion equation (18.4.2), deduced for longitudinal waves in an isotropic plasma, except for the fact the directions x and z are interchanged (here $\vec{k} \parallel \vec{B}_0 \parallel \hat{z}$). Thus, the characteristic behavior of the longitudinal plasma wave, for propagation along the magnetostatic field, is identical to the case of a plasma with no external magnetostatic field. The magnetostatic field, therefore, has no influence on the longitudinal plasma wave. This result is due to the fact that the magnetostatic field exerts no force on the charged particles moving in the direction parallel to it, and therefore it does not influence the distribution of the electrons in the longitudinal direction. It is the perturbation in the distribution of the velocities of the electrons in the longitudinal direction that accounts for the characteristics of the longitudinal plasma wave. Recall that the longitudinal plasma wave separates out as an independent mode of propagation.

2.6 Transverse electromagnetic waves

Consider now the two circularly polarized transverse waves ($\vec{E}$ normal to the direction of propagation). To deduce the dispersion relation for both waves, we first eliminate $B_{\pm}$ from Eqs. (2.61) and (2.63), and express $J_{\pm}$ in terms of $E_{\pm}$ as

$$J_{\pm} = (i\varepsilon_0/\omega)\,(\omega^2 - k^2 c^2)\,E_{\pm} \tag{2.67}$$

Substituting $J_{\pm}$, from (2.48) and (2.49), with $f_{1+}(v_{\shortparallel},v_{\perp})$ and $f_{1-}(v_{\shortparallel},v_{\perp})$ given by (2.31) and (2.33), respectively, yields

$$- e\,\pi\,\sqrt{2}\,\omega_{ce} \int_0^{+\infty} v_{\perp}^2\,dv_{\perp} \int_{-\infty}^{+\infty} \frac{F_{\pm}(v_{\shortparallel},v_{\perp})}{(\omega - kv_{\shortparallel} \mp \omega_{ce})}\,dv_{\shortparallel} =$$

$$= (\varepsilon_0/\omega)\,(\omega^2 - k^2 c^2)\,E_{\pm} \tag{2.68}$$

If (2.23) and (2.24) are used to replace $F_{\pm}(v_{\shortparallel},v_{\perp})$, we find the following *dispersion relation* for the transverse electromagnetic waves ($E_{\pm} \neq 0$)

$$k^2 c^2 = \omega^2 + \omega_{pe}^2 \frac{\pi}{n_0} \int_0^{\infty} v_{\perp}^2\,dv_{\perp} \int_{-\infty}^{+\infty} \frac{(\omega - kv_{\shortparallel})(\partial f_0/\partial v_{\perp}) + kv_{\perp}(\partial f_0/\partial v_{\shortparallel})}{(\omega - kv_{\shortparallel} \mp \omega_{ce})}\,dv_{\shortparallel} \tag{2.69}$$

where the upper sign refers to the right circularly polarized wave, and the lower sign to the left circularly polarized wave.

An alternative form of this equation can be obtained by integrating the right-hand side by parts. First, integrating over $v_{\shortparallel}$ by parts, we have

$$\int_0^{\infty} v_{\perp}^2\,dv_{\perp} \int_{-\infty}^{+\infty} \frac{kv_{\perp}\,(\partial f_0/\partial v_{\shortparallel})}{(\omega - kv_{\shortparallel} \mp \omega_{ce})}\,dv_{\shortparallel} =$$

$$= - \int_0^{\infty} \int_{-\infty}^{+\infty} \frac{(k^2 v_{\perp}^2)\,f_0}{(\omega - kv_{\shortparallel} \mp \omega_{ce})}\,v_{\perp}\,dv_{\perp}\,dv_{\shortparallel} \tag{2.70}$$

and integrating over $v_\perp$ by parts, we have

$$\int_{-\infty}^{+\infty} \frac{(\omega - kv_\parallel)\, dv_\parallel}{(\omega - kv_\parallel \mp \omega_{ce})} \int_0^\infty v_\perp^2 (\partial f_0/\partial v_\perp)\, dv_\perp =$$

$$= -2 \int_0^\infty \int_{-\infty}^{+\infty} \frac{(\omega - kv_\parallel)\, f_0}{(\omega - kv_\parallel \mp \omega_{ce})} v_\perp\, dv_\perp\, dv_\parallel \tag{2.71}$$

Since $2\pi = \int_0^{2\pi} d\phi$ and $d^3v = v_\perp\, dv_\perp\, dv_\parallel\, d\phi$, we obtain the following alternative form of the dispersion relation (2.69)

$$k^2 c^2 = \omega^2 - \frac{\omega_{pe}^2}{n_0} \int_V \left[\frac{(\omega - kv_\parallel)}{(\omega - kv_\parallel \mp \omega_{ce})} + \frac{(k^2 v_\perp^2)/2}{(\omega - kv_\parallel \mp \omega_{ce})^2} \right] f_0\, d^3v \tag{2.72}$$

Let us investigate first the plasma behavior for the case of an *isotropic* equilibrium distribution function. Thus, we choose $f_0(v)$ to be the Maxwell-Boltzmann distribution function (18.4.22). Note that, in this case, the vector $\vec{\nabla}_v f_0(v)$ is parallel to $\vec{v}$, so that the magnetic force term $[\vec{v} \times \vec{B}(\vec{r},t)] \cdot \vec{\nabla}_v f_0(v)$, in the linearized Vlasov equation (2.4), vanishes. Consequently, for an *isotropic* equilibrium function, the magnetic field $\vec{B}(\vec{r},t)$ of the wave has no influence on the plasma behavior in the linear approximation. Also, it is easy to verify that, in the isotropic case, all factors in the numerator of the integrands in Eqs. (2.69) and (2.72), which contain the propagation coefficient k, vanish. The dispersion equation (2.69) then reduces to

$$k^2 c^2 = \omega^2 + \frac{\omega_{pe}^2 \pi}{n_0} \int_0^\infty v_\perp^2\, dv_\perp \int_{-\infty}^{+\infty} \frac{\omega (\partial f_0/\partial v_\perp)}{(\omega - kv_\parallel \mp \omega_{ce})}\, dv_\parallel \tag{2.73}$$

or, equivalently,

$$k^2 c^2 = \omega^2 + \frac{\omega_{pe}^2 \omega}{2n_0} \int_V \frac{v_\perp (\partial f_0/\partial v_\perp)}{(\omega \mp \omega_{ce}) - kv_\parallel}\, d^3v \tag{2.74}$$

The alternative form of this equation, corresponding to (2.72) for the isotropic case, is

$$k^2 c^2 = \omega^2 - \frac{\omega_{pe}^2 \omega}{n_0} \int_v \frac{f_0}{(\omega \mp \omega_{ce}) - kv_{\shortparallel}} d^3v \tag{2.75}$$

Substituting $f_0(v)$ from (18.4.22) and performing the integration over $v_\perp$ and ϕ, the dispersion relation (2.75) becomes

$$k^2 c^2 = \omega^2 - \omega_{pe}^2 \omega \left(\frac{m_e}{2\pi k_B T_e}\right)^{1/2} \int_{-\infty}^{+\infty} \frac{\exp(-m_e v_{\shortparallel}^2 / 2k_B T_e)}{(\omega \mp \omega_{ce}) - kv_{\shortparallel}} dv_{\shortparallel} \tag{2.76}$$

where the upper and the lower signs correspond to the right and the left circularly polarized waves, respectively.

At this point it is convenient to introduce the following dimensionless parameters

$$\alpha_\pm = \frac{(\omega \mp \omega_{ce})/k_\pm}{(2 k_B T_e/m_e)^{1/2}} \tag{2.77}$$

$$\beta_\pm = \frac{\omega/k_\pm}{(2 k_B T_e/m_e)^{1/2}} \tag{2.78}$$

The subscripts + and - are used in k to denote that it corresponds either to the right or to the left circularly polarized wave, respectively. Thus, $\beta_\pm$ represents the phase velocity of the wave normalized to the most probable speed of the electrons $(2 k_B T_e/m_e)^{1/2}$. Setting, as in (18.4.25),

$$q = \frac{v_{\shortparallel}}{(2 k_B T_e/m_e)^{1/2}} \tag{2.79}$$

the dispersion relation (2.76) can be rewritten in the following simplified form

$$k_\pm^2 c^2 = \omega^2 + \omega_{pe}^2 \beta_\pm I(\alpha_\pm) \tag{2.80}$$

where $I(\alpha_\pm)$ denotes the integral

$$I(\alpha_\pm) = \frac{1}{\sqrt{\pi}} \int_{-\infty}^{+\infty} \frac{\exp(-q^2)}{(q - \alpha_\pm)} dq \tag{2.81}$$

This integral is the same as that defined by (18.4.32), with s = 1, and has been calculated in section 4, Chapter 18. Hence, with the help of (18.4.32) and (18.4.45), (2.80) can be rewritten as

$$k_{\pm}^2 c^2 = \omega^2 + i\sqrt{\pi}\ \omega_{pe}^2 \beta_{\pm} \exp(-\alpha_{\pm}^2) - 2\,\omega_{pe}^2 \beta_{\pm} \int_0^{\alpha_{\pm}} \exp(W^2 - \alpha_{\pm}^2)\, dW \tag{2.82}$$

This is the dispersion equation for the *right* (upper sign) and the *left* (lower sign) *circularly polarized transverse electromagnetic waves* propagating along the magnetostatic field in a hot plasma, whose equilibrium state is characterized by the isotropic Maxwell-Boltzmann distribution function.

2.7 Temporal damping of the transverse electromagnetic waves

A careful examination of (2.82) reveals that, for $k_{\pm}$ real, ω has a negative imaginary part, indicating that the amplitude of the waves are damped with time.

To establish if this temporal damping is significant or not, let us evaluate the asymptotic series expansion of the integral in (2.82) for the case when $|\alpha_{\pm}| >> 1$. For this purpose, we expand the integral in (2.82) in inverse powers of $\alpha_{\pm}$. According to (18.4.51), it is found that, as the first approximation (retaining only the leading term),

$$\int_0^{\alpha_{\pm}} \exp(W^2 - \alpha_{\pm})\, dW = \frac{1}{2\alpha_{\pm}} \tag{2.83}$$

With this result, and making use of the definitions (2.77) and (2.78), the dispersion equation (2.82) simplifies to

$$k_{\pm}^2\ c^2 = \omega^2 - \omega_{pe}^2 \frac{\omega}{(\omega \mp \omega_{ce})} + i\sqrt{\pi}\ \omega_{pe}^2\ \beta_{\pm} \exp(-\alpha_{\pm}^2) \tag{2.84}$$

Furthermore, for $|\alpha_{\pm}| >> 1$ the exponential damping term may be omitted in the first approximation, so that Eq. (2.84) becomes

$$k_{\pm}^2\ c^2 = \omega^2 - \omega_{pe}^2\ \omega/(\omega \mp \omega_{ce}) \tag{2.85}$$

This dispersion equation corresponds to the results obtained using the *cold plasma* model, with the upper sign for the right circularly polarized wave and the lower sign for the left circularly polarized wave. Consequently, it follows that the results of the cold plasma model are valid only if $|\alpha_{\pm}| >> 1$. In the case of the *left circularly polarized wave*, for a given real propagation coefficient, k_-, we find, from Eq. (2.85), that ω is real and satisfies the condition

$$\omega > -(\omega_{ce}/2) + (\omega_{ce}^2/4 + \omega_{pe}^2)^{1/2} \tag{2.86}$$

The phase velocity (ω/k_-) of the left circularly polarized wave is greater than the velocity of light, c, for all k_- and, therefore, β_- is a large number of the order of the ratio of c to the thermal velocity of the electrons. Since $\alpha_-/\beta_- = (\omega + \omega_{ce})/\omega$ is positive and greater than unity, it follows that $\alpha_+ >> 1$ for all k_-. Consequently, the Landau damping of the left circularly polarized wave propagating along the magnetostatic field in a hot plasma is always negligible. This result was also obtained for the case of transverse electromagnetic waves in a hot isotropic plasma. Further, as far as the characteristics of the left circularly polarized waves are concerned, the cold plasma model is a very good approximation for all real propagation coefficients.

In the case of the *right circularly polarized wave*, for a given real propagation coefficient, k_+, it is seen, from Eq. (2.85), that ω is real and satisfies the conditions

$$0 < \omega < \omega_{ce} \tag{2.87}$$

$$\omega > (\omega_{ce}/2) + (\omega_{ce}^2/4 + \omega_{pe}^2)^{1/2} \tag{2.88}$$

An important feature associated with the right circularly polarized wave is the existence of two natural frequency ranges of propagation, whereas for the left circularly polarized wave there is only one natural frequency range of propagation. However, the results for ω, in the range specified in (2.87), do not strictly hold for frequencies of the order of the ion plasma frequency and lower , since at these frequencies the motion of the ions cannot be neglected. For this reason we omit, in the following discussion, the very low frequency region ($\omega < \omega_{ci}$) of (2.87). In the frequency range (2.88), it is found that the phase velocity (ω/k_+) of the right circularly polarized wave is always greater than the velocity of light c, whereas in the frequency range (2.87) the phase velocity (ω/k_+) is less than, but of the order of c, except in the close neighborhood of ω_{ce}. Therefore, we see that β_+ is a large number and, since $|\alpha_+/\beta_+| = |(\omega - \omega_{ce})/\omega|$ is of the order of unity, we conclude that $|\alpha_+| >> 1$, except for ω close to ω_{ce}. Thus, the temporal damping of the right circularly polarized wave is also negligibly small and the cold plasma model is a very good approximation for ω not close to ω_{ce}.

2.8 Cyclotron damping of the right circularly polarized transverse wave

For ω in the close neighborhood of ω_{ce}, the phase velocity (ω/k_+) of the right circularly polarized wave is of the order of the thermal velocity of the electrons or lower, so that $\beta_+ \lesssim 1$. Consequently, since $|\alpha_+/\beta_+| = |(\omega - \omega_{ce})/\omega|$ is much less than unity, it follows that $|\alpha_+| << 1$. This implies that the asymptotic series expansion in inverse powers of $\alpha_\pm$, Eq. (2.83), valid for $|\alpha_\pm| >> 1$, is not applicable for $\omega \approx \omega_{ce}$.

As a first approximation to the dispersion relation (2.82) for the limiting case of $|\alpha_+| << 1$, we can set α_+ equal to zero in Eq. (2.82), to obtain

$$\left(\frac{k_+ c}{\omega}\right)^3 - \left(\frac{k_+ c}{\omega}\right) = i \frac{c\,\pi^{1/2}}{(2 k_B T_e/m_e)^{1/2}} \left(\frac{\omega_{pe}}{\omega_{ce}}\right)^2 \tag{2.89}$$

The second term in the left hand side of (2.89) can be omitted in a first approximation, as compared to the first term, since $(\omega/k)/c << 1$. Hence, (2.89) simplifies to

$$\left(\frac{\omega}{k_+ c}\right)^3 = -i \frac{(2 k_B T_e/m_e)^{1/2}}{c\pi^{1/2}} \left(\frac{\omega_{ce}}{\omega_{pe}}\right)^2 \tag{2.90}$$

Solving this equation explicitly for ω, gives

$$\omega = \omega_r + i\,\omega_i \tag{2.91}$$

where

$$\omega_r = \frac{\sqrt{3}}{2} k_+ \left[\frac{(2 k_B T_e/m_e)^{1/2}}{\pi^{1/2}} c^2 \left(\frac{\omega_{ce}}{\omega_{pe}}\right)^2\right]^{1/3} \tag{2.92}$$

$$\omega_i = -\frac{1}{2} k_+ \left[\frac{(2 k_B T_e/m_e)^{1/2}}{\pi^{1/2}} c^2 \left(\frac{\omega_{ce}}{pe}\right)^2\right]^{1/3} \tag{2.93}$$

Since ω has a negative imaginary part, it follows that the right circularly polarized wave, which is initially set to propagate along the magnetostatic field, is *temporally damped* for ω_r close to ω_{ce}. This temporal damping is usually called *cyclotron damping* and is similar to the Landau damping of the longitudinal plasma wave.

The cyclotron damping, however, differs from the Landau damping in some aspects. The most important one is the fact that the acceleration is perpendicular to the drift motion of the particles and, since the perpendicular electric acceleration does not, in the first approximation, modify the parallel drift velocity, there is no tendency toward trapping. Therefore, trapping is insignificant in cyclotron damping. The charged particles moving along lines of force will feel the oscillations of the perpendicular electric field at a frequency which differs from the plasma rest-frame frequency by the Doppler shift. Since the electrons rotate about $\vec{B}_0$ in the same direction as the electric field of the right circularly polarized wave (Fig. 2), some of them will feel the oscillations at their own cyclotron frequency and they will absorb energy from the field. As a consequence of this wave-particle interaction at the resonance frequency $\omega = \omega_{ce}$, the electrons absorb energy

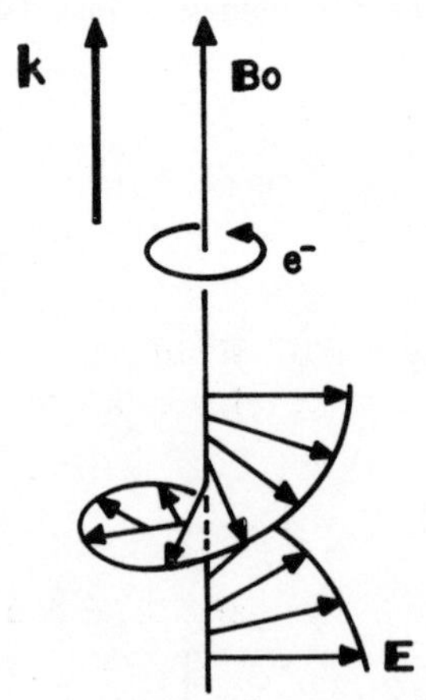

Fig. 2 - Illustrating the resonance which occurs at $\omega = \omega_{ce}$ between the electrons and the electric field of the right circularly polarized wave propagating along $\vec{B}_0$.

from the wave electric field, causing the plasma wave to damp out with time. In the absence of resonant particles, there is no energy exchange between the electric field and the particles, and hence ω is real.

As a final point note that, in the limiting case of $\omega_{ce} \to 0$, that is, in the absence of the magnetostatic field, we have $\alpha_{\pm} = \beta_{\pm} = C$ and (2.82) becomes identical to the dispersion relation (18.5.8) for transverse waves in an isotropic plasma.

2.9 Instabilities in the right circularly polarized transverse wave

We have seen that for an *isotropic* equilibrium distribution function the resonance at ω_{ce}, between the electrons and the right circularly polarized wave, leads to a temporal damping of the wave amplitude. However, depending on the characteristics of the distribution function, resonance can also lead to instabilities (which are associated with a *positive* imaginary part of ω).

Recall that for the case of an isotropic velocity distribution function the magnetic field of the wave has no effect on the plasma behavior in the linear approximation, since $\vec{\nabla}_v f_0(v)$ is parallel to $\vec{v}$ and, consequently, the magnetic force term $[\vec{v} \times \vec{B}(\vec{r},t)] \cdot \vec{\nabla}_v f_0(v)$ vanishes. However, when the condition of velocity isotropy is dropped, the effects that arise from the magnetic field associated with the wave become important and may lead to instabilities. Although the wave magnetic field itself does not exchange energy with the particles, it exerts a force in the parallel (z) direction on the particles, which destroys the isotropy of the velocity distribution function in the plane perpendicular to $\vec{B}_0$. This effect can lead to instabilities depending on the particle distribution function.

For the purpose of demonstrating such an instability, consider the following simple *anisotropic* equilibrium distribution function

$$f_0(v_\parallel, v_\perp) = \delta(v_\parallel)\, f_0(v_\perp) \tag{2.94}$$

which represents cold electrons in the parallel (z) direction, but with a Maxwellian velocity distribution function in the plane normal to $\vec{B}_0$. Inserting (2.94) into the dispersion relation (2.72) for the right circularly polarized wave (upper sign), gives

$$k^2c^2 = \omega^2 - \frac{\omega_{pe}^2}{n_0}\left[\int_{-\infty}^{+\infty}\frac{(\omega - kv_\parallel)\,\delta(v_\parallel)}{(\omega - kv_\parallel - \omega_{ce})}\,dv_\parallel \int_0^{\infty} f_0(v_\perp)\,dv_\perp \int_0^{2\pi} d\phi\; + \right.$$

$$+ \int_{-\infty}^{+\infty} \frac{\delta(v_{\parallel})}{(\omega - kv_{\parallel} - \omega_{ce})^2} dv_{\parallel} \int_0^{\infty} \frac{1}{2} k^2 v_\perp^2 f_0(v_\perp) v_\perp dv_\perp \int_0^{2\pi} d\phi \Bigg] \tag{2.95}$$

Using the following property of the Dirac delta function

$$\int_{-\infty}^{+\infty} f(x) \delta(x - x_0) dx = f(x_0), \tag{2.96}$$

substituting $f_0(v_\perp)$ by

$$f_0(v_\perp) = n_0 \left(\frac{m_e}{2\pi k_B T_e}\right) \exp\left(- \frac{m_e v_\perp^2}{2k_B T_e}\right) \tag{2.97}$$

and performing the integrals, yields

$$k^2 c^2 = \omega^2 - \omega_{pe}^2 \left[\frac{\omega}{(\omega - \omega_{ce})} + \frac{(k^2/2)(2k_B T_e/m_e)}{(\omega - \omega_{ce})^2} \right] \tag{2.98}$$

This equation can be rearranged in the form

$$k^2 = \frac{\omega^2 (\omega - \omega_{ce})^2 - \omega_{pe}^2 \omega(\omega - \omega_{ce})}{c^2 (\omega - \omega_{ce})^2 + (\omega_{pe}^2/2)(2 k_B T_e/m_e)} \tag{2.99}$$

It is a simple matter to verify that, for large values of k^2, ω becomes complex. Thus, in the limit of $k^2 \to \infty$, the denominator of (2.99) vanishes,and we obtain

$$\omega^2 - 2 \omega_{ce}\omega + \omega_{ce}^2 + \frac{\omega_{pe}^2 (2k_B T_e/m_e)}{2 c^2} = 0 \tag{2.100}$$

The solution of this second degree equation is

$$\omega = \omega_{ce} \pm i \frac{\omega_{pe} (2k_B T_e/m_e)^{1/2}}{\sqrt{2} c} \tag{2.101}$$

which shows that *growing modes* (instabilities) can occur for $\omega_r \simeq \omega_{ce}$.

Choosing an anisotropic equilibrium distribution function with some velocity spread along the parallel (z) direction, instead of (2.94), we expect this instability to diminish, while turning into damping for an isotropic distribution function. The analysis of this statement is left as an exercise for the reader.

3. WAVE PROPAGATION ACROSS THE MAGNETOSTATIC FIELD IN A HOT PLASMA

We consider now the problem of wave propagation in a direction perpendicular to an externally applied uniform magnetostatic field, $\vec{B}_0$. As before, we choose the z axis along the magnetostatic field, that is, $\vec{B}_0 = B_0\,\hat{z}$. The propagation coefficient, $\vec{k}$, is normal to $\vec{B}_0$ and along the x axis, $\vec{k} = k\,\hat{x}$ (Fig. 3), with $\vec{k}$ considered to be real. All field quantities are assumed to vary harmonically in space and time, with the phase factor exp $(i\vec{k}.\vec{r} - i\omega t)$. As in the previous cases, we take

$$f(\vec{r},\vec{v},t) = f_0(v_\parallel,v_\perp) + f_1(\vec{r},\vec{v},t) \quad (|f_1| \ll f_0) \tag{3.1}$$

where $f_0(v_\parallel,v_\perp)$ is the equilibrium distribution function of the electrons under the presence of the magnetostatic field, satisfying (2.1), $v_\parallel = v_z$ is the velocity component of the electrons in the direction parallel to $\vec{B}_0$, and $v_\perp$ the velocity component of the electrons in the plane (x,y) normal to $\vec{B}_0$.

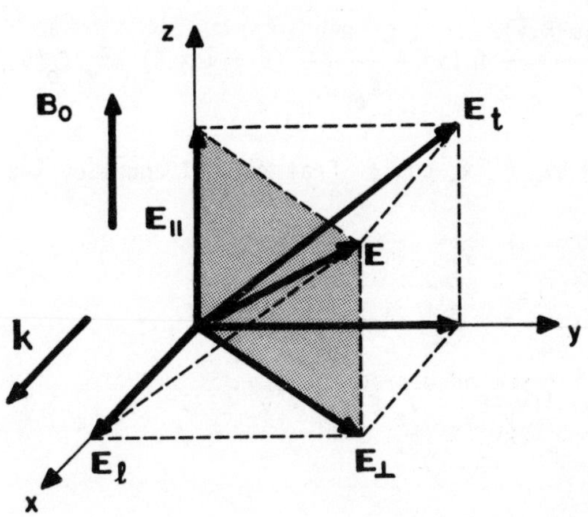

Fig. 3 - Decomposition of the wave electric field vector into components parallel and perpendicular to $\vec{B}_0$, or in components longitudinal and transverse with respect to $\vec{k}$.

For the perturbation distribution function we have

$$f_1(\vec{r},\vec{v},t) = f_1(\vec{v}) \exp(i\,kx - i\omega t) \tag{3.2}$$

and for the *wave* electric and magnetic fields

$$\vec{E}(\vec{r},t) = \vec{E} \exp(ikx - i\omega t) \tag{3.3}$$

$$\vec{B}(\vec{r},t) = \vec{B} \exp(ikx - i\omega t) \tag{3.4}$$

where $f_1(\vec{v})$, $\vec{E}$ and $\vec{B}$ are the phasor amplitudes.

As in the previous section, the purpose is to deduce the dispersion equation giving the functional relationship between k and ω, and, from an analysis of the dispersion relation, determine the intrinsic behavior of the plasma for the case under consideration.

3.1 Solution of the linearized Vlasov equation

From the linearized Vlasov equation (2.4), replacing the differential operators $\partial/\partial t$ and $\vec{\nabla}$ by $-i\omega$ and $ik\hat{x}$, respectively, and making use of relation (2.10), we obtain

$$\frac{df_1(\vec{v})}{d\phi} - i\,\frac{(\omega - \vec{k}.\vec{v})}{\omega_{ce}}\, f_1(v) = \frac{e}{m_e\,\omega_{ce}}\,(\vec{E} + \vec{v} \times \vec{B}).\,\vec{\nabla}_v\, f_0(v_{\parallel}, v_{\perp}) \tag{3.5}$$

where, now, $\vec{k}.\vec{v} = kv_x = kv_{\perp}\cos\phi$. From Maxwell equation $\vec{k} \times \vec{E} = \omega\,\vec{B}$ we have

$$\vec{B} = (k/\omega)(E_y\,\hat{z} - E_z\,\hat{y}) \tag{3.6}$$

Using this expression for $\vec{B}$, we get

$$\vec{v} \times \vec{B} = (k/\omega)\,[(v_y\,E_y + v_z\,E_z)\,\hat{x} - v_x\,E_y\,\hat{y} - v_x\,E_z\,\hat{z}\,] \tag{3.7}$$

Noting that

$$\partial f_0/\partial v_x = \cos\phi\,\partial f_0/\partial v_{\perp} \tag{3.8}$$

$$\partial f_0/\partial v_y = \sin\phi\,\partial f_0/\partial v_{\perp} \tag{3.9}$$

$$\partial f_0/\partial v_z = \partial f_0/\partial v_{\parallel} \tag{3.10}$$

$$(\vec{E} + \vec{v} \times \vec{B}) . \vec{\nabla}_v f_0 = \left[E_x + (\frac{k}{\omega})(v_y E_y + v_z E_z)\right] \cos \phi \frac{\partial f_0}{\partial v_\perp} +$$

$$+ (1 - \frac{kv_x}{\omega}) E_y \sin \phi \frac{\partial f_0}{\partial v_\perp} + (1 - \frac{kv_x}{\omega}) E_z \frac{\partial f_0}{\partial v_z} = \frac{\partial f_0}{\partial v_\perp} (\cos \phi E_x +$$

$$+ \sin \phi E_y) + \left[(\frac{k}{\omega})(v_z \frac{\partial f_0}{\partial v_\perp} - v_\perp \frac{\partial f_0}{\partial v_z}) \cos \quad + \frac{\partial f_0}{\partial v_z}\right] E_z \tag{3.11}$$

The linearized Vlasov equation (3.5) becomes, therefore,

$$\frac{df_1(\vec{v})}{d\phi} - i \frac{(\omega - kv_\perp \cos \phi)}{\omega_{ce}} f_1(\vec{v}) = \frac{e}{m_e \omega_{ce}} \left\{ \frac{\partial f_0}{\partial v_\perp} (\cos \phi E_x + \sin \phi E_y) + \right.$$

$$\left. + \left[(\frac{k}{\omega})(v_z \frac{\partial f_0}{\partial v_\perp} - v_\perp \frac{\partial f_0}{\partial v_z}) \cos \phi + \frac{\partial f_0}{\partial v_z}\right] E_z \right\} \tag{3.12}$$

The integrating factor for this first order differential equation is found to be

$$h(\phi) = \exp \left[- \int_0^\phi i \frac{(\omega - kv_\perp \cos \phi)}{\omega_{ce}} d\phi \right]$$

$$= \exp \left[- i (\frac{\omega}{\omega_{ce}}) \phi + i (\frac{kv_\perp}{\omega_{ce}}) \sin \phi \right] \tag{3.13}$$

Multiplying both sides of (3.12) by the integrating factor (3.13), gives

$$\frac{d}{d\phi} \left\{ f_1(\vec{v}) \exp \left[- i (\frac{\omega}{\omega_{ce}}) \phi + i (\frac{kv_\perp}{\omega_{ce}}) \sin \phi \right] \right\} =$$

$$= \frac{e}{m_e \omega_{ce}} \left\{ \frac{\partial f_0}{\partial v_\perp} (\cos \phi E_x + \sin \phi E_y) + \left[(\frac{k}{\omega})(v_z \frac{\partial f_0}{\partial v_\perp} - \right. \right.$$

$$\left. \left. - v_\perp \frac{f_0}{\partial v_z}) \cos \phi + \frac{f_0}{\partial v_z} \right] E_z \right\} \exp \left[- i (\frac{\omega}{\omega_{ce}}) \phi + i (\frac{kv_\perp}{\omega_{ce}}) \sin \phi \right] \tag{3.14}$$

The solution for $f_1(\vec{v})$ is obtained by integrating this equation over ϕ,

$$f_1(\vec{v}) = \frac{e}{m_e \omega_{ce}} \exp\left[i \left(\frac{\omega}{\omega_{ce}}\right) \phi - i \left(\frac{kv_\perp}{\omega_{ce}}\right) \sin \phi\right] \int_{-\infty}^{\phi} \left\{ \frac{\partial f_0}{\partial v_\perp} (\cos \phi'' \, E_x + \sin \phi'' \, E_y) + \left[\left(\frac{k}{\omega}\right)\left(v_z \frac{\partial f_0}{\partial v_\perp} - v_\perp \frac{\partial f_0}{\partial v_z}\right) \cos \phi'' + \frac{\partial f_0}{\partial v_z}\right] E_z \right\} .$$

$$\exp\left[- i \left(\frac{\omega}{\omega_{ce}}\right) \phi'' + i \left(\frac{kv_\perp}{\omega_{ce}}\right) \sin \phi''\right] d\phi'' \tag{3.15}$$

If the variable of integration is changed to $\phi' = \phi - \phi''$, (3.15) becomes

$$f_1(\vec{v}) = \frac{e}{m_e \omega_{ce}} \exp\left[-i \left(\frac{kv_\perp}{\omega_{ce}}\right) \sin \phi\right] \int_0^{\infty} \left\{ \frac{\partial f_0}{\partial v_\perp} \left[\cos (\phi - \phi') \, E_x + \sin (\phi - \phi') \, E_y\right] + \left[\left(\frac{k}{\omega}\right)\left(v_z \frac{\partial f_0}{\partial v_\perp} - v_\perp \frac{\partial f_0}{\partial v_z}\right) \cos (\phi - \phi') + \frac{\partial f_0}{\partial v_z}\right] E_z \right\} .$$

$$\exp\left[+ i \left(\frac{\omega}{\omega_{ce}}\right) \phi' + i \left(\frac{kv_\perp}{\omega_{ce}}\right) \sin (\phi - \phi')\right] d\phi' \tag{3.16}$$

Note that ϕ occurs only as the argument of periodic functions of period 2π, which is in agreement with the physical requirement that $f_1(\vec{v})$ be a single valued function of ϕ.

3.2 Current density and the conductivity tensor

The current density is given by

$$\vec{J}(\vec{r},t) = \vec{J} \exp (i \, kx - i\omega t) \tag{3.17}$$

where the phasor amplitude, $\vec{J}$, is

$$\vec{J} = e - \int_V f_1(\vec{v}) \, \vec{v} \, d^3v \tag{3.18}$$

or

$$J = -e \int_0^\infty v_\perp \, dv_\perp \int_0^{2\pi} d\phi \int_{-\infty}^{+\infty} dv_z \; f_1(\vec{v}) (v_\perp \cos\phi \, \hat{x} +$$

$$+ v_\perp \sin\phi \, \hat{y} + v_z \, \hat{z}) \tag{3.19}$$

For the purpose of calculating the components of $\vec{J}$, it is appropriate to express $\vec{J}$ as

$$\vec{J} = \overline{\overline{\sigma}} \cdot \vec{E} \tag{3.20}$$

or, in explicit form,

$$J_x = \sigma_{xx} E_x + \sigma_{xy} E_y + \sigma_{xz} E_z \tag{3.21}$$

$$J_y = \sigma_{yx} E_x + \sigma_{yy} E_y + \sigma_{yz} E_z \tag{3.22}$$

$$J_z = \sigma_{zx} E_x + \sigma_{zy} E_y + \sigma_{zz} E_z \tag{3.23}$$

where $\overline{\overline{\sigma}}$ is the conductivity tensor, whose components can be arranged in matrix form as

$$\overline{\overline{\sigma}} = \begin{pmatrix} \sigma_{xx} & \sigma_{xy} & \sigma_{xz} \\ \sigma_{yx} & \sigma_{yy} & \sigma_{yz} \\ \sigma_{zx} & \sigma_{zy} & \sigma_{zz} \end{pmatrix} \tag{3.24}$$

If $f_1(\vec{v})$, from (3.16), is substituted into (3.19), and the resulting expression is compared with Eqs.(3.21) to (3.23), we identify the components of the conductivity tensor as

$$\sigma_{xx} = - \frac{e^2}{m_e \, \omega_{ce}} \int_0^\infty v_\perp^2 \, dv_\perp \int_0^\pi \cos\phi \, d\phi \int_{-\infty}^\infty dv_z \exp - \left[i \left(\frac{k v_\perp}{\omega_{ce}}\right) \sin\phi \right] .$$

$$. \int_0^\infty d\phi' \left(\frac{\partial f_0}{\partial v_\perp}\right) \cos(\phi - \phi') \exp\left[g_1(\phi')\right] \tag{3.25}$$

$$\sigma_{xy} = - \frac{e^2}{m_e \omega_{ce}} \int_0^{\infty} v_\perp^2 \, dv_\perp \int_0^{2\pi} \cos\phi \, d\phi \int_{-\infty}^{+\infty} dv_z \exp\left[- i \left(\frac{kv_\perp}{\omega_{ce}}\right) \sin\phi\right] .$$

$$\cdot \int_0^{\infty} d\phi' \left(\frac{\partial f_0}{\partial v_\perp}\right) \sin(\phi - \phi') \exp[g_1(\phi')] \tag{3.26}$$

$$\sigma_{xz} = - \frac{e^2}{m_e \omega_{ce}} \int_0^{\infty} v_\perp^2 \, dv_\perp \int_0^{2\pi} \cos\phi \, d\phi \int_{-\infty}^{+\infty} dv_z \exp\left[-i \left(\frac{kv_\perp}{\omega_{ce}}\right) \sin\phi\right] .$$

$$\cdot \int_0^{\infty} d\phi' \left[\left(\frac{k}{\omega}\right)\left(v_z \frac{\partial f_0}{\partial v_\perp} - v_\perp \frac{\partial f_0}{\partial v_z}\right) \cos(\phi - \phi') + \frac{\partial f_0}{\partial v_z}\right] . \exp[g_1(\phi')] \tag{3.27}$$

$$\sigma_{yx} = \frac{e^2}{m_e \omega_{ce}} \int_0^{\infty} v_\perp^2 \, dv_\perp \int_0^{2\pi} \sin\phi \, d\phi \int_{-\infty}^{+\infty} dv_z \exp\left[- i \left(\frac{kv_\perp}{\omega_{ce}}\right) \sin\phi\right] .$$

$$\cdot \int_0^{\infty} d\phi' \left(\frac{\partial f_0}{\partial v_\perp}\right) \cos(\phi - \phi') \exp[g_1(\phi')] \tag{3.28}$$

$$\sigma_{yy} = - \frac{e^2}{m_e \omega_{ce}} \int_0^{\infty} v_\perp^2 \, dv_\perp \int_0^{2\pi} \sin\phi \, d\phi \int_{-\infty}^{+\infty} dv_z \exp\left[- i \left(\frac{kv_\perp}{\omega_{ce}}\right) \sin\phi\right] .$$

$$\cdot \int_0^{\infty} d\phi' \left(\frac{\partial f_0}{\partial v_\perp}\right) \sin(\phi - \phi') \exp[g_1(\phi')] \tag{3.29}$$

$$\sigma_{yz} = - \frac{e^2}{m_e \omega_{ce}} \int_0^{\infty} v_\perp^2 \, dv_\perp \int_0^{2\pi} \sin\phi \, d\phi \int_{-\infty}^{+\infty} dv_z \exp\left[- i \left(\frac{kv_\perp}{\omega_{ce}}\right) \sin\phi\right] .$$

$$\cdot \int_0^{\infty} d\phi' \left[\left(\frac{k}{\omega}\right)\left(v_z \frac{\partial f_0}{\partial v_\perp} - v_\perp \frac{\partial f_0}{\partial v_z}\right) \cos(\phi - \phi') + \frac{\partial f_0}{\partial v_z}\right] .$$

$$\exp[g_1(\phi')] \tag{3.30}$$

$$\sigma_{zx} = - \frac{e^2}{m_e\, \omega_{ce}} \int_0^{\infty} v_\perp\, dv_\perp \int_0^{2\pi} d\phi \int_{-\infty}^{+\infty} v_z\, dv_z \exp\left[- i \left(\frac{kv_\perp}{\omega_{ce}}\right) \sin \phi \right] .$$

$$\cdot \int_0^{\infty} d\phi' \left(\frac{\partial f_0}{\partial v_\perp}\right) \cos (\phi - \phi') \exp [g_1(\phi')] \qquad (3.31)$$

$$\sigma_{zy} = - \frac{e^2}{m_e\, \omega_{ce}} \int_0^{\infty} v_\perp dv_\perp \int_0^{2\pi} d\phi \int_{-\infty}^{+\infty} v_z\, dv_z \exp\left[-i \left(\frac{kv_\perp}{\omega_{ce}}\right) \sin \phi \right] .$$

$$\cdot \int_0^{\infty} d\phi' \left(\frac{\partial f_0}{\partial v_\perp}\right) \sin (\phi - \phi') \exp [g_1(\phi')] \qquad (3.32)$$

$$\sigma_{zz} = - \frac{e^2}{m_e\, \omega_{ce}} \int_0^{\infty} v_\perp\, dv_\perp \int_0^{2\pi} d\phi \int_{-\infty}^{+\infty} v_z\, dv_z \exp \left[- i \left(\frac{kv_\perp}{\omega_{ce}}\right) \sin \phi \right] .$$

$$\cdot \int_0^{\infty} d\phi' \left[\left(\frac{k}{\omega}\right)\left(v_z \frac{\partial f_0}{\partial v_\perp} - v_\perp \frac{\partial f_0}{\partial v_z}\right) \cos (\phi - \phi') + \frac{\partial f_0}{\partial v_z}\right] . \exp [g_1(\phi')] \qquad (3.33)$$

where we have used the notation

$$g_1(\phi') = i\, (\omega/\omega_{ce})\, \phi' + i\, (kv_\perp/\omega_{ce}) \sin (\phi - \phi') \qquad (3.34)$$

3.3 Evaluation of the integrals

In simplifying the expression for σ_{xx}, it is advantageous to calculate first the integral with respect to ϕ'. From (3.25) consider, therefore, the integral

$$I_1 = \int_0^{\infty} d\phi' \cos (\phi - \phi') \exp [g_1(\phi')] \qquad (3.35)$$

Differentiating (3.34) with respect to ϕ', we find

$$\cos (\phi - \phi') = \left(\frac{\omega}{kv_\perp}\right) + i \left(\frac{\omega_{ce}}{kv_\perp}\right) \frac{dg_1(\phi')}{d\phi'} \qquad (3.36)$$

Thus, Eq. (3.35) becomes

$$I_1 = \left(\frac{\omega}{kv_\perp}\right)\int_0^\infty d\phi' \exp[g_1(\phi')] + i\left(\frac{\omega_{ce}}{kv_\perp}\right)\int_0^\infty d\left\{\exp[g_1(\phi')]\right\} \tag{3.37}$$

since $d\{\exp[g_1(\phi')]\} = \exp[g_1(\phi')]\, dg_1(\phi')$. Therefore,

$$I_1 = \left(\frac{\omega}{kv_\perp}\right)\int_0^\infty d\phi' \exp[g_1(\phi')] - i\left(\frac{\omega_{ce}}{kv_\perp}\right)\exp\left[i\left(\frac{kv_\perp}{\omega_{ce}}\right)\sin\phi\right] \tag{3.38}$$

In order to evaluate the integral in Eq. (3.38), let

$$\xi = kv_\perp/\omega_{ce} \tag{3.39}$$

and express the term $\exp[g_1(\phi')]$ in a infinite series expansion in terms of the Bessel functions, $J_n(\xi)$,

$$\exp[i(\omega/\omega_{ce})\phi']\exp[i\xi\sin(\phi-\phi')] =$$

$$\exp[i(\omega/\omega_{ce})\phi']\sum_{n=-\infty}^{+\infty} J_n(\xi)\exp[in(\phi-\phi')] =$$

$$= \sum_{n=-\infty}^{+\infty} J_n(\xi)\exp(in\phi)\exp[i(\omega/\omega_{ce}-n)\phi'] \tag{3.40}$$

where $J_n(\xi)$ is the Bessel function of the first kind of order n, and where the factor $\exp[i\xi\sin(\phi-\phi')]$ is identified with the so-called generating function of the Bessel functions. Substituting (3.40) into (3.38), gives

$$I_1 = -\frac{i}{\xi}\exp(i\xi\sin\phi) + \frac{\omega}{kv_\perp}\sum_{n=-\infty}^{+\infty} J_n(\xi)\exp(in\phi)\int_0^\infty d\phi'$$

$$\exp[i(\omega/\omega_{ce}-n)\phi'] \tag{3.41}$$

$$= -\frac{i}{\xi}\exp(i\xi\sin\phi) + \frac{i\omega}{kv_\perp}\sum_{n=-\infty}^{+\infty} J_n(\xi)\exp(in\phi)\frac{1}{(\omega/\omega_{ce}-n)} \tag{3.42}$$

As the next step in evaluating σ_{xx}, we calculate the integral with respect to ϕ. Substituting (3.42) into the expression (3.25) for σ_{xx}, we find the integral with respect to ϕ to be

$$I_2 = \int_0^{2\pi} d\phi \cos\phi \left[-\frac{i}{\xi} + \frac{i\omega}{kv_\perp} \sum_{n=-\infty}^{+\infty} \frac{J_n(\xi) \exp(in\phi - i\xi \sin\phi)}{(\omega/\omega_{ce} - n)} \right] \quad (3.43)$$

The first term within the square brackets in this equation integrates to zero. For the remaining terms note first that we can write

$$\cos\phi = \frac{n}{\xi} + \frac{i}{\xi} \frac{d}{d\phi} (i\, n\, \phi - i\, \xi \sin\phi) \quad (3.44)$$

so that the integral I_2 becomes

$$I_2 = \frac{i\omega}{kv_\perp} \sum_{n=-\infty}^{+\infty} \frac{J_n(\xi)}{(\omega/\omega_{ce} - n)} \left\{ \frac{n}{\xi} \int_0^{2\pi} d\phi \exp(in\phi - i\phi \sin\phi) + \right.$$

$$\left. + \frac{i}{\xi} \int_0^{2\pi} d\left[\exp(in\phi - i\phi \sin\phi)\right] \right\} \quad (3.45)$$

The second integral within brackets in this equation vanishes and the first integral can be expressed in terms of a Bessel function, according to the relation,

$$\int_0^{2\pi} d\phi \exp(i\, n\, \phi - i\xi \sin\phi) = 2\pi\, J_n(\xi) \quad (3.46)$$

which is known as the Bessel integral. Therefore, (3.45) becomes

$$I_2 = \frac{2\pi i\omega}{\xi^2 \omega_{ce}} \sum_{n=-\infty}^{+\infty} \frac{n\, J_n^2(\xi)}{(\omega/\omega_{ce} - n)} \quad (3.47)$$

This result can be written in a slightly different form by noting that

$$I_2 = \frac{2\pi i}{\xi^2} \sum_{n=-\infty}^{+\infty} J_n(\xi) \frac{n\,(\omega/\omega_{ce} - n + n)}{(\omega/\omega_{ce} - n)}$$

$$= \frac{2\pi i}{\varepsilon^2} \sum_{n=-\infty}^{+\infty} \left[n\, J_n^2(\xi) + \frac{n^2\, J_n^2(\xi)}{(\omega/\omega_{ce} - n)} \right] \quad (3.48)$$

Now, since $J_{-n}(\xi) = (-1)^n J_n(\xi)$, we have

$$\sum_{n=-\infty}^{+\infty} n\, J_n^2(\xi) = 0 \tag{3.49}$$

and the integral (3.48) simplifies to

$$I_2 = \frac{2\pi i}{\xi^2} \sum_{n=-\infty}^{+\infty} \frac{n^2 J_n^2(\xi)}{(\omega/\omega_{ce} - n)} \tag{3.50}$$

From (3.25), (3.35) and (3.43), we see that the expression for σ_{xx} can be written as

$$\sigma_{xx} = -\frac{e^2}{m_e\, \omega_{ce}} \int_0^{\infty} v_\perp^2\, dv_\perp \int_{-\infty}^{+\infty} dv_z \frac{\partial f_0}{\partial v_\perp} I_2 \tag{3.51}$$

Thus, the substitution of (3.50) into (3.51), yields

$$\sigma_{xx} = -\frac{2\pi i\, e^2}{m_e\, \omega_{ce}} \int_0^{\infty} v_\perp^2\, dv_\perp \int_{-\infty}^{+\infty} dv_z \left(\frac{\omega_{ce}}{k v_\perp}\right) \frac{\partial f_0}{\partial v_\perp} \sum_{n=-\infty}^{+\infty} \frac{n^2 J_n^2(k v_\perp/\omega_{ce})}{(\omega/\omega_{ce} - n)} \tag{3.52}$$

This expression for σ_{xx} is valid for any cylindrically symmetric equilibrium distribution function $f_0(v_\parallel, v_\perp)$. However, in what follows, all the details will be restricted to the case in which the equilibrium state is characterized by the isotropic Maxwell-Boltzmann distribution function $f_0(v)$, for simplicity. Thus, we take

$$f_0(v) = n_0 \left(\frac{m_e}{2\pi k_B T_e}\right)^{3/2} \exp\left[-\frac{m_e (v_\perp^2 + v_z^2)}{2 k_B T_e}\right] \tag{3.53}$$

for the evaluation of the integrals over $v_\perp$ and v_z in (3.52). In order to perform the integral over $v_\perp$, it is convenient to introduce the following parameter

$$\tilde{\nu} = (k_B T_e/m_e)(k^2/\omega_{ce}^2) \tag{3.54}$$

Performing the differentiation $\partial f_0/\partial v_\perp$ and using (3.54), the expression (3.52) for σ_{xx} simplifies to

$$\sigma_{xx} = \frac{i\, n_0\, e^2}{m_e\, \omega_{ce}\, \tilde{\nu}^2} \sum_{n=-\infty}^{+\infty} \frac{n^2}{(\omega/\omega_{ce} - n)} \int_0^{\infty} \xi\, d\xi\, J_n^2(\xi) \exp(-\xi^2/2\tilde{\nu}) \tag{3.55}$$

From the theory of Bessel functions we have the following Weber's second exponential integral

$$\int_0^\infty \exp(-p^2 t^2) J_n(at) J_n(bt)\, t\, dt =$$

$$= \frac{1}{2p^2} \exp\left(-\frac{a^2+b^2}{4p^2}\right) I_n(ab/2p^2) \tag{3.56}$$

where $I_n(x)$ is the Bessel function of the second kind, which is related to the Bessel function of the first kind with an imaginary argument, $J_n(ix)$, by

$$I_n(x) = (-i)^n J_n(ix) \tag{3.57}$$

Substituting (3.56) into (3.55), yields

$$\sigma_{xx} = \frac{i\, n_0\, e^2}{m_e\, \omega_{ce}} \frac{e^{-\tilde{\nu}}}{\tilde{\nu}} \sum_{n=-\infty}^{+\infty} \frac{n^2 I_n(\tilde{\nu})}{(\omega/\omega_{ce} - n)} \tag{3.58}$$

The components σ_{xz}, σ_{yz}, σ_{zx} and σ_{zy} of the conductivity tensor vanish, since the integrands in (3.27), (3.30), (3.31) and (3.32) are found to be odd functions of v_z. Thus, performing the integrations with respect to v_z first, we find

$$\sigma_{xz} = \sigma_{yz} = \sigma_{zx} = \sigma_{zy} = 0 \tag{3.59}$$

The component σ_{zz} of the conductivity tensor, for the case of the isotropic Maxwell-Boltzmann distribution function, simplifies to

$$\sigma_{zz} = -\frac{e^2}{m_e\, \omega_{ce}} \int_0^\infty v_\perp\, dv_\perp \int_0^{2\pi} d\phi \int_{-\infty}^{+\infty} v_z\, dv_z \exp(-i\,\xi \sin\phi).$$

$$\cdot \int_0^\infty d\phi' \frac{\partial f_0}{\partial v_z} \exp[g_1(\phi')] \tag{3.60}$$

The integrals appearing here are evaluated as in the case of σ_{xx}, yielding the result

$$\sigma_{zz} = \frac{i\, n_o\, e^2}{m_e\, \omega_{ce}}\, e^{-\tilde{\nu}} \sum_{n=-\infty}^{+\infty} \frac{I_n(\tilde{\nu})}{(\omega/\omega_{ce} - n)} \tag{3.61}$$

The components σ_{xy}, σ_{yx} and σ_{yy} of the conductivity tensor will not be needed here, in order to investigate the characteristics of waves propagating across the magnetostatic field in a hot plasma. Therefore, explicit expressions for these components of $\bar{\bar{\sigma}}$ will not be presented.

3.4 Separation into the various modes

With the time and space dependence of the fields, as given by (3.3) and (3.4), and expressing $\vec{J}$ as $\bar{\bar{\sigma}} \cdot \vec{E}$, Maxwell curl equations reduce to

$$k\hat{x} \times \vec{E} = \omega \vec{B} \tag{3.62}$$

$$ik\hat{x} \times \vec{B} = (\mu_o\, \bar{\bar{\sigma}} - \bar{\bar{1}}\, i\omega/c^2) \cdot \vec{E} = -(i\omega/c^2)\, \bar{\bar{\varepsilon}} \cdot \vec{E} \tag{3.63}$$

where $\bar{\bar{1}}$ denotes the unit dyad and

$$\bar{\bar{\varepsilon}} = \bar{\bar{1}} + (i/\omega\varepsilon_o)\, \bar{\bar{\sigma}} \tag{3.64}$$

is the relative permittivity dyad. In component form, (3.62) and (3.63) become, respectively,

$$B_x = 0 \tag{3.65}$$

$$E_z = -(\omega/k)\, B_y \tag{3.66}$$

$$E_y = (\omega/k)\, B_z \tag{3.67}$$

and

$$-(\omega/kc^2)(\varepsilon_{xx}\, E_x + \varepsilon_{xy}\, E_y) = 0 \tag{3.68}$$

$$-(\omega/kc^2)(\varepsilon_{yx}\, E_x + \varepsilon_{yy}\, E_y) = -B_z \tag{3.69}$$

$$-(\omega/kc^2)\, \varepsilon_{zz}\, E_z = B_y \tag{3.70}$$

From Eqs. (3.64), (3.58) and (3.61), it follows that

$$\varepsilon_{xx} = 1 - \frac{\omega_{pe}^2}{\omega\,\omega_{ce}} \frac{e^{-\tilde{\nu}}}{\tilde{\nu}} \sum_{n=-\infty}^{+\infty} \frac{n^2 I_n(\tilde{\nu})}{(\omega/\omega_{ce} - n)} \tag{3.71}$$

$$\varepsilon_{zz} = 1 - \frac{\omega_{pe}^2}{\omega\,\omega_{ce}} e^{-\tilde{\nu}} \sum_{n=-\infty}^{+\infty} \frac{I_n(\tilde{\nu})}{(\omega/\omega_{ce} - n)} \tag{3.72}$$

and, from (3.59),

$$\varepsilon_{xz} = \varepsilon_{yz} = \varepsilon_{zx} = \varepsilon_{zy} = 0 \tag{3.73}$$

The expressions for the other components of $\bar{\bar{\varepsilon}}$ will not be needed in what follows.

An analysis of Eqs. (3.65) to (3.70) shows that the waves are *transverse magnetic* (TM) with respect to the direction (x) of propagation, since $B_x = 0$. Also, we see that the remaining field components can be separated into two independent groups, involving the following variables each:

1. E_x, E_y, B_z [Eqs. (3.67), (3.68), (3.69)] (TM mode).
2. E_z, B_y [Eqs. (3.66) and (3.70)] (TEM mode).

The first group represents the TM (Transverse Magnetic) mode, since there is no component of the wave magnetic field in the direction of propagtion (x). The second group represents the TEM (Transverse Electric-Magnetic) mode, since it has no component of either the electric or the magnetic field in the direction of propagation (Fig. 3). It is a degenerate case of the TM mode. Since the electric field is in the direction of $\vec{B}_0$, the TEM mode is called (in magnetoionic theory) the *ordinary wave,* and it is not affected by $\vec{B}_0$.

3.5 Dispersion relations

To deduce the dispersion equation for the TM mode, we first combine (3.68) and (3.69) to eliminate E_x, obtaining

$$(kc^2/\omega)\, B_z = (\varepsilon_{yy} - \varepsilon_{xy}\, \varepsilon_{yx}/\varepsilon_{xx})\, E_y \tag{3.74}$$

Substituting E_y from (3.67), into (3.74), yields

$$(k^2c^2/\omega^2 - \varepsilon_{yy} + \varepsilon_{xy}\, \varepsilon_{yx}/\varepsilon_{xx})\, B_z = 0 \tag{3.75}$$

For a nontrivial solution for B_z, and also for E_x and E_y, the term within parenthesis in (3.75) must vanish, which results in the following *dispersion relation* for the *TM mode*,

$$k^2c^2/\omega^2 = (\varepsilon_{xx}\,\varepsilon_{yy} - \varepsilon_{xy}\,\varepsilon_{yx})/\varepsilon_{xx} \tag{3.76}$$

To obtain the dispersion equation for the TEM mode, we substitute E_z from (3.66), into (3.70), to find

$$(k^2c^2/\omega^2 - \varepsilon_{zz})\,B_y = 0 \tag{3.77}$$

For a nontrivial solution for B_y, and therefore E_z, we must require that

$$k^2c^2/\omega^2 = \varepsilon_{zz} \tag{3.78}$$

which is the *dispersion relation* for the *TEM mode.*

3.6 The quasistatic mode

Since the dispersion relation (3.76), for the TM mode, is very complicated, in what follows we shall analyze this dispersion relation only for the limiting case of kc/ω tending to infinity. This limiting condition defines the *resonance* condition.

From (3.69) we see that, for finite values of E_x and E_y, B_z must be equal to zero in the limiting case of kc/ω equal to infinity. From (3.67) it follows therefore that E_y vanishes. Consequently, for a nontrivial solution for E_x, the dispersion relation for $kc/\omega \to \infty$ becomes

$$\varepsilon_{xx} = 0 \tag{3.79}$$

This equation is known as the dispersion relation for the *quasistatic wave* propagating across the magnetostatic field, since the magnetic field is negligible and the electric field is essentially in the direction of propagation. In the limit $kc/\omega \to \infty$, the longitudinal wave is already uncoupled from the transverse wave and the dispersion relation $\varepsilon_{xx} = 0$ refers to the *longitudinal* wave ($E_x \neq 0$). The TM mode, in the zero-temperature limit, corresponds to the *extraordinary* wave of magnetoionic theory. As a matter of fact, the dispersion relation (3.79) can be derived directly from the laws of electrostatics, instead of using Maxwell equations. Thus, since the magnetic field can be omitted at the outset, (3.79) is also called the dispersion relation for the *electrostatic wave*. Although (3.79) is strictly correct only for $kc/\omega = \infty$, it can be considered to be a reasonably good approximation for $kc/\omega >> 1$.

From (3.71) the explicit expression for the dispersion relation (3.79), for the quasistatic wave, is found to be

$$\frac{\omega_{pe}^2}{\omega\,\omega_{ce}} \frac{e^{-\tilde{\nu}}}{\tilde{\nu}} \sum_{n=-\infty}^{+\infty} \frac{n^2 I_n(\tilde{\nu})}{(\omega/\omega_{ce} - n)} = 1 \tag{3.80}$$

Since $I_{-n}(\tilde{\nu}) = I_n(\tilde{\nu})$, we have,

$$\sum_{n=-\infty}^{+\infty} n\, I_n(\tilde{\nu}) = 0 \tag{3.81}$$

so that multiplying (3.81) by $(\omega_{pe}^2/\omega\omega_{ce})\,(e^{-\tilde{\nu}}/\tilde{\nu})$ and adding it to (3.80), we find

$$\tilde{\nu}\,\frac{\omega_{ce}^2}{\omega_{pe}^2} = e^{-\tilde{\nu}} \sum_{n=-\infty}^{+\infty} \frac{n\, I_n(\tilde{\nu})}{(\omega/\omega_{ce} - n)} \tag{3.82}$$

This equation was extensively investigated by Bernstein, who showed that is has solutions for both ω and k real. For this reason, these solutions are often called the *Bernstein modes*.

In order to show the *absence of complex solutions for* ω, let us first write the dispersion equation (3.82) in a more convenient form. Making use of the expansion

$$\exp\,(\tilde{\nu}\cos y) = \sum_{n=-\infty}^{+\infty} I_n(\tilde{\nu})\,\exp\,(i\,n\,y) \tag{3.83}$$

and setting $y = 0$ in this expansion, we obtain

$$1 = e^{-\tilde{\nu}} \sum_{n=-\infty}^{+\infty} I_n(\tilde{\nu}) \tag{3.84}$$

Adding (3.84) and (3.82), gives

$$1 + \tilde{\nu}\,\frac{\omega_{ce}^2}{\omega_{pe}^2} = \omega e^{-\tilde{\nu}} \sum_{n=-\infty}^{+\infty} \frac{I_n(\tilde{\nu})}{(\omega - n\,\omega_{ce})} \tag{3.85}$$

From (3.54) it is seen that $\tilde{\nu}$ is real and positive, and therefore $I_n(\tilde{\nu})$ is also real and positive. Hence, writing the angular frequency as

$$\omega = \omega_r + i\,\omega_i \tag{3.86}$$

where ω_r and ω_i are the real and imaginary parts of ω, respectively, we can separate Eq. (3.85) into its real and imaginary parts, as follows:

$$\textit{real part:}\quad 1+\tilde{\nu}\,\frac{\omega_{ce}^2}{\omega_{pe}^2} = e^{-\tilde{\nu}} \sum_{n=-\infty}^{+\infty} I_n(\tilde{\nu}) \left[1+\frac{\omega_{ce}+n(\omega_r - n\omega_{ce})}{(\omega_r - n\omega_{ce})^2 + \omega_i^2}\right] \tag{3.87}$$

$$\textit{imaginary part:}\quad 0=-\omega_i\, e^{-\tilde{\nu}} \sum_{n=-\infty}^{+\infty} I_n(\tilde{\nu})\, \frac{n\,\omega_{ce}}{(\omega_r - n\omega_{ce})^2 + \omega_i^2} \tag{3.88}$$

It can be shown that (3.88) can be satisfied only if $\omega_i = 0$. This result means that the dispersion equation for the quasistatic wave has only real solutions for ω and, therefore, there is neither temporal damping nor instability of the quasistatic waves.

Next, we obtain explicit *real solutions for* ω , for two limiting cases. First, we consider the special case $\tilde{\nu} \ll 1$ which, as seen from (3.54), correponds to the *zero-temperature limit,* and afterwards we analyze the case $\tilde{\nu} \gg 1$, which corresponds to the *high-temperature limit.*

For $\tilde{\nu} \ll 1$ (zero-temperature limit), we have $I_{\pm 1}(\tilde{\nu}) \simeq \tilde{\nu}/2$, while $I_{\pm n}(\tilde{\nu}) = O(\tilde{\nu}^n)$. If ω/ω_{ce} is not close to n only the terms correponding to n = ± 1, in the infinite series on the right-hand side of (3.83), contribute significantly, whereas the other terms are small and can be neglected. Thus, Eq. (3.82) becomes for $\tilde{\nu} \ll 1$,

$$\tilde{\nu}\,\frac{\omega_{ce}^2}{\omega_{pe}^2} = -\frac{I_{-1}(\tilde{\nu})}{(\omega/\omega_{ce} + 1)} + \frac{I_1(\tilde{\nu})}{(\omega/\omega_{ce} - 1)} \tag{3.89}$$

which simplifies to

$$\omega = (\omega_{pe}^2 + \omega_{ce}^2)^{1/2} \tag{3.90}$$

This is known as the *upper hybrid resonant frequency.* This resonant frequency is also predicted in the cold plasma model treatment of waves propagating across a magnetostatic field. Thus, we find that the hot plasma theory confirms the results predicted by the cold plasma model in the zero-temperature limit.

In addition, the hot plasma theory establishes the existence of other resonant frequencies not predicted by the cold plasma model. The dispersion equation (3.82) can also be satisfied by taking $\omega = n\omega_{ce}$, for $n \geq 2$, and arranging such that only the nth term contributes, which it will if $\omega/\omega_{ce} - n = O(\tilde{\nu}^{n-1})$. Hence, in the zero-temperature limit ($\tilde{\nu} \ll 1$), the hot plasma theory predicts resonant frequencies at each harmonic of the electron cyclotron frequency,

$$\omega = n\omega_{ce} \qquad n \geq 2 \quad (\text{for } \tilde{\nu} \ll 1) \tag{3.91}$$

These resonant frequencies are not predicted by the cold plasma model.

For the high-temperature case ($\tilde{\nu} \gg 1$), we have $e^{-\tilde{\nu}} \cdot I_n(\tilde{\nu}) = O(\tilde{\nu}^{-1/2})$ and it is found that the dispersion relation (3.82) is satisfied for

$$\omega = n\omega_{ce} \qquad n \geq 1 \quad (\text{for } \tilde{\nu} \gg 1) \tag{3.92}$$

Therefore, in the limit $\tilde{\nu} \gg 1$, the resonances occur at the fundamental, as well as at all the harmonics of the electron cyclotron frequency.

To obtain the resonant frequencies for intermediate values of $\tilde{\nu}$, (3.82) needs to be solved numerically. It is convenient, for numerical purposes, to rewrite (3.82) in the form

$$\tilde{\nu}\,\frac{\omega_{ce}^2}{\omega_{pe}^2} = F(\omega/\omega_{ce}, \tilde{\nu}); \qquad F(\omega/\omega_{ce}, \tilde{\nu}) = 2\,e^{-\tilde{\nu}} \sum_{n=1}^{\infty} \frac{n^2\, I_n(\tilde{\nu})}{(\omega/\omega_{ce})^2 - n^2} \tag{3.93}$$

In Fig. 4 it is plotted the function $F(\omega/\omega_{ce}, \tilde{\nu})$ in terms of ω/ω_{ce}, for $\tilde{\nu} = 0.1$. The intersection points of this curve with the horizontal line correponding to $\tilde{\nu}\,(\omega_{ce}^2/\omega_{pe}^2)$ give the resonant frequencies in the normalized form ω/ω_{ce}.

In Fig. 5 it is shown the normalized resonant frequency, ω/ω_{ce}, as a function of $(\tilde{\nu})^{1/2}$, for a specified value of $(\omega_{ce}/\omega_{pe})$. Note, from this figure, that below each resonant frequency curve corresponding to frequencies greater than the upper hybrid resonant frequency, there is a range ω in which resonance does not occur for any value of $\tilde{\nu}$. Also, for $\tilde{\nu} \ll 1$ it is verified, from Fig. 5, that the first harmonic of the electron cyclotron frequency is not a solution of the dispersion equation (3.82).

An important difference between the quasistatic waves treated here, and the longitudinal plasma waves analyzed previously, is the absence of Landau damping for the quasistatic waves. The treatment of propagation of quasistatic waves at an arbitrary direction with respect to $\vec{B}_0$ is left as an exercise for the student.

3.7 The TEM mode

From (3.78) and (3.72), the dispersion relation for the TEM mode propagating across the magnetostatic field in a hot plasma is given explicitly by

$$\frac{k^2 c^2}{\omega^2} = 1 - \frac{{}_{pe}}{\omega\,\omega_{ce}}\; e^{-\tilde{\nu}} \sum_{n=-\infty}^{+\infty} \frac{I_n(\tilde{\nu})}{(\omega/\omega_{ce} - n)} \tag{3.94}$$

This equation has to be analyzed numerically. However, some useful results can be obtained directly, without resorting to numerical work, for some special limiting cases.

For the limiting case $\tilde{\nu} \ll 1$, only the term corresponding to $n = 0$ is

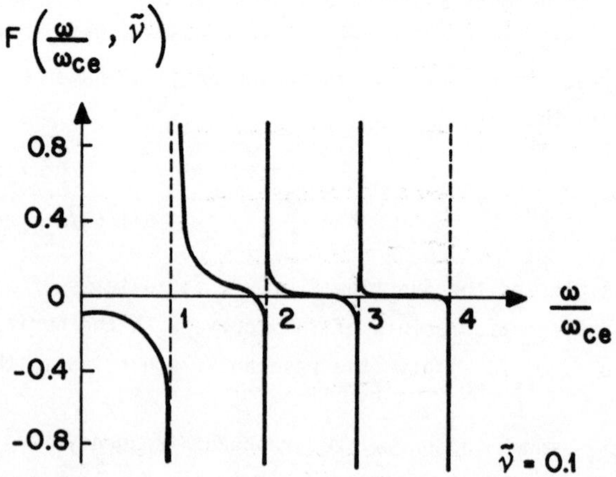

Fig. 4 - Dependence of the function $F(\omega/\omega_{ce}, \tilde{\nu})$, given by (3.93), in terms of ω/ω_{ce} for a fixed value of $\tilde{\nu}$ (here $\tilde{\nu} = 0.1$), for the quasistatic waves.

significant, while all other terms are small and can be neglected. Therefore, for $\tilde{\nu} \ll 1$, (3.94) simplifies to

$$k^2c^2/\omega^2 = 1 - \omega_{pe}^2/\omega^2 \tag{3.95}$$

where we have used the relation $I_0(0) = 1$. This result is the same as the dispersion relation for the TEM (ordinary) mode deduced from the cold plasma model. Thus we find that, in the limit of zero-temperature, the hot plasma theory agrees with the results of the cold plasma model, for the characteristics of the TEM mode propagating across the magnetostatic field.

For the limiting case $\tilde{\nu} \gg 1$, we have $e^{-\tilde{\nu}}I_n(\tilde{\nu}) = O(\tilde{\nu}^{-1/2})$ and (3.94) reduces to

$$kc/\omega = 1 \tag{3.96}$$

which is the dispersion relation for electromagnetic waves propagating in free space. Note that the condition $\tilde{\nu} \gg 1$, together with (3.54) and (3.96), is equivalent to

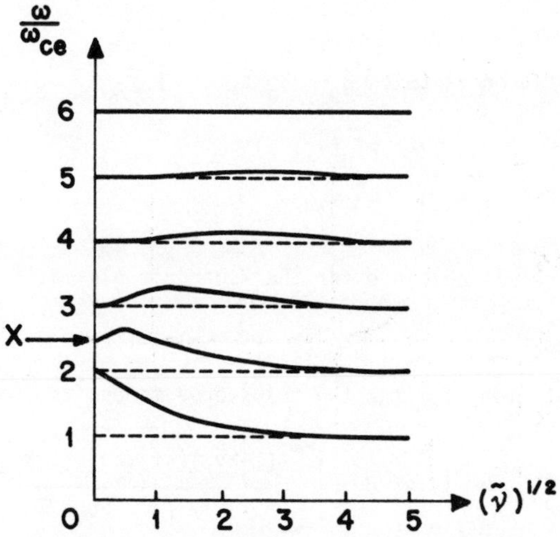

Fig. 5 - Curves of resonant frequencies for the quasistatic waves propagating across the magnetostatic field, as a function of $(\tilde{\nu})^{1/2}$, when $(\omega_{ce}/\omega_{pe})^2 = 0.2$. The resonant frequency, denoted by X, is the normalized upper hybrid frequency, $X = (\omega_{pe}^2 + \omega_{ce}^2)^{1/2}/\omega_{ce}$.

$$\omega \gg \omega_{ce} \left(\frac{m_e c^2}{k T_e}\right)^{1/2} \tag{3.97}$$

showing that the frequency must be very high. Hence, for $\tilde{\nu} \gg 1$, or equivalently, for very high frequencies, the results of the hot plasma theory are also in agreement with those predicted by the cold plasma model.

Furthermore, according to the hot plasma theory, the TEM mode has resonances at the electron cyclotron frequency and all its harmonics, since (3.94) shows that $kc/\omega = \infty$ for

$$\omega = n\,\omega_{ce}; \qquad n \geqslant 1 \tag{3.98}$$

where n is an integer. The cold plasma model does not predict the existence of these harmonic resonances.

4. SUMMARY

4.1 Propagation along B_0 in hot magnetoplasmas

(a) Longitudinal mode:

The dispersion relation is (with $\vec{B}_0 = B_0\,\hat{z}$; $\vec{k} = k\,\hat{z}$)

$$1 = -\frac{\omega_{pe}^2}{n_0\,\omega} \int_V \frac{v_\parallel}{(\omega - kv_\parallel)} \frac{\partial f_0(v_\perp, v_\parallel)}{\partial v_\parallel} d^3v \tag{2.66}$$

which is the same result obtained for the isotropic plasma.

(b) Transverse modes:

The dispersion relation, for the two transverse modes, is

$$k_\pm^2 c^2 - \omega^2 = \omega_{pe}^2 \frac{\pi}{n_0} \int_0^\infty v_\perp^2\, dv_\perp \int_{-\infty}^{+\infty} \frac{(\omega - kv_\parallel)(\partial f_o/\partial v_\perp) + kv_\perp(\partial f_0/\partial v_\parallel)}{(\omega - kv_\parallel \mp \omega_{ce})} dv_\parallel \tag{2.69}$$

The *upper* sign gives the right circularly polarized wave, and the *lower* sign gives the *left* circularly polarized wave. An alternative form for this dispersion relation is

$$k_\pm^2 c^2 - \omega^2 = -\frac{\omega_{pe}^2}{n_0}\int_v \left[\frac{(\omega - kv_{\shortparallel})}{(\omega - kv_{\shortparallel} \mp \omega_{ce})} + \frac{k^2 v_\perp^2/2}{(\omega - kv_{\shortparallel} \mp \omega_{ce})^2}\right] f_0(v_\perp, v_{\shortparallel})\, d^3v \tag{2.72}$$

When f_0 is the isotropic Maxwell-Bolzmann distribution function,

$$k_\pm^2 \; c^2 - \omega^2 = \omega_{pe}^2 \left[i\sqrt{\pi}\, \beta_\pm(-\alpha_\pm^2) - 2\beta_\pm \int_0^{\alpha_\pm} \exp\,(W^2 - \alpha_\pm^2)\, dW \right] \tag{2.82}$$

In the limits of cold and warm plasma models,

$$k_\pm^2 \; c^2 = \omega^2 - \omega_{pe}^2\, \omega/(\omega \mp \omega_{ce}) \tag{2.85}$$

The Landau (temporal) damping is negligible, since $v_{ph} \geq c$. But the right circularly polarized wave has temporal damping (cyclotron damping) for $\omega_r = \omega_{ce}$. The cyclotron damping constant (with $\omega = \omega_r + i\omega_i$), is

$$\omega_i = -\frac{1}{2} k_+ \left[\frac{(2k_B\, T_e/m_e)^{1/2}}{\pi^{1/2}} c^2 \left(\frac{\omega_{ce}}{\omega_{pe}}\right)^2 \right]^{1/3} \tag{2.93}$$

4.2 Propagation across B_0 in hot magnetoplasmas

(a) Transverse magnetic (TM) modes:

The longitudinal and transverse modes are coupled. The dispersion relation (with $\vec{B}_0 = B_0\, \hat{z}$; $\vec{k} = k\, \hat{x}$), when f_0 is the isotropic Maxwell-Boltzmann distribution function, is

$$k^2\; c^2/\omega^2 = (\varepsilon_{xx}\, \varepsilon_{yy} - \varepsilon_{xy}\, \varepsilon_{yx})/\varepsilon_{xx} \tag{3.76}$$

The TM mode corresponds to the extraordinary wave in magnetoionic theory (cold plasma).

In the limit $kc/\omega \to \infty$ (resonance condition),

$$\varepsilon_{xx} = 1 - \frac{\omega_{pe}}{\omega\, \omega_{ce}} \frac{e^{-\tilde{\nu}}}{\tilde{\nu}} \sum_{n=-\infty}^{+\infty} \frac{n^2\, I_n(\tilde{\nu})}{(\omega/\omega_{ce} - n)} = 0 \tag{3.79}$$

which is called the dispersion relation for the quasistatic mode (longitudinal mode; $E_x \neq 0$). In the limit $kc/\omega \to \infty$ the two TM modes are uncoupled, and the equation $\varepsilon_{xx} = 0$ applies to the longitudinal mode. The resonances are given by

$$\omega = (\omega_{pe}^2 + \omega_{ce}^2)^{1/2} \quad ; \text{ cold plasma result} \tag{3.90}$$

$$\omega = n\,\omega_{ce} \; ; \quad n \geq 2 \quad \text{for} \quad \tilde{\nu} \ll 1 \quad \text{Bernstein modes} \tag{3.91}$$

$$\omega = n\,\omega_{ce} \; ; \quad n \geq 1 \quad \text{for} \quad \tilde{\nu} \gg 1 \tag{3.92}$$

(b) Transverse eletric magnetic (TEM) mode:

It corresponds to the ordinary mode in magnetoinic theory (cold plasma). The dispersion relation, when f_0 is the Maxwell-Boltzmann distribution function, is

$$\frac{k^2 c^2}{\omega^2} = 1 - \frac{\omega_{pe}^2}{\omega\,\omega_{ce}}\, e^{-\tilde{\nu}} \sum_{n=-\infty}^{+\infty} \frac{I_n(\tilde{\nu})}{(\omega/\omega_{ce} - n)} \tag{3.94}$$

In the limit of cold plasma model ($\tilde{\nu} \ll 1$),

$$k^2 c^2 = \omega^2 - \omega_{pe}^2 \tag{3.95}$$

In hot plasma theory the resonances are given by

$$\omega = n\,\omega_{ce}; \qquad n \geq 1 \tag{3.98}$$

PROBLEMS

19.1 - Show that the first and second terms in the right-hand side of (2.16) represent, respectively, right and leftcircularly polarized wave fields.

19.2 - Derive expression (3.61) for σ_{zz}, starting from (3.60).

19.3 - Consider plane wave disturbances propagating along the magnetostatic field $\vec{B}_0$ in a hot electron gas, whose equilibrium distribution function is homogeneous and isotropic. In spherical coordinates in velocity space (v,Θ,ϕ) with $\vec{B}_0 = B_0\,\hat{z}$ and $\vec{k} \parallel \vec{B}_0$ (Fig. P 19.1), show that the linearized Vlasov equation reduces to

$$i\omega_{ce} \frac{df_1(\vec{v})}{d\phi} + (\omega - \vec{k}.\vec{v})\, f_1(\vec{v}) = \frac{ie}{m_e}\; \vec{E} \,.\, \vec{\nabla}_v\, f_0(v)$$

Verify that this differential equation has the formal solution

$$f_1(\vec{v}) = \frac{e}{m_e\,\omega_{ce}} \int_{-\infty}^{\phi} \vec{E} \,.\, \vec{\nabla}_{v'}\, f_0(v)\, \exp\left[\frac{i}{\omega_{ce}}\, (\omega - \vec{k} \,.\, \vec{v})(\phi - \phi')\right] d\phi'$$

where $\vec{v}'$ is the velocity vector with components $(v,\Theta,\ \phi')$. Note that $\vec{\nabla}_{v'}\, f_0(v) = (\vec{v}'/v)\; d\, f_0(v)/dv$. Perform the integral in this expression for $f_1(\vec{v})$ to obtain

$$f_1(\vec{v}) = -\frac{ie}{m_e\omega_{ce}} \left\{ \frac{1}{(A^2-1)} \left[E_x \left(A \frac{\partial f_0}{\partial v_x} - i \frac{\partial f_0}{\partial v_y}\right) + E_y\left(A \frac{\partial f_0}{\partial v_y} + i \frac{\partial f_0}{\partial v_x}\right)\right] + \right.$$

$$\left. + \frac{E_z}{A}\; \frac{\partial f_0}{\partial v_z} \right\}$$

where $A = -\,(\omega - \vec{k}.\vec{v})/\omega_{ce}$. From Maxwell equations obtain the relation

$$\vec{E} - \left(\frac{kc}{\omega}\right)^2 \vec{E}_t = \frac{ie}{\omega\varepsilon_0} \int_v \vec{v}\; f_1(\vec{v})\; d^3v$$

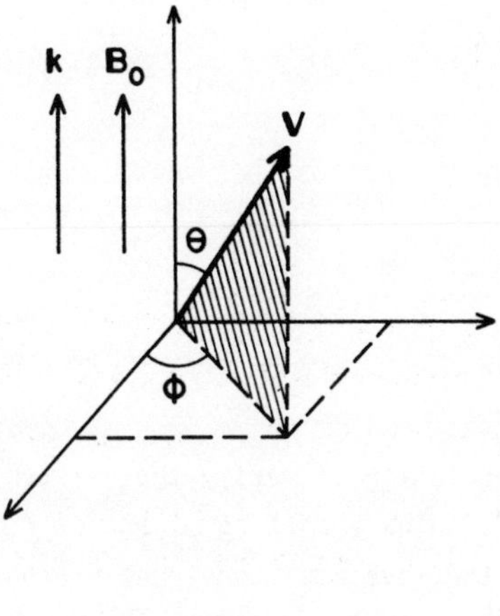

Fig. 19.1

where $\vec{E}_t = \vec{E} - E_z \hat{z}$ is the transverse part of the electric field $\vec{E}$. Using the expression for $f_1(\vec{v})$ in this equation, show that we obtain a dispersion relation with three wave solutions: the usual Landau damped longitudinal waves, and the left and the right circularly polarized waves (with $E_x = \pm i E_y$).

19.4 - An electron gas, immersed in a uniform magnetostatic field $\vec{B}_0$, is characterized by the following modified Maxwellian distribution function

$$f_0(v_\parallel, v_\perp) = n_0 \left(\frac{m_e}{2\pi k_B T_\parallel}\right)^{1/2} \left(\frac{m_e}{2\pi k_B T_\perp}\right) \exp\left(-\frac{m_e v_\parallel^2}{2 k_B T_\parallel} - \frac{m_e v_\perp^2}{2 k_B T_\perp}\right)$$

Use this distribution function in the dispersion relation for the *right circularly polarized transverse wave* propagating along $\vec{B}_0$ Eq.(2.69) ,

$$k^2c^2 = \omega^2 + \frac{\omega_{pe}^2 \pi}{n_0} \int_0^\infty v_\perp^2 dv_\perp \int_{-\infty}^{+\infty} \frac{(\omega - kv_\parallel)(\partial f_0/\partial v_\perp) + kv_\perp(\partial f_0/\partial v_\parallel)}{(\omega - kv_\parallel - \omega_{ce})} dv_\parallel$$

and evaluate the integrals to obtain the following dispersion relation

$$k^2c^2 = \omega^2 - \tau\omega_{pe}^2 - i\sqrt{\pi}\ \omega_{pe}^2\ \tau(\alpha - \beta) \exp(-\alpha^2) +$$

$$+ 2\ \omega_{pe}^2\ \tau(\alpha - \beta) \int_0^\alpha \exp(W^2 - \alpha^2)\, dW$$

where

$$\tau = 1 - (T_\perp/T_\parallel)$$

$$\alpha = \frac{(\omega - \omega_{ce})/k}{(2 k_B T_\parallel/m_e)^{1/2}}$$

$$\beta = \frac{\omega/(k\tau)}{(2 k_B T_\parallel/m_e)^{1/2}}$$

Analyze this dispersion relation to verify the existence or not of instabilities (positive imaginary part of ω) and/or damping (negative imaginary part of ω) of the wave amplitude considering the propagation coefficient $\vec{k} = k\hat{z}$ to be real. Determine the cyclotron damping coefficient. Analyze the results considering the isotropic case for which $T_\perp = T_\parallel$.

19.5 - In Problem 19.4 suppose that in the equilibrium state the velocity distribution function of the electrons is given by

$$f_0(\vec{v}) = n_0 \left(\frac{m_e}{2\pi k_B T_e}\right)^{3/2} \exp\left\{ - \frac{m_e}{2 k_B T_e} \left[v_\perp^2 + (v_\parallel - u_0)^2 \right] \right\}$$

which corresponds to an isotropic distribution but with the electrons drifting with speed u_0 along $\vec{B}_0$. Show that, with this choice of $f_0(\vec{v})$, the dispersion relation for the right circularly polarized wave reduces to

$$k^2 c^2 = \omega^2 - \frac{\omega_{pe}^2}{n_0} \int_v \frac{(\omega - k u_0)}{\omega - k v_\parallel - \omega_{ce}} f_0(\vec{v})\, d^3v$$

For the limiting case of $T_e = 0$, find the form of the distribution function $f_0(\vec{v})$ and show that the dispersion relation becomes

$$k^2 c^2 = \omega^2 - \frac{\omega_{pe}^2 (\omega - k\, u_0)}{(\omega - k\, u_0 - \omega_{ce})}$$

19.6 - For an unbounded homogeneous electron gas, characterized by the following velocity distribution function

$$f_0(\vec{v}) = n_0\, (a_0/\pi^2)\, (v^2 + a_0^2)^{-2}$$

where a_0 is a constant, show that the dispersion relation for the right circularly polarized wave, propagating along the magnetostatic field $\vec{B}_0 = B_0\, \hat{z}$, is given by

$$k^2 c^2 = \omega^2 - \frac{\omega_{pe}^2\ \omega}{\omega + i\, k\, a_0 - \omega_{ce}}$$

From this result show that the cyclotron damping coefficient is given approximately by

$$\omega_i = - (k/2) \left[a_0 (c\, \omega_{ce}/\omega_{pe})^2 \right]^{1/3}$$

19.7 - (a) Show that starting from the Vlasov equation and the laws of electrostatics it is obtained the following dispersion relation for the *quasistatic wave* propagating at an arbitrary direction with respect to a magnetostatic field $\vec{B}_0$ in a hot plasma:

$$\frac{\omega_{ce}^2 \ \tilde{\nu}}{\omega_{pe}^2 \sin^2 \Theta} = \frac{k_B T_e}{m_e} \left(\frac{k}{\omega_{pe}}\right)^2 = - \exp(-\tilde{\nu}) \sum_{n=-\infty}^{+\infty} (1 + \tilde{\nu} H_n) I_n(\tilde{\nu})$$

where

$$\tilde{\nu} = \frac{k_B T_e}{m_e} \frac{k^2 \sin^2 \Theta}{\omega_{ce}^2}$$

$$H_n = \frac{1}{\sqrt{\pi}} \int_{-\infty}^{+\infty} d\tilde{v}_z \frac{\exp(-\tilde{v}_z^2)}{(\tilde{v}_z + n\tilde{\omega}_{ce} - \tilde{\omega})}$$

$$\tilde{v}_z = \frac{v_z}{V_e} \ ; \qquad V_e = \left(\frac{2 k_B T_e}{m_e}\right)^{1/2}$$

$$\tilde{\omega} = \omega/(k \cos \Theta \ V_e)$$

$$\tilde{\omega}_{ce} = \omega_{ce}/(k \cos \Theta V_e)$$

$I_n(\tilde{\nu})$ is the Bessel function of the second kind, and Θ is the angle between $\vec{k}$ and $\vec{B}_o$, as indicated in Fig. P.19.2.

(b) Rewrite this dispersion relation in the form

$$1 + \frac{k_B T_e}{m_e} \left(\frac{k}{\omega_{pe}}\right)^2 = - i\omega \int_0^{\infty} dt \exp \Big\{ i\omega t -$$

$$- [1 - \cos(\omega_{ce} t)] \ k^2 V_e^2 \sin^2 \Theta/2\omega_{ce}^2 - k^2 V_e^2 t^2 \cos^2 \Theta/4 \Big\}$$

(c) Simplify this expression for the case of a very weak magnetostatic field to obtain the following approximate expression for the frequency of oscillation

$$\omega^2 = \omega_{pe}^2 + (3k_B T_e/m_e) k^2 + \omega_{ce}^2 \sin^2 \Theta$$

Compare this result with the cold and the warm plasma results for both the cases of $\vec{k} \parallel \vec{B}_o$ and $\vec{k} \perp \vec{B}_o$.

19.8 - Deduce the dispersion relation for small amplitude waves propagating at an *arbitrary direction* with respect to an externally applied magnetostatic field $\vec{B}_0 = B_0 \hat{z}$ in a hot plasma. Carry through the derivation as far as possible for an arbitrary value of the strength of the magnetostatic field. Then, particularize for the special case of a very weak magnetostatic field. For simplicity, assume the equilibrium distribution function to be the isotropic Maxwell-Boltzmann distribution (you may refer to the article "Waves in a Plasma in a Magnetic Field, by Ira B. Bernstein, *Physical Review*, 109(1), 10-21, 1958).

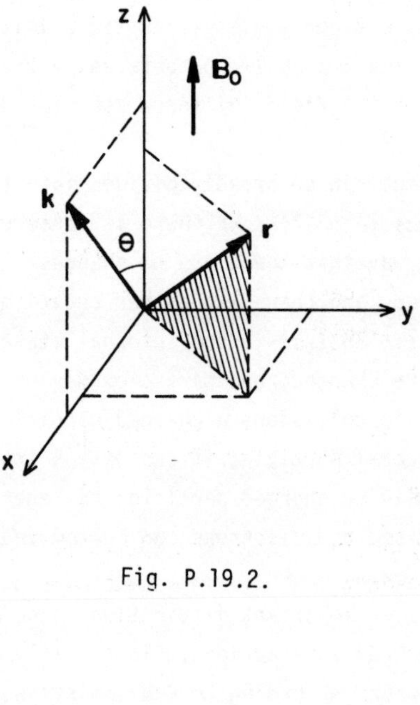

Fig. P.19.2.

CHAPTER 20

Particle Interactions in Plasmas

1. INTRODUCTION

The fundamental properties of a plasma depend upon the interactions of the plasma particles with the force fields existing inside it. These fields may be externally applied, or they can be internal fields associated with the nature and motion of the particles themselves. In this text the words *collision* and *interaction* are used synonymously. The notion of a collision as a physical contact between bodies loses its utility in the microscopic world. In the atomic level a *collision* between particles must be regarded as an *interaction* between the fields of force associated with each of the interacting particles.

Collisional phenomena can be broadly divided into two categories: *elastic* and *inelastic*. In *elastic collisions* there is conservation of mass, momentum and energy in such a way that there are no changes in the internal states of the particles involved, and there is neither creation nor annihilation of particles. In *inelastic collisions* the internal states of some or all of the particles involved are changed, and particles may be created, as well as destroyed. In inelastic collisions a charged particle may *recombine* with another to form a neutral particle; it can *attach* itself with a neutral particle to form a heavier charged particle; the energy state of an electron in an atom may be raised and electrons can be removed from their atoms resulting in *ionization*.

In plasmas there is an important distinction to be made between charge-charge and charge-neutral interactions. Electrically charged particles interact with one another according to Coulomb's law. This Coulomb interaction, in view of its $1/r^2$ dependence, is a *long-range interaction*, so that the field of one particle interacts simultaneously with a large number of other particles. Therefore, it involves *multiple* interactions. In contrast, the fields associated with neutral particles are significantly strong only within the electronic shells of the particles. Thus, they are *short-range fields* and a neutral particle only occasionally interacts with another particle, and very rarely it interacts simultaneously with more than one particle. Therefore, these short-range fields result primarily in *binary* interactions.

The multiple particle Coulomb interaction, however, can be thought of as a number of simultaneous binary interactions. In fact, one way of dealing with multiple interactions is to consider that a series of consecutive small-angle binary interactions describes the situation. The multiple interactions, which result from the Coulomb force, are of essential importance in understanding the behavior of plasmas and underlines the validity of describing a plasma as the fourth state of matter. Nevertheless, binary collisions adequately describe plasma phenomena in the case of *weakly ionized plasmas*. In fact, we use the term weakly ionized plasma to mean a plasma in which multiple particle interactions can be ignored. In these plasmas the electrons tend to dominate the situation, since they respond quickly to the influence of electric and magnetic fields, in view of their low inertia.

In this chapter we deal with the collisional processes that are of importance in plasmas, form the point of view of classical dynamics. The results are valid to a good approximation, even though the internal structure of the particles is ignored. More important, however, the procedures to be developed are useful whether the mechanics is classical or quantum.

2. BINARY COLLISIONS

Consider an elastic collision between two particles of mass m and m_1, having velocities $\vec{v}$ and $\vec{v}_1$ before collision, and $\vec{v}'$ and $\vec{v}_1'$ after collision. This binary interaction is illustrated in Fig. 1, as seen from the laboratory system. In what follows, the variables indicated with a *prime* are *after-collision* variables.

It is convenient to adopt a coordinate system in which the particle having mass m is at rest, and the particle having mass m_1 approaches with the *relative velocity*

$$\vec{g} = \vec{v}_1 - \vec{v} \tag{2.1}$$

After collision, the relative velocity is

$$\vec{g}' = \vec{v}_1' - \vec{v}' \tag{2.2}$$

The geometry of the interaction is shown in Fig. 2. The *impact parameter*, defined as the minimum distance of approach if there were no interaction, is denoted by b, the *scattering angle* by χ, and the orientation of the *orbital plane* (or *collision plane*), with respect to some given direction in a plane normal to the orbital plane, is denoted by ε.

The velocity of the *center of mass* of the colliding particles, before collision, is defined by

$$\vec{c}_0 = (m\vec{v} + m_1\,\vec{v}_1)/(m + m_1) \tag{2.3}$$

and, after collision, by

$$\vec{c}_0' = (m\vec{v}' + m_1\,\vec{v}_1')/(m + m_1) \tag{2.4}$$

We can express the initial velocities in terms of $\vec{c}_0$ and $\vec{g}$. From (2.3) and (2.1), which define $\vec{c}_0$ and $\vec{g}$, respectively, we find

$$\vec{v} = \vec{c}_0 - (\mu/m)\,\vec{g} \tag{2.5}$$

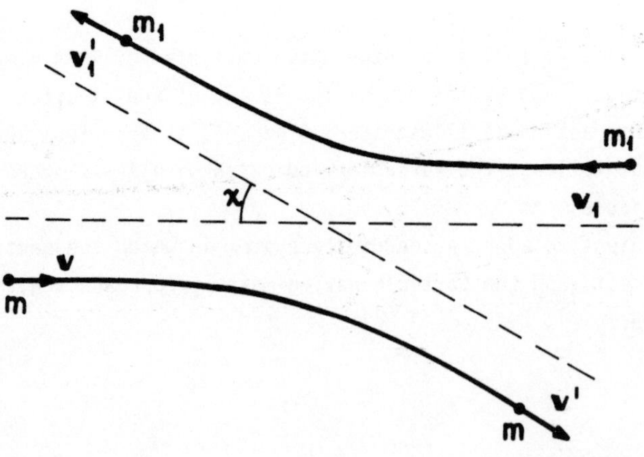

Fig. 1 - Binary interaction of two particles of masses m and m_1, with velocities $\vec{v}$ and $\vec{v}_1$, respectively, as viewed from the laboratory system.

$$\vec{v}_1 = \vec{c}_0 + (\mu/m_1)\,\vec{g} \tag{2.6}$$

where μ denotes the reduced mass, defined by

$$\mu = m\,m_1/(m + m_1) \tag{2.7}$$

Similarly, from (2.4) and (2.2) we obtain, for the final velocities,

$$\vec{v}\,' = \vec{c}_0' - (\mu/m)\,\vec{g}\,' \tag{2.8}$$

$$\vec{v}_1' = \vec{c}_0' + (\mu/m_1)\,\vec{g}\,' \tag{2.9}$$

From the law of *conservation of momentum* for the collision event, we have

$$m\,\vec{v} + m_1\,\vec{v}_1 = m\vec{v}\,' + m_1\,\vec{v}_1\,' \tag{2.10}$$

or, using (2.3) and (2.4),

$$(m + m_1)\,\vec{c}_0 = (m + m_1)\,\vec{c}_0' \tag{2.11}$$

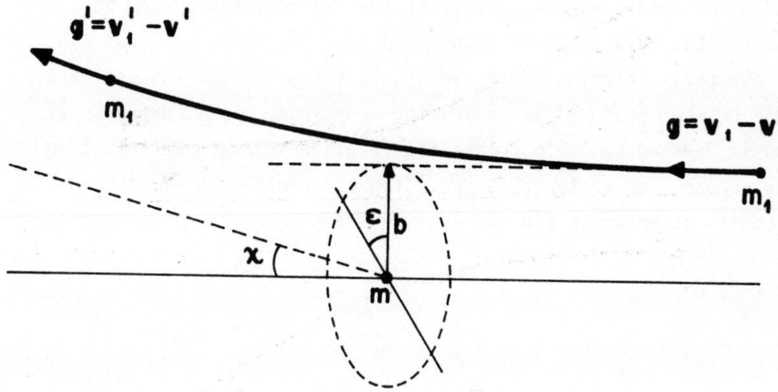

Fig. 2 - Geometry of a collision between a particle of mass m and velocity $\vec{v}$, and a particle of mass m_1 and velocity $\vec{v}_1$, viewed from a coordinate system in which the first particle is at rest.

Thus,

$$\vec{c}_0 = \vec{c}_0' \tag{2.12}$$

that is, the velocity of the center of mass is the same before and after the interaction event.

From the law of *conservation of energy* for elastic collisions, we have

$$(mv^2 + m_1 v_1^2)/2 = \left[m (v')^2 + m_1 (v_1')^2 \right] /2 \tag{2.13}$$

and using (2.5), (2.6), (2.8) and (2.9) we find, by direct calculation,

$$(mv^2 + m_1 v_1^2)/2 = (m + m_1) c_0^2/2 + \mu g^2/2 \tag{2.14}$$

$$\left[m (v')^2 + m_1 (v_1')^2 \right] /2 = (m + m_1) (c_0')^2/2 + \mu (g')^2/2 \tag{2.15}$$

Now, since $\vec{c}_0 = \vec{c}_0'$, we conclude that

$$g = g' \tag{2.16}$$

Thus, the *magnitude*, but not the direction of the relative velocity is conserved in a binary elastic collision. Eqs. (2.14) and (2.15) show that the total instantaneous kinetic energy of the two-particle system is equivalent to that associated with the motion of the center of mass plus the motion of one particle relative to the other, but using the reduced mass.

The angle between $\vec{g}$ and $\vec{g}'$ is called the *scattering angle*, or *deflection angle*, and is denoted by χ. To relate the relative velocity vectors $\vec{g}$ and $\vec{g}'$, we can choose, for instance, a Cartesian coordinate system with the z axis along $\vec{g}$, as shown in Fig. 3. Thus, we have

$$g_x = g_y = 0 \tag{2.17}$$

$$g_z = g = g' \tag{2.18}$$

$$g_x' = g \sin (\chi) \cos (\varepsilon) \tag{2.19}$$

$$g_y' = g \sin (\chi) \sin (\varepsilon) \tag{2.20}$$

$$g'_z = g \cos(\chi) \tag{2.21}$$

where ε defines the relative orientation of the *collision plane*. Therefore, knowing the initial velocites and the scattering angle χ we can determine the after-collision velocities. The opposite is also true, that is, if we know the final velocities and the scattering angle we can find the initial velocities.

It is of interest to consider the *inverse collision* (see Fig. 1, Chapter 7), in which a particle with initial velocity $\vec{v}'$ collides with another particle having initial velocity $\vec{v}_1'$, the velocities *after* collision being $\vec{v}$ and $\vec{v}_1$, respectively. For the inverse collision the scattering angle χ is the same as that for the direct collision, since the impact parameter b, the interparticle force law and the relative speed $\vec{g}$ are all the same.

The scattering angle is the only quantity appearing in the analysis presented in this section that depends on the details of the collision process. For interparticle force laws which depend only on the distance between the interacting particles, χ depends on the following quantities:

(a) Interparticle force law;

(b) Magnitude of the relative velocity g;

(c) The value of the impact parameter b.

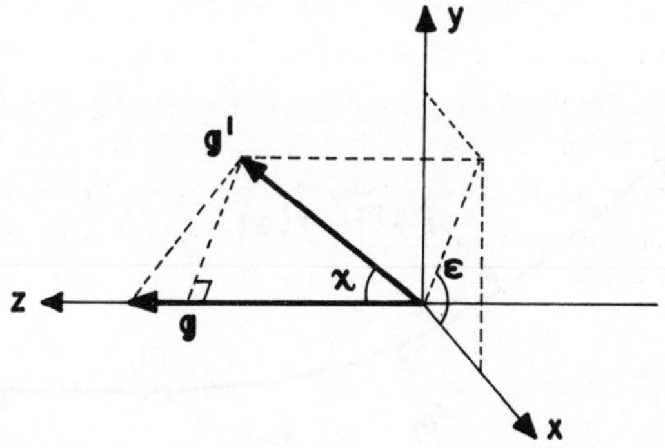

Fig. 3 - Relationship between the relative velocities $\vec{g}$ and $\vec{g}'$, in a Cartesian coordinate system for which $\vec{g} = g\,\hat{z}$. The angle ε defines the relative orientation of the plane containing the particle trajectory.

Therefore, in order to determine χ we must analyze the classical dynamics of binary collisions.

3. DYNAMICS OF BINARY COLLISIONS

The dynamics of a binary collision is governed by the interparticle force law. For each impact parameter b there will be associated a given scattering angle χ, the relation being dependent on the interparticle force law. This information is contained in the *differential cross section*, which is defined in section 5.

Consider the collision of two particles of masses m and m_1, viewed from a system of reference in which the first particle is at rest. Let $\vec{r}$ be the position vector of the particle of mass m_1 with respect to that of mass m (Fig. 4). The force of interaction between the two particles is assumed to be a central force, which acts along the straight line joining the two particles, that is,

$$\vec{F}(\vec{r}) = F(r)\,\hat{r} \tag{3.1}$$

This force is related to the potential energy $U(\vec{r})$ of the interaction by the condition

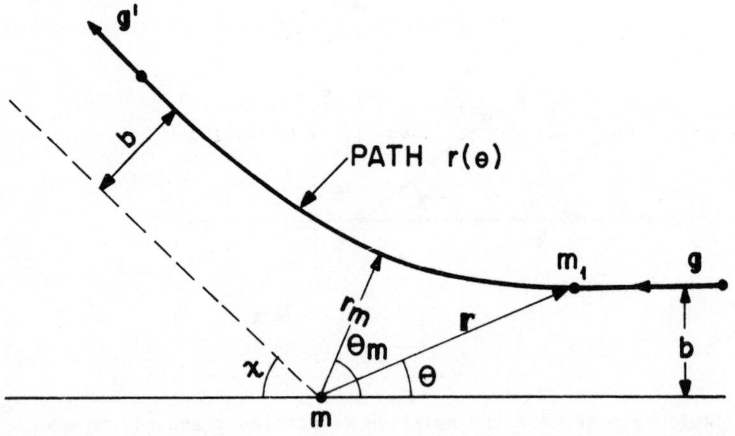

Fig. 4 - Path $r(\theta)$ of the particle of mass m_1, relative to the particle of mass m.

$$\vec{F}(\vec{r}) = -\ \vec{\nabla}U(\vec{r}) = \ -\ \hat{r}\partial U(r)/\partial r \tag{3.2}$$

For a central force the torque $\vec{N} = \vec{r} \times \vec{F}(\vec{r})$ vanishes, because $\vec{F}(\vec{r})$ is parallel to $\vec{r}$. Since the torque is the time rate of change of the angular momentum $\vec{L}$,

$$\vec{N} = d\vec{L}/dt \tag{3.3}$$

we conclude that the angular momentum is a constant of the motion. Furthermore, since $\vec{L} = \vec{r} \times \vec{p}$, we see that $\vec{r}$ is always normal to the constant direction of $\vec{L}$ in space, and the motion lies therefore in a plane.

Using polar coordinates (r, θ) and noting that the unit vectors $\hat{r}$ and $\hat{\theta}$ depend on θ (Fig. 5), we have for the instantaneous relative velocity,

$$\frac{d\vec{r}}{dt} = \frac{dr}{dt}\hat{r} + r\frac{d\hat{r}}{dt} = \frac{dr}{dt}\hat{r} + r\frac{d\hat{r}}{d\theta}\frac{d\theta}{dt} \tag{3.4}$$

Since it can be show that $d\hat{r}/d\theta = \hat{\theta}$, we obtain

$$d\vec{r}/dt = (dr/dt)\ \hat{r} + r(d\theta/dt)\ \hat{\theta} \tag{3.5}$$

or, using a dot over the variable to denote the time derivative,

$$\dot{\vec{r}} = \dot{r}\ \hat{r} + r\ \dot{\theta}\ \hat{\theta} \tag{3.6}$$

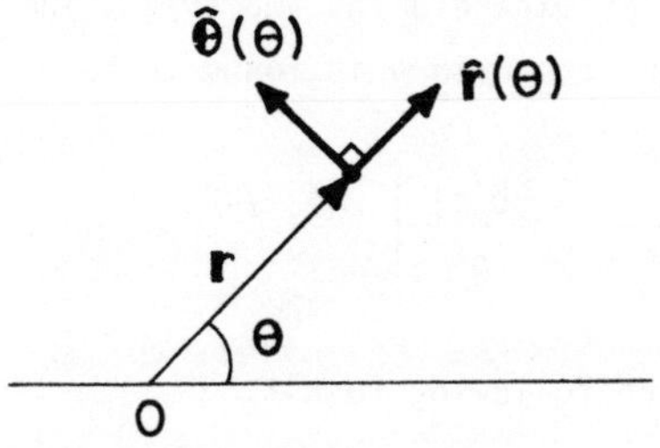

Fig. 5 - Polar coordinates (r, θ), showing that the directions of the unit vectors $\hat{r}$ and $\hat{\theta}$ depend on θ.

The trajectory of the particle in the equivalent one body problem can be easily found by using the laws of conservation of *energy* and *angular momentum*. The kinetic energy of the relative motion is given by

$$K = \mu \dot{\vec{r}} \cdot \dot{\vec{r}}/2 = \mu (\dot{r}^2 + r^2 \dot{\theta}^2)/2 \tag{3.7}$$

where μ is the reduced mass, defined by $m\, m_1/(m + m_1)$. From the law of conservation of energy we can equate the kinetic plus potential energy, at any point, to the initial kinetic energy ($\mu g^2/2$), since the initial potential energy is zero. Thus, we have

$$\mu (\dot{r}^2 + r^2 \dot{\theta}^2)/2 + (U(r)) = \mu g^2/2 \tag{3.8}$$

The angular momentum relative to the origin is given by

$$\vec{L} = \vec{r} \times (\mu \dot{\vec{r}}) = \mu r^2 \dot{\theta} (\hat{r} \times \hat{\theta}) \tag{3.9}$$

Setting the angular momentum, at any point, equal to its initial value, $b\mu g (\hat{r} \times \hat{\theta})$, we get

$$r^2 \dot{\theta} = bg \tag{3.10}$$

From (3.8) and (3.1) we can easily obtain a differential equation for the orbit $r(\theta)$. First, we write

$$dr/dt = (dr/d\theta)(d\theta/dt) \tag{3.11}$$

and use (3.10) and (3.8) to eliminate $d\theta/dt$ and dr/dt. The resulting differential equation for the trajectory is found to be

$$\left(\frac{dr}{d\theta}\right)^2 = \frac{r^4}{b^2}\left[1 - \frac{b^2}{r^2} - \frac{2U(r)}{\mu g^2}\right] \tag{3.12}$$

Rearranging (3.12) yields the following result

$$d\theta = \pm \frac{b}{r^2}\left[1 - \frac{b^2}{r^2} - \frac{2U(r)}{\mu g^2}\right]^{-1/2} dr \tag{3.13}$$

The choice of sign must be made on physical grounds. The coordinates of the position of the particle, when it is at the distance of closest approach, are denoted by r_m and Θ_m (Fig. 4). This position is called the *vertex* of the trajectory, and the line connecting the origin to the vertex is called the *apse line*. Thus, Θ_m specifies the orientation of the apse line. The plus sign in (3.13) must be used when Θ is greater than Θ_m, since for $\Theta > \Theta_m$ we see that r increases with Θ. On the other hand, for $\Theta < \Theta_m$ we see that r decreases as Θ increases, so that the minus sign in (3.13) must be used when Θ is less than Θ_m. This also shows that the trajectory is *symmetrical* about the apse line.

The *distance of closest approach* r_m can be obtained from (3.12) by noting that $dr/d\Theta = 0$ when $r = r_m$. Thus, we have

$$1 - b^2/r_m^2 - 2U(r_m)/\mu g^2 = 0 \tag{3.14}$$

or

$$r_m = b\,[1 - 2U(r_m)/\mu g^2]^{-1/2} \tag{3.15}$$

To compute the *scattering angle* χ, we first note, from Fig. 4, that

$$\chi = \pi - 2\Theta_m \tag{3.16}$$

In order to determine Θ_m we integrate (3.13) from Θ_m to some other angle Θ, obtaining

$$\Theta - \Theta_m = \pm \int_{r_m}^{r} \frac{b}{r'^2}\left[1 - \frac{b^2}{r'^2} - \frac{2U(r')}{\mu g^2}\right]^{-1/2} dr' \tag{3.17}$$

where the plus sign is to be used when $\Theta > \Theta_m$ and the minus sign when $\Theta < \Theta_m$. When $r \to \infty$ we have $\Theta_{(-)} \to 0$, while $\Theta_{(+)} \to 2\Theta_m$, so that (3.17) gives, for the orientation of the apse line,

$$\Theta_m = \int_{r_m}^{\infty} \frac{b}{r^2}\left[1 - \frac{b^2}{r^2} - \frac{2U(r)}{\mu g^2}\right]^{-1/2} dr \tag{3.18}$$

The scattering angle is therefore given by

$$\chi\ (b,\ g) = \pi - 2 \int_{r_m}^{\infty} \frac{b}{r^2} \left[1 - \frac{b^2}{r^2} - \frac{2U(r)}{\mu g^2} \right]^{-1/2} dr \tag{3.19}$$

To compute χ from this equation we must know the impact parameter b, the magnitude of the initial relative velocity g, and the interparticle potential energy U(r).

4. EVALUATION OF χ FOR SOME SPECIAL CASES

In this section we present two examples of the use of (3.19) to determine the scattering angle in terms of the impact parameter b, and the initial relative velocity g. First, we consider the collision between two perfectly elastic hard spheres and afterwards the case of the Coulomb interaction.

4.1 Two perfectly elastic hard spheres

Consider the collision between two perfectly elastic hard spheres of radii R_1 and R_2 (Fig. 6). The potential energy of interaction is given by

$$U(r) = \begin{cases} 0 \text{ for } r > R_1 + R_2 \\ \infty \text{ for } r < R_1 + R_2 \end{cases} \tag{4.1}$$

For $b > R_1 + R_2$ there is no interaction and we must have $r_m = b$, whereas for $b < R_1 + R_2$ the particles collide and we have $r_m = R_1 + R_2$. In either case, however, since the spheres are impenetrable, we have $r > R_1 + R_2$, so that (3.19) becomes

$$\chi = \pi - 2 \int_{r_m}^{\infty} \frac{b}{r^2} \left(1 - \frac{b^2}{r^2}\right)^{-1/2} dr \tag{4.2}$$

To solve this integral it is convenient to define a new variable by $y = b/r$ and write Eq. (4.2) in the form

$$\chi = \pi - 2 \int_0^{b/r_m} (1 - y^2)^{-1/2} dy \tag{4.3}$$

which gives

$$\chi = \pi - 2 \sin^{-1} (b/r_m) \tag{4.4}$$

Therefore, we find that

$$\chi = \begin{cases} \pi - 2 \sin^{-1} [b/(R_1 + R_2)] & \text{for } b \leqslant R_1 + R_2 \\ 0 & \text{for } b \geqslant R_1 + R_2 \end{cases} \tag{4.5}$$

4.2 Coulomb interaction potential

Let us consider now the important case of the Coulomb potential field, whose potential energy of interaction is given by

$$U(r) = \frac{1}{4\pi\varepsilon_0} \frac{q\, q_1}{r} \tag{4.6}$$

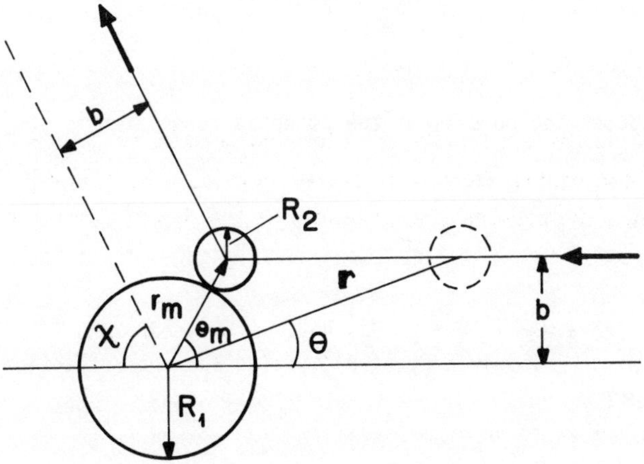

Fig. 6 - Collision between two perfectly elastic impenetrable spheres.

where q and q_1 denote the electric charge of the particles of mass m and m_1, respectively. Substituting (4.6) into (3.19), gives

$$\chi (b, g) = \pi - 2 \int_{r_m}^{\infty} \frac{b}{r^2} \left(1 - \frac{b^2}{r^2} - \frac{2 q q_1}{\pi g^2 4\pi\varepsilon_0 r}\right)^{-1/2} dr \qquad (4.7)$$

The distance of closest approach r_m is obtained from (3.15) and (4.6), and is found to be

$$r_m = b^2/[-b_0 + (b_0^2 + b^2)^{1/2}] \qquad (4.8)$$

where, for convenience, we have introduced the notation

$$b_0 = q\, q_1/(4\pi\varepsilon_0\, \mu g^2) \qquad (4.9)$$

Thus, b_0 represents the distance at which the electric potential energy of interaction is twice the relative kinetic energy at infinity. Making the change of variable $y = 1/r$ and inserting the value for b_0, given by (4.9), into (4.7), gives for the deflection angle

$$\chi(b,g) = \pi - 2b \int_0^{1/r_m} (1 - 2b_0 y - b^2 y^2)^{-1/2}\, dy \qquad (4.10)$$

The integral appearing here is of the standard form

$$\int (\alpha + \beta x + \gamma x^2)^{-1/2}\, dx = \frac{1}{\sqrt{-\gamma}} \sin^{-1} \left[\frac{(-2\gamma x - \beta)}{(\beta^2 - 4\alpha\gamma)^{1/2}}\right] \qquad (4.11)$$

where, in our case, $\alpha = 1$, $\beta = -2b_0$, and $\gamma = -b^2$. Applying the limits of integration, with r_m as given by (4.8), yields for the deflection angle

$$\chi (b, g) = 2 \sin^{-1} [b_0/(b_0^2 + b^2)^{1/2}] \qquad (4.12)$$

This equation for χ (b, g) can be written in the alternative form

$$\tan(\chi/2) = b_0/b \tag{4.13}$$

Note that for $\chi = \pi/2$ we have $b = b_0$, that is, b_0 is the value of the impact parameter for a 90^0 deflection angle. If the sign of the two charged particles are the same, then b_0 and χ will both be positive. On the other hand, if the sign of the two charged particles are opposite, then b_0 and χ will be negative. The two situations are illustrated in Fig. 7 for a deflection angle of 90^0. It is also noted, from (4.13), that $\chi = \pi$ for $b = 0$,

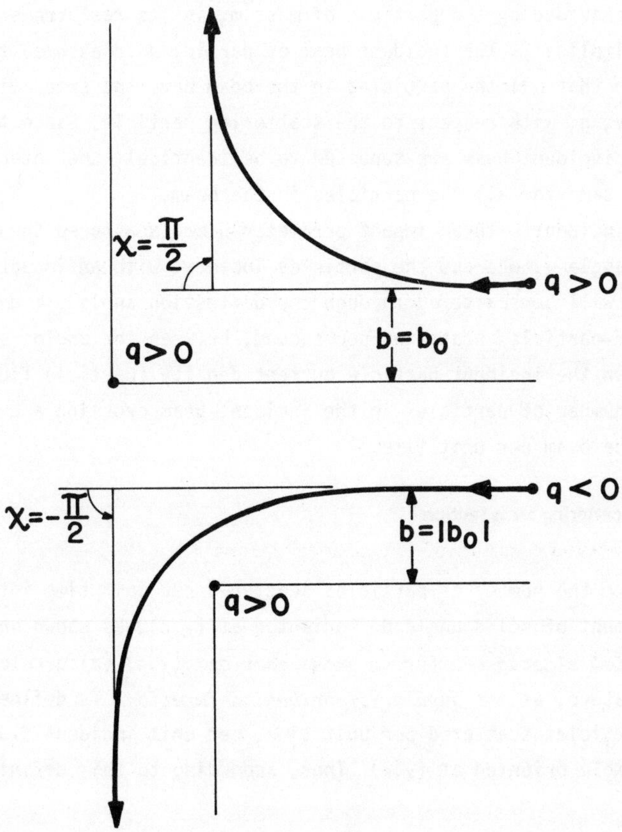

Fig. 7 - Scattering in a Coulomb potential field for a deflection angle, χ, of 90^0.

χ decreases as b increases, and $\chi = 0$ only in the limit of $b \to \infty$. Therefore, scattering occurs for all (finite) values of the impact parameter b, so that there is no cut-off value for b.

5. CROSS SECTIONS

So far we have considered specifically only the interaction between two particles. Cross sections are usually defined in terms of a beam of identical particles incident on a center of force (target particle). Therefore, let us imagine a steady beam of identical particles of mass m_1, uniformly spread out in space, incident with velocity $\vec{g} = \vec{v}_1 - \vec{v}$ upon the center of force provided by the particle of mass m, in its rest frame of reference. For simplicity, the incident beam of particles is assumed to be monoenergetic, so that all the particles in the beam have the same initial relative velocity, $\vec{g}$, with respect to the scattering particle. Since the particles in the incident beam are supposed to be identical, the interaction potential is the same for all the particles in the beam.

The particles incident with an impact parameter b are scattered through some deflection angle χ, whereas the particles incident with an impact parameter b + db will be scattered through the deflection angle $\chi + d\chi$ (Fig. 8). The number of particles scattered per second, between the angles χ and $\chi + d\chi$, depends on the incident particle current density (particle flux) $\vec{\Gamma}$, that is, on the number of particles in the incident beam crossing a unit area normal to the beam per unit time.

5.1 Differential scattering cross section

Let dN/dt denote the number of particles scattered per unit time into the differential element of solid angle $d\Omega$, oriented at (χ,ε), as shown in Fig. 8. The *differential elastic scattering cross section* $\sigma(\chi,\varepsilon)$ (also referred to, in the literature, as the *angular distribution function*) is defined as the number of particles scattered per unit time, per unit incident flux and per unit solid angle oriented at (χ,ε). Thus, according to this definition we have

$$dN/dt = \sigma(\chi,\varepsilon)\ \Gamma\ d\Omega \qquad (5.1)$$

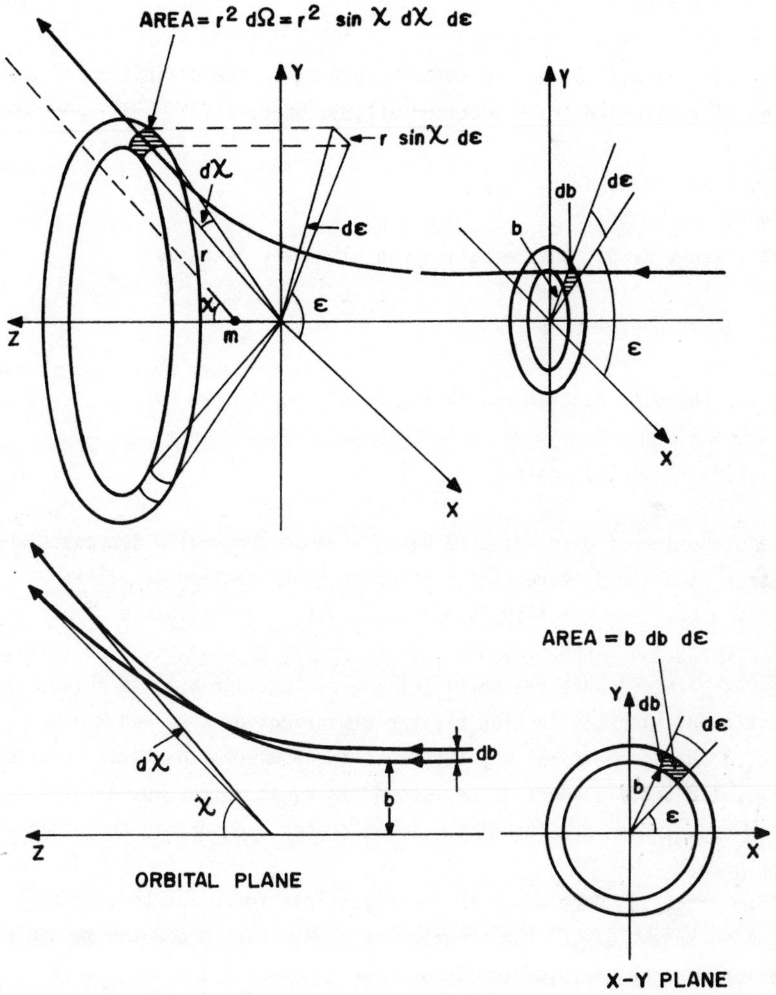

Fig. 8 - Scattering in a central field of force.

The number of particles incident per unit time, with impact parameter between b and b + db, and with the orbital plane oriented between ε and $\varepsilon + d\varepsilon$, is Γ bdb $d\varepsilon$. These same particles are scattered per unit time into the differential element of solid angle $d\Omega$ contained between χ and $\chi + d\chi$ and between ε and $\varepsilon + d\varepsilon$. Thus, we have

$$dN/dt = \Gamma\, b\, db\, d\varepsilon \tag{5.2}$$

Comparing (5.1) and (5.2) we see that, according to the definition of the differential scattering cross section $\sigma\,(\chi,\varepsilon)$, we have

$$\sigma(\chi,\varepsilon)\, d\Omega = b\, db\, d\varepsilon \tag{5.3}$$

Since $d\Omega = \sin\chi\, d\chi\, d\varepsilon$, this equation can also be written as

$$\sigma(\chi,\varepsilon)\, \sin\chi\, d\chi = b\, db \tag{5.4}$$

Solving for the differential scattering cross section, we obtain

$$\sigma(\chi,\varepsilon) = (b/\sin\chi)\, |db/d\chi| \tag{5.5}$$

The absolute value of $db/d\chi$ is used here, because χ normally decreases when b increases, and the differential scattering cross section $\sigma(\chi,\varepsilon)$ is inherently a positive quantity, since it is associated with the number of particles being scattered. The quantity $db/d\chi$ can be obtained from (3.19), which gives χ (b, g) once the potential energy function $\phi(r)$ is known.

The differential cross section has the dimensions of area and may be interpreted in a geometrical way such that the number of particles scattered into the solid angle element $d\Omega$ per second is equal to the number of particles crossing an area equal to $\sigma(\chi,\varepsilon)\, d\Omega$ (or b db $d\varepsilon$) in the incident beam per second.

The form of $\sigma(\chi,\varepsilon)$ depends on the interparticle force law and, if this force law is known, $\sigma(\chi,\varepsilon)$ can be calculated. However, a quantum mechanical calculation must be employed for this purpose, since the quantum wave packets of the colliding particles necessarily overlap and the problem is no longer a classical one. For a collection of atoms or molecules to be regarded classically, with each particle having a rather well-defined position and momentum, it is necessary that the particles be localized wave packets whose extensions are small compared to the average interparticle distance. For a classical treatment the average de Broglie wavelength of each particle must, therefore, be much smaller than the average interparticle separation.

The differential cross section is also a directly measurable quantity and can be obtained experimentally. For our purposes, we will consider the differential cross section $\sigma(\chi,\varepsilon)$, which contains the nature of the collisional interaction, as a known quantity.

5.2 Total scattering cross section

The *total scattering cross section* σ_t is defined as the number of particles scattered per unit time and per unit incident flux, in *all directions* from the scattering center. It is obtained by integrating $\sigma(\chi,\varepsilon)$ $d\Omega$ over the entire solid angle,

$$\sigma_t = \int_\Omega \sigma(\chi,\varepsilon)\, d\Omega = \int_0^{2\pi} d\varepsilon \int_0^{\pi} \sigma(\chi,\varepsilon) \sin\chi \, d\chi \tag{5.6}$$

Both $\sigma(\chi,\varepsilon)$ and σ_t depend on the magnitude of the relative particle velocity g.

In the special case when the interaction potential is *isotropic*, that is, when the differential scattering cross section is independent of ε, we can immediately perform the integral over ε in (5.6), to get

$$\sigma_t = 2\pi \int_0^{\pi} \sigma(\chi) \sin\chi \, d\chi \tag{5.7}$$

This is the case, for example, of the Coulomb interaction potential.

5.3 Momentum transfer cross section

A cross section can be defined for various processes of interaction. It will be seen later that the transfer of momentum, during a collision, is the basic microscopic event in the transport phenomena of diffusion and mobility.

Hence, it is appropriate to define a cross section for the rate of transfer of momentum σ_m as the total momentum transferred per unit time to the scattering center, per unit incident momentum flux (momentum per normal unit area per unit time),

$$\sigma_m = \frac{\text{(momentum transferred per second)}}{\text{(incident momentum flux)}} \tag{5.8}$$

The momentum of a particle in the beam, before interaction, is μg, where μ is the reduced mass and g is the initial relative velocity. The incident momentum flux is therefore $\Gamma\mu g$. After interaction, the momentum of a particle in the beam, in the direction of incidence, and which is scattered at an angle χ, is $\mu g \cos\chi$. Therefore, the momentum transferred by this particle to the scattering center is $\mu g\,(1 - \cos\chi)$. The total momentum transferred per second to the scattering center, by all the particles scattered in all directions in space, is given by

$$\Gamma\ \mu g \int_{\Omega} (1 - \cos\chi)\ \sigma(\chi,\varepsilon)\ d\Omega \tag{5.9}$$

Recall that $\sigma(\chi,\varepsilon)$ can be considered as an angular distribution function. Since the total incident momentum flux is $\Gamma\ \mu g$, we obtain for the *momentum transfer cross section*,

$$\sigma_m = \int_{\Omega} (1 - \cos\chi)\ \sigma(\chi,\varepsilon)\ d\Omega \tag{5.10}$$

For the special case of an *isotropic* interaction potential, and noting that $d\Omega = \sin\chi\ d\chi\ d\varepsilon$, we can perform the integral over ε, in (5.10), obtaining

$$\sigma_m = 2\pi \int_0^{\pi} (1 - \cos\chi)\ \sigma(\chi)\ \sin\chi\ d\chi \tag{5.11}$$

Since $\sigma(\chi)$ is an angular distribution function, it can be used as a weight function to calculate the *mean value* of any function $F(\chi)$ of the scattering angle. The contribution to the total value of $F(\chi)$, resulting from the particles scattered into $d\Omega$, is $F(\chi)\ \sigma(\chi)\ d\Omega$. Since the total number of particles scattered is $\int \sigma(\chi)\ d\Omega$, it follows that the *mean value* of $F(\chi)$, averaged over all values of χ, is given by

$$\langle F(\chi) \rangle = \frac{\int_\Omega F(\chi)\ \sigma(\chi)\ d\Omega}{\int_\Omega \sigma(\chi)\ d\Omega} \tag{5.12}$$

which may be written as

$$\langle F(\chi) \rangle = \frac{2\pi}{\sigma_t} \int^\pi F(\chi)\ \sigma(\chi)\ \sin\chi\ d\chi \tag{5.13}$$

According to this definition of mean values we see that (5.11) can be written in the form

$$\sigma_m = \sigma_t \langle 1 - \cos\chi \rangle \tag{5.14}$$

Thus, the cross section for momentum transfer is a weighted cross section in which scattering angles of zero degrees do not count at all, scatterings of 90^0 count as one, and scatterings of 180^0 count as two. This weighting is proportional to the amount of momentum transferred from the incident beam to the scattering center.

6. CROSS SECTIONS FOR THE HARD SPHERE MODEL

6.1 Differential scattering cross section

To calculate $\sigma(\chi,\varepsilon)$, as given in (5.5), we first obtain from (4.5), for $b < R_1 + R_2$,

$$b = (R_1 + R_2)\cos(\chi/2) \tag{6.1}$$

and

$$|db/d\chi| = (1/2)\ (R_1 + R_2) \sin (\chi/2) \tag{6.2}$$

Substituting these last two expressions into (5.5), yields for the differential scattering cross section

$$\sigma = (R_1 + R_2)^2/4 \tag{6.3}$$

6.2 Total scattering cross section

Integrating (6.3) over the whole solid angle, we obtain

$$\sigma_t = 2\pi \int_0^\pi \frac{(R_1 + R_2)^2}{4} \sin \chi \, d\chi = \pi\ (R_1 + R_2)^2 \tag{6.4}$$

Two special simple cases may be mentioned here. For the collision between an electron and a molecule of radius R, we have $\sigma = R^2/4$ and $\sigma_t = \pi R^2$. For molecules colliding with themselves, their diameter being D, we have $\sigma = D^2/4$ and $\sigma_t = \pi D^2$.

Note that in this case there is a cut-off value for the impact parameter b, beyond which collisions do not occur. It is the existence of this cut-off value for b that leads to a finite value for the total scattering cross section σ_t. This conclusion is made clear in the next section.

6.3 Momentum transfer cross section

From (6.3) and (5.11) we obtain

$$\sigma_m = 2\pi \int_0^\pi \frac{(R_1 + R_2)^2}{4} (1 - \cos \chi) \sin \chi \, d\chi \tag{6.5}$$

$$= \frac{\pi\ (R_1 + R_2)^2}{2} \left(\int_0^\pi \sin \chi \, d\chi - \int_0^\pi \cos \chi \sin \chi \, d\chi \right)$$

Performing the integrals, yields

$$\sigma_m = \pi\ (R_1 + R_2)^2 \tag{6.6}$$

The average value of momentum loss per particle is found, from (5.13), to be given by

$$< \mu g\ (1 - \cos\chi) > = \frac{2\pi}{\sigma_t} \int_0^{\pi} \mu g\ (1 - \cos\chi)\ \sigma(\chi)\ \sin\chi\ d\chi \tag{6.7}$$

Using (5.11), we get

$$< \mu g\ (1 - \cos\chi) > = \mu g\ \sigma_m/\sigma_t \tag{6.8}$$

Thus, from (6.4), (6.7) and (6.8) we deduce that the average value of momentum loss, per particle, for the hard sphere model, is

$$< \mu g\ (1 - \cos\chi) > = \mu g \tag{6.9}$$

For collisions between electrons and neutral particles, for example, in a weakly ionized plasma, the mass of the electron can be neglected as compared to the mass of the neutral particle, so that the reduced mass becomes equal to the electron mass. From (6.9) it is seen, in the first approximation, that the entire momentum of an electron is lost in a collision with a neutral particle. Assuming that the motion of the heavy particles can be ignored, and if ν is the collision frequency, that is, the number of collisions between electrons and neutral particles per second, then the rate of loss of momentum of an average electron is $\nu\ m_e\ \vec{u}$, where $\vec{u}$ denotes the electron velocity. However, in general an electron does not lose its entire momentum on a collision with a neutral particle, and also the perfectly elastic hard sphere model is not a very good representation for the interaction of an electron with a neutral particle. Consequently, the rate of loss of momentum is written as $\nu_{en}\ m_e\ \vec{u}$, where ν_{en} is an *effective collision frequency* for momentum transfer between electrons and neutral particles. This term is used in the *Langevin equation*, introduced in Chapter 10, to represent the time rate of transfer of momentum due to collisions.

7. CROSS SECTIONS FOR THE COULOMB POTENTIAL FIELD

7.1 Differential scattering cross section

Differentiating (4.13) we find

$$|db/d\chi| = b^2/[2b_0 \cos^2 (\chi/2)] \tag{7.1}$$

Thus, the differential scattering cross section, given in (5.5), becomes

$$\sigma(\chi) = b^3/[2b_0 \sin (\chi) \cos^2 (\chi/2)] \tag{7.2}$$

Using (4.13) this equation can be rearranged as

$$\sigma(\chi) = b_0^2/[4 \sin^4 (\chi/2)] \tag{7.3}$$

This equation is known as the *Rutherford scattering formula*. Since $2 \sin^2 (\chi/2) = (1 - \cos \chi)$, it can also be written as

$$\sigma(\chi) = b_0^2/(1 - \cos \chi)^2 \tag{7.4}$$

The Rutherford scattering formula shows that the differential scattering cross section is equal to $b_0^2/4$ for the deflection angle $\chi = \pi$, increases monotonically as χ is decreased and tends to infinity as χ tends to zero.

7.2 Total scattering cross section

Since the differential scattering cross section increases rapidly to infinity as χ goes to zero, it turns out that the total scattering cross section σ_t becomes infinite. From (5.7) and (7.4), we obtain

$$\sigma_t = 2\pi \int_{\chi_{min}}^{\pi} \sigma(\chi) \sin \chi \, d\chi = 2\pi b_0^2 \int_{\chi_{min}}^{\pi} \frac{\sin \chi}{(1 - \cos \chi)^2} d\chi \tag{7.5}$$

where $\chi_{min} = 0$. The lower limit has been written implicity for reasons that will become apparent in what follows. Evaluating the integral in (7.5), yields

$$\sigma_t = \pi b_0^2 \, [1/\sin^2 (\chi_{min}/2) - 1] \tag{7.6}$$

which clearly gives $\sigma_t = \infty$ for $\chi_{min} = 0$. The particles with very small deflection angles contribute to make σ_t infinite.

7.3 Momentum transfer cross section

The substitution of (7.4) into (5.11) gives the following expression for the momentum transfer cross section

$$\sigma_m = 2\pi \int_{\chi_{min}}^{\pi} (1 - \cos \chi) \, \sigma (\chi) \sin \chi \, d\chi = 2\pi \, b_0^2 \int_{\chi_{min}}^{\pi} \frac{\sin \chi}{(1 - \cos \chi)} \, d\chi \tag{7.7}$$

where, again, $\chi_{min} = 0$. Evaluating the integral we find that

$$\sigma_m = 4\pi \, b_0^2 \, \ell n \, [1/\sin (\chi_{min}/2)] \tag{7.8}$$

Setting $\chi_{min} = 0$ we also find that $\sigma_m = \infty$. Thus, the Coulomb potential gives infinite values for both σ_t and σ_m, the particles with very small deflection angles being responsible for this infinite result.

8. EFFECT OF SCREENING OF THE COULOMB POTENTIAL

The infinite results for σ_t and σ_m, obtained in the previous section, may be interpreted as due to the absence of a cut-off value for the impact parameter b. Note that small values of χ correspond to large values of b, and that for $\chi_{min} = 0$ we must have, from (4.13), $b_{max} = \infty$. In order to obtain finite and meaningful values for σ_t and σ_m, it is necessary to modify the basis of the treatment of interactions between individual charged particles and introduce, on some plausible grounds, a cut-off value $b = b_c$ for the impact parameter.

From (5.3) and (5.6) we have for the total scattering cross section (considering σ independent of ε)

$$\sigma_t = 2\pi \int_0^{b_c} b \, db \tag{8.1}$$

where a cut-off value $b = b_c$ has been introduced for the impact parameter. With this cut-off, σ_t for the Coulomb potential is found to be

$$\sigma_t = \pi\, b_c^2 \tag{8.2}$$

The introduction of a cut-off value for b corresponds to the assumption that, for the charged particles incident with b greater than b_c there will be no interaction, whereas for the charged particles incident with b less than b_c there will be a Coulomb type interaction with the target particle.

The deflections that yield scattering angles between $\pi/2$ and π, and which are associated with values of b between 0 and b_0, are usually called *large-angle deflections*, or *close encounters*. If only the large-angle deflections are taken into account, we obtain

$$\sigma_{t,\text{large}} = \pi b_0^2 \qquad (\pi/2 < \chi < \pi) \tag{8.3}$$

with b_0 as given by (4.9).

If the charged particle is located inside a plasma we know that it will be surrounded by a shielding cloud of particles of opposite sign. The scale length for an effective shielding of the charged particle under consideration is the *Debye length* defined by

$$\lambda_D = (\varepsilon_0\, kT/n_0\, e^2)^{1/2} \tag{8.4}$$

The sphere of radius λ_D, surrounding the charged particle under consideration, is called the *Debye sphere*. We have seen (Chapter 11) that the charged particles lying within the Debye sphere shield the Coulomb potential due to the charged particle under consideration, reducing significantly its effect on the particles lying outside its Debye sphere. Taking this screening effect into account, we find that the interaction potential energy is of the form

$$U(r) = (q\, q_1/4\pi\varepsilon_0\, r)\, \exp\,(-r/\lambda_D) \tag{8.5}$$

Thus, when $r \ll \lambda_D$ the Debye potential, as given by (8.5), is very nearly equal to the Coulomb potential, whereas when $r \gg \lambda_D$ the Debye potential is nearly equal to zero. The analysis required for calculating σ_t, using the

Debye potential, is excessively complicated and it must be done numerically. However, an alternative simple approach can be used that leads to results in very good agreement with those evaluated numerically using the Debye potential. It consists in assuming that the interaction potential is exactly equal to the Coulomb potential for $r < \lambda_D$ and is equal to zero for $r > \lambda_D$, as illustrated in Fig. 9. Therefore, it is convenient and more legitimate to introduce the cut-off in the impact parameter at $b_c = \lambda_D$ and not at $b_c = b_o$. In general we have

$$\lambda_D >> b_o \tag{8.6}$$

It is usual to denominate the deflections corresponding to $b_o < b < \lambda_D$, leading to $\chi < \pi/2$, as *small-angle deflections*. The contribution to the total scattering cross section from the small-angle deflections is deduced to be given by

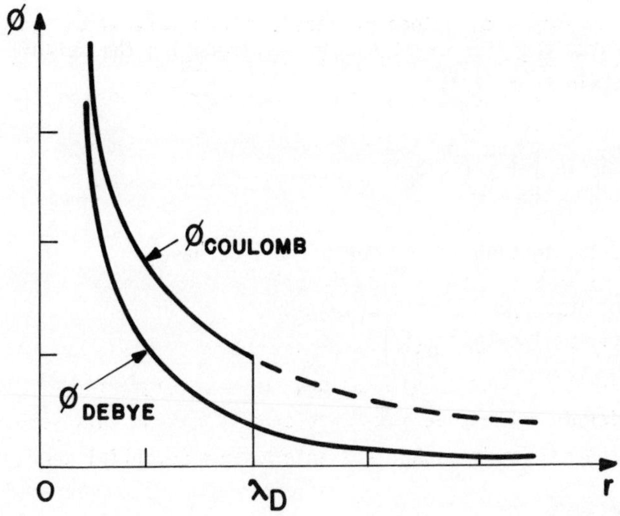

Fig. 9 - The approximation used to obtain $\sigma_t = \pi\lambda_D^2$ consists in assuming that for $r < \lambda_D$ the shielding effect is completely omitted and the particles interact according to the Coulomb potential, whereas for $r > \lambda_D$ the shielding of the target particle is assumed to be complete and there is no interaction.

$$\sigma_{t,small} = 2\pi \int_{b_0}^{\lambda_D} b \, db = \pi \, (\lambda_D^2 - b_0^2) \qquad (\chi < \pi/2) \tag{8.7}$$

Therefore, it follows from (8.3) and (8.7) that

$$\sigma_{t,small}/\sigma_{t,\,large} = (\lambda_D/b_0)^2 - 1 \simeq (\lambda_D/b_0)^2 \tag{8.8}$$

since $\lambda_D >> b_0$. This result shows that the large number of particles interacting mildly with the target particle, and therefore producing only small-angle deflections, are much more important than the small number of particles interacting strongly with the target particle and producing large-angle deflections. Therefore, if the impact parameter is cut-off at $b_c = \lambda_D$ we obtain, from (8.1), the following value for the total scattering cross section

$$\sigma_t = \pi \, \lambda_D^2 \tag{8.9}$$

For the momentum transfer cross section, introducing the cut-off at $b_c = \lambda_D$, we obtain from (7.8)

$$\sigma_m = 2 \, \pi \, b_0^2 \, \ell n \left[1 + (\lambda_D/b_0)^2 \right] \tag{8.10}$$

since from (4.12), letting $\chi = \chi_c$ for $b = b_c$, we have

$$\sin \, (\chi_c/2) = \left[1 + (b_c/b_0)^2 \right]^{-1/2} \tag{8.11}$$

Using the notation

$$\Lambda = \lambda_D/b_0 \tag{8.12}$$

and noting that in general $\Lambda >> 1$, (8.10) becomes

$$\sigma_m = 4\pi \, b_0^2 \, \ell n \, \Lambda \tag{8.13}$$

It can be shown that for σ_m, as for the case of σ_t, the large number of particles producing small-angle deflections are much more important than the small number of particles producing large-angle deflections.

The function $\ell n\ \Lambda$ varies very slowly over a large range of variation of the parameters on which Λ depends. For most laboratory plasmas $\ell n\ \Lambda$ lies between 10 and 20. In order to calculate Λ it is usual to make some approximations. For this purpose consider the interaction between an electron gas (charge $q = -e$) and a singly charged ion gas ($q_1 = e$, $Z = 1$). Let n_0 denote the number density of both electrons and ions in the gas, which we assume to constitute a plasma, and let the temperature of the electrons and ions be equal to T. If we further assume that the electron and ion velocities have a Maxwellian equilibrium distribution function with no drift velocity, then we find, by direct calculation,

$$< g^2 > = \frac{1}{n_0^2} \int_v \int_{v_1} f_e\, f_{i1} (\vec{v}_1 - \vec{v})^2\, d^3v\, d^3v_1 = \frac{1}{n_0} \int_v f_e \left(\frac{3kT}{m_i} + v^2\right) d^3v$$

$$= 3kT/\mu \tag{8.14}$$

where k is Boltzmann's constant and μ is the reduced mass. Replacing g^2, in (4.9), by its average value, we find (for $q_1 = -q = e$)

$$b_0 = e^2/(4\pi\varepsilon_0\, \mu < g^2 >) = e^2/(12\pi\varepsilon_0 kT) \tag{8.15}$$

Substituting this result into the expression (8.12), with λ_D as given by (8.4), gives

$$\Lambda = (12\ \pi\varepsilon_0\ kT/e^2)\ \lambda_D = 12\ \pi n_0\ \lambda_D^3 = 9\ N_D \tag{8.16}$$

where N_D is the number of electrons in a Debye sphere. Table 1 presents values for $\ell n\ \Lambda$, for various values of the electron number density, n_e, and the electron temperature, T.

TABLE 1

VALUES OF $\ell n\ \Lambda$, FOR Z = 1

T (^oK)	Electron Density n_e (cm^{-3})								
	1	10^3	10^6	10^9	10^{12}	10^{15}	10^{18}	10^{21}	10^{24}
10^2	16.3	12.8	9.43	5.97					
10^3	19.7	16.3	12.8	9.43	5.97				
10^4	23.2	19.7	16.3	12.8	9.43	5.97			
10^5	26.7	23.2	19.7	16.3	12.8	9.43	5.97		
10^6	29.7	26.3	22.8	19.3	15.9	12.4	8.96	5.54	
10^7	32.0	28.5	25.1	21.6	18.1	14.7	11.2	7.85	4.39
10^8	43.3	30.9	27.4	24.0	20.5	17.0	13.6	10.1	6.69

PROBLEMS

20.1 - For a differential scattering cross section with an angular dependence given by

$$\sigma(\chi) = \frac{\sigma_0}{2} \mid 3 \cos^2 \chi - 1 \mid$$

where σ_0 is a constant, calculate the total cross section and the momentum transfer cross section.

20.2 - Consider a collision between two particles of mass m and m_1, in which the particle of mass m_1 is initially at rest. Denote the scattering angle in the *center of mass* coordinate system by χ, and in the *laboratory* coordinate system (as seen by an observer at rest) by χ_L.

(a) Show that

$$\tan \chi_L = \sin (\chi)/[\cos(\chi) + m/m_1]$$

(b) Show that the relationship between the differential scattering cross section in the laboratory system $\sigma_L(\chi_L)$ and in the center of mass coordinate system $\sigma(\chi)$ is given by

$$\sigma_L(\chi_L) = \sigma(\chi) \frac{[1 + 2(m/m_1) \cos \chi + (m/m_1)^2]^{3/2}}{[1 + (m/m_1) \cos \chi]}$$

Note that when $m_1 = \infty$ we have $\chi_L = \chi$ and $\sigma_L(\chi_L) = \sigma(\chi)$.
(c) Verify that when $m = m_1$ we obtain $\chi_L = \chi/2$ and $\sigma_L(\chi_L) = 4 \cos(\chi/2)\, \sigma(\chi)$.

20.3 - Consider two particles whose interaction is governed by the following retangular-well potential,

$$\phi(r) = 0 \quad \text{for} \quad r > a$$

$$\phi(r) = -\phi_0 \ \text{for} \quad r \leqslant a$$

(a) Calculate the differential scattering cross section $\sigma(\chi)$, and show that it is given by (for $b < a$)

$$\sigma(\chi) = \frac{p^2\, a^2\, [p \cos(\chi/2) - 1]\, [p - \cos(\chi/2)]}{4 \cos(\chi/2)\, [1 - 2p \cos(\chi/2) + p^2]^2}$$

where

$$p = (1 + 2\phi_0/\mu g^2)^{1/2}$$

(b) Show that the total scattering cross section is given by

$$\sigma_t = 2\pi \int_0^a b\, db = \pi\, a^2$$

20.4 - Consider a general inverse-power interparticle force of the form

$$F(r) = K/r^p$$

where K is a constant and p is a positive integer number.
(a) Determine expressions for the scattering angle χ, the differential scattering cross section $\sigma(\chi, \varepsilon)$, the total scattering cross section σ_t and

the momentum transfer cross section σ_m.
(b) Calculate χ, $\sigma(\chi,\varepsilon)$, σ_t and σ_m for the case of Maxwell molecules, for which p = 5.

20.5 - From the expression for σ_m, obtained in part (a) of Problem 20.4, verify that for p = 2 the momentum transfer cross section is given by

$$\sigma_m = 2\pi\ (K/\mu)^2\ A_1(2)\ g^{-4}$$

where $A_1(2)$ is given by (with ℓ = 1 and p = 2)

$$A_\ell(p) = \int_0^\infty (1 - \cos^\ell \chi)\ v_0\ dv_0$$

$$v_0 = b\ (\mu g^2/K)^{1/(p-1)}$$

Consequently, the velocity-dependent collision frequency, defined by

$$\nu_r(g) = n\sigma_m\ g$$

varies as g^{-3}. This inverse dependence on g accounts for the *electron runaway effect*. (In the presence of a sufficiently large electric field $\vec{E}$, some electrons will gain enough energy between collisions so as to decrease their cross section and collision frequency, which in turn allow them to pick up more energy from the field and decrease their cross section and collision frequency even further. If $\vec{E}$ is large enough, the collision frequency will fall so fast that these electrons will form an accelerated beam of runaway electrons).

20.6 - Show that for the case of Coulomb interactions (p = 2) we have

$$A_1(2) = (\mu g^2/K)^2\ b_0^2\ \ell n\ (1 + \Lambda^2)$$

$$A_2(2) = (\mu g^2/K)^2\ 2\ b_0^2\ [\ell n\ (1 + \Lambda^2) - \Lambda^2/(1 + \Lambda^2)]$$

where $b_0 = qq_1/(4\ \pi\ \varepsilon_0\ \mu g^2)$, $A_\ell(p)$ is as defined in Problem 20.5, and $\Lambda = \lambda_D/b_0$. For $\Lambda \gg 1$ verify that

$$A_1(2) = (\mu g^2/K)^2\ 2\ b_0^2\ \ell n\Lambda = 2\ \ell n\ \Lambda$$

$$A_2(2) = (\mu g^2/K)^2\ 2\ b_0^2\ (2\ \ell n\Lambda - 1) = 2\ (2\ \ell n\Lambda - 1)$$

Note that, since $K = qq_1/(4\pi\varepsilon_0)$ for Coulomb interactions, we have $(\mu g^2/K)^2\ b_0^2 = 1$.

CHAPTER 21

The Boltzmann and the Fokker-Planck Equations

1. INTRODUCTION

When the Boltzmann equation was first introduced in Chapter 5, the effects of collisions were incorporated in its right-hand side [see Eq. (5.5.27)] through a general collision term $(\delta f_\alpha/\delta t)_{coll}$, still to be specified. We present now a derivation of the Boltzmann collision term, which takes into account only binary collisions. This collision term involves integrals over the particle velocities, so that the Boltzmann equation turns out to be an *integro-differential equation*. The fact that the Boltzmann collision integral takes into account only binary collisions limits considerably its applicability for a plasma, where each charged particle interacts simultaneously with a large number of neighboring charged particles. Although these multiple Coulomb collisions are very important for a plasma, there are some cases, however, as in weakly ionized plasmas, where the binary charged-neutral collisions play a dominant role.

The collision term originally proposed by Boltzmann applies to a gas of low density, in which only binary elastic collisions are important. These binary collisions may involve neutral atoms or molecules in a dilute gas, or charged and neutral particles in a plasma. We have seen that in a plasma these are not the only particle interactions of importance. The multiple Coulomb interactions need to be taken into account and in most cases are much more important than the binary collisions. Nevertheless, the Boltzmann collision term can in some cases be used for a plasma, but the results obtained have to be interpreted cautiously. Furthermore, the Fokker-Planck collision term, which applies to charged particle interactions, can be derived from the Boltzmann collision term by considering the charged particle encounters as a series of consecutive weak (small deflection angle) binary collisions.

2. THE BOLTZMANN EQUATION

2.1 Derivation of the Boltzmann collision integral

The collision term, $(\delta f_\alpha/\delta t)_{coll}$, represents the rate of change of the distribution function, $f_\alpha(\vec{r},\vec{v},t)$, as a result of particle collisions. Some of the particles of type α originally situated inside the volume element $d^3r\, d^3v$ at $(\vec{r},\vec{v})$ in phase space may leave this volume element, whereas some particles of type α originally outside the volume element may enter it, as a result of collisions during the time interval dt. Let ΔN_α denote this net gain or loss of particles of type α in $d^3r\, d^3v$ at $(\vec{r},\vec{v})$ during dt, that is,

$$\Delta N_\alpha = (\delta f_\alpha/\delta t)_{coll}\, d^3r\, d^3v\, dt \tag{2.1}$$

It is convenient to separate ΔN_α into two parts

$$\Delta N_\alpha = \Delta N_\alpha^+ - \Delta N_\alpha^- \tag{2.2}$$

where ΔN_α^+ represents the gain term due to collisions in which a particle of type α situated inside d^3r about r has, *after* collision, a velocity lying within d^3v at $\vec{v}$, and ΔN_α^- represents the loss term due to collisions in which a particle of type α situated inside d^3r about $\vec{r}$ has, *before* collision, a velocity lying within d^3v at $\vec{v}$.

We proceed now to determine ΔN_α, defined in (2.1), by calculating initially ΔN_α^- and afterwards ΔN_α^+.

To calculate ΔN_α^- we consider the particles of type α situated within the volume element d^3r at $\vec{r}$, whose velocities lie within d^3v about $\vec{v}$, and which are scattered out of this velocity range as a result of collisions with particles of some type β (which may or may not be type α particles) lying in the *same* volume element d^3r at $\vec{r}$, and having some velocity within d^3v_1 about $\vec{v}_1$. Let us focus attention on a single particle of type α situated within the volume element of phase space $d^3r\, d^3v$ at the coordinates $(\vec{r},\vec{v})$. The particles of type β inside $d^3r\, d^3v_1$ at $(\vec{r},\vec{v}_1)$ may be viewed as constituting a particle flux incident on this particle of type α. Noting that $f_\beta(\vec{r},\vec{v}_1,t)\, d^3v_1$ is the number of type β particles per unit volume, with velocities within d^3v_1 about $\vec{v}_1$, the flux of this incident beam can be written as

$$\Gamma_\beta = f_\beta(\vec{r},\vec{v}_1,t)\, d^3v_1\, |\vec{v}_1 - \vec{v}| = f_\beta(\vec{r},\vec{v}_1,t)\, d^3v_1\, g \tag{2.3}$$

Consider the type β particles which approach with an impact parameter between b and b + db, in a collision plane lying between the angles ε and $\varepsilon + d\varepsilon$. The average number of interactions of this part of the type β particles with the type α particle, occurring in the time interval dt, is equal to the number of particles crossing the element of area b db $d\varepsilon$ during dt. This number can be obtained by multiplying the flux of type β particles, given in (2.3), by the element of area b db $d\varepsilon$ and by the time interval dt,

$$\Gamma_\beta \, b \, db \, d\varepsilon \, dt = f_\beta(\vec{r},\vec{v}_1,t) \, d^3v_1 \, g \, b \, db \, d\varepsilon \, dt \tag{2.4}$$

This expression is just the number of particles of type β with velocities within d^3v_1 about $\vec{v}_1$, lying inside the elementary cylindrical volume of length g dt and cross sectional area b db $d\varepsilon$, shown in Fig. 1, and whose volume is g b db $d\varepsilon$ dt. It is assumed here that dt is large compared to the time of interaction between the colliding particles. To determine the number of collisions between the indicated part of the type β particles with all the type α particles lying within $d^3r \, d^3v$ at $(\vec{r},\vec{v})$, during dt, we multiply (2.4) by $f_\alpha(\vec{r},\vec{v},t) \, d^3r \, d^3v$, the number of particles of type α lying within the volume element of phase space $d^3r \, d^3v$ at $(\vec{r},\vec{v})$,

$$f_\alpha(\vec{r},\vec{v},t) \, d^3r \, d^3v \, f_\beta(\vec{r},\vec{v}_1,t) \, d^3v_1 \, g \, b \, db \, d\varepsilon \, dt \tag{2.5}$$

In deducing this expression it has been assumed that the number of particles of the types α and β, with velocities in d^3v about $\vec{v}$, and d^3v_1 abaut $\vec{v}_1$, respectively, lying in the same volume element d^3r about $\vec{r}$, is proportional to the product $f_\alpha(\vec{r},\vec{v},t) \, f_\beta(\vec{r},\vec{v}_1,t)$. However, in a system of interacting particles the existence of a particle within a given volume element d^3r at $\vec{r}$, with a given velocity $\vec{v}$, affects the probability that another particle be found with a specified velocity $\vec{v}_1$ in the same volume element d^3r, at the same instant of time. Thus, in expression (2.5) we are neglecting any possible correlation that may exist between the velocity of a particle and its position. This approximation, known as the *molecular chaos* assumption, is introduced as a mathematical convenience, but although it may represent a possible condition for a system of particles, it is not a general condition.

The total number of particles of type α in d^3r about $\vec{r}$ that are scattered *out* of the velocity space element d^3v about $\vec{v}$, during dt, is obtained integrating expression (2.5) over all possible values of b, ε and $\vec{v}_1$, and summing over all species β

$$\Delta N_\alpha^- = f_\alpha(\vec{r},\vec{v},t)\ d^3r\ d^3v\ dt \sum_\beta \int_{v_1}\int_b\int_\varepsilon f_\beta(\vec{r},\vec{v}_1,t)\ d^3v_1 g\ b\ db\ d\varepsilon \qquad (2.6)$$

where the triple integral over $\vec{v}_1$ is represented again by a single integral sign.

To determine the *gain* term, ΔN_α^+, we proceed in a way similar to the determination of ΔN_α^-, by considering the *inverse collision*, in which a particle of type α with initial velocity in d^3v' about $\vec{v}'$ collides with a particle of type β having initial velocity in d^3v_1' about $\vec{v}_1'$, resulting in the particle of type α scattered into the velocity element d^3v about $\vec{v}$, the event occurring inside the volume element d^3r about $\vec{r}$. The average number of interactions between a single particle of type α, inside the volume element of phase space $d^3r\ d^3v'$ at $(\vec{r},\vec{v}')$, with the particles of type β inside $d^3r\ d^3v_1'$ at $(\vec{r},\vec{v}_1')$, which approach with an impact parameter between b and $b + db$, and with the collision plane oriented between the angles ε and $\varepsilon + d\varepsilon$, is given by

$$f_\beta(\vec{r},\vec{v}_1',t)\ d^3v_1'\ g'\ b\ db\ d\varepsilon\ dt \qquad (2.7)$$

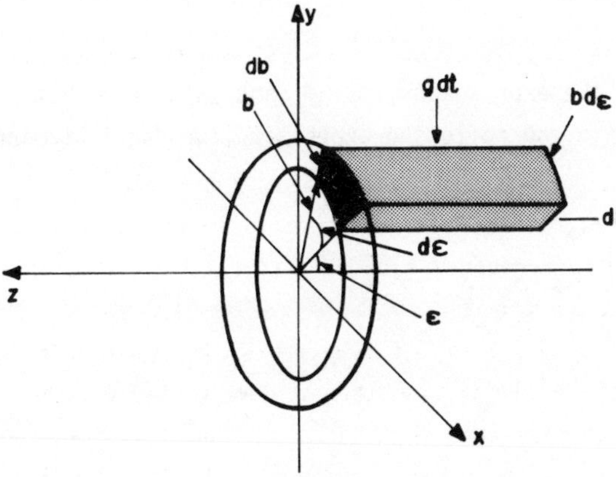

Fig. 1 - The volume element of height g dt and cross sectional area b db dε, with sides lying between b and b + db, and between ε and ε + dε

To take into account all collisions occurring within d^3r, at $\vec{r}$, during the time interval dt, between the particles of type α and type β which scatter the particles of type α into the volume element d^3v, about $\vec{v}$, we must multiply (2.7) by the number of particles of type α which lie initially inside $d^3r d^3v'$, at $(\vec{r},\vec{v}')$, that is $f(\vec{r},\vec{v}',t)\, d^3r d^3v'$, integrate the result over all possible values of b, ε and $\vec{v}_1'$, and sum over all species β,

$$\Delta N_\alpha^+ = f_\alpha(\vec{r},\vec{v}',t) d^3r\, d^3v'\, dt \sum_\beta \int_{v_1'} \int_b \int_\varepsilon f_\beta(\vec{r},\vec{v}_1',t)\, d^3v_1' g'\, b\, db\, d\varepsilon \tag{2.8}$$

We have seen that $g' = g = |\vec{v}_1 - \vec{v}|$, and from the theory of Jacobians

$$d^3v' d^3v_1' = |J|\, d^3v\, d^3v_1 \tag{2.9}$$

It is shown in the following subsection that for this transformation of velocities we have $|J| = 1$, so that

$$d^3v'\, d^3v_1' = d^3v\, d^3v_1 \tag{2.10}$$

and (2.8) becomes

$$\Delta N_\alpha^+ = f_\alpha(\vec{r},\vec{v}',t)\, d^3r\, d^3v\, dt \sum_\beta \int_{v_1} \int_b \int_\varepsilon f_\beta(\vec{r},\vec{v}_1,t)\, d^3v_1\, g\, b\, db\, d\varepsilon \tag{2.11}$$

If we now combine the expressions for ΔN_α^- and ΔN_α^+, and substitute b db $d\varepsilon = \sigma(\Omega)\, d\Omega$, we obtain the following expression for the Boltzmann collision integral

$$\left(\frac{\delta f_\alpha}{\delta t}\right)_{coll} = \left(\frac{\Delta N_\alpha^+ - \Delta N_\alpha^-}{d^3r\, d^3v\, dt}\right)$$

$$= \sum_\beta \int_{v_1} \int_\Omega (f_\alpha' f_{\beta 1}' - f_\alpha f_{\beta 1})\, d^3v_1\, g\, \sigma(\Omega)\, d\Omega \tag{2.12}$$

where we have used the notation

$$f_\alpha' = f_\alpha(\vec{r},\vec{v}',t)$$

$$f_{\beta 1}' = f_\beta(\vec{r},\vec{v}_1',t)$$

$$f_\alpha = f_\alpha(\vec{r},\vec{v},t)$$

$$f_{\beta 1} = f_\beta(\vec{r},\vec{v}_1,t)$$

In explicit form, the Boltzmann equation can finally be written as

$$\frac{\partial f_\alpha}{\partial t} + \vec{v} \cdot \vec{\nabla} f_\alpha + \vec{a}.\vec{\nabla}_v f_\alpha = \sum_\beta \int_{v_1} \int_\Omega (f'_\alpha f'_{\beta_1} - f_\alpha f_{\beta_1})\, d^3v_1 g\sigma(\Omega)\, d\Omega \qquad (2.14)$$

The Boltzmann equation is therefore an *integro-differential* equation, involving integrals and partial derivatives of the distribution function. The external force $\vec{F} = m_\alpha \vec{a}$, in the case of a plasma, includes also the electromagnetic Lorentz force $\vec{F} = q_\alpha(\vec{E}+\vec{v} \times \vec{B})$ due to externally applied fields.

For a system consisting of various different species of particles, there is one equation for each species. For an ionized gas composed of electrons, one type of positive ions and one type of neutral particles, for example, we have a system of three Boltzmann equations coupled through the collision term. In the Boltzmann equation for the electrons, for example, the collision term contains the distribution function for the electrons f_e, the distribution function for the ions f_i and the distribution function for the neutral particles f_n. Since the collision term involves products of the distribution functions, the Boltzmann equation is also *nonlinear*. For a system consisting of only one type of particles the summation over the type β particles disappears and the collision term involves only the product of distribution functions of the same particle species.

2.2 Jacobian of the transformation

The relation between the velocity elements $d^3v\, d^3v_1$ and $d^3v'\, d^3v'_1$ is given by

$$d^3v'\, d^3v'_1 = |J|\, d^3v\, d^3v_1 \qquad (2.15)$$

where J is the Jacobian of the transformation from the variables $(\vec{v},\vec{v}_1)$ to $(\vec{v}',\vec{v}'_1)$,

$$J = \frac{\partial(\vec{v}',\vec{v}'_1)}{\partial(\vec{v},\vec{v}_1)} = \frac{\partial(v'_x, v'_y, v'_z, v'_{1x}, v'_{1y}, v'_{1z})}{\partial(v_x, v_y, v_z, v_{1x}, v_{1y}, v_{1z})} \qquad (2.16)$$

which corresponds to the determinant

$$J = \begin{vmatrix} \dfrac{\partial v'_x}{\partial v_x} & \dfrac{\partial v'_y}{\partial v_x} & \cdots & \dfrac{\partial v'_{1z}}{\partial v_x} \\ \dfrac{\partial v'_x}{\partial v_y} & \dfrac{\partial v'_y}{\partial v_y} & \cdots & \dfrac{\partial v'_{1z}}{\partial v_y} \\ \cdots & \cdots & \cdots & \cdots \\ \dfrac{\partial v'_x}{\partial v_{1z}} & \dfrac{\partial v'_y}{\partial v_{1z}} & \cdots & \dfrac{\partial v'_{1z}}{\partial v_{1z}} \end{vmatrix} \tag{2.17}$$

Using (20.2.25) and (20.2.6) we can express d^3v and d^3v_1 in terms of d^3c_o and d^3g,

$$d^3v\, d^3v_1 = |J_c|\, d^3c_o\, d^3g \tag{2.18}$$

where J_c denotes the Jacobian of the transformation indicated in Eqs. (20.2.5) and (20.2.6). Let us consider initially only the x component of (2.18),

$$dv_x\, dv_{1x} = \left| \frac{\partial(v_x, v_{1x})}{\partial(c_{ox}, g_x)} \right| dc_{ox}\, dg_x \tag{2.19}$$

From (20.2.5) and (20.2.6) we obtain

$$dv_x\, dv_{1x} = \begin{vmatrix} 1 & 1 \\ -\mu/m & \mu/m_1 \end{vmatrix} dc_{ox}\, dg_x$$

$$= \mu \left(\frac{1}{m_1} + \frac{1}{m} \right) dc_{ox}\, dg_x = dc_{ox}\, dg_x \tag{2.20}$$

Taking the product of three such terms, corresponding to the components x, y, z, gives

$$d^3v \; d^3v_1 = d^3c_0 \; d^3g \tag{2.21}$$

In a similar way, using (20.2.8) and (20.2.9) we find

$$d^3v' \; d^3\vec{v}_1' = d^3c_0' \; d^3g' \tag{2.22}$$

Whe have seen that $\vec{c}_0 = \vec{c}_0'$. Furthermore, $\vec{g}$ and $\vec{g}'$ differ only in direction, having the same magnitude, and since volume elements are not changed by a simple rotation of coordinates, we must have $d^3g = d^3g'$. Consequently, (2.21) and (2.22) imply

$$d^3v \; d^3v_1 = d^3v' \; d^3v_1' \tag{2.23}$$

2.3 Assumptions in the derivation of the Boltzmann collision integral

The derivation of the Boltzmann collision integral involves four basic assumptions:

(a) The distribution function does not vary appreciably over a distance of the order of the range of the interparticle force law, as well as over time scales of the order of the interaction time.

(b) Effects of the external force, on the magnitude of the collision cross section, are ignored.

(c) Only binary collisions are taken into account.

(d) The velocities of the interacting particles, before collision, are assumed to be uncorrelated.

The first assumption is quite reasonable and is incorporated in the calculation of $(\delta f_\alpha/\delta t)_{coll}$ when we evaluate all the distribution functions at the position $\vec{r}$ and at the instant t. The element of volume d^3r is considered to be large compared to the range of the interparticle forces, and the time interval dt is taken to be large compared to the time of interaction. On the other hand, as far as variations in the distribution functions are concerned, the elements d^3r and dt must be infinitesimally small quantities.

Next, the external force was assumed to have a negligible effect on the two-body collission problem. This is valid if the external force is negligibly small compared to the force of interaction between the particles. When external forces of magnitude comparable to the short-range interparticle forces are present, the collision process is modified. The constancy of the relative spped $g = |\vec{v}_1 - \vec{v}|$ is valid only in the absence of external forces.

The assumption of binary collisions is justified for a dilute gas, whose molecules interact through short-range forces. However, it is not strictly valid for a plasma, where the Coulomb force between the charged particles is a long range force. In a plasma a charged particle interacts simultaneously with all the charged particles inside its Debye sphere. Since there is a large number of charged particles inside a Debye sphere, each charged particle in the plasma does not move freely, as does a neutral particle between collisions, but is permanently interacting with a large number of charged particles. Each long-range individual interaction, however, results only in a small deflection in the particle trajectory. Since each individual interaction is relatively weak, the collective effect of many simultaneous interactions can be considered as a cumulative succession of weak binary collisions. Thus, in general, the Boltzmann collision term is not strictly valid for a plasma and the results obtained for the case of charged-neutral particle interactions in weakly ionized plasmas must be interpreted cautiously.

Assumption (d) is known as the *molecular chaos* assumption. It is justified for a gas in which the density is sufficiently small, so that the mean free path is much larger than the characteristic range of the interparticle forces. This is certainly not a general situation for a plasma, in view of the long-range characteristic of the Coulomb force. Generally, the *joint* probability of having, at the position $\vec{r}$ and at the instant t, a particle of type α with velocity $\vec{v}$ and a particle of type β with velocity $\vec{v}_1$, is proportional to

$$f_\alpha(\vec{r},\vec{v},t)\ f_\beta(\vec{r},\vec{v}_1,t)\ [1+\Psi_{\alpha\beta}(\vec{v},\vec{v}_1,\vec{r},t)] \tag{2.24}$$

where $\Psi_{\alpha\beta}(\vec{v},\vec{v}_1,\vec{r},t)$ is known as the *correlation function*. In the derivation of the Boltzmann collision integral we have neglected the correlation effects between the particles and we have taken this joint probability as being proportional to $f_\alpha(\vec{r},\vec{v},t)\ f_\beta(\vec{r},\vec{v}_1,t)$. The *irreversible* character of the Boltzmann equation, to be discussed in the next section, is a consequence of the molecular chaos assumption. In order to avoid this approximation the only alternative is to work with the reversible equations of the BBGKY (Bogoliubov, Born, Green, Kirkwood and Yvon) hierarchy. This treatment, however, is beyond the scope of this text.

For gaseous systems in which the characteristic interaction length is much less than the average interparticle distance, and where temporal and spatial gradients are not very large, the Boltzmann equation is nevertheless very well verified experimentally and, in this respect, constitutes one of the basic relations of the kinetic theory of gases.

2.4 Rate of change of a physical quantity as a result of collisions

In section 2, Chapter 8, we have represented the time rate of change of a physical quantity $\chi(\vec{v})$ per unit volume, for the particles of type α, due to collisions with the other particles in the plasma, by

$$\left[\frac{\delta}{\delta t}(n_\alpha < \chi >_\alpha)\right]_{coll} = \int_v \chi \left(\frac{\delta f_\alpha}{\delta t}\right)_{coll} d^3v \tag{2.25}$$

Using the Boltzmann collision integral,

$$\left(\frac{\delta f_\alpha}{\delta t}\right)_{coll} = \sum_\beta \int_\Omega \int_{v_1} (f'_\alpha f'_{\beta 1} - f_\alpha f_{\beta 1})\, g\, \sigma(\Omega)\, d\Omega\, d^3v_1 \tag{2.26}$$

we obtain the following expression for (2.25),

$$\left[\frac{\delta}{\delta t}(n_\alpha < \chi >_\alpha)\right]_{coll} =$$

$$= \sum_\beta \int_\Omega \int_{v_1} \int_v (f'_\alpha f'_{\beta_1} - f_\alpha f_{\beta_1})\, \chi\, g\, \sigma(\Omega)\, d\Omega\, d^3v\, d^3v_1 \tag{2.27}$$

Recall that for each direct collision there is a corresponding inverse collision with the same cross section. Hence, the integrals over $\vec{v}$ and $\vec{v}_1$ can be replaced by integrals over $\vec{v}\,'$ and $\vec{v}_1'$, respectively, without altering the result. Therefore, the first group of integrals in (2.27) may be written as

$$\sum_\beta \int_\Omega \int_{v_1} \int_v f'_\alpha f'_{\beta_1}\, \chi\, g\, \sigma(\Omega)\, d\Omega\, d^3v_1\, d^3v =$$

$$= \sum_{\beta} \int_{\Omega} \int_{v_1} \int_{v} f_\alpha \, f_{\beta 1} \, \chi' \, g \, \sigma(\Omega) \, d\Omega \, d^3v_1 \, d^3v \tag{2.28}$$

where we have replaced $d^3v' \, d^3v_1'$ by $d^3v \, d^3v_1$. Using this expression, results in the following alternative form for the collision term in (2.27),

$$\left[\frac{\delta}{\delta t}(n_\alpha < \chi >_\alpha)\right]_{coll} = \sum_{\beta} \int_{\Omega} \int_{v_1} \int_{v} f_\alpha \, f_{\beta_1} (\chi' - \chi) g \, \sigma(\Omega) \, d\Omega \, d^3v_1 \, d^3v \tag{2.29}$$

Note that the property $\chi(\vec{v})$ is associated with the particles of type α and that χ' denotes $\chi(\vec{v}')$. Note also that only the quantity χ' on the right-hand side of (2.29) is a function of the *after* collision velocity $\vec{v}'$.

The result just derived applies to the special case of binary collisions in a dilute plasma (or gas), when processes of creation and disappearance of particles, as well as radiation losses are unimportant.

3. THE BOLTZMANN'S H FUNCTION

An important characteristic of the Boltzmann collision term is that it drives the distribution functions towards the equilibrium state in an *irreversible* way. This irreversible character of the Boltzmann collision term, as mentioned before, is a consequence of the molecular chaos assumption, which neglects the correlation effects between the particles.

In order to place in evidence this aspect of the Boltzmann collision term, we introduce now the Boltzmann's function H(t). For simplicity, we will consider the particles to be uniformly distributed in space (having no density gradients) and isolated from external forces. The distribution function is therefore independent of $\vec{r}$ and can be denoted $f(\vec{v},t)$. We define, according to Boltzmann, the function H(t) by

$$H(t) = \int_{v} f(\vec{v},t) \, \ln f(\vec{v},t) \, d^3v \tag{3.1}$$

For problems involving spatial gradients the function H(t), defined in (3.1), is $H_{total}(t)$ per unit volume, where

$$H_{total}(t) = \int_r \int_v f(\vec{r},\vec{v},t)\ \ell n\ f(\vec{r},\vec{v},t)\ d^3v\ d^3r \tag{3.2}$$

The function H(t) is proportional to the *entropy* per unit volume of the system according to

$$S/V = -\ kH \tag{3.3}$$

where S denotes the total entropy, V is the volume of the system and k is Boltzmann's constant. More generally, for systems in which spatial gradients are present we have

$$S = -\ kH_{total} \tag{3.4}$$

3.1 Boltzmann's H theorem

The Boltzmann's H theorem states that if $f(\vec{v},t)$ satisfies the Boltzmann equation, that is, if

$$\frac{\partial f(\vec{v},t)}{\partial t} = \int_\Omega \int_{v_1} [f(\vec{v}\,',t)\ f(\vec{v}_1',t) - f(\vec{v},t)\ f(\vec{v}_1,t)]\ g\ \sigma(\Omega)\ d\Omega\ d^3v_1 \tag{3.5}$$

then

$$\partial H(t)/\partial t \leqq 0 \tag{3.6}$$

To prove this theorem let us take the derivate of (3.1) with respect to time, which gives

$$\frac{\partial H}{\partial t} = \int_v (1 + \ell nf)\ \frac{\partial f}{\partial t}\ d^3v \tag{3.7}$$

Substituting (3.5) into (3.7) gives

$$\frac{\partial H}{\partial t} = \int_\Omega \int_v \int_{v_1} (1+\ell nf)\ (f'\ f_1' - f\ f_1)g\ \ \sigma(\Omega)\ d\Omega\ d^3v_1\ d^3v \tag{3.8}$$

where the notation $f_1' = f(\vec{v}_1',t)$, and so on, has been used. The variables of integration $\vec{v}$ and $\vec{v}_1$ are dummy variables, and can be interchanged in the integrand of (3.8) without changing the value of the integral, since $\sigma(\Omega)$ and $g = |\vec{v}_1 - \vec{v}|$ are also invariants. Thus, (3.8) can be written as

$$\frac{\partial H}{\partial t} = \int_\Omega \int_{v_1} \int_v (1+\ln f_1)\,(f_1'\, f' - f_1\, f)\, g\, \sigma(\Omega)\, d\Omega\, d^3v\, d^3v_1 \tag{3.9}$$

Adding Eqs. (3.8) and (3.9), and dividing by 2, gives

$$\frac{\partial H}{\partial t} = \frac{1}{2}\int_\Omega \int_{v} \int_{v_1} [2+\ln (f\, f_1)]\,(f' f_1' - f\, f_1)\, g\, \sigma(\Omega)\, d\Omega\, d^3v_1\, d^3v \tag{3.10}$$

In this equation we can replace the velocities *before* collision, $\vec{v}$ and $\vec{v}_1$, by the velocities *after* collision, $\vec{v}\,'$ and $\vec{v}_1'$, respectively, without altering the value of the integral, since for each collision there exists an inverse collision with the same differential cross section $\sigma(\Omega)$. We have already seen that $d^3v'\, d^3v_1' = d^3v\, d^3v_1$ and $g = g'$. Consequently, (3.10) may be written as

$$\frac{\partial H}{\partial t} = \frac{1}{2}\int_\Omega \int_{v} \int_{v_1} [2+\ln(f'\, f_1')]\,(f\, f_1 - f'\, f_1')\, g\, \sigma(\Omega)\, d\Omega\, d^3v_1\, d^3v \tag{3.11}$$

We now combine Eqs. (3.10) and (3.11) to obtain

$$\frac{\partial H}{\partial t} = \frac{1}{4}\int_\Omega \int_{v} \int_{v_1} \left[\ln\left(\frac{f\, f_1}{f' f_1'}\right)\right] (f' f_1' - f f_1) g\, \sigma(\Omega)\, d\Omega\, d^3v_1\, d^3v \tag{3.12}$$

In this expression it is clear that if $f' f_1' > f f_1$ then $\ln(f\, f_1/f'\, f_1') < 0$, and consequently $\partial H/\partial t < 0$, since all other factors appearing in the right-hand side of (3.12) are positive. On the other hand, if $f' f_1' < f\, f_1$ then $\ln\, (f\, f_1/f' f_1') > 0$ and, again, $\partial H/\partial t < 0$. When $f' f_1' = f\, f_1$, both factors are zero and $\partial H/\partial t = 0$.

This proves the H theorem and shows that, when f satifies the Boltzmann equation, the functional H(t) always decreases monotonically with time until it reaches a limiting value, which occurs when there is no further change with time in the system. This limiting value is reached *only* when

$$f' f'_1 = f\, f_1 \tag{3.13}$$

so that this condition is *necessary* for $\partial H/\partial t = 0$ and, consequently, it is also a necessary condition for the equilibrium state. According to the Boltzmann equation (3.5), the *equilibrium* distribution function satifies the following integral equation

$$\int_\Omega \int_{v_1} [f(\vec{v}\,')\, f(\vec{v}_1') - f(\vec{v})\, f(\vec{v}_1)]\, g\, \sigma(\Omega)\, d\Omega\, d^3v_1 = 0 \tag{3.14}$$

so that the condition (3.13) is also a *sufficient* condition for the equilibrium state.

It is instructive to note that (3.13) can be considered as an example of the *general principle of detailed balance* of statistical mechanics, as discussed in section 1, Chapter 7, where it was used to derive the Maxwell-Boltzmann equilibrium distribution function. An important conclusion that can be drawn from (3.13) is that the equilibrium distribution function is independent of the differential collision cross section $\sigma(\Omega)$, considered to be nonzero. The Maxwell-Boltzmann distribution function is therefore the only distribution for the equilibrium state that can exist in a uniform gas in the absence of external forces.

3.2 Analysis of Boltzmann's H theorem

According to (3.3), the H theorem states that the entropy of a given isolated system always increases with time until it reaches the equilibrium state.

Although this *irreversible* behavior is compatible with the laws of thermodynamics it is, nevertheless, in disagreement with the laws of mechanics, which are *reversible*. If, at a given instant of time, the velocities of all the particles in the system were reversed, the laws of mechanics predict that each particle would describe, in the opposite sense, its previous trajectory. However, we have seen that the Boltzmann collision term leads to a irreversible temporal evolution of the distribution function and of the function H. The existence of this paradox has its origin in the molecular chaos assumption that was used in the derivation of the Boltzmann collision term.

Recall that the molecular chaos assumption admits that, if $f(\vec{r},\vec{v},t)$ is proportional to the probability of finding in a given volume element d^3r about $\vec{r}$ a particle with velocity $\vec{v}$, at the instant t, then the joint probability of simultaneously finding in the same volume element d^3r about $\vec{r}$ a particle with velocity $\vec{v}$ and another particle with velocity $\vec{v}_1$, at the instant t, is proportional to the product $f(\vec{r},\vec{v},t)\, f(\vec{r},\vec{v}_1,t)$. Thus, it neglects any possible correlation that may exist between the particles. Generally, the state of the gas may or may not satisfy the molecular chaos assumption and consequently the distribution function describing the gas may or may not satisfy the Boltzmann equation. The distribution function, which characterizes the state of the gas, will obey the Boltzmann equation only at the instants of time when the molecular chaos assumption holds true for the gas. The H theorem, therefore, is also valid only when this condition is satisfied.

We shall show that at the instants of time when the state of the gas satisfies the molecular chaos assumption, the function H(t) is at a local maximum. For this purpose consider a gas not in equilibrium, which is in the state of molecular chaos at the instant $t = t_0$. The H theorem implies, therefore, that at the instant $t_0 + dt$ we have $dH/dt \leq 0$. Consider a second gas which at the instant $t = t_0$ is exactly identical to the first one, except that the velocities of all the particles have directions opposite to the velocities of the first one, has the same function H(t) as the first one, and is in a state of molecular chaos at $t = t_0$. Consequently, at the instant $t_0 + dt$ we must have $dH/dt \leq 0$, according to the H theorem. On the other hand, due to the invariance of the equations of motion under time reversal, the time evolution of the second gas corresponds to the past of the first. This means that for the first gas we must have

$$dH(t)/dt \leq 0 \qquad \text{at} \qquad t = t_0 + dt \tag{3.15a}$$

$$dH(t)/dt \geq 0 \qquad \text{at} \qquad t = t_0 - dt \tag{3.15b}$$

which shows that at the instant when the condition of molecular chaos is satisfied, the function H(t) is at a local maximum. This is illustrated in Fig. 2 at the instant $t = t_0$ indicated by the number (2). At the instants when H(t) does not present a local maximum, as for example at the instants indicated by the numbers (1) and (3) in Fig. 2, the gas is not in a state of molecular chaos. Note that dH/dt need not necessarily be a continuous function of time and may change abruptly as a result of collisions.

The time evolution of H(t) is governed by the collisional interactions between the particles, which occur at random, and which can establish as well as destroy the state of molecular chaos as time passes. Fig. 3 illustrates how H(t) may vary with time. Some of the instants when the condition of molecular chaos is satisfied are indicated by dots in the curve of H(t). If the condition of molecular chaos prevails during most of the time, as in a dilute gas, for example, H(t) will be at a local maximum most of the time. Due to the random character of the sequence of collisions, these instants of molecular chaos will probably be distributed in time in an almost uniform way. On the other hand, the time variation of H(t) obtained using the distribution function that satisfies the Boltzmann equation is represented by a smooth curve of negative slope which tries to fit, with a minimum deviation, all the points (instants) of the real curve of H(t) in which the condition of molecular chaos is satisfied, as shown by the dashed line in Fig. 3. The state of molecular chaos can therefore be considered as a convenient mathematical model to describe a state not in equilibrium.

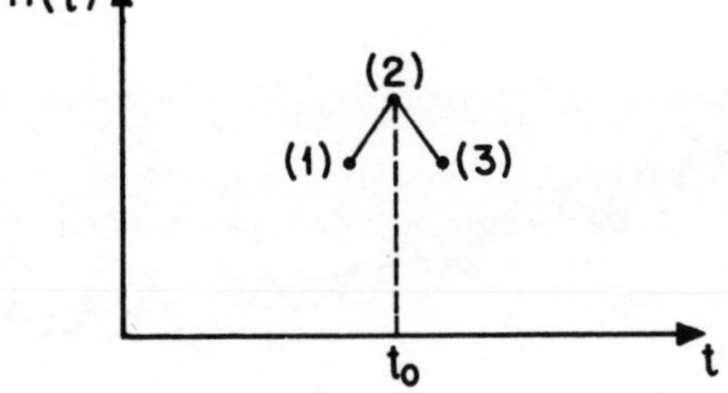

Fig. 2 - When the gas satisfies the condition of molecular chaos, the function H(t) is at a local maximum, indicated here by the point denoted (2).

The Boltzmann equation, although strictly valid only at the instants when the gas is in the state of molecular chaos, can, nevertheless, be considered generally valid in a *statistical* sense at any instant. Similarly, the H theorem is also valid only in a statistical sense.

3.3 Maximum entropy or minimum H approach for deriving the equilibrium distribution function

The Maxwell-Boltzmann equilibrium distribution function can also be derived by performing a variational calculation on the function H(t). We have seen that at equilibrium H is a minimum, so that for a one-component uniform gas we must have at *equilibrium*

$$\delta H = \delta \int_V f \ \ln f \ d^3v = 0 \tag{3.16}$$

The symbol δ, before a given quantity, denotes a variation in that quantity as a result of a small change in the distribution function f. Carrying out the variation indicated in Eq. (3.16) in a formal way, we have

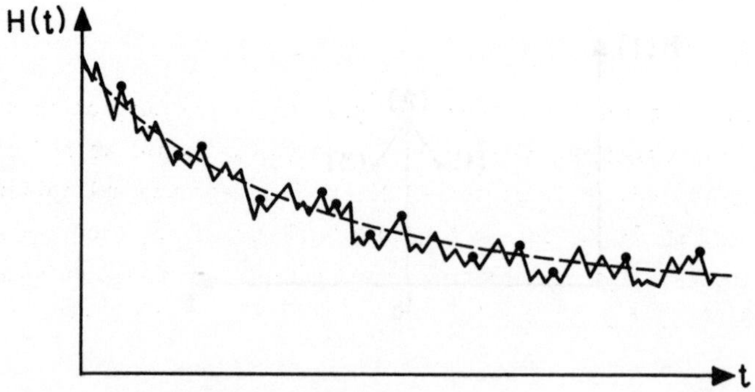

Fig. 3 - The time evolution of H(t) for a gas initially not in an equilibrium state is indicated by the solid curve. The dashed curve is the time variation of H(t) predicted by the Boltzmann equation. The dots indicate some of the instants when the condition of molecular chaos is satisfied.

$$\delta H = \int_V (1+\ell n f)\ \delta f\ d^3v = 0 \tag{3.17}$$

There are, however, certain macroscopic constraints imposed on the system. When we vary f slightly, we cannot violate the conservation of mass, momentum and energy for the system as a whole. Therefore, the variational integral (3.17) is subjected to the constraints that the total mass, momentum and energy densities of the uniform gas remain constant under the variation δf. The constancy of the mass density under a small change δf in f requires that

$$\delta(\rho_m) = m \int_V \delta f\ d^3v = 0 \tag{3.18}$$

Similarly, for the constancy of the momentum density,

$$\delta(\rho_m < \vec{v} >) = m \int_V \vec{v}\ \delta f\ d^3v = 0 \tag{3.19}$$

and for the energy density,

$$\delta\left(\frac{1}{2}\ \rho_m < v^2 > \right) = \frac{1}{2}\ m \int_V v^2\ \delta f\ d^3v = 0 \tag{3.20}$$

We can now solve the variational integral in (3.17) subject to the constraints expressed in (3.18), (3.19) and (3.20) by the method of the *Lagrange multipliers*. Multiplying (3.18) by the Lagrange multiplier a_1, the i'th component of (3.19) by the Lagrange multiplier a_{2i} (for i = x,y,z), (3.20) by the Lagrange multiplier a_3, and adding the resulting equations together with (3.17), gives

$$m \int_V (1 + \ell n f + a_1 + \vec{a}_2 \cdot \vec{v} + \frac{1}{2}\ a_3 v^2)\ \delta f\ d^3v = 0 \tag{3.21}$$

where we used the notation $\vec{a}_2 \cdot \vec{v} = a_{2x} v_x + a_{2y} v_y + a_{2z} v_z$. The variation δf is now completely *arbitrary*, since all the constraints imposed on the system have been taken into account in (3.21). Thus, this integral can be equal to zero if and only if

$$\ell n\ f = -\ (1 + a_1 + \vec{a}_2 \ .\ \vec{v} + \frac{1}{2}\ a_3 v^2) \tag{3.22}$$

The form of this equation is identical to (7.1.19), which we solved in Chapter 7 to obtain the Maxwell-Boltzmann distribution function. Hence, it leads in identical fashion to the equilibrium distribution function

$$f = n \left(\frac{m}{2\pi kT} \right)^{3/2} \exp\ [-\ m\ (\vec{v} - \vec{u})^2/2kT] \tag{3.23}$$

with $\vec{v} - \vec{u} = \vec{c}$. The Maxwellian distribution function, besides being the *equilibrium* solution of the Boltzmann equation is, therefore, also the *most probable* distribution consistent with the specified macroscopic parameters n, $\vec{u}$ and T of the system.

3.4 Mixture of various species of particles

For the case of a mixture containing different species of particles, *each* species having a given number density n_α, average velocity $\vec{u}_\alpha$, and temperature T_α, we can still perform a variational calculation to determine the *most probable* distribution subject to the constraints provided by the set of macroscopic parameters n_α, $\vec{u}_\alpha$, T_α, for each species of particles. Note that this is *not* an equilibrium situation, unless the temperatures and mean velocities of all species are equal. To determine the most probable distribution function for this nonequilibrium gas mixture (*each* species having their *own* number density, mean velocity and temperature), we independently minimize each H_α,

$$H_\alpha = \int_V f_\alpha\ \ell n f_\alpha\ d^3v \tag{3.24}$$

This also minimizes H for the mixture, since

$$H = \sum_\alpha H_\alpha \tag{3.25}$$

For the species of type α, when H_α is at its minimum, we must have $\delta H_\alpha = 0$ for a small variation δf_α in f_α. The macroscopic parameters n_α, $\vec{u}_\alpha$ and T_α must all remain fixed when f_α is varied. The problem is completely analogous to the one we solved in the previous subsection for a one-component gas and leads, in identical fashion, to Eq. (3.23) for each species. Therefore, each particle species has a Maxwellian distribution function, but with its own number density, mean velocity and temperature. Although this is not an equilibrium situation for the whole gas (unless the mean velocities and temperatures of all species are the same), it is nevertheless the most probable distribution function for this system subject to the specified constraints.

4. BOLTZMANN COLLISION TERM FOR A WEAKLY IONIZED PLASMA

In this section we derive from the Boltzmann equation an approximate expression for the collision term for a weakly ionized plasma in which only the collisions between electrons and neutral particles are important. The distribution function for the neutral particles is assumed to be homogeneous and isotropic. The external forces acting on the electrons are assumed to be small, so that the electrons are not very far from the equilibrium state. Consequently, the spatial inhomogeneity and the anisotropy of the nonequilibrium distribution function for the electrons are very small, since the nonequilibrium state is only slightly perturbed from the equilibrium state. In the equilibrium state the electrons are assumed to have no drift velocity and their distribution function is isotropic and homogeneous.

4.1 Spherical harmonic expansion of the distribution function

Let (v,θ,ϕ) denote spherical coordinates in velocity space, as shown in Fig. 4. Since the anisotropy of the nonequilibrium distribution function is very small, the dependence of $f(\vec{r},\vec{v},t)$ on θ and ϕ is very weak. Hence, it is appropriate to expand $f(\vec{r},\vec{v},t)$ in terms of the velocity space angular variables θ and ϕ, and retain only the first few terms of this expansion. Since ϕ varies between 0 and 2π, we can expand $f(\vec{r},\vec{v},t)$ in a Fourier series in ϕ. Furthermore, θ varies between 0 and π and, consequently, $\cos(\theta)$ varies between + 1 abd -1, which means that we can expand $f(\vec{r},\vec{v},t)$ in a series of Legendre polynomials in $\cos(\theta)$. Therefore, we can make a spherical harmonic expansion of the distribution function as follows

$$f(\vec{r},\vec{v},t) = \sum_{m=0}^{\infty} \sum_{n=0}^{\infty} P_n^m (\cos\theta)[f_{mn}(\vec{r},v,t)\cos(m\phi)+g_{mn}(\vec{r},v,t)\sin(m\phi)] \tag{4.1}$$

where the functions $P_n^m(\cos\theta)$ represent the associated Legendre polynomials, and the functions f_{mn} and g_{mn} can be considered as coefficients of the expansion.

The first term in the expansion (4.1) corresponds to $m = 0$, $n = 0$, and since $P_0^0(\cos\theta) = 1$, it follows that it is given by $f_{00}(\vec{r},v,t)$. This leading term is the isotropic distribution function corresponding to the equilibrium state of the electrons. The term corresponding to m=1, n=0, vanishes, since $P_0^1(\cos\theta) = 0$. The next highest order term in (4.1) corresponds to $m = 0$, $n = 1$, and since $P_1^0(\cos\theta) = \cos\theta$, it is given by $f_{01}(\vec{r},v,t)\cos\theta$. Therefore, retaining only the first two nonzero terms of the spherical harmonic expansion (4.1), in view of the fact that the anisotropy is assumed to be small, we obtain

$$f(\vec{r},\vec{v},t) = f_{00}(\vec{r},v,t) + \frac{\vec{v}\cdot\hat{v}_z}{v} f_{01}(\vec{r},v,t) \tag{4.2}$$

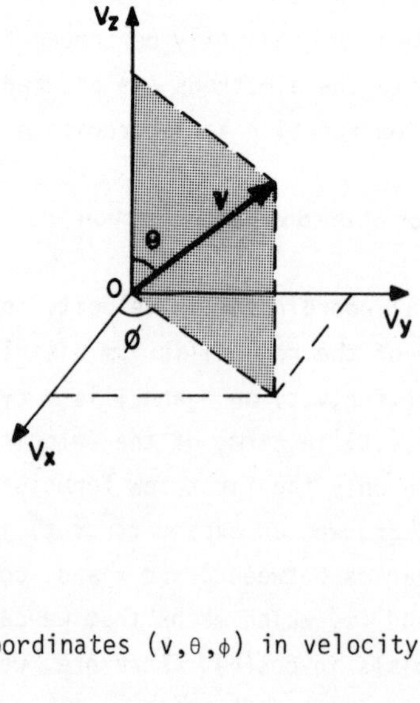

Fig. 4 - Spherical coordinates (v,θ,ϕ) in velocity space.

where we have replaced cos θ by $\vec{v}.\hat{v}_z/v$ (see Fig. 4). The second term in the right-hand side of (4.2) corresponds to the small anisotropy due to the spatial inhomogeneity and the external forces on the electrons.

4.2 Approximate expression for the Boltzmann collision term

The Boltzmann collision integral, given in Eq. (2.12), can be written for the case of binary electron-neutral collisions as

$$\left(\frac{\delta f_e}{\delta t}\right)_{coll} = \int_b \int_\varepsilon \int_{v_1} (f_e' f_{n1}' - f_e f_{n1})\, g\, b\, db\, d\varepsilon\, d^3v_1 \tag{4.3}$$

where we have replaced $\sigma(\Omega)\, d\Omega$ by $b\, db\, d\varepsilon$. Here f_e represents the nonequilibrium distribution function for the electrons and f_n is the isotropic equilibrium distribution function for the neutral particles.

In a first approximation we may assume the neutral particles to be stationary and not affected by collisions with electrons, since the mass of a neutral particle is much larger than that of an electron. Hence, we assume that

$$\vec{v}_1 = \vec{v}_1' \cong 0 \tag{4.4}$$

$$f_{n1} = f_{n1}' \tag{4.5}$$

Therefore, (4.3) becomes

$$\left(\frac{\delta f_e}{\delta t}\right)_{coll} = \int_{v_1} f_n\, d^3v_1 \int_0^{2\pi} d\varepsilon \int_0^{\infty} (f_e' - f_e)\, g\, b\, db \tag{4.6}$$

Since the number density of the neutral particles is given by

$$n_n = \int_{v_1} f_n\, d^3v_1 \tag{4.7}$$

we can write (4.6) as

$$\left(\frac{\delta f_e}{\delta t}\right)_{coll} = n_n \int_0^{2\pi} d\varepsilon \int_0^{\infty} (f_e' - f_e)\, g\, b\, db \tag{4.8}$$

Further, from (4.2), the distribution function for the electrons, *before collision*, is given by

$$f_e \equiv f_e(\vec{r},\vec{v},t) = f_{00}(\vec{r},v,t) + \frac{\vec{v}\cdot\hat{v}_z}{v} f_{01}(\vec{r},v,t) \tag{4.9}$$

and, *after collision*, by

$$f'_e \equiv f_e(\vec{r},\vec{v}',t) = f_{00}(\vec{r},v',t) + \frac{\vec{v}'\cdot\hat{v}_z}{v'} f_{01}(\vec{r},v',t)$$

$$= f_{00}(\vec{r},v,t) + \frac{\vec{v}'\cdot\hat{v}_z}{v} f_{01}(\vec{r},v,t) \tag{4.10}$$

In this last equation we have considered $v' = v$, in view of the fact that the electrons do not loose energy on collisions, since the neutrals are much more massive and considered at rest in a first approximation. This means that $\vec{v} = \vec{g}$ and $\vec{v}' = \vec{g}'$ [see (4.4)], and since $g = g'$ (see (20.2.16)] we have $v = v'$. Note, however, that $\vec{v} \neq \vec{v}'$. Therefore, from (4.9) and (4.10), we have

$$f_e' - f_e = \frac{(\vec{v}' - \vec{v})\cdot\hat{v}_z}{v} f_0(\vec{r},v,t) \tag{4.11}$$

Without any loss of generality we can choose the v_z axis as being parallel to the initial relative velocity g of the electron. Therefore,

$$(\vec{v}' - \vec{v})\cdot\hat{v}_z = \vec{g}'\cdot\hat{v}_z - \vec{g}\cdot\hat{v}_z$$

$$= g\cos\chi - g = v\,(\cos\chi - 1) \tag{4.12}$$

where χ denotes the scattering angle, that is, the angle between $\vec{g}$ and $\vec{g}'$ (see Fig. 3, Chapter 20). Substituting (4.12) into (4.11), we obtain

$$f'_e - f_e = -(1 - \cos\chi)\, f_{01}(\vec{r},v,t) \tag{4.13}$$

The substitution of (4.13) into (4.8), yields

$$\left(\frac{\delta f_e}{\delta t}\right)_{coll} = - n_n\, g\, f_{01}(\vec{r},v,t) \int_0^{2\pi} d\varepsilon \int_0^{\infty} (1 - \cos\chi)\, b\, db \tag{4.14}$$

Since the momentum transfer cross section σ_m for collisions between electrons and neutral particles, is defined by [see (20.5.10)]

$$\sigma_m = \int_\Omega (1-\cos\chi)\,\sigma(\Omega)\, d\Omega = \int_0^{2\pi} d\varepsilon \int_0^{\infty} (1 - \cos\chi)\, d\, db \tag{4.15}$$

we can write (4.14) as

$$\left(\frac{\delta f_e}{\delta t}\right)_{coll} = - n_n\, g\, \sigma_m\, f_{01}(\vec{r},v,t) \tag{4.16}$$

If we substitute $f_{01}(\vec{r},v,t)$ in (4.16) using (4.9), and noting that $\vec{v}.\hat{v}_z/v = \vec{g}.\hat{v}_z/g = 1$, we obtain

$$\left(\frac{\delta f_e}{\delta t}\right)_{coll} = - n_n\, v\, \sigma_m (f_e - f_{oe}) = - \nu_r (v)\, (f_e - f_{oe}) \tag{4.17}$$

where we have introduced the velocity-dependent relaxation collision frequency $\nu_r(v) = n_n\, v\, \sigma_m$, and where f_{oo}, which characterizes the isotropic equilibrium state of the electrons, has been replaced by f_{oe}, in accordance with the notation used previously. Expression (4.17) is similar to the relaxation model (or Krook model) for the collision term introduced in section 6, Chapter 5, except for the velocity-dependent collision frequency. Once the force of interaction between the electrons and the neutral particles has been specified, the momentum transfer cross section σ_m, and consequently the relaxation collision frequency $\nu_r(v)$, can be determined as functions of velocity.

4.3 Rate of change of momentum due to collisions

The time rate of change of momentum per unit volume of the electron gas due to collisions with neutral particles is given, from (8.4.3), by

$$\vec{A}_e = \left[\frac{\delta(\rho_{me}\,\vec{u}_e)}{\delta t}\right]_{coll} = m_e \int_v \vec{v} \left(\frac{\delta f_e}{\delta t}\right)_{coll} d^3v \tag{4.18}$$

Substituting (4.17) into (4.18), we obtain

$$\vec{A}_e = - m_e \int_v \nu_r(v)\, \vec{v}\, f_e\, d^3v + m_e \int_v \nu_r(v)\, \vec{v}\, f_{oe}\, d^3v \tag{4.19}$$

If we assume that the relaxation collision frequency ν_r does not depend on velocity, and if we consider that the electron gas has no drift velocity in the equilibrium state, that is,

$$\vec{u}_{oe} = \frac{1}{n_e}\int_v \vec{v}\, f_{oe}\, d^3v = 0 \tag{4.20}$$

then (4.19) becomes

$$\vec{A}_e = - n_e\, m_e\, \nu_r\, u_e = -\rho_{me}\, \nu_r\, \vec{u}_e \tag{4.21}$$

where $\vec{u}_e$ is the average velocity of the electrons in the nonequilibrium state. (4.21) corresponds to the expression used in the Langevin equation for the time rate of change of momentum per unit volume as a result of collisions, in which a constant collision frequency ν_c was introduced phenomenologically.

5. THE FOKKER–PLANCK EQUATION

In this section we present a derivation of the Fokker-Planck equation, in which the collision term takes into account the simultaneous Coulomb interactions of a charged particle with the other charged particles in its Debye sphere. For this purpose we assume that the large-angle deflection of a charged particle, in a multiple Coulomb interaction, can be considered as a series of consecutive weak binary collisions (or grazing collisions), that is, as a succession of small-angle scatterings. Therefore, the Fokker--Planck collision term can be derived directly from the Boltzmann collision integral, which is valid for binary collisions, under the assumption that a series of consecutive weak (small-angle deflection) binary collisions is a good representation for the multiple Coulomb interaction. Only collisions between species of particles represented by the indices α and β will be considered.

5.1 Derivation of the Fokker-Planck collision term

If $\chi(\vec{v})$ is some arbitrary function of velocity, associated with the particles of type α, then, according to (2.27) and (2.29), the time rate of change of the quantity $\chi(\vec{v})$ per unit volume, as a result of collisions between particles of type α and those of type β, can be expressed as

$$\int_V \chi(\vec{v}) \left(\frac{\delta f_\alpha}{\delta t}\right)_{coll} d^3v = \int_\Omega \int_{V_1} \int_V (f'_\alpha f'_{\beta 1} - f_\alpha f_{\beta 1})\, \chi\, g\, \sigma(\Omega)\, d\Omega\, d^3v_1\, d^3v$$

$$= \int_\Omega \int_{V_1} \int_V f_\alpha f_{\beta 1} (\chi' - \chi)\, g\, \sigma(\Omega)\, d\Omega\, d^3v_1\, d^3v \tag{5.1}$$

where χ' denotes $\chi(\vec{v}')$. In this last expression only the quantity χ' is a function of the velocity after collision $\vec{v}'$. For weak binary collisions (or grazing collisions), we can write

$$\vec{v}' = \vec{v} + \Delta\vec{v} \tag{5.2}$$

where the change $\Delta\vec{v}$, due to collision, is assumed to be small. Since

$$\chi' \equiv \chi(\vec{v}') = \chi(\vec{v} + \Delta\vec{v}) \tag{5.3}$$

we can expand χ' in a Taylor series about the velocity $\vec{v}$, as

$$\chi(\vec{v} + \Delta\vec{v}) = \chi(\vec{v}) + \sum_i \frac{\partial \chi}{\partial v_i} \Delta v_i + \frac{1}{2} \sum_{i,j} \frac{\partial^2 \chi}{\partial v_i\, \partial v_j} \Delta v_i\, \Delta v_j + \ldots \tag{5.4}$$

Substituting (5.4) into (5.1), and neglecting higher order terms, we obtain

$$\int_V \chi \left(\frac{\delta f_\alpha}{\delta t}\right)_{coll} d^3v = \int_\Omega \int_{V_1} \int_V f_\alpha \; f_{\beta_1} \; (\sum_i \frac{\delta \chi}{\partial v_i} \Delta v_i +$$

$$+ \frac{1}{2} \sum_{i,j} \frac{\partial^2 \chi}{\partial v_i \; \partial v_j} \Delta v_i \; \Delta v_j) \; g \; \sigma(\Omega) \; d\Omega \; d^3v_1 \; d^3v \tag{5.5}$$

The next step is to factor out the arbitrary function $\chi(\vec{v})$ from (5.5). This can be accomplished by integrating the first group of integrals involving $\partial\chi/\partial v_i$, in (5.5), by parts once, and the second group of integrals involving $\partial^2\chi/(\partial v_i \; \partial v_j)$ by parts twice. For example, for the x component of the first group of integrals involving $\partial\chi/\partial v_i$, we have

$$\int_\Omega \int_{V_1} \int_V \frac{\partial \chi(\vec{v})}{\partial v_x} \Delta v_x \; f_\alpha(\vec{v}) \;\; f_\beta(\vec{v}_1) \; g \; \sigma(\Omega) \; d\Omega \; d^3v_1 \; d^3v =$$

$$= \int_\Omega \int_{V_1} \left[\int_V dv_y \; dv_z \frac{\partial \chi(\vec{v})}{\partial v_x} dv_x \; (v_x' - v_x) \; f_\alpha(\vec{v}) \; g \; \sigma(\Omega) \; d\Omega \right] f_\beta(\vec{v}_1) \; d^3v_1 \tag{5.6}$$

For the term within brackets we can take

$$dV = [\partial\chi(\vec{v})/\partial v_x] \; dv_x \tag{5.7}$$

$$U = (v_x' - v_x) \; f_\alpha(\vec{v}) \; g \; \sigma(\Omega) \; d\Omega \tag{5.8}$$

and perform the integral over v_x by parts to obtain

$$\int_{v_x} \frac{\partial \chi(\vec{v})}{\partial v_x} dv_x \; (v_x' - v_x) \; f_\alpha(\vec{v}) \; g \; \sigma(\Omega) \; d\Omega =$$

$$= - \int_{v_x} \chi(\vec{v}) \; \frac{\partial}{\partial v_x} \left[(v_x' - v_x) \; f_\alpha(\vec{v}) \; g \; \sigma(\Omega) \; d\Omega \right] dv_x \tag{5.9}$$

where the integrated term vanishes since f must be zero at ± ∞. Therefore, we find for the integral in (5.6)

$$\int_\Omega \int_{v_1} \int_v \frac{\partial \chi(\vec{v})}{\partial v_x} \Delta v_x \, f_\alpha(\vec{v}) \, f_\beta(\vec{v}_1) \, g \, \sigma(\Omega) \, d\Omega \, d^3v_1 \, d^3v =$$

$$= \int_\Omega \int_{v_1} \int_v - \chi(\vec{v}) \frac{\partial}{\partial v_x} \left[\Delta v_x \, f_\alpha(\vec{v}) \, g \, \sigma(\Omega) \, d\Omega\right] f_\beta(\vec{v}_1) \, d^3v_1 \, d^3v \tag{5.10}$$

Performing the other integrals in (5.5) by parts, in a similar way, yields for the collision term

$$\int_v \chi \left(\frac{\delta f_\alpha}{\delta t}\right)_{coll} d^3v = \int_\Omega \int_{v_1} \int_v - \chi \sum_i \frac{\partial}{\partial v_i} [\Delta v_i \, f_\alpha \, g \, \sigma(\Omega) \, d\Omega] \, f_{\beta_1} d^3v_1 \, d^3v \, +$$

$$+ \int_\Omega \int_{v_1} \int_v \frac{1}{2} \chi \sum_{i,j} \frac{\partial^2}{\partial v_i \, \partial v_j} [\Delta v_i \, \Delta v_j \, f_\alpha \, g \, \sigma(\Omega) \, d\Omega] \, f_{\beta_1} d^3v_1 \, d^3v =$$

$$= \int_v \chi \left[- \sum_i \frac{\partial}{\partial v_i} \left(f_\alpha \int_\Omega \int_{v_1} \Delta v_i \, g \, \sigma(\Omega) \, d\Omega \, f_{\beta_1} \, d^3v_1\right) d^3v \right] +$$

$$+ \int_v \chi \left[\frac{1}{2} \sum_{i,j} \frac{\partial^2}{\partial v_i \, \partial v_j} \left(f_\alpha \int_\Omega \int_{v_1} \Delta v_i \, \Delta v_j \, g \, \sigma(\Omega) \, d\Omega \, f_{\beta_1} \, d^3v_1\right) d^3v \right] \tag{5.11}$$

We now define the quantities

$$< \Delta v_i >_{av} = \int_\Omega \int_{v_1} \Delta v_i \, g \, \sigma(\Omega) \, d\Omega \, f_{\beta_1} \, d^3v_1 \tag{5.12}$$

and

$$< \Delta v_i \, \Delta v_j >_{av} = \int_\Omega \int_{v_1} \Delta v_i \, \Delta v_j \, g \, \sigma(\Omega) \, d\Omega \, f_{\beta_1} \, d^3v_1 \tag{5.13}$$

which are modified averages over the scattering angle and the velocity distribution of the scatterers. Using this notation (5.11) becomes

$$\int_V \chi \left(\frac{\delta f_\alpha}{\delta t}\right)_{coll} d^3v = \int_V \chi \left[- \sum_i \frac{\partial}{\partial v_i} (f_\alpha < \Delta v_i >_{av}) + \right.$$

$$\left. + \frac{1}{2} \sum_{i,j} \frac{\partial^2}{\partial v_i \partial v_j} (f_\alpha < \Delta v_i \Delta v_j >_{av})\right] d^3v \qquad (5.14)$$

Since this equation must hold for any arbitrary function of velocity $\chi(v)$, it follows that we must have (taking $\chi = 1$)

$$\left(\frac{\delta f_\alpha}{\delta t}\right)_{coll} = - \sum_i \frac{\partial}{\partial v_i} (f_\alpha < \Delta v_i >_{av}) +$$

$$+ \frac{1}{2} \sum_{i,j} \frac{\partial^2}{\partial v_i \partial v_j} (f_\alpha < \Delta v_i \Delta v_j >_{av}) \qquad (5.15)$$

This is the *Fokker-Planck collision term* and the quantities $< v_i >_{av}$ and $< \Delta v_i \Delta v_j >_{av}$ are known, respectively, as the Fokker-Planck coefficients of *dynamical friction* and of *diffusion in velocity space*. They give the mean rate at which Δv_i and $\Delta v_i \Delta v_j$, respectively, are changed due to many consecutive weak Coulomb collisions.

Note that the Fokker-Planck collision term (5.15) has terms of opposite sign, which may result in no net change in f_α as a result of collisions. A dimensional analysis of (5.12) shows that the Fokker-Planck coefficient $< \Delta v_i >_{av}$ has dimensions of force per unit mass and tends to accelerate or deccelerate the particles until they reach the average equilibrium velocity. This process is called dynamical friction. On the other hand, the Fokker-Planck coefficient $< \Delta v_i \Delta v_j >_{av}$ represents diffusion in velocity space until equilibrium is reached. Under equilibrium conditions, diffusion in velocity space is balanced by dynamical friction and there is no net change in f_α as a result of collisions, so that $(\delta f_\alpha/\delta t)_{coll} = 0$. This process is illustrated schematically in Fig. 5.

In principle, the expansion procedure used to obtain the Fokker-Planck collision term can be extended to any number of terms. However, in practice only the first two terms of the expansion, shown in (5.15), are ever used, so that (5.15) can be considered as a reasonable approximation to the collision term $(\delta f_\alpha/\delta t)_{coll}$ when $\Delta\vec{v} = \vec{v}' - \vec{v}$ is small for most collisions. This is generally supposed to be the case for long-range forces such as the Coulomb force.

5.2 The Fokker-Planck coefficients for Coulomb interactions

We now evaluate the coefficients of dynamical friction $<\Delta v_i>_{av}$ and diffusion in velocity $<\Delta v_i \, \Delta v_j>_{av}$, which appear in the Fokker-Planck collision term (5.15), for the case of the Coulomb interaction. It is convenient to perform first the integral over the solid angle Ω, since it does not require a knowledge of the velocity distribution function $f_\beta(\vec{v}_1)$. For this purpose let us write

$$<\Delta v_i>_{av} = \int_{v_1} \{\Delta v_i\} \, f_{\beta 1} \, d^3 v_1 \tag{5.16}$$

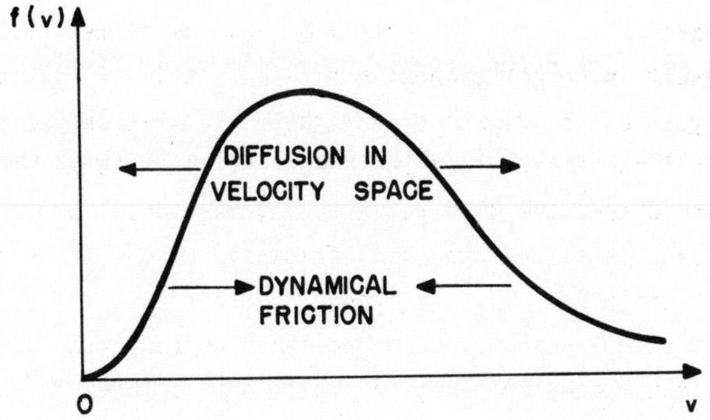

Fig. 5 . Illustrating the processes of dynamical friction and diffusion in velocity space.

and

$$< \Delta v_i \, \Delta v_j >_{av} = \int_{v_1} \{\Delta v_i \; \Delta v_j\} \, f_{\beta_1} \, d^3 v_1 \tag{5.17}$$

where the curly bracket notation has been introduced to represent the following integrals over solid angle

$$\{\Delta v_i\} = \int_{\Omega} \Delta v_i \; g \; \sigma(\Omega) \; d\Omega \tag{5.18}$$

and

$$\{\Delta v_i \; \Delta v_j\} = \int_{\Omega} \Delta v_i \; \Delta v_j \; g \; \sigma(\Omega) \; d\Omega \tag{5.19}$$

In order to calculate $\{\Delta v_i\}$ and $\{\Delta v_i \; \Delta v_j\}$, we recall first that in the center of mass coordinate system we have, from (20.2.5) and (20.2.8),

$$\vec{v} = \vec{c}_0 - \vec{g} \; m_\beta/(m_\alpha + m_\beta) \tag{5.20}$$

$$\vec{v}\,' = \vec{c}_0 - \vec{g}\,' \; m_\beta/(m_\alpha + m_\beta) \tag{5.21}$$

so that

$$\Delta\vec{v} = \vec{v}\,' - \vec{v} = (\vec{g} - \vec{g}\,') \; m_\beta/(m_\alpha + m_\beta) \tag{5.22}$$

In a Cartesian coordinate system in which the vector g is along the z axis (see Fig. 3, of Chapter 20), we have

$$g_z = g \qquad ; \qquad g_x = g_y = 0 \tag{5.23}$$

and

$$g_x' = g \sin \chi \cos\varepsilon \tag{5.24}$$

$$g_y' = g \sin \chi \sin\varepsilon \tag{5.25}$$

$$g'_z = g \cos \chi \tag{5.26}$$

In the next sub-section we consider electrons deflected by the field of a stationary group of positive ions, and in this case there is no difference between the center of mass system and the laboratory system. Using Eqs. (5.23) to (5.26), in (5.22), gives

$$\Delta\vec{v} = \left(\frac{m_\beta}{m_\alpha+m_\beta}\right) g\,[(1-\cos\chi)\,\hat{z} - \sin\chi\,(\cos\varepsilon\,\hat{x} + \sin\varepsilon\,\hat{y})] \tag{5.27}$$

The differential scattering cross section for the Coulomb potential was calculated in section 7, Chapter 20, and is given by

$$\sigma(\chi) = \frac{b_0^2}{4\sin^4(\chi/2)} = \frac{b_0^2}{(1-\cos\chi)^2} \tag{5.28}$$

[see (20.7.3) and (20.7.4)] where b_0 is defined in (20.4.9).

Proceeding in the evaluation of $\{\Delta v_i\} = \{v'_i - v_i\}$ for i = x, y, z, let us first calculate $\{\Delta v_z\}$. From (5.18), (5.27) and (5.28), we have

$$\{\Delta v_z\} = \int_\Omega \Delta v_z\, g\, \sigma(\Omega)\, d\Omega = \left(\frac{m_\beta}{m_\alpha+m_\beta}\right) g^2 b_0^2 \int_0^{2\pi} d\varepsilon \int_{\chi_{min}}^{\pi} \frac{\sin\chi}{(1-\cos\chi)}\, d\chi \tag{5.29}$$

where the lower limit of the integral in χ was taken to be χ_{min}, in order to avoid the divergence of the integral that would result if we take $\chi_{min} = 0$. As we have seen, the charged particles in the plasma that are separated by distances greater than λ_D are effectively shielded from one another. Therefore, in order to avoid an infinite result for the integral in (5.29), we take the lower limit χ_{min} to be value of the scattering angle that corresponds to an impact parameter equal to λ_D.

With reference to (20.4.13) let us introduce the new variable

$$u = (b/b_0) = \cot(\chi/2) \tag{5.30}$$

from which we obtain

$$du = -\, d\chi/(1-\cos\chi) \tag{5.31}$$

and

$$\sin \chi = 2u/(1+u^2) \tag{5.32}$$

With this change of variables and introducing the cut-off value for the impact parameter at $b_c = \lambda_D$, that is, at

$$u_c = \lambda_D/b_0 = \Lambda \tag{5.33}$$

we obtain for (5.29),

$$\{\Delta v_z\} = 2\pi \left(\frac{m_\beta}{m_\alpha+m_\beta}\right) g^2\, b_0^2 \int_\Lambda^0 \frac{2u}{(1+u^2)}\,(-du)$$

$$= 4\pi \left(\frac{m_\beta}{m_\alpha+m_\beta}\right) g^2\, b_0^2 \int_0^\Lambda \frac{u}{(1+u^2)}\,du$$

$$= 2\pi \left(\frac{m_\beta}{m_\alpha+m_\beta}\right) g^2\, b_0^2\, \ln\,(1+\Lambda^2) \tag{5.34}$$

In general $\Lambda >> 1$, so that $\ln\,(1+\Lambda^2) \simeq 2\,\ln\,\Lambda$ and (5.34) simplifies to

$$\{\Delta v_z\} = \left(\frac{m_\beta}{m_\alpha+m_\beta}\right) \frac{Z^2\, e^4\, \ln\,\Lambda}{4\pi\, \varepsilon_0^2\, \mu^2\, g^2} \tag{5.35}$$

where we have substituted b_0 by the expression given in (20.4.9). Introducing the notation

$$\Theta = \frac{Z^2\, e^4 \ln\,\Lambda}{4\pi\, \varepsilon_0^{\,2}\, \mu^2} \tag{5.36}$$

we can write (5.35) as

$$\{\Delta v_z\} = \left(\frac{m_\beta}{m_\alpha+m_\beta}\right) \frac{\Theta}{g^2} \tag{5.37}$$

Next let us consider the quantities $\{\Delta v_x\}$ and $\{\Delta v_y\}$. From (5.18) and (5.27) we see that these quantities involve integrals of either cos ε or sin ε from 0 to 2π, which are clearly equal to zero. Therefore,

$$\{\Delta v_x\} = \{\Delta v_y\} = 0 \tag{5.38}$$

In a similar way, it can be shown from (5.19) and (5.27) that

$$\{\Delta v_i \; \Delta v_j\} = 0 \qquad \text{for} \quad i \neq j \tag{5.39}$$

since the integrals over ε from 0 to 2π vanish.

To evaluate $\{\Delta v_z \; \Delta v_z\} = \{\Delta v_z^2\}$ we use (5.19), (5.27) and (5.28), which give

$$\{\Delta v_z^2\} = 2\pi \left(\frac{m_\beta}{m_\alpha + m_\beta}\right)^2 g^3 \, b_0^2 \int_{\chi_{min}}^{\pi} \sin\chi \, d\chi \tag{5.40}$$

Changing variables according to (5.30), we obtain

$$\{\Delta v_z\} = 2\pi \left(\frac{m_\beta}{m_\alpha + m_\beta}\right)^2 g^3 \, b_0^2 \int_{\Lambda}^{0} \frac{4u}{(1+u^2)^2} (-du)$$

$$= 4\pi \left(\frac{m_\beta}{m_\alpha + m_\beta}\right)^2 g^3 \, b_0^2 \, \frac{\Lambda^2}{(1+\Lambda^2)} \tag{5.41}$$

Since $\Lambda >> 1$, (5.41) simplifies to

$$\{\Delta v_z^2\} = 4\pi \left(\frac{m_\beta}{m_\alpha + m_\beta}\right)^2 g^3 \, b_0^2 = \left(\frac{m_\beta}{m_\alpha + m_\beta}\right)^2 \frac{Z^2 \, e^4}{4\pi \, \varepsilon_0^2 \, \mu^2 \, g} \tag{5.42}$$

or, using the notation introduced in (5.36),

$$\{\Delta v_z^2\} = \left(\frac{m_\beta}{m_\alpha + m_\beta}\right)^2 \frac{\Theta}{g \, \ell n \, \Lambda} \tag{5.43}$$

In a similar way, we can calculate $\{\Delta v_x^2\}$ and $\{\Delta v_y^2\}$ from (5.19), (5.27) and (5.28), which give

$$\{\Delta v_x^2\} = \left(\frac{m_\beta}{m_\alpha+m_\beta}\right)^2 g^3 b_0^2 \int_0^{2\pi} \cos^2\varepsilon\, d\varepsilon \int_{\chi_{min}}^{\pi} \frac{\sin^3\chi}{(1-\cos\chi)^2}\, d\chi \tag{5.44}$$

$$\{\Delta v_y^2\} = \left(\frac{m_\beta}{m_\alpha+m_\beta}\right)^2 g^3 b_0^2 \int_0^{2\pi} \sin^2\varepsilon\, d\varepsilon \int_{\chi_{min}}^{\pi} \frac{\sin^3\chi}{(1-\cos\chi)^2}\, d\chi \tag{5.45}$$

Therefore, evaluating the integral over ε we find

$$\{\Delta v_x^2\} = \{\Delta v_y^2\} = \pi \left(\frac{m_\beta}{m_\alpha+m_\beta}\right)^2 g^3 b_0^2 \int_{\chi_{min}}^{\pi} \frac{\sin^3\chi}{(1-\cos\chi)^2}\, d\chi \tag{5.46}$$

If we change variables according to (5.30), we readily find that (5.46) can be written as

$$\{\Delta v_x^2\} = \{\Delta v_y^2\} = 4\pi \left(\frac{m_\beta}{m_\alpha+m_\beta}\right)^2 g^3 b_0^2 \int_0^{\Lambda} \frac{u^3}{(1+u^2)^2}\, du$$

$$= 2\pi \left(\frac{m_\beta}{m_\alpha+m_\beta}\right)^2 g^3 b_0^2 \left[\ln(1+\Lambda^2) - \frac{\Lambda^2}{(1+\Lambda^2)}\right] \tag{5.47}$$

Since $\Lambda >> 1$, and replacing b_0 by (20.4.9), we can write

$$\{\Delta v_x^2\} = \{\Delta v_y^2\} = \left(\frac{m_\beta}{m_\alpha+m_\beta}\right)^2 \frac{Z^2 e^4 \ln\Lambda}{4\pi\, \varepsilon_0^2\, \mu^2\, g} = \left(\frac{m_\beta}{m_\alpha+m_\beta}\right)^2 \frac{\Theta}{g} \tag{5.48}$$

The next step in the evaluation of the Fokker-Planck coefficients consists in integrating the values of $\{\Delta v_i\}$ and $\{\Delta v_i\, \Delta v_j\}$, for i, j = x, y, z, over the distribution function of the particles which constitute the scattering center. Thus, using the results we have just calculated we find

$$< \Delta v_z >_{av} = \left(\frac{m_\beta}{m_\alpha + m_\beta}\right) \int_{v_1} \frac{\Theta}{g^2} f_{\beta_1} \, d^3 v_1 \tag{5.49}$$

$$< \Delta v_z^2 >_{av} = \left(\frac{m_\beta}{m_\alpha + m_\beta}\right)^2 \int_{v_1} \frac{\Theta}{g \, \ell n \, \Lambda} f_{\beta_1} \, d^3 v_1 \tag{5.50}$$

$$< \Delta v_x^2 >_{av} = < \Delta v_y^2 >_{av} = \left(\frac{m_\beta}{m_\alpha + m_\beta}\right)^2 \int_{v_1} \frac{\Theta}{g} f_{\beta_1} \, d^3 v_1 \tag{5.51}$$

All other coefficients vanish.

5.3 Application to electron-ion collisions

Let us calculate the Fokker-Planck coefficients for the case of electron-ion collisions. For simplicity we assume that the electron is colliding with a field of heavy *stationary* positive ions. This assumption is reasonable, since on the average the electron velocities are much larger than the ion velocities ($< v^2_i > = 3kT_i/m_i$ while $< v_e^2 > = 3kT_e/m_e$ and generally $T_e/m_e \gg T_i/m_i$). Thus, assuming that the positive ions are motionless we can set their velocity distribution function equal to the Dirac delta function,

$$f_{\beta_1} = n_0 \, \delta(v_{1x}) \, \delta(v_{1y}) \, \delta(v_{1z}) \tag{5.52}$$

In addition, because $m_i \gg m_e$, we can take $(m_e + m_i) \simeq m_i$ and $\mu \simeq m_e$.
Substituting(5.52) into Eqs. (5.49) through (5.51), we obtain at once

$$< \Delta v_z >_{av} \equiv < \Delta v_{\parallel} >_{av} = n_0 \Theta / g^2 \tag{5.53}$$

$$< \Delta v_z^2 >_{av} \equiv < \Delta v_{\parallel}^2 >_{av} = n_0 \Theta / g \, \ell n \, \Lambda \tag{5.54}$$

$$< \Delta v_x^2 >_{av} = < \Delta v_y^2 >_{av} \equiv < \Delta v_{\perp}^2 >_{av} = n_0 \Theta / g \tag{5.55}$$

where Θ is defined according to (5.36). Since v_z is in the direction of the initial relative velocities, we have used in (5.53) and (5.54) the notation $\Delta v_z \equiv \Delta v_{\parallel}$, whereas in (5.55) we have used $\Delta v_x = \Delta v_y = \Delta v_{\perp}$, to denote the change in velocity in the directions parallel and perpendicular to the initial relative velocities, respectively.

PROBLEMS

21.1 - Consider a system consisting of a mixture of *two types* of particles having masses m and M, and subjected to an external force $\vec{F}$. Denote the corresponding distribution functions by f and g, respectively, and write down the set of coupled Boltzmann transport equations for the system.

21.2 - Consider a plasma in which the electrons and ions are characterized, respectively, by the following distribution functions:

$$f_e = n_0 \left(\frac{m_e}{2\pi k T_e}\right)^{3/2} \exp\left[-\frac{m_e(\vec{v}-\vec{u}_e)^2}{2k\, T_e}\right]$$

$$f_i = n_0 \left(\frac{m_i}{2\pi k T_i}\right)^{3/2} \exp\left[-\frac{m_i(\vec{v}-\vec{u}_i)^2}{2k\, T_i}\right]$$

(a) Calculate the difference $(f_e' \, f_{i1}' - f_e \, f_{i1})$.

(b) Show that this plasma of electrons and ions are in the equilibrium state, that is, the difference $(f_e' \, f_{i1}' - f_e \, f_{i1})$ vanishes, if and only if $\vec{u}_e = \vec{u}_i$ and $T_e = T_i$.

21.3 - Use a Lagrange multiplier technique to show that for a system characterized by the following modified Maxwell-Boltzmann distribution

$$f(\vec{r},v) = n\,(\vec{r}) \left(\frac{m}{2\pi kT}\right)^{3/2} \exp\left(-\frac{mv^2}{2kT}\right)$$

where T is constant, the entropy S defined by

$$S = -k \int_r \int_v f \,\ell n\, f \, d^3v \, d^3r$$

is a maximum when the density n is a constant, independent of r. Consider that the system has a total of N particles in a fixed volume V at a temperature T.

21.4 - Consider the case of *Maxwell molecules*, for which the interparticle force is of the form

$$F(r) = K/r^5$$

where K is a constant.

(a) Without specifying the form of the distribution functions $f_\alpha(\vec{v})$ and $f_{\beta 1}(\vec{v}\,)$ for the particles of type α and β, show that the time rate of change of momentum for the particles of type α per unit volume, due to collisions, is given by

$$\left(\frac{\delta \vec{P}_\alpha}{\delta t}\right)_{coll} \equiv \int_V m_\alpha \vec{c}_\alpha \left(\frac{\delta f_\alpha}{\delta t}\right)_{coll} d^3v = \sum_\beta n_\alpha m_\alpha \nu_{\alpha\beta} (\vec{u}_\beta - \vec{u}_\alpha)$$

where $\nu_{\alpha\beta}$ is the collision frequency for momentum transfer given explicitly by

$$\nu_{\alpha\beta} = 2\pi \, (K\mu)^{1/2} \, (n_\beta/m_\alpha) \, A_1(5)$$

where $A_1(5)$ is a dimensionless number (of order unity) defined by (with $p = 5$ and $\ell = 1$)

$$A_\ell(p) = \int_0^\infty (1 - \cos^\ell \chi) \, v_0 \, dv_0$$

with

$$v_0 = b \, (\mu g^2/K)^{1/(p-1)}$$

Also,

$$n_\alpha = \int_V f_\alpha \, d^3v \quad ; \quad \vec{u}_\alpha = \frac{1}{n_\alpha} \int_V \vec{v} f_\alpha \, d^3v$$

$$n_\beta = \int_{V_1} f_{\beta 1} \, d^3v_1 \quad ; \quad \vec{u}_\beta = \frac{1}{n_\beta} \int_{V_1} \vec{v}_1 f_{\beta 1} \, d^3v_1$$

and $(\delta f_\alpha/\delta t)_{coll}$ denotes the Boltzmann collision term.

(b) For the same case, show that the time rate of change of the energy for the particles of type α per unit volume, due to collisions, is given by

$$\left(\frac{\delta E_\alpha}{\delta t}\right)_{coll} \equiv \int_V \frac{1}{2} m_\alpha c_\alpha^2 \left(\frac{\delta f_\alpha}{\delta t}\right) d^3v = \frac{n_\alpha m_\alpha \nu_{\alpha\beta}}{(m_\alpha + m_\beta)}.$$

$$\cdot \, [\, 3k\,(T_\beta - T_\alpha) + m_\beta\,(\vec{u}_\beta - \vec{u}_\alpha)^2]$$

where

$$T_\alpha = \frac{m_\alpha}{3k} < c_\alpha^2 > = \frac{m_\alpha}{3k}\frac{1}{n_\alpha}\int_V c_\alpha^2\, f_\alpha\, d^3v$$

$$T_\beta = \frac{m_\beta}{3k} < c_{\beta 1}^2 > = \frac{m_\beta}{3k}\frac{1}{n_\beta}\int_{V_1} c_{\beta 1}^2\, f_{\beta 1}\, d^3v_1$$

21.5 - Consider a gas mixture of two types of particles ($\alpha = 1,2$), each one characterized by a Maxwellian distribution function

$$f_\alpha(\vec{v}_\alpha) = n_\alpha \left(\frac{m_\alpha}{2\pi kT_\alpha}\right)^{3/2} \exp\left(-\frac{m_\alpha v_\alpha^2}{2kT_\alpha}\right); \qquad (\alpha = 1,2)$$

with its own mass, density and temperature.

(a) Make the following transformation of velocity variables

$$\vec{v}_1 = \vec{v}_c^{\,\prime} + \overline{M}_2\,\vec{g}$$

$$\vec{v}_2 = \vec{v}_c^{\,\prime} - \overline{M}_1\,\vec{g}$$

where $\vec{v}_c^{\,\prime}$ is a velocity similar to the center of mass velocity, $\vec{g}$ is the relative velocity between the two species ($\vec{g} = \vec{v}_1 - \vec{v}_2$) and

$$\overline{M}_1 = (m_1/T_1)/[(m_1/T_1) + (m_2/T_2)]$$

$$\overline{M}_2 = (m_2/T_2)/[(m_1/T_1) + (m_2/T_2)]$$

Show that the Jacobian of this transformation satisfies

$$|J| = \left|\frac{\partial(\vec{v}_c, \vec{g})}{\partial(\vec{v}_1, \vec{v}_2)}\right| = 1$$

so that $d^3v_c\, d^3g = d^3v_1\, d^3v_2$.

(b) The relative speed between the two species, $g = |\vec{v}_1 - \vec{v}_2|$, when averaged over both their velocity distribution functions, is given by

$$\langle g \rangle = \frac{1}{n_1 n_2} \int_{v_1} \int_{v_2} g\, f_1(\vec{v}_1)\, f_2(\vec{v}_2)\, d^3v_1\, d^3v_2$$

Transform the variables of integration $\vec{v}_1$ and $\vec{v}_2$ to $\vec{v}_c'$ and $\vec{g}$, and perform the integrals over $\vec{v}_c'$ and $\vec{g}$, to show that

$$\langle g \rangle = [(8k/\pi)(T_1/m_1 + T_2/m_2)]^{1/2}$$

(c) If only one kind of particles is present, so that $m_1 = m_2 = m$, $T_1 = T_2 = T$, and $n_1 = n_2 = n$, show that

$$\langle g \rangle = \sqrt{2} \langle v \rangle = (8kT/\pi\mu)^{1/2}$$

where $\langle v \rangle = (8kT/\pi m)^{1/2}$ is the average speed and $\mu = m/2$ is the reduced mass. If the mutual scattering cross section is σ, show that the collision frequency in a homogeneous Maxwellian gas is given by

$$\nu = n \sigma \langle g \rangle = 4 n \sigma (kT/\pi m)^{1/2}$$

21.6 - Consider the following expressions which define the Fokker-Planck coefficients of dynamical friction and of diffusion in velocity,

$$\langle \Delta v_i \rangle_{av} = \int_{\Omega} \int_{v_1} \Delta v_i\, g\, \sigma(\Omega)\, d\Omega\, f_{\beta 1}\, d^3v_1$$

$$\langle \Delta v_i \Delta v_j \rangle_{av} = \int_{\Omega} \int_{v_1} \Delta v_i\, \Delta v_j\, g\, \sigma(\Omega)\, d\Omega\, f_{\beta 1}\, d^3v_1$$

(a) With reference to Fig. P.21.1, verify that

$$\Delta v_x = -[m_\beta/(m_\alpha + m_\beta)]\, g \sin\chi \cos\varepsilon$$

$$\Delta v_y = -[m_\beta/(m_\alpha + m_\beta)]\, g \sin\chi \sin\varepsilon$$

$$\Delta v_z = [m_\beta/(m_\alpha + m_\beta)]\, g\, (1 - \cos\chi)$$

For a general inverse-power interparticle force of the form $F(r) = K/r^p$, where K is a constant and p is a positive integer number, show that [see (5.18)]

$$\{\Delta v_x\} = \{\Delta v_y\} = 0$$

$$\{\Delta v_z\} = \mu g^2 \sigma_m / m_\alpha$$

where $\mu = m_\alpha m_\beta/(m_\alpha + m_\beta)$ is the reduced mass and σ_m is the momentum transfer cross section given by

$$\sigma_m = 2\pi (K/\mu g^2)^{2/(p-1)} A_1(p)$$

where

$$A_\ell(p) = \int_0^\infty (1 - \cos^\ell \chi) \, v_0 \, dv_0$$

$$v_0 = b \, (\mu g^2/K)^{1/(p-1)}$$

Verify also that [see Eq. (5.19)]

$$\{\Delta v_i \, \Delta v_j\} = 0 \quad \text{for} \quad i \neq j$$

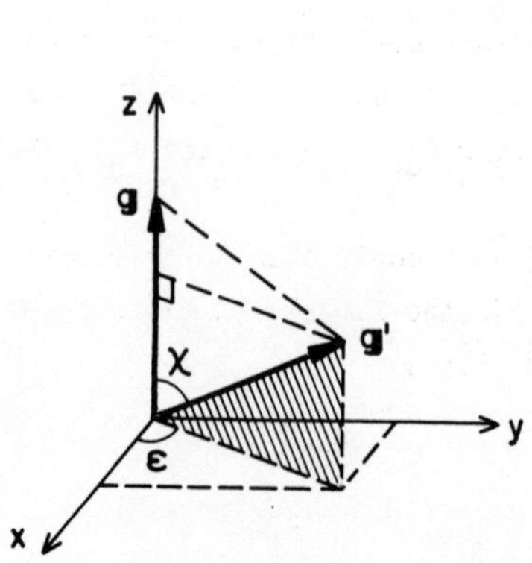

Fig. P.21.1

$$\{\Delta v_x^2\} = \{\Delta v_y^2\} = \pi \frac{\mu^2 g^3}{m_\alpha^2} \left(\frac{K}{\mu g^2}\right)^{2/(p-1)} A_2(p)$$

$$\{\Delta v_z^2\} = 2\pi \frac{\mu^2 g^3}{m_\alpha^2} \left(\frac{K}{\mu g^2}\right)^{2/(p-1)} [2 A_1(p) - A_2(p)]$$

(b) For the case of *Maxwell molecules* (p = 5), where the results are independent of $f_{\beta 1}$, show that the Fokker-Planck coefficients are given by

$$< \Delta v_x >_{av} = < \Delta v_y >_{av} = 0$$

$$< \Delta v_z >_{av} = \nu_{\alpha\beta} < g >_\beta$$

$$< \Delta v_i \ \Delta v_j >_{av} = 0 \quad \text{for} \quad i \neq j$$

$$< \Delta v_x^2 >_{av} = < \Delta v_y^2 >_{av} = \frac{\mu}{m_\alpha} \frac{A_2(5)}{2A_1(5)} \nu_{\alpha\beta} < g^2 >_\beta$$

$$< \Delta v_z^2 >_{av} = \frac{\mu}{m_\alpha} \left[2 - \frac{A_2(5)}{A_1(5)}\right] \nu_{\alpha\beta} < g^2 >_\beta$$

where

$$\nu_{\alpha\beta} = 2\pi \ (K\mu)^{1/2} (n_\beta/m_\alpha) \ A_1(5)$$

$$< g^i >_\beta = \frac{1}{n_\beta} \int_{v_1} g^i \ f_{\beta 1} \ d^3v_1; \quad (i = 1,2)$$

(c) Calculate the Fokker-Planck coefficients for the case of *Coulomb interactions* (p=2) using the resuls of part (a) and of Problem 20.6, in terms of integrals over f_{β_1}, and compare with the results derived in sub-section 5.2.

(d) Calculate the Fokker-Planck coefficients for electron-electron interactions, when $f_{\beta 1}$ is the Maxwellian distribution function. Refer to (5.49), (5.50), and (5.51).

CHAPTER 22
Transport Processes in Plasmas

1. INTRODUCTION

In this chapter we analyze some basic transport processes in weakly ionized plasmas using the Boltzmann equation with the relaxation model for the collision term, considering a velocity-dependent collision frequency.

Transport phenomena in plasmas can be promoted by external and internal forces. In a *spatially homogeneous* plasma under the influence of external forces, a drifting of the electrons can occur. This motion induced by external forces is referred to as mobility. Since the electrons have mass and electric charge, their motion implies in transport of mass and in conduction of electricity when acted upon by an external electric field. On the other hand, in a *spatially inhomogeneous* plasma, collisions cause the electrons to drift from the high-pressure to the low-pressure regions. The existence of pressure gradients is associated with the existence of either density gradients or temperature gradients, or both. This motion of the electrons, induced by internal pressure gradients, is called diffusion. Since the electrons also have kinetic energy associated with their random thermal motion, their drift implies in the transport of thermal energy and therefore in heat conduction. When the plasma is spatially inhomogeneous and is also acted upon by external forces, then the particle flux is due to both diffusion and mobility. The basic transport phenomena which we analyze in this chapter using the Boltzmann equation with the relaxation model are electric conduction, particle diffusion and thermal energy flux.

2. ELECTRIC CONDUCTIVITY IN A NONMAGNETIZED PLASMA

Initially we derive an expression for the AC conductivity of a *weakly ionized plasma*, in which only the electron-neutral collisions are important. [illegible] consider that the spatial inhomogeneity and the anisotropy of the electron [illegible]ibrium distribution function are very small, so that we can apply the [illegible]ed in section 4, Chapter 21. Thus, according to (21.4.17) we have

$$[\delta f(\vec{r},\vec{v},t) / \delta t]_{coll} = - \nu_r(v) \quad [f(\vec{r},\vec{v},t) - f_0(v)] \tag{2.1}$$

where $f_0(v)$ denotes the homogeneous isotropic equilibrium distribution function of the electrons, and $\nu_r(v)$ is a velocity-dependent relaxation collision frequency. Expression (2.1) assumes that the neutral particles are stationary and do not recoil as they collide with electrons, in view of their much larger mass.

2.1 Solution of Boltzmann equation

We assume that the electron distribution function $f(\vec{r},\vec{v},t)$ deviates only slightly from the equilibrium function $f_0(v)$, so that

$$f(\vec{r},\vec{v},t) = f_0(v) + f_1(\vec{r},\vec{v},t) \; ; \qquad |f_1| \ll f_0 \tag{2.2}$$

where $f_1(\vec{r},\vec{v},t)$ corresponds to the small anisotropy and spatial inhomogeneity of the electrons in the nonequilibrium state. Using (2.2) and the relaxation model (2.1), the collision term in the Boltzmann equation becomes

$$(\delta f/\delta t)_{coll} = - \nu_r(v) \quad f_1(\vec{r},\vec{v},t) \tag{2.3}$$

Substituting (2.2) and (2.3) into the Boltzmann equation, and neglecting second order quantities, we obtain

$$\partial f_1(\vec{r},\vec{v},t) / \partial t + (\vec{v} \cdot \vec{\nabla}) \; f_1(\vec{r},\vec{v},t) - (e/m_e) \; \vec{E}(\vec{r},t) \cdot \vec{\nabla}_v \; f_0(v) =$$

$$= - \nu_r(v) \; f_1(\vec{r},\vec{v},t) \tag{2.4}$$

where we have considered the electric field $\vec{E}(\vec{r},t)$ as the only field externally applied to the plasma. For the purpose of evaluating the conductivity, the perturbation $f_1(\vec{r},\vec{v},t)$ in the velocity distribution function can be assumed to be essentially independent of the position coordinate $\vec{r}$, and therefore denoted by $f_1(\vec{v},t)$, since the main effect associated with a spatial variation is the diffusion of particles and, at the moment, we are interested primarily in the charged particle current density induced by an electric fiel The electric field is considered to vary harmonically in time at a frequen ω, according to

$$\vec{E}(\vec{r},t) = \vec{E}(\vec{r}) \exp(-i\omega t) \tag{2.5}$$

Therefore, we assume that $f_1(\vec{v},t)$ has also the same time variation,

$$f_1(\vec{v},t) = f_1(\vec{v}) \exp(-i\omega t) \tag{2.6}$$

Consequently, for the phasor amplitudes, the Boltzmann equation (2.4) simplifies to

$$-i\omega f_1(\vec{v}) - (e/m_e)\, \vec{E}(\vec{r}) \cdot \vec{\nabla}_v f_0(v) = -\nu_r(v)\, f_1(\vec{v}) \tag{2.7}$$

Using the following identity, given in (18.3.17),

$$\vec{\nabla}_v f_0(v) = (\vec{v}/v)\, df_0(v)/dv \tag{2.8}$$

we obtain, from (2.7),

$$f_1(\vec{v}) = \frac{ie}{m_e} \frac{\vec{E}(\vec{r}) \cdot \vec{v}}{v[\omega + i\nu_r(v)]} \frac{df_0(v)}{dv} \tag{2.9}$$

2.2 Electric current density and conductivity

The electric current density is given by

$$\vec{J}(\vec{r},t) = -e\, n_e \langle\vec{v}\rangle_e = -e \int_v \vec{v}\, f(\vec{r},\vec{v},t)\, d^3v \tag{2.10}$$

Using (2.2), (2.6) and (2.9) we find that

$$\vec{J}(\vec{r},t) = \vec{J}(\vec{r}) \exp(-i\omega t) \tag{2.11}$$

ere

$$= -e \int_v \vec{v}\, f_1(\vec{v})\, d^3v$$

$$\int_v \frac{\vec{v}\, [\vec{E}(\vec{r}) \cdot \vec{v}]}{v[\omega + i\nu_r(v)]} \frac{df_0(v)}{dv}\, d^3v \tag{2.12}$$

In this result we have assumed that the electrons have no average flow velocity in the equilibrium state, that is,

$$\vec{u}_0 = \frac{1}{n} \int_V \vec{v} \; f_0(v) \; d^3v \; = 0 \tag{2.13}$$

In spherical coordinates (v, θ, ϕ) in velocity space (Fig. 1), we have $d^3v = v^2 \, dv \sin\theta \, d\theta \, d\phi$, so that (2.12) can be rewritten as

$$\vec{J}(\vec{r}) = - \frac{ie^2}{me} \int_0^\infty \frac{v \, dv}{[\omega + i\nu_r(v)]} \frac{df_0(v)}{dv} \int_0^\pi d\theta \sin\theta \int_0^{2\pi} \vec{v} [\vec{E}(\vec{r}) \cdot \vec{v}] \, d\phi \tag{2.14}$$

Using the following orthogonality relation

$$\int_0^\pi \int_0^{2\pi} v_i \, v_j \sin\theta \, d\theta \, d\phi = \frac{4\pi}{3} v^2 \, \delta_{ij} \tag{2.15}$$

with i, j = x, y, z, we find that

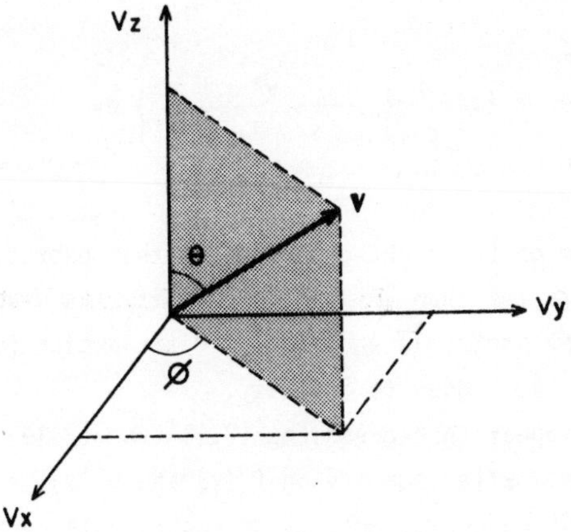

Fig. 1 - Spherical coordinates (v, θ, ϕ) in velocity space.

$$\int_0^{\pi} d\theta \sin\theta \int_0^{2\pi} d\phi \, \vec{v} \, (\vec{E} \cdot \vec{v}) = \frac{4\pi}{3} v^2 \vec{E} \tag{2.16}$$

Consequently, (2.14) becomes

$$\vec{J}(\vec{r}) = -\frac{i\, 4\pi\, e^2}{3\, m_e} \vec{E}(\vec{r}) \int_0^{\infty} \frac{v^3}{[\omega + i\nu_r(v)]} \frac{df_0(v)}{dv} dv \tag{2.17}$$

From the relation $\vec{J} = \sigma\vec{E}$ we identify the following expression for the electric conductivity

$$\sigma = -\frac{i\, 4\pi\, e^2}{3\, m_e} \int_0^{\infty} \frac{v^3}{[\omega + i\nu_r(v)]} \frac{df_0(v)}{dv} dv \tag{2.18}$$

An alternative expression for the electric conductivity can be obtained by integrating (2.18) by parts,

$$\sigma = -\frac{i\, 4\pi\, e^2}{3\, m_e} \left\{ \frac{v^3 \, f_0(v)}{[\omega + i\nu_r(v)]} \right\} \Bigg|_0^{\infty} +$$

$$+ \frac{i\, 4\pi\, e^2}{3\, m_e} \int_0^{\infty} f_0(v) \frac{d}{dv} \left\{ \frac{v^3}{[\omega + i\nu_r(v)]} \right\} dv \tag{2.19}$$

The integrated out term on the right-hand side of this expression vanishes, since f_0 goes to zero faster than v^3 goes to infinity, as v approaches infinity. In general the isotropic equilibrium distribution function, $f_0(v)$, decreases exponentially as v goes to infinity.

The integrals which appear in expressions (2.18) and (2.19) can be calculated explicitly only after specifying $f_0(v)$ and $\nu_r(v)$. The functional dependence of ν_r on v is generally determined experimentally from cross section measurements.

If we assume that ν_r is independent of v, then we obtain, from (2.19), for any $f_0(v)$,

$$\sigma = \frac{i\, 4\pi\, e^2}{3\, m_e\, (\omega + i\, \nu_r)} \int_0^\infty f_0(v)\, 3v^2\, dv = \frac{i\, n_0\, e^2}{m_e\, (\omega + i\, \nu_r)}$$

$$= \frac{\nu_r\, n_0\, e^2}{m_e\, (\omega^2 + \nu_r^2)} + i\, \frac{\omega\, n_0\, e^2}{m_e\, (\omega^2 + \nu_r^2)} \tag{2.20}$$

where n_0 denotes the electron number density at equilibrium,

$$n_0 = 4\pi \int_0^\infty f_0(v)\, v^2\, dv \tag{2.21}$$

The result (2.20) is identical to the one obtained in section 5, Chapter 10 [see Eq. (10.5.5)], for the longitudinal conductivity.

2.3 Conductivity for Maxwellian distribution function

Let us consider that $f_0(v)$ is given by the Maxwellian distribution function,

$$f_0(v) = n_0 \left(\frac{m_e}{2\pi\, k\, T} \right)^{3/2} \exp\left(- \frac{m_e\, v^2}{2\, k\, T} \right) \tag{2.22}$$

Defining a dimensionless variable by

$$\xi = (m_e / 2\, k\, T)^{1/2}\, v \tag{2.23}$$

it can be verified that

$$v^3\, \frac{df_0(v)}{dv}\, dv = - \frac{2\, n_0}{\pi^{3/2}}\, \xi^4\, \exp(-\xi^2)\, d\xi \tag{2.24}$$

Substituting this expression into (2.18) and rationalizing, we find

$$\sigma = \frac{8 n_0 e^2}{3 m_e \pi^{1/2}} \left[\int_0^\infty \frac{\nu_r(\xi)\, \xi^4 \exp(-\xi^2)}{\nu_r^2(\xi) + \omega^2} d\xi + \right.$$

$$\left. + i\omega \int_0^\infty \frac{\xi^4 \exp(-\xi^2)}{\nu_r^2(\xi) + \omega^2} d\xi \right] \tag{2.25}$$

This equation can be used to calculate the electric conductivity of a weakly ionized plasma when the equilibrium distribution function of the electrons is the Maxwell-Boltzmann distribution, for any dependence of the collision frequency ν_r on speed v. In particular, if ν_r is independent of v, then (2.25) reduces directly to the result (2.20).

3. ELECTRIC CONDUCTIVITY IN A MAGNETIZED PLASMA

We consider now a weakly ionized plasma immersed in an externally applied magnetostatic field, $\vec{B}_0$. As in the previous section, we assume that the distribution function of the electrons in the nonequilibrium state is only slightly perturbed from the equilibrium value. For purposes of calculating the conductivity, it can also be assumed that the plasma is homogeneous in space. Therefore, we can write

$$f(\vec{v},t) = f_0(v) + f_1(\vec{v},t) \tag{3.1}$$

where $|f_1| << f_0$. Suppose that an AC electric field is applied to the plasma, having a harmonic time dependence according to

$$\vec{E}(\vec{r},t) = \vec{E}(\vec{r}) \exp(-i\omega t) \tag{3.2}$$

Consequently, we also have

$$f_1(\vec{v},t) = f_1(\vec{v}) \exp(-i\omega t) \tag{3.3}$$

The total magnetic field will be denoted by

$$\vec{B}_t(\vec{r},t) = \vec{B}_0 + \vec{B}(\vec{r},t) \tag{3.4}$$

where $\vec{B}_0$ is the externally applied field and $\vec{B}(\vec{r},t)$ is a first order quantity which has the same harmonic time dependence as the electric field.

3.1 Solution of Boltzmann equation

The Boltzmann equation satisfied by the homogeneous distribution function of the electrons, and with the relaxation model (2.3) for the collision term, can be written as

$$\partial f_1(\vec{v},t)/\partial t - (e/m_e) \; [\vec{E}(\vec{r},t) + \vec{v} \times \vec{B}_t(\vec{r},t)] \cdot \vec{\nabla}_v \; [f_0(v) + f_1(\vec{v},t)] =$$

$$= - \; \nu_r(v) \; f_1(\vec{v},t) \tag{3.5}$$

From the identity (2.8) we see that the term $(\vec{v} \times \vec{B}_t) \cdot \vec{\nabla}_v \; f_0(v)$ vanishes, since it involves the dot product of two mutually orthogonal vector functions. Neglecting second order terms, the Boltzmann equation for the phasor amplitudes becomes

$$[\nu_r(v) - i\omega] \; f_1(\vec{v}) - (e/m_e) \; (\vec{v} \times \vec{B}_0) \cdot \vec{\nabla}_v \; f_1(\vec{v}) =$$

$$= (e/m_e) \; \vec{E} \cdot \vec{\nabla}_v \; f_0(v) \tag{3.6}$$

In cylindrical coordinates $(v_\perp, \phi, v_\parallel)$ in velocity space (Fig. 2), with the $\vec{v}_\parallel$ vector along the magnetostatic field $\vec{B}_0$, we have, from Eq. (19.2.10),

$$(\vec{v} \times \hat{B}_0) \cdot \vec{\nabla}_v \; f_1(\vec{v}) = - \; df_1(\vec{v}) \,/\, d\phi \tag{3.7}$$

Substituting (3.7) into (3.6) and using the identity (2.8), we obtain

$$\frac{df_1(\vec{v})}{d\phi} + \frac{\nu_r(v) - i\omega}{\omega_{ce}} \; f_1(\vec{v}) = \frac{e}{m_e \; \omega_{ce}} \; \vec{E} \cdot \frac{\vec{v}}{v} \; \frac{df_0(v)}{dv} \tag{3.8}$$

where we have used the notation $\omega_{ce} = e \; B_0/m_e$, which represents the electron cyclotron frequency. Notice that the speed v does not depend on ϕ, since $v^2 = v_\perp^2 + v_\parallel^2$.

It is convenient to decompose the electric field vector into right circularly polarized (E_+), left circularly polarized (E_-) and longitudinal ($E_\parallel$) components, that is,

$$\vec{E} = E_+ \frac{(\hat{x} + i\hat{y})}{\sqrt{2}} + E_- \frac{(\hat{x} - i\hat{y})}{\sqrt{2}} + E_\parallel \hat{z} \tag{3.9}$$

where

$$E_\pm = (E_x \mp iE_y) / \sqrt{2} \tag{3.10}$$

Similarly, we can also decompose the electron velocity as

$$\vec{v} = v_+ \frac{(\hat{x} + i\hat{y})}{\sqrt{2}} + v_- \frac{(\hat{x} - i\hat{y})}{\sqrt{2}} + v_\parallel \hat{z} \tag{3.11}$$

where

$$v_\pm = (v_x \mp iv_y) / \sqrt{2}$$

$$= (v_\perp / \sqrt{2}) \exp(\mp i\phi) \tag{3.12}$$

since $v_x = v_\perp \cos\phi$, $v_y = v_\perp \sin\phi$ and $\exp(\pm i\phi) = \cos\phi \pm i \sin\phi$. Thus, using (3.9) and (3.11),

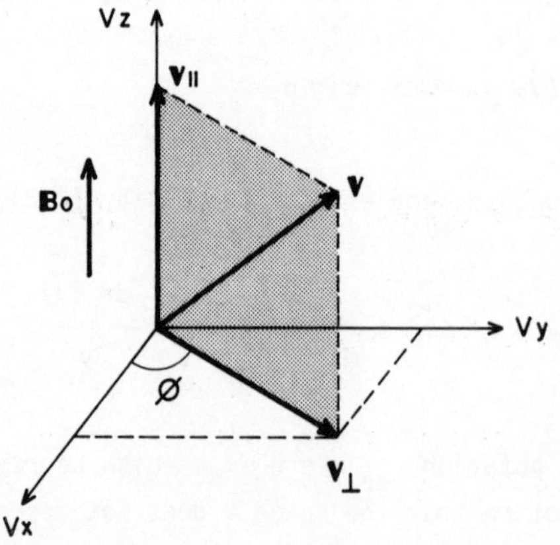

Fig. 2 - Cylindrical coordinates ($v_\perp$, ϕ, $v_\parallel$) in velocity space.

$$\vec{E} \cdot \vec{v} = E_+ v_- + E_- v_+ + E_\parallel v_\parallel$$

$$= (v_\perp / \sqrt{2})\, (E_+ e^{i\phi} + E_- e^{-i\phi}) + E_\parallel v_\parallel \tag{3.13}$$

Substituting this expression into the Boltzmann equation (3.8), we obtain

$$\frac{df_1(\vec{v})}{d\phi} + \frac{\nu_r(v) - i\omega}{\omega_{ce}} f_1(\vec{v}) = \frac{e}{m_e \omega_{ce}} \left[\frac{v_\perp}{\sqrt{2}} (E_+ e^{i\phi} + E_- e^{-i\phi}) + E_\parallel v_\parallel \right] \frac{1}{v} \frac{df_0(v)}{dv} \tag{3.14}$$

As in sub-section 2.2, of Chapter 19, we now introduce the notation

$$F_+(\vec{v}) = \frac{e}{m_e \omega_{ce}} E_+ \frac{v_\perp}{v} \frac{e^{i\phi}}{\sqrt{2}} \frac{df_0(v)}{dv} \tag{3.15}$$

$$F_-(\vec{v}) = \frac{e}{m_e \omega_{ce}} E_- \frac{v_\perp}{v} \frac{e^{-i\phi}}{\sqrt{2}} \frac{df_0(v)}{dv} \tag{3.16}$$

$$F_\parallel(\vec{v}) = \frac{e}{m_e \omega_{ce}} E_\parallel \frac{v_\parallel}{v} \frac{df_0(v)}{dv} \tag{3.17}$$

which allows (3.14) to be written as

$$\frac{df_1(\vec{v})}{d\phi} + \frac{\nu_r(v) - i\omega}{\omega_{ce}} f_1(\vec{v}) = F_+(\vec{v}) + F_-(\vec{v}) + F_\parallel(\vec{v}) \tag{3.18}$$

This differential equation is similar to Eq. (19.2.26), replacing the term $-kv_\parallel$ by $i\nu_r(v)$. Therefore, its solution can be obtained by inspection of the corresponding results contained in sub-section 2.2, Chapter 19. Hence, using Eqs. (19.2.27) to (19.2.34), we obtain

$$f_1(\vec{v}) = \frac{i\omega_{ce}}{\omega + i\nu_r(v) - \omega_{ce}} F_+(\vec{v}) + \frac{i\omega_{ce}}{\omega + i\nu_r(v) + \omega_{ce}} F_-(\vec{v}) +$$

$$+ \frac{i\omega_{ce}}{\omega + i\nu_r(v)} F_{\parallel}(\vec{v}) \tag{3.19}$$

or substituting (3.15), (3.16) and (3.17), into (3.19),

$$f_1(\vec{v}) = \frac{ie}{m_e} \frac{1}{v} \frac{df_0(v)}{dv} \left\{ \frac{v_\perp}{\sqrt{2}} \left[\frac{E_+ e^{i\phi}}{\omega + i\nu_r(v) - \omega_{ce}} + \right.\right.$$

$$\left.\left. + \frac{E_- e^{-i\phi}}{\omega + i\nu_r(v) + \omega_{ce}} \right] + \frac{v_\parallel E_\parallel}{\omega + i\nu_r(v)} \right\} \tag{3.20}$$

3.2 Electric current density and conductivity

Assuming that the electron gas has no average flow velocity in the equilibrium state ($\vec{u}_0 = 0$), we can write for the electric current density,

$$\vec{J} = - e \int_V \vec{v}\, f_1(\vec{v})\, d^3v \tag{3.21}$$

As in Eqs. (3.9) to (3.12), we can also decompose the current density into three components, according to

$$J_+ = - e \int_V v_+ f_1(\vec{v})\, d^3v \tag{3.22}$$

$$J_- = - e \int_V v_- f_1(\vec{v})\, d^3v \tag{3.23}$$

$$J_\parallel = - e \int_V v_\parallel f_1(\vec{v})\, d^3v \tag{3.24}$$

For purposes of calculating the conductivity, it is convenient to use spherical coordinates (v, θ, ϕ) in velocity space (Fig. 1), so that $v_\perp = v \sin\theta$, $v_\parallel =$ $= v \cos\theta$ and $d^3v = v^2 dv \sin\theta\, d\theta\, d\phi$. Plugging $f_1(\vec{v})$, from (3.20), into the expressions for J_+, J_- and $J_\parallel$, given in (3.22), (3.23) and (3.24), respectively, transforming to spherical coordinates, and performing the integrals over ϕ [making use of Eq. (19.2.51)], we find

$$J_{\pm} = -\frac{i\,\pi\, e^2}{m_e} E_{\pm} \int_0^{\pi} \sin^3\theta\, d\theta \int_0^{\infty} \frac{v^3}{\omega + i\nu_r(v) \mp \omega_{ce}} \frac{df_0(v)}{dv}\, dv \tag{3.25}$$

$$J_\parallel = -\frac{i\,2\pi\, e^2}{m_e} E_\parallel \int_0^{\pi} \cos^2\theta \sin\theta\, d\theta \int_0^{\infty} \frac{v^3}{\omega + i\nu_r(v)} \frac{df_0(v)}{dv}\, dv \tag{3.26}$$

Note that, in (3.25), either upper signs or lower signs are to be used. Carrying out the integrations over θ in these last two equations, yields

$$J_{\pm} = -\frac{i\,4\pi\, e^2}{3m_e} E_{\pm} \int_0^{\infty} \frac{v^3}{\omega + i\nu_r(v) \mp \omega_{ce}} \frac{df_0(v)}{dv}\, dv \tag{3.27}$$

$$J_\parallel = -\frac{i\,4\pi\, e^2}{3m_e} E_\parallel \int_0^{\infty} \frac{v^3}{\omega + i\nu_r(v)} \frac{df_0(v)}{dv}\, dv \tag{3.28}$$

The advantage of using the right and left circularly polarized components in the plane normal to $\vec{B}_0$ is that the corresponding equations for J_+ and J_- are uncoupled, so that J_+ depends only on E_+, whereas J_- depends only on E_-. Therefore, writing $\vec{J} = \bar{\bar{\sigma}} \cdot \vec{E}$, where $\bar{\bar{\sigma}}$ is the conductivity tensor, we obtain, from (3.27) and (3.28),

$$\begin{pmatrix} J_+ \\ J_- \\ J_\parallel \end{pmatrix} = \begin{pmatrix} \sigma_+ & 0 & 0 \\ 0 & \sigma_- & 0 \\ 0 & 0 & \sigma_\parallel \end{pmatrix} \begin{pmatrix} E_+ \\ E_- \\ E_\parallel \end{pmatrix} \tag{3.29}$$

with the following expressions for the elements of the conductivity tensor

$$\sigma_{\pm} = -\frac{i\,4\pi\,e^2}{3m_e}\int_0^\infty \frac{v^3}{\omega + i\nu_r(v) \mp \omega_{ce}}\,\frac{df_0(v)}{dv}\,dv \qquad (3.30)$$

$$\sigma_{\parallel} = -\frac{i\,4\pi\,e^2}{3m_e}\int_0^\infty \frac{v^3}{\omega + i\nu_r(v)}\,\frac{df_0(v)}{dv}\,dv \qquad (3.31)$$

Note that the longitudinal conductivity $\sigma_{\parallel}$ is the same as that for the case of a nonmagnetized plasma, deduced in the previous section.

The elements of the conductivity tensor, in *Cartesian coordinates*, can be obtained as follows. From (3.9) and (3.10) we can write in matrix form

$$\begin{pmatrix} E_+ \\ E_- \\ E_{\parallel} \end{pmatrix} = \begin{pmatrix} \frac{1}{\sqrt{2}} & -\frac{i}{\sqrt{2}} & 0 \\ \frac{1}{\sqrt{2}} & \frac{i}{\sqrt{2}} & 0 \\ 0 & 0 & 1 \end{pmatrix} \begin{pmatrix} E_x \\ E_y \\ E_z \end{pmatrix} \qquad (3.32)$$

Using a relation analogous to (3.32) for the current density $\vec{J}$, and inverting it, we obtain

$$\begin{pmatrix} J_x \\ J_y \\ J_z \end{pmatrix} = \begin{pmatrix} \frac{1}{\sqrt{2}} & \frac{1}{\sqrt{2}} & 0 \\ \frac{i}{\sqrt{2}} & -\frac{i}{\sqrt{2}} & 0 \\ 0 & 0 & 1 \end{pmatrix} \begin{pmatrix} J_+ \\ J_- \\ J_{\parallel} \end{pmatrix} \qquad (3.33)$$

Substituting (3.29) into (3.33), and combining the resulting expression with (3.32), we find that

$$\begin{pmatrix} J_x \\ J_y \\ J_z \end{pmatrix} = \begin{pmatrix} \sigma_{\perp} & -\sigma_H & 0 \\ \sigma_H & \sigma_{\perp} & 0 \\ 0 & 0 & \sigma_{\parallel} \end{pmatrix} \begin{pmatrix} E_x \\ E_y \\ E_z \end{pmatrix} \qquad (3.34)$$

where

$$\sigma_\perp = (\sigma_+ + \sigma_-)/2 \tag{3.35}$$

$$\sigma_H = i(\sigma_+ - \sigma_-)/2 \tag{3.36}$$

with σ_+, σ_- and $\sigma_\parallel$ as given in (3.30) and (3.31).

The integrals over v can only be evaluated after specifying the dependence of ν_r on v. In general, when ν_r is an arbitrary function of v, the elements of the conductivity tensor have to be determined by a numerical procedure. In cases when ν_r can be expressed as a polynominal in v, it is possible to obtain simple expressions for the conductivities in the limiting cases of very high and very low collision frequencies. In particular, for the special case when ν_r is independent of v, the integrals over v in (3.30) and (3.31) can be explicitly evaluated, yielding

$$\sigma_\pm = \frac{i\, n_0\, e^2}{m_e\, (\omega + i\nu_r \mp \omega_{ce})} \tag{3.37}$$

$$\sigma_\parallel = \frac{i\, n_0\, e^2}{m_e\, (\omega + i\nu_r)} \tag{3.38}$$

If these expressions are substituted into (3.35) and (3.36), we obtain the following results for the Cartesian components $\sigma_\perp$ and σ_H of the conductivity tensor:

$$\sigma_\perp = \frac{i\, n_0\, e^2\, (\omega + i\nu_r)}{m_e\, [(\omega + i\nu_r)^2 - \omega_{ce}^2]} \tag{3.39}$$

$$\sigma_H = \frac{n_0\, e^2\, \omega_{ce}}{m_e\, [(\omega + i\nu_r)^2 - \omega_{ce}^2]} \tag{3.40}$$

These are the same results deduced in section 5, of Chapter 10, which were calculated using the macroscopic transport equations with a constant collision frequency.

4. FREE DIFFUSION

In this section we derive an expression for the free diffusion coefficient of a weakly ionized plasma, considering that the relaxation collision frequency is a function of the speed of the electrons. For the analysis of diffusion phenomena we must consider specifically a spatial inhomogeneity in the electron density. Hence, we assume that the equilibrium velocity distribution function of the electrons has a spatial inhomogeneity, but is isotropic in velocity space, and will be denoted as $f_0(\vec{r},v)$. Since we are interested in calculating the electron flux due to diffusion only, we also assume that there are no external electromagnetic fields applied to the plasma. Furthermore, we study the free diffusion problem only under steady state conditions, in which all physical parameters are time-independent.

4.1 Perturbation distribution function

We assume that, under diffusion, the actual distribution function of the electrons, $f(\vec{r},\vec{v})$, deviates only slightly from the equilibrium value $f_0(\vec{r},v)$, so that we can write

$$f(\vec{r},\vec{v}) = f_0(\vec{r},v) + f_1(\vec{r},\vec{v}) \tag{4.1}$$

where $f_1(\vec{r},\vec{v})$ is a first order quantity, $|f_1| << f_0$.

Under steady state conditions, in the absence of external forces, and using the relaxation model for the collision term, the Boltzmann equation simplifies to

$$\vec{v} \cdot \vec{\nabla} f_0(\vec{r},v) = - \nu_r(v)\ f_1(\vec{r},\vec{v}) \tag{4.2}$$

where only the first order terms have been retained. Thus, we obtain directly for the perturbation distribution function,

$$f_1(\vec{r},\vec{v}) = [-1/\nu_r(v)]\ \vec{v} \cdot \vec{\nabla} f_0(\vec{r},v) \tag{4.3}$$

4.2 Particle flux

The expression for the particle current density (or flux) for the electrons, considering $\vec{u}_0 = 0$, is

$$\vec{\Gamma}_e = n_e \langle\vec{v}\rangle_e = \int_V f_1(\vec{r},\vec{v})\, \vec{v}\, d^3v \tag{4.4}$$

Substituting (4.3) into (4.4), gives

$$\vec{\Gamma}_e = - \int_V \frac{1}{\nu_r(v)} \vec{v}\, [\vec{v} \cdot \vec{\nabla} f_0(\vec{r},v)]\, d^3v \tag{4.5}$$

In spherical coordinates in velocity space (v, θ, ϕ) we have $d^3v = v^2 \sin\theta\, dv\, d\theta\, d\phi$, and using the result contained in (2.16) we obtain

$$\int_0^{\pi} d\theta \sin\theta \int_0^{2\pi} d\phi\, \vec{v}\, [\vec{v} \cdot \vec{\nabla} f_0(\vec{r},v)] = \frac{4\pi}{3} v^2\, \vec{\nabla} f_0(\vec{r},v) \tag{4.6}$$

Therefore, the electron flux vector (4.5) can be written as

$$\vec{\Gamma}_e = - \frac{4\pi}{3} \int_0^{\infty} \frac{v^4}{\nu_r(v)} \vec{\nabla} f_0(\vec{r},v)\, dv \tag{4.7}$$

4.3 Free diffusion coefficient

The distribution function $f_0(\vec{r},v)$ is in general a function of the electron number density n_e, the electron speed v, and the electron temperature T_e, so that it can generally be written in the form

$$f_0(\vec{r},v) = n_e\, F(v,T_e) \tag{4.8}$$

since the number density appears only as a result of normalization of the distribution function. The function $f_0(\vec{r},v)$ could be, for example, a local Maxwellian distribution.

For the purpose of calculating the free electron diffusion coefficient, we assume that the electron temperature has no spatial variation, so that

$$\vec{\nabla} f_0(\vec{r},v) = \vec{\nabla} n_e(\vec{r})\ F(v,T_e) \tag{4.9}$$

or, using (4.8),

$$\vec{\nabla} f_0(\vec{r},v) = \vec{\nabla} n_e(\vec{r})\ \frac{f_0(\vec{r},v)}{n_e(\vec{r})} \tag{4.10}$$

Substituting (4.10) into (4.7), we obtain

$$\vec{\Gamma}_e = -\ \frac{4\pi}{3}\ \frac{\vec{\nabla} n_e(\vec{r})}{n_e(\vec{r})} \int_0^\infty \frac{v^4}{\nu_r(v)}\ f_0(\vec{r},v)\ dv \tag{4.11}$$

Defining the free electron diffusion coefficient D_e by the relation

$$\vec{\Gamma}_e = -\ D_e\ \vec{\nabla} n_e(\vec{r}) \tag{4.12}$$

we deduce the following expression for D_e, by inspection of (4.11),

$$D_e = \frac{4\pi}{3\ n_e(\vec{r})} \int_0^\infty \frac{v^4}{\nu_r(v)}\ f_0(\vec{r},v)\ dv \tag{4.13}$$

Note that this expression for D_e is constant, independent of $\vec{r}$ and v, in view of (4.8) and (4.9).

If we consider $f_0(\vec{r},v)$ as being a modified (or local) Maxwellian distribution function given by

$$f_0(\vec{r},v) = n_e(\vec{r}) \left(\frac{m_e}{2\pi k T_e} \right)^{3/2} \exp \left(-\ \frac{m_e v^2}{2kT_e} \right) \tag{4.14}$$

then (4.13) becomes

$$D_e = \frac{4\pi}{3} \left(\frac{m_e}{2\pi k T_e} \right)^{3/2} \int_0^\infty \frac{v^4}{\nu_r(v)}\ \exp \left(-\ \frac{m_e v^2}{2kT_e} \right)\ dv \tag{4.15}$$

Furthermore, if the relaxation collision frequency ν_r is taken to be constant, independent of v, then the integral in (4.15) can be explicitly evaluated [see Eq. (7.4.22)], which gives

$$D_e = k\, T_e / m_e\, \nu_r \tag{4.16}$$

This is the same result obtained in section 8, Chapter 10 [see Eq.(10.8.9)], which was deduced using the macroscopic transport equations with a constant collision frequency.

5. DIFFUSION IN A MAGNETIC FIELD

In this section we want to include the effects of an externally applied magnetostatic field, $\vec{B}_0$, on the problem of electron diffusion in a weakly ionized plasma. We consider the same assumptions made in the previous section, except for the inclusion of the external magnetic field.

5.1 Solution of Boltzmann equation

Retaining only the first order terms, the linearized Boltzmann equation is now

$$\vec{v} \cdot \vec{\nabla} f_0(\vec{r},v) - (e/m_e)\, (\vec{v} \times \vec{B}_0) \cdot \vec{\nabla}_v f_1(\vec{r},\vec{v}) = -\, \nu_r(v)\, f_1(\vec{r},\vec{v}) \tag{5.1}$$

Note that in view of the isotropy of $f_0(\vec{r},v)$ we can use the identity (2.8), so that

$$(\vec{v} \times \vec{B}_0) \cdot \vec{\nabla}_v\, f_0(\vec{r},v) = 0 \tag{5.2}$$

In cylindrical coordinates ($v_\perp$, ϕ, $v_\parallel$) in velocity space (Fig. 2) we have, from (3.7),

$$(\vec{v} \times \hat{B}_0) \cdot \vec{\nabla}_v\, f_1(\vec{r},\vec{v}) = -\, df_1(\vec{r},\vec{v}) / d\phi \tag{5.3}$$

Choosing the unit vector $\hat{z}$ along the magnetic field $\vec{B}_0$, we can write

$$\vec{v} \cdot \vec{\nabla} f_0(\vec{r},v) = \left(v_\perp \cos\phi \frac{\partial}{\partial x} + v_\perp \sin\phi \frac{\partial}{\partial y} + v_\parallel \frac{\partial}{\partial z} \right) f_0(\vec{r},v) \tag{5.4}$$

Substituting (5.4) and (5.3) into (5.1), and rearranging, yields

$$\left[\frac{d}{d\phi} + \frac{\nu_r(v)}{\omega_{ce}} \right] f_1(\vec{r},\vec{v}) = - \frac{1}{\omega_{ce}} \left(v_\perp \cos\phi \frac{\partial}{\partial x} + \right.$$

$$\left. + v_\perp \sin\phi \frac{\partial}{\partial y} + v_\parallel \frac{\partial}{\partial z} \right) f_0(\vec{r},v) \tag{5.5}$$

In order to solve this linear differential equation let

$$f_1(\vec{r},\vec{v}) = F_1(\vec{r},\vec{v}) + F_2(\vec{r},\vec{v}) + F_3(\vec{r},\vec{v}) \tag{5.6}$$

where F_1, F_2 and F_3 are the solutions of (5.5) corresponding, respectively, to only the first, the second and the third terms within parenthesis in the right-hand side of (5.5), that is,

$$\left[\frac{d}{d\phi} + \frac{\nu_r(v)}{\omega_{ce}} \right] F_1(\vec{r},\vec{v}) = - \frac{1}{\omega_{ce}} v_\perp \cos\phi \frac{\partial f_0(\vec{r},v)}{\partial x} \tag{5.7}$$

$$\left[\frac{d}{d\phi} + \frac{\nu_r(v)}{\omega_{ce}} \right] F_2(\vec{r},\vec{v}) = - \frac{1}{\omega_{ce}} v_\perp \sin\phi \frac{\partial f_0(\vec{r},v)}{\partial y} \tag{5.8}$$

$$\left[\frac{d}{d\phi} + \frac{\nu_r(v)}{\omega_{ce}} \right] F_3(\vec{r},\vec{v}) = - \frac{1}{\omega_{ce}} v_\parallel \frac{\partial f_0(\vec{r},v)}{\partial z} \tag{5.9}$$

To solve (5.7) let us first rewrite it in the form

$$\left[\frac{d}{d\phi} + \frac{\nu_r(v)}{\omega_{ce}} \right] F_1(\vec{r},\vec{v}) \equiv \exp\left[- \frac{\nu_r(v)}{\omega_{ce}} \phi \right] \cdot$$

$$\cdot \frac{d}{d\phi} \left\{ F_1(\vec{r},\vec{v}) \exp\left[\frac{\nu_r(v)}{\omega_{ce}} \phi \right] \right\} =$$

$$= - \frac{1}{\omega_{ce}} v_\perp \cos\phi \frac{\partial f_0(\vec{r},v)}{\partial x} \tag{5.10}$$

The solution of this differential equation is given by

$$F_1(\vec{r},\vec{v}) = - \frac{1}{\omega_{ce}} v_\perp \frac{\partial f_0(\vec{r},v)}{\partial x} \exp\left[- \frac{\nu_r(v)}{\omega_{ce}} \phi \right] \cdot$$

$$\cdot \int_{-\infty}^{\phi} \cos\phi' \exp\left[\frac{\nu_r(v)}{\omega_{ce}} \phi' \right] d\phi'$$

$$= - v_\perp \frac{\partial f_0(\vec{r},v)}{\partial x} \frac{[\nu_r(v)\cos\phi + \omega_{ce}\sin\phi]}{[\nu_r^2(v) + \omega_{ce}^2]} \tag{5.11}$$

Notice that $F_1(\vec{r},\vec{v})$ is a periodic function of ϕ, with period 2π.

In a similar way, the solution of (5.8) and (5.9) are given, respectively, by

$$F_2(\vec{r},\vec{v}) = - v_\perp \frac{\partial f_0(\vec{r},v)}{\partial y} \frac{[\nu_r(v)\sin\phi - \omega_{ce}\cos\phi]}{[\nu_r^2(v) + \omega_{ce}^2]} \tag{5.12}$$

$$F_3(\vec{r},\vec{v}) = - v_\parallel \frac{\partial f_0(\vec{r},v)}{\partial z} \frac{1}{\nu_r(v)} \tag{5.13}$$

Adding (5.11), (5.12) and (5.13), gives the solution for $f_1(\vec{r},\vec{v})$ in terms of $f_0(\vec{r},v)$ and $\nu_r(v)$.

5.2 Particle flux and diffusion coefficients

From (4.4), the expression for the x component of the electron flux vector is found to be

$$\Gamma_{ex} = \int_V v_x f_1(\vec{r},\vec{v}) \, d^3v \tag{5.14}$$

In cylindrical coordinates (Fig. 2) we have $d^3v = v_\perp \, dv_\perp \, dv_\parallel \, d\phi$ and $v_x = v_\perp \cos\phi$. Therefore,

$$\Gamma_{ex} = \int_0^\infty dv_\perp \int_0^{2\pi} d\phi \int_{-\infty}^{+\infty} dv_\parallel \; v_\perp^2 \cos\phi \; f_1(\vec{r},\vec{v}) \tag{5.15}$$

Using (5.6), (5.11), (5.12) and (5.13), and performing the integration over ϕ, we obtain

$$\Gamma_{ex} = -\pi \int_0^\infty dv_\perp \int_{-\infty}^{+\infty} dv_\parallel \; \frac{v_\perp^3 \, [\nu_r(v) \; \partial f_0(\vec{r},v)/\partial x - \omega_{ce} \; \partial f_0(\vec{r},v)/\partial y]}{[\nu_r^2(v) + \omega_{ce}^2]} \tag{5.16}$$

To perform the integrals in (5.16) it is convenient to use spherical coordinates (v, θ, ϕ) in velocity space (Fig. 1). Transforming to spherical coordinates, Eq. (5.16) becomes

$$\Gamma_{ex} = -\pi \int_0^\infty dv \int_0^\pi d\theta \; \cdot$$

$$\cdot \; \frac{v^4 \sin^3\theta \; [\nu_r(v) \; \partial f_0(\vec{r},v)/\partial x - \omega_{ce} \; \partial f_0(\vec{r},v)/\partial y]}{[\nu_r^2(v) + \omega_{ce}^2]} \tag{5.17}$$

Carrying out the integration over θ, we obtain

$$\Gamma_{ex} = -\frac{4\pi}{3} \int_0^\infty \frac{v^4 \, \nu_r(v)}{\nu_r^2(v) + \omega_{ce}^2} \; \frac{\partial f_0(\vec{r},v)}{\partial x} \, dv \; -$$

$$- \frac{4\pi}{3} \int_0^\infty \frac{v^4 \; \omega_{ce}}{\nu_r^2(v) + \omega_{ce}^2} \; \frac{\partial f_0(\vec{r},v)}{\partial y} \, dv \tag{5.18}$$

This equation can be written in the form

$$\Gamma_{ex} = - \frac{\partial}{\partial x} [D_\perp n_e(\vec{r})] - \frac{\partial}{\partial y} [- D_H n_e(\vec{r})] \tag{5.19}$$

where the electron diffusion coefficients $D_\perp$ and D_H are given by

$$D_\perp = \frac{4\pi}{3n_e(\vec{r})} \int_0^\infty \frac{v^4 \nu_r(v)}{\nu_r^2(v) + \omega_{ce}^2} f_0(\vec{r},v) \, dv \tag{5.20}$$

$$D_H = \frac{4\pi}{3n_e(\vec{r})} \int_0^\infty \frac{v^4 \omega_{ce}}{\nu_r^2(v) + \omega_{ce}^2} f_0(\vec{r},v) \, dv \tag{5.21}$$

Along similar lines, we obtain for the y component of the electron flux vector,

$$\Gamma_{ey} = - \frac{\partial}{\partial x} [D_H n_e(\vec{r})] - \frac{\partial}{\partial y} [D_\perp n_e(\vec{r})] \tag{5.22}$$

and for the z component,

$$\Gamma_{ez} = - \frac{\partial}{\partial z} [D_\parallel n_e(\vec{r})] \tag{5.23}$$

where

$$D_\parallel = \frac{4\pi}{3n_e(\vec{r})} \int_0^\infty \frac{v^4}{\nu_r(v)} f_0(\vec{r},v) \, dv \tag{5.24}$$

Eqs. (5.19), (5.22) and (5.23) can be written in a succint vector form as

$$\vec{\Gamma}_e = - \vec{\nabla} \cdot [\bar{\bar{D}} \, n_e(\vec{r})] \tag{5.25}$$

where $\bar{\bar{D}}$ denotes the dyadic coefficient for electron diffusion in a magnetic field, given in matrix form by

$$\bar{\bar{D}} = \begin{pmatrix} D_\perp & D_H & 0 \\ -D_H & D_\perp & 0 \\ 0 & 0 & D_\parallel \end{pmatrix} \tag{5.26}$$

The diffusion coefficient $D_\parallel$ is the same as that obtained in the absence of a magnetostatic field ($D_\parallel = D_e$). Therefore, the diffusion of particles along the magnetic field is the same as if there were no field present, whereas the diffusion in the perpendicular direction is inhibited by the magnetic field since $D_\perp < D_\parallel$, as can be verified from (5.20) and (5.24).

For the special case in which $f_0(\vec{r},v)$ is given by a local Maxwellian distribution function, as in (4.14), and ν_r is independent of v, the integrals in (5.20), (5.21) and (5.24) can be evaluated directly, yielding

$$D_\perp = [\nu_r^2/(\nu_r^2 + \omega_{ce}^2)]\ D_e \tag{5.27}$$

$$D_H = [\nu_r\ \omega_{ce}/(\nu_r^2 + \omega_{ce}^2)]\ D_e \tag{5.28}$$

$$D_\parallel = D_e = k\ T_e/m_e\ \nu_r \tag{5.29}$$

which are the same results obtained in section 9, Chapter 10, deduced from the macroscopic transport equations [see Eqs. (10.9.4) to (10.9.7)].

6. HEAT FLOW

We shall now derive expressions for the heat flow vector, $\vec{q}_e$, and for the thermal conductivity, K_e, due to the random motion of the electrons in a weakly ionized plasma. As in the previous sections, we shall determine the nonequilibrium distribution function $f(\vec{r},\vec{v})$, under steady state conditions, by applying a perturbation technique to the Boltzmann equation, using the relaxation model for the collision term. To simplify matters we assume that there are no externally applied electromagnetic fields.

Using (4.1), we find that the Boltzmann equation for this case is the same as that given by (4.2). Therefore, as in sub-section 4.1,

$$f_1(\vec{r},\vec{v}) = -\ [1/\nu_r(v)]\ \ \vec{v}\cdot\vec{\nabla}f_0(\vec{r},v) \tag{6.1}$$

6.1 General expression for the heat flow vector

The expression for the heat flow vector due to the thermal motion of the electrons, and considering $\vec{u}_0 = 0$, is

$$\vec{q}_e = \frac{1}{2} m_e \int_v v^2 \, \vec{v} \, f_1(\vec{r},\vec{v}) \, d^3v \tag{6.2}$$

Substituting (6.1) into (6.2), yields

$$\vec{q}_e = - \frac{1}{2} m_e \int_v \frac{v^2}{\nu_r(v)} \, \vec{v} \, [\vec{v} \cdot \vec{\nabla} f_0(\vec{r},v)] \, d^3v \tag{6.3}$$

In spherical coordinates in velocity space and using (4.6), we obtain, from (6.3),

$$\vec{q}_e = - \frac{2\pi m_e}{3} \int_0^\infty \frac{v^6}{\nu_r(v)} \, \vec{\nabla} f_0(\vec{r},v) \, dv \tag{6.4}$$

This expression gives the electron heat flow vector $\vec{q}_e$ in terms of the distribution function $f_0(\vec{r},v)$ and the relaxation collision frequency $\nu_r(v)$.

6.2 Thermal conductivity for a constant kinetic pressure

Next, we evaluate (6.4) for the case when $f_0(\vec{r},v)$ is given by a local Maxwellian distribution function,

$$f_0(\vec{r},v) = n_e(\vec{r}) \left[\frac{m_e}{2\pi k T_e(\vec{r})} \right]^{3/2} \exp \left[- \frac{m_e \, v^2}{2kT_e(\vec{r})} \right] \tag{6.5}$$

in which both n_e and T_e may have a spatial variation, but such that the electron kinetic pressure stays constant, that is,

$$p_e = n_e(\vec{r}) \, k \, T_e(\vec{r}) = \text{constant.} \tag{6.6}$$

From (6.6) we have

$$k\, T_e(\vec{r})\, \vec{\nabla} n_e(\vec{r}) = -\, n_e(\vec{r})\, k\, \vec{\nabla} T_e(\vec{r}) \tag{6.7}$$

and calculating the gradient of (6.5) we find

$$\vec{\nabla} f_0(\vec{r},v) = \left[-\frac{5}{2} + \frac{m_e v^2}{2kT_e(\vec{r})} \right] \frac{\vec{\nabla} T_e(\vec{r})}{T_e(\vec{r})} f_0(\vec{r},v) \tag{6.8}$$

Substituting (6.8) into (6.4) gives

$$\vec{q}_e = -\frac{2\pi m_e}{3} \frac{\vec{\nabla} T_e(\vec{r})}{T_e(\vec{r})} \int_0^\infty \frac{v^6}{\nu_r(v)} \left[-\frac{5}{2} + \frac{m_e v^2}{2kT_e(\vec{r})} \right] f_0(\vec{r},v)\, dv \tag{6.9}$$

This equation can be written in the form

$$\vec{q}_e = -\, K_e\, \vec{\nabla} T_e(\vec{r}) \tag{6.10}$$

where K_e is the thermal conductivity coefficient given by

$$K_e = \frac{2\pi m_e}{3T_e(\vec{r})} \int_0^\infty \frac{v^6}{\nu_r(v)} \left[-\frac{5}{2} + \frac{m_e v^2}{2kT_e(r)} \right] f_0(\vec{r},v)\, dv \tag{6.11}$$

In the special case when ν_r is independent of v we can write (6.11) as

$$K_e = \frac{2\pi m_e}{3T_e(\vec{r})\, \nu_r} \left[-\frac{5}{2} \int_0^\infty v^6 f_0(\vec{r},v)\, dv + \right.$$

$$\left. + \frac{m_e}{2kT_e(\vec{r})} \int_0^\infty v^8 f_0(\vec{r},v)\, dv \right] \tag{6.12}$$

Now,

$$\int_0^\infty v^6 \; f_0(\vec{r},v) \; dv = \frac{15 \; k \; T_e(\vec{r}) \; p_e}{4\pi \; m_e^2} \tag{6.13}$$

$$\int_0^\infty v^8 \; f_0(\vec{r},v) \; dv = \frac{105 \; k^2 \; T_e^2(\vec{r}) \; p_e}{4\pi \; m_e^3} \tag{6.14}$$

so that, substituting (6.13) and (6.14) into (6.12) and simplifying, we obtain the following expression for the thermal conductivity, when ν_r = constant,

$$K_e = (5/2) \; k \; p_e/m_e \; \nu_r \tag{6.15}$$

6.3 Thermal conductivity for the adiabatic case

We consider now the case when the electron kinetic pressure is not constant, but follows the adiabatic law

$$p_e(\vec{r}) \; n_e^{-\gamma}(\vec{r}) = \text{constant} \tag{6.16}$$

where γ is the adiabatic constant, defined as the ratio of the specific heats at constant pressure and at constant volume, which may be expressed as

$$\gamma = (2 + N)/N \tag{6.17}$$

where N denotes the number of degrees of freedom. Eq. (6.16) can also be written as

$$n_e(\vec{r}) \; T_e(\vec{r})^{1/(1 - \gamma)} = \text{constant} \tag{6.18}$$

Taking the gradient of the local Maxwellian distribution function (6.5), and making use of (6.18), we obtain

$$\vec{\nabla} f_0(\vec{r},v) = \left[\frac{1}{\gamma - 1} - \frac{3}{2} + \frac{m_e \; v^2}{2k \; T_e(\vec{r})} \right] \frac{\vec{\nabla} T_e(\vec{r})}{T_e(\vec{r})} \; f_0(\vec{r},v) \tag{6.19}$$

Now we substitute (6.19) into (6.4), which gives for the heat flow vector

$$\vec{q}_e = - \frac{2\pi m_e}{3} \frac{\vec{\nabla} T_e(\vec{r})}{T_e(\vec{r})} \int_0^\infty \frac{v^6}{\nu_r(v)} \left[\frac{1}{\gamma - 1} - \frac{3}{2} + \frac{m_e v^2}{2kT_e(\vec{r})} \right] f_0(\vec{r},v)\, dv \tag{6.20}$$

With reference to (6.10) we identify the following expression for the thermal conductivity

$$K_e = \frac{2\pi m_e}{3T_e(\vec{r})} \int_0^\infty \frac{v^6}{\nu_r(v)} \left[\frac{1}{\gamma - 1} - \frac{3}{2} + \frac{m_e v^2}{2kT_e(\vec{r})} \right] f_0(\vec{r},v)\, dv \tag{6.21}$$

For the special case in which ν_r does not depend on v, we can use the results given in (6.13) and (6.14), so that (6.21) simplifies to

$$K_e = (5/2)\ [2 + 1/(\gamma - 1)]\ k\ p_e/m_e\ \nu_r \tag{6.22}$$

If three degrees of freedom corresponding to the three-dimensional translational motion are considered, we have $\gamma = 5/3$, so that

$$K_e = (35/4)\ k\ p_e/m_e\ \nu_r \tag{6.23}$$

When the plasma is immersed in an externally applied magnetostatic field $\vec{B}_0$, an anisotropy is introduced in the thermal energy flux, so that the thermal conductivity coefficient is replaced by a thermal conductivity dyad $\bar{\bar{K}}$. Expressions for the components of the thermal conductivity dyad can be deduced along lines similar to the calculations presented in section 5 for the diffusion coefficient dyad. The derivation of explicit expressions for the components of $\bar{\bar{K}}$ in a magnetized plasma will be left as an exercise for the reader.

PROBLEMS

22.1 - In Cartesian coordinates in velocity space (refer to Fig. 1), with the components expressed in spherical coordinates (v, θ, ϕ), we have

$$\vec{v} = v\ \hat{v} = v(\sin\theta \cos\phi\ \hat{v}_x + \sin\theta \sin\phi\ \hat{v}_y + \cos\theta\ \hat{v}_z)$$

(a) Show that the dyad $\vec{v}\,\vec{v}$ can be written in matrix form as

$$\vec{v}\,\vec{v} = v^2 \begin{pmatrix} (\sin^2\Theta\cos^2\phi) & (\sin^2\Theta\sin\phi\cos\phi) & (\sin\Theta\cos\Theta\cos\phi) \\ (\sin^2\Theta\sin\phi\cos\phi) & (\sin^2\Theta\sin^2\phi) & (\sin\Theta\cos\Theta\sin\phi) \\ (\sin\Theta\cos\Theta\cos\phi) & (\sin\Theta\cos\Theta\sin\phi) & (\cos^2\Theta) \end{pmatrix}$$

(b) Prove the following orthogonality relations

$$\int_0^{\pi}\int_0^{2\pi} \sin\Theta \, d\Theta \, d\phi = 4\pi$$

$$\int_0^{\pi}\int_0^{2\pi} v_i \sin\Theta \, d\Theta \, d\phi = 0$$

$$\int_0^{\pi}\int_0^{2\pi} v_i v_j \sin\Theta \, d\Theta \, d\phi = \frac{4\pi}{3} v^2 \delta_{ij}$$

$$\int_0^{\pi}\int_0^{2\pi} v_i v_j v_k \sin\Theta \, d\Theta \, d\phi = 0$$

with i, j, k = x, y, z, and where δ_{ij} is the Kronecker delta.

22.2 - Using the Maxwell-Boltzmann distribution function (2.22) and the definition (2.23), verify (2.24).

22.3 - Show that, when ν_r is independent of v, (2.25) reduces to (2.20).

22.4 - Consider (2.25), which gives the AC electric conductivity of a weakly ionized plasma for a velocity-dependent collision frequency $\nu_r(v)$.

(a) Show that in the high-frequency limit, $\omega^2 >> \nu_r^2$, we have

$$\sigma = (\nu_c + i\omega)\, n_0\, e^2/m_e\, \omega^2$$

where

$$\nu_c = \frac{8}{3\sqrt{\Pi}} \int_0^{\infty} \nu_r(\xi)\, \xi^4 \exp(-\xi^2)\, d\xi$$

(b) Show that in the low-frequency limit, $\omega^2 << \nu_r^2$, we have

$$\sigma = \frac{n_0 e^2}{m_e} \left[\frac{1}{\nu_c'} + i \frac{\omega}{(\nu_c'')^2} \right]$$

where

$$\frac{1}{\nu_c'} = \frac{8}{3\pi^{1/2}} \int_0^\infty \frac{1}{\nu_r(\xi)} \xi^4 \exp(-\xi^2) \, d\xi$$

$$\frac{1}{(\nu_c'')^2} = \frac{8}{3\pi^{1/2}} \int_0^\infty \frac{1}{\nu_r(\xi)^2} \xi^4 \exp(-\xi^2) \, d\xi$$

(c) For intermediate frequencies, show that

$$\sigma = (n_0 e^2/m_e) (\nu_c K_1 + i \omega K_2)$$

where

$$K_1 = \frac{8}{3\pi^{1/2}} \frac{1}{\nu_c \omega^2} \int_0^\infty \left[1 - \frac{\nu_r(\xi)^2}{\omega^2} + \frac{\nu_r(\xi)^4}{\omega^4} + \dots \right] \nu_r(\xi) \xi^4 \exp(-\xi^2) \, d\xi$$

$$K_2 = \frac{8}{3\pi^{1/2}} \frac{1}{\omega^2} \int_0^\infty \left[1 - \frac{\nu_r(\xi)^2}{\omega^2} + \frac{r(\xi)^4}{\omega^4} + \dots \right] \xi^4 \exp(-\xi^2) \, d\xi$$

and where ν_c is the same quantity defined in part (a) for the high-frequency limit.

22.5 - If we define an *effective collision frequency*, $\nu_{eff}(\omega)$, such that the longitudinal electric conductivity is given by

$$\sigma = \frac{i n_0 e^2}{m_e [\omega + i \nu_{eff}(\omega)]}$$

then, by comparison with Eq. (2.18), we find that

$$[\omega + i\,\nu_{eff}(\omega)]^{-1} = -\frac{4\pi}{3n_0}\int_0^{\infty} \frac{v^3}{[\omega + i\,\nu_r(v)]}\,\frac{df_0(v)}{dv}\,dv$$

(a) Show that in the low-frequency limit, $\omega \ll \nu_{eff}(\omega)$, we have

$$\frac{1}{\nu_{eff}} = -\frac{4\pi}{3n_0}\int_0^{\infty} \frac{v^3}{\nu_r(v)}\,\frac{df_0(v)}{dv}\,dv$$

(b) Show that in the high-frequency limit, $\omega \gg \nu_{eff}(\omega)$, we have

$$\nu_{eff} = -\frac{4\pi}{3n_0}\int_0^{\infty} v^3\,\nu_r(v)\,\frac{df_0(v)}{dv}\,dv$$

Thus, in both limits ν_{eff} is independent of ω.

22.6 - In the expression deduced for ν_{eff} in part (b) of the previous problem (high-frequency limit), consider that f_0 is the Maxwell-Boltzmann distribution function and that $\nu_r(v) = \nu_0\, v^n$, where ν_0 is a constant and n is an integer.

(a) Show that in this case we have

$$\nu_{eff} = \frac{4\nu_0}{3\pi^{1/2}}\left(\frac{2kT}{m}\right)^{n/2} \Gamma\left(\frac{n+5}{2}\right)$$

where $\Gamma(z)$ is the gamma function defined by

$$\Gamma(z) = \int_0^{\infty} t^{(z-1)}\,e^{-t}\,dt$$

(b) Calculate the average value of the collision frequency, $\langle \nu_r(v) \rangle_0$, using the Maxwell-Boltzmann distribution function, and show that

$$\langle \nu_r(v) \rangle_0 = \nu_{eff}/(1 + n/3)$$

22.7 - Derive (3.34), from Eqs. (3.29) to (3.33).

22.8 - Show that (3.30) and (3.31) yield, respectively, (3.37) and (3.38), when ν_r is independent of v for any $f_0(v)$.

22.9 - Deduce (5.22) and (5.23) starting from the definition of the electron flux vector, and the expression for $f_1(\vec{r},\vec{v})$ given by (5.6), (5.11), (5.12) and (5.13).

22.10 - Analyze the problem of heat flow in a weakly ionized plasma immersed in an externally applied magnetostatic field, $\vec{B}_0$, and derive expressions for the heat flow vector, $\vec{q}_e$, and for the components of the thermal conductivity dyad, $\bar{\bar{K}}$, considering a velocity-dependent collision frequency, $\nu_r(v)$. Analyze the problem for the adiabatic case and for the case of a constant kinetic pressure.

APPENDIX I

Useful Vector Relations

(1) $\vec{A}.\vec{B} = \vec{B}.\vec{A} = A_xB_x + A_yB_y + A_zB_z$

(2) $\vec{A}x\vec{B} = -\vec{B}x\vec{A} = \begin{vmatrix} \hat{x} & \hat{y} & \hat{z} \\ A_x & A_y & A_z \\ B_x & B_y & B_z \end{vmatrix}$

(3) $\vec{A}\,.\,(\vec{B}x\vec{C}) = (\vec{A}x\vec{B})\,.\,\vec{C} = (\vec{C}x\vec{A})\,.\,\vec{B}$

(4) $\vec{A}\,x\,(\vec{B}x\vec{C}) = (\vec{A}.\vec{C})\,\vec{B} - (\vec{A}.\vec{B})\,\vec{C}$

(5) $(\vec{A}x\vec{B})\,x\,\vec{C} = (\vec{A}.\vec{C})\,\vec{B} - (\vec{B}.\vec{C})\,\vec{A}$

(6) $(\vec{A}x\vec{B})\,.\,(\vec{C}x\vec{D}) = (\vec{A}.\vec{C})\,(\vec{B}.\vec{D}) - (\vec{A}.\vec{D})\,(\vec{B}.\vec{C})$

(7) $(\vec{A}x\vec{B})\,x\,(\vec{C}x\vec{D}) = [\vec{A}.(\vec{B}x\vec{D})]\,\vec{C} - [\vec{A}.(\vec{B}x\vec{C})]\,\vec{D}$

(8) $\vec{\nabla}(\phi\psi) = \phi\vec{\nabla}\psi + \psi\vec{\nabla}\phi$

(9) $\vec{\nabla}.(\phi\vec{A}) = \phi\vec{\nabla}.\vec{A} + \vec{A}.\vec{\nabla}\phi$

(10) $\vec{\nabla}x(\phi\vec{A}) = \phi\vec{\nabla}x\vec{A} + (\vec{\nabla}\phi)\,x\,\vec{A}$

(11) $\vec{\nabla}\,.\,(\vec{A}x\vec{B}) = \vec{B}.(\vec{\nabla}x\vec{A}) - \vec{A}\,.\,(\vec{\nabla}x\vec{B})$

(12) $\vec{\nabla}\,(\vec{A}.\vec{B}) = (\vec{A}.\vec{\nabla})\,\vec{B} + (\vec{B}.\vec{\nabla})\,\vec{A} + \vec{A}x(\vec{\nabla}x\vec{B}) + \vec{B}x(\vec{\nabla}x\vec{A})$

(13) $\vec{\nabla}x(\vec{A}x\vec{B}) = \vec{A}(\vec{\nabla}.\vec{B}) + (\vec{B}.\vec{\nabla})\,\vec{A} - \vec{B}(\vec{\nabla}.\vec{A}) - (\vec{A}.\vec{\nabla})\,\vec{B}$

(14) $\vec{\nabla}\,x(\vec{\nabla}x\vec{A}) = \vec{\nabla}(\vec{\nabla}.\vec{A}) - (\vec{\nabla}.\vec{\nabla})\,\vec{A}$

(15) $\vec{\nabla}\,.\,(\vec{\nabla}x\vec{A}) = 0$

(16) $\vec{\nabla}\,x\,(\vec{\nabla}\phi) = 0$

(17) $(\vec{\nabla}.\vec{\nabla})\ \phi = \nabla^2\phi$

If $\vec{r}$ is the radius vector, of magnitude r, drawn from the origin to a general point x,y,z, then

(18) $\vec{\nabla}.\vec{r} = 3$

(19) $\vec{\nabla}\times\vec{r} = 0$

(20) $\vec{\nabla}r = \vec{r}/r$

(21) $\vec{\nabla}(1/r) = -\ \vec{r}/r^3$

(22) $\vec{\nabla}.(\vec{r}/r^3) = -\nabla^2(1/r) = 4\pi\ \delta(\vec{r})$

In the following integral relations, V is the volume bounded by the closed surface S and $\hat{n}$ is a unit normal vector drawn outwardly to the closed surface S:

(23) $\oint_S \phi\ \hat{n}\ dS = \int_V (\vec{\nabla}\phi)\ dV$

(24) $\oint_S \vec{A}\ .\ \hat{n}\ dS = \int_V (\vec{\nabla}.\vec{A})\ dV$ (Gauss' theorem)

(25) $\oint_S (\hat{n}\times\vec{A})\ dS = \int_V (\vec{\nabla}\times\vec{A})\ dV$

(26) $\oint_S \phi\ (\vec{\nabla}\psi)\ .\ \hat{n}\ dS = \int_V [\phi\nabla^2\psi + (\vec{\nabla}\phi)\ .\ (\vec{\nabla}\psi)]\ dV$ (Green's first identity)

(27) $\oint_S (\phi\vec{\nabla}\psi - \psi\vec{\nabla}\phi).\ \hat{n}\ dS = \int_V (\phi\nabla^2\psi - \psi\nabla^2\phi)\ dV$ (Green's second identity or Green's theorem)

(28) $\oint_S [\vec{B}\times(\vec{\nabla}\times\vec{A}) - \vec{A}\times(\vec{\nabla}\times\vec{B})]\ .\hat{n}\ dS =$

$$= \int_V \{\vec{A}.\,[\vec{\nabla}\times(\vec{\nabla}\times B)] - \vec{B}.[\vec{\nabla}\times(\vec{\nabla}\times\vec{A})]\}\,dV$$ (Vector version of Green's theorem)

If S is an open surface bounded by the contour C, of which the line element is $d\vec{\ell}$, then

(29) $$\oint_C \phi\, d\vec{\ell} = \int_S \hat{n} \times (\vec{\nabla}\phi)\, dS$$

(30) $$\oint_C \vec{A}.\, d\vec{\ell} = \int_S (\vec{\nabla}\times\vec{A})\, .\, \hat{n}\, dS$$ (Stoke's theorem)

If $\bar{\bar{T}}$ is a tensor, then

(31) $$\vec{\nabla}.\,(\phi\bar{\bar{T}}) = \phi(\vec{\nabla}.\bar{\bar{T}}) + (\vec{\nabla}\phi)\, .\, \bar{\bar{T}}$$

(32) $$\oint_S \bar{\bar{T}}\, .\, \hat{n}\, dS = \int_V (\vec{\nabla}.\bar{\bar{T}})\, dV$$

APPENDIX II

Useful Relations in Cartesian and in Curvilinear Coordinates

1. CARTESIAN COORDINATES

Orthogonal unit vectors:

$\hat{x}, \hat{y}, \hat{z}$

Orthogonal line elements:

dx, dy, dz

Components of gradient:

$$(\vec{\nabla}\psi)_x = \frac{\partial \psi}{\partial x}$$

$$(\vec{\nabla}\psi)_y = \frac{\partial \psi}{\partial y}$$

$$(\vec{\nabla}\psi)_z = \frac{\partial \psi}{\partial z}$$

Divergence:

$$\vec{\nabla}.\vec{A} = \frac{\partial}{\partial x} A_x + \frac{\partial}{\partial y} A_y + \frac{\partial}{\partial z} A_z$$

Components of curl:

$$(\vec{\nabla}x\vec{A})_x = \left(\frac{\partial}{\partial y} A_z - \frac{\partial}{\partial z} A_y \right)$$

$$(\vec{\nabla}x\vec{A})_y = \left(\frac{\partial}{\partial z} A_x - \frac{\partial}{\partial x} A_z \right)$$

$$(\vec{\nabla}\times\vec{A})_z = \left(\frac{\partial}{\partial x} A_y - \frac{\partial}{\partial y} A_x \right)$$

Laplacian:

$$\nabla^2\psi = \frac{\partial^2\psi}{\partial x^2} + \frac{\partial^2\psi}{\partial y^2} + \frac{\partial^2\psi}{\partial z^2}$$

Components of divergence of a tensor:

$$(\vec{\nabla}.\bar{\bar{T}})_x = \frac{\partial}{\partial x} T_{xx} + \frac{\partial}{\partial y} T_{yx} + \frac{\partial}{\partial z} T_{zx}$$

$$(\vec{\nabla}.\bar{\bar{T}})_y = \frac{\partial}{\partial x} T_{xy} + \frac{\partial}{\partial y} T_{yy} + \frac{\partial}{\partial z} T_{zy}$$

$$(\vec{\nabla}.\bar{\bar{T}})_z = \frac{\partial}{\partial x} T_{xz} + \frac{\partial}{\partial y} T_{yz} + \frac{\partial}{\partial z} T_{zz}$$

2. CYLINDRICAL COORDINATES

Orthogonal unit vectors:

$\hat{\rho}$, $\hat{\phi}$, $\hat{z}$

Orthogonal line elements:

$d\rho$, $\rho d\phi$, dz

Components of gradient:

$$(\vec{\nabla}\psi)_\rho = \frac{\partial\psi}{\partial\rho}$$

$$(\vec{\nabla}\psi)_\phi = \frac{1}{\rho} \frac{\partial\psi}{\partial\phi}$$

$$(\vec{\nabla}\psi)_z = \frac{\partial\psi}{\partial z}$$

Divergence:

$$\vec{\nabla}.\vec{A} = \frac{1}{\rho}\frac{\partial}{\partial\rho}(\rho A_\rho) + \frac{1}{\rho}\frac{\partial}{\partial\phi}A_\phi + \frac{\partial}{\partial z}A_z$$

Components of curl:

$$(\vec{\nabla}x\vec{A})_\rho = \frac{1}{\rho}\frac{\partial}{\partial\phi}A_z - \frac{\partial}{\partial z}A_\phi$$

$$(\vec{\nabla}x\vec{A})_\phi = \frac{\partial}{\partial z}A_\rho - \frac{\partial}{\partial\rho}A_z$$

$$(\vec{\nabla}x\vec{A})_z = \frac{1}{\rho}\frac{\partial}{\partial\rho}(\rho A_\phi) - \frac{1}{\rho}\frac{\partial}{\partial\phi}A_\rho$$

Laplacian:

$$\nabla^2\psi = \frac{1}{\rho}\frac{\partial}{\partial\rho}\left(\rho\frac{\partial\psi}{\partial\rho}\right) + \frac{1}{\rho^2}\frac{\partial^2\psi}{\partial\phi^2} + \frac{\partial^2\psi}{\partial z^2}$$

Components of divergence of a tensor:

$$(\vec{\nabla}.\bar{\bar{T}})_\rho = \frac{1}{\rho}\frac{\partial}{\partial\rho}(\rho T_{\rho\rho}) + \frac{1}{\rho}\frac{\partial}{\partial\phi}(T_{\phi\rho}) + \frac{\partial}{\partial z}T_{z\rho} - \frac{1}{\rho}T_{\phi\phi}$$

$$(\vec{\nabla}.\bar{\bar{T}})_\phi = \frac{1}{\rho}\frac{\partial}{\partial\rho}(\rho T_{\rho\phi}) + \frac{1}{\rho}\frac{\partial}{\partial\phi}T_{\phi\phi} + \frac{\partial}{\partial z}T_{z\phi} + \frac{1}{\rho}T_{\phi\rho}$$

$$(\vec{\nabla}.\bar{\bar{T}})_z = \frac{1}{\rho}\frac{\partial}{\partial\rho}(\rho T_{\rho z}) + \frac{1}{\rho}\frac{\partial}{\partial\phi}T_{\phi z} + \frac{\partial}{\partial z}T_{zz}$$

3. SPHERICAL COORDINATES

Orthogonal unit vectors:

$\hat{r}$, $\hat{\theta}$, $\hat{\phi}$

Orthogonal line elements:

$$dr,\ r d\theta,\ r \sin\theta\, d\phi$$

Components of gradient:

$$(\vec{\nabla}\psi)_r = \frac{\partial\psi}{\partial r}$$

$$(\vec{\nabla}\psi)_\theta = \frac{1}{r}\frac{\partial\psi}{\partial\theta}$$

$$(\vec{\nabla}\psi)_\phi = \frac{1}{r\sin\theta}\frac{\partial\psi}{\partial\phi}$$

Divergence:

$$\vec{\nabla}.\vec{A} = \frac{1}{r^2}\frac{\partial}{\partial r}(r^2 A_r) + \frac{1}{r\sin\theta}\frac{\partial}{\partial\theta}(A_\theta \sin\theta) + \frac{1}{r\sin\theta}\frac{\partial}{\partial\phi}A_\phi$$

Components of curl:

$$(\vec{\nabla}\mathrm{x}\vec{A})_r = \frac{1}{r\sin\theta}\frac{\partial}{\partial\theta}(A_\phi \sin\theta) - \frac{1}{r\sin\theta}\frac{\partial}{\partial\phi}A_\theta$$

$$(\vec{\nabla}\mathrm{x}\vec{A})_\theta = \frac{1}{r\sin\theta}\frac{\partial}{\partial\phi}A_r - \frac{1}{r}\frac{\partial}{\partial r}(r\, A_\phi)$$

$$(\vec{\nabla}\mathrm{x}\vec{A})_\phi = \frac{1}{r}\frac{\partial}{\partial r}(r\, A_\theta) - \frac{1}{r}\frac{\partial}{\partial\theta}A_r$$

Laplacian:

$$\nabla^2\psi = \frac{1}{r^2}\frac{\partial}{\partial r}\left(r^2\frac{\partial\psi}{\partial r}\right) + \frac{1}{r^2\sin\theta}\frac{\partial}{\partial\theta}\left(\sin\theta\frac{\partial\psi}{\partial\theta}\right) + \frac{1}{r^2\sin^2\theta}\frac{\partial^2\psi}{\partial\phi^2}$$

Components of divergence of a tensor:

$$(\vec{\nabla}.\bar{\bar{T}})_r = \frac{1}{r^2}\frac{\partial}{\partial r}(r^2 T_{rr}) + \frac{1}{r \sin\theta}\frac{\partial}{\partial\theta}(T_{\theta r}\sin\theta) +$$

$$+ \frac{1}{r \sin\theta}\frac{\partial}{\partial\phi}T_{\phi r} - \frac{1}{r}(T_{\theta\theta} + T_{\phi\phi})$$

$$(\vec{\nabla}.\bar{\bar{T}})_\theta = \frac{1}{r^2}\frac{\partial}{\partial r}(r^2 T_{r\theta}) + \frac{1}{r \sin\theta}\frac{\partial}{\partial\theta}(T_{\theta\theta}\sin\theta) +$$

$$+ \frac{1}{r \sin\theta}\frac{\partial}{\partial\phi}T_{\phi\theta} + \frac{1}{r}(T_{\theta r} - \cot\theta\, T_{\phi\phi})$$

$$(\vec{\nabla}.\bar{\bar{T}})_\phi = \frac{1}{r^2}\frac{\partial}{\partial r}(r^2 T_{r\phi}) + \frac{1}{r \sin\theta}\frac{\partial}{\partial\theta}(T_{\theta\phi}\sin\theta) +$$

$$+ \frac{1}{r \sin\theta}\frac{\partial}{\partial\phi}T_{\phi\phi} + \frac{1}{r}(T_{\phi r} + \cot\theta\, T_{\phi\theta})$$

APPENDIX III

Physical Constants (MKSA)

c	Speed of light in vacuum	2.998×10^{8} m/sec
ε_o	Permittivity of vacuum	8.854×10^{-12} farad/m
μ_o	Permeability of vacuum	$4\pi \times 10^{-7}$ henry/m
h	Planck's constant	6.626×10^{-34} joule . sec
k	Boltzmann's constant	1.381×10^{-23} joule/K
G	Gravitational constant	6.671×10^{-11} Nm^2/kg^2
e	Charge of proton	1.602×10^{-19} coul
m_e	Rest mass of proton	1.673×10^{-27} kg
m_p	Rest mass of electron	9.109×10^{-31} kg
m_n	Rest mass of neutron	1.675×10^{-27} kg
m_p/m_e	Proton/electron mass ratio	1.836×10^{3}
amu	Unified atomic mass unit	1.661×10^{-27} kg
a_o	Bohr radius	5.292×10^{-11} m
r_e	Classical electron radius	2.818×10^{-15} m
N_A	Avogadro's number	6.022×10^{23} mol^{-1}
N_L	Loschmidt's number	2.687×10^{25} m^{-3}
V_o	Molar volume at STP	22.4×10^{-3} m^3/mol
R	Gas constant ($N_A k$)	8.314 joule/(K mol)
g	Standard acceleration of gravity	9.807 m/sec^2

APPENDIX IV

Conversion Factors for Units

Charge:	1 coulomb = 2.998×10^9 statcoulomb
Current:	1 ampere ≡ 1 coul/sec = 2.998×10^9 statampere
Potential:	1 volt = $(2.998 \times 10^2)^{-1}$ statvolt
Electric field:	1 volt/m = $(2.998 \times 10^4)^{-1}$ statvolt/cm
Magnetic induction:	1 weber/m^2 ≡ 1 tesla = 10^4 gauss
Resistance:	1 ohm = $(2.998)^{-2} \times 10^{-11}$ sec/cm
Conductivity:	1 mho/m = $(2.998)^2 \times 10^9$ sec^{-1}
Capacitance:	1 farad = $(2.998)^2 \times 10^{11}$ cm
Magnetic flux:	1 weber = 10^8 gauss . cm^2 (or maxwells)
Magnetic field:	1 ampere-turn/m = $4\pi \times 10^{-3}$ oersted
Force:	1 newton = 10^5 dyne
Energy:	1 joule = 10^7 erg
	1 electron volt (ev) = 1.602×10^{-19} joule
	1 ev = kT, where k is Boltzmann's constant, for T = 1.160×10^4 K
	1 Rydberg = 13.61 ev
Power:	1 watt ≡ 1 joule/sec = 10^2 erg/sec
Pressure:	1 newton/m^2 = 10 dyne/cm^2
	1 atm = 760 mm Hg = 1.013×10^5 newton/m^2
	1 torr = 1 mm Hg

APPENDIX V

Some Important Plasma Parameters

1. Electron plasma frequency:

$$\omega_{pe} = \left(\frac{n_e e^2}{m_e \varepsilon_o} \right)^{1/2} = 56.5\, n_e^{1/2} \text{ rad/sec } (n_e \text{ in } m^{-3})$$

2. Ion plasma frequency:

$$\omega_{pi} = \left(\frac{n_i Z^2 e^2}{m_i \varepsilon_o} \right)^{1/2}$$

3. Debye length:

$$\lambda_D = \left(\frac{\varepsilon_o k T}{n_e e^2} \right)^{1/2} = 69.0 \left(\frac{T}{n_e} \right)^{1/2} \text{ m}$$

(T in degrees K and n_e in m^{-3})

4. Electron cyclotron frequency:

$$\omega_{ce} = \frac{eB}{m_e} = 1.76 \times 10^{11} \text{ B rad/sec} \quad \text{(B in tesla)}$$

5. Ion cyclotron frequency:

$$\omega_{ci} = \frac{Z e B}{m_i}$$

6. Particle magnetic moment:

$$\vec{m} = - \frac{W_\perp}{B^2} \vec{B} = - \frac{m v_\perp^2/2}{B^2} \vec{B}$$

7. Electron cyclotron radius:

$$r_{ce} = \frac{v_{\perp e}}{\omega_{ce}} = \frac{m_e v_{\perp e}}{e B}$$

8. Ion cyclotron radius:

$$r_{ci} = \frac{v_{\perp i}}{\omega_{ci}} = \frac{m_i v_{\perp i}}{Z e B}$$

9. Number of electrons in Debye sphere:

$$N_D = \frac{4}{3} \pi \lambda_D^3 n_e = 1.37 \times 10^6 \frac{T^{3/2}}{n_e^{1/2}}$$

(T in degrees K and n_e in m^{-3})

10. Alfvēn velocity:

$$\vec{V}_A = \frac{\vec{B}}{(\mu_o \rho_m)^{1/2}}$$

11. DC conductivity:

$$\sigma_o = \frac{n_e e^2}{m_e \nu_e}$$

12. Free electron diffusion coefficient:

$$D_e = \frac{k T_e}{m_e \nu_e}$$

13. Ambipolar diffusion coefficient:

$$D_a = \frac{k(T_e + T_i)}{m_e \nu_{en} + m_i \nu_{in}}$$

14. Magnetic pressure:

$$p_m = \frac{B^2}{2\mu_o}$$

15. Magnetic viscosity:

$$\eta_m = \frac{1}{\mu_o \sigma_o}$$

16. Magnetic Reynolds number:

$$R_m = \frac{uL}{\eta_m}$$

17. Coulomb cutoff parameter:

$$\Lambda = 12 \pi n_e \lambda_D^3 = 9 N_D$$

$$= 1.23 \times 10^7 \frac{T^{3/2}}{n_e^{1/2}}$$

(T in degrees K and n_e in m^{-3})

18. Electron collision frequencies for momentum transfer:

$$\nu_{ei} = 3.62 \times 10^{-6} n_i T_e^{-3/2} \ln \Lambda \text{ sec}^{-1}$$

$$\nu_{en} = 2.60 \times 10^4 \sigma^2 n_n T_e^{1/2} \text{ sec}^{-1}$$

(T in degrees K, $n_{i,n}$ in m^{-3}; σ is the sum of the radii of the colliding particles and is of the order of 10^{-10} m, and $\ln \Lambda$ is typically about 10).

APPENDIX VI

Approximate Magnitudes in Some Typical Plasmas

PLASMA TYPE	n_o (m^{-3})	T (ev)	ω_{pe} (sec^{-1})	λ_D (m)	$n_o \lambda_D^3$
Interstellar gas	10^6	10^{-1}	6×10^4	1	10^6
Interplanetary gas	10^8	1	6×10^5	1	10^8
Solar corona	10^{12}	10^2	6×10^7	10^{-1}	10^9
Solar atmosphere	10^{20}	1	6×10^{11}	10^{-6}	10^2
Ionosphere	10^{12}	10^{-1}	6×10^7	10^{-3}	10^4
Gas discharge	10^{20}	1	6×10^{11}	10^{-6}	10^2
Hot plasma	10^{20}	10^2	6×10^{11}	10^{-5}	10^5
Diffuse hot plasma	10^{18}	10^2	6×10^{10}	10^{-4}	10^6
Dense hot plasma	10^{22}	10^2	6×10^{12}	10^{-6}	10^4
Thermonuclear plasma	10^{22}	10^4	6×10^{12}	10^{-5}	10^7

Index